Water-Resources Engineering

Water-Resources Engineering

David A. Chin

Professor of Civil and
Environmental Engineering
University of Miami

Prentice Hall
Upper Saddle River, New Jersey 07458

Library of Congress Cataloging-in-Publication Data

Chin, David A.

    Water-resources engineering / David A. Chin.

       p.   cm.

    Includes bibliographical references and index.

    ISBN 0-201-35091-2

    1. Hydraulics.   2. Hydraulic engineering.   3. Hydrology.

  I. Title.

  TC160.C52   2000

627—dc21   99-21054

Acquisition Editor: *Michael Slaughter*
Editor-in-Chief: *Marcia Horton*
Production Editor: *The Book Company*
Executive Managing Editor: *Vince O'Brien*
Assistant Managing Editor: *Eileen Clark*
Art Director: *Jayne Conte*
Cover Design: *Bruce Kenselaar*
Manufacturing Manager: *Trudy Pisciotti*
Assistant vice president of production and marketing: *David W. Riccardi*

© 2000 by Prentice Hall
Prentice-Hall Inc.
Upper Saddle River, New Jersey 07458

Printed in the United States of America
10   9   8   7   6   5   4   3   2

ISBN 0-201-35091-2

Prentice-Hall International (UK) Limited, *London*
Prentice-Hall of Australia Pty. Limited, *Sydney*
Prentice-Hall Canada, Inc., *Toronto*
Prentice-Hall Hispanoamericana, S.A., *Mexico*
Prentice-Hall of India Private Limited, *New Delhi*
Prentice-Hall of Japan, Inc., *Tokyo*
Prentice-Hall (Singapore) Pt. Ltd., *Singapore*
Editora Prentice-Hall do Brazil, Ltda., *Rio de Janeiro*

To my wife Linda Sue for her love and support

Contents

4 Flow in Open Channels   138

Appendices

Preface

Water-resources engineering is concerned with the design of systems that control the quantity, quality, timing, and distribution of water to support both human habitation and the needs of the environment. Water-resources engineers are typically trained in civil or environmental engineering programs and specialize in a variety of areas, including the design of water-supply systems, water and wastewater treatment facilities, irrigation and drainage systems, hydropower systems, and flood-control systems.

The technical and scientific bases for most water-resources specializations are found in the areas of fluid mechanics, hydraulics, hydrology, contaminant fate and transport processes, and water-treatment processes. Many engineering schools classify water-treatment processes as a subject that belongs to environmental engineering rather than water-resources engineering; however, a holistic view of the practice of water-resources engineering supports the study of water-treatment processes as part of both water-resources and environmental engineering specialties. The pigeonholing of fluid mechanics, hydraulics, hydrology, and contaminant fate and transport into discrete subjects, usually taught in separate courses using different textbooks, has resulted in large part from the extensive knowledge base that has developed in each of these areas and the commensurate specialization of engineers involved in research and academic practice. Engineering students are consequently left with a sense of compartmentalization and intimidation. Typically, they fail to see the complete picture of water-resources engineering and view each specialty as so vast that mastery at the undergraduate level is impossible. To address this misperception, an integrated treatment of water-resources engineering must necessarily present the fundamental aspects of the field while providing sufficient detail for the student to feel comfortable and competent in all the areas covered. Such an integrated approach has been taken in preparing this text, resulting in a book that covers the topics most fundamental to the practicing water-resources engineer—and does so with sufficient rigor that further instruction, whether at the graduate level or in professional journals, can be assimilated at a high technical level.

A course in fluid mechanics is generally regarded as the first step in a water-resources engineering track, and criteria for accrediting civil and environmental engineering programs in the United States (ABET Engineering Criteria 2000) require at most that engineering students demonstrate a proficiency in fluid mechanics relevant to their program of study. This book covers the elements of fluid mechanics relevant to a water-resources engineering track as well as the fundamentals of fluid mechanics covered on the Fundamentals of Engineering (FE) exam. The majority of this book provides detailed treatment of hydraulics, surface-water hydrology, ground-water hydrology, and hydrologic fate and transport processes, and it features practical design applications in all of these areas. The text incorporates, and explains in detail, the design of water distribution systems, sanitary sewer systems, stormwater management systems, and water-quality control systems in rivers, lakes, ground waters, and coastal waters. Care has been taken that all the design protocols presented in this book are consistent with the relevant American Society of Civil Engineers (ASCE), Water Environment Federation (WEF), and American Water Works Association (AWWA) Manuals of Practice.

The topics covered in this book constitute much of the technical background expected of water-resources engineers and part of the core requirements for environmental engineering students. This text is appropriate for undergraduate and first-year graduate courses in hydraulics, hydrology, and contaminant fate and transport processes. It also incorporates enough fluid mechanics background to rigorously cover the fundamentals of hydraulics and hydrology. Prerequisites for courses that use this text should include calculus up to differential equations.

The book begins with an introduction to water-resources engineering (Chapter 1) that orients the reader to the depth and breadth of the field. Chapter 2 covers the fundamentals of classical fluid mechanics relevant to water-resources engineering, and Chapter 3 presents the fundamentals of flow in closed conduits, including a detailed exposition on the design of water-supply systems. Chapter 4 covers flow in open channels from basic principles, including the computation of water-surface profiles and the performance of hydraulic structures. Applications of this material to the design of lined, unlined, and grassed drainage channels are presented along with the design of sanitary sewer systems. Computer models commonly used in practice to apply the principles of open-channel hydraulics are reviewed at the end of the chapter.

Many of the analytical methods used by water-resources engineers are based in the theory of probability and statistics, and Chapter 5 presents elements of probability and statistics relevant to the practice of water-resources engineering. Useful probability distributions, hydrologic data analysis, and frequency analysis are all covered, and the applications of these techniques to risk analysis in engineering design are illustrated by examples. Chapter 6 covers surface-water hydrology and focuses mostly on urban design applications. The ASCE *Manuals of Practice* on the design of surface-water management systems (ASCE, 1992) and urban runoff quality management (ASCE, 1998) were used as bases for much of the material presented. Coverage includes the specification of design rainfall, runoff models, routing models, and water-quality models. Applications of this material to the design of both minor and major components of stormwater management systems are presented, along with computer models widely used in practice to implement these techniques in complex stormwater management systems.

Chapter 7 covers ground-water hydrology, including the basic equations of ground-water flow, analytic solutions describing flow in aquifers, saltwater intrusion, and ground-water flow in the unsaturated zone. Applications to the design of municipal wellfields and individual water-supply wells, the delineation of wellhead protection areas, the design of aquifer pumping tests, and the design of exfiltration trenches are presented. Numerical models of ground-water flow used in practice are also reviewed. Chapter 8, finally, covers hydrologic fate and transport processes, including water-quality regulations, and quantitative analyses of fate and transport processes in rivers, lakes, ground waters, and coastal waters. The applications of these analyses to the design of water-quality management systems are presented. Seven appendices at the end of the book include conversion factors between SI and U.S. Customary units, fluid properties, geometric properties of plane surfaces, statistical tables, special functions, and drinking water standards.

This book can be used in a variety of ways, depending on the needs of students and instructors. As a guideline, the material in this text could be substantially covered in a two-course sequence. The first course could cover the material in Chapters 1 through 5 (Introduction, Fundamentals of Fluid Mechanics, Flow in Closed Conduits, Flow in Open Channels, Probability and Statistics in Water-Resources Engineering); the second, Chapters 6 through 8 (Surface-Water Hydrology, Ground-Water Hydrology, Hydrologic

Fate and Transport Processes). A course plan that complements other required courses in the engineering curriculum is generally recommended.

In summary, this book is a reflection of the author's belief that water-resources engineers must have a firm understanding of the depth and breadth of the technical areas fundamental to their discipline. This knowledge will allow them to be more innovative, view water-resource systems holistically, and be technically prepared for a lifetime of learning. On the basis of this vision, the material contained in this book is presented mostly from first principles, is rigorous, is relevant to the practice of water-resources engineering, and is reinforced by detailed presentations of design applications.

Even though the United States is squarely on the road to adopting International Standard (SI) units, most textbooks in hydraulics and hydrology published in this country continue to use the system of U.S. Customary units. Providing a mix of units can sometimes be confusing and usually forces the reader to adopt one set and ignore the other. Unfortunately, many engineering students tend to adopt the U.S. Customary unit system and disregard the SI system. If they are to be competitive in the future, American engineers cannot afford this luxury. Therefore, this textbook preferentially uses SI units.

Many people have contributed both directly and indirectly to the creation of this book. I acknowledge the many inspirational teachers who kindled my interest in water-resources engineering and whose philosophical ideas have contributed to development of my present view of the field. To name only a few people would be a disservice to many, but the faculty I studied under at Caltech and Georgia Tech during my graduate school days certainly deserve special recognition. My students in the civil and environmental engineering programs at the University of Miami provided valuable feedback in the development of this book, and Michael Slaughter of Addison-Wesley was a source of advice and help. I would like to join with the publisher in thanking the following reviewers for their comments and suggestions during the development of the manuscript: Mary Bergs, University of Toledo; Paul C. Chan, New Jersey Institute of Technology; Alexander Cheng, University of Delaware; Steven Chiesa, Santa Clara University; Bruce DeVantier, Southern Illinois University–Carbondale; Robert Kersten, University of Central Florida; Jay Lund, University of California, Davis; Joe Middlebrooks, University of Nevada, Reno; Paul Trotta, Northern Arizona University; and Ralph Wurbs, Texas A&M University. A special thanks to Bob Liu, who drafted most of the figures, and whose dedication to this project was beyond the call of duty.

David A. Chin

Introduction

1.1 Water-Resources Engineering

Water-resources engineering is concerned with the analysis and design of systems to control the quantity, quality, timing, and distribution of water to meet the needs of human habitation and the environment. The design of water-resource systems requires an understanding of the importance of water to various ecosystems, the behavior of engineered systems, and the fate and transport of contaminants contained in water. Aside from the engineering and environmental aspects of water-resource systems, their feasibility from legal, economic, financial, political, and social viewpoints must also be considered in the development process. In fact, the successful operation of an engineered system usually depends as much on nonengineering analyses (e.g., economic and social analyses) as on sound engineering design. This is particularly true in developing countries (Brookshire and Whittington, 1993). Examples of water-resource systems include domestic and industrial water supply, wastewater treatment, irrigation, drainage, flood control, salinity control, sediment control, pollution abatement, and hydropower-generation systems.

The waters of the earth are found on land, in the oceans, and in the atmosphere. The core science of water-resources engineering is *hydrology*, which deals with the occurrence, distribution, movement, and properties of water on earth. Engineering hydrologists are primarily concerned with water on land and in the atmosphere, from its deposition as atmospheric precipitation, such as rainfall and snowfall, to its inflow into the oceans and its vaporization into the atmosphere. The technical areas of expertise that are fundamental to water-resources engineering can be grouped into the following four categories:

1. Subsurface hydrology
2. Surface water and climate studies
3. Erosion and sedimentation studies; geomorphology
4. Water policy, economics, and systems analyses

Subsurface hydrology is concerned with the occurrence and movement of water below the surface of the earth; surface water and climate studies are concerned with the occurrence and movement of water above the surface of the earth; erosion and sedimentation studies and geomorphology deal with the effects of sediment transport on landforms; and water policy, economics, and systems analyses are concerned with the political, economic, and environmental constraints in the design and operation of water-resource systems. The quantity and quality of water are inseparable issues in design, and the modern practice of water-resources engineering demands that practitioners be technically competent in understanding both the physical processes that govern the movement of water, and the chemical and biological processes that affect the quality of water.

The design of water-resource systems usually involves interaction with government agencies. Collection of hydrologic and geologic data, granting of development permits, specification of design criteria, and use of government-developed computer models of water-resource systems are some of the many areas in which water-resource engineers interact with government agencies. The following are some of the key water-resources agencies in the United States:

■ **U.S. Geological Survey (USGS).** The primary source for streamflow and ground-water data in the United States, USGS maintains a network of thousands of stream gages and ground-water monitoring wells. USGS constructs and distributes $7\frac{1}{2}$-minute quadrangle topographic maps, which are useful in hydrologic studies; the agency also produces geological maps of subsurface formations.

■ **National Climatic Data Center (NCDC).** NCDC, the world's largest active source of weather data, produces numerous climate publications and responds to data requests from all over the world. Most of the data available from NCDC is collected and analyzed by the National Weather Service (NWS), formerly the United States Weather Bureau, which is a division of the National Oceanic and Atmospheric Administration (NOAA). Data collected by NWS include rainfall and evaporation measurements at over 10,000 locations in the United States.

■ **U.S. Bureau of Reclamation (USBR).** USBR manages water-related resources west of the Mississippi river. Besides being the largest wholesale supplier of water in the United States, USBR is the sixth largest hydroelectric supplier in the United States. Water management in the western United States includes water conservation, treatment, quality control, and supply.

■ **U.S. Environmental Protection Agency (USEPA).** USEPA is responsible for the implementation and enforcement of federal environmental laws. The agency's mission is to protect public health and to safeguard and improve the natural environment—air, water, and land—upon which human life depends.

■ **U.S. Natural Resources Conservation Service (NRCS).** NRCS, formerly the Soil Conservation Service, works with landowners on private lands to conserve natural resources. NRCS provides technical assistance to farmers and ranchers to develop conservation systems and to rural and urban communities to reduce erosion, conserve and protect water, and solve other resource problems. The NRCS publishes general soil maps for each state and detailed soil maps for each county in the United States.

■ **U.S. Army Corps of Engineers (USACE).** USACE is responsible for the planning, construction, operation, and maintenance of a variety of water-resource facilities whose objectives include navigation, flood control, water supply, water quality control, and other purposes.

These government agencies provide a wealth of information on water resources, relevant government regulations, and useful computer software that can be found on the Internet. Several of the more useful websites currently in use (and likely to be around for the foreseeable future) are listed in Table 1.1.

■ Table 1.1	Organization	Web address
Selected Internet Sites Relevant to Water-Resources Engineering	U.S. Geological Survey (USGS)	www.usgs.gov
	National Climatic Data Center (NCDC)	www.ncdc.noaa.gov
	U.S. Bureau of Reclamation (USBR)	www.usbr.gov
	U.S. Environmental Protection Agency (EPA)	www.epa.gov
	U.S. Natural Resources Conservation Service (NRCS)	www.ncg.nrcs.gov
	U.S. Army Corps of Engineers (USACE)	www.usace.army.mil

1.2 The Hydrologic Cycle

The *hydrologic cycle* is defined as the pathway of water as it moves in its various phases through the atmosphere, to the earth, over and through the land, to the ocean, and back to the atmosphere (National Research Council, 1991). The movement of water in the hydrologic cycle is illustrated in Figure 1.1, where the relative magnitudes of various hydrologic processes are given in units relative to a value of 100 for the rate of precipitation on land. The relative magnitudes are based on global annual averages (Chow et al., 1988).

A description of the hydrologic cycle can start with the evaporation of water from the oceans driven by energy from the sun. The evaporated water, in the form of water vapor,

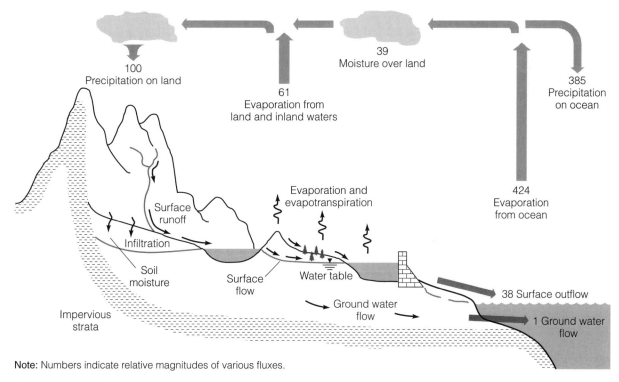

Note: Numbers indicate relative magnitudes of various fluxes.

Figure 1.1 ■ Hydrologic Cycle
Source: *Applied Hydrology.* Ven Te Chow, David R. Maidment, Larry W. Mays. Copyright © 1988 The McGraw-Hill Companies. Reprinted by permission of The McGraw-Hill Companies.

rises by convection; condenses in the atmosphere to form clouds; and precipitates onto land and ocean surfaces, predominantly as rain or snow. Precipitation on land surfaces is partially intercepted by surface vegetation, partially stored in surface depressions, partially infiltrated into the ground, and partially flows over land into drainage channels and rivers that ultimately lead back to the ocean. Precipitation that is intercepted by surface vegetation is eventually evaporated into the atmosphere; water held in depression storage either evaporates or infiltrates into the ground; and water that infiltrates into the ground contributes to the recharge of ground water, which is either utilized by plants or becomes subsurface flow that ultimately emerges as recharge to streams or directly to the ocean. *Ground water* is defined as the water below the land surface; water above the land surface (in liquid form) is called *surface water*. In urban areas, the ground surface is typically much less pervious than in rural areas, and surface runoff is mostly controlled by constructed drainage systems. Surface waters and ground waters in urban areas also tend to be significantly influenced by the water supply and wastewater removal systems that are an integral part of urban developments. Since humanmade systems are part of the hydrologic cycle, it is the responsibility of the water-resources engineer to ensure that systems constructed for water use and control are in harmony with the needs of the natural environment.

The quality of water varies considerably as it moves through the hydrologic cycle, with contamination resulting from several sources. Classes of contaminants commonly found in water, along with some examples, are listed in Table 1.2. The importance of the quantity and quality of water on the health of terrestrial ecosystems and the value of these ecosystems in the hydrologic cycle are often overlooked. For example, the modification of free-flowing rivers for energy or water supply, and the drainage of wetlands, can have a variety of adverse effects on aquatic ecosystems, including losses in species diversity, floodplain fertility, and biofiltration capability (Gleick, 1993).

On a global scale, the distribution of the water resources of the earth is given in Table 1.3, where it is clear that the vast majority of the earth's water resources is contained in the oceans, with most of the fresh water contained in ground water and polar ice. The amount of water stored in the atmosphere is relatively small, however, the flux of water into and out of the atmosphere dominates the hydrologic cycle. Wood (1991) estimated that the typical residence time for atmospheric water is on the order of a week, while soil moisture typical residence times go from weeks to months. The estimated fluxes of precipitation, evaporation, and runoff within the global hydrologic cycle are given in Table 1.4. These data indicate that the global average precipitation over land is on the order of

■ Table 1.2	Contaminant class	Example
Classes of Water Contaminants	Oxygen-demanding wastes	plant and animal material
	Infectious agents	bacteria and viruses
	Plant nutrients	fertilizers, such as nitrates and phosphates
	Organic chemicals	pesticides, such as DDT, detergent molecules
	Inorganic chemicals	acids from coal mine drainage, inorganic chemicals such as iron from steel plants
	Sediment from land erosion	clay silt on streambed, which may reduce or even destroy life forms living at the solid-liquid interface
	Radioactive substances	waste products from mining and processing of radioactive material, radioactive isotopes after use
	Heat from industry	cooling water used in steam generation of electricity

Source: U.S. Public Health Service.

■ **Table 1.3**
Estimated World
Water Quantities

Item	Volume ($\times 10^3$ km^3)	Percent total water (%)	Percent fresh water (%)
Oceans	1,338,000.	96.5	
Ground water			
Fresh	10,530.	0.76	30.1
Saline	12,870.	0.93	
Soil moisture	16.5	0.0012	0.05
Polar ice	24,023.5	1.7	68.6
Other ice and snow	340.6	0.025	1.0
Lakes			
Fresh	91.	0.007	0.26
Saline	85.4	0.006	
Marshes	11.47	0.0008	0.03
Rivers	2.12	0.0002	0.006
Biological water	1.12	0.0001	0.003
Atmospheric water	12.9	0.001	0.04
Total water	1,385,984.61	100.	
Fresh water	35,029.21	2.5	100.

Source: USSR National Committee for the International Hydrological Decade (1978).

■ **Table 1.4**
Fluxes in Global
Hydrologic Cycle

Component	Oceanic flux (mm/year)	Terrestrial flux (mm/year)
Precipitation	1,270	800
Evaporation	1,400	484
Runoff to ocean (rivers plus ground water)	—	316

Source: USSR National Committee for the International Hydrological Decade (1978).

800 mm/yr (31 in./yr), of which 484 mm/yr (19 in./yr) is returned to the atmosphere as evaporation and 316 mm/yr (12 in./yr) is returned to the ocean via surface runoff. On a global scale, large variations from these average values are observed. In the United States, for example, the highest annual rainfall is found at Mount Wai'ale'ale on the Hawaiian island of Kauai with an annual rainfall of 1,168 cm (460 in.), while Greenland Ranch in Death Valley, California, has the lowest annual average rainfall—4.5 cm (1.78 in.)—in the United States.

On regional scales, water resources are managed within topographically defined areas called *watersheds* or *basins*. These areas are typically enclosed by topographic high points in the land surface, and within these bounded areas the path of the surface runoff can usually be controlled with a reasonable degree of coordination.

1.3 Design of Water-Resource Systems

The uncertainty and natural variability of hydrologic processes require that most water-resource systems be designed with some degree of *risk*. Approaches to designing water-resource systems are classified as either *frequency-based design*, *risk-based design*, or *critical-event design*. In frequency-based design, the exceedance probability of the design event is selected a priori and the water-resource system is designed to accommodate all lesser events up to and including an event with the selected exceedance probability. The water-resource system will then be expected to fail with a probability equal to the exceedance probability of the design event. The frequency-based design approach is commonly used in designing the minor structures of urban drainage systems. For example, urban storm-drainage systems are typically designed for precipitation events with return periods of 10 years or less, where the *return period* of an event is defined as the reciprocal of the (annual) exceedance probability of the event. In risk-based design, systems are designed such that the sum of the capital cost and the cost of failure is minimized. Capital costs tend to increase and the cost of failure tends to decrease with increasing system capacity. Because any threats to human life are generally assigned extremely high failure costs, structures such as large dams are usually designed for rare hydrologic events with long return periods and commensurate small failure risks. In some extreme cases, where the consequences of failure are truly catastrophic, water-resource systems are designed for the largest possible magnitude of a hydrologic event. This approach is called critical-event design, and the value of the design (hydrologic) variable in this case is referred to as the *estimated limiting value* (ELV).

Water-resource systems can be broadly categorized as *water-control systems* or *water-use systems*, as shown in Table 1.5, but it should be noted that these systems are not mutually exclusive. A third category of *environmental restoration* has been suggested by Mays (1996). The following sections present a brief overview of the design objectives in these systems.

1.3.1 ■ Water-Control Systems

Water-control systems are primarily designed to control the spatial and temporal distribution of surface runoff resulting from rainfall events. Flood-control structures and storage impoundments reduce the peak flows in streams, rivers, and drainage channels, thereby reducing the occurrence of floods. A *flood* is defined as a high flow that exceeds the capacity of a stream or drainage channel, and the elevation at which the flood overflows the embankments is called the *flood stage*. A *floodplain* is the normally dry land adjoining rivers, streams, lakes, bays, or oceans that is inundated during flood events. Typically, flows with return periods from 1.5 to 3 years represent bankfull conditions, with larger flows causing inundation of the floodplain (McCuen, 1989). The 100-year flood has been adopted by the U.S. Federal Emergency Management Agency (FEMA) as the base flood for delineating floodplains, and the area inundated by the 500-year flood is sometimes delineated to indicate areas of additional risk. Encroachment onto floodplains reduces the capacity of the watercourse and increases the extent of the floodplain. Approximately 7%–10% of the land in the United States is in a floodplain, and in the 1970s flood-related deaths were 200 per year, with another 80,000 per year forced from their homes (Wanielista and Yousef, 1993). The largest floodplain areas in the United

■ Table 1.5	Water-control systems	Water-use systems
Water-Resource Systems	Drainage	Domestic and industrial water supply
	Flood control	Wastewater treatment
	Salinity control	Irrigation
	Sediment control	Hydropower generation
	Pollution abatement	

States are in the South; the most populated floodplains are along the north Atlantic coast, the Great Lakes region, and in California (Viessman and Lewis, 1996).

In urban settings, water-control systems include storm-sewer systems for collecting and transporting surface runoff, and storage reservoirs that attenuate peak runoff rates and reduce pollutant loads in drainage channels. Urban stormwater control systems are typically designed to prevent flooding from runoff events with return periods of 10 years or less. For larger runoff events, the capacity of these systems are exceeded and surface (street) flooding usually results.

1.3.2 ■ Water-Use Systems

Water-use systems are designed to support human habitation and include water-treatment systems, water-distribution systems, wastewater-collection systems, and wastewater-treatment systems. The design capacity of these systems are generally dictated by the population of the service area, commercial and industrial requirements, and the economic design life of the system. Water-use systems are designed to provide specified levels of service: water-treatment systems, for example, must produce water of sufficient quality to meet drinking water standards, water-distribution systems must deliver peak demands while sustaining adequate water pressures, wastewater-collection systems must have sufficient capacity to transport wastes without overflowing into the streets, and wastewater-treatment systems must provide a sufficient level of treatment that effluent discharges will not degrade the aquatic environment. In agricultural areas, the water requirements of plants are met by a combination of rainfall and irrigation. The design of irrigation systems requires the estimation of crop evapotranspiration rates and leaching requirements in agricultural areas, with the portion of these requirements that are not met by rainfall being met by irrigation systems. In rivers where there is sufficient available energy, such as behind large dams or in rapidly flowing rivers, hydroelectric power generation may be economically feasible.

1.4   About This Book

The previous sections have indicated the breadth of applications in water-resources engineering. The fundamental technical aspects of water-resources engineering derive mostly from the subject areas of fluid mechanics, hydraulics, probability and statistics, hydrology, and contaminant fate and transport processes. A good understanding of these subjects is the foundation on which engineers produce sound designs and operational protocols for water-resources systems. This book addresses each of these fundamental subjects in detail, beginning with the basics of fluid mechanics relevant to the field of water-resources engineering. The hydraulics of flow in closed conduits is presented along with a detailed presentation of the application of closed-conduit flow principles to the design of water-supply systems. The estimation of design flows and service pressures and the design of storage reservoirs are all covered. The hydraulics of flow in open channels is rigorously presented, from basic principles to the computation of water-surface profiles, hydraulic structures, the design of channels, and sanitary-sewer systems.

A chapter on probability and statistics reviews probability distribution functions, the estimation of probability distributions from measured data, and applications of probability and statistics to risk analysis in water-resource systems. Surface-water hydrology is presented from an urban design perspective through the specification of design rainfall and abstraction models and the computation of the quantity and quality of surface runoff resulting from a specified rainfall. The design of urban stormwater management systems is presented in detail. The chapter on ground-water hydrology covers the basic principles of flow in porous media through analytic descriptions of ground-water flow. The design of municipal wellfields, the design of water-supply wells, the delineation of wellhead-protection areas, the design of aquifer pumping tests, and the design of exfiltration trenches for ground-water injection are all presented in detail to demonstrate the application of ground-water flow principles in design.

Hydrologic fate and transport processes are covered in the context of water-quality standards and the mixing of polluted effluents in natural water bodies. The properties of the general diffusion equation are derived from basic principles, and the fate and transport of contaminants in rivers, lakes, ground waters, and oceans are all examined, with an emphasis on the application of fate and transport models to the design of water-quality control systems.

2 Fundamentals of Fluid Mechanics

2.1 Introduction

Fluid mechanics is an engineering science on which much of water-resources engineering is based. As a field, fluid mechanics covers many different types of fluids and many phenomena of fluid behavior. Those areas of fluid mechanics particularly relevant to water-resources engineering are presented in this chapter, including fluid properties, forces on plane and curved surfaces, basic equations of fluid dynamics, and the theory of dimensional analysis and similitude.

2.2 Physical Properties of Water

Fluid properties commonly encountered in studying the behavior of static and flowing fluids are density, viscosity, compressibility, surface tension, saturation vapor pressure, and latent heat of vaporization. The definitions of these properties, along with their importance in various areas of water-resources engineering are presented here.

The *density* of a fluid, ρ, is defined as the mass of fluid per unit volume. The density of pure water varies nonlinearly between 999.8 kg/m^3 at 0°C and 958.4 kg/m^3 at 100°C, with a density of 998.2 kg/m^3 at 20°C. The maximum density of water is 1,000.0 kg/m^3 at 4°C. The densities of water at temperatures between 0°C and 100°C are given in Table B.1 of Appendix B. Other properties that are derived from the density are *specific weight* and *specific gravity*. The specific weight of a fluid, γ, is defined as the weight per unit volume and is related to the density by

$$\gamma = \rho g \tag{2.1}$$

where g is the acceleration due to gravity. By international agreement, standard gravity, g, at sea level is 9.80665 m/s^2, and variation in g on the earth's surface is small and is usually neglected. The specific gravity of a fluid, SG, is defined as the ratio of the density of the fluid to the density of pure water at some specified temperature, usually 15°C (Wandmacher and Johnson, 1995). Some fluid mechanics textbooks cite a reference temperature of 4°C (Munson et al., 1994). The definition of the specific gravity can be expressed as

$$SG = \frac{\rho}{\rho_{H_2O @ 15°C}} \tag{2.2}$$

where the density of pure water at 15°C is equal to 999.10 kg/m^3. The addition of salt to water increases the density of the water, supresses the temperature at which the maximum density occurs, and suppresses the freezing point of the water. These properties are of particular interest in using salt to prevent ice formation on roads and in controlling the intrusion of seawater in coastal regions.

The *viscosity* of a fluid is the proportionality constant between the shear stress and the strain rate of a fluid element. Consider the rectangular fluid element illustrated in Figure 2.1, where the shape of the element at time t is ABCD, and the shape of the element at time $t + \Delta t$ is ABC'D'. The element shown in Figure 2.1 is acted upon by shear stresses, τ, on the upper and lower surfaces, and the angle $\Delta\theta$ is the shear strain in the time interval Δt resulting from the applied shear stress. If the velocity of the bottom surface of the fluid element is u, and the velocity of the top surface is $u + \Delta u$, then for small values of $\Delta\theta$,

$$\Delta\theta \approx \tan \Delta\theta = \frac{CC'}{\Delta y} = \frac{(\Delta u)(\Delta t)}{\Delta y} \tag{2.3}$$

Defining the viscosity, μ, as the proportionality constant between the shear stress, τ, and the strain rate, $\Delta\theta/\Delta t$,

$$\tau = \mu \frac{\Delta\theta}{\Delta t} \tag{2.4}$$

and combining Equations 2.3 and 2.4 gives

$$\tau = \mu \frac{\Delta u}{\Delta y} \tag{2.5}$$

Taking the limit as $\Delta y \to 0$ yields

$$\boxed{\tau = \mu \frac{du}{dy}} \tag{2.6}$$

This proportionality only exists for a class of fluids called *Newtonian fluids*, including water, and so Equation 2.6 is sometimes referred to as *Newton's equation of viscosity*. The relationship between shear stress and strain rate in *non-Newtonian fluids* takes the form

$$\tau = K \left(\frac{du}{dy} \right)^n$$

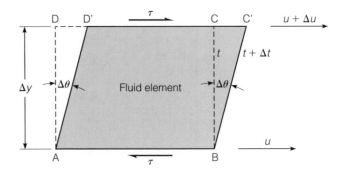

Figure 2.1 ■
Viscous Effect

where K is a consistency index and n is a power-law index. The viscosity, μ, of Newtonian fluids (such as water) in Equation 2.6 is also called the *absolute viscosity* or the *dynamic viscosity*, and is quite sensitive to temperature changes. The viscosity of water varies nonlinearly between 1.78 mPa·s at 0°C, and 0.282 mPa·s at 100°C, with a viscosity of 1.00 mPa·s at 20°C. The (dynamic) viscosity of water as a function of temperature is given in Table B.1 of Appendix B. A fluid property that is closely related to the (dynamic) viscosity, μ, is the *kinematic viscosity*, ν, which is defined by the relation

$$\nu = \frac{\mu}{\rho} \qquad (2.7)$$

where ρ is the density of the fluid. The values of μ and ρ as a function of temperature given in Table B.1 can be combined to yield the kinematic viscosity, ν, as a function of temperature.

■ Example 2.1

The velocity distribution, $u(y)$, in an open channel of large width is given by

$$u(y) = 5.75\sqrt{gy_oS_o}\,\log\left(\frac{30y}{k}\right)$$

where y is the distance from the bottom of the channel, y_o is the depth of flow, S_o is the slope of the channel, and k is the height of the roughness elements on the bottom of the channel. Estimate the shear stress 1 cm above the bottom of a channel on a slope of 0.5% where the depth of flow is 2.5 m and the characteristic roughness height is 1 cm. If the critical shear stress required to move the bed material in the channel is 1 N/m², determine whether the channel is in a stable condition. Assume that the temperature of the water is 20°C.

Solution

Taking $y_o = 2.5$ m, $S_o = 0.005$, and $k = 0.01$ m, then the velocity distribution is given by

$$u(y) = 5.75\sqrt{gy_oS_o}\,\log\left(\frac{30y}{k}\right)$$

$$= 5.75\sqrt{(9.81)(2.5)(0.005)}\,\log\left(\frac{30y}{0.01}\right) \text{ m/s}$$

$$= 2.01(3.48 + \log y) \text{ m/s}$$

$$= 6.99 + 0.873\ln y \text{ m/s}$$

At 20°C, $\mu = 0.00100$ Pa·s $= 0.00100$ N·s/m², and the shear stress, τ_o, 1 cm from the bottom of the channel is given by

$$\tau_o = \mu\left.\frac{du}{dy}\right|_{y=0.01\text{m}}$$

$$= \mu\left.\frac{d}{dy}(6.99 + 0.873\ln y)\right|_{y=0.01\text{m}} \text{ N/m}^2$$

$$= 0.00100 \left[\frac{0.873}{y} \right]_{y=0.01\text{m}} \text{N/m}^2$$

$$= 0.0873 \text{ N/m}^2$$

Since the shear stress required to move the bed material is 1 N/m^2 and the actual shear stress caused by the flowing water is 0.0873 N/m^2, the channel is stable. ▪

The *compressibility* of a fluid is characterized by the *bulk modulus*, E_v, defined by the relation

$$E_v = -\frac{dp}{dV/V} \approx -\frac{\Delta p}{\Delta V/V} \tag{2.8}$$

where Δp is the pressure increment that causes a volume V of fluid to change by ΔV. In the limit as $\Delta p \rightarrow 0$, Δp becomes the differential dp, and ΔV becomes dV. The bulk modulus is also known as the *volume modulus of elasticity*, and is sometimes expressed in terms of density, ρ, and density differential, $d\rho$, by the relation

$$\boxed{E_v = \frac{dp}{d\rho/\rho}} \tag{2.9}$$

The bulk modulus of water is a function of both temperature and pressure, and values of E_v for conditions close to atmospheric pressure are shown in Table B.1 of Appendix B. For most practical purposes, water can be taken as incompressible, with a bulk modulus of 2.15×10^6 kPa. The compressibility of water is taken into consideration primarily in calculating the pressure rise in closed conduits resulting from the sudden closure of valves, an effect that is called *water hammer*.

▪ Example 2.2

Water in municipal supply systems typically experience pressures in the range of 140 kPa–700 kPa, yet the density is usually taken as a constant in engineering calculations. Estimate the maximum percentage error that is expected by assuming a constant density.

Solution

The compressibility of water, E_v, can be taken as 2.15×10^6 kPa and, according to Equation 2.9:

$$E_v \approx \frac{\Delta p}{\Delta \rho/\rho}$$

which can be rearranged to yield

$$\frac{\Delta \rho}{\rho} \approx \frac{\Delta p}{E_v} = \frac{700 - 140}{2.15 \times 10^6} = 0.000260$$

The percentage change in density corresponding to a pressure change of 560 kPa ($= 700$ kPa $- 140$ kPa) is 0.026%. Therefore, the pressure variation has a negligible effect on the density. ▪

Surface tension results from the unbalanced cohesive forces acting on liquid molecules at the surface of a liquid. The molecular attraction per unit length along any line in the liquid surface is called the surface tension, σ, which can be expressed as

$$\sigma = \frac{F}{L} \tag{2.10}$$

where F is the surface tension force over a distance L in the surface of the fluid. Surface tension depends on both the temperature and the fluid and/or the nature of the material that contacts the liquid. The surface tension of pure water in contact with air varies from 75.6 mN/m at 0°C to 58.9 mN/m at 100°C, with a surface tension of 72.8 mN/m at 20°C. The surface tension of water in contact with air as a function of temperature between 0°C and 100°C is given in Table B.1 of Appendix B. Surface tension has an important influence on the movement and distribution of ground water in unsaturated soils.

The *saturation vapor pressure* of a fluid is defined as the partial pressure of the gaseous phase of the fluid that is in contact with the liquid phase of the fluid when there is no net exchange of mass between the two phases. This state of equilibrium is reached when the mass flux of the fluid from the liquid phase to the gas phase is equal to the mass flux from the gas phase to the liquid phase. If the actual vapor pressure exceeds the saturation vapor pressure, there is a net flux of mass into the liquid phase until equilibrium is restored. If the vapor pressure in the gas phase is less than the saturation vapor pressure, there is a net flux of mass from the liquid to the gas phase. This cycle of mass transfer occurs almost daily on earth, where, during the cooler nighttime hours, the saturation vapor pressure of water is reduced below the actual vapor pressure, causing liquid precipitation called *dew*, while during the warmer daytime hours the saturation vapor pressure typically rises above the actual vapor pressure causing evaporation.

The process of evaporation of water from the surface of the earth and from the stomatae of plants is a key component of the hydrologic cycle. The rate of evaporation is an important quantity in many agricultural operations, where plant transpiration rates determine irrigation requirements and the rate of transpiration is directly proportional to the difference between the saturation vapor pressure and the actual vapor pressure. The saturation vapor pressure of pure water varies between 0.611 kPa at 0°C and 101.3 kPa at 100°C, with a value of 2.34 kPa at 20°C. The saturation vapor pressure of water as a function of temperature is given in Table B.1 of Appendix B. In hydrologic applications, the saturation vapor pressure is commonly denoted by e_s, and the actual vapor pressure is commonly denoted by e. A property that is related to both e_s and e is the *relative humidity*, RH, which is defined by the relation

$$RH = \frac{e}{e_s} \times 100 \tag{2.11}$$

where RH is in percent. Clearly, evaporation can only occcur when the relative humidity is less than 100%.

■ Example 2.3

The air temperature in an agricultural area during a typical summer day is shown in the following table. If the relative humidity at 12:00 noon is 70%, estimate the time of day that the relative humidity becomes 100%.

Time	Midnight	3 a.m.	6 a.m.	9:00 a.m.	Noon	3:00 p.m.	6:00 p.m.	9:00 p.m.
Temperature (°C)	20	18.3	20	23.3	32.2	33.9	26.7	22.7

Solution

Saturation vapor pressures as a function of temperature are given in Table B.1 of Appendix B. At 12:00 noon, the temperature is 32.2°C and the saturation vapor pressure is 4.93 kPa. Since the relative humidity is 70%, the actual vapor pressure, e, is given by

$$e = (0.7)(4.93) = 3.45 \text{ kPa}$$

Assuming that the actual vapor pressure of water in the atmosphere remains constant, then at other times of the day the relative humidity, RH, can be taken as

$$\text{RH} = \frac{3.45}{e_s} \times 100$$

where e_s is the saturation vapor pressure at the ambient air temperature. Reading the saturation vapor pressures as a function of temperature from Table B.1 yields the following variation in relative humidity, beginning at noon:

Time	Noon	3:00 p.m.	6:00 p.m.	9:00 p.m.
Temperature (°C)	32.2	33.9	26.7	22.7
e_s (kPa)	4.93	5.47	3.53	2.79
RH (%)	70	63.1	96.3	100

The relative humidity at 9:00 p.m. is calculated to be more than 100%, which will result in condensation to maintain the relative humidity at 100%. Therefore the relative humidity becomes equal to 100% at some time between 6:00 p.m. and 9:00 p.m. Linearly interpolating the vapor pressure between 6:00 p.m. and 9:00 p.m. indicates that at approximately 6:20 p.m. the saturation vapor pressure is 3.45 kPa and the humidity equals 100%. ■

The vapor pressure of water in the atmosphere normally contributes no more than about 2 kPa to the total atmospheric pressure of about 101 kPa, with the remainder of the atmospheric pressure contributed by other gasses, primarily nitrogen (N_2) and oxygen (O_2). The saturation vapor pressure of water increases with temperature. Whenever the temperature of the water is such that the saturation vapor pressure equals the pressure in the water body, vapor cavities are formed within the water body to maintain equilibrium between the liquid and vapor phases. The process by which vapor cavities are formed is called *cavitation*. These cavities contain water vapor at a pressure equal to the saturation vapor pressure at the water temperature, and the water is said to be *boiling*. This boiling phenomenon can also be achieved by reducing the pressure in the water body so that the water pressure becomes equal to the saturation vapor pressure. This mechanism for the formation of vapor cavities is frequently of concern when water is pumped through low-pressure regions in closed conduits. Severe damage to the conduit and hydromachinery can occur when these vapor-filled cavities collapse in regions of higher pressure; the rotors of pumps and turbines are particularly susceptible to damage.

■ Example 2.4

Atmospheric pressure in a mountain region is equal to 82 kPa. Estimate the boiling point of water. Compare your result with the boiling point at sea level, where the atmospheric pressure can be taken as 101 kPa.

Solution

At a vapor pressure of 82 kPa, interpolation of Table B.1 in Appendix B gives the corresponding water temperature as 93.8°C. Hence, assuming that the pressure in the water body is approximately equal to the atmospheric pressure, the water will boil at 93.8°C.

At sea level, the pressure in a water body is approximately equal to 101 kPa, and the corresponding boiling point is 100°C. ■

■ Example 2.5

Water at 20°C is pumped through a closed conduit and experiences a range of pressures. What is the minimum allowable pressure to prevent cavitation within the conduit?

Solution

The saturation vapor pressure at 20°C is 2.337 kPa (Table B.1, Appendix B). Hence, the (absolute) water pressure in the conduit should be maintained above 2.337 kPa. ■

The *latent heat of vaporization*, λ, is the amount of heat required to convert a unit mass of a fluid from the liquid to the vapor phase at a given temperature. This quantity is also called the *heat of vaporization*. For water it ranges from 2.499 MJ/kg at 0°C to 2.256 MJ/kg at 100°C, with a value of 2.452 MJ/kg at 20°C. Values of the latent heat of vaporization as a function of temperature are given in Table B.1 in Appendix B. The heat of vaporization is most commonly used in applications relating to evaporation and plant transpiration, where the net solar and long-wave energy fluxes at the surface of the earth, in J/(m²·d), result in a mass flux of liquid terrestrial water to water vapor in the atmosphere.

■ Example 2.6

The net solar and long-wave energy flux measured during July in Miami averages 19 MJ/(m²·d). Estimate the resulting evaporation from an open-water body.

Solution

$$\text{energy flux, } E = 19 \text{ MJ/(m}^2\text{·d)}$$

$$\text{heat of vaporization @20°C, } \lambda = 2.452 \text{ MJ/kg}$$

$$\text{density of water @20°C, } \rho = 998 \text{ kg/m}^3$$

The energy required to evaporate 1 m³ of water at 20°C is equal to $\rho\lambda$, where

$$\rho\lambda = (998)(2.452) = 2450 \text{ MJ/m}^3$$

The amount of evaporation resulting from the energy flux, E, is therefore given by $E/\rho\lambda$, where

$$\frac{E}{\rho\lambda} = \frac{19}{2450} \text{ m/d} = 7.76 \text{ mm/d}$$

The total evaporation in July from an open-water body in Miami can therefore be estimated as 7.76 mm/d × 31 d = 241 mm. ■

Summary of Water Properties. Most design applications in water-resources engineering require specification of water properties and a corresponding specification of water

■ **Table 2.1**
Physical Properties of
Water at 20°C

Property	Symbol	Value
Density	ρ	998 kg/m^3
Specific weight	γ	9.79 kN/m^3
(Dynamic) viscosity	μ	1.00×10^{-3} N·s/m^2
(Kinematic) viscosity	ν	1.00×10^{-6} m^2/s
Bulk modulus	E_v	2.15×10^6 kPa
Surface tension	σ	72.8×10^{-3} N/m
Saturation vapor pressure	e_s	2.34 kPa
Heat of vaporization	λ	2.453 MJ/kg

temperature. In cases where temperature changes are not a significant factor, a temperature of 20°C is commonly assumed. The corresponding properties of water are given in Table 2.1.

2.3 Fluid Statics

2.3.1 ■ Pressure Distribution in Static Fluids

Fluid statics is the study of the forces exerted by fluids at rest. Consider the control volume within a static fluid illustrated in Figure 2.2. If the dimensions of the control volume are $\Delta x \times \Delta y \times \Delta z$, the pressure at the center of the control volume is p, and the specific weight of the fluid is γ, then a force balance in the x-direction yields

$$\left(p - \frac{\partial p}{\partial x}\frac{\Delta x}{2}\right)\Delta y \Delta z - \left(p + \frac{\partial p}{\partial x}\frac{\Delta x}{2}\right)\Delta y \Delta z = 0 \qquad (2.12)$$

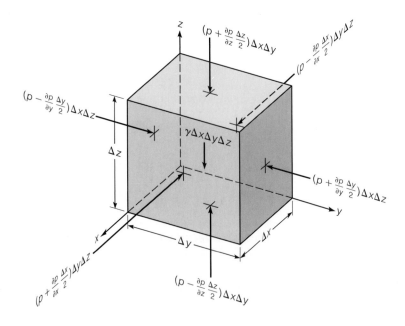

Figure 2.2 ■ Control
Volume in a Static Fluid

which simplifies to

$$\frac{\partial p}{\partial x} = 0 \tag{2.13}$$

indicating that there is no variation in pressure in the (horizontal) x-direction. Similarly, a force balance in the y-direction yields

$$\left(p - \frac{\partial p}{\partial y}\frac{\Delta y}{2}\right)\Delta x \Delta z - \left(p + \frac{\partial p}{\partial y}\frac{\Delta y}{2}\right)\Delta x \Delta z = 0 \tag{2.14}$$

which simplifies to

$$\frac{\partial p}{\partial y} = 0 \tag{2.15}$$

indicating that there is no variation in pressure in the (horizontal) y-direction. Equations 2.13 and 2.15 collectively indicate that *there are no pressure variations in the horizontal plane in static fluids.* A force balance in the (vertical) z-direction yields

$$\left(p - \frac{\partial p}{\partial z}\frac{\Delta z}{2}\right)\Delta x \Delta y - \left(p + \frac{\partial p}{\partial z}\frac{\Delta z}{2}\right)\Delta x \Delta y - \gamma \Delta x \Delta y \Delta z = 0 \tag{2.16}$$

which simplifies to

$$-\frac{\partial p}{\partial z} - \gamma = 0 \tag{2.17}$$

or

$$\frac{\partial p}{\partial z} = -\gamma \tag{2.18}$$

According to Equations 2.13 and 2.15, the pressure, p, is independent of x and y and therefore depends only on the vertical coordinate z. Consquently, the partial derivative in Equation 2.18 can be replaced by the total derivative and written as

$$\boxed{\frac{dp}{dz} = -\gamma} \tag{2.19}$$

This equation describes the pressure variation within all static fluids. In cases where the fluid density is constant, Equation 2.19 can be integrated to yield

$$p = -\gamma z + \text{constant} \tag{2.20}$$

or

$$\boxed{p + \gamma z = \text{constant}} \tag{2.21}$$

Pressure distributions described by Equation 2.21 are called *hydrostatic* pressure distributions. Pressure measured relative to absolute zero is called *absolute pressure*, and pressure measured relative to atmospheric pressure is called *gage pressure*. Standard atmospheric pressure is usually taken as 101.3 kPa.

Consider the case illustrated in Figure 2.3, where the point P is located a distance h below a fluid surface and a distance z above a fixed datum. If the pressure at the fluid

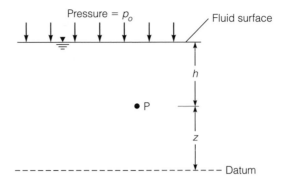

Figure 2.3 ▪ Pressure
Distribution in a Static Fluid

surface is equal to p_o, and the pressure at point P is equal to p, then applying Equation 2.21 at both the fluid surface and P yields

$$p_o + \gamma(z + h) = p + \gamma z \qquad (2.22)$$

which simplifies to

$$\boxed{p = p_o + \gamma h} \qquad (2.23)$$

This commonly used equation states that the pressure at any point located a distance h below another point in a static fluid is equal to the pressure at the higher point plus γh. In cases where the higher point corresponds to the surface of a fluid exposed to atmospheric pressure, p_{atm}, then Equation 2.23 becomes

$$p = p_{atm} + \gamma h \qquad (2.24)$$

or

$$p - p_{atm} = \gamma h \qquad (2.25)$$

which means that if p is taken as a gage pressure, then Equation 2.25 can be written simply as

$$p = \gamma h \qquad (2.26)$$

Pressure is sometimes expressed as the height of a static column of fluid that would cause the same pressure. For a pressure p, the corresponding height of the fluid column, h, is given by

$$h = \frac{p}{\gamma} \qquad (2.27)$$

where γ is the specific weight of the fluid. The height h is called the *pressure head* and, unless stated otherwise, the fluid is generally taken as water at 20°C.

▪ Example 2.7

A pipeline leads from a reservoir to a closed valve, as illustrated in Figure 2.4. Calculate the gage and absolute pressures at the valve.

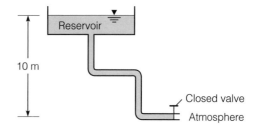

Figure 2.4 ■
Pipeline System

Solution

The gage pressure, p, at the valve is given by Equation 2.26 as

$$p = \gamma h$$

where $\gamma = 9.79$ kN/m^3 and $h = 10$ m. Therefore

$$p = (9.79)(10) = 97.9 \text{ kN/m}^2 = 97.9 \text{ kPa}$$

The absolute pressure, p_{abs}, at the valve location is given by

$$p_{abs} = p + p_{atm}$$

where $p_{atm} = 101.3$ kPa, and therefore

$$p_{abs} = 97.9 + 101.3 = 199.2 \text{ kPa}$$ ■

■ Example 2.8

Water pressure at a pipeline junction measures 450 kPa. What is the corresponding pressure head?

Solution

For water at 20°C, Table 2.1 gives $\gamma = 9.79$ kN/m^3. The pressure head, h, corresponding to $p = 450$ kPa is therefore

$$h = \frac{p}{\gamma} = \frac{450}{9.79} = 46.0 \text{ m}$$ ■

2.3.2 ■ Pressure Measurements

The performance of many hydraulic and water-transmission systems is routinely monitored by measuring the pressure at strategic locations in the system. There are a variety of devices for measuring pressure, but mechanical and electrical pressure gages* are most common in practice. Manual pressure gages based on measuring fluid levels are primarily used in research applications.

The most widely used mechanical device for measuring pressure is the Bourdon-tube pressure gage, illustrated in Figure 2.5. The Bourdon-tube gage is directly connected to the fluid system, and the fluid fills a metal tube having an elliptical cross-section. The elliptical tube is mechanically attached to a pointer that registers zero (gage) pressure

---

*The spellings *gage* and *gauge* are used interchangeably, and both are acceptable.

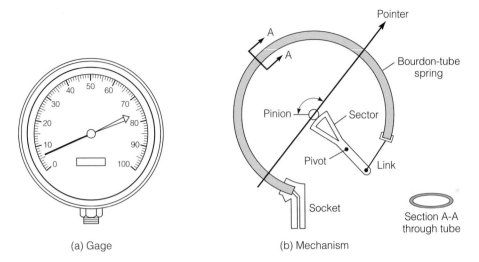

Figure 2.5 ■ Bourdon-
Tube Pressure Gage

(a) Gage   (b) Mechanism

when the tube is empty. When the fluid fills the tube and the pressure is increased, the elliptical cross-section moves toward becoming circular, the tube straightens, and the attached pointer moves to indicate the pressure. The pressure scale on Bourdon-tube pressure gages is generally calibrated by the manufacturer, and the gage is quite reliable as long as it is not subjected to excessive pressure pulses.

Pressure transducers are devices that measure the variations in electrical signals caused by pressure variations. In a typical pressure transducer, one side of a small diaphragm is exposed to the fluid system and the flexure of the diaphragm caused by pressure variations is measured by a sensing element connected to the other side of the diaphragm. There are several types of pressure transducers (Mott, 1994), with a variety of designs for the sensing element. A common type of sensing element is one in which a resistance-wire strain gage is attached to the diaphragm. As the diaphragm flexes, the resistance wire changes length, and the resulting change in resistance is manifested as a measurable change in voltage across the resistance wire. The transducer is calibrated to read pressures rather than voltage.

Manual pressure measurements typically use *manometers*, which relate fluid elevations in attached tubes to pressures in connected fluid systems. The principles of manometry are based on the hydrostatic pressure distribution and are illustrated in the following examples.

■ Example 2.9

The water pressure in a pipeline is to be measured by a U-tube manometer illustrated in Figure 2.6. If the specific weight of the water is 9.79 kN/m³ and the specific weight of the gage fluid is 29.6 kN/m³, determine the water pressure in the pipeline.

Solution

Since the fluids in the manometer tube are not moving, the pressure distribution in the manometer is hydrostatic in accordance with Equation 2.23. From the given data, $\gamma_w = 9.79$ kN/m³, $\gamma_f = 29.6$ kN/m³, and since the pressure at D, p_D, is equal to atmospheric pressure, p_{atm}, the following relationships are derived using the hydrostatic pressure distribution

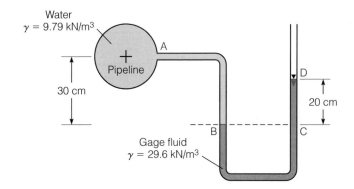

Figure 2.6 ■
U-Tube Manometer

$$p_C = p_D + \gamma_f(0.20) = p_{atm} + \gamma_f(0.20)$$

$$p_B = p_C = p_{atm} + \gamma_f(0.20)$$

$$p_A = p_B - \gamma_w(0.30)$$

$$= p_{atm} + \gamma_f(0.20) - \gamma_w(0.30)$$

Hence the gage pressure in the pipeline at A, $p_A - p_{atm}$, is given by

$$p_A - p_{atm} = \gamma_f(0.20) - \gamma_w(0.30) = 29.6(0.2) - 9.79(0.3) = 2.98 \text{ kPa} \qquad ■$$

■ **Example 2.10**

A differential manometer, which is used to measure the difference in pressure between two water-transmission lines, is illustrated in Figure 2.7. If the specific weight of the water is 9.79 kN/m³ and the specific weight of the gage fluid is 29.6 kN/m³, what is the difference in pressure between the two pipelines?

Solution

From the given data, $\gamma_w = 9.79$ kN/m³, $\gamma_f = 29.6$ kN/m³. The following relationships are derived using the hydrostatic pressure distribution

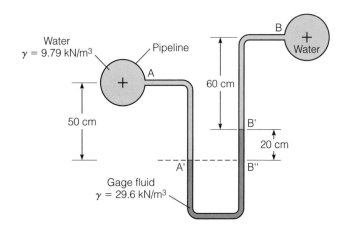

Figure 2.7 ■
Differential Manometer

$$p_{B'} = p_B + \gamma_w(0.60)$$

$$p_{B''} = p_{B'} + \gamma_f(0.20) = p_B + \gamma_w(0.60) + \gamma_f(0.20)$$

$$p_{A'} = p_{B''} = p_B + \gamma_w(0.60) + \gamma_f(0.20)$$

$$p_A = p_{A'} - \gamma_w(0.50) = p_B + \gamma_w(0.60) + \gamma_f(0.20) - \gamma_w(0.50)$$

Hence the difference in pressure between the two pipelines, $p_A - p_B$, is given by

$$p_A - p_B = \gamma_w(0.60) + \gamma_f(0.20) - \gamma_w(0.50)$$

$$= \gamma_w(0.60 - 0.50) + \gamma_f(0.20)$$

$$= (9.79)(0.60 - 0.50) + (29.6)(0.20) = 6.90 \text{ kPa} \qquad ■$$

Manometers are often used to measure the pressure in closed conduits (such as pipes) with flowing fluids. To ensure that the fluid motion in the conduit does not influence the pressure measurement, the manometer tube must be installed in a wall that is parallel to the flow and in such a way that the flow pattern in the conduit is not disturbed. When the attached manometer contains only the fluid in the conduit, the manometer is commonly referred to as a *piezometer*.

2.3.3 ■ Hydrostatic Forces on Plane Surfaces

Consider the general case of a plane (flat) surface within a static fluid, as illustrated in Figure 2.8. The (gage) pressure, p, on the surface area dA located at a vertical distance h below the fluid surface is given by

$$p = \gamma h \qquad (2.28)$$

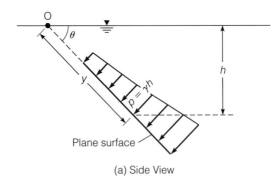

(a) Side View

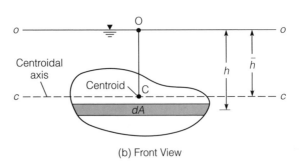

Figure 2.8 ■ Force on a Plane Surface

(b) Front View

It is convenient to utilize the y-coordinate measured along the intersection of the plane surface and the vertical plane as illustrated in Figure 2.8(a). If the y-axis intersects the fluid surface at an angle θ, then the resultant hydrostatic pressure force, F, acting normal to the plane surface area, A, is given by

$$F = \int_A p \, dA$$

$$= \int_A \gamma h \, dA = \int_A \gamma y \sin\theta \, dA = \gamma \sin\theta \int_A y \, dA$$

which can be written as

$$F = \gamma A \bar{y} \sin\theta \tag{2.29}$$

where $\bar{y}$ is the distance to the centroid of the plane surface measured along the y-axis and given by

$$\bar{y} = \frac{1}{A} \int_A y \, dA \tag{2.30}$$

The location of the resultant force, F, that yields the same total moment as the distributed pressure force can be calculated by taking moments about the point O, where the y-axis intersects the fluid surface. The location of F is called the *center of pressure*, y_{cp}, and satisfies the moment equation

$$F y_{cp} = \int_A y p \, dA$$

$$= \int_A y(\gamma y \sin\theta) \, dA$$

$$= \gamma \sin\theta \int_A y^2 \, dA \tag{2.31}$$

Substituting Equation 2.29 for F yields the following expression for the center of pressure,

$$y_{cp} = \frac{1}{A\bar{y}} \int_A y^2 \, dA \tag{2.32}$$

The integral in Equation 2.32 is equal to the *moment of inertia*, I_{oo}, of the plane surface about an axis through O, where

$$I_{oo} = \int_A y^2 \, dA \tag{2.33}$$

The moment of inertia of a plane area is commonly stated relative to an axis passing through the centroid of the area, in which case the moment of inertia depends only on the size and shape of the area. A centroidal axis is illustrated in Figure 2.8(b). The moment of inertia about the centroidal axis, I_{cc}, is related to the moment of inertia, I_{oo}, about another axis by the relation

$$I_{oo} = I_{cc} + Ad^2 \tag{2.34}$$

where d is the distance between the centroidal axis and the other axis. The relation given by Equation 2.34 is commonly called the *parallel axis theorem*. In the present case, $d = \bar{y}$ and Equation 2.34 can be written as

$$I_{oo} = I_{cc} + A\bar{y}^2 \qquad (2.35)$$

Combining Equations 2.32, 2.33, and 2.35 leads to

$$\boxed{y_{cp} = \bar{y} + \frac{I_{cc}}{A\bar{y}}} \qquad (2.36)$$

The moments of inertia of several plane areas about their centroidal axes are given in Appendix C.

■ Example 2.11

A 10-m wide salinity-control gate shown in Figure 2.9 is used to control the movement of saltwater into inland areas. Determine the magnitude and location of the net hydrostatic force on the gate. If the gate is mounted on rollers with a coefficient of friction equal to 0.2, and the gate weighs 10 kN, calculate the force required to lift the gate. The density of seawater at 20°C can be taken as 1,025 kg/m³.

Solution

On the freshwater side of the gate, the total hydrostatic force, F_f, is given by Equation 2.29 as

$$F_f = \gamma_f A_f \bar{y}_f \sin\theta$$

where $\gamma_f = 9.79$ kN/m³, $A_f = (6)(10) = 60$ m², $\bar{y}_f = 0.5(6) = 3$ m, and $\theta = 90°$. Therefore,

$$F_f = (9.79)(60)(3)\sin 90° = 1{,}760 \text{ kN}$$

On the saltwater side, the total hydrostatic force, F_s, is given by

$$F_s = \gamma_s A_s \bar{y}_s \sin\theta$$

where $\gamma_s = (1025)(9.807) = 10050$ N/m³ $= 10.1$ kN/m³, $A_s = (3)(10) = 30$ m², $\bar{y}_s = 0.5(3) = 1.5$ m, and $\theta = 90°$, and therefore

$$F_s = (10.1)(30)(1.5)\sin 90° = 455 \text{ kN}$$

The net hydrostatic force, F, on the gate is given by

$$F = F_f - F_s = 1760 - 455 = 1{,}305 \text{ kN}$$

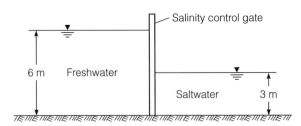

Figure 2.9 ■ Salinity Control Gate

The location, y_{cp}, of the resultant hydrostatic force on each side of the gate is calculated using Equation 2.36, where

$$y_{cp} = \bar{y} + \frac{I_{cc}}{A\bar{y}}$$

On the freshwater side of the gate, $\bar{y} = 3$ m, $I_{cc} = (10)(6)^3/12 = 180$ m^4, $A = 60$ m^2, and therefore the center of pressure, y_{cp}, is given by

$$y_{cp}|_{\text{fresh}} = 3 + \frac{180}{(60)(3)} = 4 \text{ m}$$

On the saltwater side of the gate, $\bar{y} = 1.5$ m, $I_{cc} = (10)(3)^3/12 = 22.5$ m^4, $A = 30$ m^2, and the center of pressure, y_{cp}, is given by to

$$y_{cp}|_{\text{salt}} = 1.5 + \frac{22.5}{(30)(1.5)} = 2 \text{ m}$$

The location of the net hydrostatic force, y_o, can be determined by taking moments about the surface of the freshwater which gives

$$\begin{aligned} Fy_o &= F_f \; y_{cp}|_{\text{fresh}} - F_s\left(3 + y_{cp}|_{\text{salt}}\right) \\ &= (1760)(4) - (455)(3 + 2) \\ &= 4{,}765 \text{ kN m} \end{aligned} \tag{2.37}$$

Since $F = F_f - F_s = 1760 - 455 = 1{,}305$ kN, then

$$y_o = \frac{4765}{1305} = 3.65 \text{ m}$$

which means that the net hydrostatic force of 1,305 kN is located 3.65 m below the freshwater surface.

Since the coefficient of friction, μ, of the rollers is equal to 0.2, the frictional force, F_μ, that must be overcome in lifting the gate is given by

$$F_\mu = \mu F = 0.2(1305) = 261 \text{ kN}$$

The total force to lift the gate is equal to the weight of the gate (10 kN) plus the frictional force, F_μ. Therefore the force required to lift the gate is 10 kN + 261 kN = 271 kN. ■

■ Example 2.12

The rectangular gate shown in Figure 2.10(a) has dimensions 3 m × 2 m and is pin-connected at B. If the surface on which the gate rests at A is frictionless, what is the reaction at A?

Solution

A free-body diagram of the gate is shown in Figure 2.10(b), where B_x and B_y are the components of the support force at B (the pin is assumed to be frictionless, hence there is no fixed moment at B), A_y is the normal reaction of the support surface at A (the surface at A is frictionless, hence the support force is normal to the surface), and F is the hydrostatic force on the gate exerted by the water. The hydrostatic force on the gate is

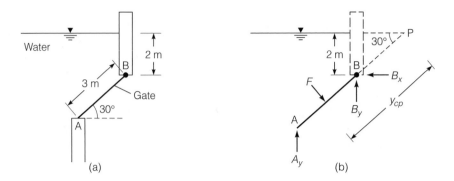

Figure 2.10 ■
Rectangular Gate

the same as if the water surface extends to the point P. Since the gate is rectangular, the distance, $\bar{y}$, of the centroid of the gate from P is given by

$$\bar{y} = PB + \frac{3}{2}\, \text{m} = \frac{2}{\sin 30°} + \frac{3}{2} = 5.5 \text{ m}$$

and the geometric properties of the gate are

$$A = (2)(3) = 6 \text{ m}^2$$

$$I_{xx} = \frac{bd^3}{12} = \frac{(2)(3)^3}{12} = 4.5 \text{ m}^4$$

$$y_{cp} = \bar{y} + \frac{I_{xx}}{A\bar{y}} = 5.5 + \frac{4.5}{(6)(5.5)} = 5.64 \text{ m}$$

The hydrostatic force, F, is therefore given by

$$F = \gamma A \bar{y} \sin \theta$$

$$= (9.79)(6)(5.5) \sin 30° = 162 \text{ kN}$$

The reaction force at A, A_y, can be determined by taking moments about B, in which case

$$F(y_{cp} - PB) = A_y(3 \cos 30°)$$

which gives

$$162 \left(5.64 - \frac{2}{\sin 30°} \right) = A_y(3 \cos 30°)$$

and hence

$$A_y = 102 \text{ kN}$$

The gate therefore exerts a force of 102 kN on the support at A.   ■

2.3.4 ■ Hydrostatic Forces on Curved Surfaces

Consider the general case of a curved surface within a static fluid, as illustrated in Figure 2.11. If the fluid between the curved surface and the fluid surface is taken as a free body, then the (horizontal) hydrostatic pressures above the plane AA on both sides of the free

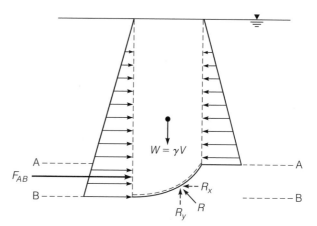

Figure 2.11 ■ Force on a Curved Surface, with Fluid Above Surface

body are equal and opposite. Consequently, the net hydrostatic force, F_{AB}, on the sides of the free body is equal to the hydrostatic force on the free body between plane AA and the bottom of the curved surface (plane BB). The reaction force at the curved surface is equal in magnitude and opposite in direction to the hydrostatic force on the curved surface. Denoting the reaction of the curved surface by R, with components R_x and R_y, then balancing forces in the x-direction yields

$$R_x = F_{AB} = \gamma A_v \bar{y}_v \qquad (2.38)$$

where A_v is the projection of the curved surface in the vertical plane, and $\bar{y}_v$ is the vertical distance from the fluid surface to the centroid of A_v. Balancing forces in the vertical direction yields

$$R_y = W = \gamma V \qquad (2.39)$$

where W is the weight of the fluid between the curved surface and the fluid surface and V is the corresponding volume of fluid. Equations 2.38 and 2.39 collectively give the magnitudes of the components of the hydrostatic force on the curved surface, and the directions of these components are opposite to the directions of the reactions shown in Figure 2.11. Equilibrium of the free body shown in Figure 2.11 requires that F_{AB}, W, and R be concurrent; therefore, the line of action of R passes through the point where F_{AB} and W intersect.

In many cases, the curved surface does not have any fluid directly above the surface, as illustrated in Figure 2.12. Consider the free body containing the volume of fluid between the horizontal and vertical projections of the curved surface. Equilibrium in the x-direction requires that the reaction force on the curved surface, R_x, be equal to the hydrostatic force on the vertical projection of the curved surface, F_v, therefore

$$R_x = F_v = \gamma A_v \bar{y}_v \qquad (2.40)$$

where A_v is the area of the vertical projection of the curved surface, and $\bar{y}_v$ is the distance of the centroid of A_v from the surface. Equilibrium in the vertical (y) direction requires that

$$F_h - W - R_y = 0 \qquad (2.41)$$

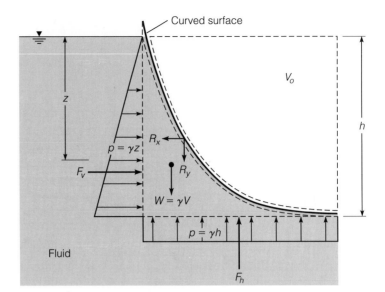

Figure 2.12 ■ Force on a Curved Surface, Without Fluid Above Surface

where F_h is the force exerted on the horizontal projection of the curved surface, and given by

$$F_h = \gamma h A_h \qquad (2.42)$$

where h is the depth of the horizontal projection from the fluid surface and A_h is the area of the horizontal projection. The weight of the fluid in the free body, W, is given by

$$W = \gamma V \qquad (2.43)$$

where V is the volume of the free body. Combining Equations 2.41 to 2.43 yields

$$R_y = F_h - W$$
$$= \gamma h A_h - \gamma V$$
$$= \gamma (h A_h - V) \qquad (2.44)$$

The term in brackets is equal to the volume between the curved surface and the fluid level, indicated in Figure 2.12, and therefore Equation 2.44 can be written as

$$\boxed{R_y = \gamma V_o} \qquad (2.45)$$

where V_o is the volume between the curved surface and the fluid level.

■ Example 2.13

Water is restrained behind a spillway using the radial (Tainter) gate illustrated in Figure 2.13. If the gate is 10 m wide and the water level is at the midpoint of the gate, calculate the hydrostatic force on the gate.

Solution

The horizontal component of the hydrostatic force on the gate, F_x, is given by

$$F_x = \gamma A_v \bar{y}_v$$

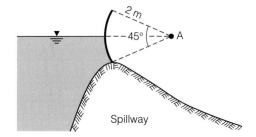

Figure 2.13 ■
Tainter Gate

where $\gamma = 9.79$ kN/m^3, $A_v = (2 \sin 22.5°)(10) = 7.65$ m^2, and $\bar{y}_v = 0.5(2 \sin 22.5°) = 0.383$ m. Therefore

$$F_x = (9.79)(7.65)(0.383) = 28.7 \text{ kN}$$

The vertical component of the hydrostatic force on the gate, F_y, is equal to the weight of the water that would occupy the volume, V_o, illustrated in Figure 2.14(a), where

$$F_y = \gamma V_o$$

Based on the geometry of a circle, illustrated in Figure 2.14(b),

$$V_o = \left[\frac{1}{4} r^2 (\theta - \sin \theta) \right] w$$

where $r = 2$ m, $w = 10$ m (= width of the gate), and $\theta = 45° = \pi/4$ radians. Therefore

$$V_o = \left[\frac{1}{4} (2^2) \left(\frac{\pi}{4} - \sin \frac{\pi}{4} \right) \right] (10) = 7.72 \text{ m}^3$$

which gives

$$F_y = \gamma V_o = (9.79)(7.72) = 75.6 \text{ kN}$$

Area of shaded portion $= \frac{1}{2} r^2 (\theta - \sin \theta)$

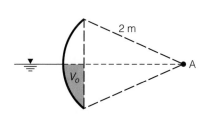

Figure 2.14 ■ Geometric
Properties of Circle

(a) Contributing Volume
(b) Geometry of Circle

The total hydrostatic force, F, is then given by

$$F = \sqrt{F_x^2 + F_y^2} = \sqrt{(28.7)^2 + (75.6)^2} = 80.9 \text{ kN}$$

The Tainter gate must therefore support a hydrostatic force of 80.9 kN.   ▪

Buoyancy.   The net hydrostatic force acting on a submerged body is called the *buoyant force*. Consider the submerged body shown in Figure 2.15, where the volume of fluid above the upper surface of the body, above AB, is equal to V_o, and the volume of the body is V. The hydrostatic force on the upper surface of the submerged body is equal to the weight of the fluid above the upper surface, F_u, hence

$$F_u = \gamma_f V_o \qquad (2.46)$$

where γ_f is the specific weight of the fluid. The hydrostatic force on the lower surface of the submerged body is equal to the weight of the fluid above the lower surface, F_l, where

$$F_l = \gamma_f (V_o + V) \qquad (2.47)$$

The net hydrostatic force, F, on the body is therefore given by

$$F = F_l - F_u = \gamma_f (V_o + V) - \gamma_f V_o = \gamma_f V \qquad (2.48)$$

This result indicates that the buoyant force on a submerged body is equal to the weight of the fluid displaced by the body. This relationship is commonly known as *Archimedes' principle*.[*] The buoyant force, F, acts vertically upward and passes through the centroid of the volume of fluid displaced by the submerged body. The buoyant force is countered by the weight, W, of the body that acts vertically downward, where

$$W = \gamma_s V \qquad (2.49)$$

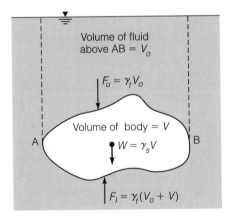

Figure 2.15 ▪ Forces on a Submerged Body

---

[*]The Greek philosopher Archimedes (287–212 B.C.E.) presented the laws of buoyancy and flotation in about 250 B.C.E., and is generally considered the father of hydrostatics.

and γ_s is the specific weight of the submerged body. The net downward force, F_{net}, acting on a submerged body is therefore given by

$$\boxed{F_{net} = (\gamma_s - \gamma_f)V} \tag{2.50}$$

Clearly, a homogeneous body will sink when its density is greater than that of the surrounding fluid ($\gamma_s > \gamma_f$) and will float when its density is less than that of the fluid ($\gamma_s < \gamma_f$). In the case of floating bodies, the buoyant force is equal to the volume of fluid displaced, which is less than the volume of the body.

■ Example 2.14

A spherical sediment particle on the bottom of a river has a diameter of 3 mm and a specific gravity of 2.65. If the coefficient of friction is 0.37, estimate the shear force on the bottom of the channel that is required to move the particle.

Solution

The buoyant force, F, on a spherical particle is given by

$$F = \gamma_f V$$

where $\gamma_f = 9.79$ kN/m^3 and V is given by

$$V = \frac{1}{6}\pi D^3 = \frac{1}{6}\pi(0.003)^3 = 1.41 \times 10^{-8} \text{ m}^3$$

Hence

$$F = (9.79)(1.41 \times 10^{-8}) = 1.38 \times 10^{-7} \text{ kN} = 1.38 \times 10^{-4} \text{ N}$$

The weight, W, of a sediment particle is

$$W = \gamma_s V = 2.65\gamma_f V = 2.65F = 2.65(1.38 \times 10^{-4}) = 3.66 \times 10^{-4} \text{ N}$$

and the net (downward) force, F_{net}, on a sediment particle is therefore

$$F_{net} = W - F = 3.66 \times 10^{-4} - 1.38 \times 10^{-4} = 2.28 \times 10^{-4} \text{ N}$$

The friction force, F_f, corresponding to a net downward force of F_{net} is

$$F_f = \mu_f F_{net}$$

where μ_f is the friction coefficient ($= 0.37$), and hence

$$F_f = (0.37)(2.28 \times 10^{-4}) = 8.44 \times 10^{-5} \text{ N}$$

The shear force needed to move a sediment particle is therefore equal to 8.44×10^{-5} N.

■

■ Example 2.15

A barge is 10 m long, 5 m wide, 3.5 m deep, and weighs 1,500 kN. How much of the barge will be below the water line?

Solution

Let V be the volume of water displaced by the barge, and hence for equilibrium

$$\gamma_f V = W \tag{2.51}$$

where $\gamma_f = 9.79$ kN/m^3, and $W = 1{,}500$ kN. Rearranging Equation 2.51 gives

$$V = \frac{W}{\gamma_f} = \frac{1500}{9.79} = 153 \text{ m}^3$$

If x is the depth of the barge below the water line, then

$$10 \times 5 \times x = 153 \text{ m}^3$$

which gives

$$x = 3.06 \text{ m}$$

Hence 3.06 m of the barge is below water and $3.5 - 3.06 = 0.44$ m is above water.   ■

2.4 Fluid Kinematics

Fluid kinematics is the study of the geometry of fluid motions without regard to the forces producing motion. Fluid motions are generally described in units of length and time, where the length measure is typically associated with location. The kinematic parameters used to describe the motion of fluid elements are position, velocity, and acceleration, where the velocity, $\mathbf{v}$, and acceleration, $\mathbf{a}$, are related to the position, $\mathbf{x}$, by the relations

$$\mathbf{v} = \frac{d\mathbf{x}}{dt} \tag{2.52}$$

$$\mathbf{a} = \frac{d\mathbf{v}}{dt} = \frac{d^2\mathbf{x}}{dt} \tag{2.53}$$

where the quantities in bold are vector quantities.

Flows in which the velocity and acceleration are independent of location are called *uniform flows*, and flows where the velocity and acceleration are a function of location are called *nonuniform flows*. If the velocity and/or acceleration at any location varies with time, then the flow is called *unsteady*, whereas if the velocity and acceleration are independent of time the flow is called *steady*. On the basis of these definitions, it is possible to have such combinations as steady nonuniform flow and unsteady uniform flow. The motion of fluid elements can be viewed from either *Eulerian** or *Lagrangian†* *reference frames*. An Eulerian reference frame is fixed in space, and changes in fluid properties are described at fixed locations. For example, the velocity of a fluid in an Eulerian reference frame is described in the form $\mathbf{v}(\mathbf{x}, t)$, where $\mathbf{x}$ is the location and t is the time. A Lagrangian reference frame moves with a fluid element that contains a fixed mass of fluid, and fluid properties within the fluid element are only observed to change with time. The velocity of a fluid element in a Lagrangian reference frame is described in the form, $\mathbf{V}(t)$, where t is the time.

From a practical viewpoint, we are usually interested in the behavior of fluids at particular locations in space, in which case an Eulerian reference frame is preferable. The

*Named after Leonhard Euler (1707–1783).
†Named after Joseph-Louis Lagrange (1736–1813).

complicating factor in working with Eulerian reference frames is that the fundamental equations of fluid motion and thermodynamics, which are so useful in understanding the behavior of fluids, are all stated for fluid elements in Lagrangian reference frames. For example, Newton's second law states that the sum of the forces on any fluid element (as it moves within the fluid continuum) is equal to the mass of fluid within the fluid element multiplied by the acceleration of the fluid element. To transform the fundamental equations of fluid motion and thermodynamics into useful equations in Eulerian reference frames, it is necessary to understand the relationship between Lagrangian equations and Eulerian equations.

2.4.1 ■ Turbulence

Turbulent flow is characterized by persistent random instabilities in the flow field called turbulence. Turbulence in fluids is a result of spatial variations in velocity called *velocity shear*. The fluid viscosity causes small eddies to form within the fluid as a result of velocity shears. If the inertia of the fluid is sufficiently high, then some of these eddies become unstable and grow to form large-scale disturbances that propagate within the flow field. Conversely, if the inertia of the fluid is relatively low, then the small-scale eddies caused by viscous effects do not become unstable, and large-scale random perturbations in the flow field are not present. Under these circumstances, the flow is called *laminar flow* or *viscous flow*. This view of turbulence in terms of the stability of small-scale eddies is known as the *Tollmien-Schlichting theory*.

The tendency of small-scale eddies to cause turbulence depends on the relative magnitudes of the inertial and viscous forces. The inertial force, F_I, is proportional to the mass times acceleration of a fluid and can be approximated (to within an order of magnitude) by

$$F_I = ma \sim \rho L^3 \frac{V^2}{L} = \rho V^2 L^2 \tag{2.54}$$

where m is the mass of fluid within a fluid system, ρ is the density of the fluid, L is the length scale of the fluid system, and V is a measure of the velocity of the system. The viscous force on the system, F_V, can be estimated from the definition of the viscosity (Equation 2.6) as

$$F_V \sim \mu \frac{V}{L} L^2 = \mu V L \tag{2.55}$$

where μ is the dynamic viscosity of the fluid. The ratio of the inertial to the viscous force is called the *Reynolds number*, Re, and is given by

$$\boxed{Re = \frac{F_I}{F_V} = \frac{\rho V L}{\mu}} \tag{2.56}$$

For higher values of the Reynolds number, inertial forces are much larger than viscous forces and turbulence tends to occur within the fluid. On the other hand, small values of the Reynolds number indicate that viscous forces are comparable to or greater than inertial forces, indicating that turbulence is less likely to occur within the fluid. Turbulence has a significant effect on energy losses within a flowing fluid, and the values of Re at which turbulent flow occurs depends on the geometry of the flow and the characteristic length scale. In the case of pipe flow, where the diameter of the pipe is the characteristic length scale, for Re $\leq$ 2,000 the flow is laminar, for 2,000 < Re < 4,000 there is a gradual change to turbulent flow, and for Re > 4,000 the flow is turbulent. In the case of open-channel flow, such as in canals and rivers, the depth of flow is used as the characteristic length scale, and open-channel flows are turbulent for Re ≥ 1,000. In most engineering applications involving closed-conduit and open-channel flow, the Reynolds number limits are far exceeded and the flows are fully turbulent.

2.4.2 ■ Reynolds Transport Theorem

Lagrangian equations describe the behavior of a fluid within a system, whereas Eulerian equations describe the behavior of fluid within a control volume. A *system* is defined as a collection of matter of fixed identity that may move, flow, and interact with its surroundings (Young et al., 1997). In other words, a fluid system always contains the same set of atoms. In contrast to a fluid system, which moves around within a fluid continuum, a *control volume* is a fixed volume in space through which fluid may flow both in and out. The *Reynolds transport theorem* relates the rate of change of some property in a fluid system to the rate of change of that same property in a control volume containing the fluid.

Consider the control volume illustrated in Figure 2.16 and define a fluid system as the fluid contained in the control volume at time t. In other words, at time t, the fluid in the control volume and the fluid in the system are the same. During a subsequent time interval, Δt, a volume V_I of fluid enters the control volume, a volume V_{II} of fluid remains in the control volume, and a volume V_{III} of fluid leaves the control volume. Defining the amount of a fluid property within the system at time t as $B(t)$, then the change in B over the time interval Δt is given by

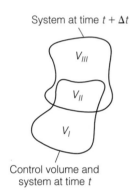

System at time $t + \Delta t$

V_{III}

V_{II}

V_I

Control volume and
system at time t

Figure 2.16 ■ System and Control Volume

$$B(t + \Delta t) - B(t) = \left(\int_{V_{II}} \rho\beta \, dV + \int_{V_{III}} \rho\beta \, dV \right)_{t+\Delta t} - \left(\int_{V_{II}} \rho\beta \, dV \right)_t \quad (2.57)$$

where ρ is the density of the fluid and β is the amount of fluid property per unit mass of the fluid. Adding and subtracting the quantity $\left(\int_{V_I} \rho\beta \, dV \right)_{t+\Delta t}$ and dividing by Δt yields

$$\frac{B(t + \Delta t) - B(t)}{\Delta t} = \frac{\left(\int_{V_{II}} \rho\beta \, dV + \int_{V_I} \rho\beta \, dV \right)_{t+\Delta t} - \left(\int_{V_{II}} \rho\beta \, dV \right)_t}{\Delta t}$$

$$+ \frac{\left(\int_{V_{III}} \rho\beta \, dV \right)_{t+\Delta t}}{\Delta t} - \frac{\left(\int_{V_I} \rho\beta \, dV \right)_{t+\Delta t}}{\Delta t} \quad (2.58)$$

Taking the limit as $\Delta t \to 0$, the term on the left hand side of Equation 2.58 becomes dB/dt, the first term on the right hand side is equal to the rate of change of the amount of fluid property in the control volume, and the last two terms are collectively equal to the net rate at which the fluid property leaves the control volume. Therefore, in the limit as $\Delta t \to 0$, Equation 2.58 becomes

$$\boxed{\frac{dB}{dt} = \frac{\partial}{\partial t} \int_V \beta\rho \, dV + \int_A \beta\rho \mathbf{v} \cdot \mathbf{n} \, dA} \quad (2.59)$$

where B is the total amount of property in the fluid system, dB/dt is the rate of change of B in the fluid system, V is the control volume, A is the surface area of the control volume, $\mathbf{v}$ is the velocity field, and $\mathbf{n}$ is the unit normal directed outward from the control

surface. Equation 2.59 is called the Reynolds transport theorem and provides a fundamental relationship between the rate of change of a fluid property in a fluid system and
the properties within a control volume.

2.5 Fluid Dynamics

The study of fluid motions within defined control volumes are typically of interest to
water-resources engineers. Equations that must be satisfied by all control volumes are
(1) the law of conservation of mass, also known as the continuity equation, (2) the law
of conservation of momemtum, also known as Newton's second law, and (3) the law of
conservation of energy, also known as the first law of thermodynamics.

2.5.1 ■ Conservation of Mass

The law of conservation of mass requires that the mass of fluid within a system remains
constant. If B is defined as the mass of fluid within a system, then the law of conservation
of mass can be written as

$$\frac{dB}{dt} = 0 \tag{2.60}$$

Combining Equations 2.59 and 2.60 and noting that $\beta = 1$ by definition (= mass of
fluid per unit mass of fluid) lead to the following expression for the law of conservation
of mass for a control volume contained within the fluid

$$\boxed{\frac{\partial}{\partial t} \int_V \rho \, dV + \int_A \rho \mathbf{v} \cdot \mathbf{n} \, dA = 0} \tag{2.61}$$

where ρ is the fluid density, $\mathbf{v}$ is the velocity field, $\mathbf{n}$ is the unit normal pointing out of
the control volume, V denotes the control volume, and A denotes the surface area of
the control volume. The terms in the continuity equation state that the rate of change
of fluid mass within the control volume plus the net flux of fluid mass out of the control
volume is equal to zero. Upon rearrangement, Equation 2.61 also states that the rate of
change of fluid mass within a control volume is equal to the net influx of mass into the
control volume. In cases where the fluid density is constant, which is usually the case in
dealing with water, then the continuity equation becomes

$$\boxed{\int_A \mathbf{v} \cdot \mathbf{n} \, dA = 0} \tag{2.62}$$

which states that the net flux of fluid out of the control volume is equal to zero, or simply
that the fluid mass does not accumulate within the control volume.

■ Example 2.16

The velocity distribution, $u(r)$, in pipes with circular diameters can be approximated by
an equation of the form

$$u(r) = V_{\max}\left(1 - \frac{r^2}{R^2}\right)$$

where r is the radial distance from the centerline of the pipe, $V_{\max}$ is the maximum (centerline) velocity, and R is the radius of the pipe. If water enters a reservoir through a 100-mm diameter pipe with a maximum velocity of 1 m/s and leaves the reservoir through a 75-mm diameter pipe with a maximum velocity of 0.8 m/s, what is the rate at which water is accumulating in the reservoir?

Solution

Defining the reservoir as the control volume, the continuity equation given by Equation 2.61 can be written as

$$\frac{\partial}{\partial t} \int_{V_{\text{res}}} \rho \, dV + \int_{A_{\text{in}}} \rho \mathbf{v} \cdot \mathbf{n} \, dA + \int_{A_{\text{out}}} \rho \mathbf{v} \cdot \mathbf{n} \, dA = 0 \tag{2.63}$$

where V_{res} is the volume of the reservoir, A_{in} and A_{out} are the cross-sectional areas of the inflow and outflow pipes respectively, and ρ is the density of water (= 998 kg/m³). Evaluating the mass fluxes into and out of the reservoir yields

$$\int_{A_{\text{in}}} \rho \mathbf{v} \cdot \mathbf{n} \, dA = -\rho \int_0^{R_{\text{in}}} V_{\text{in}} \left(1 - \frac{r^2}{R_{\text{in}}^2}\right) 2\pi r \, dr = -\pi \rho V_{\text{in}} R_{\text{in}}^2 \tag{2.64}$$

and

$$\int_{A_{\text{out}}} \rho \mathbf{v} \cdot \mathbf{n} \, dA = \rho \int_0^{R_{\text{out}}} V_{\text{out}} \left(1 - \frac{r^2}{R_{\text{out}}^2}\right) 2\pi r \, dr = \pi \rho V_{\text{out}} R_{\text{out}}^2 \tag{2.65}$$

where V_{in} and V_{out} are the maximum velocities in the inflow and outflow pipes respectively, and R_{in} and R_{out} are the inflow and outflow pipe radii. To understand the negative sign in Equation 2.64, it should be noted that where the inflow pipe enters the reservoir, the velocity $\mathbf{v}$ is directed into the reservoir and the outward normal $\mathbf{n}$ is directed outward, and hence $\mathbf{v} \cdot \mathbf{n} = -v$, where v is the magnitude of the inflow velocity. Similarly, in Equation 2.65, it should be noted that where the outflow pipe leaves the reservoir, $\mathbf{v}$ is directed outward from the reservoir and $\mathbf{n}$ is also directed outward, hence $\mathbf{v} \cdot \mathbf{n} = +v$, where v is the magnitude of the outflow velocity. Substituting Equations 2.64 and 2.65 into Equation 2.63 yields the following expression for the continuity equation

$$\frac{\partial}{\partial t} \int_{V_{\text{res}}} \rho \, dV = \pi \rho \left[V_{\text{in}} R_{\text{in}}^2 - V_{\text{out}} R_{\text{out}}^2\right]$$

$$= \pi (998) \left[(1)(0.05)^2 - (0.8)(0.0375)^2\right]$$

$$= 4.31 \text{ kg/s}$$

Water is therefore accumulating in the reservoir at the rate of 4.31 kg/s. ■

2.5.2 ■ Conservation of Momentum

The law of conservation of momentum requires that the sum of the forces acting on a fluid system be equal to the rate of change of momentum of the system. If B is defined as the total momentum of the system, then Newton's second law can be stated as

$$\sum \mathbf{F} = \frac{dB}{dt} \tag{2.66}$$

where $\sum \mathbf{F}$ is the sum of the forces acting on the system. Since the momentum per unit mass, β, is equal to the fluid velocity, $\mathbf{v}$, then Equations 2.59 and 2.66 can be combined to the following expression for the law of conservation of momentum

$$\boxed{\sum \mathbf{F} = \frac{\partial}{\partial t} \int_V \rho \mathbf{v}\, dV + \int_A \rho \mathbf{v}\, \mathbf{v} \cdot \mathbf{n}\, dA}$$ (2.67)

which can be written in the component form

$$\sum F_x = \frac{\partial}{\partial t} \int_V \rho v_x\, dV + \int_A \rho v_x \mathbf{v} \cdot \mathbf{n}\, dA$$

$$\sum F_y = \frac{\partial}{\partial t} \int_V \rho v_y\, dV + \int_A \rho v_y \mathbf{v} \cdot \mathbf{n}\, dA$$

$$\sum F_z = \frac{\partial}{\partial t} \int_V \rho v_z\, dV + \int_A \rho v_z \mathbf{v} \cdot \mathbf{n}\, dA$$

where v_x, v_y, and v_z are components of the fluid velocity.

■ Example 2.17

Water flowing at 30 m³/s in a 10-m wide rectangular channel passes under an open gate as shown in Figure 2.17(a). If the shear stress on the bottom of the channel can be neglected, calculate the force exerted by the water on the gate.

Solution

A control volume between the upstream section (Section 1) and the downstream section (Section 2) is illustrated in Figure 2.17(b), where the pressure force on Section 1 is P_1, the pressure force on Section 2 is P_2, and the reaction of the gate to the force exerted by the water is F. Applying the steady-state momentum equation in the x-direction (= flow direction), and assuming a uniform velocity distribution at the inflow and

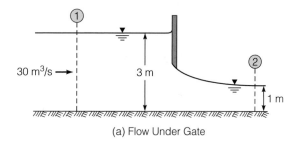

(a) Flow Under Gate

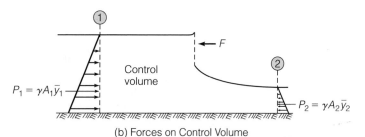

Figure 2.17 ■ Flow Under an Open Gate

(b) Forces on Control Volume

outflow sections, yields

$$P_1 - P_2 - F = \int_{A_1} \rho v_x \mathbf{v} \cdot \mathbf{n} \, dA + \int_{A_2} \rho v_x \mathbf{v} \cdot \mathbf{n} \, dA$$

$$= \rho V_1(-V_1)A_1 + \rho V_2(V_2)A_2$$

$$= -\rho V_1^2 A_1 + \rho V_2^2 A_2$$

where V_1 and V_2 are the average velocities at Sections 1 and 2 and A_1 and A_2 are the surface areas of Sections 1 and 2. Note that since the normal vector $\mathbf{n}$ points out of the control volume, $\mathbf{v} \cdot \mathbf{n} = -V_1$ on A_1, and $\mathbf{v} \cdot \mathbf{n} = V_2$ on A_2. The continuity equation requires that

$$V_1 = \frac{Q}{A_1} = \frac{30}{(3)(10)} = 1 \text{ m/s} \quad \text{and} \quad V_2 = \frac{Q}{A_2} = \frac{30}{(1)(10)} = 3 \text{ m/s}$$

where

$$A_1 = (3)(10) = 30 \text{ m}^2 \quad \text{and} \quad A_2 = (1)(10) = 10 \text{ m}^2$$

Because the upstream and downstream surfaces (Sections 1 and 2) are vertical plane surfaces, the resultant forces are given by

$$P_1 = \gamma A_1 \bar{y}_1 \quad \text{and} \quad P_2 = \gamma A_1 \bar{y}_2$$

where $\bar{y}_1$ and $\bar{y}_2$ are the depths to the centroids at Sections 1 and 2 respectively. In the present case, $\bar{y}_1 = 1.5$ m and $\bar{y}_2 = 0.5$ m. The momentum equation can now be written as

$$\gamma A_1 \bar{y}_1 - \gamma A_2 \bar{y}_2 - F = \rho V_2^2 A_2 - \rho V_1^2 A_1$$

or

$$(9790)(30)(1.5) - (9790)(10)(0.5) - F = (998)(3^2)(10) - (998)(1^2)(30)$$

which yields

$$F = 332 \times 10^3 \text{ N} = 332 \text{ kN}$$

Therefore the force exerted by the water on the gate is 332 kN.   ■

The momentum equation is commonly applied to determine the force on various components of hydraulic systems, such as deflectors and blades in turbomachines and pipe bends, enlargements, and contractions in pipeline systems. In most cases of practical interest, the reaction force of the hydraulic structure balances the rate of change of momentum as the fluid passes over the surface.

2.5.2.1 Forces on Deflectors and Blades

Deflectors and blades are commonly used to adjust the flow direction (deflectors) and convert hydraulic energy to mechanical energy (blades). The forces exerted on deflectors and blades are of interest for a variety of reasons, such as to ensure that adequate structural support is provided or to determine the mechanical energy that can be extracted from the flow.

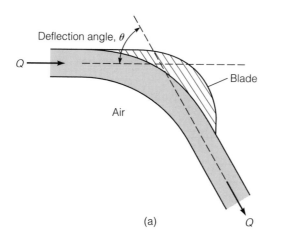

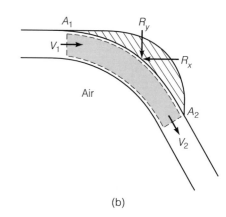

(a)   (b)

Figure 2.18 ■ Force on Stationary Blade

Stationary Blade. Consider the case of a fluid jet deflected by a stationary blade or vane as illustrated in Figure 2.18(a). Using the control volume shown in Figure 2.18(b), the x- and y-components of the steady state momentum equation can be written as

$$\sum F_x = \int_A \rho v_x \mathbf{v} \cdot \mathbf{n} \, dA \tag{2.68}$$

$$\sum F_y = \int_A \rho v_y \mathbf{v} \cdot \mathbf{n} \, dA \tag{2.69}$$

Applying the x-component of the momentum equation (Equation 2.68) to the control volume shown in Figure 2.18(b) gives

$$-R_x = \int_{A_1} \rho V_1(-V_1) \, dA + \int_{A_2} \rho (V_2 \cos\theta)(V_2) \, dA = -\rho A_1 V_1^2 + \rho A_2 V_2^2 \cos\theta \tag{2.70}$$

where R_x is the x-component of the reaction force on the control volume, A_1 and A_2 are the inflow and outflow areas of the control volume respectively, V_1 and V_2 are the inflow and outflow velocities, and θ is the deflection angle. Equation 2.70 assumes that the blade is in the horizontal plane, in which case the weight of the fluid in the control volume does not contribute to the force balance. External pressures on the control volume are also neglected, since the fluid is surrounded by air at atmospheric pressure. The continuity equation requires that the volumetric flowrate, Q, into the control volume is equal to the volumetric flowrate out of the control volume, hence

$$Q = A_1 V_1 = A_2 V_2 \tag{2.71}$$

Combining Equations 2.70 and 2.71 and simplifying yields the following relation

$$\boxed{R_x = \rho Q(V_1 - V_2 \cos\theta)} \tag{2.72}$$

Applying the y-component of the momentum equation (Equation 2.69) gives

$$-R_y = \int_{A_2} \rho(-V_2 \sin\theta)(V_2) \, dA = \rho A_2 (-V_2 \sin\theta) V_2 \tag{2.73}$$

which simplifies to

$$\boxed{R_y = \rho Q V_2 \sin\theta}$$

(2.74)

Since R_x and R_y are the reactions of the blade, the force exerted by the fluid on the blade are equal and opposite to R_x and R_y.

Moving Blade.   Consider the case of a fluid jet deflected by a moving blade as illustrated in Figure 2.19(a), where the blade moves with a velocity V_b. As long as the blade moves with a constant velocity, the momentum equation is applicable to the moving reference frame. The control volume viewed relative to the moving blade, is shown in Figure 2.19(b), and the x-component of the momentum equation (Equation 2.68) can be written as

$$-R_x = \int_{A_1} \rho V_{1r}(-V_{1r})\, dA + \int_{A_2} \rho(V_{2r}\cos\theta)(V_{2r})\, dA = -\rho A_1 V_{1r}^2 + \rho A_2 V_{2r}^2 \cos\theta$$

(2.75)

where the V_{1r} and V_{2r} are the relative fluid velocities across the inflow and outflow boundaries of the control volume, and A_1 and A_2 are the areas of the inflow and outflow surfaces. In terms of the absolute velocity of the impinging jet, V_1, and the velocity of the blade, V_b, the relative velocities can be written as

$$V_{1r} = V_1 - V_b$$

(2.76)

$$V_{2r} = V_{1r}\frac{A_1}{A_2}$$

(2.77)

where Equation 2.77 is derived by applying the continuity equation to the control volume. Combining Equations 2.75 to 2.77 yields

$$\boxed{R_x = \rho Q'(V_{1r} - V_{2r}\cos\theta)}$$

(2.78)

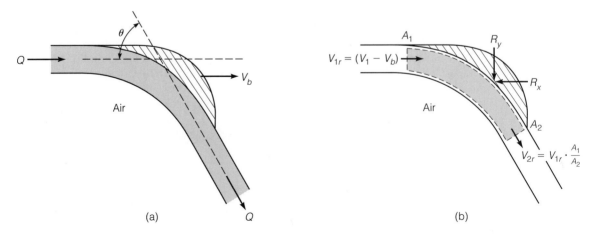

(a)

(b)

Figure 2.19 ■ Force on Moving Blade

where

$$Q' = A_1 V_{1r} \tag{2.79}$$

Applying the y-component of the momentum equation (Equation 2.69) to the moving control volume gives

$$-R_y = \int_{A_2} \rho(-V_{2r}\sin\theta)V_{2r}\, dA = \rho A_2(-V_{2r}\sin\theta)V_{2r} \tag{2.80}$$

which simplifies to

$$\boxed{R_y = \rho Q' V_{2r}\sin\theta} \tag{2.81}$$

Again, since R_x and R_y are the reactions of the blade, the force exerted by the fluid on the blade are equal and opposite to R_x and R_y.

■ Example 2.18

A jet of water having a velocity of 15 m/s impinges on a stationary vane whose section is in the form of a circular arc. The vane deflects the jet through an angle of 120°, as illustrated in Figure 2.20. (a) Find the magnitude and direction of the force on the vane when the jet discharge is 0.45 kg/s. (b) If the vane moves with a velocity of 6 m/s in the direction of the jet and the velocity relative to the moving vane remains constant, determine the power delivered by the jet to the moving blade.

Solution

(a) In the case of a stationary vane, the x-component of the reaction, R_x, is given by Equation 2.72 as

$$R_x = \rho Q(V_1 - V_2\cos\theta)$$

where $\rho Q = 0.45$ kg/s, $V_1 = V_2 = 15$ m/s, and $\theta = 120°$. Substituting these values gives

$$R_x = (0.45)(15 - 15\cos 120°) = 10.1 \text{ N}$$

The y-component of the reaction is given by Equation 2.74 as

$$R_y = \rho Q V_2\sin\theta = (0.45)(15)\sin 120° = 5.85 \text{ N}$$

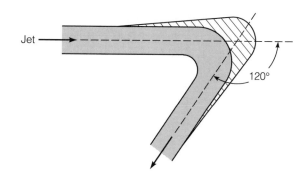

Figure 2.20 ■ Force on Vane

The force exerted by the water on the vane is equal and opposite to the reaction on the vane. The magnitude of the force on the vane, R, is given by

$$R = \sqrt{R_x^2 + R_y^2} = \sqrt{10.1^2 + 5.85^2} = 11.7 \text{ N}$$

and the angle with the horizontal, α, is given by

$$\alpha = \tan^{-1}\left(\frac{5.85}{10.1}\right) = 30.1°$$

(b) If the vane moves at 6 m/s, the x-component of the reaction, R_x, is given by Equation 2.78 as

$$R_x = \rho Q'(V_{1r} - V_{2r}\cos\theta)$$

where the mass flowrate over the moving vane is

$$\rho Q' = 0.45 \times \frac{15 - 6}{15} = 0.27 \text{ kg/s}$$

and $V_{1r} = 15 - 6 = 9$ m/s, $V_{2r} = V_{1r} = 9$ m/s, and $\theta = 120°$. Substituting these values gives

$$R_x = 0.27(9 - 9\cos 120°) = 3.65 \text{ N}$$

The power, P, delivered by the jet to the moving blade is given by

$$P = R_x V_b$$

where $V_b = 6$ m/s, and therefore

$$P = (3.65)(6) = 21.9 \text{ N·m} = 21.9 \text{ W}$$ ■

2.5.2.2 Forces on Pressure Conduits

Pressure conduits are commonplace in water-distribution systems. Components of these systems that change the velocity or flow direction must be designed with sufficient structural support to withstand dynamic forces. Two common components of water-distribution systems are reducers and bends.

Forces on Reducers.   Consider the case of a fluid flowing through the reducer shown in Figure 2.21(a). Upstream of the reducer, the cross-sectional area is A_1, the average velocity is V_1, the pressure is p_1, and the corresponding variables downstream of the reducer are A_2, V_2, and p_2. A control volume within the reducer section is shown in

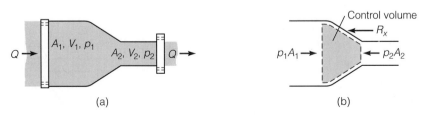

Figure 2.21 ■ Force on Reducer

(a)   (b)

Figure 2.21(b), where the forces acting on the fluid in the control volume are the reaction, R_x, of the reducer, and the pressure forces $p_1 A_1$ and $p_2 A_2$. The x-component of the steady-state momentum equation is

$$\sum F_x = \int_A \rho v_x \mathbf{v} \cdot \mathbf{n} \, dA \qquad (2.82)$$

which, in this case, becomes

$$p_1 A_1 - p_2 A_2 - R_x = \rho V_2^2 A_2 - \rho V_1^2 A_1 \qquad (2.83)$$

Equation 2.83 assumes that the reducer is in the horizontal plane, in which case the weight of the fluid in the control volume does not have a component on the flow direction. Equation 2.83 simplifies to

$$\boxed{R_x = p_1 A_1 - p_2 A_2 - \rho Q (V_2 - V_1)} \qquad (2.84)$$

where Q is the volumetric flowrate given by

$$Q = A_1 V_1 = A_2 V_2 \qquad (2.85)$$

The reaction force, R_x, is equal and opposite to the force exerted by the fluid on the reducer, and a sufficient support force must be provided to keep the reducer in place. Lateral forces on the reducer, normal to R_x, are small because of the symmetry of the reducer.

Forces on Bends.   Consider a fluid flowing through the bend illustrated in Figure 2.22(a). Upstream of the bend, the cross-sectional area is A_1, the average velocity is V_1, the pressure is p_1, and the corresponding variables downstream of the bend are A_2, V_2, and p_2. A control volume within the bend is shown in Figure 2.22(b), where the forces acting on the fluid in the control volume are the reaction components, R_x and R_z, the pressure forces, $p_1 A_1$ and $p_2 A_2$, and the weight, W, of fluid within the control volume. This assumes that the bend is in the vertical plane. The x-component of the steady-state momentum equation is

$$\sum F_x = \int_A \rho v_x \mathbf{v} \cdot \mathbf{n} \, dA \qquad (2.86)$$

which, in this case, becomes

$$p_1 A_1 - p_2 A_2 \cos \theta - R_x = \rho (V_2 \cos \theta) V_2 A_2 - \rho V_1^2 A_1 \qquad (2.87)$$

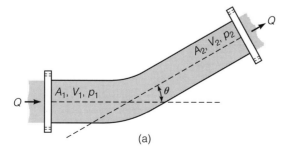

(a)   (b)

Figure 2.22 ■ Force on Bend

which simplifies to

$$\boxed{R_x = p_1 A_1 - p_2 A_2 \cos\theta - \rho Q(V_2 \cos\theta - V_1)} \tag{2.88}$$

where Q is the volumetric flowrate given by

$$Q = V_1 A_1 = V_2 A_2 \tag{2.89}$$

The z-component of the steady-state momentum equation is

$$\sum F_z = \int_A \rho v_z \mathbf{v} \cdot \mathbf{n}\, dA \tag{2.90}$$

which in this case becomes

$$R_z - p_2 A_2 \sin\theta - W = \rho(V_2 \sin\theta) V_2 A_2 \tag{2.91}$$

and simplifies to

$$\boxed{R_z = p_2 A_2 \sin\theta + W + \rho Q V_2 \sin\theta} \tag{2.92}$$

As noted in the case of a reducer, the reaction force is equal and opposite to the force exerted on the bend by the fluid, and sufficient supporting force must be provided to keep the bend in place.

▪ Example 2.19

Water under a pressure of 350 kPa flows with a velocity of 3 m/s through a 90° bend in the horizontal plane. If the bend has a uniform diameter of 300 mm, and assuming no drop in pressure, calculate the force required to keep the bend in place.

Solution

The components of the reaction force can be calculated using Equations 2.88 and 2.92, where the weight component of the fluid in the horizontal plane is zero. Equation 2.88 gives

$$R_x = p_1 A_1 - p_2 A_2 \cos\theta - \rho Q(V_2 \cos\theta - V_1)$$

where $p_1 = p_2 = 350$ kPa $= 3.5 \times 10^5$ Pa, $A_1 = A_2 = \pi(0.3)^2/4 = 0.0707$ m^2, $\theta = 90°$, $\rho = 998$ kg/m^3, $V_1 = V_2 = 3$ m/s, and $Q = A_1 V_1 = (0.0707)(3) = 0.212$ m^3/s. Substituting these values gives

$$R_x = (3.50 \times 10^5)(0.0707) - (350 \times 10^5)(0.0707)\cos 90°$$

$$- (998)(0.212)(3\cos 90° - 3)$$

$$= 25400\text{ N} = 25.4\text{ kN}$$

The y-component of the reaction force is given by Equation 2.92 as

$$R_z = p_2 A_2 \sin\theta + \rho Q V_2 \sin\theta$$

$$= (3.50 \times 10^5)(0.0707)\sin 90° + (998)(0.212)(3)\sin 90°$$

$$= 25400\text{ N} = 25.4\text{ kN}$$

The magnitude, R, of the force required to keep the bend in place is

$$R = \sqrt{R_x^2 + R_y^2} = \sqrt{25.4^2 + 25.4^2} = 35.9 \text{ kN}$$

■

Forces on Junctions. In the case of junctions, where multiple flow conduits intersect, it is usually easier to determine the reaction forces directly from the steady-state momentum equation, rather than using generalized formulae. This approach is illustrated by the following example.

■ Example 2.20

Several water pipes intersect at the junction box illustrated in Figure 2.23. Determine the force required to keep the junction in place.

Solution

The x-component of the steady-state momentum equation is

$$\sum F_x = \int_A \rho v_x \mathbf{v} \cdot \mathbf{n} \, dA$$

which in this case can be written as

$$R_x + p_A A_A \sin 40° - p_B A_B \cos 30° - p_C A_C \sin 50° - p_D A_D \cos 60°$$
$$= \rho(V_A \sin 40°)(-Q_A) + \rho(-V_B \cos 30°)(-Q_B)$$
$$+ \rho(V_C \sin 50°)(Q_C) + \rho(V_D \cos 60°)(Q_D) \qquad (2.93)$$

The velocities in each pipe are calculated using the relation $V = Q/A = Q/(\pi D^2/4)$. Using the flowrates, Q, and diameters, D, given in Figure 2.23 yields

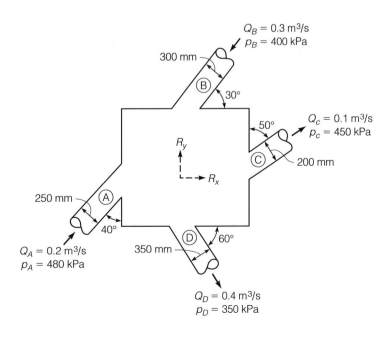

Figure 2.23 ■ Force on Junction

Pipe	Q (m³/s)	D (m)	A (m²)	V (m/s)
A	0.2	0.25	0.0491	4.07
B	0.3	0.30	0.0707	4.24
C	0.1	0.20	0.0314	3.18
D	0.4	0.35	0.0962	4.16

Substituting into Equation 2.93 gives

$$R_x + (480)(0.0491)\sin 40° - (400)(0.0707)\cos 30°$$
$$- (450)(0.0314)\sin 50° - (350)(0.0962)\cos 60°$$
$$= (0.998)[(4.07)\sin 40°](-0.2) + (0.998)[(-4.24)\cos 30°](-0.3)$$
$$+ (0.998)[(3.18)\sin 50°](0.1) + (0.998)[(4.16)\cos 60°](0.4)$$

which leads to

$$R_x = 38.7 \text{ kN}$$

The y-component of the steady-state momentum equation is

$$\sum F_y = \int_A \rho v_y \mathbf{v} \cdot \mathbf{n} \, dA$$

which becomes

$$R_y + p_A A_A \cos 40° - p_B A_B \sin 30° - p_C A_C \cos 50° + p_D A_D \sin 60°$$
$$= \rho(V_A \cos 40°)(-Q_A) + \rho(-V_B \sin 30°)(-Q_B)$$
$$+ \rho(V_C \cos 50°)(Q_C) + \rho(-V_D \sin 60°)(Q_D) \tag{2.94}$$

Substituting the given parameters into Equation 2.94 gives

$$R_y + (480)(0.0491)\cos 40° - (400)(0.0707)\sin 30°$$
$$- (450)(0.0314)\cos 50° + (350)(0.0962)\sin 60°$$
$$= (0.998)[(4.07)\cos 40°](-0.2) + (0.998)[(-4.24)\sin 30°](-0.3)$$
$$+ (0.998)[(3.18)\cos 50°](0.1) + (0.998)[(-4.16)\sin 60°](0.4)$$

which leads to

$$R_y = -25.3 \text{ kN}$$

The force, R, required to keep the junction in place is therefore given by

$$R = \sqrt{R_x^2 + R_y^2} = \sqrt{38.7^2 + 25.3^2} = 46.2 \text{ kN} \qquad ■$$

2.5.2.3   Water Hammer

The sudden closure of valves in pipeline systems can result in significant pressure increases through a process called *water hammer*. Consider the situation in Figure 2.24,

Figure 2.24 ■
Pressure Wave

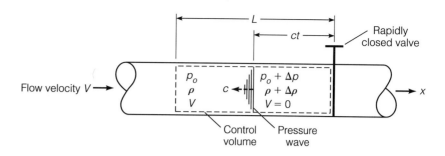

where the flow of a fluid in a pipe is halted by the rapid closure of a valve and the pipe walls are *rigid*, indicating that the pipe walls do not flex in response to pressure changes. Before the valve is closed, the velocity in the pipe is V, the fluid pressure is p_o, and the density of the fluid is ρ. After the valve is closed, the fluid adjacent to the valve is immediately halted and the effect of valve closure is propagated upstream by a pressure wave that moves with a velocity c. Behind the pressure wave, the velocity is equal to zero, the fluid pressure is $p_o + \Delta p$, and the fluid density is $\rho + \Delta \rho$, whereas in front of the pressure wave the velocity is V, the pressure is p_o, and the density is ρ. For the control volume illustrated in Figure 2.24, the momentum equation in the flow direction is

$$\sum F_x = \frac{\partial}{\partial t} \int_{\mathcal{V}} \rho v_x \, d\mathcal{V} + \int_A \rho v_x \mathbf{v} \cdot \mathbf{n} \, dA \tag{2.95}$$

Neglecting the shear resistance on the pipe boundary, Equation 2.95 can be written as

$$p_o A - (p_o + \Delta p)A = \frac{\partial}{\partial t}[\rho V(L - ct)A] + \rho V(-VA) \tag{2.96}$$

where A is the cross-sectional area of the pipe and L is the length of the control volume. Simplifying Equation 2.96 gives

$$(-\Delta p)A = \frac{\partial}{\partial t}(\rho LAV - \rho ctAV) - \rho V^2 A$$

$$\Delta p A = \rho c\,AV + \rho V^2 A$$

$$\Delta p = \rho c\,V + \rho V^2 \tag{2.97}$$

which relates the pressure increase, Δp, associated with sudden valve closure to the fluid and flow properties. The pressure change Δp is commonly called the *water-hammer pressure*. In most cases, the velocity of the pressure wave, c, is much higher than the fluid velocity, V, and Equation 2.97 can be approximated as

$$\boxed{\Delta p = \rho c\,V} \tag{2.98}$$

Applying the continuity equation to the control volume shown in Figure 2.24 gives

$$\frac{\partial}{\partial t} \int_{\mathcal{V}} \rho \, d\mathcal{V} + \int_A \rho \mathbf{v} \cdot \mathbf{n} \, dA = 0 \tag{2.99}$$

which in this case yields

$$\frac{\partial}{\partial t}[\rho(L-ct)A + (\rho+\Delta\rho)(ct)A] + \rho(-V)A = 0$$

$$\frac{\partial}{\partial t}[\rho LA + \Delta\rho ctA] - \rho VA = 0$$

$$\Delta\rho cA - \rho VA = 0 \qquad (2.100)$$

which simplifies to

$$\frac{\Delta\rho}{\rho} = \frac{V}{c} \qquad (2.101)$$

Recalling the definition of the bulk modulus, E_v, as

$$E_v = \frac{\Delta p}{\Delta\rho/\rho} \qquad (2.102)$$

then Equations 2.101 and 2.102 combine to give

$$c = \frac{VE_v}{\Delta p} \qquad (2.103)$$

Substituting Equation 2.98 into Equation 2.103 yields wave speed, c, in terms of the fluid properties as

$$\boxed{c = \sqrt{\frac{E_v}{\rho}}} \qquad (2.104)$$

The wave speed, c, is also equal to the speed of sound in the fluid, since sound waves are in fact pressure waves. In the case of water at 20°C, $E_v = 2.15 \times 10^6$ kPa, $\rho = 998$ kg/m^3, and therefore $c = 1{,}470$ m/s. Clearly, the approximation that $V \ll c$ is reasonable for water in all practical cases. The foregoing analysis has assumed that the pipe walls are rigid. If the pipe walls are slightly deformable, the wave speed, c, is given by (Roberson et al., 1998)

$$c = \sqrt{\frac{E_v/\rho}{1 + (KD/eE)}}$$

where D is the diameter of the pipe, e is the wall thickness, and E is the modulus of elasticity of the conduit-wall material.

The propagation of a pressure wave generated by sudden valve closure is illustrated in Figure 2.25, where a source reservoir is located at a distance L upstream of the valve. The initial condition, at the instant of valve closure, is shown in Figure 2.25(a), where the pressure head in the pipeline is p_o/γ and is equal to the height of the source reservoir above the pipe. This approximation neglects the frictional resistance of the pipe. During the time interval $0 < t < L/c$, the pressure wave propagates upstream, as illustrated in Figure 2.25(b), and at $t = L/c$ the pressure wave reaches the reservoir. At this instant, the velocity in the pipe is zero, the pressure in the pipe is $p_o + \Delta p$, and the pressure in the reservoir is p_o. This abrupt pressure change is the same as that which occurred at the valve when the valve was suddenly closed. To equalize the pressure difference

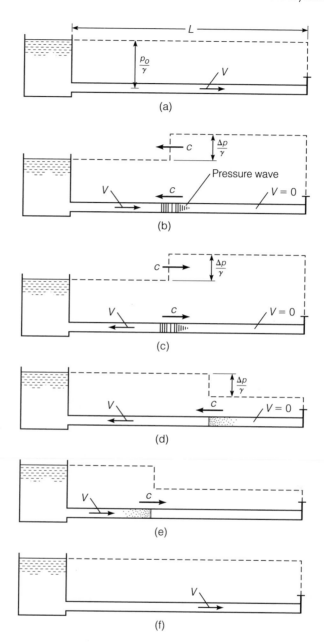

Figure 2.25 ■ Propagation of Pressure Wave

between the reservoir and the pipe, the fluid flows with a velocity, V, from the pipe into the reservoir. This causes a pressure wave to reflect back in the direction of the valve $(L/c < t < 2L/c)$, as illustrated in Figure 2.25(c). At $t = 2L/c$, the wave arrives at the valve, where the velocity must equal zero. Since the velocity, V, in the pipe is directed toward the reservoir, this causes a sudden pressure drop of Δp, creating another pressure wave where the pressure on the valve side of the wave is Δp less than the pressure on the reservoir side of the wave. The pressure drop, Δp, has the same magnitude as the pressure rise when the valve was closed, since the abrupt change in flow velocity, V, is the same in both cases. During time $2L/c < t < 3L/c$, this wave travels toward the reservoir, as illustrated in Figure 2.25(d). At $t = 3L/c$, the pressure wave reaches the reservoir, at which time the pressure in the pipe is Δp less than the pressure in the reservoir. This pressure difference is the same that occurred at the valve when the valve was closed. To equalize the pressures, fluid enters the pipe with velocity V. During the time interval $3L/c < t < 4L/c$, a pressure wave propagates back in the direction of the closed valve, as illustrated in Figure 2.25(e). At $t = 4L/c$, the pressure wave arrives back at the valve, Figure 2.25(f), and conditions are now exactly the same as immediately after the valve was closed. The cycle then repeats itself until the pressure wave is dissipated by frictional effects.

If the valve is closed gradually rather than instantaneously, the pressure increase also occurs gradually. As long as the valve is completely closed in a time less than $2L/c$, however, the maximum pressure occurring as a result of valve closure is still equal to Δp. The *critical time of closure*, t_c, is therefore equal to

$$t_c = \frac{2L}{c} \tag{2.105}$$

If the valve is closed in a time longer than t_c, then any pressure increases generated in the pipe are damped by the open valve.

■ Example 2.21

Estimate the maximum water-hammer pressure generated in a rigid pipe where the initial water velocity is 2.5 m/s, the pipe is 5 km long, and a valve at the downstream end of the pipe is closed in 4 seconds. Assume a water temperature of 25°C.

Solution

At 25°C, $E_v = 2.22 \times 10^6$ kPa, and $\rho = 997.0$ kg/m³ (Table B.1, Appendix B). The speed of the pressure wave, c, is given by Equation 2.104 as

$$c = \sqrt{\frac{E_v}{\rho}} = \sqrt{\frac{2.22 \times 10^9}{997}} = 1{,}490 \text{ m/s}$$

The critical time of closure, t_c, is given by Equation 2.105 as

$$t_c = \frac{2L}{c} = \frac{2(5000)}{1490} = 6.71 \text{ s}$$

Since the closure time of 4 seconds is less than t_c, then the maximum pressure increase in the pipe, Δp, is given by Equation 2.98 as

$$\Delta p = \rho c V = (997)(1490)(2.5) = 3.71 \times 10^6 \text{ Pa} = 3{,}710 \text{ kPa}$$

This pressure increase is significantly higher than the pressures normally encountered in water-distribution systems, which are typically less than 1,000 kPa.   ■

A variety of approaches can be taken to mitigate the effects of water hammer in pipeline systems. In some cases, valves that prevent rapid closure can be used, while in other cases pressure-relief valves or *surge tanks* are more practical. Surge tanks are commonly used in hydropower systems, and consist of large open tanks that are directly connected to pipelines.

2.5.3 ■ Conservation of Energy

The first law of thermodynamics, also called the *law of conservation of energy*, states that within any defined system, the heat added to the system, ΔQ_h, minus the work done by the system, ΔW, is equal to the change in internal energy within the system, ΔE. This statement can be put in the form

$$\Delta Q_h - \Delta W = \Delta E \qquad (2.106)$$

Dividing Equation 2.106 by a time interval, Δt, and taking the limit as Δt approaches zero leads to the differential form of the first law of thermodynamics,

$$\lim_{\Delta t \to 0} \frac{\Delta Q_h}{\Delta t} - \lim_{\Delta t \to 0} \frac{\Delta W}{\Delta t} = \lim_{\Delta t \to 0} \frac{\Delta E}{\Delta t} \qquad (2.107)$$

or

$$\frac{dQ_h}{dt} - \frac{dW}{dt} = \frac{dE}{dt} \qquad (2.108)$$

The energy per unit mass of fluid, e, can be written in the form

$$e = gz + \frac{v^2}{2} + u \qquad (2.109)$$

where g is the acceleration due to gravity, z is the elevation of the fluid mass, v is the magnitude of the fluid velocity, and u is the internal energy per unit mass. Equation 2.109 expresses the energy per unit mass in terms of three components: potential energy (gz), kinetic energy ($v^2/2$), and internal energy (u), which accounts for the chemical energy, electrical energy, and molecular activity per unit mass of the fluid. The internal energy, u, depends primarily on the temperature and phase of the fluid. Combining Equations 2.59 and 2.108 leads to the following expression for the first law of thermodynamics that is applicable to control volumes

$$\frac{dQ_h}{dt} - \frac{dW}{dt} = \frac{\partial}{\partial t} \int_V \rho e \, dV + \int_A \rho e \, \mathbf{v} \cdot \mathbf{n} \, dA \qquad (2.110)$$

where e is given by Equation 2.109. The energy equation is fundamental to describing the flow in closed conduits and open channels.

■ Example 2.22

Water flows under an open gate, as illustrated in Figure 2.17. Estimate the heat loss between the upstream and downstream sections. Neglect the internal energy of the water.

Solution

For the control volume shown in Figure 2.17(b), conditions are steady state and the energy equation (Equation 2.110) is given by

$$\frac{dQ_h}{dt} - \frac{dW}{dt} = \int_A \rho e \, \mathbf{v} \cdot \mathbf{n} \, dA$$

or, taking the velocity distribution as uniform across the upstream and downstream sections,

$$\frac{dQ_h}{dt} = \frac{dW}{dt} - \rho V_1 \int_{A_1} e \, dA + \rho V_2 \int_{A_2} e \, dA \qquad (2.111)$$

where A_1 and A_2 are the cross-sectional areas at Sections 1 and 2, respectively, and V_1 and V_2 are the corresponding velocities. Each of the terms on the right-hand side of Equation 2.111 can now be addressed separately.

First, dW/dt is the rate at which work is being done by the fluid in the control volume. Since the fluid in the control volume only does work against the external pressure forces, then

$$\frac{dW}{dt} = \int_{A_1} p_1 \mathbf{v} \cdot \mathbf{n} \, dA + \int_{A_2} p_2 \mathbf{v} \cdot \mathbf{n} \, dA$$

where p_1 and p_2 are the pressures on the upstream and downstream surfaces of the control volume respectively. Since

$$p_1 = \gamma y_1 \qquad \text{and} \qquad p_2 = \gamma y_2$$

where y_1 and y_2 are measured vertically downward from the upstream and downstream water surfaces, respectively, then

$$\frac{dW}{dt} = -\int_0^3 \gamma y_1 V_1 b \, dy_1 + \int_0^1 \gamma y_2 V_2 b \, dy_2$$

where b is the channel width. Substituting $\gamma = 9{,}790$ N/m^3, $V_1 = 1$ m/s, $b = 10$ m, $V_2 = 3$ m/s, and integrating leads to

$$\frac{dW}{dt} = -\frac{(9790)(1)(10)}{2} y_1^2 \Big|_0^3 + \frac{(9790)(3)(10)}{2} y_2^2 \Big|_0^1$$

$$= -440550 + 146850$$

$$= -293700 \text{ W} = -294 \text{ kW}$$

The next terms in the energy equation (Equation 2.111) to be evaluated are:

$$-\rho V_1 \int_{A_1} e \, dA \qquad \text{and} \qquad \rho V_2 \int_{A_2} e \, dA$$

where e is defined by

$$e = gz + \frac{V^2}{2} + u$$

Neglecting the internal energy, u, and taking $\rho = 998$ kg/m^3 leads to

$$-\rho V_1 \int_{A_1} e \, dA = -\rho V_1 \int_0^3 \left(gy_1 + \frac{V_1^2}{2} \right) w \, dy_1$$

$$= -(998)(1)(10) \int_0^3 \left(9.81 y_1 + \frac{1^2}{2} \right) dy_1$$

$$= -9980 \left[9.81 \frac{y_1^2}{2} + \frac{1}{2} y_1 \right]_0^3$$

$$= -9980[44.15 + 1.5] = -455587 \text{ W}$$

$$= -456 \text{ kW}$$

and

$$\rho V_2 \int_{A_2} e \, dA = \rho V_2 \int_0^1 \left(g y_2 + \frac{V_2^2}{2} \right) w \, dy_2$$

$$= (998)(3)(10) \int_0^1 \left(9.81 y_2 + \frac{3^2}{2} \right) dy_2$$

$$= 29940 \left[9.81 \frac{y_2^2}{2} + 4.5 y_2 \right]_0^1$$

$$= 29940[4.905 + 4.5] = 281586 \text{ W}$$

$$= 282 \text{ kW}$$

Substituting the computed terms into the energy equation (Equation 2.111) leads to

$$\frac{dQ_h}{dt} = -294 - 456 + 282 = -468 \text{ kW}$$

Therefore, heat losses from the control volume shown in Figure 2.17 amount to 468 kW.

■

2.6   Dimensional Analysis and Similitude

Dimensional analysis is the process by which functional relationships are formulated in terms of nondimensional groups. For example, consider the case where there is an unknown relationship between the N variables $x_1, x_2, \ldots, x_N$, which can be written in the form

$$f_1(x_1, x_2, \ldots, x_N) = 0 \tag{2.112}$$

where f_1 is an unknown function. In cases where the relationship given by Equation 2.112 must be determined experimentally, this can become a daunting task as N becomes large, since all possible combinations of values of the variables must be considered. Fortunately, the complexity of this problem can be reduced by combining the variables into a lesser number of nondimensional groups, thereby reducing the number of combinations of variables that must be considered in designing the experiments. This process of transforming a functional relationship between N variables into a functional relationship between a lesser number of nondimensional groups is called *dimensional analysis*. The foundation of dimensional analysis is the *Buckingham pi* (Π) *theorem* (Buckingham, 1915), which can be stated as follows

■ **Theorem 2.1**   *If there are* N *dimensional variables in a dimensionally homogeneous equation, described by* m *fundamental dimensions, they may be grouped in* N − m *dimensionless groups.*

The SI system of units has seven fundamental dimensions: meter, kilogram, second, ampere, Kelvin, candela, and mole. To illustrate the application of the Buckingham pi theorem, consider again the functional relationship between the N variables given in

Equation 2.112, and assume that there are m fundamental dimensions involved in the units of these variables. Then, according to the Buckingham pi theorem, Equation 2.112 can also be written in the form

$$f_2(\Pi_1, \Pi_2, \ldots, \Pi_{N-m}) = 0 \qquad (2.113)$$

where $\Pi_1, \Pi_2, \ldots, \Pi_{N-m}$ are independent nondimensional groups of the original dimensional variables $x_1, x_2, \ldots, x_N$. Clearly the experimental effort required to determine the relationship between the $N - m$ nondimensional variables is less complex than the effort required to determine the relationship between the N dimensional variables.

■ Example 2.23

The motion of an object falling under the influence of gravity in a frictionless environment is described by the relation

$$s = u_o t + \frac{1}{2}gt^2 \qquad (2.114)$$

where s is the distance fallen in time t, u_o is the initial velocity, and g is the acceleration due to gravity. Demonstrate the validity of the Buckingham pi theorem.

Solution

The equation describing the motion of the falling object is given by

$$s - u_o t - \frac{1}{2}gt^2 = 0$$

which can be put in the functional form

$$f_1(s, u_o, t, g) = 0 \qquad (2.115)$$

where the dimensions of the variables are:

Variable	Dimension
s	$[L]$
u_o	$[LT^{-1}]$
t	$[T]$
g	$[LT^{-2}]$

In this case, there are four variables ($N = 4$) and two dimensions ($m = 2$). The Buckingham pi theorem indicates that Equation 2.115 can be expressed as a relation between $N - m = 2$ dimensionless groups. Defining the nondimensional groups, Π_1 and Π_2, as

$$\Pi_1 = \frac{s}{u_o t} \qquad \text{and} \qquad \Pi_2 = \frac{gt}{u_o}$$

then, according to the Buckingham pi theorem, Equation 2.115 can be written as

$$f_2\left(\frac{s}{u_o t}, \frac{gt}{u_o}\right) = 0 \qquad (2.116)$$

This can be verified by rearranging Equation 2.114 as

$$\left(\frac{s}{u_o t}\right) - 1 - \frac{1}{2}\left(\frac{gt}{u_o}\right) = 0$$

which is the actual functional relation between the nondimensional variables and verifies the result (Equation 2.116) derived using the Buckingham pi theorem.

In this example, the analytic relationship between variables is known and it is therefore possible to verify the result of the dimensional analysis, given by Equation 2.116. In the more usual case, the analytic relationship between variables is unknown and the result of a dimensional analysis is the functional relationship between dimensionless groups. The actual relationship is then determined by experimentation. This example also demonstrates that the Buckingham pi theorem is applicable to any functional relationship, not just those in water-resources engineering. ■

The two key steps in applying dimensional analysis to any problem are (1) selection of (dimensional) variables, and (2) formulation of nondimensional groups. The selection of a complete set of relevant dimensional variables is required for a correct dimensional analysis. Selected variables must describe (a) the geometry of the flow system, (b) the properties of the fluid, (c) the external effects driving the fluid flow, and (d) the internal property of the fluid flow that is of interest. The dimensional variables selected must be independent of each other, which means that none of the variables can be obtained by combining the other variables. After selecting the N dimensional variables that describe the flow system, the next step is to identify the m fundamental dimensions and form the $N - m$ dimensionless groups. A variety of algebraic methodologies for formulating dimensionless groups are available, and the groups that result from these methods are not necessarily unique. Of course, if there is only one dimensionless group, then that group is certainly unique. Regardless of the method used to form the dimensionless groups, the total number of groups remains fixed by the number of variables involved, and any combination of dimensionless groups can be converted to any other combination of dimensionless groups by multiplying and/or dividing the dimensionless groups by each other. The ideal method of formulating dimensionless groups is one that forms groups to which physical significance can be attached. Under these circumstances, it is usually possible to infer whether the dependent group will be sensitive to variations in a specific dimensionless group.

Some of the dimensionless groups used in water-resources engineering are listed in Table 2.2. These dimensionless groups are all measures of the ratios of various forces to

■ Table 2.2 Common Dimensionless Groups	Name	Symbol	Formula	Physical meaning
	Reynolds number	Re	$\frac{\rho V L}{\mu}$	ratio of inertial force to viscous force
	Froude number	Fr	$\frac{V}{\sqrt{gL}}$	ratio of inertial force to gravitational force
	Euler number	Eu	$\frac{V}{\sqrt{2\Delta p/\rho}}$	ratio of inertial force to pressure force
	Weber number	We	$\frac{V}{\sqrt{\sigma/\rho L}}$	ratio of inertial force to surface tension force
	Cauchy number	Ca	$\frac{\rho V^2}{E_v}$	ratio of inertial force to compressibility force
	Strouhal number	St	$\frac{\omega L}{V}$	ratio of local inertial force to convective inertia force

the inertial force, and can be used as a basis for neglecting the variables associated with forces that have a negligible influence on the motion of the fluid. For example, high values of the Reynolds number would indicate that viscous forces are small relative to inertial forces; therefore, the fluid viscosity could possibly be neglected as a variable. The most common applications of dimensional analysis in hydraulics include flow in closed conduits, open-channel flow, and turbomachinery.

■ Example 2.24

Show that the ratio of the inertial force to the gravitational force is measured by the Froude number.

Solution

If L is the length scale of a fluid element, then the inertial force, F_I, is given by

$$F_I = ma \sim \rho L^3 \frac{L}{T^2} = \rho L^2 \left(\frac{L}{T}\right)^2 \sim \rho L^2 V^2$$

where ρ is the fluid density and V is the velocity scale. The gravitational force, F_G, on a fluid element is given by

$$F_G = mg \sim \rho L^3 g$$

The ratio of the inertial force to the gravitational force is therefore given by

$$\frac{F_I}{F_G} \sim \frac{\rho L^2 V^2}{\rho L^3 g} = \frac{V^2}{gL}$$

Since the Froude number, Fr, is defined as

$$\text{Fr} = \frac{V}{\sqrt{gL}}$$

then

$$\frac{F_I}{F_G} \sim \text{Fr}^2$$

Therefore the Froude number, Fr, is a measure of F_I/F_G.   ■

Modeling and Similitude.   In cases where large flow systems are to be either constructed or modified, it is prudent to study the expected behavior of these systems by constructing smaller-scale models and inferring the behavior of the larger system from observations on the smaller system. Model tests are commonly used as part of the design process for large dams and for studying the impact of major modifications to surface-water systems. However, model tests should be used to supplement but not replace the theoretical knowledge, good judgment, and experience of the design engineer.

A fundamental question is how the magnitudes of various measured parameters in the model are related to the corresponding parameters in the full-scale flow system, commonly referred to as the *prototype*. The answer to this question is derived from a dimensional analysis of the flow system, the results of which can generally be written in the form

$$\Pi_o = f(\Pi_1, \Pi_2, \ldots, \Pi_n) \tag{2.117}$$

where Π_o is a nondimensional group containing the variable of interest, and $\Pi_1, \Pi_2, \ldots, \Pi_n$ are nondimensional groups containing other variables that affect the flow. Based on Equation 2.117, it is clear that if the model and prototype both represent the same physical system, and if the model is constructed such that $\Pi_1, \Pi_2, \ldots, \Pi_n$ are the same in both the model and prototype, then Π_o must also be the same in both the model and prototype. Therefore, by measuring Π_o in the model, the parameter of interest in the prototype can be inferred from the relation

$$(\Pi_o)_{\text{prototype}} = (\Pi_o)_{\text{model}} \tag{2.118}$$

Usually, several of the independent Π groups on the lefthand side of Equation 2.117 involve the ratio of length scales, and equality of these Π groups in the model and prototype require that the model be *geometrically similar* to the prototype. Also, several of the independent Π groups on the lefthand side of Equation 2.117 involve the ratio of forces, and equality of these Π groups in the model and prototype require that the model be *dynamically similar* to the prototype. In most models, dynamic similarity is only required for the dominant force. For example, in most open-channel models the dominant force is gravity, and similitude is only achieved with respect to gravity forces while neglecting the effects of viscosity, surface tension, and other forces. Surface tension effects in models are minimized by using model flow depths greater than 5 cm, and viscous effects are minimized by requiring that the flow be turbulent in the model, assuming that the flow is turbulent in the prototype. Distortions in model performance that result from neglecting certain similarity requirements are called *scale effects*.

■ Example 2.25

The impact of major channel modifications on the water surface profile in a river are to be studied in the laboratory using a scale model. Dimensional analysis shows that the water surface profile in the river can be written in the functional form

$$\frac{y}{L} = f\left(\frac{x}{L}, \frac{V}{\sqrt{gL}}\right)$$

where y is the depth of flow in the river at a distance x downstream from a reference location, L is the depth at the reference location, and V is the velocity at the reference location. If the model is to be designed based on Froude number similarity, which would be required since the dominant force is gravity, the model scale is 1:20, and the flowrate in the prototype is 50 m^3/s, what flowrate should be used in the model? If the reference depth, L, in the prototype is 5 m, then what is the corresponding depth in the model? If the measured water surface profile in the model is given in the following table, tabulate the water surface profile in the prototype.

x/L	0	1	10	100	1000
y/L	1.00	0.96	0.90	0.71	0.83

Solution

Since the model and prototype are geometrically similar, the theory of models requires that when x/L and $V/\sqrt{gL}$ are the same in the model and prototype, then y/L is also the same in the model and the prototype. That is, using subscripts m and p to denote the model and prototype respectively, then

$$\left(\frac{V}{\sqrt{gL}}\right)_m = \left(\frac{V}{\sqrt{gL}}\right)_p$$

and

$$\left(\frac{x}{L}\right)_m = \left(\frac{x}{L}\right)_p$$

requires that

$$\left(\frac{y}{L}\right)_m = \left(\frac{y}{L}\right)_p$$

Also, since

$$Q \propto VL^2$$

the Froude similarity requirement can be written as

$$\left(\frac{VL^2}{\sqrt{gL^5}}\right)_m = \left(\frac{VL^2}{\sqrt{gL^5}}\right)_p$$

or

$$\left(\frac{Q}{\sqrt{gL^5}}\right)_m = \left(\frac{Q}{\sqrt{gL^5}}\right)_p$$

which leads to

$$\frac{Q_m}{Q_p} = \left(\frac{L_m}{L_p}\right)^{\frac{5}{2}}$$

Since $L_m/L_p = 1/20$ and $Q_p = 50$ m³/s, then the required flowrate in the model, Q_m, is given by

$$Q_m = Q_p \left(\frac{L_m}{L_p}\right)^{\frac{5}{2}} = 50\left(\frac{1}{20}\right)^{\frac{5}{2}} = 0.028 \text{ m}^3/\text{s} = 28 \text{ L/s}$$

Since the reference depth, L_p, in the prototype is 5 m, then the corresponding reference depth in the model, L_m, is given by

$$L_m = L_p\left(\frac{1}{20}\right) = 5\left(\frac{1}{20}\right) = 0.25 \text{ m} = 25 \text{ cm}$$

The nondimensional water surface profile is the same in both the model and prototype. Therefore the water surface profile in the prototype can be calculated using the relations

$$x_p = \left(\frac{x}{L}\right)_m L_p = 5\left(\frac{x}{L}\right)_m$$

and

$$y_p = \left(\frac{y}{L}\right)_m L_p = 5\left(\frac{y}{L}\right)_m$$

The following water-surface profile in the prototype can be derived from the given water-surface profile in the model:

x (m)	0	5	50	500	5,000
y (m)	5	4.8	4.5	3.6	4.2

■

Models of spillways, conduits, and other hydraulic structures typically have scale ratios that range from 1:50 to 1:15. Models of rivers, harbors, estuaries, and reservoirs are usually distorted and have horizontal-scale ratios that range from 1:2,000 to 1:100 and vertical-scale ratios that range from 1:150 to 1:50 (French, 1985).

Summary

This chapter covered the elements of fluid mechanics that are fundamental to water-resources engineering. Physical properties of water commonly used in engineering design include density, viscosity (dynamic and kinematic), bulk modulus, surface tension, saturation vapor pressure, and heat of vaporization. The values of these properties at 20°C are listed in Table 2.1. The pressure distribution in static fluids is derived from first principles. Specific applications to pressure measurement and the calculation of hydrostatic forces on both plane and curved surfaces are presented. Fluids in motion must satisfy the conservation laws of mass, momentum, and energy; these fundamental relationships are converted to forms that are useful in engineering applications via the Reynolds transport theorem. Included are specific applications of the momentum equation to calculate the forces exerted by fluid jets on deflectors and blades; the forces on reducers, bends, and junctions in pressure conduits; and water-hammer pressures.

In cases where solutions of fluid-flow problems are not possible analytically, engineers frequently resort to laboratory-scale models of the phenomena. The design of laboratory-scale experiments and the relationship between model and prototype parameters are derived using dimensional analysis and similitude. Dimensionless groups encountered in engineering practice include the Reynolds, Froude, Euler, Weber, Cauchy, and Strouhal numbers. Dimensional analysis and similitude form much of the basis for the analysis of closed-conduit and open-channel flow, which are covered in subsequent chapters.

Problems

2.1. What is the largest percentage error in the density of water that can be made if the density is always assumed to be equal to 1,000 kg/m³?

2.2. What is the specific weight of water at 0°C, 20°C, and 100°C?

2.3. What is the specific gravity of water at 0°C, 20°C, and 100°C?

2.4. Tabulate the kinematic viscosity of water, ν, as a function of temperature.

2.5. If the velocity distribution in a pipe of radius R is given by

$$V(r) = V_o\left(1 - \frac{r^2}{R^2}\right)$$

where V_o is the velocity at the pipe centerline and r is the distance from the center of the pipe, then find: (a) the shear stress as a function of r, and (b) the shear force on the pipe boundary per unit length of pipe.

2.6. Derive Equation 2.9 from Equation 2.8.

2.7. What pressure change would be necessary to change the density of water by 1% at 20°C?

2.8. Would you expect that the surface tension of water in contact with air is the same as the surface tension of water in contact with pure oxygen? Explain.

2.9. What would be the boiling point of water in a high mountain area where the atmospheric pressure is 90 kPa?

2.10. What is the minimum allowable pressure to prevent cavitation of water in a pipeline system that transmits water at 35°C?

2.11. Estimate the evaporation rate of water that results from a net solar radiation of 10 MJ/(m²·d). Assume that the temperature of the water body is 15°C.

2.12. If water in a lake evaporates at a rate of 5 mm/d without any change in temperature, estimate the net incident radiation. The lake temperature is 20°C.

2.13. At what depth below the surface of a water body will the (gage) pressure be equal to 101.3 kPa?

2.14. The pressure in the air space above an oil ($SG = 0.80$) surface in a tank is 14 kPa. Find the pressure 1.5 m below the surface of the oil.

2.15. What is the pressure head corresponding to a pressure of 800 kPa?

2.16. Find the pressure head in millimeters of mercury (Hg) equivalent to 80 mm of water plus 60 mm of a fluid whose specific gravity is 2.90. The specific weight of mercury can be taken as 133 kN/m³.

2.17. Find the pressure head corresponding to an atmospheric pressure of 101.3 kPa. Give your answer in terms of millimeters of mercury.

2.18. It is common practice to use elevated reservoirs to maintain the pressures in municipal water-supply systems. Such a system is illustrated in Figure 2.26. If the pressure in the pipeline is to be maintained in the range of 350 kPa to 500 kPa

and the reservoir is not to be less than half full, then estimate the height of the midpoint of the reservoir above the pipeline and the minimum required space between the midpoint and the top of the reservoir.

2.19. The U-tube manometer shown in Figure 2.27 is used to measure the pressure in the pipeline at A. The water is at 20°C and the density of the gage fluid is 40 kN/m³. For the given fluid heights, determine the pressure in the pipeline.

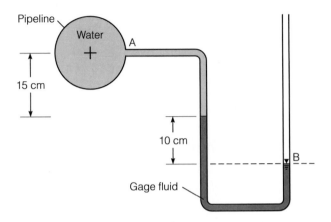

Figure 2.27 ■ Problem 2.19

2.20. Consider the differential manometer illustrated in Figure 2.28. Express the pressure difference between points A and B in terms of γ_w, γ_f, h_1, h_2, and h_3.

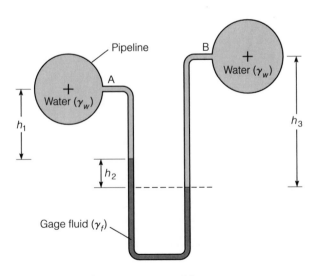

Figure 2.28 ■ Problem 2.20

2.21. The 2-m diameter pipe illustrated in Figure 2.29 drains water from a reservoir, where the crown of the pipe is 3 m be-

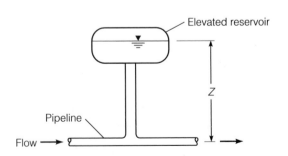

Figure 2.26 ■ Problem 2.18

low the surface of the reservoir. If the pipe entrance is covered with a 2-m diameter gate hinged at A, determine the magnitude and location of the net hydrostatic force on the gate. What moment would need to be applied at A to open the gate?

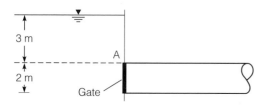

Figure 2.29 ■ Problem 2.21

2.22. A circular hatch is located in a sloping wall of a water-storage reservoir, where the wall slopes at 35° to the horizontal, the radius of the hatch is 420 mm, and the center of the hatch is 3 m below the water surface (measured along the sloping wall). Find the magnitude and location of the resultant hydrostatic force on the hatch.

2.23. The water in a reservoir is contained by an elliptical gate illustrated in Figure 2.30, where D is the diameter of the pipe leading to the gate and D and $D/\sin\theta$ are the lengths of the minor and major principal axes of the elliptical gate, respectively. If $D = 1.2$ m, $\theta = 30°$, and the water surface in the reservoir is 9 m above the centerline of the gate, then determine the resultant hydrostatic force on the gate and the center of pressure. What moment at point P would be required to keep the gate closed? Neglect the weight of the gate.

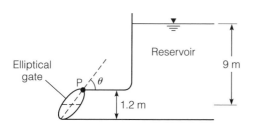

Figure 2.30 ■ Problem 2.23

2.24. An 8-m deep by 20-m long aquarium is to be designed with a glass viewing area at the bottom. If the viewing section

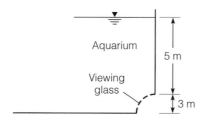

Figure 2.31 ■ Problem 2.24

is shaped like the quadrant of a circle (Figure 2.31), calculate the hydrostatic force on the viewing glass.

2.25. Water is contained within a reservoir by the Tainter gate illustrated in Figure 2.32. The gate has the shape of a quadrant of a circle, is 5 m wide, weighs 10 kN, and the center of gravity of the gate is at point G. Determine the net hydrostatic force on the gate and the magnitude and direction of the moment that must be applied at P to open the gate.

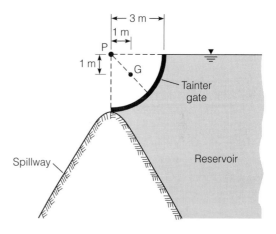

Figure 2.32 ■ Problem 2.25

2.26. The drag force, F_D, on a spherical particle settling with velocity, v, in a fluid can be approximated by (Franzini and Finnemore, 1997)

$$F_D = 3\pi\mu v D$$

where μ is the viscosity of the fluid and D is the particle diameter. If a 2-mm diameter particle with a specific gravity of 2.65 is stirred up from the bottom of a river, estimate the sedimentation velocity, v. [*Hint*: A particle settles with a constant velocity when the sum of the forces on the particle is equal to zero.]

2.27. It is conventional wisdom that in areas where the ground water is close to the land surface, swimming pools should not be completely emptied. The dimensions of a swimming pool are 10 m long by 5 m wide by 2.5 m deep, the weight of the pool is 500 kN, and the ground water is 1.25 m below the top of the pool. Determine the minimum depth of water that must be maintained in the pool. [*Hint*: The net force on the pool structure must remain downward.]

2.28. The flow in a pipeline is divided as shown in Figure 2.33. The diameter of the pipe at Sections 1, 2, and 3 are 100 mm, 75 mm, and 50 mm, respectively, and the flowrate at Section 1 is 10 L/s. Calculate the volumetric flowrate and velocity at Section 2.

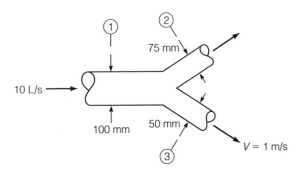

Figure 2.33 ■ Problem 2.28

2.29. The flowrate on the intake side of a pump is equal to Q, as illustrated in Figure 2.34. If the power delivered by the pump is 100 kW, how does the flowrate on the discharge side of the pump compare with Q?

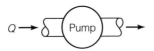

Figure 2.34 ■ Problem 2.29

2.30. An empirical equation for the velocity distribution in a horizontal open channel is given by

$$u = u_{max} \left(\frac{y}{d}\right)^{\frac{1}{7}}$$

where u is the velocity at a distance y above the floor of the channel, u_{max} is the maximum velocity, and d is the depth of flow. If $d = 1$ m and $u_{max} = 3$ m/s, what is the volumetric flowrate (i.e., discharge) is m³/s per meter of width of channel?

2.31. Water enters the cylindrical reservoir shown in Figure 2.35 through a pipe at point A at a rate of 2.80 L/s, and exits

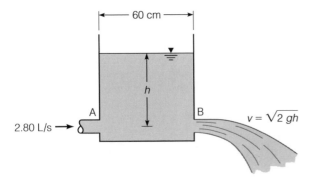

Figure 2.35 ■ Problem 2.31

through a 5-cm diameter orifice at B. The radius of the cylindrical reservoir is 60 cm, and the velocity, v, of water leaving the orifice is given by

$$v = \sqrt{2gh}$$

where h is the height of the water surface (in the reservoir) above the orifice. How long will it take for the water surface in the reservoir to drop from $h = 2$ m to $h = 1$ m? [*Hint:* You may need to use integral tables.]

2.32. Water flows over a 0.2-m high step in a 5-m wide channel, as illustrated in Figure 2.36. If the flowrate in the channel is 15 m³/s and the upstream and downstream depths are 3.00 m and 2.79 m, respectively, calculate the force on the step.

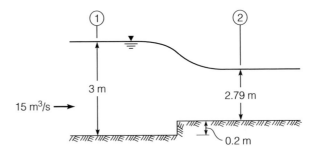

Figure 2.36 ■ Problem 2.32

2.33. A jet of water at 10 m³/s is impinging on a stationary deflector that changes the flow direction of the jet by 45°. The velocity of the impinging jet is 10 m/s and the velocity of the deflected jet is 9 m/s. What force is required to keep the deflector in place?

2.34. A blade attached to a turbine rotor is driven by a stream of water that has a velocity of 18 m/s. The blade moves with a velocity of 8 m/s and deflects the stream of water through an angle of 85°; the entrance and exit flow areas are each equal to 1.5 m². Estimate the force on the moving blade and the power transferred to the turbine rotor.

2.35. A reducer is to be used to attach a 400-mm diameter pipe to a 300-mm diameter pipe. For any flowrate, the pressures in the pipes upstream and downstream of the reducer are expected to be related by

$$\frac{p_1}{\gamma} + \frac{V_1^2}{2g} = \frac{p_2}{\gamma} + \frac{V_2^2}{2g}$$

where p_1 and p_2 are the upstream and downstream pressures and V_1 and V_2 are the upstream and downstream velocities. Estimate the force on the reducer when water flows through the reducer at 200 L/s and the upstream pressure is 400 kPa.

2.36. Water flows at 100 L/s through a 200-mm diameter vertical bend, as shown in Figure 2.37. If the pressure at Section 1 is 500 kPa and the pressure at Section 2 is 450 kPa, then determine the horizontal and vertical thrust on the support structure. The volume of the bend is 0.16 m³.

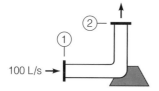

Figure 2.37 ■ Problem 2.36

2.37. A 30° bend connects a 250-mm diameter pipe (inflow) to a 400-mm pipe (outflow). The volume of the bend is 0.2 m³, the weight of the bend is 400 N, and the pressures on the inflow outflow sections are related by

$$\frac{p_1}{\gamma} + \frac{V_1^2}{2g} = \frac{p_2}{\gamma} + \frac{V_2^2}{2g}$$

The pressure at the inflow section is 500 kPa, the bend is in the vertical plane, and the bend-support structure has a maximum allowable load of 18 kN in the horizontal direction and 40 kN in the vertical direction. Determine the maximum allowable flowrate in the bend.

2.38. Determine the force required to restrain the pipe junction illustrated in Figure 2.38. Assume that the junction is in the horizontal plane and explain how your answer would differ if the junction were in the vertical plane.

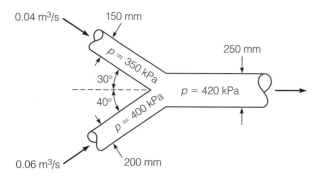

Figure 2.38 ■ Problem 2.38

2.39. Plot the water-hammer pressure versus time at the midpoint of the pipeline shown in Figure 2.25. Assume that the valve is closed instantaneously.

2.40. Water flows in a 100-m long pipe at 3 m/s. If the water temperature is 20°C, determine the minimum valve-closure time to avoid creating water-hammer pressures. What is the maximum water-hammer pressure that can occur? How is this pressure affected if the water temperature drops to 10°C?

2.41. Based on your result in Problem 2.40, do you think that water hammer can be a serious problem in household plumbing?

2.42. Calculate the heat loss between the upstream and downstream sections in Figure 2.36. Neglect the internal energy of the water.

2.43. Calculate the heat loss in the reducer described in Problem 2.35. Neglect the internal energy of the water, and neglect changes in elevation. What role does the internal energy play in the heat balance?

2.44. Calculate the heat loss in the junction described in Problem 2.38. Neglect the internal energy of the water, and neglect changes in elevation.

2.45. The velocity distribution, $V(r)$, in a pipe with fully developed laminar flow is given by (Shames, 1992)

$$V = \frac{\Delta p}{4\mu L}\left(\frac{D^2}{4} - r^2\right)$$

where Δp is the pressure drop over a distance L, μ is the dynamic viscosity, D is the diameter of the pipe, and r is the distance from the pipe centerline. Derive this functional relationship using the Buckingham pi theorem.

2.46. The force F on a reducer is related to the flow conditions and pipe geometry by

$$F = p_1A_1 - p_2A_2 - \rho A_1 V_1(V_2 - V_1)$$

where p_1 and p_2 are the pressures upstream and downstream of the reducer, respectively; A_1 and A_2 are the corresponding flow areas; and V_1 and V_2 are the velocities. Derive this functional relationship using the Buckingham pi theorem.

2.47. Show that the ratio of the inertial force to the pressure force is measured by the Euler number.

2.48. Show that the ratio of the inertial force to the surface tension force is measured by the Weber number.

2.49. Show that the ratio of the inertial force to the compressibility force is measured by the Cauchy number.

2.50. The energy per unit mass, e, added by a pump of a given shape depends on the pump size, D, flowrate, Q, speed of the rotor, ω, density of the fluid, ρ, and dynamic viscosity of the fluid, μ. This functional relation can be stated as:

$$e = f(D, Q, \omega, \rho, \mu)$$

Express this as a relationship between dimensionless groups. What is gained by expressing the pump performance as an empirical relationship between dimensionless groups versus expressing the pump performance as a relationship between the given dimensional variables?

2.51. The height of rise, h, of a fluid in a capillary tube depends on the specific weight, γ, the surface tension, σ, of the fluid, and the diameter, D, of the capillary tube. In other words,

$$h = f(\gamma, \sigma, D)$$

Express this relationship in terms of dimensionless groups. How does your result compare with the theoretical expression

$$h = \frac{4\sigma}{\gamma D}$$

2.52. The flow over a sharp-crested weir, Q, is commonly expressed in the form

$$Q = f\left(\text{We, Re, } \frac{H}{H_w}\right) b\sqrt{g}H^{\frac{3}{2}} \qquad (2.119)$$

where We is the Weber number defined by

$$\text{We} = \frac{V}{\sqrt{\sigma/\rho H}}$$

and V is the flow velocity over the weir, σ is the surface tension of the fluid, ρ is the fluid density, H is the depth of fluid over the crest of the weir, and Re is the Reynolds number defined by

$$\text{Re} = \frac{\rho VH}{\mu}$$

where μ is the viscosity of the fluid, H_w is the height of the weir, and b is the length of the weir crest. Derive Equation 2.119 using dimensional analysis.

2.53. The shear stress, τ_o, exerted on a pipeline of diameter, D, by a fluid of density, ρ, and viscosity, μ, moving with a velocity, V, is given by the following dimensionless relation

$$\frac{\tau_o}{\rho V^2} = f\left(\frac{\rho VD}{\mu}, \frac{\epsilon}{D}\right)$$

where ϵ is the height of the roughness elements on the boundary of the pipe. In a prototype pipeline to be modeled, the velocity of flow is 2 m/s, the diameter of the pipeline is 3 m, and the height of the roughness elements are 2 mm. If a model of the pipeline is to be constructed based on Reynolds number similarity (since viscous force is important), a model scale of 1:20, and the same fluid is used in both the model and prototype, what velocity and roughness height should be used in the model? If the shear stress on the pipe boundary in the model is measured at 2.25 kPa, then what is the shear stress in the prototype?

2.54. A sluice gate is to be installed to control the flow in a coastal canal. The flow through the gate, Q, depends on the upstream water depth, h_1, the downstream water depth, h_2, the gate opening, s, the width of the gate, b, and the acceleration due to gravity, g. Use dimensional analysis to determine a functional relationship between nondimensional groups that can be used to guide laboratory model experiments on the gate. Identify the physical significance of each nondimensional group. A $\frac{1}{7}$ scale model having a width of 1 m is constructed in the laboratory, and the following operating condition is observed in the model:

h_1 (m)	h_2 (m)	s (m)	Q (m³/s)
0.57	0.50	0.16	0.14

Find the corresponding operating condition in the prototype. How would you justify neglecting the viscosity of the water in your analysis?

3 Flow in Closed Conduits

3.1 ■ Introduction

Flow in closed conduits includes all cases where the flowing fluid completely fills the conduit. The cross-sections of closed conduits can be of any shape or size and can be made of a variety of materials. Engineering applications of the principles of flow in closed conduits include the design of municipal water-supply systems and plumbing systems in buildings. The basic equations governing the flow of fluids in closed conduits are the continuity, momentum, and energy equations. The most useful forms of these equations for application to pipe flow problems are derived in this chapter. The governing equations are presented in forms that are applicable to any fluid flowing in a closed conduit, but particular attention is given to the flow of water.

The computation of flows in pipe networks is a natural extension of the flows in single pipelines, and methods of calculating flows and pressure distributions in pipeline systems are also described here. These methods are particularly applicable to the analysis and design of municipal water distribution systems, where the engineer is frequently interested in assessing the effects of various modifications to the system. Because transmission of water in closed conduits is typically accomplished using pumps, the fundamentals of pump operation and performance are also presented in this chapter. A sound understanding of pumps is important in selecting the appropriate pump to achieve the desired operational characteristics in water transmission systems. The design protocol for municipal water distribution systems is presented as an example of the application of the principles of flow in closed conduits. Methods for estimating water demand, design of the functional components of distribution systems, network analysis, and the operational criteria for municipal water distribution systems are all covered.

3.2 ■ Single Pipelines

The governing equations for flows in pipelines are derived from the conservation laws of mass, momentum, and energy. The control-volume forms of these equations were derived in Section 5 of Chapter 2. The forms of these equations that are most useful for application to closed-conduit flow are derived in the following sections.

3.2.1 ■ Continuity Equation

Consider the application of the continuity equation to the control volume illustrated in Figure 3.1. Fluid enters and leaves the control volume normal to the control surfaces,

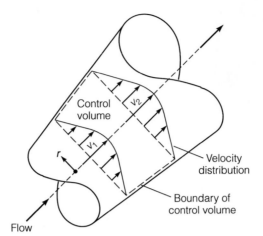

Figure 3.1 ▪ Flow
Through Closed Conduit

Control
volume

v_2

v_1

r

Velocity
distribution

Boundary of
control volume

Flow

with the inflow velocity denoted by $v_1(\mathbf{r})$ and the outflow velocity by $v_2(\mathbf{r})$. Both the inflow and outflow velocities vary across the control surface. The steady-state continuity equation for an incompressible fluid (Equation 2.62) can thus be written as

$$\int_{A_1} v_1 \, dA = \int_{A_2} v_2 \, dA \tag{3.1}$$

Defining V_1 and V_2 as the average velocities across A_1 and A_2, respectively, where

$$V_1 = \frac{1}{A_1} \int_{A_1} v_1 \, dA \tag{3.2}$$

and

$$V_2 = \frac{1}{A_2} \int_{A_2} v_2 \, dA \tag{3.3}$$

the steady-state continuity equation becomes

$$\boxed{V_1 A_1 = V_2 A_2 (= Q)} \tag{3.4}$$

The terms on each side of Equation 3.4 are equal to the volumetric flowrate, Q. The steady-state continuity equation simply states that the volumetric flowrate across any surface normal to the flow is a constant.

▪ Example 3.1

Water enters a pump through a 150-mm diameter intake pipe and leaves the pump through a 200-mm diameter discharge pipe. If the average velocity in the intake pipeline is 1 m/s, calculate the average velocity in the discharge pipeline. What is the flowrate through the pump?

Solution

In the intake pipeline, $V_1 = 1$ m/s, $D_1 = 0.15$ m and

$$A_1 = \frac{\pi}{4} D_1^2 = \frac{\pi}{4} (0.15)^2 = 0.0177 \text{ m}^2$$

In the discharge pipeline, $D_2 = 0.20$ m and

$$A_2 = \frac{\pi}{4}D_2^2 = \frac{\pi}{4}(0.20)^2 = 0.0314 \text{ m}^2$$

According to the continuity equation,

$$V_1 A_1 = V_2 A_2$$

Therefore,

$$V_2 = V_1\left(\frac{A_1}{A_2}\right) = (1)\left(\frac{0.0177}{0.0314}\right) = 0.56 \text{ m/s}$$

The flowrate, Q, is given by

$$Q = A_1 V_1 = (0.0177)(1) = 0.0177 \text{ m}^3/\text{s}$$

The average velocity in the discharge pipeline is 0.56 m/s, and the flowrate through the pump is 0.0177 m^3/s. ■

3.2.2 ■ Momentum Equation

Consider the application of the momentum equation to the control volume illustrated in Figure 3.1. Under steady-state conditions, the component of the momentum equation (Equation 2.67) in the direction of flow (x-direction) can be written as

$$\sum F_x = \int_A \rho v_x \mathbf{v} \cdot \mathbf{n} \, dA \tag{3.5}$$

where $\sum F_x$ is the sum of the x-components of the forces acting on the fluid in the control volume, ρ is the density of the fluid, v_x is the flow velocity in the x-direction, and $\mathbf{v} \cdot \mathbf{n}$ is the component of the flow velocity normal to the control surface. Since the unit normal vector, $\mathbf{n}$, in Equation 3.5 is directed outward from the control volume, then the momentum equation for an incompressible fluid ($\rho = $ constant) can be written as

$$\sum F_x = \rho \int_{A_2} v_2^2 \, dA - \rho \int_{A_1} v_1^2 \, dA \tag{3.6}$$

where the integral terms depend on the velocity distributions across the inflow and outflow control surfaces. The velocity distribution across each control surface is generally accounted for by the *momentum correction coefficient*, β, defined by the relation

$$\boxed{\beta = \frac{1}{AV^2} \int_A v^2 \, dA} \tag{3.7}$$

where A is the area of the control surface and V is the average velocity over the control surface. The momentum coefficients for the inflow and outflow control surfaces, A_1 and A_2, are then given by β_1 and β_2, where

$$\beta_1 = \frac{1}{A_1 V_1^2} \int_{A_1} v_1^2 \, dA \tag{3.8}$$

$$\beta_2 = \frac{1}{A_2 V_2^2} \int_{A_2} v_2^2 \, dA \tag{3.9}$$

Substituting Equations 3.8 and 3.9 into Equation 3.6 leads to the following form of the momentum equation

$$\sum F_x = \rho \beta_2 V_2^2 A_2 - \rho \beta_1 V_1^2 A_1 \tag{3.10}$$

Recalling that the continuity equation states that the volumetric flowrate, Q, is the same across both the inflow and outflow control surfaces, where

$$Q = V_1 A_1 = V_2 A_2 \tag{3.11}$$

then combining Equations 3.10 and 3.11 leads to the following form of the momentum equation

$$\sum F_x = \rho \beta_2 Q V_2 - \rho \beta_1 Q V_1 \tag{3.12}$$

or

$$\sum F_x = \rho Q (\beta_2 V_2 - \beta_1 V_1) \tag{3.13}$$

In many cases of practical interest, the velocity distribution across the cross-section of the closed conduit is approximately uniform, in which case the momentum coefficients, β_1 and β_2, are approximately equal to unity and the momentum equation becomes

$$\sum F_x = \rho Q (V_2 - V_1) \tag{3.14}$$

Consider the common case of flow in a straight pipe with a uniform circular cross-section illustrated in Figure 3.2, where the average velocity remains constant at each cross section,

$$V_1 = V_2 = V \tag{3.15}$$

then the momentum equation becomes

$$\sum F_x = 0 \tag{3.16}$$

The forces that act on the fluid in a control volume of uniform cross-section are illustrated in Figure 3.2. At Section 1, the average pressure over the control surface is equal to p_1 and the elevation of the midpoint of the section relative to a defined datum is equal to z_1, at Section 2, located a distance L downstream from Section 1, the pressure is p_2, and the elevation of the midpoint of the section is z_2. The average shear stress exerted on the

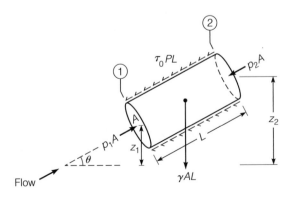

Figure 3.2 ■ Forces on Flow in Closed Conduit

fluid by the pipe surface is equal to τ_o, and the total shear force opposing flow is $\tau_o PL$, where P is the perimeter of the pipe. The fluid weight acts vertically downward and is equal to γAL, where γ is the specific weight of the fluid and A is the cross-sectional area of the pipe. The forces acting on the fluid system that have components in the direction of flow are the shear force, $\tau_o PL$; the weight of the fluid in the control volume, γAL; and the pressure forces on the upstream and downstream faces, $p_1 A$ and $p_2 A$, respectively. Substituting the expressions for the forces into the momentum equation, Equation 3.16, yields

$$p_1 A - p_2 A - \tau_o PL - \gamma AL \sin \theta = 0 \tag{3.17}$$

where θ is the angle that the pipe makes with the horizontal and is given by the relation

$$\sin \theta = \frac{z_2 - z_1}{L} \tag{3.18}$$

Combining Equations 3.17 and 3.18 yields

$$\frac{p_1}{\gamma} - \frac{p_2}{\gamma} - z_2 + z_1 = \frac{\tau_o PL}{\gamma A} \tag{3.19}$$

Defining the *total head*, or energy per unit weight, at Sections 1 and 2 as h_1 and h_2, where

$$h_1 = \frac{p_1}{\gamma} + \frac{V^2}{2g} + z_1 \tag{3.20}$$

and

$$h_2 = \frac{p_2}{\gamma} + \frac{V^2}{2g} + z_2 \tag{3.21}$$

then the *head loss* between Sections 1 and 2, Δh, is given by

$$\Delta h = h_1 - h_2 = \left(\frac{p_1}{\gamma} + z_1 \right) - \left(\frac{p_2}{\gamma} + z_2 \right) \tag{3.22}$$

Combining Equations 3.19 and 3.22 leads to the following expression for head loss

$$\Delta h = \frac{\tau_o PL}{\gamma A} \tag{3.23}$$

In this case, the head loss, Δh, is entirely due to pipe friction and is commonly denoted by h_f. In the case of pipes with circular cross-sections, Equation 3.23 can be written as

$$h_f = \frac{\tau_o(\pi D)L}{\gamma (\pi D^2/4)} = \frac{4\tau_o L}{\gamma D} \tag{3.24}$$

where D is the diameter of the pipe. The ratio of the cross-sectional area, A, to the perimeter, P, is defined as the *hydraulic radius*, R, where

$$R = \frac{A}{P} \tag{3.25}$$

and the head loss can be written in terms of the hydraulic radius as

$$h_f = \frac{\tau_o L}{\gamma R} \tag{3.26}$$

The form of the momentum equation given by Equation 3.26 is of limited utility in that the head loss, h_f, is expressed in terms of the boundary shear stress, τ_o, which is not a measurable quantity. However, the boundary shear stress, τ_o, can be expressed in terms of measurable flow variables using dimensional analysis, where τ_o can be taken as a function of the mean flow velocity, V; density of the fluid, ρ; dynamic viscosity of the fluid, μ; diameter of the pipe, D; characteristic size of roughness projections, ϵ; characteristic spacing of the roughness projections, ϵ'; and a (dimensionless) form factor, m, that depends on the shape of the roughness elements on the surface of the conduit. This functional relationship can be expressed as

$$\tau_o = f_1(V, \rho, \mu, D, \epsilon, \epsilon', m) \tag{3.27}$$

According to the Buckingham pi theorem, this relationship between eight variables in three fundamental dimensions can also be expressed as a relationship between five nondimensional groups. The following relation is proposed

$$\frac{\tau_o}{\rho V^2} = f_2\left(\text{Re}, \frac{\epsilon}{D}, \frac{\epsilon'}{D}, m\right) \tag{3.28}$$

where Re is the Reynolds number defined by

$$\text{Re} = \frac{\rho V D}{\mu} \tag{3.29}$$

The relationship given by Equation 3.28 is as far as dimensional analysis goes, and experiments are necessary to determine an empirical relationship between the nondimensional groups. Nikuradse (1932, 1933) conducted a series of experiments in pipes in which the inner surfaces were roughened with sand grains of uniform diameter, ϵ. In these experiments, the spacing, ϵ', and shape, m, of the roughness elements (sand grains) were constant and Nikuradse's experimental data fitted to the following functional relation

$$\frac{\tau_o}{\rho V^2} = f_3\left(\text{Re}, \frac{\epsilon}{D}\right) \tag{3.30}$$

It is convenient for subsequent analysis to introduce a factor of 8 into this relationship, which can then be written as

$$\frac{\tau_o}{\rho V^2} = \frac{1}{8} f\left(\text{Re}, \frac{\epsilon}{D}\right) \tag{3.31}$$

or simply

$$\frac{\tau_o}{\rho V^2} = \frac{f}{8} \tag{3.32}$$

where the dependence of the *friction factor*, f, on the Reynolds number, Re, and relative roughness, ϵ/D, is understood. Combining Equations 3.32 and 3.24 leads to the following form of the momentum equation for flows in circular pipes

$$\boxed{h_f = \frac{f L}{D} \frac{V^2}{2g}} \tag{3.33}$$

This equation, called the *Darcy-Weisbach equation*,* expresses the frictional head loss, h_f, of the fluid over a length L of pipe in terms of measurable parameters, including the pipe diameter (D), average flow velocity (V), and the friction factor (f) that characterizes the shear stress of the fluid on the pipe.

Based on Nikuradse's (1932, 1933) experiments on sand-roughened pipes, Prandl and von Kármán established the following empirical formulae for estimating the friction factor in turbulent pipe flows

$$\text{Smooth pipe } \left(\frac{k}{D} \approx 0\right): \quad \frac{1}{\sqrt{f}} = -2\log\left(\frac{2.51}{\text{Re}\sqrt{f}}\right)$$

$$\text{Rough pipe } \left(\frac{k}{D} \gg 0\right): \quad \frac{1}{\sqrt{f}} = -2\log\left(\frac{k/D}{3.7}\right)$$

$$(3.34)$$

where k is the roughness height of the sand grains on the surface of the pipe. Turbulent flow in pipes is generally present when Re > 4,000; transition to turbulent flow begins at about Re = 2,300. The pipe behaves like a *smooth pipe* when the friction factor does not depend on the height of the roughness projections on the wall of the pipe and therefore depends only on the Reynolds number. In *rough pipes*, the friction factor is determined by the relative roughness, k/D, and becomes independent of the Reynolds number. The smooth pipe case generally occurs at lower Reynolds numbers, when the roughness projections are submerged within the viscous boundary layer. At higher values of the Reynolds number, the thickness of the viscous boundary layer decreases and eventually the roughness projections protrude sufficiently far outside the viscous boundary layer that the shear stress of the pipe boundary is dominated by the hydrodynamic drag associated with the roughness projections into the main body of the flow. Under these circumstances, the flow in the pipe becomes *fully turbulent*, the friction factor is independent of the Reynolds number, and the pipe is considered to be (hydraulically) rough. The flow is actually turbulent under both smooth-pipe and rough-pipe conditions, but the flow is termed *fully turbulent* when the friction factor is independent of the Reynolds number. Between the smooth- and rough-pipe conditions, there is a transition region in which the friction factor depends on both the Reynolds number and the relative roughness. Colebrook (1939) developed the following relationship that asymptotes to the Prandl and von Kármán relations

$$\frac{1}{\sqrt{f}} = -2\log\left(\frac{k/D}{3.7} + \frac{2.51}{\text{Re}\sqrt{f}}\right) \tag{3.35}$$

This equation is commonly referred to as the *Colebrook equation* or *Colebrook formula*. Equation 3.35 can be applied in the transition region between smooth-pipe and rough-pipe conditions.

Commercial pipes differ from Nikuradse's experimental pipes in that the heights of the roughness projections are not uniform and are not uniformly distributed. In commercial pipes, an *equivalent sand roughness*, k_s, is defined as the diameter of Nikuradse's sand grains that would cause the same head loss as in the commercial pipe. The equivalent sand roughness, k_s, of several commercial pipe materials are given in Table 3.1.

*Henry Darcy (1803–1858) was a nineteenth-century French engineer; Julius Weisbach was a German engineer of the same era. Weisbach proposed the use of a dimensionless resistance coefficient, and Darcy carried out the tests on water pipes.

■ **Table 3.1**
Equivalent Sand Roughness
in New Commercial Pipes

Material	Equivalent sand roughness, k_s (mm)
Riveted steel	0.9–9.0
Concrete	0.3–3.0
Ductile and cast iron	0.26
Galvanized iron	0.15
Asphalt-dipped ductile/cast iron	0.12
Commercial steel or wrought iron	0.046
Copper or brass tubing	0.0015
Glass, plastic (PVC)	≈ 0

Source: Moody (1944).

These values of k_s apply to clean new pipe only; pipe that has been in service for a long time usually experiences corrosion or scale buildup that results in values of k_s that are orders of magnitude larger than the values given in Table 3.1 (Echávez, 1997; Gerhart et al., 1992). The rate of increase of k_s with time depends primarily on the quality of the water being transported, and the roughness coefficients for older water mains are usually determined through field testing (AWWA, 1992). The expression for the friction factor derived by Colebrook (Equation 3.35) was plotted by Moody (1944) in what is today commonly referred to as the *Moody diagram*,* reproduced in Figure 3.3. The Moody diagram indicates that for Re $\leq 2,000$, the flow is laminar and the friction factor is given by

$$f = \frac{64}{\text{Re}} \qquad (3.36)$$

which can be derived theoretically based on the assumption of laminar flow of a Newtonian fluid (Daily and Harleman, 1966). Beyond a Reynolds number of 2,000–4,000, the flow is turbulent and the friction factor is controlled by the thickness of the laminar boundary layer relative to the height of the roughness projections on the surface of the pipe. The dashed line in Figure 3.3 indicates the boundary between the fully turbulent flow regime, where f is independent of Re, and the transition regime, where f depends on both Re and the relative roughness, k_s/D. The equation of this dashed line is given by Mott (1994) as

$$\frac{1}{\sqrt{f}} = \frac{\text{Re}}{200(D/k_s)} \qquad (3.37)$$

The line in the Moody diagram corresponding to a relative roughness of zero describes the friction factor for pipes that are hydraulically smooth.

Although the Colebrook equation (Equation 3.35) can be used to calculate the friction factor in lieu of the Moody diagram, this equation has the drawback that it is an *implicit equation* for the friction factor and must be solved iteratively. This minor inconvenience was circumvented by Jain (1976), who suggested the following explicit equation

---

*This type of diagram was originally suggested by Blasius in 1913 and Stanton in 1914 (Stanton and Pannell, 1914).

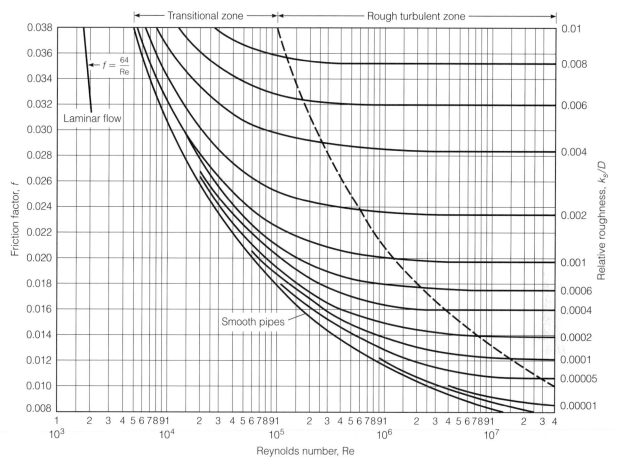

Figure 3.3 ■ Moody Diagram
Source: Moody, L. F. "Friction Factors for Pipe Flow" 66(8), 1944, ASME, New York.

for the friction factor

$$\frac{1}{\sqrt{f}} = -2\log\left(\frac{k_s/D}{3.7} + \frac{5.74}{\text{Re}^{0.9}}\right), \qquad 10^{-6} \le \frac{k_s}{D} \le 10^{-2}, \ 5{,}000 \le \text{Re} \le 10^8 \quad (3.38)$$

where, according to Jain (1976), Equation 3.38 deviates by less than 1% from the Colebrook equation within the entire turbulent flow regime, provided that the restrictions on k_s/D and Re are honored. According to Franzini and Finnemore (1997) and Granger (1985), values of the friction factor calculated using the Colebrook equation are generally accurate to within 10% to 15% of experimental data, while Potter and Wiggert (1991) put the accuracy of the Moody diagram at no more than 5%. The Jain equation (Equation 3.38) can be more conveniently written as

$$f = \frac{1.325}{[\ln(k_s/3.7D + 5.74/\text{Re}^{0.9})]^2}, \qquad 10^{-6} \le \frac{k_s}{D} \le 10^{-2}, \ 5{,}000 \le \text{Re} \le 10^8 \quad (3.39)$$

Uncertainties in relative roughness and in the data used to produce the Colebrook equation make the use of several-place accuracy in pipe flow problems unjustified. As a rule of thumb, an accuracy of 10% in calculating friction losses in pipes is to be expected (Munson et al., 1994; Gerhart et al., 1992).

■ Example 3.2

Water from a treatment plant is pumped into a distribution system at a rate of 4.38 m³/s, a pressure of 480 kPa, and a temperature of 20°C. The diameter of the pipe is 750 mm and is made of ductile iron. Estimate the pressure 200 m downstream of the treatment plant if the pipeline remains horizontal. Compare the friction factor estimated using the Colebrook equation to the friction factor estimated using the Jain equation. After 20 years in operation, scale buildup is expected to cause the equivalent sand roughness of the pipe to increase by a factor of 10. Determine the effect on the water pressure 200 m downstream of the treatment plant.

Solution

According to the Darcy-Weisbach equation, the difference in total head, Δh, between the upstream section (at exit from treatment plant) and the downstream section (200 m downstream from the upstream section) is given by

$$\Delta h = \frac{fL}{D}\frac{V^2}{2g}$$

where f is the friction factor, L is the pipe length between the upstream and downstream sections ($= 200$ m), D is the pipe diameter ($= 750$ mm), and V is the velocity in the pipe. The velocity, V, is given by

$$V = \frac{Q}{A}$$

where Q is the flowrate in the pipe ($= 4.38$ m³/s) and A is the area of the pipe cross-section given by

$$A = \frac{\pi}{4}D^2 = \frac{\pi}{4}(0.75)^2 = 0.442 \text{ m}^2$$

The pipeline velocity is therefore

$$V = \frac{Q}{A} = \frac{4.38}{0.442} = 9.91 \text{ m/s}$$

The friction factor, f, in the Darcy-Weisbach equation is calculated using the Colebrook equation

$$\frac{1}{\sqrt{f}} = -2\log\left[\frac{k_s}{3.7D} + \frac{2.51}{\text{Re}\sqrt{f}}\right]$$

where Re is the Reynolds number and k_s is the equivalent sand roughness of ductile iron ($= 0.26$ mm). The Reynolds number is given by

$$\text{Re} = \frac{VD}{\nu}$$

where v is the kinematic viscosity of water at 20°C, which is equal to 1.00×10^{-6} m²/s. Therefore

$$\text{Re} = \frac{VD}{v} = \frac{(9.91)(0.75)}{1.00 \times 10^{-6}} = 7.43 \times 10^6$$

Substituting into the Colebrook equation leads to

$$\frac{1}{\sqrt{f}} = -2 \log \left[\frac{0.26}{(3.7)(750)} + \frac{2.51}{7.43 \times 10^6 \sqrt{f}} \right]$$

or

$$\frac{1}{\sqrt{f}} = -2 \log \left[9.37 \times 10^{-5} + \frac{3.38 \times 10^{-7}}{\sqrt{f}} \right]$$

This is an implicit equation for f, and by trial and error the solution is

$$f = 0.016$$

The head loss, Δh, between the upstream and downstream sections can now be calculated using the Darcy-Weisbach equation as

$$\Delta h = \frac{fL}{D} \frac{V^2}{2g} = \frac{(0.016)(200)}{0.75} \frac{(9.91)^2}{(2)(9.81)} = 21.4 \text{ m}$$

Using the definition of head loss, Δh,

$$\Delta h = \left(\frac{p_1}{\gamma} + z_1 \right) - \left(\frac{p_2}{\gamma} + z_2 \right)$$

where p_1 and p_2 are the upstream and downstream pressures, γ is the specific weight of water, and z_1 and z_2 are the upstream and downstream pipe elevations. Since the pipe is horizontal, $z_1 = z_2$ and Δh can be written in terms of the pressures at the upstream and downstream sections as

$$\Delta h = \frac{p_1}{\gamma} - \frac{p_2}{\gamma}$$

In this case, $p_1 = 480$ kPa, $\gamma = 9.79$ kN/m³, and therefore

$$21.4 = \frac{480}{9.79} - \frac{p_2}{9.79}$$

which yields

$$p_2 = 270 \text{ kPa}$$

Therefore, the pressure 200 m downstream of the treatment plant is 270 kPa. The Colebrook equation required that f be determined iteratively, but the explicit Jain approximation for f is given by

$$\frac{1}{\sqrt{f}} = -2 \log \left[\frac{k_s}{3.7D} + \frac{5.74}{\text{Re}^{0.9}} \right]$$

Substituting for k_s, D, and Re gives

$$\frac{1}{\sqrt{f}} = -2\log\left[\frac{0.26}{(3.7)(750)} + \frac{5.74}{(7.43 \times 10^6)^{0.9}}\right]$$

which leads to

$$f = 0.016$$

This is the same friction factor obtained using the Colebrook equation within an accuracy of two significant digits.

After 20 years, the equivalent sand roughness, k_s, of the pipe is 2.6 mm, the (previously calculated) Reynolds number is 7.43×10^6, and the Colebrook equation gives

$$\frac{1}{\sqrt{f}} = -2\log\left[\frac{2.6}{(3.7)(750)} + \frac{2.51}{7.43 \times 10^6 \sqrt{f}}\right]$$

or

$$\frac{1}{\sqrt{f}} = -2\log\left[9.37 \times 10^{-4} + \frac{3.38 \times 10^{-7}}{\sqrt{f}}\right]$$

which yields

$$f = 0.027$$

The head loss, Δh, between the upstream and downstream sections is given by the Darcy-Weisbach equation as

$$\Delta h = \frac{f L}{D}\frac{V^2}{2g} = \frac{(0.027)(200)}{0.75}\frac{(9.91)^2}{(2)(9.81)} = 36.0 \text{ m}$$

Hence the pressure, p_2, 200 m downstream of the treatment plant is given by the relation

$$\Delta h = \frac{p_1}{\gamma} - \frac{p_2}{\gamma}$$

where $p_1 = 480$ kPa, $\gamma = 9.79$ kN/m³, and therefore

$$36.0 = \frac{480}{9.79} - \frac{p_2}{9.79}$$

which yields

$$p_2 = 128 \text{ kPa}$$

Therefore, pipe aging over 20 years will cause the pressure 200 m downstream of the treatment plant to decrease from 270 kPa to 128 kPa. This is quite a significant drop and shows why velocities of 9.91 m/s are not used in these pipelines, even for short lengths of pipe.   ■

The problem in Example 3.2 illustrates the case where the flowrate through a pipe is known and the objective is to calculate the head loss and pressure drop over a given length of pipe. The approach is summarized as follows: (1) calculate the Reynolds number, Re, and the relative roughness, k_s/D, from the given data; (2) use the Cole-brook equation (Equation 3.35) or Jain equation (Equation 3.38) to calculate f; and (3) use the calculated value of f to calculate the head loss from the Darcy-Weisbach equation (Equation 3.33), and the corresponding pressure drop from Equation 3.22.

Flowrate for a Given Head Loss. In many cases, the flowrate through a pipe is not controlled but attains a level that matches the pressure drop available. For example, the flowrate through faucets in home plumbing is determined by the gage pressure in the water main, which is relatively insensitive to the flow through the faucet. A useful ap-proach to this problem that utilizes the Colebrook equation has been suggested by Fay (1994), where the first step is to calculate $\mathrm{Re}\sqrt{f}$ using the rearranged Darcy-Weisbach equation

$$\mathrm{Re}\sqrt{f} = \left(\frac{2gh_f D^3}{\nu^2 L}\right)^{\frac{1}{2}} \qquad (3.40)$$

Using this value of $\mathrm{Re}\sqrt{f}$, solve for Re using the rearranged Colebrook equation

$$\mathrm{Re} = -2.0(\mathrm{Re}\sqrt{f}) \log\left(\frac{k_s/D}{3.7} + \frac{2.51}{\mathrm{Re}\sqrt{f}}\right) \qquad (3.41)$$

Using this value of Re, the flowrate, Q, can then be calculated by

$$Q = \frac{1}{4}\pi D^2 V = \frac{1}{4}\pi D\nu\mathrm{Re} \qquad (3.42)$$

This approach must necessarily be validated by verifying that $\mathrm{Re} > 2{,}300$, which is re-quired for the application of the Colebrook equation. Swamee and Jain (1976) combine Equations 3.40 to 3.42 to yield

$$Q = -0.965D^2\sqrt{\frac{gDh_f}{L}}\ln\left(\frac{k_s/D}{3.7} + \frac{1.784\nu}{D\sqrt{gDh_f/L}}\right) \qquad (3.43)$$

▪ Example 3.3

A 50-mm diameter galvanized iron service pipe is connected to a water main in which the pressure is 450 kPa gage. If the length of the service pipe to a faucet is 40 m and the faucet is 1.2 m above the main, estimate the flowrate when the faucet is fully open.

Solution

The head loss, h_f, in the pipe is estimated by

$$h_f = \left(\frac{p_{\mathrm{main}}}{\gamma} + z_{\mathrm{main}}\right) - \left(\frac{p_{\mathrm{outlet}}}{\gamma} + z_{\mathrm{outlet}}\right)$$

where $p_{\mathrm{main}} = 450$ kPa, $z_{\mathrm{main}} = 0$ m, $p_{\mathrm{outlet}} = 0$ kPa, and $z_{\mathrm{outlet}} = 1.2$ m. Therefore, taking $\gamma = 9.79$ kn/m^3 (at 20°C) gives

$$h_f = \left(\frac{450}{9.79} + 0\right) - (0 + 1.2) = 44.8 \text{ m}$$

Also, since $D = 50$ mm, $L = 40$ m, $k_s = 0.15$ mm (from Table 3.1), $\nu = 1.00 \times 10^{-6}$ m²/s (at 20°C), the Swamee-Jain equation (Equation 3.43) yields

$$Q = -0.965 D^2 \sqrt{\frac{gDh_f}{L}} \ln\left(\frac{k_s/D}{3.7} + \frac{1.784\nu}{D\sqrt{gDh_f/L}}\right)$$

$$= -0.965(0.05)^2 \sqrt{\frac{(9.81)(0.05)(44.8)}{40}} \ln\left[\frac{0.15/50}{3.7} + \frac{1.784(1.00 \times 10^{-6})}{(0.05)\sqrt{(9.81)(0.05)(44.8)/40}}\right]$$

$$= 0.0126 \text{ m}^3/\text{s} = 12.6 \text{ L/s}$$

The faucet can therefore be expected to deliver 12.6 L/s when fully open.   ■

Diameter for a Given Flowrate and Head Loss.   In many cases, an engineer must select a size of pipe to provide a given level of service. For example, the maximum flowrate and maximum allowable pressure drop may be specified for a water delivery pipe, and the engineer is required to calculate the minimum diameter pipe that will satisfy these design constraints. Solution of this problem necessarily requires an iterative procedure. Streeter and Wylie (1985) have suggested the following steps

1. Assume a value of f.
2. Calculate D from the rearranged Darcy-Weisbach equation,

$$D = \sqrt[5]{\left(\frac{8LQ^2}{h_f g\pi^2}f\right)} \tag{3.44}$$

   where the term in parentheses can be calculated from given data.
3. Calculate Re from

$$\text{Re} = \frac{VD}{\nu} = \left(\frac{4Q}{\pi\nu}\right)\frac{1}{D} \tag{3.45}$$

   where the term in parentheses can be calculated from given data.
4. Calculate k_s/D.
5. Use Re and k_s/D to calculate f from the Colebrook equation.
6. Using the new f, repeat the procedure until the new f agrees with the old f to the first two significant digits.

■ Example 3.4

A galvanized iron service pipe from a water main is required to deliver 200 L/s during a fire. If the length of the service pipe is 35 m and the head loss in the pipe is not to exceed 50 m, calculate the minimum pipe diameter that can be used.

Solution

Step 1: Assume $f = 0.03$
Step 2: Since $Q = 0.2$ m³/s, $L = 35$ m, and $h_f = 50$ m, then

$$D = \sqrt[5]{\left[\frac{8LQ^2}{h_f g\pi^2}\right]f} = \sqrt[5]{\left[\frac{8(35)(0.2)^2}{(50)(9.81)\pi^2}\right](0.03)} = 0.147 \text{ m}$$

Step 3: Since $v = 1.00 \times 10^{-6}$ m²/s (at 20°C), then

$$\text{Re} = \left[\frac{4Q}{\pi v}\right]\frac{1}{D} = \left[\frac{4(0.2)}{\pi(1.00 \times 10^{-6})}\right]\frac{1}{0.147} = 1.73 \times 10^6$$

Step 4: Since $k_s = 0.15$ mm (from Table 3.1, for new pipe), then

$$\frac{k_s}{D} = \frac{1.5 \times 10^{-4}}{0.147} = 0.00102$$

Step 5: Using the Colebrook equation (Equation 3.35) gives

$$\frac{1}{\sqrt{f}} = -2\log\left(\frac{k_s/D}{3.7} + \frac{2.51}{\text{Re}\sqrt{f}}\right) = -2\log\left(\frac{0.00102}{3.7} + \frac{2.51}{1.73 \times 10^6\sqrt{f}}\right)$$

which leads to

$$f = 0.020$$

Step 6: $f = 0.020$ differs from the assumed f (= 0.03), so repeat the procedure with $f = 0.020$.
Step 2: For $f = 0.020$, $D = 0.136$ m
Step 3: For $D = 0.136$, Re $= 1.87 \times 10^6$
Step 4: For $D = 0.136$, $k_s/D = 0.00110$
Step 5: $f = 0.020$
Step 6: The calculated f (= 0.020) is equal to the assumed f. The required pipe diameter is therefore equal to 0.136 m or 136 mm. A commercially available pipe with the closest diameter larger than 136 mm should be used. ■

The iterative procedure demonstrated in the previous example converges fairly quickly, and does not pose any computational difficulty. Swamee and Jain (1976) have suggested the following explicit formula for calculating the pipe diameter, D,

$$D = 0.66\left[k_s^{1.25}\left(\frac{LQ^2}{gh_f}\right)^{4.75} + vQ^{9.4}\left(\frac{L}{gh_f}\right)^{5.2}\right]^{0.04},$$

$$3{,}000 \le \text{Re} \le 3 \times 10^8, \quad 10^{-6} \le \frac{k_s}{D} \le 2 \times 10^{-2} \tag{3.46}$$

Equation 3.46 will yield a D within 5% of the value obtained by the method using the Colebrook equation. This method is illustrated by repeating the previous example.

■ Example 3.5

A galvanized iron service pipe from a water main is required to deliver 200 L/s during a fire. If the length of the service pipe is 35 m, and the head loss in the pipe is not to exceed 50 m, use the Swamee-Jain equation to calculate the minimum pipe diameter that can be used.

Solution

Since $k_s = 0.15$ mm, $L = 35$ m, $Q = 0.2$ m³/s, $h_f = 50$ m, $v = 1.00 \times 10^{-6}$ m²/s, the Swamee-Jain equation gives

$$D = 0.66 \left[k_s^{1.25} \left(\frac{LQ^2}{gh_f} \right)^{4.75} + \nu Q^{9.4} \left(\frac{L}{gh_f} \right)^{5.2} \right]^{0.04}$$

$$= 0.66 \left\{ (0.00015)^{1.25} \left[\frac{(35)(0.2)^2}{(9.81)(50)} \right]^{4.75} + (1.00 \times 10^{-6})(0.2)^{9.4} \left[\frac{35}{(9.81)(50)} \right]^{5.2} \right\}^{0.04}$$

$$= 0.140 \text{ m}$$

The calculated pipe diameter (140 mm) is about 3% higher than calculated by the Cole-brook equation (136 mm). ■

3.2.3 ■ Energy Equation

The steady-state energy equation for the control volume illustrated in Figure 3.4 is given by

$$\frac{dQ_h}{dt} - \frac{dW}{dt} = \int_A \rho e \, \mathbf{v} \cdot \mathbf{n} \, dA \tag{3.47}$$

where Q_h is the heat added to the fluid in the control volume, W is the work done by the fluid in the control volume, A is the surface area of the control volume, ρ is the density of the fluid in the control volume, and e is the internal energy per unit mass of fluid in the control volume given by

$$e = gz + \frac{v^2}{2} + u \tag{3.48}$$

where z is the elevation of the fluid mass having a velocity v and internal energy u. By convention, the heat added to a system and the work done by a system are positive quantities. The normal stresses on the inflow and outflow boundaries of the control volume are equal to the pressure, p, with shear stresses tangential to the boundaries of the control volume. As the fluid moves across the control surface with velocity $\mathbf{v}$, the power (= rate of doing work) expended by the fluid against the external pressure forces is given by

$$\frac{dW_p}{dt} = \int_A p \mathbf{v} \cdot \mathbf{n} \, dA \tag{3.49}$$

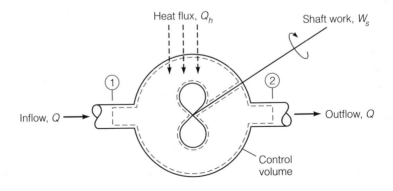

Figure 3.4 ■ Energy Balance in Closed Conduit

where W_p is the work done against external pressure forces. The work done by a fluid in the control volume is typically separated into work done against external pressure forces, W_p, plus work done against rotating surfaces, W_s, commonly referred to as the *shaft work*. The rotating element is called a *rotor* in a gas or steam turbine, an *impeller* in a pump, and a *runner* in a hydraulic turbine. The rate at which work is done by a fluid system, dW/dt, can therefore be written as

$$\frac{dW}{dt} = \frac{dW_p}{dt} + \frac{dW_s}{dt} = \int_A p\mathbf{v} \cdot \mathbf{n}\, dA + \frac{dW_s}{dt} \tag{3.50}$$

Combining Equation 3.50 with the steady-state energy equation (Equation 3.47) leads to

$$\frac{dQ_h}{dt} - \frac{dW_s}{dt} = \int_A \rho\left(\frac{p}{\rho} + e\right)\mathbf{v} \cdot \mathbf{n}\, dA \tag{3.51}$$

Substituting the definition of the internal energy, e, given by Equation 3.48 into Equation 3.51 yields

$$\frac{dQ_h}{dt} - \frac{dW_s}{dt} = \int_A \rho\left(h + gz + \frac{v^2}{2}\right)\mathbf{v} \cdot \mathbf{n}\, dA \tag{3.52}$$

where h is the enthalpy of the fluid defined by

$$h = \frac{p}{\rho} + u \tag{3.53}$$

Denoting the rate at which heat is being added to the fluid system by $\dot{Q}$, and the rate at which work is being done against moving impervious boundaries (shaft work) by $\dot{W}_s$, then the energy equation can be written in the form

$$\dot{Q} - \dot{W}_s = \int_A \rho\left(h + gz + \frac{v^2}{2}\right)\mathbf{v} \cdot \mathbf{n}\, dA \tag{3.54}$$

Considering the terms $h + gz$, where

$$h + gz = \frac{p}{\rho} + u + gz = g\left(\frac{p}{\gamma} + z\right) + u \tag{3.55}$$

and γ is the specific weight of the fluid, then Equation 3.55 indicates that $h + gz$ can be assumed to be constant across the inflow and outflow openings illustrated in Figure 3.4, since a hydrostatic pressure distribution across the inflow/outflow boundaries guarantees that $p/\gamma + z$ is constant across the inflow/outflow boundaries normal to the flow direction, and the internal energy, u, depends only on the temperature, which can be assumed constant across each boundary. Since $\mathbf{v} \cdot \mathbf{n}$ is equal to zero over the impervious boundaries in contact with the fluid system, Equation 3.54 can be integrated to yield

$$\dot{Q} - \dot{W}_s = (h_1 + gz_1)\int_{A_1} \rho\mathbf{v} \cdot \mathbf{n}\, dA + \int_{A_1} \rho\frac{v^2}{2}\mathbf{v} \cdot \mathbf{n}\, dA + (h_2 + gz_2)\int_{A_2} \rho\mathbf{v} \cdot \mathbf{n}\, dA$$
$$+ \int_{A_2} \rho\frac{v^2}{2}\mathbf{v} \cdot \mathbf{n}\, dA$$

$$= -(h_1 + gz_1) \int_{A_1} \rho v_1 \, dA - \int_{A_1} \rho \frac{v_1^3}{2} \, dA + (h_2 + gz_2) \int_{A_2} \rho v_2 \, dA$$

$$+ \int_{A_2} \rho \frac{v_2^3}{2} \, dA \tag{3.56}$$

where the subscripts 1 and 2 refer to the inflow and outflow boundaries, respectively, and the negative signs result from the fact that the unit normal points out of the control volume, causing $\mathbf{v} \cdot \mathbf{n}$ to be negative on the inflow boundary and positive on the outflow boundary.

Equation 3.56 can be simplified by noting that the assumption of steady flow requires that rate of mass inflow to the control volume is equal to the mass outflow rate and, denoting the mass flow rate by $\dot{m}$, the continuity equation requires that

$$\dot{m} = \int_{A_1} \rho v_1 \, dA = \int_{A_2} \rho v_2 \, dA \tag{3.57}$$

Furthermore, the constants α_1 and α_2 can be defined by the equations

$$\int_{A_1} \rho \frac{v^3}{2} \, dA = \alpha_1 \rho \frac{V_1^3}{2} A_1 \tag{3.58}$$

$$\int_{A_2} \rho \frac{v^3}{2} \, dA = \alpha_2 \rho \frac{V_2^3}{2} A_2 \tag{3.59}$$

where A_1 and A_2 are the areas of the inflow and outflow boundaries, respectively, and V_1 and V_2 are the corresponding mean velocities across these boundaries. The constants α_1 and α_2 are determined by the velocity profile across the flow boundaries, and these constants are called *kinetic energy correction factors*. If the velocity is constant across a flow boundary, then it is clear from Equation 3.58 that the kinetic energy correction factor for that boundary is equal to unity; for any other velocity distribution, the kinetic energy factor is greater than unity. Combining Equations 3.56 to 3.59 leads to

$$\dot{Q} - \dot{W}_s = -(h_1 + gz_1)\dot{m} - \alpha_1 \rho \frac{V_1^3}{2} A_1 + (h_2 + gz_2)\dot{m} + \alpha_2 \rho \frac{V_2^3}{2} A_2 \tag{3.60}$$

Invoking the continuity equation requires that

$$\rho V_1 A_1 = \rho V_2 A_2 = \dot{m} \tag{3.61}$$

and combining Equations 3.60 and 3.61 leads to

$$\dot{Q} - \dot{W}_s = \dot{m} \left[\left(h_2 + gz_2 + \alpha_2 \frac{V_2^2}{2} \right) - \left(h_1 + gz_1 + \alpha_1 \frac{V_1^2}{2} \right) \right] \tag{3.62}$$

which can be put in the form

$$\frac{\dot{Q}}{\dot{m}g} - \frac{\dot{W}_s}{\dot{m}g} = \left(\frac{p_2}{\gamma} + \frac{u_2}{g} + z_2 + \alpha_2 \frac{V_2^2}{2g} \right) - \left(\frac{p_1}{\gamma} + \frac{u_1}{g} + z_1 + \alpha_1 \frac{V_1^2}{2g} \right) \tag{3.63}$$

and can be further rearranged into the useful form

$$\left(\frac{p_1}{\gamma} + \alpha_1 \frac{V_1^2}{2g} + z_1 \right) = \left(\frac{p_2}{\gamma} + \alpha_2 \frac{V_2^2}{2g} + z_2 \right) + \left[\frac{1}{g}(u_2 - u_1) - \frac{\dot{Q}}{\dot{m}g} \right] + \left[\frac{\dot{W}_s}{\dot{m}g} \right] \tag{3.64}$$

Two key terms can be identified in Equation 3.64: the (shaft) work done by the fluid per unit weight, h_s, defined by the relation

$$h_s = \frac{\dot{W}_s}{\dot{m}g} \tag{3.65}$$

and the energy loss per unit weight, commonly called the head loss, h_L, defined by the relation

$$h_L = \frac{1}{g}(u_2 - u_1) - \frac{\dot{Q}}{\dot{m}g} \tag{3.66}$$

Combining Equations 3.64 to 3.66 leads to the most common form of the *energy equation*

$$\boxed{\left(\frac{p_1}{\gamma} + \alpha_1 \frac{V_1^2}{2g} + z_1\right) = \left(\frac{p_2}{\gamma} + \alpha_2 \frac{V_2^2}{2g} + z_2\right) + h_L + h_s} \tag{3.67}$$

where a positive head loss indicates an increase in internal energy, manifested by an increase in temperature or a loss of heat, and a positive value of h_s is associated with work being done by the fluid, such as in moving a turbine runner. Many practitioners incorrectly refer to Equation 3.67 as the *Bernoulli equation*, which bears some resemblance to Equation 3.67 but is different in several important respects. Fundamental differences between the energy equation and the Bernoulli equation are that the Bernoulli equation is derived from the momentum equation, which is independent of the energy equation, and the Bernoulli equation does not account for fluid friction.

Energy and Hydraulic Grade Lines.   The *total head*, h, of a fluid at any cross-section of a pipe is defined by

$$\boxed{h = \frac{p}{\gamma} + \alpha \frac{V^2}{2g} + z} \tag{3.68}$$

where p is the pressure in the fluid at the centroid of the cross-section, γ is the specific weight of the fluid, α is the kinetic energy correction factor, V is the average velocity across the pipe, and z is the elevation of the centroid of the pipe. The total head measures the average energy per unit weight of the fluid flowing across a pipe cross-section. The energy equation, Equation 3.67, states that changes in the total head along the pipe are described by

$$h(x + \Delta x) = h(x) - (h_L + h_s) \tag{3.69}$$

where x is the coordinate measured along the pipe centerline, Δx is the distance between two cross-sections in the pipe, h_L is the head loss, and h_s is the shaft work done by the fluid over the distance Δx. The practical application of Equation 3.69 is illustrated in Figure 3.5, where the head loss, h_L, between two sections a distance Δx apart is indicated. At each cross-section, the total energy, h, is plotted relative to a defined datum, and the locus of these points is called the *energy grade line*. The energy grade line at each pipe cross-section is located a distance $p/\gamma + \alpha V^2/2g$ vertically above the centroid of the cross-section, and between any two cross-sections the elevation of the energy grade line falls by a vertical distance equal to the head loss caused by pipe friction, h_L, plus the shaft work, h_s, done by the fluid. The *hydraulic grade line* measures the hydraulic head

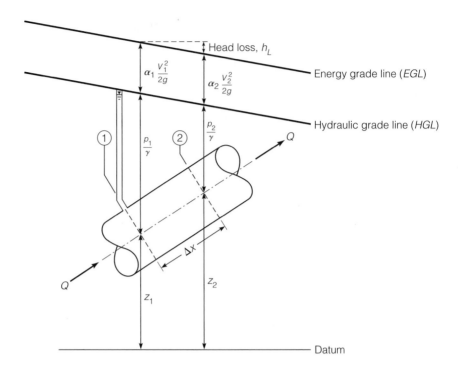

Figure 3.5 ■ Head Loss
Along Pipe

$p/\gamma + z$ at each pipe cross-section. It is located a distance p/γ above the pipe centerline and indicates the elevation to which the fluid would rise in an open tube connected to the wall of the pipe section. The hydraulic grade line is therefore located a distance $\alpha V^2/2g$ below the energy grade line.

Both the hydraulic grade line and the energy grade line are useful in visualizing the state of the fluid as it flows along the pipe and are frequently used in assessing the performance of fluid delivery systems. Most fluid delivery systems, for example, require that the fluid pressure remain positive, in which case the hydraulic grade line must remain above the pipe. In circumstances where additional energy is required to maintain acceptable pressures in pipelines, a pump is installed along the pipeline to elevate the energy grade line by an amount h_s, which also elevates the hydraulic grade line by the same amount. This condition is illustrated in Figure 3.6. In cases where the pipeline upstream and downstream of the pump are of the same diameter, then the velocity heads $\alpha V^2/2g$ both upstream and downstream of the pump are the same, and the head added by the pump, h_s, goes entirely to increase the pressure head, p/γ, of the fluid. It should also be clear from Figure 3.5 that the pressure head in a pipeline can be increased by simply increasing the pipeline diameter, which reduces the velocity head, $\alpha V^2/2g$, and thereby increases the pressure head, p/γ, to maintain the same total energy at the pipe section.

Velocity Profile. The momentum and energy correction factors, α and β, depend on the cross-sectional velocity distribution. The velocity profile in both smooth and rough pipes of circular cross-section can be estimated by the semi-empirical equation

$$v(r) = \left[(1 + 1.326\sqrt{f}) - 2.04\sqrt{f} \log\left(\frac{R}{R-r}\right) \right] V \qquad (3.70)$$

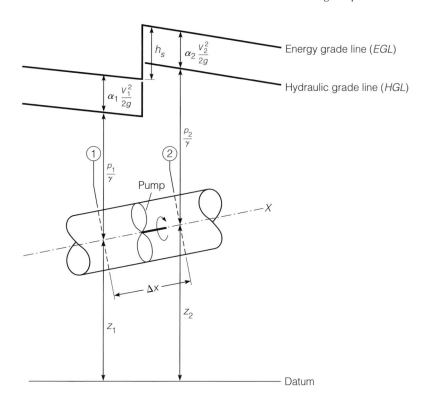

Figure 3.6 ■ Pump Effect on Flow in Pipeline

where $v(r)$ is the velocity at a radial distance r from the centerline of the pipe, R is the radius of the pipe, f is the friction factor, and V is the average velocity across the pipe.

The velocity distribution given by Equation 3.70 agrees well with velocity measurements in both smooth and rough pipes. This equation, however, is not applicable within the small region close to the centerline of the pipe and is also not applicable in the small region close to the pipe boundary. This is apparent since at the axis of the pipe dv/dr must be equal to zero, but Equation 3.70 does not have a zero slope at $r = 0$. The pipe boundary v must also be equal to zero, but Equation 3.70 gives a velocity of zero at a small distance from the wall, with a velocity of $-\infty$ at $r = R$. The energy and momentum correction factors, α and β, derived from the velocity profile are (Moody, 1950)

$$\alpha = 1 + 2.7f \tag{3.71}$$

$$\beta = 1 + 0.98f \tag{3.72}$$

Another commonly used equation to describe the velocity distribution in turbulent pipe flow is the empirical *power law* equation given by

$$v(r) = V_o \left(1 - \frac{r}{R}\right)^{\frac{1}{n}} \tag{3.73}$$

where V_o is the centerline velocity and n is a function of the Reynolds number, Re. Values of n typically range between 6 and 10 and can be approximated by (Fox and McDonald, 1992)

$$n = 1.83 \log \text{Re} - 1.86 \tag{3.74}$$

The power law is not applicable within $0.04R$ of the wall, since the power law gives an infinite velocity gradient at the wall. Although the profile fits the data close to the centerline of the pipe, it does not give zero slope at the centerline. The kinetic energy coefficient, α, derived from the power law equation is given by

$$\alpha = \frac{(1+n)^3(1+2n)^3}{4n^4(3+n)(3+2n)} \qquad (3.75)$$

For n between 6 and 10, α varies from 1.08 to 1.03.

In most engineering applications, α and β are taken as unity (see Problem 3.14).

Head Losses in Transitions and Fittings.   The head losses in straight pipes of constant diameter are caused by friction between the moving fluid and the pipe boundary and are estimated using the Darcy-Weisbach equation. Flow through pipe fittings, around bends, and through changes in pipeline geometry cause additional head losses, h_o, that are quantified by an equation of the form

$$\boxed{h_o = \sum K \frac{V^2}{2g}} \qquad (3.76)$$

where K is a loss coefficient that is specific to each fitting and transition and V is the average velocity at a defined location within the transition or fitting. The loss coefficients for several fittings and transitions are shown in Figure 3.7. Head losses in transitions and fittings are also called *local head losses* or *minor head losses*. The latter term should be avoided, however, since in some cases these head losses are a significant portion of the total head loss in a pipe. Detailed descriptions of local head losses in various valve geometries can be found in Mott (1994), and additional data on local head losses in pipeline systems can be found in Brater and colleagues (1996).

■ Example 3.6

A pump is to be selected that will pump water from a well into a storage reservoir. In order to fill the reservoir in a timely manner, the pump is required to deliver 5 L/s when the water level in the reservoir is 5 m above the water level in the well. Find the head that must be added by the pump. The pipeline system is shown in Figure 3.8. Assume that the minor loss coefficient for each of the bends is equal to 0.25 and that the temperature of the water is 20°C.

Solution

Taking the elevation of the water surface in the well to be equal to 0 m, and proceeding from the well to the storage reservoir (where the head is equal to 5 m), the energy equation (Equation 3.67) can be written as

$$0 - \frac{V_1^2}{2g} - \frac{f_1 L_1}{D_1}\frac{V_1^2}{2g} - K_1\frac{V_1^2}{2g} + h_p - \frac{f_2 L_2}{D_2}\frac{V_2^2}{2g} - (K_2 + K_3)\frac{V_2^2}{2g} - \frac{V_2^2}{2g} = 5$$

where V_1 and V_2 are the velocities in the 50-mm ($= D_1$) and 100-mm ($= D_2$) pipes, respectively; L_1 and L_2 are the corresponding pipe lengths; f_1 and f_2 are the corresponding friction factors; K_1, K_2, and K_3 are the loss coefficients for each of the three bends; and

Description	Sketch	Additional Data		K	
Pipe entrance	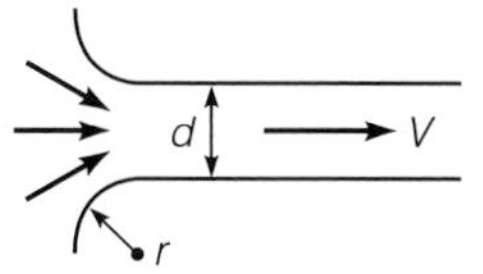	r/d		K	
		0.0		0.50	
		0.1		0.12	
		>0.2		0.03	
Contraction	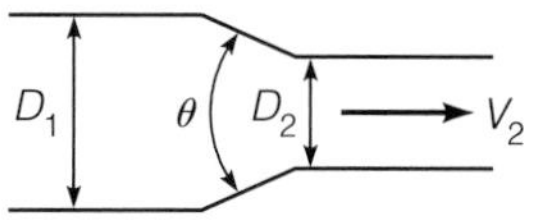	D_2/D_1		K $\theta = 60°$	K $\theta = 180°$
		0.0		0.08	0.50
		0.20		0.08	0.49
		0.40		0.07	0.42
		0.60		0.06	0.27
		0.80		0.06	0.20
		0.90		0.06	0.10
Expansion	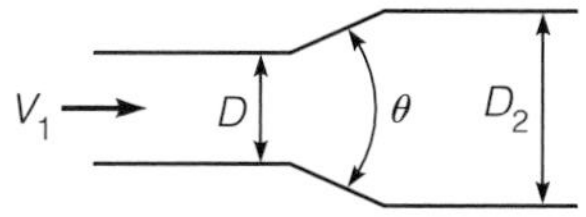	D_1/D_2		K $\theta = 20°$	K $\theta = 180°$
		0.0			1.00
		0.20		0.30	0.87
		0.40		0.25	0.70
		0.60		0.15	0.41
		0.80		0.10	0.15
90° miter bend	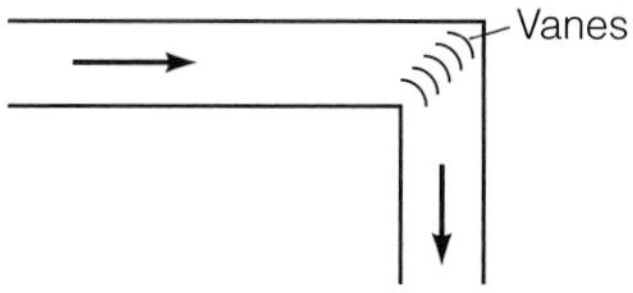	Without vanes		$K = 1.1$	
		With vanes		$K = 0.2$	
90° smooth bend	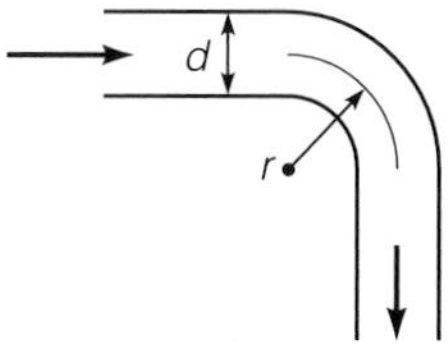	r/d		K $\theta = 90°$	
		1		0.35	
		2		0.19	
		4		0.16	
		6		0.21	
Threaded pipe fittings	Globe valve — wide open			K 10.0	
	Angle valve — wide open			5.0	
	Gate valve — wide open			0.2	
	Gate valve — half open			5.6	
	Return bend			2.2	
	Tee				
	straight-through flow			0.4	
	side-outlet flow			1.8	
	90° elbow			0.9	
	45° elbow			0.4	

Figure 3.7 ■ Loss Coefficients for Transitions and Fittings

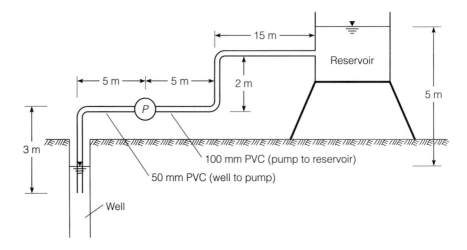

Figure 3.8 ■
Pipeline System

h_p is the head added by the pump. The cross-sectional areas of each of the pipes, A_1 and A_2, are given by

$$A_1 = \frac{\pi}{4}D_1^2 = \frac{\pi}{4}(0.05)^2 = 0.001963 \text{ m}^2$$

$$A_2 = \frac{\pi}{4}D_2^2 = \frac{\pi}{4}(0.10)^2 = 0.007854 \text{ m}^2$$

When the flowrate, Q, is 5 L/s, the velocities V_1 and V_2 are given by

$$V_1 = \frac{Q}{A_1} = \frac{0.005}{0.001963} = 2.54 \text{ m/s}$$

$$V_2 = \frac{Q}{A_2} = \frac{0.005}{0.007854} = 0.637 \text{ m/s}$$

PVC pipe is considered smooth ($k_s \approx 0$) and therefore the friction factor, f, can be estimated using the Jain equation

$$f = \frac{0.25}{\left[\log_{10}\dfrac{5.74}{\text{Re}^{0.9}}\right]^2}$$

where Re is the Reynolds number. At 20°C, the kinematic viscosity, ν, is equal to 1.00×10^{-6} m^2/s and for the 50-mm pipe

$$\text{Re}_1 = \frac{V_1 D_1}{\nu} = \frac{(2.51)(0.05)}{1.00 \times 10^{-6}} = 1.27 \times 10^5$$

which leads to

$$f_1 = \frac{0.25}{\left[\log_{10}\dfrac{5.74}{(1.27 \times 10^5)^{0.9}}\right]^2} = 0.0170$$

and for the 100-mm pipe

$$\text{Re}_2 = \frac{V_2 D_2}{\nu} = \frac{(0.637)(0.10)}{1.00 \times 10^{-6}} = 6.37 \times 10^4$$

which leads to

$$f_2 = \frac{0.25}{\left[\log_{10}\dfrac{5.74}{(6.37 \times 10^4)^{0.9}}\right]^2} = 0.0197$$

Substituting the values of these parameters into the energy equation yields

$$0 - \left[1 + \frac{(0.0170)(8)}{0.05} + 0.25\right]\frac{2.54^2}{(2)(9.81)} + h_p$$

$$- \left[\frac{(0.0197)(22)}{0.10} + 0.25 + 0.25 + 1\right]\frac{0.637^2}{(2)(9.81)} = 5$$

which leads to

$$h_p = 6.43 \text{ m}$$

Therefore the head to be added by the pump is 6.43 m. ■

Minor losses are frequently neglected in the analysis of pipeline systems. As a general rule, neglecting minor losses is justified when, on average, there is a length of 1,000 diameters between each minor loss (Streeter et al., 1998).

Head Losses in Noncircular Conduits. Most pipelines are of circular cross-section, but flow of water in noncircular conduits is commonly encountered in cases such as rectangular culverts flowing full. The hydraulic radius, R, of a conduit of any shape is defined by the relation

$$R = \frac{A}{P} \tag{3.77}$$

where A is the cross-sectional area of the conduit and P is the wetted perimeter. For circular conduits of diameter D, the hydraulic radius is given by

$$R = \frac{\pi D^2/4}{\pi D} = \frac{D}{4} \tag{3.78}$$

or

$$D = 4R \tag{3.79}$$

Using the hydraulic radius, R, as the length scale of a closed conduit instead of D, the frictional head losses, h_f, in noncircular conduits can be estimated using the Darcy-Weisbach equation for circular conduits by simply replacing D by $4R$, which yields

$$\boxed{h_f = \frac{f L}{4R}\frac{V^2}{2g}} \tag{3.80}$$

where the friction factor, f, is calculated using a Reynolds number, Re, defined by

$$\text{Re} = \frac{\rho V(4R)}{\mu} \tag{3.81}$$

and a relative roughness defined by $k_s/4R$.

Characterizing a noncircular conduit by the hydraulic radius, R, is necessarily approximate since conduits of arbitrary cross-section cannot be described with a single parameter. Secondary currents that are generated across a noncircular conduit cross-section to redistribute the shears are another reason why noncircular conduits cannot be completely characterized by the hydraulic radius (Liggett, 1994). However, according to Munson and colleagues (1994) and White (1994), using the hydraulic radius as a basis for calculating frictional head losses in noncircular conduits is usually acccurate to within 15% for turbulent flow. This approximation is much less accurate for laminar flows, where the accuracy is on the order of $\pm 40\%$ (White, 1994). Olson and Wright (1990) state that this approach can be used for rectangular conduits where the ratio of sides, called the *aspect ratio*, does not exceed about 8. Potter and Wiggert (1991) state that aspect ratios must be less than 4:1.

■ Example 3.7

Water flows through a rectangular concrete culvert of width 2 m and depth 1 m. If the length of the culvert is 10 m and the flowrate is 6 m^3/s, estimate the head loss through the culvert. Assume that the culvert flows full.

Solution

The head loss can be calculated using Equation 3.80. The hydraulic radius, R, is given by

$$R = \frac{A}{P} = \frac{(2)(1)}{2(2+1)} = 0.333 \text{ m}$$

and the mean velocity, V, is given by

$$V = \frac{Q}{A} = \frac{6}{(2)(1)} = 3 \text{ m/s}$$

At 20°C, $\nu = 1.00 \times 10^{-6}$ m^2/s, and therefore the Reynolds number, Re, is given by

$$\text{Re} = \frac{V(4R)}{\nu} = \frac{(3)(4 \times 0.333)}{1.00 \times 10^{-6}} = 4.00 \times 10^6$$

A median equivalent sand roughness for concrete can be taken as $k_s = 1.6$ mm (Table 3.1), and therefore the relative roughness, $k_s/4R$, is given by

$$\frac{k_s}{4R} = \frac{1.6 \times 10^{-3}}{4(0.333)} = 0.00120$$

Substituting Re and $k_s/4R$ into the Jain equation (Equation 3.38) for the friction factor yields

$$\frac{1}{\sqrt{f}} = -2\log\left[\frac{k_s/4R}{3.7} + \frac{5.74}{\text{Re}^{0.9}}\right] = -2\log\left[\frac{0.00120}{3.7} + \frac{5.74}{(4.00 \times 10^6)^{0.9}}\right] = 6.96$$

which yields

$$f = 0.0206$$

The frictional head loss in the culvert, h_f, is therefore given by the Darcy-Weisbach equation as

$$h_f = \frac{fL}{4R}\frac{V^2}{2g} = \frac{(0.0206)(10)}{(4 \times 0.333)}\frac{3^2}{2(9.81)} = 0.0709 \text{ m}$$

The head loss in the culvert can therefore be estimated as 7.1 cm. ■

Empirical Friction-Loss Formulae. Friction losses in pipelines should generally be calculated using the Darcy-Weisbach equation. However, a minor inconvenience in using the Darcy-Weisbach equation to relate the friction loss to the flow velocity results from the dependence of the friction factor on the flow velocity; therefore, the Darcy-Weisbach equation must be solved simultaneously with the Colebrook equation. In modern engineering practice, computer hardware and software make this a very minor inconvenience. In earlier years, however, this was considered a real problem, and various empirical head-loss formulae were developed to relate the head loss directly to the flow velocity. The most commonly used empirical formulae are the *Hazen-Williams formula* and the *Manning formula*.

The *Hazen-Williams formula* (Williams and Hazen, 1933) is applicable only to the flow of water in pipes and is given by

$$V = 0.849C_H R^{0.63} S_f^{0.54} \tag{3.82}$$

where V is the flow velocity (in m/s), C_H is the Hazen-Williams roughness coefficient, R is the hydraulic radius (in m), and S_f is the slope of the energy grade line, defined by

$$S_f = \frac{h_f}{L} \tag{3.83}$$

where h_f is the head loss due to friction over a length L of pipe. Values of C_H for a variety of commonly used pipe materials are given in Table 3.2. Solving Equations 3.82 and 3.83 yields the following expression for the frictional head loss,

$$h_f = 6.82\frac{L}{D^{1.17}}\left(\frac{V}{C_H}\right)^{1.85} \tag{3.84}$$

where D is the diameter of the pipe. The Hazen-Williams equation is applicable to the flow of water at 16°C in pipes with diameters between 50 mm and 1850 mm, and flow velocities less than 3 m/s. Application of the Hazen-Williams formula to temper-

■ Table 3.2
Pipe Roughness Coefficients

Pipe material	C_H Range	C_H Typical	n Range	n Typical
Ductile and cast iron:				
New, unlined	120–140	130	—	0.013
Old, unlined	40–100	80	—	—
Cement lined and seal coated	100–140	120	0.011–0.015	0.013
Steel:				
Welded and seamless	80–150	120	—	—
Riveted	—	110	0.012–0.018	0.015
Concrete pipe	100–140	110	0.011–0.015	0.015
Vitrified clay pipe (VCP)	—	—	0.012–0.014	—
Polyvinyl chloride (PVC)	—	—	0.007–0.011	—

Source: Velon and Johnson (1993). Reprinted by permission of The McGraw-Hill Companies.

atures much different than 16°C results in some error (Mott, 1994). Street and colleagues (1996) and Liou (1998) have shown that the Hazen-Williams coefficient has a strong Reynolds number dependence and is mostly applicable where the pipe is relatively smooth and in the early part of its transition to rough flow. Jain and colleagues (1978) have shown that an error of up to 39% can be expected in the evaluation of the velocity by the Hazen-Williams formula over a wide range of diameters and slopes. The Hazen-Williams formula is generally not recommended, since the roughness coefficient, C_H, is strictly limited to a particular flow condition, pipe size, and water temperature, and application outside this range can lead to significant errors (Liou, 1998).

A second empirical formula that is sometimes used to describe flow in pipes is the Manning formula, which is given by

$$V = \frac{1}{n} R^{\frac{2}{3}} S_f^{\frac{1}{2}} \tag{3.85}$$

where V, R, and S_f have the same meaning and units as in the Hazen-Williams formula, and n is the Manning roughness coefficient. Values of n for a variety of commonly used pipe materials are given in Table 3.2 (Velon and Johnson, 1993). Solving Equations 3.85 and 3.83 yields the following expression for the frictional head loss

$$h_f = 6.35 \frac{n^2 L V^2}{D^{\frac{4}{3}}} \tag{3.86}$$

The Manning formula applies to fully turbulent flows, where the frictional head losses depend primarily on the relative roughness.

■ Example 3.8

Water flows at a velocity of 1 m/s in a 150 mm new ductile iron pipe. Estimate the head loss over 500 m using: (a) the Hazen-Williams formula, (b) the Manning formula, and (c) the Darcy-Weisbach equation. Compare your results.

Solution

(a) The Hazen-Williams roughness coefficient, C_H, can be taken as 130 (Table 3.2), $L = 500$ m, $D = 0.150$ m, $V = 1$ m/s, and therefore the head loss, h_f, is given by Equation 3.84 as

$$h_f = 6.82 \frac{L}{D^{1.17}} \left(\frac{V}{C_H}\right)^{1.85} = 6.82 \frac{500}{(0.15)^{1.17}} \left(\frac{1}{130}\right)^{1.85} = 3.85 \text{ m}$$

(b) The Manning roughness coefficient, n, can be taken as 0.013 (approximation from Table 3.2), and therefore the head loss, h_f, is given by Equation 3.86 as

$$h_f = 6.35 \frac{n^2 L V^2}{D^{\frac{4}{3}}} = 6.35 \frac{(0.013)^2 (500)(1)^2}{(0.15)^{\frac{4}{3}}} = 6.73 \text{ m}$$

(c) The equivalent sand roughness, k_s, can be taken as 0.26 mm (Table 3.1), and the Reynolds number, Re, is given by

$$\text{Re} = \frac{VD}{\nu} = \frac{(1)(0.15)}{1.00 \times 10^{-6}} = 1.5 \times 10^5$$

where $\nu = 1.00 \times 10^{-6}$ m²/s at 20°C. Substituting k_s, D, and Re into the Colebrook equation yields the friction factor, f, where

$$\frac{1}{\sqrt{f}} = -2\log\left[\frac{k_s}{3.7D} + \frac{2.51}{\text{Re}\sqrt{f}}\right] = -2\log\left[\frac{0.26}{3.7(150)} + \frac{2.51}{1.5 \times 10^5\sqrt{f}}\right]$$

Solving by trial and error leads to

$$f = 0.0238$$

The head loss, h_f, is therefore given by the Darcy-Weisbach equation as

$$h_f = f\frac{L}{D}\frac{V^2}{2g} = 0.0238\frac{500}{0.15}\frac{1^2}{2(9.81)} = 4.04 \text{ m}$$

It is reasonable to assume that the Darcy-Weisbach equation yields the most accurate estimate of the head loss. In this case, the Hazen-Williams formula gives a head loss that is 5% too low and the Manning formula yields a head loss that is 67% too high. ■

3.3 Multiple Pipelines

Multiple-pipeline systems are commonly encountered in the context of water-supply systems. The performance criteria of these systems are typically specified in terms of minimum flow rates and pressure heads that must be maintained at the specified points in the network. Analyses of pipe networks are usually within the context of (1) designing a new network, (2) designing a modification to an existing network, and/or (3) evaluating the reliability of a network. The procedure for analyzing a pipe network usually aims at finding the flow distribution within the network, with the pressure distribution being derived from the flow distribution using the energy equation. A typical pipe network is illustrated in Figure 3.9, where the boundary conditions consist of inflows, outflows, and constant-head boundaries such as storage reservoirs. Inflows are typically

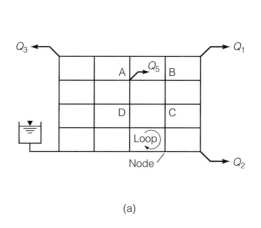

(a)

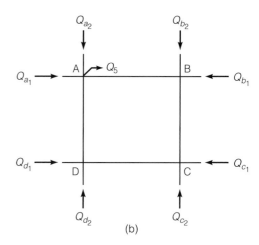

(b)

Figure 3.9 ■ Typical Pipe Network

from water-treatment facilities, outflows from consumer withdrawals or fires. All out-flows are assumed to occur at network junctions.

The basic equations to be satisfied in pipe networks are the continuity and energy equations. The continuity equation requires that, at each node in the network, the sum of the outflows is equal to the sum of the inflows. This requirement is expressed by the relation

$$\sum_{i=1}^{NP(j)} Q_{ij} - F_j = 0, \qquad j = 1, NJ \qquad (3.87)$$

where Q_{ij} is the flowrate in pipe i at junction j (inflows positive); $NP(j)$ is the number of pipes meeting at junction j; F_j is the external flow rate (outflows positive) at junction j; and NJ is the total number of junctions in the network. The energy equation requires that the heads at each of the nodes in the pipe network be consistent with the head losses in the pipelines connecting the nodes. There are two principal methods of calculating the flows in pipe networks: the nodal method and the loop method.

3.3.1 ■ Nodal Method

In the nodal method, the energy equation is written for each pipeline in the network as

$$h_2 = h_1 - \left(\frac{fL}{D} + \sum k_m\right)\frac{Q|Q|}{2gA^2} + \frac{Q}{|Q|}h_p \qquad (3.88)$$

where h_2 and h_1 are the heads at the upstream and downstream ends of a pipe; the terms in parentheses measure the friction loss and minor losses, respectively; and h_p is the head added by pumps in the pipeline. The energy equation stated in Equation 3.88 has been modified to account for the fact that the flow direction is in many cases unknown, in which case a positive flow direction in each pipeline must be assumed, and a consistent set of energy equations stated for the entire network. Equation 3.88 assumes that the positive flow direction is from node 1 to node 2. Application of the nodal method in practice is usually limited to relatively simple networks.

■ Example 3.9

The high-pressure ductile-iron pipeline shown in Figure 3.10 becomes divided at point B and rejoins at point C. The pipeline characteristics are given in the following tables.

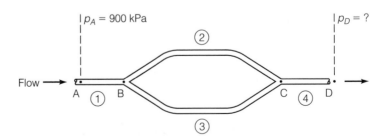

Figure 3.10 ■
Pipe Network

Pipe	Diameter (mm)	Length (m)
1	750	500
2	400	600
3	500	650
4	700	400

Location	Elevation (m)
A	5.0
B	4.5
C	4.0
D	3.5

If the flowrate in Pipe 1 is 2 m³/s and the pressure at point A is 900 kPa, calculate the pressure at point D. Assume that the flows are fully turbulent in all pipes.

Solution

The equivalent sand roughness, k_s, of ductile-iron pipe is 0.26 mm, and the pipe and flow characteristics are as follows:

Pipe	Area (m²)	Velocity (m/s)	k_s/D	f
1	0.442	4.53	0.000347	0.0154
2	0.126	—	0.000650	0.0177
3	0.196	—	0.000520	0.0168
4	0.385	5.20	0.000371	0.0156

where it has been assumed that the flows are fully turbulent. Taking $\gamma = 9.79$ kN/m³, the head at location A, h_A, is given by

$$h_A = \frac{p_A}{\gamma} + \frac{V_1^2}{2g} + z_A = \frac{900}{9.79} + \frac{4.53^2}{(2)(9.81)} + 5 = 98.0 \text{ m}$$

and the energy equations for each pipe are as follows

$$\text{Pipe 1: } \quad h_B = h_A - \frac{f_1 L_1}{D_1} \frac{V_1^2}{2g} = 98.0 - \frac{(0.0154)(500)}{0.75} \frac{4.53^2}{(2)(9.81)}$$

$$= 87.3 \text{ m} \tag{3.89}$$

$$\text{Pipe 2: } \quad h_C = h_B - \frac{f_2 L_2}{D_2} \frac{Q_2^2}{2g A_2^2} = 87.3 - \frac{(0.0177)(600)}{0.40} \frac{Q_2^2}{(2)(9.81)(0.126)^2}$$

$$= 87.3 - 85.2 Q_2^2 \tag{3.90}$$

$$\text{Pipe 3: } \quad h_C = h_B - \frac{f_3 L_3}{D_3} \frac{Q_3^2}{2g A_3^2} = 87.3 - \frac{(0.0168)(650)}{0.50} \frac{Q_3^2}{(2)(9.81)(0.196)^2}$$

$$= 87.3 - 29.0 Q_3^2 \tag{3.91}$$

$$\text{Pipe 4:} \quad h_D = h_C - \frac{f_4 L_4}{D_4} \frac{Q_4^2}{2gA_4^2} = h_C - \frac{(0.0156)(400)}{0.70} \frac{Q_4^2}{(2)(9.81)(0.385)^2}$$

$$= h_C - 3.07 Q_4^2 \tag{3.92}$$

and the continuity equations at the two pipe junctions are

$$\text{Junction B:} \quad Q_2 + Q_3 = 2 \text{ m}^3/\text{s} \tag{3.93}$$

$$\text{Junction C:} \quad Q_2 + Q_3 = Q_4 \tag{3.94}$$

Equations 3.90 to 3.94 are five equations in five unknowns: h_C, h_D, Q_2, Q_3, and Q_4. Equations 3.93 and 3.94 indicate that

$$Q_4 = 2 \text{ m}^3/\text{s}$$

Combining Equations 3.90 and 3.91 leads to

$$87.3 - 85.2 Q_2^2 = 87.3 - 29.0 Q_3^2$$

and therefore

$$Q_2 = 0.583 Q_3 \tag{3.95}$$

Substituting Equation 3.95 into Equation 3.93 yields

$$2 = (0.583 + 1) Q_3$$

or

$$Q_3 = 1.26 \text{ m}^3/\text{s}$$

and from Equation 3.95

$$Q_2 = 0.74 \text{ m}^3/\text{s}$$

According to Equation 3.91

$$h_C = 87.3 - 29.0 Q_3^2 = 87.3 - 29.0(1.26)^2 = 41.3 \text{ m}$$

and Equation 3.92 gives

$$h_D = h_C - 3.07 Q_4^2 = 41.3 - 3.07(2)^2 = 29.0 \text{ m}$$

Therefore, since the total head at D, h_D, is equal to 29.0 m, then

$$29.0 = \frac{p_D}{\gamma} + \frac{V_4^2}{2g} + z_D = \frac{p_D}{9.79} + \frac{5.20^2}{(2)(9.81)} + 3.5$$

which yields

$$p_D = 236 \text{ kPa}$$

Therefore, the pressure at location D is 236 kPa.

This problem has been solved by assuming that the flows in all pipes are fully turbulent. This is generally not known for sure a priori, and therefore a complete solution would require repeating the calculations until the assumed friction factors are consistent with the calculated flowrates. ■

3.3.2 ■ Loop Method

In the loop method, the energy equation is written for each loop of the network, in which case the algebraic sum of the head losses within each loop is equal to zero. This requirement is expressed by the relation

$$\sum_{j=1}^{NP(i)} (h_{L,ij} - h_{p,ij}) = 0, \qquad i = 1, NL \tag{3.96}$$

where $h_{L,ij}$ is the head loss in pipe j of loop i, and $h_{p,ij}$ is the head added by any pumps that may exist in line ij. Combining Equations 3.87 and 3.96 with an expression for calculating the head losses in pipes, such as the Darcy-Weisbach equation, and the pump characteristic curves, which relate the head added by the pump to the flowrate through the pump, yields a complete mathematical description of the flow problem. Solution of this system of flow equations is complicated by the fact that the equations are nonlinear, and numerical methods such as the Newton-Raphson or Hardy Cross methods must be used to solve for the flow distribution in the pipe network.

Hardy Cross Method.   A simple and popular technique for solving the system of equations governing the flow in pipe networks is called the Hardy Cross method (Cross, 1936). This iterative method assumes that the head loss, h_L, in each pipe is proportional to the discharge, Q, raised to some power n, in which case

$$h_L = rQ^n \tag{3.97}$$

The proportionality constant, r, depends on which head loss equation is used and the types of losses in the pipe. Clearly, if all head losses are due to friction and the Darcy-Weisbach equation is used to calculate the head losses, then r is given by

$$r = \frac{fL}{2gA^2D} \tag{3.98}$$

If the flow in each pipe is approximated as $\hat{Q}$, and ΔQ is the error in this estimate, then the actual flowrate, Q, is related to $\hat{Q}$ and ΔQ by

$$Q = \hat{Q} + \Delta Q \tag{3.99}$$

and the head loss in each pipe is given by

$$
\begin{aligned}
h_L &= rQ^n \\
&= r(\hat{Q} + \Delta Q)^n \\
&= r\left[\hat{Q}^n + n\hat{Q}^{n-1}\Delta Q + \frac{n(n-1)}{2}\hat{Q}^{n-2}(\Delta Q)^2 + \cdots + (\Delta Q)^n \right]
\end{aligned}
\tag{3.100}
$$

If the error in the flow estimate, ΔQ, is small, then the higher order terms in ΔQ can be neglected and the head loss in each pipe can be approximated by the relation

$$h_L \approx r\hat{Q}^n + rn\hat{Q}^{n-1}\Delta Q \tag{3.101}$$

This relation approximates the head loss in the flow direction. However, in working with pipe networks, it is required that the algebraic sum of the head losses in any loop of the network (see Figure 3.9) must be equal to zero. We must therefore define a positive flow direction (such as clockwise), and count head losses as positive in pipes when the flow is in the positive direction and negative when the flow is opposite to the selected positive direction. Under these circumstances, the sign of the head loss must be the same as the sign of the flow direction. Further, when the flow is in the positive direction, positive values of ΔQ require a positive correction to the head loss; when the flow is in the negative direction, positive values in ΔQ also require a positive correction to the calculated head loss. To preserve the algebraic relation among head loss, flow direction, and flow error (ΔQ), Equation 3.101 for each pipe can be written as

$$h_L = r\hat{Q}|\hat{Q}|^{n-1} + rn|\hat{Q}|^{n-1}\Delta Q \tag{3.102}$$

where the approximation has been replaced by an equal sign. On the basis of Equation 3.102, the requirement that the algebraic sum of the head losses around each loop be equal to zero can be written as

$$\sum_{j=1}^{NP(i)} r_{ij}Q_j|Q_j|^{n-1} + \Delta Q_i \sum_{j=1}^{NP(i)} r_{ij}n|Q_j|^{n-1} = 0, \qquad i = 1, NL \tag{3.103}$$

where $NP(i)$ is the number of pipes in loop i, r_{ij} is the head-loss coefficient in pipe j (in loop i), Q_j is the estimated flow in pipe j, ΔQ_i is the flow correction for the pipes in loop i, and NL is the number of loops in the entire network. The approximation given by Equation 3.103 assumes that there are no pumps in the loop, and that the flow correction, ΔQ_i, is the same for each pipe in each loop. Solving Equation 3.103 for ΔQ_i leads to

$$\Delta Q_i = -\frac{\sum_{j=1}^{NP(i)} r_{ij}Q_j|Q_j|^{n-1}}{\sum_{j=1}^{NP(i)} nr_{ij}|Q_j|^{n-1}} \tag{3.104}$$

This equation forms the basis of the Hardy Cross method.

The steps to be followed in using the Hardy Cross method to calculate the flow distribution in pipe networks are:

1. Assume a reasonable distribution of flows in the pipe network. This assumed flow distribution must satisfy continuity.
2. For each loop, i, in the network, calculate the quantities $r_{ij}Q_j|Q_j|^{n-1}$ and $nr_{ij}|Q_j|^{n-1}$ for each pipe in the loop. Calculate the flow correction, ΔQ_i, using Equation 3.104. Add the correction algebraically to the estimated flow in each pipe. [*Note:* Values of r_{ij} occur in both the numerator and denominator of Equation 3.104; therefore, values proportional to the actual r_{ij} may be used to calculate ΔQ_i.]
3. Proceed to another circuit and repeat step 2.
4. Repeat steps 2 and 3 until the corrections (ΔQ_i) are small.

The application of the Hardy Cross method is best demonstrated by an example.

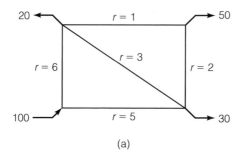

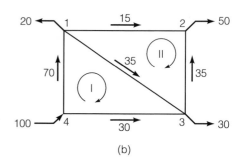

(a)   (b)

Figure 3.11 ■ Flows in Pipe Network

■ **Example 3.10**

Compute the distribution of flows in the pipe network shown in Figure 3.11(a), where the head loss in each pipe is given by

$$h_L = rQ^2$$

and the values of r are shown in Figure 3.11(a). The flows are taken as dimensionless for the sake of illustration.

Solution

The first step is to assume a distribution of flows in the pipe network that satisfies continuity. The assumed distribution of flows is shown in Figure 3.11(b), along with the positive-flow directions in each of the two loops. The flow correction for each loop is calculated using Equation 3.104. Since $n = 2$ in this case, the flow correction formula becomes

$$\Delta Q_i = -\frac{\sum_{j=1}^{NP(i)} r_{ij} Q_j |Q_j|}{\sum_{j=1}^{NP(i)} 2 r_{ij} |Q_j|}$$

The calculation of the numerator and denominator of this flow correction formula for loop I is tabulated as follows

Loop	Pipe	Q	$r\,Q|Q|$	$2r|Q|$
I	4–1	70	29,400	840
	1–3	35	3,675	210
	3–4	−30	−4,500	300
			28,575	1,350

The flow correction for loop I, ΔQ_I, is therefore given by

$$\Delta Q_I = -\frac{28575}{1350} = -21.2$$

and the corrected flows are

Loop	Pipe	Q
I	4–1	48.8
	1–3	13.8
	3–4	−51.2

Moving to loop II, the calculation of the numerator and denominator of the flow correction formula for loop II is given by

Loop	Pipe	Q	$r\,Q\lvert Q\rvert$	$2r\lvert Q\rvert$
II	1–2	15	225	30
	2–3	−35	−2,450	140
	3–1	−13.8	−574	83
			−2,799	253

The flow correction for loop II, ΔQ_{II}, is therefore given by

$$\Delta Q_{II} = -\frac{-2799}{253} = 11.1$$

and the corrected flows are

Loop	Pipe	Q
II	1–2	26.1
	2–3	−23.9
	3–1	−2.7

This procedure is repeated in the following table until the calculated flow corrections do not affect the calculated flows, to the level of significant digits retained in the calculations.

Iteration	Loop	Pipe	Q	$r\,Q\lvert Q\rvert$	$2r\lvert Q\rvert$	ΔQ	Corrected Q
2	I	4–1	48.8	14,289	586		47.7
		1–3	2.7	22	16		1.6
		3–4	−51.2	−13,107	512		−52.3
				1,204	1,114	−1.1	
	II	1–2	26.1	681	52		29.1
		2–3	−23.9	−1,142	96		−20.9
		3–1	−1.6	−8	10		1.4
				−469	157	3.0	
3	I	4–1	47.7	13,663	573		47.7
		1–3	1.4	6	8		1.4
		3–4	−52.3	−13,666	523		−52.3
				3	1,104	0.0	
	II	1–2	29.1	847	58		29.2
		2–3	−20.9	−874	84		−20.8
		3–1	1.4	6	8		1.5
				−21	150	0.1	

Iteration	Loop	Pipe	Q	$rQ\lvert Q\rvert$	$2r\lvert Q\rvert$	ΔQ	Corrected Q
4	I	4–1	47.7	13,662	573		47.7
		1–3	1.5	7	9		1.5
		3–4	−52.3	−13,668	523		−52.3
				1	1,104	0.0	
	II	1–2	29.2	853	58		29.2
		2–3	−20.8	−865	83		−20.8
		3–1	1.5	7	9		1.5
				−5	150	0.0	

The final flow distribution, after four iterations, is given by

Pipe	Q
1–2	29.2
2–3	−20.8
3–4	−52.3
4–1	47.7
1–3	−1.5

It is clear that the final results are fairly close to the flow estimates after only one iteration.

■

As Example 3.10 illustrates, complex pipe networks can generally be treated as a combination of simple circuits or loops, with each balanced in turn until compatible flow conditions exist in all loops.

3.4 Pumps

Pumps are hydraulic machines that convert mechanical energy to fluid energy. They can be classified into two main categories: (1) positive displacement pumps, and (2) kinetic pumps. Positive displacement pumps deliver a fixed quantity of fluid with each revolution of the pump rotor, such as with a piston or cylinder, while kinetic pumps add energy to the fluid by accelerating it through the action of a rotating impeller. Kinetic pumps are far more common in practice and will be the focus of this section.

Three types of kinetic pumps commonly encountered are centrifugal pumps, axial-flow pumps, and mixed-flow pumps. In centrifugal pumps, the flow enters the pump chamber along the axis of the impeller and is discharged radially by centrifugal action, as illustrated in Figure 3.12(a). In axial-flow pumps, the flow enters and leaves the pump chamber along the axis of the impeller, as shown in Figure 3.12(b). In mixed-flow pumps, outflows have both radial and axial components. Typical centrifugal and axial-flow pump installations are illustrated in Figure 3.13. Key components of the centrifugal pump are a foot valve installed in the suction pipe to prevent water from leaving the pump when it is stopped and a check valve in the discharge pipe to prevent backflow if there is a power failure. If the suction line is empty prior to starting the pump, then the suction line must be *primed* prior to startup. Unless the water is known to be very clean, a strainer should be installed at the inlet to the suction piping. The pipe size of

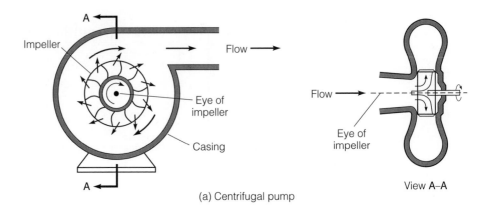

(a) Centrifugal pump

View A–A

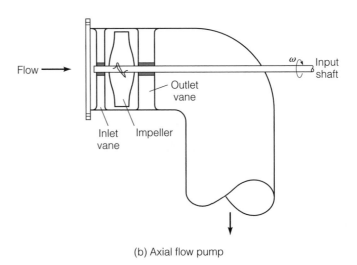

(b) Axial flow pump

Figure 3.12 ■ Types of Pumps

the suction line should never be smaller than the inlet connection on the pump; if a reducer is required, it should be of the eccentric type since concentric reducers place part of the supply pipe above the pump inlet where an air pocket could form. The discharge line from the pump should contain a valve close to the pump to allow service or pump replacement.

The pumps illustrated in Figure 3.13 are both *single-stage* pumps, which means that they have only one impeller. In *multistage* pumps, two or more impellers are arranged in series in such a way that the discharge from one impeller enters the eye of the next impeller. These types of pumps are typically used when large pumping heads are required.

The performance of a pump is measured by the head added by the pump, h_p, and the pump efficiency, η, defined by

$$\eta = \frac{\text{power delivered to the fluid}}{\text{power supplied to the shaft}} \qquad (3.105)$$

The pump-performance parameters, h_p and η, can be expressed in terms of the fluid properties and the physical characteristics of the pump by the functional relation

$$gh_p \quad \text{or} \quad \eta = f_1(\rho, \mu, D, \omega, Q) \qquad (3.106)$$

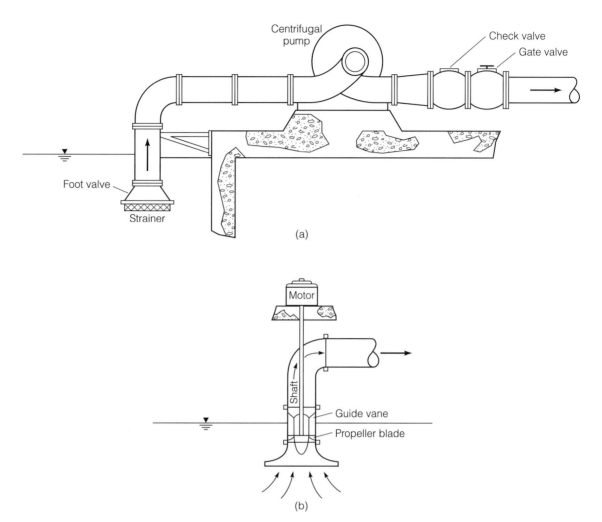

Figure 3.13 ■ Centrifugal and Axial Flow Pump Installations
Source: Franzini, Joseph B. and Finnemore, E. John, *Fluid Mechanics*. Copyright © 1997 by The McGraw-Hill Companies.

where the energy added per unit mass of fluid, gh_p, is used instead of h_p (to account for the effect of gravity), f_1 is an unknown function, ρ and μ are the density and dynamic viscosity of the fluid respectively, Q is the flowrate through the pump, D is a characteristic dimension of the pump (usually the inlet diameter), and ω is the speed of the pump impeller. Equation 3.106 is a functional relationship between six variables in three dimensions. According to the Buckingham pi theorem, this relationship can be expressed as a relation between three dimensionless groups as follows

$$\frac{gh_p}{\omega^2 D^2} \quad \text{or} \quad \eta = f_2\left(\frac{Q}{\omega D^3}, \frac{\rho\omega D^2}{\mu}\right) \tag{3.107}$$

In most cases, the flow through the pump is fully turbulent and viscous forces are negligible relative to the inertial forces. Under these circumstances, the viscosity of the fluid is neglected and Equation 3.107 becomes

$$\frac{gh_p}{\omega^2 D^2} \quad \text{or} \quad \eta = f_3\left(\frac{Q}{\omega D^3}\right) \tag{3.108}$$

This relationship describes the performance of all pumps in which viscous effects are negligible, but the exact form of the function in Equation 3.108 depends on the geometry of the pump. A series of pumps having the same shape (but different sizes) are expected to have the same functional relationships between $gh_p/(\omega^2 D^2)$ and $Q/(\omega D^3)$ as well as η and $Q/(\omega D^3)$. A class of pumps that have the same shape is called a *homologous series*, and the performance characteristics of a homologous series of pumps are described by curves such as those in Figure 3.14. It is clear that the efficiency of the pump varies with the operating condition, and it is usually desirable to select a pump that operates at or near the point of maximum efficiency, indicated by the point P in Figure 3.14. At the point of maximum efficiency,

$$\frac{gh_p}{\omega^2 D^2} = K_1, \quad \text{and} \quad \frac{Q}{\omega D^3} = K_2 \tag{3.109}$$

Eliminating D from these equations yields

$$\frac{\omega Q^{\frac{1}{2}}}{(gh_p)^{\frac{3}{4}}} = \sqrt{\frac{K_2}{K_1^{\frac{3}{2}}}} \tag{3.110}$$

The term on the righthand side of this equation is a constant for a homologous series of pumps and is denoted by the *specific speed* or *shape number*, n_s, defined by

$$n_s = \frac{\omega Q^{\frac{1}{2}}}{(gh_p)^{\frac{3}{4}}} \tag{3.111}$$

where any consistent set of units can be used. In SI units, ω is in rad/s, Q in m³/s, g in m/s², and h_p in meters. The most efficient operating point for a homologous series of pumps is therefore specified by the specific speed. This nomenclature is somewhat unfortunate since the specific speed is dimensionless and hence does not have units of speed. It is common practice in the United States to define the specific speed by N_s, as

$$N_s = \frac{\omega Q^{\frac{1}{2}}}{h_p^{\frac{3}{4}}} \tag{3.112}$$

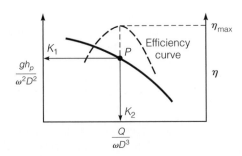

Figure 3.14 ■ Performance Curves of a Homologous Series of Pumps

■ Table 3.3
Pump Selection Guidelines

Type of pump	Range of specific speeds, n_s*	Typical flowrates (L/s)	Typical efficiencies (%)
Centrifugal	0.15–1.5 (400–4,000)	< 60	70–94
Mixed flow	1.5–3.7 (4,000–10,000)	60–300	90–94
Axial flow	3.7–5.5 (10,000–15,000)	> 300	84–90

*The specific speeds in parentheses correspond to N_s given by Equation 3.112, with ω in rpm, Q in gpm, and h_p in ft.

where N_s is not dimensionless, ω is in revolutions per minute (rpm), Q is in gallons per minute (gpm), and h_p is in feet (ft). Although N_s has dimensions, the units are seldom stated in practice. The required pump performance characteristics usually dictate the flowrate, Q, and head, h_p, required from the pump; the rotational speed, ω, is determined by available motors for the size of the pump; and the specific speed calculated from the required performance characteristics is usually the basis for selecting the appropriate pump.

The types of pump that give the maximum efficiency for given specific speeds, n_s, are listed in Table 3.3 along with typical flowrates delivered by the pumps. Table 3.3 indicates that centrifugal pumps have low specific speeds, $n_s < 1.5$; mixed-flow pumps have medium specific speeds, $1.5 < n_s < 3.7$; and axial-flow pumps have high specific speeds, $n_s > 3.7$. The efficiencies of radial-flow (centrifugal) pumps tend to increase with increasing specific speed, while the efficiencies of mixed-flow and axial-flow pumps tend to decrease with increasing specific speed. Pumps with specific speed less than 0.3 tend to be inefficient (Franzini and Finnemore, 1997). Most pumps are driven by standard electric motors. The standard speed of AC synchronous induction motors at 60 cycles and 220 to 440 volts is given by

$$\text{Synchronous speed (rpm)} = \frac{3600}{\text{no. of pairs of poles}} \qquad (3.113)$$

3.4.1 ■ Affinity Laws

The performance curves for a homologous series of pumps is illustrated in Figure 3.14. Any two pumps in the homologous series are expected to operate at the same efficiency when

$$\left(\frac{Q}{\omega D^3}\right)_1 = \left(\frac{Q}{\omega D^3}\right)_2, \text{ and } \left(\frac{h_p}{\omega^2 D^2}\right)_1 = \left(\frac{h_p}{\omega^2 D^2}\right)_2 \qquad (3.114)$$

These relationships are sometimes called the *affinity laws for homologous pumps*. An affinity law for the power delivered to the fluid, P, can be derived from the affinity relations given in Equation 3.114, since P is defined by

$$P = \gamma Q h_p \qquad (3.115)$$

which leads to the following derived affinity relation

$$\left(\frac{P}{\omega^3 D^5}\right)_1 = \left(\frac{P}{\omega^3 D^5}\right)_2 \qquad (3.116)$$

In accordance with the dimensional analysis of pump performance, Equation 3.107, the affinity laws for scaling pump performance are valid as long as viscous effects are negligible. The effect of viscosity is measured by the Reynolds number, Re, defined by

$$\text{Re} = \frac{\rho \omega D^2}{\mu} \tag{3.117}$$

and scale effects are negligible when Re > 3×10^5 (Gerhart et al., 1992). In lieu of stating a Reynolds number criterion for scale effects to be negligible, it is sometimes stated that larger pumps are more efficient than smaller pumps and that the scale effect on efficiency is given by (Potter and Wiggert, 1991)

$$\frac{1 - \eta_2}{1 - \eta_1} = \left(\frac{D_1}{D_2}\right)^{\frac{1}{4}} \tag{3.118}$$

where η_1 and η_2 are the efficiencies of homologous pumps of diameters D_1 and D_2, respectively. The effect of changes in flowrate on efficiency can be estimated using the relation

$$\frac{0.94 - \eta_2}{0.94 - \eta_1} = \left(\frac{Q_1}{Q_2}\right)^{0.32} \tag{3.119}$$

where Q_1 and Q_2 are corresponding homologous flowrates.

▪ Example 3.11

A pump with a 1,200 rpm motor has a performance curve of

$$h_p = 12 - 0.1Q^2$$

If the motor is changed to 2,400 rpm, estimate the new performance curve.

Solution

The performance characteristics of a homologous series of pumps is given by

$$\frac{gh_p}{\omega^2 D^2} = f\left(\frac{Q}{\omega D^3}\right)$$

For a fixed pump size, D, for two different motor speeds, ω_1 and ω_2, the general performance curve can be written as

$$\frac{h_1}{\omega_1^2} = f\left(\frac{Q_1}{\omega_1}\right)$$

and

$$\frac{h_2}{\omega_2^2} = f\left(\frac{Q_2}{\omega_2}\right)$$

where h_1 and Q_1 are the head added by the pump and the flowrate, respectively, when the speed is ω_1; and h_2 and Q_2 are the head and flowrate when the speed is ω_2. Since the pumps are part of a homologous series, then, neglecting scale effects, the function f is fixed and therefore when

$$\frac{Q_1}{\omega_1} = \frac{Q_2}{\omega_2}$$

then

$$\frac{h_1}{\omega_1^2} = \frac{h_2}{\omega_2^2}$$

These relations are simply statements of the affinity laws. In the present case, $\omega_1 = 1{,}200$ rpm and $\omega_2 = 2{,}400$ rpm and the affinity laws give that

$$Q_1 = \frac{\omega_1}{\omega_2} Q_2 = \frac{1200}{2400} Q_2 = 0.5 Q_2$$

$$h_1 = \frac{\omega_1^2}{\omega_2^2} h_2 = \frac{1200^2}{2400^2} h_2 = 0.25 h_2$$

Since the performance curve of the pump at speed ω_1 is given by

$$h_1 = 12 - 0.1 Q_1^2$$

then the performance curve at speed ω_2 is given by

$$0.25 h_2 = 12 - 0.1 (0.5 Q_2)^2$$

which leads to

$$h_2 = 48 - 0.1 Q_2^2$$

The performance curve of the pump with a 2,400 rpm motor is therefore given by

$$h_p = 48 - 0.1 Q^2 \qquad\blacksquare$$

3.4.2 ■ Operating Point

Selection of a pump from a homologous series requires specification of the characteristic size, D, and rotational speed, ω. For given values of D and ω, the *performance curve* of a pump can generally be expressed as a relationship between h_p and Q as shown in Figure 3.15. This curve is sometimes referred to as the *characteristic curve* of the pump. Pumps

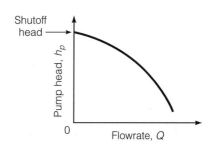

Figure 3.15 ■ Pump Performance Curve

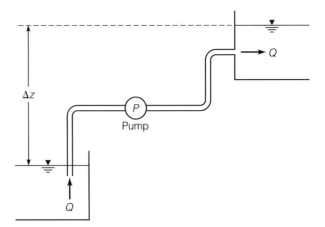

Figure 3.16 ■
Pipeline System

are placed in pipeline systems such as that illustrated in Figure 3.16, in which case the energy equation requires that the head, h_p, added by the pump is given by

$$h_p = \Delta z + Q^2 \left[\sum \frac{fL}{2gA^2D} + \sum \frac{K_m}{2gA^2} \right] \qquad (3.120)$$

where Δz is the difference in elevation between the water surfaces of the inflow and outflow reservoirs, the first term in the square brackets is the sum of the head losses due to friction, and the second term is the sum of the minor head losses. Equation 3.120 gives a relationship between h_p and Q for the pipeline system. This relationship is commonly called the *system curve*. Because the flowrate and head added by the pump must satisfy both the system curve and the pump characteristic curve, Q and h_p are determined by simultaneous solution of Equation 3.120 and the pump characteristic curve. The resulting values of Q and h_p identify the *operating point* of the pump. The location of the operating point on the performance curve is shown in Figure 3.17.

■ **Example 3.12**

Water is being pumped from a lower to an upper reservoir through a pipeline system similar to the one shown in Figure 3.16. The reservoirs differ in elevation by 15.2 m, and the length of the steel pipe ($k_s = 0.046$ mm) connecting the reservoirs is 21.3 m. The pipe is 50-mm in diameter and the performance curve of the pump is given by

$$h_p = 24.4 - 7.65Q^2 \qquad (3.121)$$

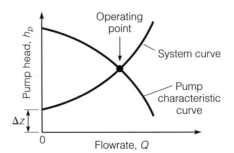

Figure 3.17 ■ Operating
Point in Pipeline System

where h_p is in meters and Q is in liters per second. Using this pump, what flow do you expect in the pipeline? If the motor on the pump rotates at 2,400 rpm, calculate the specific speed of the pump in U.S. Customary units and state the type of pump that should be used.

Solution

Neglecting minor losses, the energy equation for the pipeline system is given by

$$h_p = 15.2 + \frac{fL}{2gA^2D}Q^2 \tag{3.122}$$

where h_p is the head added by the pump, f is the friction factor, L is the pipe length, A is the cross-sectional area of the pipe, and D is the diameter of the pipe. The area, A, is given by

$$A = \frac{\pi}{4}D^2 = \frac{\pi}{4}(0.05)^2 = 0.00196 \text{ m}^2$$

In general, f is a function of both the Reynolds number and the relative roughness. However, if the flow is fully turbulent, then the friction factor depends only on the relative roughness according to Equation 3.34

$$\frac{1}{\sqrt{f}} = -2\log\left(\frac{\epsilon}{3.7}\right) \tag{3.123}$$

where ϵ is the relative roughness of the pipe given by

$$\epsilon = \frac{k_s}{D}$$

and k_s is the equivalent sand roughness. In this case, $k_s = 0.046$ mm and $D = 50$ mm. Therefore

$$\epsilon = \frac{0.046}{50} = 0.000920$$

Substituting for ϵ into Equation 3.123 yields

$$\frac{1}{\sqrt{f}} = -2\log\left(\frac{0.000920}{3.7}\right) = 7.21$$

Therefore

$$f = 0.0192$$

Substituting this value of f into the system equation (Equation 3.122) yields

$$h_p = 15.2 + \frac{fL}{2gA^2D}Q^2 = 15.2 + \frac{(0.0192)(21.3)}{(2)(9.81)(0.00196)^2(0.05)}Q^2 = 15.2 + 108500Q^2$$

This equation is for Q in m^3/s, and the corresponding equation for Q in L/s is

$$h_p = 15.2 + 0.109Q^2 \tag{3.124}$$

Combining the system curve, Equation 3.124, with the pump characteristic curve, Equation 3.121, leads to

$$15.2 + 0.109Q^2 = 24.4 - 7.65Q^2$$

$$Q = 1.09 \text{ L/s}$$

This flowrate was derived by assuming that the flow in the pipeline is fully turbulent. This assumption can now be verified by recalculating the friction factor with the Reynolds number corresponding to the calculated flowrate. The velocity in the pipeline, V, is given by

$$V = \frac{Q}{A} = \frac{1.09 \times 10^{-3}}{0.00196} = 0.556 \text{ m/s}$$

Since the kinematic viscosity, ν, at 20°C is 1.00×10^{-6} m²/s, the Reynolds number, Re, is

$$Re = \frac{VD}{\nu} = \frac{(0.556)(0.05)}{1.00 \times 10^{-6}} = 2.78 \times 10^4$$

The friction factor can now be recalculated using the Jain equation (Equation 3.38), where

$$\frac{1}{\sqrt{f}} = -2\log\left[\frac{\epsilon}{3.7} + \frac{5.74}{Re^{0.9}}\right] = -2\log\left[\frac{0.000920}{3.7} + \frac{5.74}{(2.78 \times 10^4)^{0.9}}\right] = 6.17$$

which leads to

$$f = 0.0263$$

Therefore, the original approximation of $f = 0.0192$ was in error and the estimation of the flow in the pipeline must be recalculated using $f = 0.0263$. After calculating the revised flow, the friction factor must again be calculated to see if it is equal to the assumed value. If not, the process is repeated until the assumed and calculated friction factors are equal. In this case, one more iteration shows that $f = 0.0263$ and

$$Q = 1.09 \text{ L/s}$$

According to the pump characteristic curve, the head added by the pump, h_p, is

$$h_p = 24.4 - 7.65Q^2 = 24.4 - 7.65(1.09)^2 = 15.3 \text{ m}$$

In U.S. Customary units, $Q = 1.09$ L/s $= 17.3$ gpm; $h_p = 15.3$ m $= 50.2$ ft; $\omega = 2,400$ rpm; and the specific speed, N_s, is given by Equation 3.112 as

$$N_s = \frac{\omega Q^{\frac{1}{2}}}{h_p^{\frac{3}{4}}} = \frac{(2400)(17.3)^{\frac{1}{2}}}{(50.2)^{\frac{3}{4}}} = 529$$

According to Table 3.3, with a pump operating at a specific speed of $N_s = 529$, the best type of pump is a centrifugal pump. ■

3.4.3 ■ Limits on Pump Location

If the absolute pressure on the suction side of a pump falls below the saturation vapor pressure of the fluid, the water will begin to vaporize. This process of vaporization is called *cavitation*. Cavitation is usually a transient phenomenon that occurs as water enters the low-pressure suction side of a pump and experiences the even lower pressures adjacent to the rotating pump impeller. As the water containing vapor cavities moves toward the high-pressure environment of the discharge side of the pump, the vapor cavities are compressed and ultimately implode, creating small localized high-velocity jets that can cause considerable damage to the pump machinery. This damage usually manifests itself as *pitting* of the metal casing and impeller, reduced pump efficiency, and excessive vibration of the pump. The noise generated by imploding vapor cavities resembles the sound of gravel going through a centrifugal pump.

The potential for cavitation is measured by the net positive suction head, NPSH, defined by

$$\text{NPSH} = \left(\frac{p_s}{\gamma} + \frac{V_s^2}{2g} + z_s \right) - \left(\frac{p_v}{\gamma} + z_s \right)$$

$$= \frac{p_s}{\gamma} + \frac{V_s^2}{2g} - \frac{p_v}{\gamma} \tag{3.125}$$

where p_s, V_s, and z_s are the pressure, velocity, and elevation of the fluid at the suction side of the pump, and p_v is the saturation vapor pressure of water at the temperature of the fluid. The net positive suction head is equal to the difference between the head in the fluid and the head at which cavitation occurs. In cases where water is being pumped from a reservoir, application of the energy equation between the reservoir and the suction side of the pump leads to the following expression for NPSH that can be used in lieu of Equation 3.125

$$\boxed{\text{NPSH} = \frac{p_o}{\gamma} - \Delta z_s - h_L - \frac{p_v}{\gamma}} \tag{3.126}$$

where p_o is the pressure at the surface of the reservoir (usually atmospheric), Δz_s is the difference in elevation between the suction side of the pump and the fluid surface in the reservoir (called the *suction lift*), h_L is the head loss in the pipeline between the reservoir and suction side of the pump (including minor losses), and p_v is the saturation vapor pressure.

In applying either Equation 3.125 or 3.126 to calculate NPSH, care must be taken to use a consistent measure of the pressures, using either gage pressures or absolute pressures. Absolute pressures are usually more convenient, since the vapor pressure is typically given as an absolute pressure. For any operating pump, a *cavitation parameter*, σ, is defined by the relation

$$\boxed{\sigma = \frac{\text{NPSH}}{h_p}} \tag{3.127}$$

where h_p is the head added by the pump. For all pumps, there is a critical value of the cavitation parameter, σ_c, below which cavitation in the pump can be expected to occur. This critical value of the cavitation parameter is usually provided by the pump manufacturer and generally places a limit on the operating range of the pump. The potential for cavitation to occur should always be considered when the operating conditions of

$$n_s = \frac{\omega Q^{1/2}}{(gh_p)^{3/4}}$$

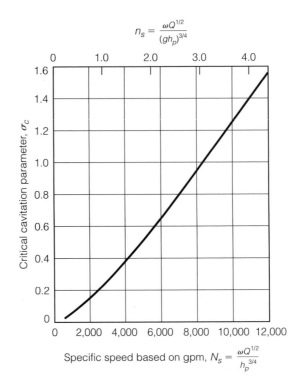

Figure 3.18 ■ Critical Value of Cavitation Parameter Versus Specific Speed

Source: Franzini, Joseph B. and Finnemore E. John, *Fluid Mechanics*. Copyright © 1997 by The McGraw-Hill Companies.

pumps are derived from the simultaneous solution pump performance curves and system curves. Values of σ_c as a function of specific speed are illustrated in Figure 3.18, where it is clear that σ_c ranges from 0.05 for pumps with a specific speed of 1,000 (in U.S. Customary units) to 1.55 for pumps with a specific speed of 12,000. In lieu of the critical cavitation parameter, some manufacturers supply the minimum allowable net positive suction head as a function of Q.

■ Example 3.13

A pump with a performance curve of $h_p = 12 - 0.01Q^2$ (h_p in m, Q in L/s) is operating 3 m above a water reservoir and pumps water at a rate of 24.5 L/s through a 102-mm diameter ductile iron pipe ($k_s = 0.26$ mm). If the length of the pipeline between the water surface and the suction side of the pump is 3.5 m, and the temperature of the water is 20°C, calculate the cavitation parameter of the pump. If the specific speed of the pump is 0.94, estimate the critical value of the cavitation parameter, and the maximum height above the surface of the reservoir that the pump can be located.

Solution

The cavitation parameter, σ, is defined by Equation 3.127 as

$$\sigma = \frac{\text{NPSH}}{h_p}$$

where NPSH is given by Equation 3.126 as

$$\text{NPSH} = \frac{p_o}{\gamma} - \Delta z_s - h_L - \frac{p_v}{\gamma}$$

and h_p is the head added by the pump. Atmospheric pressure, p_o, can be taken as 101 kPa; the specific weight of water, γ, is 9.79 kN/m³; the suction lift, Δz_s, is 3 m; and at 20°C, and the saturated vapor pressure of water, p_v, is 2.34 kPa. The head loss, h_L, must be estimated using the Darcy-Weisbach and minor loss coefficients. The flowrate, Q, is 0.0245 m³/s, and the average velocity of flow in the pipe, V, is given by

$$V = \frac{Q}{A} = \frac{0.0245}{\frac{\pi}{4}(0.102)^2} = 3.0 \text{ m/s}$$

At 20°C the kinematic viscosity, ν, is 1.00×10^{-6} m²/s and therefore the Reynolds number of the flow, Re, is given by

$$\text{Re} = \frac{VD}{\nu} = \frac{(3)(0.102)}{1.00 \times 10^{-6}} = 3.06 \times 10^5$$

Using Re $= 3.06 \times 10^5$, $k_s = 0.26$ mm and $D = 0.102$ m, the Colebrook equation gives $f = 0.0257$. Assuming an inlet head loss of $V^2/2g$ (for a projecting inlet), then the head loss, h_L, in the pipeline between the reservoir and the suction side of the pump is given by

$$h_L = \left(1 + \frac{fL}{D}\right)\frac{V^2}{2g} = \left[1 + \frac{(0.0257)(3.5)}{(0.102)}\right]\frac{(3.0)^2}{2(9.81)} = 0.863 \text{ m}$$

Now that all the parameters necessary to calculate the NPSH have been determined,

$$\text{NPSH} = \frac{p_o}{\gamma} - z_s - h_L - \frac{p_v}{\gamma} = \frac{101}{9.79} - 3 - 0.863 - \frac{2.34}{9.79} = 6.21 \text{ m}$$

The head added by the pump, h_p, can be calculated from the pump performance curve and the given flowrate (24.5 L/s) as

$$h_p = 12 - 0.01Q^2 = 12 - 0.01(24.5)^2 = 6 \text{ m}$$

The cavitation parameter of the pump, σ, is therefore given by

$$\sigma = \frac{\text{NPSH}}{h_p} = \frac{6.21}{6} = 1.04$$

If the specific speed of the pump is 0.94, then Figure 3.18 indicates that the critical cavitation parameter, σ_c, can be estimated as 0.20. Since the cavitation parameter in this case was found to be 1.04, then cavitation should not be a problem with this pump. When cavitation is imminent, the cavitation parameter is equal to 0.20 and the NPSH is given by

$$\text{NPSH} = \sigma_c h_p = 0.20(6) = 1.2 \text{ m}$$

or

$$1.2 = \frac{p_o}{\gamma} - z_s - h_L - \frac{p_v}{\gamma}$$

where $p_o = 101$ kPa, and $p_v = 2.34$ kPa. The head loss, h_L, between the pump and the reservoir can be estimated by the relation

$$h_L = \left[1 + \frac{f(z_s + 0.5)}{D}\right]\frac{V^2}{2g}$$

where the length of the pipe from the reservoir to the pump is estimated as $z_s + 0.5$ m, $D = 0.102$ m, $f = 0.0257$, and $V = 3.0$ m/s. Therefore,

$$h_L = \left[1 + \frac{(0.0257)(z_s + 0.5)}{0.102}\right]\frac{3^2}{2(9.81)} = 0.517 + 0.116z_s$$

Substituting this equation into the expression for NPSH gives

$$1.2 = \frac{p_o}{\gamma} - z_s - h_L - \frac{p_v}{\gamma}$$

$$= \frac{101}{9.79} - z_s - (0.517 + 0.116z_s) - \frac{2.34}{9.79}$$

$$= 9.56 - 1.116z_s$$

Solving for z_s yields

$$z_s = 7.49 \text{ m}$$

Therefore, the pump should be no more than 7.49 m above the reservoir. ■

3.4.4 ■ Multiple-Pump Systems

In cases where a single pump is inadequate to achieve a desired operating condition, multiple pumps can be used. Combinations of pumps are referred to as *pump systems*, and the pumps within these systems are typically arranged either in series or parallel. The characteristic curve of a pump system is determined by the arrangement of pumps within the system. Consider the case of two identical pumps in series, illustrated in Figure 3.19(a). The flow through each pump is equal to Q, and the head added by each pump is h_p. For the two-pump system, the flow through the system is equal to Q and the head added by the system is $2h_p$. Consequently, the characteristic curve of the two-pump (in series) system is related to the characteristic curve of each pump in that for any flow Q the head added by the system is twice the head added by a single pump, and the relationship between the single-pump characteristic curve and the two-pump

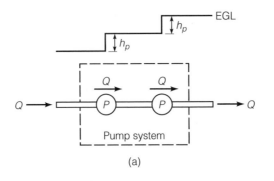

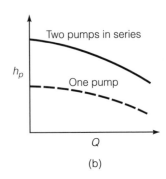

Figure 3.19 ■ Pumps in Series

(a)   (b)

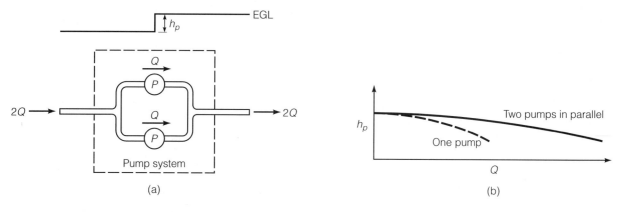

Figure 3.20 ■ Pumps in Parallel

characteristic curve is illustrated in Figure 3.19(b). This analysis can be extended to cases where the pump system contains n identical pumps in series, in which case the n-pump characteristic curve is derived from the single-pump characteristic curve by multiplying the ordinate of the single-pump characteristic curve (h_p) by n. Pumps in series are used in applications involving unusually high heads.

The case of two identical pumps arranged in parallel is illustrated in Figure 3.20. In this case, the flow through each pump is Q and the head added is h_p; therefore, the flow through the two-pump system is equal to $2Q$, while the head added is h_p. Consequently, the characteristic curve of the two-pump system is derived from the characteristic curve of the individual pumps by multiplying the abcissa (Q) by two. This is illustrated in Figure 3.20(b). In a similar manner, the characteristic curves of systems containing n identical pumps in parallel can be derived from the single-pump characteristic curve by multiplying the abcissa (Q) by n. Pumps in parallel are used in cases where the desired flowrate is beyond the range of a single pump and also to provide flexibility in pump operations, since some pumps in the system can be shut down during low-demand conditions or for service. This arrangement is common in sewage pump stations and water-distribution systems, where flowrates vary significantly during the course of a day.

When pumps are placed either in series or parallel, it is usually desirable that these pumps be identical; otherwise, the pumps will be loaded unequally and the efficiency of the pump system will be less than optimal. In cases where nonidentical pumps are placed in series, the characteristic curve of the pump system is obtained by summing the heads added by the individual pumps for a given flowrate. In cases where nonidentical pumps are placed in parallel, the characteristic curve of the pump system is obtained by summing the flowrates through the individual pumps for a given head.

■ **Example 3.14**

If a pump has a performance curve given by

$$h_p = 12 - 0.1Q^2$$

then what is the performance curve for: (a) a system having three of these pumps in series; and (b) a system having three of these pumps in parallel?

Solution

(a) For a system with three pumps in series, the same flow, Q, goes through each pump, and each pump adds one-third of the head, H_p, added by the pump system. Therefore,

$$\frac{H_p}{3} = 12 - 0.1Q^2$$

and the characteristic curve of the pump system is

$$H_p = 36 - 0.3Q^2$$

(b) For a system consisting of three pumps in parallel, one-third of the total flow, Q, goes through each pump, and the head added by each pump is the same as the total head, H_p added by the pump system. Therefore

$$H_p = 12 - 0.1\left(\frac{Q}{3}\right)^2$$

and the characteristic curve of the pump system is

$$H_p = 12 - 0.011Q^2 \qquad \blacksquare$$

3.5 Design of Water Distribution Systems

3.5.1 ■ Components of a Distribution System

The major components of a water distribution system are pipelines, pumps, storage facilities, valves, and meters. Water usually enters the system at a fairly constant rate from the treatment plant. To accommodate fluctuations in demand, a storage reservoir is typically located at the head of the system to store the excess water during periods of low demand and provide water from storage during periods of high demand. In addition to the operational storage required to accommodate diurnal (24-hour cycle) variations in water demand, storage facilities are also used to provide storage to fight fires, to provide storage for emergency conditions, and to equalize pressures in water-distribution systems. Storage facilities are classified as either ground storage or elevated storage. *Ground-storage* facilities are constructed at about ground level and discharge water to the distribution system through a pump station. *Elevated storage* facilities are constructed such that the height of the water in the elevated storage tank is sufficient to deliver water to the distribution system at the required pressure. Since the water level in a storage tank is equal to the elevation of the hydraulic grade line in the distribution system (at the outlet of the storage tank), elevated storage tanks are sometimes said to *float* on the system. Elevated storage is useful in the case of fires and emergency conditions since pumping of water from elevated tanks is not necessary, although the water must generally be pumped into elevated storage tanks.

Service pressures are typically maintained by pumps, with head losses and increases in pipeline elevations acting to reduce pressures and decreases in pipeline elevations acting to increase pressures. When portions of the distribution system are separated by long distances or significant changes in elevation, *booster pumps* are sometimes used to maintain acceptable service pressures. In some cases, *fire-service pumps* are used to provide additional capacity for emergency fire protection. Pumps operate at the intersection of the pump performance curve and the system curve. Since the system curve is significantly affected by variations in water demand, there is a significant variation in pump operating conditions. In most cases, the range of operating conditions is too wide to be met by a single pump, and multiple-pump installations or variable speed pumps are required (Velon and Johnson, 1993).

3.5.2 ■ Water Demand

Major considerations in designing water-supply systems are the water demands of the population being served, the fire flows needed to protect life and property, and the proximity of the service area to sources of water. There are usually several categories of water demand within any populated area, and these sources of demand can be broadly grouped into residential, commercial, industrial, and public. Residential water use is associated with houses and apartments where people live; commercial water use is associated with retail businesses, offices, hotels, and restaurants; industrial water use is associated with manufacturing and processing operations; and public water use includes governmental facilities that use water. Large industrial requirements are typically satisfied by sources other than the public water supply.

A typical distribution of water use for an average city is given in Table 3.4. These rates, however, vary from city to city as a result of differences in local conditions that are unrelated to the efficiency of water use. Water consumption is frequently stated in terms of the average amount of water delivered per day (to all categories of water use) divided by the population served, which is called the *average per capita demand*. The distribution of average per capita rates among 392 water-supply systems serving approximately 95 million people in the United States is shown in Table 3.5. The average per capita water

■ Table 3.4
Typical Distribution of Water Demand

Category	Average use (liters/day)/person	Percent of total
Residential	260	40
Commercial	90	14
Industrial	190	29
Public	70	10
Loss	50	7
Total	660	100

Source: Portions reprinted by permission of Waveland Press, Inc. from R. S. Gupta, *Hydrology and Hydraulic Systems* (Prospect Heights, IL; Waveland Press, Inc., 1989 [reissued 1995]). All rights reserved.

■ Table 3.5
Distribution of Per Capita Water Demand

Range (liters/day)/person	Number of systems	Percent of systems
190–370	30	7.7
380–560	132	33.7
570–750	133	33.9
760–940	51	13.0
950–1130	19	4.8
> 1140	27	6.9

Source: Reprinted from *1984 Water Utility Operating Data*, by permission. Copyright © 1986 American Water Works Association.

usage in this sample was 660 L/d, with a standard deviation of 270 L/d. Generally, high per capita rates are found in water-supply systems servicing large industrial or commercial sectors (Dziegielewski et al., 1996).

In the planning of municipal water-supply projects, the water demand at the end of the design life of the project is usually the basis for design. For existing water-supply systems, the American Water Works Association (AWWA, 1992) recommends that every 5 or 10 years, as a minimum, water-distribution systems be thoroughly reevaluated for requirements that would be placed on it by development and reconstruction over a 20-year period into the future. The estimation of the design flowrates for components of the water-supply system typically requires prediction of the population of the service area at the end of the design life, which is then multiplied by the per capita water demand to yield the design flowrate. Whereas the per capita water demand can usually be assumed to be fairly constant, the estimation of the future population typically involves a nonlinear extrapolation of past population trends.

A variety of methods are used in population forecasting. The simplest models treat the population as a whole, forecast future populations based on past trends, and fit empirical growth functions to historical population data. The most complex models disaggregate the population into various groups and forecast the growth of each group separately. A popular approach that segregates the population by age and gender is *cohort analysis* (Sykes, 1995). High levels of disaggregation have the advantage of making forecast assumptions very explicit, but these models tend to be complex and require more data than the empirical models that treat the population as a whole. Over relatively short time horizons, on the order of 10 years, detailed disaggregation models may not be any more accurate than using empirical extrapolation models of the population as a whole. Several conventional extrapolation models are illustrated in the following paragraphs.

Populated areas tend to grow in at varying rates, as illustrated in Figure 3.21. In the early stages of growth, there are wide open spaces. Population, P, tends to grow geometrically according to the relation

$$\frac{dP}{dt} = k_1 P \tag{3.128}$$

where k_1 is a growth constant. Integrating Equation 3.128 gives the following expression for the population as a function of time

$$P(t) = P_o e^{k_1 t} \tag{3.129}$$

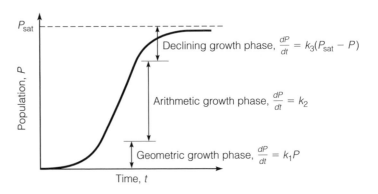

Figure 3.21 ■ Growth Phases in Populated Areas

where P_o is the population at some initial time designated as $t = 0$. Beyond the initial geometric growth phase, the rate of growth begins to level off and the following arithmetic growth relation may be more appropriate

$$\frac{dP}{dt} = k_2 \tag{3.130}$$

where k_2 is an arithmetic growth constant. Integrating Equation 3.130 gives the following expression for the population as a function of time

$$\boxed{P(t) = P_o + k_2 t} \tag{3.131}$$

where P_o is the population at $t = 0$. Ultimately, the growth of population centers becomes limited by the resources available to support the population, and further growth is influenced by the saturation population of the area, P_{sat}, and the population growth is described by a relation such as

$$\frac{dP}{dt} = k_3(P_{\text{sat}} - P) \tag{3.132}$$

where k_3 is a constant. This phase of growth is called the *declining growth* phase. Almost all communities have zoning regulations that control the use of both developed and undeveloped areas within their jurisdiction (sometimes called a master plan), and a review of these regulations will yield an estimate of the saturation population of the undeveloped areas. Integrating Equation 3.132 gives the following expression for the population as a function of time

$$\boxed{P(t) = P_{\text{sat}} - (P_{\text{sat}} - P_o)e^{-k_3 t}} \tag{3.133}$$

where P_o is the population at $t = 0$.

The time scale associated with each growth phase is typically on the order of 10 years, although the actual duration of each phase can deviate significantly from this number. The duration of each phase is important in that population extrapolation using a single-phase equation can only be justified for the duration of that growth phase. Consequently, single-phase extrapolations are typically limited to 10 years or less, and these population predictions are termed *short-term projections* (Viessman and Welty, 1985). Extrapolation beyond 10 years, called *long-term projections*, involve fitting an S-shaped curve to the historical population trends and then extrapolating using the fitted equation. The most commonly fitted S-curve is the so-called *logistic curve*, which is described by the equation

$$\boxed{P(t) = \frac{P_{\text{sat}}}{1 + ae^{bt}}} \tag{3.134}$$

where a and b are constants. The conventional methodology to fit the population equations to historical data is to plot the historical data, observe the trend in the data, and fit the curve that best matches the population trend. Regardless of which method is used to forecast the population, errors less than 10% can be expected for planning periods shorter than 10 years, and errors greater than 50% can be expected for planning periods longer than 20 years (Sykes, 1995).

■ **Example 3.15**

You are in the process of designing a water-supply system for a town, and the design life of your system is to end in the year 2020. The population in the town has been measured every 10 years since 1920 by the U.S. Census Bureau, and the reported populations are tabulated here. Estimate the population in the town using (a) graphical extension, (b) arithmetic growth projection, (c) geometric growth projection, (d) declining growth projection (assuming a saturation concentration of 600,000 people), and (e) logistic curve projection.

Year	Population
1920	125,000
1930	150,000
1940	150,000
1950	185,000
1960	185,000
1970	210,000
1980	280,000
1990	320,000

Solution

The population trend is plotted in Figure 3.22, where a geometric growth rate approaching an arithmetic growth rate is indicated.

(a) A growth curve matching the trend in the measured populations is indicated in Figure 3.22. Graphical extension to the year 2020 leads to a population estimate of 530,000 people.

(b) Arithmetic growth is described by Equation 3.131 as

$$P(t) = P_o + k_2 t \tag{3.135}$$

where P_o and k_2 are constants. Consider the arithmetic projection of a line passing through points B and C on the approximate growth curve shown in Figure 3.22. At point B, $t = 0$ (year 1980) and $P = 270,000$; at point C, $t = 10$ (year 1990) and $P = 330,000$. Applying these conditions to Equation 3.135 yields

$$P = 270000 + 6000t \tag{3.136}$$

In the year 2020, $t = 40$ years and the population estimate given by Equation 3.136 is

$$P = 510,000 \text{ people}$$

(c) Geometric growth is described by Equation 3.129 as

$$P = P_o e^{k_1 t} \tag{3.137}$$

where k_1 and P_o are constants. Using points A and C in Figure 3.22 as a basis for projection, then at $t = 0$ (year 1970), $P = 225,000$, and at $t = 20$ (year 1990), $P = 330,000$. Applying these conditions to Equation 3.137 yields

$$P = 225000 e^{0.0195t} \tag{3.138}$$

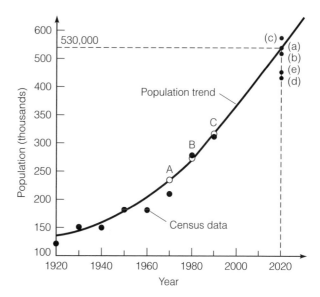

Figure 3.22 ■
Population Trend

In the year 2020, $t = 50$ years and the population estimate given by Equation 3.138 is

$$P = 597,000 \text{ people}$$

(d) Declining growth is described by Equation 3.133 as

$$P(t) = P_{sat} - (P_{sat} - P_o)e^{-k_3 t} \tag{3.139}$$

where P_o and k_3 are constants. Using points A and C in Figure 3.22, then at $t = 0$ (year 1970), $P = 225,000$, at $t = 20$ (year 1990), $P = 330,000$, and $P_{sat} = 600,000$. Applying these conditions to Equation 3.139 yields

$$P = 600000 - 375000e^{-0.0164t} \tag{3.140}$$

In the year 2020, $t = 50$ years and the population given by Equation 3.140 is given by

$$P = 434,800 \text{ people}$$

(e) The logistic curve is described by Equation 3.134. Using points A and C in Figure 3.22 to evaluate the constants in Equation 3.134 ($t = 0$ in 1970) yields

$$P = \frac{600000}{1 + 1.67e^{-0.0357t}} \tag{3.141}$$

In 2020, $t = 50$ years and the population given by Equation 3.141 is

$$P = 469,000 \text{ people}$$

These results indicate that the population projection in 2020 is quite uncertain, with estimates ranging from 597,000 for geometric growth to 434,800 for declining growth. The projected results are compared graphically in Figure 3.22. Closer inspection of the predictions indicate that the declining and logistic growth models

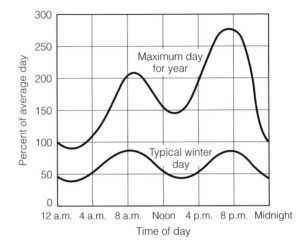

Figure 3.23 ■ Typical
Daily Cycles in Water
Demand
Source: Linsley, Ray K. et al,
Water-Resources Engineering.
Copyright © 1992 by The
McGraw-Hill Companies.

are limited by the specified saturation population of 600,000, while the geometric
growth model is not limited by saturation conditions and produces the highest projected population.   ■

The multiplication of the population projection by the per capita water demand is
used to estimate the *average daily demand* for a municipal water-supply system. The
(annual) average daily demand is equal to the average of the daily demands over one
year and is typically given in m^3/s.

Variations in Demand.   Water demand generally fluctuates between being below the
average daily demand in the early morning hours (before sunrise) and above the average
daily demand during the midday hours. Typical daily cycles in water demand are shown
in Figure 3.23. Water-use patterns within a typical 24-hour period are characterized by
demands that are 25% to 40% of the average daily demand during the hours between
midnight and 6:00 a.m. and 150% to 175% of the average daily demand during the
morning or evening peak periods (Velon and Johnson, 1993).

The range of demand conditions that are to be expected in water-distribution systems
are specified by *demand factors* or *peaking factors* that express the ratio of the demand
under certain conditions to the average daily demand. Typical demand factors for various conditions are given in Table 3.6, where the *maximum daily demand* is defined as the
demand on the day of the year that uses the most volume of water, and the *maximum
hourly demand* is defined as the demand during the hour that uses the most volume of

■ Table 3.6 Typical Demand Factors	Condition	Range of demand factors	Typical value
	Daily average in maximum month	1.10–1.50	1.20
	Daily average in maximum week	1.20–1.60	1.40
	Maximum daily demand	1.50–3.00	1.80
	Maximum hourly demand	2.00–4.00	3.25
	Minimum hourly demand	0.20–0.60	0.30

Source: Velon and Johnson (1993). Reprinted by permission of The McGraw-Hill Companies.

water. The demand factors in Table 3.6 should serve only as guidelines, with the actual demand factors in any one distribution system being best estimated from local measurements. In small water systems, demand factors may be significantly higher than those shown in Table 3.6.

Fire Demand.   Besides the fluctuations in demand that occur under normal operating conditions, water-distribution systems are usually designed to accommodate the large (short-term) water demands associated with fighting fires. Although there is no legal requirement that a governing body must size its water-distribution system to provide fire protection, the governing bodies of most communities provide water for fire protection for reasons that include protection of the tax base from destruction by fire, preservation of jobs, preservation of human life, and reduction of human suffering. Flowrates required to fight fires can significantly exceed the maximum flowrates in the absence of fires, particularly in small water systems.

Numerous methods have been proposed for estimating fire flows (AWWA, 1992), the most popular of which was proposed by the Insurance Services Office, Inc. (ISO, 1980). The required fire flow for individual buildings can be estimated using the formula (ISO, 1980)

$$\boxed{\text{NFF}_i = C_i O_i (X + P)_i} \tag{3.142}$$

where NFF_i is the *needed fire flow* at location i, C_i is the *construction factor* based on the size of the building and its construction, O_i is the *occupancy factor* reflecting the kinds of materials stored in the building (values range from 0.75 to 1.25), and $(X + P)_i$ is the sum of the *exposure factor* (X_i) and *communication factor* (P_i) that reflect the proximity and exposure of other buildings (values range from 1.0 to 1.75). The construction factor, C_i, is the portion of the NFF attributed to the size of the building and its construction and is given by

$$C_i = 220F\sqrt{A_i} \tag{3.143}$$

where C_i is in L/min; A_i (m^2) is the effective floor area, typically equal to the area of the largest floor in the building plus 50% of the area of all other floors; and F is a coefficient based on the class of construction, given in Table 3.7.

The maximum value of C_i calculated using Equation 3.143 is limited by the following: 30,000 L/min for construction classes 1 and 2; 23,000 L/min for construction classes

■ Table 3.7	Class of construction	Description	F
Construction Coefficient, F	1	frame	1.5
	2	joisted masonry	1.0
	3	noncombustible	0.8
	4	masonry, noncombustible	0.8
	5	modified fire resistive	0.6
	6	fire resistive	0.6

Source: AWWA (1992).

■ **Table 3.8**
Occupancy Factors, O_i

Combustibility class	Examples	O_i
C-1 Noncombustible	steel or concrete products storage	0.75
C-2 Limited combustible	apartments, churches, offices	0.85
C-3 Combustible	department stores, supermarkets	1.00
C-4 Free burning	auditoriums, warehouses	1.15
C-5 Rapid burning	paint shops, upholstering shops	1.25

Source: Reprinted from *Distribution System Requirements for Fire Protection* (M31), by permission. Copyright © 1992 American Water Works Association.

3, 4, 5, and 6; and 23,000 L/min for a one-story building of any class of construction. The minimum value of C_i is 2,000 L/min, and the calculated value of C_i should be rounded to the nearest 1,000 L/min. The occupancy factors, O_i, for various classes of buildings are given in Table 3.8. Detailed tables for estimating the exposure and communication factors, $(X+P)_i$, can be found in AWWA (1992), and values of $(X+P)_i$ are typically on the order of 1.4. The NFF calculated using Equation 3.142 should not exceed 45,000 L/min, nor be less than 2,000 L/min. According to AWWA (1992), 2,000 L/min is the minimum amount of water with which any fire can be controlled and suppressed safely and effectively. The NFF should be rounded to the nearest 1,000 L/min if less than 9,000 L/min, and to the nearest 2,000 L/min if greater than 9,000 L/min. For one- and two-family dwellings not exceeding two stories in height, the NFF listed in Table 3.9 should be used. For other habitable buildings not listed in Table 3.9, the NFF should not exceed 13,000 L/min maximum.

Usually the local water utility will have a policy on the upper limit of fire protection that it will provide to individual buildings. Those wanting higher fire flows need to either provide their own system or reduce fire-flow requirements by installing sprinkler systems, fire walls, or fire-retardant materials (Walski, 1996; AWWA, 1992). Estimates of the needed fire flow calculated using Equation 3.142 are used to determine the fire-flow requirements of the water-supply system, where the needed fire flow is calculated at several representative locations in the service area, and it is assumed that only one building is on fire at any time (Sykes, 1995). The design duration of the fire should follow the guidelines in Table 3.10. If these durations cannot be maintained, insurance rates are typically increased accordingly. A more detailed discussion of the requirements for fire protection has been published by the American Water Works Association (AWWA, 1992).

■ **Table 3.9**
Needed Fire Flow for One- and Two-Family Dwellings

Distance between buildings (m)	Needed fire flow (L/min)
> 30	2,000
9.5–30	3,000
3.5–9.5	4,000
< 3.5	6,000

Source: Reprinted from *Distribution System Requirements for Fire Protection* (M31), by permission. Copyright © 1992 American Water Works Association.

	Required fire flow (L/min)	Duration (h)
■ Table 3.10 Required Fire Flow Durations	< 9000	2
	11,000–13,000	3
	15,000–17,000	4
	19,000–21,000	5
	23,000–26,000	6
	26,000–30,000	7
	30,000–34,000	8
	34,000–38,000	9
	38,000–45,000	10

Source: Reprinted from *Distribution System Requirements for Fire Protection* (M31), by permission. Copyright © 1992 American Water Works Association.

■ Example 3.16

Estimate the flowrate and volume of water required to provide adequate fire protection to a 10-story noncombustible building with an effective floor area of 8,000 m^2.

Solution

The NFF can be estimated by Equation 3.142 as

$$\text{NFF}_i = C_i O_i (X + P)_i$$

where the construction factor, C_i, is given by

$$C_i = 220F\sqrt{A_i}$$

For the 10-story building, $F = 0.8$ (Table 3.7, noncombustible, Class 3 construction), and $A_i = 8,000$ m^2, hence

$$C_i = 220(0.8)\sqrt{8000} = 16000 \text{ L/min}$$

where C_i has been rounded to the nearest 1,000 L/min. The occupancy factor, O_i, is given by Table 3.8 as 0.75 (C-1 Noncombustible); $(X + P)_i$ can be estimated by the median value of 1.4; and hence the needed fire flow, NFF, is given by

$$\text{NFF}_i = (16000)(0.75)(1.4) = 17000 \text{ L/min}$$

This flow must be maintained for a duration of four hours (Table 3.10). Hence the required volume, V, of water is given by

$$V = 17000 \times 4 \times 60 = 4.08 \times 10^6 \text{ L} = 4080 \text{ m}^3 \qquad ■$$

Fire hydrants are placed throughout the service area to provide either direct hose connections for firefighting or connections to pumper trucks (also known as fire engines). A single-hose stream is generally taken as 950 L/min, and hydrants are typically located at street intersections or spaced 60–150 m apart (McGhee, 1991). In high-value districts, additional hydrants may be necessary in the middle of long blocks. Fire hydrants may

■ **Table 3.11**
Design Periods and
Capacities in Water-Supply
Systems

Component	Design period (years)	Design capacity
Sources of supply:		
River	indefinite	maximum daily demand
Wellfield	10–25	maximum daily demand
Reservoir	25–50	average annual demand
Pumps:		
Low-lift	10	maximum daily demand, one reserve unit
High-lift	10	maximum hourly demand, one reserve unit
Water-treatment plant	10–15	maximum daily demand
Service reservoir	20–25	working storage plus fire demand plus emergency storage
Distribution system:		
Supply pipe or conduit	25–50	greater of (1) maximum daily demand plus fire demand, or (2) maximum hourly demand
Distribution grid	full development	same as for supply pipes

Source: Reprinted by permission of Waveland Press, Inc. from R. S. Gupta, *Hydrology and Hydraulic Systems*. (Prospect Heights, IL; Waveland Press, Inc., 1989 [reissued 1995]). All rights reserved.

also be used to release air at high points in the water-distribution system and blow off sediments at low points in the system.

Design Flows.   The design capacity of various components of the water-supply system are given in Table 3.11, where *low-lift pumps* refer to low-head, high-rate units that convey the raw-water supply to the treatment facility and *high-lift pumps* deliver finished water from the treatment facility into the distribution network at suitable pressures. The required capacities consist of various combinations of the maximum daily demand, maximum hourly demand, and the fire demand. Typically, the delivery pipelines from the water source to the treatment plant, as well as the treatment plant itself, are designed with a capacity equal to the maximum daily demand. The flowrates and pressures in the distribution system are analyzed under both maximum daily plus fire demand and the maximum hourly demand, and the larger flowrate governs the design. Pumps are sized for a variety of conditions from maximum daily to maximum hourly demand, depending on their function in the distribution system. Additional reserve capacity is usually installed in water-supply systems to allow for redundancy and maintenance requirements.

■ **Example 3.17**

A metropolitan area has a population of 130,000 people with an average daily demand of 600 L/d/person. If the needed fire flow is 20,000 L/min, estimate: (a) the design capacities for the wellfield and the water-treatment plant; (b) the duration that the fire flow must be sustained and the volume of water that must be kept in the service reservoir in case of a fire; and (c) the design capacity of the main supply pipeline to the distribution system.

Solution

(a) The design capacity of the wellfield should be equal to the maximum daily demand (Table 3.11). With a demand factor of 1.8 (Table 3.6), the per capita demand on the maximum day is equal to $1.8 \times 600 = 1,080$ L/day/person. Since the population served is 130,000 people, the design capacity of the wellfield, Q_{well}, is given by

$$Q_{well} = 1080 \times 130000 = 1.4 \times 10^8 \text{ L/d} = 1.62 \text{ m}^3/\text{s}$$

The design capacity of the water-treatment plant is also equal to the maximum daily demand, and therefore should also be taken as 1.62 m³/s.

(b) The needed fire flow, Q_{fire}, is 20000 L/min = 0.33 m³/s. According to Table 3.10, the fire flow must be sustained for 5 hours. The volume, V_{fire}, required for the fire flow will be stored in the service reservoir and is given by

$$V_{fire} = 0.33 \times 5 \times 3600 = 5,940 \text{ m}^3$$

(c) The required flowrate in the main supply pipeline is equal to the maximum daily demand plus fire demand or the maximum hourly demand, whichever is greater.

$$\text{Maximum daily demand + fire demand} = 1.62 + 0.33 = 1.95 \text{ m}^3/\text{s}$$
$$\text{Maximum hourly demand} = \frac{3.25}{1.80} \times 1.62 = 2.92 \text{ m}^3/\text{s}$$

where a demand factor of 3.25 has been assumed for the maximum hourly demand. The main supply pipe to the distribution system should therefore be designed with a capacity of 2.92 m³/s. The water pressure within the distribution system must be above acceptable levels when the system demand is 2.92 m³/s. ■

3.5.3 ■ Pipelines

Water-distribution systems typically consist of connected pipe loops throughout the service area. The principal arteries of distribution systems are the *transmission* and *feeder mains* that carry flow from the water-treatment plant to the service area. Transmission mains typically have a looped configuration, with diameters greater than 600 mm, and are usually on the order of 3 km apart. Feeder mains are connected to transmission mains and are laid out in interlocking loops with the pipelines not more than 1 km apart and diameters in the range of 400–500 mm. Smaller *distribution mains* form a grid over the entire service area, with diameters in the range of 150–300 mm, and supply water to every user. Pipelines in distribution systems are collectively called *water mains*, and a pipe that carries water from a main to a building or property is called a *service line*. Water mains are normally installed within the rights-of-way of streets and should be buried to a depth below the frost line in northern climates and at a depth sufficient to cushion the pipe against traffic loads in warmer climates (Clark, 1990). Generally, a cover of 1.2 m to 1.5 m is used for large mains and 0.75 m to 1.0 m for smaller mains. In areas where frost penetration is a significant factor, mains can have as much as 2.5 m of cover. Trenches for water mains should be as narrow as possible and still be wide enough to allow for proper joining and compaction around the pipe. The suggested trench width is the nominal pipe diameter plus 0.6 m; in deep trenches, sloping may be necessary to keep the bank from caving in. Trench bottoms should be undercut 15 to 25 cm, and sand, clean fill, or crushed stone installed to provide a cushion against the bottom of the excavation, which is usually rock (Clark, 1990).

Ductile iron is the most widely used material for pipe diameters up to 760 mm, and it has largely replaced *cast iron* in new construction. Ductile iron pipe is manufactured in diameters from 100 to 1200 mm (4–48 in.). For diameters from 100 to 500 mm (4–20 in.), standard commercial sizes are available in 50-mm (2 in.) increments, while for diameters from 600 to 1200 mm (24–48 in.), the size increments are 150 mm (6 in.). The standard lengths of ductile iron pipe are 5.5 m (18 ft) and 6.1 m (20 ft). Ductile iron pipes are usually coated (outside and inside) with an asphaltic coating to minimize corrosion. For ductile iron pipes larger than 760 mm, steel pipe and prestressed reinforced concrete pipe compare favorably with ductile iron.

Standards for pipe construction, installation, and performance are published by the American Water Works Association in its C-series standards, which are continuously being updated. For fire protection, insurance underwriters typically require a minimum main size of 150 mm for residential areas and 200 mm for high-value districts if cross-connecting mains are not more than 180 m apart. On principal streets, and for all long lines not connected at frequent intervals, 300-mm and larger mains are required. Dead ends should be avoided whenever possible, since the lack of flow in such lines may contribute to water-quality problems. Allowable velocities in pipeline systems are governed by the characteristics of the water and the magnitude of hydraulic transients. Velocities of 0.9 to 1.8 m/s are common in water-distribution pipes, and the upper limit lies between 3 and 6 m/s for most types of conduit materials.

3.5.4 ■ Operating Criteria for Water-Distribution Systems

The primary functions of water-distribution systems (Zipparro and Hasen, 1993) are to (1) meet the water demands of users while maintaining acceptable pressures in the system; (2) supply water for fire protection at specific locations within the system, while maintaining acceptable pressures for normal service throughout the remainder of the system; and (3) provide a sufficient level of redundancy to support a minimum level of reliable service during emergency conditions, such as an extended loss of power or a major water-main failure.

Service Pressures.   The requirement that adequate pressures be maintained in the distribution system while supplying the service demands requires that the system be analyzed on the basis of specified minimum allowable pressures. Criteria for minimum acceptable service pressures recommended by the Great Lakes Upper Mississippi River Board of State Public Health and Environmental Managers (GLUMB, 1987) are typical of most water-distribution systems, and they are listed in Table 3.12. There are several considerations in assessing the adequacy of service pressures, including (1) the pressure required at street level for excellent flow to a 3-story building is about 290 kPa; (2) flow is adequate for residential areas if the pressure is not reduced below 240 kPa; (3) the pressure required for adequate flow to a 20-story building is about 830 kPa, which is not desirable because of the associated leakage and waste; (4) very tall buildings are usually served with their own pumping equipment; and (5) it is usually desirable to maintain normal pressures of 410–520 kPa since these pressures are adequate for the following purposes:

■ Table 3.12
Minimum Acceptable
Pressures in Distribution
Systems

Demand condition	Minimum acceptable pressure (kPa)
Average daily demand	240–410
Maximum daily demand	240–410
Maximum hourly demand	240–410
Fire situation	> 140
Emergency conditions	> 140

Source: GLUMB (1987).

- To supply ordinary consumption for buildings up to 10 stories.
- To provide adequate sprinkler service in buildings of 4 to 5 stories.
- To provide direct hydrant service for quick response.
- To allow larger margin for fluctuations in pressure caused by clogged pipes and excessive length of service pipes.

Pressures higher than 650 kPa should be avoided if possible because of the added burden of installing and maintaining pressure-reducing valves and other specialized equipment (Clark, 1990).

Storage Facilities.   Storage facilities in a distribution system are required to have sufficient volume to meet the following criteria (Velon and Johnson, 1993): (1) adequate volume to supply peak demands in excess of the maximum daily demand using no more than 50% of the available storage capacity; (2) adequate volume to supply the critical fire demand in addition to the volume required for meeting the maximum daily demand fluctuations; and (3) adequate volume to supply the average daily demand of the system for the estimated duration of a possible emergency. Conventional design practice is to rely on pumping to meet the daily operational demands up to the maximum daily demand; where detailed demand data are not available, the storage available to supply the peak demands should equal 20% to 25% of the maximum daily demand volume. Sizing the storage volume for fire protection is based on the product of the critical fire flow and duration for the service area. Emergency storage volumes for most municipal water-supply systems vary from one to two days of supply capacity at the average daily demand. The recommended standards for water works developed by the Great Lakes Upper Mississippi River Board of State Public Health and Environmental Managers suggest a minimum emergency storage capacity equal to the average daily system demand.

The minimum acceptable height of water in an elevated storage tank is determined by computing the minimum acceptable piezometric head in the service area and then adding to that figure an estimate of the head losses between the critical service location and the location of the elevated service tank, under the condition of average daily demand. The maximum height of water in the elevated tank is then determined by adding the minimum acceptable piezometric head to the head loss between the tank location and the critical service location under the condition of maximum hourly demand. The difference between the calculated minimum and maximum heights of water in the elevated storage tank is then specified as the normal operating range within the tank. The normal operating range for water in elevated tanks is usually limited to 4.5 to 6 meters, so that fluctuations in pressure are limited to 35 to 70 kPa. In most cases, the operating range is located in the upper half of the storage tank, with storage in the lower half of the tank reserved for firefighting and emergency storage. A typical elevated storage

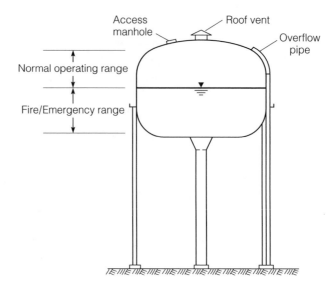

Figure 3.24 ■ Typical Elevated Storage Tank

Source: Zipparo, Vincent J. and Hans Hasen/Velon and Johnson. *Davis' Handbook of Applied Hydraulics.* Copyright © 1993 by The McGraw-Hill Companies.

tank is illustrated in Figure 3.24. These types of storage facilities generally have only a single pipe connection to the distribution system, and this single pipe handles both inflows and outflows from the storage tank. This piping arrangement is in contrast to ground storage reservoirs, which have separate inflow and outflow piping. The inflow piping delivers the outflow from the water-treatment facility to the reservoir, while the outflow piping delivers the water from the reservoir to the pumps that input water into the distribution system. Ground-level storage reservoirs can be made of either concrete or steel, while elevated tanks are generally made of steel (Walski, 1996).

■ Example 3.18

A service reservoir is to be designed for a water-supply system serving 250,000 people with an average demand of 600 L/d/capita, and a needed fire flow of 37,000 L/min. Estimate the required volume of service storage.

Solution

The required storage is the sum of three components: (1) volume to supply the demand in excess of the maximum daily demand, (2) fire storage, and (3) emergency storage.

The volume to supply the peak demand can be taken as 25% of the maximum daily demand volume. Taking the maximum daily demand factor as 1.8 (Table 3.6), then the maximum daily flowrate, Q_m, is given by

$$Q_m = (1.8)(600)(250000) = 2.7 \times 10^8 \text{ L/d} = 2.7 \times 10^5 \text{ m}^3/\text{d}$$

The storage volume to supply the peak demand, V_{peak}, is therefore given by

$$V_{\text{peak}} = (0.25)(2.7 \times 10^5) = 67500 \text{ m}^3$$

According to Table 3.10, the 37,000 L/min ($= 0.62 \text{ m}^3/\text{s}$) fire flow must be maintained for at least 9 hours. The volume to supply the fire demand, V_{fire}, is therefore given by

$$V_{\text{fire}} = 0.62 \times 9 \times 3600 = 20100 \text{ m}^3$$

The emergency storage, V_{emer}, can be taken as the average daily demand, in which case

$$V_{emer} = 250000 \times 600 = 150 \times 10^6 \text{ L} = 150{,}000 \text{ m}^3$$

The required volume, V, of the service reservoir is therefore given by

$$V = V_{peak} + V_{fire} + V_{emer}$$
$$= 67500 + 20100 + 150000$$
$$= 237{,}600 \text{ m}^3$$

The service reservoir should be designed to store 238,000 m³ of water. It is interesting to note that most of the storage in the service reservoir is reserved for emergencies.   ■

■ Example 3.19

A water-supply system is to be designed in an area where the minimum allowable pressure in the distribution system is 300 kPa. A hydraulic analysis of the distribution network under average daily demand conditions indicates that the head loss between the low-pressure service location, which has a pipeline elevation of 5.40 m, and the location of the elevated storage tank is 10 m. Under maximum hourly demand conditions, the head loss between the low-pressure service location and the elevated storage tank is 12 m. Determine the normal operating range for the water stored in the elevated tank.

Solution

Under average demand conditions, the elevation z_o of the hydraulic grade line (HGL) at the reservoir location is given by

$$z_o = \frac{p_{min}}{\gamma} + z_{min} + h_L$$

where $p_{min} = 300$ kPa, $\gamma = 9.79$ kN/m³, $z_{min} = 5.4$ m, and $h_L = 10$ m, which yields

$$z_o = \frac{300}{9.79} + 5.4 + 10 = 46.0 \text{ m}$$

Under maximum hourly demand conditions, the elevation z_1 of the HGL at the service reservoir is given by

$$z_1 = \frac{300}{9.79} + 5.4 + 12 = 48.0 \text{ m}$$

Therefore, the operating range in the storage tank should be between elevations 46.0 m and 48.0 m.   ■

3.5.5 ■ Network Analysis

Methodologies for analyzing pipe networks were discussed in Section 3.3, and these methods can be applied to any given pipe network to calculate the pressure and flow distribution under a variety of demand conditions. In complex pipe networks, the application of computer programs to implement these methodologies is standard practice (Haestad Methods, 1997a; 1997b). Computer programs allow engineers to easily calculate the hydraulic performance of complex networks and such parameters as the age of water delivered to consumers and also to trace the origin of the delivered water. Water age is measured from the time the water enters the system and gives an indication of the overall quality of the delivered water. Steady-state analyses are usually adequate for assessing the performance of various components of the distribution system, including the pipelines, storage tanks, and pumping systems, while time-dependent simulations are useful in evaluating the operation of pumping stations and variable-level storage tanks (Velon and Johnson, 1993). Modelers frequently refer to time-dependent simulations as *extended-period simulations*.

An important part of analyzing large water-distribution systems is the *skeletonizing* of the system, which consists of representing the full water-distribution system by a subset of the system that includes only the most important elements. For example, consider the case of a water supply to the subdivision shown in Figure 3.25(a), where the system shown includes the service connections to the houses. A slight degree of skeletonization could be achieved by omitting the household service pipes (and their associated head losses) from consideration and accounting for the water demands at the tie-ins and as shown in Figure 3.25(b). This reduces the number of junctions from 48 to 19. Further skeletonization can be achieved by modeling just 4 junctions, consisting of the ends of the main piping and the major intersections shown in Figure 3.25(c). In this case, the water demands are associated with the nearest junctions to each of the service connections, and the dashed lines in Figure 3.25(c) indicate the service areas for each junction.

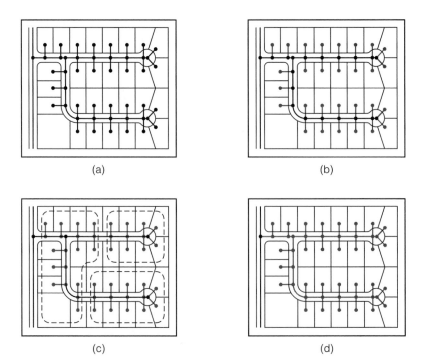

(a)

(b)

(c)

(d)

Figure 3.25 ■ Skeletonizing a Water-Distribution System

Source: Haestad Methods, 1997 *Practical Guide: Hydraulics and Hydrology* p. 61–62. Copyright © 1997 by Haestad Methods, Inc. Reprinted by permission.

A further level of skeletonization is shown in Figure 3.25(d), where the water supply to the entire subdivision is represented by a single node, at which the water demand of the subdivision is attributed.

Clearly, further levels of skeletonization could be possible in large water-distribution systems. As a general guideline, larger systems permit more degrees of skeletonization without introducing significant error in the flow conditions of main distribution pipes.

The results of a pipe-network analysis should generally include pressures and/or hydraulic grade line elevations at all nodes, flow, velocity, and head loss through all pipes as well as rates of flow into and out of all storage facilities. These results are used to assess the hydraulic performance and reliability of the network, and they are to be compared with the guidelines and specifications required for acceptable performance.

Summary

The hydraulics of flow in closed conduits is the basis for designing water-supply systems and other systems that involve the transport of water under pressure. The fundamental relationships governing flow in closed conduits are the conservation laws of mass, momentum, and energy; the forms of these equations that are most useful in engineering applications are derived from first principles. Of particular note is the momentum equation, the most useful form of which is the Darcy-Weisbach equation. Techniques for analyzing flows in both single and multiple pipelines, using the nodal and loop methods, are presented. Flows in closed conduits are usually driven by pumps, and the fundamentals of pump performance using dimensional analysis and similitude are presented. Considerations in selecting a pump include the specific speed under design conditions, the application of affinity laws in adjusting pump performance curves, the computation of operating points in pump-pipeline systems, practical limits on pump location based on the critical cavitation parameter, and the performance of pump systems containing multiple units.

Water-supply systems are designed to meet service-area demands during the design life of the system. Projection of water demand involves the estimation of per capita demands and population projections. Over short time scales, populations can be expected to follow either geometric, arithmetic, or declining growth models, while over longer time scales a logistic growth curve may be more appropriate. Components of water-supply systems must be designed to accommodate daily fluctuations in water demand plus potential fire flows. The design periods and capacities of various components of water-supply systems are listed in Table 3.11. Other key considerations in designing water distribution systems include required service pressures (Table 3.12), pipeline selection and installation, and provision of adequate storage capacity to meet fire demands and emergency conditions.

Problems

3.1. Water at 20°C is flowing in a 100-mm diameter pipe at an average velocity of 2 m/s. If the diameter of the pipe is suddenly expanded to 150 mm, what is the new velocity in the pipe? What are the volumetric and mass flowrates in the pipe?

3.2. A 200-mm diameter pipe divides into two smaller pipes each of diameter 100 mm. If the flow divides equally between the two smaller pipes and the velocity in the 200-mm pipe is 1 m/s, calculate the velocity and flowrate in each of the smaller pipes.

3.3. The velocity distribution in a pipe is given by the equation

$$v(r) = V_o \left[1 - \left(\frac{r}{R} \right)^2 \right] \tag{3.144}$$

where $v(r)$ is the velocity at a distance r from the centerline of the pipe, V_o is the centerline velocity, and R is the radius of the pipe. Calculate the average velocity and flowrate in the pipe in terms of V_o.

3.4. Calculate the momentum correction coefficient, β, for the velocity distribution given in Equation 3.144.

3.5. Water is flowing in a horizontal 200-mm diameter pipe at a rate of 0.06 m^3/s, and the pressures at sections 100 m apart are equal to 500 kPa at the upstream section and 400 kPa at the downstream section. Estimate the average shear stress on the pipe and the friction factor, f. [*Hint:* Use Equation 3.26 to calculate the shear stress and Equation 3.32 to calculate the friction factor.]

3.6. Water at 20°C flows at a velocity of 2 m/s in a 250-mm diameter horizontal ductile iron pipe. Estimate the friction factor in the pipe, and state whether the pipe is hydraulically smooth or rough. Compare the friction factors derived from the Moody diagram, the Colebrook equation, and the Jain equation. Estimate the change in pressure over 100 m of pipeline. How would the friction factor and pressure change be affected if the pipe is not horizontal but 1 m lower at the downstream section?

3.7. Show that the Colebrook equation can be written in the (slightly) more convenient form:

$$f = \frac{0.25}{\{\log[(k_s/D)/3.7 + 2.51/(\mathrm{Re}\sqrt{f})]\}^2}$$

Why is this equation termed "(slightly) more convenient"?

3.8. If you had your choice of estimating the friction factor either from the Moody diagram or from the Colebrook equation, which one would you pick? Explain your reasons.

3.9. Water leaves a treatment plant in a 500-mm diameter ductile iron pipeline at a pressure of 600 kPa and at a flowrate of 0.50 m^3/s. If the elevation of the pipeline at the treatment plant is 120 m, then estimate the pressure in the pipeline 1 km downstream where the elevation is 100 m. Assess whether the pressure in the pipeline would be sufficient to serve the top floor of a 10-story (approximately 30 m high) building.

3.10. A 25-mm diameter galvanized iron service pipe is connected to a water main in which the pressure is 400 kPa. If the length of the service pipe to a faucet is 20 m and the faucet is 2.0 m above the main, estimate the flowrate when the faucet is fully open.

3.11. A galvanized iron service pipe from a water main is required to deliver 300 L/s during a fire. If the length of the service pipe is 40 m and the head loss in the pipe is not to exceed 45 m, calculate the minimum pipe diameter that can be used. Use the Colebrook equation in your calculations.

3.12. Repeat Problem 3.11 using the Swamee-Jain equation.

3.13. Use the velocity distribution given in Problem 3.3 to estimate the kinetic energy correction factor, α, for turbulent pipe flow.

3.14. The velocity profile, $v(r)$, for turbulent flow in smooth pipes is sometimes estimated by the seventh-root law, originally proposed by Blasius (1911)

$$v(r) = V_o \left(1 - \frac{r}{R}\right)^{\frac{1}{7}}$$

where V_o is the maximum (centerline) velocity and R is the radius of the pipe. Estimate the energy and momentum correction factors corresponding to the seventh-root law.

3.15. Show that the kinetic energy correction factor, α, corresponding to the power-law velocity profile is given by Equation 3.75. Use this result to confirm your answer to Problem 3.14.

3.16. Water enters and leaves a pump in pipelines of the same diameter and approximately the same elevation. If the pressure on the inlet side of the pump is 30 kPa and a pressure of 500 kPa is desired for the water leaving the pump, what is the head that must be added by the pump, and what is the power delivered to the fluid?

3.17. Water leaves a reservoir at 0.06 m^3/s through a 200-mm riveted steel pipeline that protrudes into the reservoir and then immediately turns a 90° bend with a minor loss coefficient equal to 0.3. Estimate the length of pipeline required for the friction losses to account for 90% of the total losses, which includes both friction losses and so-called "minor losses". Would it be fair to say that for pipe lengths shorter than the length calculated in this problem that the word "minor" should not be used?

3.18. Water is pumped from a supply reservoir to a ductile iron water transmission line, as shown in Figure 3.26. The high point of the transmission line is at point A, 1 km downstream of the supply reservoir, and the low point of the transmission line is at point B, 1 km downstream of A. If the flowrate through the pipeline is 1 m^3/s, the diameter of the pipe is 750 mm, and the pressure at A is to be 350 kPa, then: (a) estimate the head that must be added by the pump; (b) estimate the power supplied by the pump; and (c) calculate the water pressure at B.

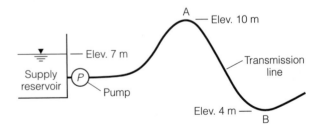

Figure 3.26 ■ Problem 3.18

3.19. A pipeline is to be run from a water-treatment plant to a major suburban development 3 km away. The average daily demand for water at the development is 0.0175 m^3/s, and the peak demand is 0.578 m³/s. Determine the required diameter of ductile iron pipe such that the flow velocity during peak demand is 2.5 m/s. Round the pipe diameter upward to the nearest 25 mm (i.e., 25 mm, 50 mm, 75 mm, ...). The water pressure at the development is to be at least 340 kPa during average demand conditions, and 140 kPa during peak demand. If the water at the treatment plant is stored in a ground-level reservoir where the level of the water is 10.00 m NGVD and the ground elevation at the suburban development is 8.80 m NGVD, determine the pump power (in kilowatts) that must be available to meet both the average daily and peak demands.

3.20. Water flows at 5 m³/s in a 1 m × 2 m rectangular concrete pipe. Calculate the head loss over a length of 100 m.

3.21. Water flows at 10 m³/s in a 2 m × 2 m square reinforced-concrete pipe. If the pipe is laid on a (downward) slope of 0.002, what is the change in pressure in the pipe over a distance of 500 m?

3.22. Derive the Hazen-Williams head-loss relation, Equation 3.84, starting from Equation 3.82.

3.23. Compare the Hazen-Williams formula for head loss (Equation 3.84) with the Darcy-Weisbach equation for head loss (Equation 3.33) to determine the expression for the friction factor that is assumed in the Hazen-Williams formula. Based on your result, identify the type of flow condition incorporated in the Hazen-Williams formula (rough, smooth, or transition).

3.24. Derive the Manning head-loss relation, Equation 3.86.

3.25. Compare the Manning formula for head loss (Equation 3.86) with the Darcy-Weisbach equation for head loss (Equation 3.33) to determine the expression for the friction factor that is assumed in the Manning formula. Based on your result, identify the type of flow condition incorporated in the Manning formula (rough, smooth, or transition).

3.26. Determine the relationship between the Hazen-Williams roughness coefficient and the Manning roughness coefficient.

3.27. Given a choice between using the Darcy-Weisbach, Hazen-Williams, or Manning equations to estimate the friction losses in a pipeline, which equation would you choose? Why?

3.28. Water flows at a velocity of 2 m/s in a 300-mm new ductile iron pipe. Estimate the head loss over 500 m using: (a) the Hazen-Williams formula; (b) the Manning formula; and (c) the Darcy-Weisbach equation. Compare your results. Calculate the Hazen-Williams roughness coefficient and the Manning coefficient that should be used to obtain the same head loss as the Darcy-Weisbach equation.

3.29. Reservoirs A, B, and C are connected as shown in Figure 3.27. The water elevations in reservoirs A, B, and C are 100 m, 80 m, and 60 m, respectively. The three pipes connecting the reservoirs meet at the junction J, with pipe AJ being 900 m long, BJ 800 m long, CJ 700 m long, and the diameter of all pipes equal to 850 mm. If all pipes are made of ductile

iron and the water temperature is 20°C, find the flow into or out of each reservoir.

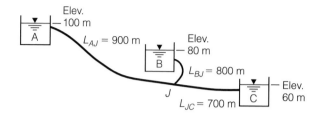

Figure 3.27 ■ Problem 3.29

3.30. The water-supply network shown in Figure 3.28 has constant-head elevated storage tanks at A and B, with inflows and withdrawals at C and D. The network is on flat terrain, and the pipeline characteristics are as follows:

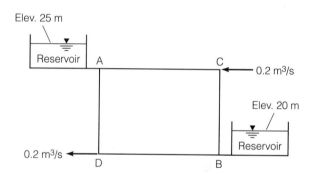

Figure 3.28 ■ Problem 3.30

Pipe	L (km)	D (mm)
AD	1.0	400
BC	0.8	300
BD	1.2	350
AC	0.7	250

If all pipes are made of ductile iron, calculate the inflows/outflows from the storage tanks. Assume that the flows in all pipes are fully turbulent.

3.31. Consider the pipe network shown in Figure 3.29. The Hardy Cross method can be used to calculate the pressure distribution in the system, where the friction loss, h_f, is estimated using the equation

$$h_f = rQ^n$$

and all pipes are made of ductile iron. What value of r and n would you use for each pipe in the system? The pipeline characteristics are as follows:

Pipe	L (m)	D (mm)
AB	1,000	300
BC	750	325
CD	800	200
DE	700	250
EF	900	300
FA	900	250
BE	950	350

You can assume that the flow in each pipe is hydraulically rough.

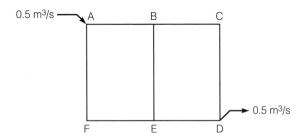

Figure 3.29 ■ Problem 3.31

3.32. A portion of a municipal water distribution network is shown in Figure 3.30, where all pipes are made of ductile iron and have diameters of 300 mm. Use the Hardy Cross method to find the flowrate in each pipe. If the pressure at point P is 500 kPa and the distribution network is on flat terrain, determine the water pressures at each pipe intersection.

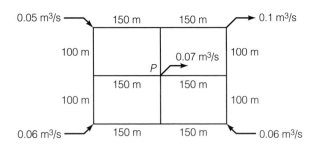

Figure 3.30 ■ Problem 3.32

3.33. What is the constant that can be used to convert the specific speed in SI units (Equation 3.111) to the specific speed in U.S. Customary units (Equation 3.112)?

3.34. What is the highest synchronous speed for a motor driving a pump?

3.35. Derive the affinity relationship for the power delivered to a fluid by two homologous pumps. [*Note*: This affinity relation is given by Equation 3.116.]

3.36. A pump is required to deliver 150 L/s ($\pm$ 10%) through a 300-mm diameter PVC pipe from a well to a reservoir. The water level in the well is 1.5 m below the ground and the water surface in the reservoir is 2 m above the ground. The delivery pipe is 300 m long, and minor losses can be neglected. A pump manufacturer suggests using a pump with a performance curve given by

$$h_p = 6 - 6.67 \times 10^{-5}Q^2$$

where h_p is in meters and Q in L/s. Is this pump adequate?

3.37. A pump is to be selected to deliver water from a well to a treatment plant through a 300-m long pipeline. The temperature of the water is 20°C, the average elevation of the water surface in the well is 5 m below ground surface, the pump is 50 cm above ground surface, and the water surface in the receiving reservoir at the water-treatment plant is 4 m above ground surface. The delivery pipe is made of ductile iron ($k_s = 0.26$ mm) with a diameter of 800 mm. If the selected pump has a performance curve of $h_p = 12 - 0.1Q^2$, where Q is in m³/s and h_p is in m, then what is the flowrate through the system? Calculate the specific speed of the required pump (in U.S. Customary units), and state what type of pump will be required when the speed of the pump motor is 1,200 rpm. Neglect minor losses.

3.38. A pump lifts water through a 100-mm diameter ductile iron pipe from a lower to an upper reservoir (Figure 3.31). If the difference in elevation between the reservoir surfaces is 10 m, and the pump performance curve is given by

$$h_p = 15 - 0.1Q^2$$

where h_p is in meters and Q in L/s, then estimate the flowrate through the system. If the pump motor rotates at 2,400 rpm, what is the maximum height above the lower reservoir that the pump can be placed?

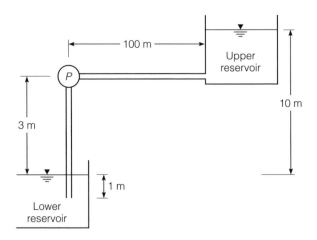

Figure 3.31 ■ Problem 3.38

3.39. Water is being pumped from reservoir A to reservoir F through a 30-m long PVC pipe of diameter 150 mm (see Figure 3.32). There is an open gate valve located at C; 90° bends (threaded) located at B, D, and E; and the pump performance curve is given by

$$h_p = 20 - 4713Q^2$$

where h_p is the head added by the pump in meters and Q is the flowrate in m³/s. The specific speed of the pump (in U.S. Customary units) is 3,000. Assuming that the flow is turbulent (in the smooth, rough, or transition range) and the temperature of the water is 20°C, then (a) write the energy equation between the upper and lower reservoirs, accounting for entrance, exit, and minor losses between A and F; (b) calculate the flowrate and velocity in the pipe; (c) calculate the cavitation parameter of the pump, and assess the potential for cavitation (for this analysis you may assume that the head loss in the pipe is negligible between the intake and the pump); and (d) use the affinity laws to estimate the pump performance curve when the motor on the pump is changed from 800 rpm to 1,600 rpm.

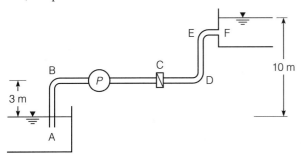

Figure 3.32 ■ Problem 3.39

3.40. If the performance curve of a certain pump model is given by

$$h_p = 30 - 0.05Q^2$$

where h_p is in meters and Q is in L/s, what is the performance curve of a pump system containing n of these pumps in series? What is the performance curve of a pump system containing n of these pumps in parallel?

3.41. A pump is placed in a pipe system in which the energy equation (system curve) is given by

$$h_p = 15 + 0.03Q^2$$

where h_p is the head added by the pump in meters and Q is the flowrate through the system in L/s. The performance curve of the pump is

$$h_p = 20 - 0.08Q^2$$

What is the flowrate through the system? If the pump is replaced by two identical pumps in parallel, what would be the flowrate in the system? If the pump is replaced by two identical pumps in series, what would be the flowrate in the system?

3.42. Derive an expression for the population, P, versus time, t, where the growth rate is: (a) geometric, (b) arithmetic, and (c) declining.

3.43. The design life of a planned water-distribution system is to end in the year 2030, and the population in the town has been measured every 10 years since 1920 by the U.S. Census Bureau. The reported populations are tabulated below. Estimate the population in the town using: (a) graphical extension, (b) arithmetic growth projection, (c) geometric growth projection, (d) declining growth projection (assuming a saturation concentration of 100,000 people), and (e) logistic curve projection.

Year	Population
1920	25,521
1930	30,208
1940	30,721
1950	37,253
1960	38,302
1970	41,983
1980	56,451
1990	64,109

3.44. A city founded in 1950 had a population of 13,000 in 1960; 125,000 in 1975; and 300,000 in 1990. Assuming that the population growth follows a logistic curve, estimate the saturation population of the city.

3.45. The average demand of a population served by a water-distribution system is 580 L/d/capita, and the population at the end of the design life is estimated to be 100,000 people. Estimate the maximum daily demand and maximum hourly demand.

3.46. Estimate the flowrate and volume of water required to provide adequate fire protection to a five-story office building constructed of joisted masonry. The effective floor area of the building is 5,000 m².

3.47. What is the maximum fire flow and corresponding duration that can be estimated for any building?

3.48. A water-supply system is being designed to serve a population of 200,000 people, with an average per capita demand of 600 L/d/person and a needed fire flow of 28,000 L/min. If the water supply is to be drawn from a river, then what should be the design capacity of the supply pumps and water-treatment plant? For what duration must the fire flow be sustained, and what volume of water must be kept in the service reservoir to accommodate a fire? What should be the design capacity of the distribution pipes be?

3.49. What is the minimum acceptable water pressure in a distribution system under average daily demand conditions?

3.50. Calculate the volume of storage required for the elevated storage reservoir in the water-supply system described in Problem 3.48.

4 Flow in Open Channels

4.1 Introduction

In open-channel flows the water surface is exposed to the atmosphere, a type of flow that is typically found in sanitary sewers, drainage conduits, canals, and rivers. Open-channel flow, sometimes referred to as *free-surface* flow, is more complicated than flow in closed conduits, since the location of the free surface is not constrained and the depth of flow depends on such factors as discharge and the shape and slope of the channel. Flows in conduits with closed sections, such as pipes, may be either open-channel or closed-conduit flow, depending on whether the conduit is flowing full. A closed pipe flowing partially full is an open-channel flow, since the water surface is exposed to the atmosphere. Open-channel flow is said to be *steady* if the depth of flow at any specified location does not change with time; if the depth of flow varies with time, the flow is called *unsteady*. Most open-channel flows are analyzed under steady-flow conditions. The flow is said to be *uniform* if the depth of flow is the same at every cross-section of the channel; if the depth of flow varies, the flow is *nonuniform* or *varied*. Uniform flow can be either steady or unsteady, depending on whether the flow depth changes with time; however, uniform flows are practically nonexistent in nature. More commonly, open-channel flows are either steady nonuniform flows or unsteady nonuniform flows. Open channels are classified as either *prismatic* or *nonprismatic*. Prismatic channels are characterized by an unvarying shape of the cross-section, constant bottom slope, and relatively straight alignment. In nonprismatic channels, the cross-section, alignment, and/or bottom slope changes along the channel.

This chapter covers the basic principles of open-channel flow and derives the most useful forms of the continuity, momentum, and energy equations. These equations are applied to the computation of water-surface profiles, predicting the performance of hydraulic structures, and designing both lined and unlined open channels.

4.2 Basic Principles

The governing equations of flow in open channels are the continuity, momentum, and energy equations. Any flow in an open channel must satisfy all three of these equations. The analysis of flow can usually be accomplished with the control-volume form of the governing equations, and the most useful forms of these equations for steady open-channel flows are derived in the following sections.

138

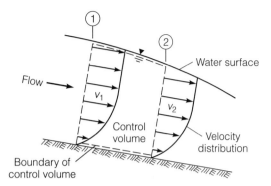

Figure 4.1 ■ Flow in an Open Channel

Water surface

Flow →

v_1

v_2

Control volume

Velocity distribution

Boundary of control volume

4.2.1 ■ Continuity Equation

Consider the case of steady flow in the open channel illustrated in Figure 4.1. The flow enters and leaves the control volume normal to the control surfaces, with the inflow velocity distribution denoted by v_1 and the outflow velocity distribution by v_2; both the inflow and outflow velocities vary across the control surfaces. The steady-state continuity equation can be written as

$$\int_{A_1} v_1 \, dA = \int_{A_2} v_2 \, dA \tag{4.1}$$

Defining V_1 and V_2 as the average velocities across A_1 and A_2, respectively, where

$$V_1 = \frac{1}{A_1} \int_{A_1} v_1 \, dA \tag{4.2}$$

and

$$V_2 = \frac{1}{A_2} \int_{A_2} v_2 \, dA \tag{4.3}$$

then the continuity equation can be written as

$$\boxed{V_1 A_1 = V_2 A_2} \tag{4.4}$$

which is the same expression that was derived for steady flow in closed conduits.

4.2.2 ■ Momentum Equation

Consider the case of steady nonuniform flow in the open channel illustrated in Figure 4.2. The steady-state momentum equation for the control volume shown in Figure 4.2 is given by

$$\sum F_x = \int_A \rho v_x \mathbf{v} \cdot \mathbf{n} \, dA \tag{4.5}$$

where F_x represents the forces in the flow direction, x; A is the surface area of the control volume; v_x is the flow velocity in the x direction, and $\mathbf{n}$ is a unit normal directed outward from the control volume. Since the velocities normal to the control surface are nonzero

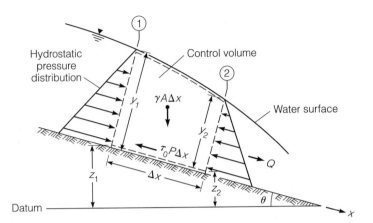

Figure 4.2 ■ Steady Nonuniform Flow in an Open Channel

only for the inflow and outflow surfaces, Equation 4.5 can be written as

$$\sum F_x = \int_{A_2} \rho v_x^2 \, dA - \int_{A_1} \rho v_x^2 \, dA \tag{4.6}$$

where A_1 and A_2 are the upstream and downstream areas of the control volume. If the velocity is uniformly distributed (i.e., constant) across the control surface, then Equation 4.6 becomes

$$\sum F_x = \rho v_2^2 A_2 - \rho v_1^2 A_1 \tag{4.7}$$

where v_1 and v_2 are the velocities on the upstream and downstream faces of the control volume, respectively. In reality, velocity distributions in open channels are never uniform, and so it is convenient to define a *momentum correction coefficient*, β, by the relation

$$\beta = \frac{\int_A v^2 \, dA}{V^2 A} \tag{4.8}$$

where V is the mean velocity across the channel section of area A. The momentum correction coefficient, β, is sometimes called the *Boussinesq coefficient*. Applying the definition of the momentum correction coefficient to Equation 4.6 leads to the following form of the momentum equation

$$\sum F_x = \rho \beta_2 V_2^2 A_2 - \rho \beta_1 V_1^2 A_1 \tag{4.9}$$

where β_1 and β_2 are the momentum correction coefficients at the upstream and downstream faces of the control volume, respectively. Values of β can be expected to be in the range 1.03–1.07 for regular channels, flumes, and spillways, and in the range 1.05–1.17 for natural streams (Chow, 1959).

Since the continuity equation requires that the discharge, Q, is the same at each cross-section, then

$$Q = A_1 V_1 = A_2 V_2 \tag{4.10}$$

and the momentum equation (Equation 4.9) can be written as

$$\sum F_x = \rho \beta_2 Q V_2 - \rho \beta_1 Q V_1 \tag{4.11}$$

By definition, values of β must be greater than or equal to unity. In practice, however, deviations of β from unity are second-order corrections that are small relative to the uncertainties in the other terms in the momentum equation. By assuming

$$\beta_1 \approx \beta_2 = 1$$

the momentum equation can be written as

$$\sum F_x = \rho Q V_2 - \rho Q V_1 = \rho Q (V_2 - V_1) \tag{4.12}$$

Considering the forces acting on the control volume shown in Figure 4.2, then Equation 4.12 can be written as

$$\gamma A \Delta x \sin \theta - \tau_o P \Delta x + \gamma A (y_1 - y_2) = \rho Q (V_2 - V_1) \tag{4.13}$$

where γ is the specific weight of the fluid; A is the average cross-sectional area of the control volume; Δx is the length of the control volume; θ is the inclination of the channel; τ_o is the mean shear stress on the control surface, P is the average (wetted) perimeter of the cross-section of the control volume; and y_1 and y_2 are the upstream and downstream depths, respectively, at the control volume. The three force terms on the lefthand side of Equation 4.13 are the component of the weight of the fluid in the direction of flow, the shear force exerted by the channel boundary on the moving fluid, and the net hydrostatic force. If z_1 and z_2 are the elevations of the bottom of the channel at the upstream and downstream faces of the control volume, then

$$\sin \theta = \frac{z_1 - z_2}{\Delta x} \tag{4.14}$$

combining Equations 4.13 and 4.14 and rearranging leads to

$$\tau_o = -\gamma \frac{A}{P} \frac{\Delta z}{\Delta x} - \gamma \frac{A}{P} \frac{\Delta y}{\Delta x} - \gamma \frac{A}{P} \frac{V}{g} \frac{\Delta V}{\Delta x} \tag{4.15}$$

where Δz, Δy, and ΔV are defined by

$$\Delta z = z_2 - z_1, \quad \Delta y = y_2 - y_1, \quad \Delta V = V_2 - V_1 \tag{4.16}$$

and V is the average velocity in the control volume. The ratio A/P is commonly called the *hydraulic radius*, R, where

$$R = \frac{A}{P} \tag{4.17}$$

Combining Equations 4.15 and 4.17 and taking the limit as $\Delta x \to 0$ yields

$$\begin{aligned}
\tau_o &= -\gamma R \left[\lim_{\Delta x \to 0} \frac{\Delta z}{\Delta x} + \lim_{\Delta x \to 0} \frac{\Delta y}{\Delta x} + \frac{V}{g} \lim_{\Delta x \to 0} \frac{\Delta V}{\Delta x} \right] \\
&= -\gamma R \left[\frac{dz}{dx} + \frac{dy}{dx} + \frac{V}{g} \frac{dV}{dx} \right] \\
&= -\gamma R \frac{d}{dx} \left[y + z + \frac{V^2}{2g} \right]
\end{aligned} \tag{4.18}$$

The term in brackets is the energy per unit weight of the fluid, E, defined as

$$E = y + z + \frac{V^2}{2g} \tag{4.19}$$

It should be noted that the energy per unit weight of a fluid element is usually defined as $p/\gamma + z' + V^2/2g$, where z' is elevation of the fluid element relative to a defined datum. If the pressure is hydrostatic across the cross-section, then $p/\gamma + z' = \text{constant} = y + z$, where y is the water depth and z is the elevation of the bottom of the channel. The energy per unit weight, E, can therefore be written as $y + z + V^2/2g$. A plot of E versus the distance along the channel is called the *energy grade line*. The momentum equation, Equation 4.18, can now be written as

$$\tau_o = -\gamma R \frac{dE}{dx} \tag{4.20}$$

or

$$\boxed{\tau_o = \gamma R S_f} \tag{4.21}$$

where S_f is equal to the slope of the energy grade line, which is positive when it slopes downward in the direction of flow.

4.2.2.1 Darcy-Weisbach Equation

A functional expression for the average shear stress, τ_o, can be derived in a similar manner to that used in pipe flow, in which case

$$\tau_o = f_o(V, R, \rho, \mu, \epsilon, \epsilon', m, s) \tag{4.22}$$

where V is the mean velocity in the channel, R is the hydraulic radius, ρ is the fluid density, μ is the (dynamic) viscosity of the fluid, ϵ is the characteristic size of the roughness projections on the channel boundary, ϵ' is the characteristic spacing of the roughness projections, m is a dimensionless form factor that describes the shape of the roughness elements, and s is a channel shape factor that describes the shape of the channel cross-section. In accordance with the Buckingham pi theorem, the functional relationship given by Equation 4.22 between nine variables in three dimensions can also be expressed as a relation between six nondimensional groups as follows

$$\frac{\tau_o}{\rho V^2} = f_1\left(\text{Re}, \frac{\epsilon}{4R}, \frac{\epsilon'}{4R}, m, s\right) \tag{4.23}$$

where Re is the Reynolds number defined by the relation

$$\text{Re} = \frac{\rho V (4R)}{\mu} \tag{4.24}$$

and the variable $4R$ is used instead of R for convenience in subsequent analyses.

The relationship given by Equation 4.23 is as far as dimensional analysis goes, and experimental data is necessary to determine an empirical relationship between the nondimensional groups. The problem of determining an empirical expression for the boundary shear stress in open-channel flow is similar to the problem faced in determining an empirical expression for the boundary shear stress in pipe flow, where $4R$ for circular conduits is equal to the pipe diameter. If the influence of the shape of the cross-section and arrangement of roughness elements on the boundary shear stress, τ_o, are small relative to the influence of the size of the roughness elements, then the shear stress can be expressed in the following functional form

$$\frac{\tau_o}{\rho V^2} = \frac{1}{8} f\left(\text{Re}, \frac{\epsilon}{4R}\right) \tag{4.25}$$

where the function f can be expected to closely approximate to the Darcy friction factor in pipes.

   In reality, the friction factor, f, in Equation 4.25 has been observed to be a function of channel shape, decreasing roughly in the order of rectangular, triangular, trapezoidal, and circular channels (Chow, 1959). According to Daily and Harleman (1966), as channels become very wide or otherwise depart radically from the circle or semicircle, the friction factors derived from pipe experiments become less applicable to open channels. Myers (1991) has shown that friction factors in wide rectangular open channels are as much as 45% greater than in narrow sections with the same Re and $\epsilon/4R$. The transition from laminar to turbulent flow in open channels occurs at a Reynolds number of about 600, but laminar free-surface flows are seldom encountered in nature.

   It is convenient to define three possible types of turbulent flow: *smooth*, *transition*, and *rough*. The flow is (hydraulically) smooth when the roughness projections on the channel boundary are submerged within a laminar sublayer, in which case the friction factor in open channels can be estimated by (Henderson, 1966)

$$f = \frac{0.316}{\text{Re}^{\frac{1}{4}}}, \quad \text{Re} < 10^5 \tag{4.26}$$

$$\frac{1}{\sqrt{f}} = -2.0 \log_{10}\left(\frac{2.51}{\text{Re}\sqrt{f}}\right), \quad \text{Re} > 10^5 \tag{4.27}$$

These relations are the same as the Blasius and Prandtl–von Kármán equations for flow in pipes. However, it should be noted that owing to the free surface and the interdependence of the hydraulic radius, discharge, and slope, the relationship between f and Re in open channel flow is not identical to that for pipe flow (Chow, 1959). The question of how to account for the shape of an open channel in estimating the friction factor remains open (Pillai, 1997). The flow is (hydraulically) rough when the roughness projections on the channel boundary extend out of the laminar sublayer, creating sufficient turbulence that the friction factor depends only on the relative roughness. Under these conditions, the friction factor can be estimated by the equation (ASCE, 1963)

$$\frac{1}{\sqrt{f}} = -2\log_{10}\left(\frac{k_s}{12R}\right) \tag{4.28}$$

where k_s is the equivalent sand roughness of the open channel. Equation 4.28 is based on the original experiments of Nikuradse, supplemented by many other experiments, and gives a higher friction factor than the Prandtl–von Kármán equation that is used in pipe flow. In the transition region between (hydraulically) smooth and rough flow, the friction factor depends on both the Reynolds number and the relative roughness and can be approximated by the relation (ASCE, 1963)

$$\frac{1}{\sqrt{f}} = -2\log_{10}\left(\frac{k_s}{12R} + \frac{2.5}{\text{Re}\sqrt{f}}\right) \tag{4.29}$$

This relation differs slightly from the Colebrook equation for transition flow and can be applied in both smooth and rough flow.

The three types of flow (smooth, transition, rough) can be characterized by the dimensionless quantity $k_s u_*/\nu$, where u_* is the *shear velocity* defined by

$$u_* = \sqrt{\frac{\tau_o}{\rho}} = \sqrt{gRS_f} \qquad (4.30)$$

and ν is the kinematic viscosity of the fluid. The transition flow region is defined by (Henderson, 1966)

$$4 < \frac{u_* k_s}{\nu} < 100 \qquad (4.31)$$

where the lower limit defines the end of smooth flow and the upper limit defines the beginning of rough flow.

Combining Equations 4.21 and 4.25 leads to the following form of the momentum equation that is most commonly used in practice

$$\boxed{V = \sqrt{\frac{8g}{f}} \sqrt{RS_f}} \qquad (4.32)$$

where f is a function of the relative roughness and Reynolds number of the flow given by Equation 4.29. In cases where the flow is uniform, the slope of the energy grade line, S_f, is equal to the slope of the channel, S_o, since under these conditions

$$S_f = -\frac{d}{dx}\left(y + z + \frac{V^2}{2g}\right) = -\frac{dz}{dx} = S_o$$

where the depth, y, and average velocity, V, are constant and independent of x under uniform flow conditions. The depth of flow under steady uniform flow conditions is called the *normal* depth of flow.

■ Example 4.1

Water flows at a depth of 1.83 m in a trapezoidal, concrete-lined section ($k_s = 1.5$ mm) with a bottom width of 3 m and side slopes of 2:1 (H:V). The slope of the channel is 0.0005 and the water temperature is 20°C. Assuming uniform flow conditions, estimate the average velocity and flowrate in the channel.

Solution

The flow in the channel is illustrated in Figure 4.3. The momentum equation gives the average velocity, V, as

$$V = \sqrt{\frac{8g}{f}} \sqrt{RS_o}$$

In this case, $S_o = 0.0005$, the flow area, A, is 12.2 m², the wetted perimeter, P, is 11.2 m, and the hydraulic radius, R, is

$$R = \frac{A}{P} = \frac{12.2}{11.2} = 1.09 \text{ m}$$

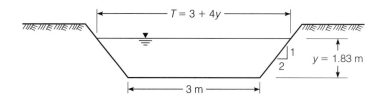

Figure 4.3 ■ Flow in a Trapezoidal Channel

For concrete, the equivalent sand roughness, k_s, is 1.5 mm. Assuming the flow is fully turbulent, the friction factor, f, can be estimated using Equation 4.28, where

$$\frac{1}{\sqrt{f}} = -2\log_{10}\left[\frac{k_s}{12R}\right] = -2\log_{10}\left[\frac{0.0015}{12(1.09)}\right] = 7.88$$

which leads to

$$f = 0.016$$

The mean velocity can now be estimated as

$$V = \sqrt{\frac{8g}{f}}\sqrt{RS_o} = \sqrt{\frac{8(9.81)}{0.016}}\sqrt{(1.09)(0.0005)} = 1.64 \text{ m/s}$$

and the corresponding flowrate, Q, is given by

$$Q = AV = (12.2)(1.64) = 20.0 \text{ m}^3/\text{s}$$

This flowrate was obtained by assuming that the flow in the channel is hydraulically rough, in which case the friction factor does not depend on the Reynolds number of the flow. This assumption can now be checked by recalculating the friction factor using the estimated flowrate. At 20°C, the density, ρ, and kinematic viscosity, μ, of water are given by $\rho = 998.2 \text{ kg/m}^3$, and $\mu = 0.001 \text{ N·s/m}^2$. The Reynold number, Re, is therefore given by

$$\text{Re} = \frac{\rho V(4R)}{\mu} = \frac{(998.2)(1.64)(4 \times 1.09)}{0.001} = 7.14 \times 10^6$$

The friction factor can now be estimated by the generalized expression for the friction factor given by Equation 4.29, where

$$\frac{1}{\sqrt{f}} = -2\log_{10}\left[\frac{k_s}{12R} + \frac{2.5}{\text{Re}\sqrt{f}}\right]$$

$$= -2\log_{10}\left[\frac{0.0015}{12(1.09)} + \frac{2.5}{7.14 \times 10^6\sqrt{f}}\right]$$

$$= -2\log_{10}\left[1.15 \times 10^{-4} + \frac{3.50 \times 10^{-7}}{\sqrt{f}}\right]$$

which by trial and error yields

$$f = 0.016$$

Since this is the same friction factor as originally estimated, the flow is indeed rough and the estimated velocity and flowrate are correct. ■

4.2.2.2   Manning Equation

To fully appreciate the advantage of using Equation 4.32 compared with other flow equations used in practice, some historical perspective is needed. Equation 4.32 is based primarily on the pipe experiments of Nikuradse and Colebrook, which were conducted between 1930 and 1940. Observations on rivers and other large open channels, however, began much earlier. In 1775, Chézy* proposed the following expression for the mean velocity in an open channel

$$V = C\sqrt{RS_f} \tag{4.33}$$

where C was referred to as the *Chézy coefficient*. Equation 4.33 has exactly the same form as Equation 4.32 and was derived in the same way, with the exception that the functional dependence of the Chézy coefficient on the Reynolds number and the relative roughness was not addressed. Comparing Equations 4.32 and 4.33, the Chézy coefficient is related to the friction factor by

$$C = \sqrt{\frac{8g}{f}} \tag{4.34}$$

In 1869, Ganguillet and Kutter (1869) published an elaborate formula for C which became widely used. In 1890, Manning (1890) demonstrated that the data used by Ganguillet and Kutter were fitted just as well by a simpler formula in which C varies as the sixth root of R, where

$$C = \frac{R^{\frac{1}{6}}}{n} \tag{4.35}$$

and n is a coefficient that is characteristic of the surface roughness alone. Since C is not a dimensionless quantity, values of n were specified to be consistent with length units measured in meters and time in seconds. If the length units are measured in feet, then Equation 4.35 becomes

$$C = 1.486\frac{R^{\frac{1}{6}}}{n} \tag{4.36}$$

When either Equation 4.35 or 4.36 is combined with the Chézy equation, the resulting expression is called the *Manning equation* or *Strickler equation* (in Europe) and is given by

$$\boxed{V = \frac{1}{n}R^{\frac{2}{3}}S^{\frac{1}{2}}} \quad \text{(SI units)} \tag{4.37}$$

and

$$\boxed{V = \frac{1.486}{n}R^{\frac{2}{3}}S^{\frac{1}{2}}} \quad \text{(U.S. Customary units)} \tag{4.38}$$

---

*Antoine de Chézy (1718–1798) was a French engineer.

where $S = S_f = S_o$ under uniform flow conditions. In reality, the coefficient 1.486 in Equation 4.38 is much too precise considering the accuracy with which n is known and should not be written any more precisely than 1.49 or even 1.5 (Henderson, 1966). Williamson (1951) investigated the consistency between the Manning equation and the momentum equation based on the friction factor (Equation 4.32). After making some minor adjustments to Nikuradse's data, Williamson found that for rough flow, the functional relation between the friction factor and the relative roughness, k_s/R, can be approximated by the relation

$$f = 0.113 \left(\frac{k_s}{R} \right)^{\frac{1}{3}} \tag{4.39}$$

Since the Chézy coefficient, C, is given by Equation 4.34 in terms of the friction factor, combining Equations 4.34 and 4.39 leads to (in U.S. Customary units)

$$C = \frac{1.49 R^{\frac{1}{6}}}{0.031 d^{\frac{1}{6}}} \tag{4.40}$$

where the equivalent sand roughness, k_s, has been replaced by d, which represents the characteristic stone or gravel size on the bed of the open channel. Equation 4.40 is identical to the Manning expression for C if the roughness coefficient, n, is given by

$$n = 0.031 d^{\frac{1}{6}} \tag{4.41}$$

where d is measured in feet. White (1994) has shown that Equation 4.41 is valid in the range $0.001 < d/4R < 0.1$ and that it is very close to the empirical relation proposed by Strickler in 1923, where

$$n = 0.034 d^{\frac{1}{6}} \tag{4.42}$$

which was derived from studies in gravel-bed streams where d was the median size of the bed material.

On the basis of Equations 4.41 and 4.42, it appears that the estimated velocities using the friction factor and Manning equation are consistent and interchangeable, provided the flow is hydraulically rough. The consistency between Nikuadse's result and experiments in open channels could not be assumed a priori, since Nikuradse's experiments were on relatively small circular pipes (the largest was 64 mm in diameter) with uniform roughness projections. It is also interesting to note that the sixth-root relationship between the roughness height, d, and the roughness coefficient, n, means that large relative errors in estimating d results in a much smaller relative error in estimating n. Typical values of of the roughness coefficient, n, are given in Table 4.1. Based on a review of the literature, Johnson (1996) indicates that Manning n values estimated from field measurements typically have errors in the range of 5% to 35%. The limit of hydraulically rough flow is given by Equation 4.31, which can be written in terms of the Manning equation parameters (French, 1985) as

$$n^6 \sqrt{R S_f} \geq 1.9 \times 10^{-13} \tag{4.43}$$

where R is the hydraulic radius in meters and S_f is the slope of the energy grade line. If this inequality is satisfied, then rough conditions exist and the Manning equation can be applied.

▪ **Table 4.1**

Manning Coefficient for Open Channels

Material	n
Lined channels:	
Asphalt	0.013–0.017
Brick	0.012–0.018
Concrete	0.011–0.020
Rubble or riprap	0.020–0.035
Vegetal	0.030–0.40
Excavated or dredged channels:	
Earth, straight and uniform	0.020–0.030
Earth, winding, fairly uniform	0.025–0.040
Rock	0.030–0.045
Unmaintained	0.050–0.14
Natural channels*:	
Fairly regular section	0.03–0.07
Irregular section with pools	0.04–0.10

▪ **Example 4.2**

Water flows at a depth of 1.83 m in a trapezoidal, concrete-lined section with a bottom width of 3 m and side slopes of 2:1 (H:V). The slope of the channel is 0.0005 and the water temperature is 20°C. Assuming uniform flow conditions, estimate the average velocity and flowrate in the channel.

Solution

The flow in the channel is illustrated in Figure 4.3. The Manning equation gives the average velocity, V, as

$$V = \frac{1}{n} R^{\frac{2}{3}} S_o^{\frac{1}{2}}$$

The channel slope, S_o, is 0.0005; the flow area, A, is 12.2 m^2; the wetted perimeter, P, is 11.2 m; and the hydraulic radius, R, is 1.09 m. Table 4.1 indicates that a midrange roughness coefficient for concrete is $n = 0.015$. According to Equation 4.43, the Manning equation is valid when

$$n^6 \sqrt{RS_o} \geq 1.9 \times 10^{-13}$$

In this case,

$$n^6 \sqrt{RS_o} = (0.015)^6 \sqrt{(1.09)(0.0005)} = 2.66 \times 10^{-13}$$

Therefore, the Manning equation can be applied. The average velocity given by the Manning equation is

$$V = \frac{1}{0.015}(1.09)^{\frac{2}{3}}(0.0005)^{\frac{1}{2}} = 1.58 \text{ m/s}$$

and the corresponding flowrate, Q, is

$$Q = AV = (12.2)(1.58) = 19.3 \text{ m}^3/\text{s}$$

This result can be compared with the flowrate computed using the friction factor in Example 4.1, which yielded $Q = 20.0$ m³/s, a difference of 3.5%. ■

In channels where the roughness varies significantly over the perimeter of the channel, the *composite roughness*, n_e, of the channel is usually computed by first subdividing the channel section into N smaller sections, where the ith section has a roughness n_i, wetted perimeter, P_i, and hydraulic radius, R_i. Several commonly-used formulae for calculating n_e are listed in Table 4.2. Motayed and Krishnamurthy (1980) used data from 36 natural-channel cross-sections to assess the performance of the various formulae in Table 4.2 and concluded that the formula proposed by Horton (1933) and Einstein (1934) performed best.

A composite roughness is usually necessary in analyzing flood flows in which a river bank is overtopped and the flow extends into the adjacent *floodplain*. The roughness elements in the floodplain, consisting of shrubs, trees, and possibly houses, are usually much larger than in the river channel.

■ Example 4.3

The floodplain shown in Figure 4.4 can be divided into sections with approximately uniform roughness characteristics. The Manning n values for each section are as follows:

Section	n
1	0.040
2	0.030
3	0.015
4	0.013
5	0.017
6	0.035
7	0.060

Use the formulae in Table 4.2 to estimate the composite roughness.

■ Table 4.2 Composite Roughness Formulae	Formula	Reference
	$n_e = \left(\dfrac{\sum_{i=1}^{N} P_i n_i^{3/2}}{\sum_{i=1}^{N} P_i} \right)^{2/3}$	Horton (1933), Einstein (1934)[*]
	$n_e = \dfrac{\left(\sum_{i=1}^{N} P_i n_i^2 \right)^{1/2}}{\left(\sum_{i=1}^{N} P_i \right)^{1/2}}$	Muhlhofer (1933), Einstein and Banks (1951)
	$n_e = \dfrac{PR^{5/3}}{\sum_{i=1}^{n} \dfrac{P_i R_i^{5/3}}{n_i}}$	Lotter (1933)[†]
	$\ln n_e = \dfrac{\sum_{i=1}^{N} P_i y_i^{3/2} \ln n_i}{\sum_{i=1}^{N} P_i y_i^{3/2}}$	Krishnamurthy and Christensen (1972)[‡]

[*]Formula assumes that the mean flow in each of the subareas is equal to the mean flow velocity.
[†]P and R are the perimeter and hydraulic radius of the entire cross-section, respectively.
[‡]y_i is the average flow depth in Section i.

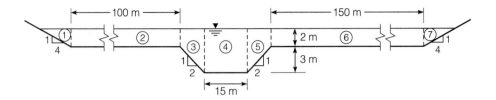

Figure 4.4 ■ Flow in a
Flood Plain

Solution

From the given shape of the floodplain (Figure 4.4), the following geometric characteristics are derived:

Section, i	P_i (m)	A_i (m²)	R_i (m)	n_i	y_i (m)
1	8.25	8.00	0.97	0.040	1.00
2	100	200	2.00	0.030	2.00
3	6.71	21	3.13	0.015	3.50
4	15.0	75	5.00	0.013	5.00
5	6.71	21	3.13	0.017	3.50
6	150	300	2.00	0.035	2.00
7	8.25	8.00	0.97	0.060	1.00
	295	633			

The total perimeter, P, of the (compound) channel is 295 m, the total area, A, is 633 m²,
and hence the hydraulic radius, R, of the compound section is given by

$$R = \frac{A}{P} = \frac{633}{295} = 2.15 \text{ m}$$

Substituting these data into the formulae listed in Table 4.2 yields the following results:

Formula	n_e
Horton/Einstein	0.033
Muhlhofer/Einstein and Banks	0.033
Lotter	0.022
Krishnamurthy and Christensen	0.026

It is apparent from this example that estimates of n_e can vary significantly.   ■

4.2.2.3 Velocity Distribution in Open Channels

In *wide channels*, lateral boundaries have negligible effects on the velocity distribution
in the central portion of the channel. According to Franzini and Finnemore (1997), a
wide channel is one in which the width of the channel exceeds 10 times the depth. The
velocity distribution, $v(y)$, in wide open channels can be approximated by the relation
(Vanoni, 1941),

$$v(y) = V + \frac{1}{\kappa}\sqrt{gdS_o}\left(1 + 2.3\log\frac{y}{d}\right) \tag{4.44}$$

where V is the depth-averaged velocity, κ is the von Kármán constant (≈ 0.4), d is the depth of flow, S_o is the slope of the channel, and y is the distance from the bottom of the channel. Equation 4.44 indicates that the average velocity, V, can be estimated by either

$$V = v(0.4d) \tag{4.45}$$

or

$$V = \frac{v(0.2d) + v(0.8d)}{2} \tag{4.46}$$

It is standard practice of the U.S. Geological Survey to use measurements at $0.2d$ and $0.8d$ with Equation 4.46 to estimate the average velocity in channel sections with depth greater than 0.6 m (2 ft) and to use measurements at $0.4d$ with Equation 4.45 to estimate the average velocity in sections with depth less than 0.6 m (2 ft).

4.2.3 ■ Energy Equation

The steady-state energy equation for the control volume shown in Figure 4.2 is

$$\frac{dQ_h}{dt} - \frac{dW}{dt} = \int_A \rho e \, \mathbf{v} \cdot \mathbf{n} \, dA \tag{4.47}$$

where Q_h is the heat added to the fluid in the control volume, W is the work done by the fluid in the control volume, A is the surface area of the control volume, ρ is the density of the fluid in the control volume, and e is the internal energy per unit mass of fluid in the control volume given by

$$e = gz + \frac{v^2}{2} + u \tag{4.48}$$

where z is the elevation of a fluid mass having a velocity v and internal energy u. The normal stresses on the inflow and outflow boundaries of the control volume are equal to the pressure, p, with shear stresses tangential to the control-volume boundaries. As the fluid moves with velocity $\mathbf{v}$, the power expended by the fluid is given by

$$\frac{dW}{dt} = \int_A p\mathbf{v} \cdot \mathbf{n} \, dA \tag{4.49}$$

No work is done by the shear forces since the velocity is equal to zero on the channel boundary and the flow direction is normal to the direction of the shear forces on the inflow and outflow boundaries. Combining Equation 4.49 with the steady-state energy equation (Equation 4.47) leads to

$$\frac{dQ_h}{dt} = \int_A \rho \left(\frac{p}{\rho} + e \right) \mathbf{v} \cdot \mathbf{n} \, dA \tag{4.50}$$

Substituting the definition of the internal energy, e, (Equation 4.48) into Equation 4.50 gives the following form of the energy equation

$$\frac{dQ_h}{dt} = \int_A \rho \left(h + gz + \frac{v^2}{2} \right) \mathbf{v} \cdot \mathbf{n} \, dA \tag{4.51}$$

where h is the enthalpy of the fluid defined by

$$h = \frac{p}{\rho} + u \tag{4.52}$$

Denoting the rate at which heat is being added to the fluid system by $\dot{Q}$, then the energy equation becomes

$$\dot{Q} = \int_A \rho \left(h + gz + \frac{v^2}{2} \right) \mathbf{v} \cdot \mathbf{n} \, dA \tag{4.53}$$

Considering the term $h + gz$, then

$$h + gz = \frac{p}{\rho} + u + gz = g \left(\frac{p}{\gamma} + z \right) + u \tag{4.54}$$

where γ is the specific weight of the fluid. Equation 4.54 indicates that $h + gz$ can be assumed to be constant across the inflow and outflow control surfaces, since the hydrostatic pressure distribution across the inflow/outflow boundaries guarantees that $p/\gamma + z$ is constant across the boundaries and since the internal energy, u, depends only on the temperature, which can be assumed constant across each boundary. Since $\mathbf{v} \cdot \mathbf{n}$ is equal to zero over the impervious boundaries in contact with the fluid system, Equation 4.53 simplifies to

$$\dot{Q} = (h_1 + gz_1) \int_{A_1} \rho \mathbf{v} \cdot \mathbf{n} \, dA + \int_{A_1} \rho \frac{v^2}{2} \mathbf{v} \cdot \mathbf{n} \, dA$$
$$+ (h_2 + gz_2) \int_{A_2} \rho \mathbf{v} \cdot \mathbf{n} \, dA + \int_{A_2} \rho \frac{v^2}{2} \mathbf{v} \cdot \mathbf{n} \, dA \tag{4.55}$$

where the subscripts 1 and 2 refer to the inflow and outflow boundaries, respectively. Equation 4.55 can be further simplified by noting that the assumption of steady state requires that rate of mass inflow, $\dot{m}$, to the control volume is equal to the mass outflow rate, where

$$\dot{m} = \int_{A_2} \rho \mathbf{v} \cdot \mathbf{n} \, dA = - \int_{A_1} \rho \mathbf{v} \cdot \mathbf{n} \, dA \tag{4.56}$$

where the negative sign comes from the fact that the unit normal points out of the control volume. Also, the kinetic energy correction factors, α_1 and α_2, can be defined by the equations

$$\int_{A_1} \rho \frac{v^3}{2} \, dA = \alpha_1 \rho \frac{V_1^3}{2} A_1 \tag{4.57}$$

$$\int_{A_2} \rho \frac{v^3}{2} \, dA = \alpha_2 \rho \frac{V_2^3}{2} A_2 \tag{4.58}$$

where A_1 and A_2 are the areas of the inflow and outflow boundaries, respectively, and V_1 and V_2 are the corresponding mean velocities across these boundaries. The kinetic energy correction factors, α_1 and α_2, are determined by the velocity profile across the flow boundaries. In regular channels, flumes, and spillways, α is typically in the range 1.1–1.2, while in natural channels α is typically in the range 1.15–1.5 (Chow, 1959).

Combining Equations 4.55 to 4.58 leads to

$$\dot{Q} = -(h_1 + gz_1)\dot{m} - \alpha_1 \rho \frac{V_1^3}{2} A_1 + (h_2 + gz_2)\dot{m} + \alpha_2 \rho \frac{V_2^3}{2} A_2 \qquad (4.59)$$

where the negative signs come from the fact that the unit normal points out of the inflow boundary, making $\mathbf{v} \cdot \mathbf{n}$ negative for the inflow boundary in Equation 4.55. Invoking the steady-state continuity equation

$$\rho V_1 A_1 = \rho V_2 A_2 = \dot{m} \qquad (4.60)$$

and combining Equations 4.59 and 4.60 leads to

$$\dot{Q} = \dot{m}\left[\left(h_2 + gz_2 + \alpha_2 \frac{V_2^2}{2}\right) - \left(h_1 + gz_1 + \alpha_1 \frac{V_1^2}{2}\right)\right] \qquad (4.61)$$

which can be put in the form

$$\frac{\dot{Q}}{\dot{m}g} = \left(\frac{p_2}{\gamma} + \frac{u_2}{g} + z_2 + \alpha_2 \frac{V_2^2}{2g}\right) - \left(\frac{p_1}{\gamma} + \frac{u_1}{g} + z_1 + \alpha_1 \frac{V_1^2}{2g}\right) \qquad (4.62)$$

where p_1 is the pressure at elevation z_1 on the inflow boundary and p_2 is the pressure at elevation z_2 on the outflow boundary. Equation 4.62 can be further rearranged into the form

$$\left(\frac{p_1}{\gamma} + \alpha_1 \frac{V_1^2}{2g} + z_1\right) = \left(\frac{p_2}{\gamma} + \alpha_2 \frac{V_2^2}{2g} + z_2\right) + \left[\frac{1}{g}(u_2 - u_1) - \frac{\dot{Q}}{\dot{m}g}\right] \qquad (4.63)$$

The energy loss per unit weight or *head loss*, h_L, is defined by the relation

$$h_L = \frac{1}{g}(u_2 - u_1) - \frac{\dot{Q}}{\dot{m}g} \qquad (4.64)$$

Combining Equations 4.63 and 4.64 leads to a useful form of the energy equation

$$\left(\frac{p_1}{\gamma} + \alpha_1 \frac{V_1^2}{2g} + z_1\right) = \left(\frac{p_2}{\gamma} + \alpha_2 \frac{V_2^2}{2g} + z_2\right) + h_L \qquad (4.65)$$

The *head*, h, of the fluid at any cross-section is defined by the relation

$$h = \frac{p}{\gamma} + \alpha \frac{V^2}{2g} + z \qquad (4.66)$$

where p is the pressure at elevation z, γ is the specific weight of the fluid, α is the kinetic energy correction factor, and V is the average velocity across the channel. The kinetic energy correction factor, α, is sometimes called the *Coriolis coefficient* or the *energy coefficient*. The head, h, measures the average energy per unit weight of the fluid flowing across a channel cross-section, where the piezometric head, $p/\gamma + z$, is taken to be constant across the section, assuming a hydrostatic pressure distribution normal to the direction of the flow. In this case, the piezometric head can be written in terms of the invert elevation of the cross-section, z_o, and the depth of flow, y, as

$$\frac{p}{\gamma} + z = y \cos\theta + z_o \qquad (4.67)$$

where θ is the angle that the channel makes with the horizontal. Combining Equations 4.66 and 4.67 leads to the following expression for the head, h, at a flow boundary of a control volume

$$h = y \cos\theta + \alpha \frac{V^2}{2g} + z_o \qquad (4.68)$$

and therefore the energy equation (Equation 4.65) can be written as

$$\left(y_1 \cos\theta + \alpha_1 \frac{V_1^2}{2g} + z_1 \right) = \left(y_2 \cos\theta + \alpha_2 \frac{V_2^2}{2g} + z_2 \right) + h_L \qquad (4.69)$$

where y_1 and y_2 are the flow depths at the upstream and downstream sections of the control volume respectively, and z_1 and z_2 are the corresponding invert elevations.

Equation 4.69 is the most widely used form of the energy equation in practice and can be written in the summary form

$$h_1 = h_2 + h_L \qquad (4.70)$$

where h_1 and h_2 are the heads at the inflow and outflow boundaries of the control volume, respectively. A rearrangement of the energy equation (Equation 4.69) gives

$$y_2 \cos\theta + \alpha_2 \frac{V_2^2}{2g} = y_1 \cos\theta + \alpha_1 \frac{V_1^2}{2g} + (z_1 - z_2) - h_L \qquad (4.71)$$

which can also be written in the more compact form

$$\left[y \cos\theta + \alpha \frac{V^2}{2g} \right]_2^1 = (S - S_o \cos\theta)L \qquad (4.72)$$

where L is the distance between the inflow and outflow sections of the control volume, S is the head loss per unit length (= slope of the energy grade line) given by

$$S = \frac{h_L}{L} \qquad (4.73)$$

and S_o is the slope of the channel defined as

$$S_o = \frac{z_1 - z_2}{L \cos\theta} \qquad (4.74)$$

In contrast to our usual definition of slopes, downward slopes are generally taken as positive in open-channel hydraulics. The relationship between S_o and $\cos\theta$ is shown in Table 4.3, where it is clear that for open-channel slopes less than 0.1 (10%), the error in assuming that $\cos\theta = 1$ is less than 0.5%. Since this error is usually less than the uncertainty in other terms in the energy equation, the energy equation (Equation 4.72)

■ Table 4.3
S_o versus $\cos\theta$

S_o	$\cos\theta$
0.001	0.9999995
0.01	0.99995
0.1	0.995
1	0.0

is frequently written as

$$\left[y + \alpha \frac{V^2}{2g} \right]_2^1 = (S - S_o)L, \qquad S_o < 0.1 \tag{4.75}$$

and the range of slopes corresponding to this approximation is frequently omitted. The slopes of rivers and canals in plain areas are usually on the order of 0.0001 to 0.01, while the slopes of mountain streams are typically on the order of 0.05 to 0.1 (Montes, 1998). In cases where the head loss is entirely due to frictional resistance, the energy equation is written as

$$\left[y + \alpha \frac{V^2}{2g} \right]_2^1 = (S_f - S_o)L, \qquad S_o < 0.1 \tag{4.76}$$

where S_f is the frictional head loss per unit length. Equation 4.76 is sometimes written in the expanded form

$$\left(y_1 + \alpha_1 \frac{V_1^2}{2g} + z_1 \right) = \left(y_2 + \alpha_2 \frac{V_2^2}{2g} + z_2 \right) + h_f, \qquad S_o < 0.1 \tag{4.77}$$

where h_f is the head loss due to friction. Equation 4.77 is superficially similar to the Bernoulli equation, but the two equations are fundamentally different because the Bernoulli equation is derived from the momentum equation (not the energy equation) and does not contain a head loss term.

4.2.3.1 Energy Grade Line

The head at each cross-section of an open channel, h, is given by

$$h = y + \alpha \frac{V^2}{2g} + z_o, \qquad S_o < 0.1 \tag{4.78}$$

where y is the depth of flow, V is the average velocity over the cross-section, z_o is the elevation of the bottom of the channel, and S_o is the slope of the channel. As stated previously, in most cases the slope restriction ($S_o < 0.1$) is met, and this restriction is not explicitly stated in the definition of the head. When the head, h, at each section is plotted versus the distance along the channel, this curve is called the *energy grade line*. The point on the energy grade line corresponding to each cross-section is located a distance $\alpha V^2/2g$ vertically above the water surface; between any two cross-sections the elevation of the energy grade drops by a distance equal to the head loss, h_L, between the two sections. The energy grade line is useful in visualizing the state of a fluid as it flows along an open channel and especially useful in visualizing the performance of hydraulic structures in open-channel systems.

4.2.3.2 Specific Energy

The *specific energy*, E, of a fluid is defined as the energy per unit weight of the fluid measured relative to the bottom of the channel and is given by

$$E = y + \alpha \frac{V^2}{2g} \tag{4.79}$$

or

$$E = y + \alpha \frac{Q^2}{2gA^2} \tag{4.80}$$

where Q is the volumetric flowrate and A is the cross-sectional area. For a given shape of the channel cross-section and flowrate, Q, the specific energy, E, depends only on the depth of flow, y. The typical relationship between E and y given by Equation 4.80, for a constant value of Q, is shown in Figure 4.5. The salient features of Figure 4.5 are (1) there is more than one possible flow depth for a given specific energy, and (2) the specific energy curve is asymptotic to the line

$$y = E \tag{4.81}$$

In accordance with the energy equation (Equation 4.77), the specific energy at any cross-section can be expressed in terms of the specific energy of the fluid at an upstream section, change in the elevation of the bottom of the channel, and energy loss due to friction. The fact that there can be more than one possible depth for a given specific energy leads to the question of which depth will exist. The specific energy diagram, Figure 4.5, indicates that there exists a depth, y_c, at which the specific energy is a minimum. At this point, Equation 4.80 indicates that

$$\frac{dE}{dy} = 0 = 1 - \frac{Q^2}{gA_c^3}\frac{dA}{dy} \tag{4.82}$$

where A_c is the flow area corresponding to $y = y_c$, and the kinetic energy correction factor, α, has been taken equal to unity.

Referring to the general open-channel cross-section shown in Figure 4.6, it is clear that for small changes in the flow depth, y,

$$dA = T\,dy \tag{4.83}$$

where dA is the increase in flow area resulting from a change in depth dy, and T is the top width of the channel when the flow depth is y. Equation 4.83 can be written as

$$\frac{dA}{dy} = T \tag{4.84}$$

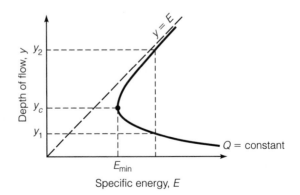

Figure 4.5 ■ Typical
Specific Energy Diagram

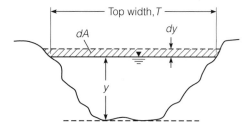

Figure 4.6 ■ Typical Channel Cross-Section

which can be combined with Equation 4.82 to yield the following relationship under *critical flow* conditions

$$\boxed{\frac{Q^2}{g} = \frac{A_c^3}{T_c}}$$
(4.85)

where T_c is the top width of the channel corresponding to $y = y_c$. Equation 4.85 forms the basis for calculating the *critical flow depth*, y_c, in a channel for a given flowrate, Q, since the lefthand side of the equation is known and the righthand side is a function of y_c for a channel of given shape. The specific energy under critical flow conditions, E_c, is then given by

$$E_c = y_c + \frac{A_c}{2T_c}$$
(4.86)

Defining the *hydraulic depth*, D, by the relation

$$D = \frac{A}{T}$$
(4.87)

then a Froude number, Fr, can be defined by

$$\mathrm{Fr}^2 = \frac{V^2}{gD} = \frac{Q^2 T}{gA^3}$$
(4.88)

Combining this definition of the Froude number with the critical flow condition given by Equation 4.85 leads to the relation

$$\boxed{\mathrm{Fr}_c = 1}$$
(4.89)

where Fr_c is the Froude number under critical flow conditions. When $y > y_c$, Equation 4.88 indicates that the Froude number is less than unity, and when $y < y_c$, the Froude number is greater than unity. Flows where $y < y_c$ are called *supercritical*; where $y > y_c$, flows are called *subcritical*. It is apparent from the specific energy diagram, Figure 4.5, that when the flow conditions are close to critical, a relatively large change of depth occurs with small variations in specific energy. Flow under these conditions is unstable and excessive wave action or undulations of the water surface may occur. Experiments in rectangular channels have shown that these instabilities can be avoided if $\mathrm{Fr} < 0.86$ or $\mathrm{Fr} > 1.13$ (U.S. Army Corps of Engineers, 1995).

■ Example 4.4

Determine the critical depth for water flowing at 10 m³/s in a trapezoidal channel with bottom width 3 m and side slopes of 2:1 (H:V).

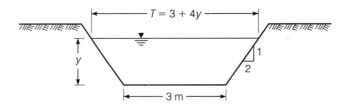

Figure 4.7 ■ Trapezoidal
Cross-Section

Solution

The channel cross-section is illustrated in Figure 4.7, where the depth of flow is y, the top width, T, is given by

$$T = 3 + 4y$$

and the flow area, A, is

$$A = 3y + 2y^2$$

Under critical flow conditions

$$\frac{Q^2}{g} = \frac{A_c^3}{T_c}$$

and since $Q = 10$ m³/s and $g = 9.81$ m/s², then under critical flow conditions

$$\frac{10^2}{9.81} = \frac{(3y_c + 2y_c^2)^3}{(3 + 4y_c)}$$

Solving for y_c by trial and error yields the critical depth

$$y_c = 0.855 \text{ m} \qquad\qquad ■$$

The critical flow condition described by Equation 4.85 can be simplified considerably in the case of flow in rectangular channels, where it is convenient to deal with the flow per unit width, q, given by

$$q = \frac{Q}{b} \tag{4.90}$$

where b is the width of the channel. The flow area, A, and top width, T, are given by

$$A = by \tag{4.91}$$

$$T = b \tag{4.92}$$

The critical flow condition given by Equation 4.85 then becomes

$$\frac{(qb)^2}{g} = \frac{(by_c)^3}{b} \tag{4.93}$$

which can be solved to yield the critical flow depth

$$\boxed{y_c = \left(\frac{q^2}{g}\right)^{\frac{1}{3}}} \tag{4.94}$$

and corresponding critical energy

$$E_c = y_c + \frac{q^2}{2gy_c^2} \tag{4.95}$$

Combining Equations 4.94 and 4.95 leads to the following simplified form of the minimum specific energy in rectangular channels

$$E_c = \frac{3}{2}y_c \tag{4.96}$$

The specific energy diagram illustrated in Figure 4.8 for a rectangular channel is similar to the nonrectangular case shown in Figure 4.5, with the exception that the specific energy curve for a rectangular channel corresponds to a fixed value of q rather than Q in a nonrectangular channel. The specific energy diagram in Figure 4.8 indicates that the specific energy curve shifts upward and to the right for increasing values of q. This form of the specific energy diagram is particularly informative in understanding what happens to the flow in a rectangular channel when there is a constriction, such as when the channel width is narrowed to accommodate a bridge. Suppose that the flow per unit width upstream of the constriction is q_1 and the depth of flow at this location is y_1. If the channel is constricted so that the flow per unit width becomes q_2, then, provided energy losses in the constriction are minimal, Figure 4.8 indicates that there are two possible flow depths in the constricted section, y_2 and y_2'. Neglecting energy losses in the constriction is reasonable if the constriction is smooth and takes place over a relatively short distance. Figure 4.8 indicates that it is physically impossible for the flow depth in the constriction to be y_2' since this would require the flow per unit width, q, to increase and then decrease to reach y_2'. Since the flow per unit width decreases monotonically in a constriction, the flow depth can only go from y_1 to y_2. A similar case arises when the flow depth upstream of the constriction is y_1' and the possible flow depths at the constriction are y_2' and y_2. In this case, only y_2' is possible, since an increase and then a decrease in q would be required to achieve a flow depth of y_2. Based on this analysis, it is clear that if the flow upstream of the constriction is subcritical ($y > y_c$ or Fr < 1), then the flow in the constriction must be either subcritical or critical, and if the flow upstream of the constriction is supercritical ($y < y_c$ or Fr > 1), then the flow in the constriction must be either supercritical or critical. If the flow upstream of the constriction is critical, then flow through the constriction is not possible under the existing flow

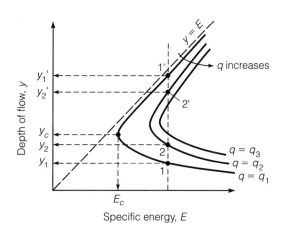

Figure 4.8 ■ Specific Energy Diagram for Rectangular Channel

conditions and the upstream flow conditions must necessarily change upon installation of the constriction in the channel.

Regardless of whether the flow upstream of the constriction is subcritical or supercritical, Figure 4.8 indicates that a maximum constriction will cause critical flow to occur at the constriction. A larger constriction (smaller opening) than that which causes critical flow to occur at the constriction is not possible based on the available specific energy upstream of the constriction. Any further constriction will result in changes in the upstream flow conditions to maintain critical flow at the constriction. Under these circumstances, the flow is said to be *choked*. Choked flows can result either from narrowing channels or from raising the invert elevation of the channel. Narrowing the channel increases the flowrate per unit width, q, while maintaining the available specific energy, E, whereas raising the channel invert maintains q and reduces E.

▪ Example 4.5

A rectangular channel 1.30 m wide carries 1.10 m³/s of water at a depth of 0.85 m. (a) If a 30-cm wide pier is placed in the middle of the channel, find the elevation of the water surface at the constriction. (b) What is the minimum width of the constriction that will not cause a rise in the upstream water surface?

Solution

(a) The cross-sections of the channel upstream of the constriction and at the constriction are shown in Figure 4.9. Neglecting the energy losses between the constriction and the upstream section, the energy equation requires that the specific energy at the constriction be equal to the specific energy at the upstream section. Therefore,

$$y_1 + \frac{V_1^2}{2g} = y_2 + \frac{V_2^2}{2g}$$

where Section 1 refers to the upstream section and Section 2 refers to the constricted section. In this case, $y_1 = 0.85$ m, and

$$V_1 = \frac{Q}{A_1} = \frac{1.10}{(0.85)(1.30)} = 1.00 \text{ m/s}$$

The specific energy at Section 1, E_1, is

$$E_1 = y_1 + \frac{V_1^2}{2g} = 0.85 + \frac{1.00^2}{2(9.81)} = 0.901 \text{ m}$$

Equating the specific energies at Sections 1 and 2 yields

$$0.901 = y_2 + \frac{Q^2}{2gA^2}$$

$$= y_2 + \frac{1.10^2}{2(9.81)[(1.30 - 0.30)y_2]^2}$$

which simplifies to

$$y_2 + \frac{0.0617}{y_2^2} = 0.901$$

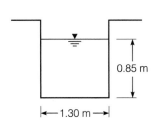

0.85 m

|←1.30 m→|

Upstream of constriction

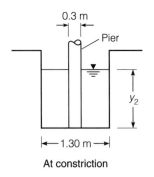

0.3 m

Pier

y_2

|←1.30 m→|

At constriction

Figure 4.9 ▪ Constriction in Rectangular Channel

There are three solutions to this cubic equation: $y_2 = 0.33$ m, 0.80 m, and -0.23 m. Of the two positive depths, we must select the depth corresponding to the same flow condition as upstream. At the upstream section, $Fr_1 = V_1/\sqrt{gy_1} = 0.35$; therefore, the upstream flow is subcritical and the flow at the constriction must also be subcritical. The flow depth must therefore be

$$y_2 = 0.80 \text{ m}$$

where $Fr_2 = V_2/\sqrt{gy_2} = 0.49$. The other depth ($y_2 = 0.33$ m, $Fr_2 = 1.9$) is supercritical and cannot be achieved.

(b) The minimum width of constriction that does not cause the upstream depth to change is associated with critical flow conditions at the constriction. Under these conditions (for a rectangular channel)

$$E_1 = E_2 = E_c = \frac{3}{2}y_c = \frac{3}{2}\left(\frac{q^2}{g}\right)^{\frac{1}{3}}$$

If b is the width of the constriction that causes critical flow, then

$$E_1 = \frac{3}{2}\left[\frac{(Q/b)^2}{g}\right]^{\frac{1}{3}}$$

From the given data: $E_1 = 0.901$ m, and $Q = 1.10$ m^3/s. Therefore

$$0.901 = \frac{3}{2}\left[\frac{1.10^2}{b^2(9.81)}\right]^{\frac{1}{3}}$$

which gives

$$b = 0.75 \text{ m}$$

If the constricted channel width is any less than 0.75 m, the flow will be choked and the upstream flow depth will increase. ∎

The specific energy analyses covered in this section have generally assumed that the kinetic energy correcion factor, α, is approximately equal to unity. Although this approximation is valid in many cases, in diverging transitions at angles exceeding 8° flows tend to separate from the wall of the transition, causing large energy losses and substantial increases in α (Montes, 1998). Under such conditions, the assumption of a constant specific energy and a value of α approximately equal to unity are not justified.

4.3 Water Surface Profiles

4.3.1 ■ Profile Equation

The equation describing the shape of the water surface in an open channel can be derived from the energy equation, Equation 4.76, which is of the form

$$S_o - S_f = \frac{\Delta\left(y + \alpha\frac{V^2}{2g}\right)}{\Delta x} \tag{4.97}$$

where S_o is the slope of the channel, S_f is the slope of the energy grade line, y is the depth of flow, α is the kinetic energy correction factor, V is the average velocity, x is a coordinate measured along the channel (the flow direction defined as positive), and Δx is the distance between the upstream and downstream sections. Equation 4.97 can be further rearranged into

$$S_o - S_f = \frac{\Delta y}{\Delta x} + \frac{\Delta\left(\alpha \frac{V^2}{2g}\right)}{\Delta x} \tag{4.98}$$

Taking the limit of Equation 4.98 as $\Delta x \to 0$, and invoking the definition of the derivative, yields

$$
\begin{aligned}
S_o - S_f &= \lim_{\Delta x \to 0} \frac{\Delta y}{\Delta x} + \lim_{\Delta x \to 0} \frac{\Delta\left(\alpha \frac{V^2}{2g}\right)}{\Delta x} \\
&= \frac{dy}{dx} + \frac{d}{dx}\left(\alpha \frac{V^2}{2g}\right) \\
&= \frac{dy}{dx} + \frac{d}{dy}\left(\alpha \frac{V^2}{2g}\right)\frac{dy}{dx} \\
&= \frac{dy}{dx}\left[1 + \frac{d}{dy}\left(\alpha \frac{Q^2}{2gA^2}\right)\right] \\
&= \frac{dy}{dx}\left[1 - \alpha \frac{Q^2}{gA^3}\frac{dA}{dy}\right]
\end{aligned}
\tag{4.99}
$$

where Q is the (constant) flowrate and A is the (variable) cross-sectional flow area in the channel. Recalling that

$$\frac{dA}{dy} = T \tag{4.100}$$

where T is the top width of the channel and the hydraulic depth, D, of the channel is defined as

$$D = \frac{A}{T} \tag{4.101}$$

then the Froude number, Fr, of the flow can be written as

$$\mathrm{Fr} = \frac{V}{\sqrt{gD}} = \frac{Q\sqrt{T}}{A\sqrt{gA}} = \frac{Q}{\sqrt{gA^3}}\sqrt{\frac{dA}{dy}} \tag{4.102}$$

or

$$\mathrm{Fr}^2 = \frac{Q^2}{gA^3}\frac{dA}{dy} \tag{4.103}$$

Combining Equations 4.99 and 4.103 and rearranging yields

$$\boxed{\frac{dy}{dx} = \frac{S_o - S_f}{1 - \alpha\mathrm{Fr}^2}} \tag{4.104}$$

This differential equation describes the water-surface profile in open channels. To appreciate the utility of Equation 4.104, consider the relative magnitudes of the channel slope, S_o, and the friction slope, S_f. According to the Manning equation, for any given flowrate,

$$\frac{S_f}{S_o} = \left(\frac{A_n R_n^{\frac{2}{3}}}{A R^{\frac{2}{3}}}\right)^2 \tag{4.105}$$

where A_n and R_n are the cross-sectional area and hydraulic radius under normal flow conditions ($S_f = S_o$), and A and R are the actual cross-sectional area and hydraulic radius of the flow. Since $A R^{\frac{2}{3}} > A_n R_n^{\frac{2}{3}}$ when $y > y_n$, then Equation 4.105 indicates that

$$S_f > S_o \quad \text{when} \quad y < y_n, \quad \text{and} \quad S_f < S_o \quad \text{when} \quad y > y_n \tag{4.106}$$

or

$$S_o - S_f < 0 \quad \text{when} \quad y < y_n, \quad \text{and} \quad S_o - S_f > 0 \quad \text{when} \quad y > y_n \tag{4.107}$$

It has already been shown that

$$\text{Fr} > 1 \quad \text{when} \quad y < y_c, \quad \text{and} \quad \text{Fr} < 1 \quad \text{when} \quad y > y_c \tag{4.108}$$

or

$$1 - \text{Fr} < 0 \quad \text{when} \quad y < y_c, \quad \text{and} \quad 1 - \text{Fr} > 0 \quad \text{when} \quad y > y_c \tag{4.109}$$

Based on Equations 4.107 and 4.109, the sign of the numerator in Equation 4.104 is determined by the magnitude of the flow depth, y, relative to the normal depth, y_n, and the sign of the denominator is determined by the magnitude of the flow depth, y, relative to the critical depth, y_c (assuming $\alpha \approx 1$). Therefore, the slope of the water surface, dy/dx, is determined by the relative magnitudes of y, y_n, and y_c.

4.3.2 ■ Classification of Water-Surface Profiles

In hydraulic engineering, channel slopes are classified based on the relative magnitudes of the normal depth, y_n, and the critical depth, y_c. The hydraulic classification of slopes is shown in Table 4.4 and is illustrated in Figure 4.10. The range of flow depths for each slope can be divided into two or more zones, delimited by the normal and critical flow depths, where the highest zone above the channel bed is called Zone 1, the intermediate zone is Zone 2, and the lowest zone is Zone 3. Water-surface profiles are classified based on both the type of slope (for example, type M) and the zone in which the water surface is located (for example, Zone 2). Therefore, each water-surface profile is classified by a letter and number: for example, an M2 profile indicates a mild slope ($y_n > y_c$) and the actual depth y is in Zone 2 ($y_n < y < y_c$). Since dy/dx is determined by the relative magnitudes of y, y_n, and y_c, and these relative magnitudes are implicit in the profile classification, then the slopes of all water surfaces are determined by their profile classification.

■ Example 4.6

Water flows in a trapezoidal channel in which the bottom width is 5 m and the side slopes are 1.5:1 (H:V). The channel lining has an estimated Manning n of 0.04, and the longitudinal slope of the channel is 1%. If the flowrate is 60 m³/s and the depth of flow at a gaging station is 4 m, classify the water-surface profile, state whether the depth

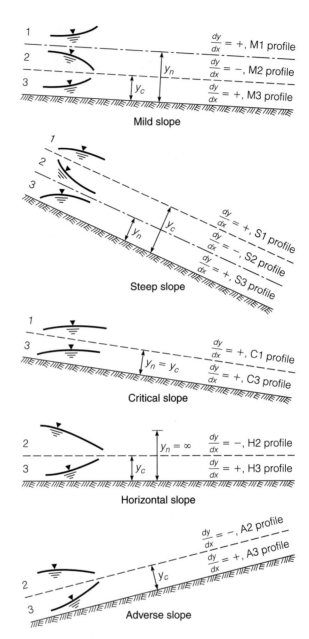

$$\frac{dy}{dx} = +, \text{M1 profile}$$

$$\frac{dy}{dx} = -, \text{M2 profile}$$

$$\frac{dy}{dx} = +, \text{M3 profile}$$

Mild slope

$$\frac{dy}{dx} = +, \text{S1 profile}$$

$$\frac{dy}{dx} = -, \text{S2 profile}$$

$$\frac{dy}{dx} = +, \text{S3 profile}$$

Steep slope

$$\frac{dy}{dx} = +, \text{C1 profile}$$

$$\frac{dy}{dx} = +, \text{C3 profile}$$

Critical slope

$$\frac{dy}{dx} = -, \text{H2 profile}$$

$$\frac{dy}{dx} = +, \text{H3 profile}$$

Horizontal slope

$$\frac{dy}{dx} = -, \text{A2 profile}$$

$$\frac{dy}{dx} = +, \text{A3 profile}$$

Adverse slope

Figure 4.10 ■
Slope Classifications

■ **Table 4.4**
Hydraulic Classification
of Slopes

Name	Type	Condition
Mild	M	$y_n > y_c$
Steep	S	$y_n < y_c$
Critical	C	$y_n = y_c$
Horizontal	H	$y_n = \infty$
Adverse	A	$S_o < 0$

increases or decreases in the downstream direction, and calculate the slope of the water surface at the gaging station. On the basis of this water-surface slope, estimate the depth of flow 100 m downstream of the gaging station.

Solution

To classify the water-surface profile, the normal and critical depths must be calculated and contrasted with the actual flow depth of 4 m. To calculate the normal flow depth, apply the Manning equation

$$Q = \frac{1}{n} A_n R_n^{2/3} S_o^{1/2} = \frac{1}{n} \frac{A_n^{5/3}}{P_n^{2/3}} S_o^{1/2}$$

where Q is the flowrate ($= 60$ m^3/s), A_n and P_n are the areas and wetted perimeters under normal flow conditions, and S_o is the longitudinal slope of the channel ($= 0.01$). The Manning equation can be written in the more useful form

$$\frac{A_n^5}{P_n^2} = \left[\frac{Qn}{\sqrt{S_o}} \right]^3$$

where the lefthand side is a function of the normal flow depth, y_n and the righthand side is in terms of given data. Substituting the given parameters leads to

$$\frac{A_n^5}{P_n^2} = \left[\frac{(60)(0.04)}{\sqrt{0.01}} \right]^3 = 13824$$

Since

$$A_n = (b + my_n)y_n = (5 + 1.5y_n)y_n$$

and

$$P_n = b + 2\sqrt{1 + m^2}\, y_n = 5 + 2\sqrt{1 + 1.5^2}\, y_n = 5 + 3.61y_n$$

then the Manning equation can be written as

$$\frac{(5 + 1.5y_n)^5 y_n^5}{(5 + 3.61y_n)^2} = 13824$$

which can be solved by trial and error to yield

$$y_n = 2.25 \text{ m}$$

When the flow conditions are critical, then

$$\frac{A_c^3}{T_c} = \frac{Q^2}{g}$$

where A_c and T_c are the area and top-width, respectively. The lefthand side of this equation is a function of the critical flow depth, y_c, and the righthand side of this equation is in terms of given data. Thus

$$\frac{A_c^3}{T_c} = \frac{60^2}{9.81} = 367$$

Since

$$A_c = (b + my_c)y_c = (5 + 1.5y_c)y_c$$

and

$$T_c = b + 2my_c = 5 + 2(1.5)y_c = 5 + 3y_c$$

then the critical flow equation can be written as

$$\frac{(5 + 1.5y_c)^3 y_c^3}{5 + 3y_c} = 367$$

which can be solved by trial and error to yield

$$y_c = 1.99 \text{ m}$$

Since $y_n (= 2.25 \text{ m}) > y_c (= 1.99 \text{ m})$, then the slope is mild. Also, since $y (= 4 \text{ m}) > y_n > y_c$ the water surface is in Zone 1 and therefore the water-surface profile is an M1 profile. This classification requires that the flow depth increases in the downstream direction.

The slope of the water surface is given by (assuming $\alpha = 1$)

$$\frac{dy}{dx} = \frac{S_o - S_f}{1 - \text{Fr}^2}$$

where S_f is the slope of the energy grade line and Fr is the Froude number. According to the Manning equation, S_f can be estimated by

$$S_f = \left[\frac{nQ}{AR^{\frac{2}{3}}} \right]^2$$

and when the depth of flow, y, is 4 m, then

$$A = (b + my)y = (5 + 1.5 \times 4)(4) = 44 \text{ m}^2$$

$$P = b + 2\sqrt{1 + m^2}y = 5 + 2\sqrt{1 + 1.5^2}(4) = 19.4 \text{ m}$$

$$R = \frac{A}{P} = \frac{44}{19.4} = 2.27 \text{ m}$$

and therefore S_f is estimated to be

$$S_f = \left[\frac{(0.04)(60)}{(44)(2.27)^{\frac{2}{3}}} \right]^2 = 0.000997$$

The Froude number, Fr, is given by

$$\text{Fr}^2 = \frac{V^2}{gD} = \frac{(Q/A)^2}{g(A/T)} = \frac{Q^2 T}{gA^3}$$

where the top width, T, is given by

$$T = b + 2my = 5 + 2(1.5)(4) = 17 \text{ m}$$

and therefore the Froude number is given by

$$\text{Fr}^2 = \frac{(60)^2(17)}{(9.81)(44)^3} = 0.0732$$

Substituting the values for S_o ($= 0.01$), S_f ($= 0.000997$), and Fr^2 ($= 0.0732$) into the profile equation yields

$$\frac{dy}{dx} = \frac{0.01 - 0.000997}{1 - 0.0732} = 0.00971$$

The depth of flow, y, at a location 100 m downstream from where the flow depth is 4 m can be estimated by

$$y = 4 + \frac{dy}{dx}(100) = 4 + (0.00971)(100) = 4.97 \text{ m}$$

The estimated flow depth 100 m downstream could be refined by recalculating dy/dx for a flow depth of 4.97 m and then using an averaged value of dy/dx to estimate the flow depth 100 m downstream. ■

4.3.3 ■ Hydraulic Jump

In some cases, supercritical flow must necessarily transition into subcritical flow, even though such a transition does not appear to be possible within the context of the water-surface profiles discussed in the previous section. An example is a case where water is discharged as supercritical flow from under a sluice gate into a body of water flowing under subcritical conditions. If the slope of the channel is mild, then the water emerging from under the gate must necessarily follow an M3 profile, where the depth increases with distance downstream. However, if the depth downstream of the gate is subcritical, then it is apparently impossible for this flow condition to be reached, since this would require a continued increase in the water depth through the M2 zone.

In reality, this transition is accomplished by an abrupt localized change in water depth called a *hydraulic jump*, which is illustrated in Figure 4.11(a), with the corresponding transition in the specific energy diagram shown in Figure 4.11(b). The supercritical (upstream) flow depth is y_1, the subcritical (downstream) flow depth is y_2, and the energy

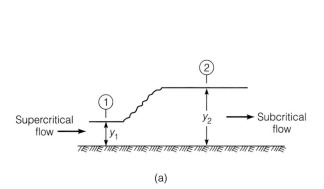

(a)   (b)

Figure 4.11 ■ Hydraulic Jump

loss between the upstream and downstream sections is ΔE. If the energy loss were equal to zero, then the downstream depth, y_2, could be calculated by equating the upstream and downstream specific energies. However, the transition between supercritical and subcritical flow is generally a turbulent process with a significant energy loss that cannot be neglected. Applying the momentum equation to the control volume between Sections 1 and 2 leads to

$$P_1 - P_2 = \rho Q(V_2 - V_1) \tag{4.110}$$

where P_1 and P_2 are the hydrostatic pressure forces at Sections 1 and 2; Q is the flowrate; and V_1 and V_2 are the average velocities at Sections 1 and 2, respectively. Equation 4.110 neglects the friction forces on the channel boundary within the control volume, which is justified by the assumption that over a short distance the friction force will be small compared with the difference in upstream and downstream hydrostatic forces. The momentum equation, Equation 4.110, can be written as

$$\gamma \bar{y}_1 A_1 - \gamma \bar{y}_2 A_2 = \rho Q\left(\frac{Q}{A_2} - \frac{Q}{A_1}\right) \tag{4.111}$$

where $\bar{y}_1$ and $\bar{y}_2$ are the distances from the water surface to the centroids of Sections 1 and 2 and A_1 and A_2 are the cross-sectional areas at Sections 1 and 2, respectively. Equation 4.111 can be rearranged as

$$\frac{Q^2}{gA_1} + A_1\bar{y}_1 = \frac{Q^2}{gA_2} + A_2\bar{y}_2 \tag{4.112}$$

or

$$\boxed{\frac{Q^2}{gA} + A\bar{y} = \text{constant}} \tag{4.113}$$

The term on the lefthand side is called the *specific momentum*, and Equation 4.113 states that the specific momentum remains constant across the hydraulic jump. In the case of a rectangular channel, Equation 4.112 becomes

$$\frac{q^2}{gy_1} + \frac{y_1^2}{2} = \frac{q^2}{gy_2} + \frac{y_2^2}{2} \tag{4.114}$$

where q is the flow per unit width. Equation 4.114 can be solved for y_2 to yield

$$y_2 = \frac{y_1}{2}\left(-1 + \sqrt{1 + \frac{8q^2}{gy_1^3}}\right) \tag{4.115}$$

which can also be written in the following nondimensional form

$$\boxed{\frac{y_2}{y_1} = \frac{1}{2}\left(-1 + \sqrt{1 + 8\text{Fr}_1^2}\right)} \tag{4.116}$$

where Fr_1 is the upstream Froude number defined by

$$\text{Fr}_1 = \frac{V_1}{\sqrt{gy_1}} = \frac{q}{y_1\sqrt{gy_1}} \tag{4.117}$$

	Fr$_1$	Jump characteristics
■ **Table 4.5** Characteristics of Hydraulic Jumps	1–1.7	standing wave (*undular jump*); kinetic-energy loss (as a percentage of the upstream kinetic energy, $V_1^2/2g$) < 5%
	1.7–2.5	smooth surface rise (*weak jump*); kinetic-energy loss 5%–15%
	2.5–4.5	unstable oscillating jump, where each irregular pulsation creates a large wave that can travel far downstream, damaging earth banks and other structures (*oscillating jump*); kinetic-energy loss 15%–45%
	4.5–9.0	stable jump, best performance and action, and insensitive to downstream conditions (*steady jump*); kinetic-energy loss 45%–70%
	> 9.0	rough, somewhat intermittent (*strong jump*); kinetic-energy loss 70%–85%

Source: Gerhart et al. (1992).

The depths upstream and downstream of a hydraulic jump, y_1 and y_2, are called the *conjugate depths* of the hydraulic jump, and experimental measurements have shown that Equation 4.116 yields values of y_2 to within 1% of observed values (Streeter et al., 1998). The theoretical relationship between conjugate depths of a hydraulic jump in a horizontal rectangular channel, Equation 4.116, can also be used for hydraulic jumps on sloping channels, provided the channel slope is less than about 5%. For larger channel slopes, the component of the weight of the fluid in the direction of flow becomes significant and must be incorporated into the momentum equation from which the hydraulic jump equation is derived. The length of hydraulic jumps are around $6y_2$ for $4.5 < $ Fr$_1 < $ 13, and somewhat smaller outside this range. The length, L, of a hydraulic jump can also be estimated by (Hager, 1991)

$$\frac{L}{y_1} = 220 \tanh \frac{\text{Fr}_1 - 1}{22} \tag{4.118}$$

and the length, L_t, of the transition region between the end of the hydraulic jump and fully developed open-channel flow can be estimated by (Wu and Rajaratnam, 1996)

$$L_t = 10y_2 \tag{4.119}$$

Physical characteristics of hydraulic jumps in relation to the upstream Froude number, Fr$_1$, are listed in Table 4.5. A steady, well-established jump, with $4.5 < $ Fr$_1 < 9.0$, is often used as an energy dissipator downstream of a dam or spillway (Potter and Wiggert, 1991) and can also be used to mix chemicals, or act as an aeration mechanism. The energy loss in the hydraulic jump, ΔE, is given by

$$\Delta E = \left(y_1 + \frac{V_1^2}{2g}\right) - \left(y_2 + \frac{V_2^2}{2g}\right) \tag{4.120}$$

Combining Equation 4.120 with the equation relating the conjugate depths of the hydraulic jump, Equation 4.116, leads to the following expression for the head loss in a hydraulic jump in a rectangular channel

$$\boxed{\Delta E = \frac{(y_2 - y_1)^3}{4y_1y_2}} \tag{4.121}$$

■ Example 4.7

Water flows down a spillway at the rate of 12 m^3/s per meter of width into a horizontal channel, where the velocity at the channel entrance is 20 m/s. Determine the (downstream) depth of flow in the channel that will cause a hydraulic jump to occur in the channel, and determine the power loss in the jump per meter of width.

Solution

In this case, $q = 12$ m²/s, and $V_1 = 20$ m/s. Therefore the initial depth of flow, y_1, is given by

$$y_1 = \frac{q}{V_1} = \frac{12}{20} = 0.60 \text{ m}$$

and the corresponding Froude number, Fr_1, is given by

$$Fr_1 = \frac{q}{y_1\sqrt{gy_1}} = \frac{12}{0.60\sqrt{(9.81)(0.60)}} = 8.24$$

which confirms that the flow is supercritical. The conjugate depth is given by Equation 4.116, where

$$\frac{y_2}{y_1} = \frac{1}{2}\left(-1 + \sqrt{1 + 8Fr_1^2}\right)$$

$$= \frac{1}{2}\left(-1 + \sqrt{1 + 8(8.24)^2}\right)$$

$$= 11.2$$

Therefore the conjugate downstream depth, y_2, is

$$y_2 = 11.2y_1 = 11.2(0.60) = 6.70 \text{ m}$$

The energy loss in the hydraulic jump, ΔE, is given by Equation 4.121 as

$$\Delta E = \frac{(y_2 - y_1)^3}{4y_1y_2} = \frac{(6.70 - 0.60)^3}{4(0.60)(6.70)} = 14.1 \text{ m}$$

and therefore the power loss, P, per unit width in the jump is given by

$$P = \gamma q\Delta E = 9790(12)(14.1) = 1.66 \times 10^6 \text{ W} = 1.7 \text{ MW} \quad ■$$

The location of a hydraulic jump is important in determining channel wall heights as well as the flow conditions (subcritical or supercritical) in the channel. The mean location of the hydraulic jump is usually estimated by computing the upstream and downstream water-surface profiles; the jump is located where the upstream and downstream water depths are equal to the conjugate depths of the hydraulic jump. In many cases, the location of the hydraulic jump is controlled by installing baffle blocks, sills, drops, or rises in the bottom of the channel to create sufficient energy loss that the hydraulic jump forms at the location of these structures. Hydraulic structures that are specifically designed to induce the formation of hydraulic jumps are called *stilling basins*.

4.3.4 ▪ Computation of Water-Surface Profiles

The differential equation describing the water-surface profile in an open channel is given by Equation 4.104, which can be written in the form

$$\boxed{\frac{dy}{dx} = F(x, y)} \qquad (4.122)$$

where y is the depth of flow in the channel, x is the distance along the channel, and $F(x, y)$ is a function defined by the relation

$$F(x, y) = \frac{S_o - S_f}{1 - \alpha \mathrm{Fr}^2} \qquad (4.123)$$

where S_o is the channel slope, S_f is the slope of the energy grade line, Fr is the Froude number of the flow, and α is the kinetic energy correction factor. The function $F(x, y)$ can be calculated using given values of Q, S_o, α, and channel geometry (which can be a function of x), using Equation 4.102 to estimate the Froude number and the Manning or Darcy-Weisbach head loss formula to estimate the slope of the energy grade line, S_f.

A basic assumption is that the head loss between upstream and downstream sections can be estimated using either the Manning or Darcy-Weisbach head loss formula, without regard to trends in depth. This approximation requires that flow conditions change gradually. Such flow conditions are called *gradually varied flow* (GVF). Conversely, flows that are not gradually varied are called *rapidly varied flow* (RVF). GVF can be further contrasted with RVF in that GVF is usually analyzed over longer distances, where friction losses due to boundary shear are significant, whereas RVF is analyzed over shorter distances, where boundary shear is less significant. Also, the pressure distribution in GVF is usually hydrostatic, whereas in RVF there is usually significant acceleration normal to the streamlines, causing a nonhydrostatic pressure distribution. An example of RVF is the hydraulic jump. Under GVF conditions, the Manning approximation (in SI units) to the friction slope, S_f, is given by

$$S_f = \left(\frac{nQ}{AR^{\frac{2}{3}}}\right)^2 \qquad (4.124)$$

where n is the Manning roughness coefficient, Q is the flowrate, and A and R are the area and hydraulic radius of the cross-section, respectively.

The solution of the water-surface profile equation (Equation 4.122) is usually obtained using numerical integration, since the function $F(x, y)$ is typically only available at discrete intervals along the channel. A technique that is used to determine the water-surface profile by numerical integration of Equation 4.122 is the *direct-integration method*. As an alternative to direct integration of Equation 4.122, water-surface profiles can also be calculated directly from the energy equation. Rearranging the form of the energy equation given by Equation 4.76 leads to

$$\boxed{\Delta L = \frac{\left[y + \alpha \frac{V^2}{2g}\right]_2^1}{\bar{S}_f - S_o}} \qquad (4.125)$$

where ΔL is the distance between the upstream section (Section 1) and the downstream section (Section 2) and $\bar{S}_f$ is the average slope of the energy grade line between Sections 1 and 2. Two commonly used techniques to calculate the water-surface profile directly from the energy equation (Equation 4.125) are the *direct-step method* and the *standard-step method*.

4.3.4.1   Direct-Integration Method

In applying the direct-integration method, the water-surface profile is expressed in the finite difference form

$$\frac{y_2 - y_1}{x_2 - x_1} = F(\bar{x}, \bar{y}) \tag{4.126}$$

where

$$\bar{x} = \frac{x_1 + x_2}{2}, \quad \text{and} \quad \bar{y} = \frac{y_1 + y_2}{2} \tag{4.127}$$

and the subscripts refer to (adjacent) cross-sections of the channel. A more convenient form of Equation 4.126 is

$$\boxed{y_2 = y_1 + F(\bar{x}, \bar{y})(x_2 - x_1)} \tag{4.128}$$

This equation is appropriate for computing the water-surface profile in the downstream direction. In most cases, however, water-surface profiles are computed in the upstream direction. A more appropriate form of Equation 4.128 is

$$\boxed{y_1 = y_2 - F(\bar{x}, \bar{y})(x_2 - x_1)} \tag{4.129}$$

In subcritical flow, calculations usually proceed in an upstream direction, while in supercritical flow calculations proceed in a downstream direction. In applying Equation 4.129, the following procedure is suggested:

1. Starting with known flow conditions at Section 2, assume a depth, y_1, at location x_1, and then calculate y_1 using Equation 4.129. On the first calculation, it is reasonable to assume that $y_1 = y_2$.
2. Repeat step 1 until the calculated value of y_1 is equal to the assumed value of y_1. This is then the depth of flow at x_1.

A similar procedure is used to apply Equation 4.128, with iterations on y_2 rather than y_1.

▪ Example 4.8

Water flows at 10 m³/s in a rectangular concrete channel of width 5 m and longitudinal slope 0.001. The Manning roughness coefficient, n, of the channel lining is 0.015, and the water depth is measured as 0.80 m at a gaging station. Use the direct-integration method to estimate the flow depth 100 m upstream of the gaging station.

Solution

From the given data: $Q = 10$ m³/s, $b = 5$ m, $S_o = 0.001$, $n = 0.015$, $y_2 = 0.80$ m, $x_1 = 0$ m, and $x_2 = 100$ m. Assuming $y_1 = y_2 = 0.80$ m, the hydraulic parameters of the flow are:

$$\bar{x} = \frac{x_1 + x_2}{2} = \frac{0 + 100}{2} = 50 \text{ m}$$

$$\bar{y} = \frac{y_1 + y_2}{2} = \frac{0.80 + 0.80}{2} = 0.80 \text{ m}$$

$$\bar{A} = b\bar{y} = (5)(0.80) = 4.0 \text{ m}^2$$

$$\bar{P} = b + 2\bar{y} = 5 + 2(0.80) = 6.60 \text{ m}$$

$$\bar{R} = \frac{\bar{A}}{\bar{P}} = \frac{4.0}{6.60} = 0.606 \text{ m}$$

$$\bar{S}_f = \left[\frac{nQ}{\bar{A}\bar{R}^{\frac{2}{3}}}\right]^2 = \left[\frac{(0.015)(10)}{(4.0)(0.606)^{\frac{2}{3}}}\right]^2 = 0.00274$$

$$\bar{D} = \frac{\bar{A}}{T} = \frac{b\bar{y}}{b} = \bar{y} = 0.80 \text{ m}$$

$$\bar{V} = \frac{Q}{\bar{A}} = \frac{10}{4} = 2.5 \text{ m/s}$$

$$\bar{Fr}^2 = \frac{\bar{V}^2}{g\bar{D}} = \frac{(2.5)^2}{(9.81)(0.80)} = 0.80$$

Using these results, and assuming $\alpha = 1$, Equation 4.123 gives

$$F(\bar{x}, \bar{y}) = \frac{S_o - \bar{S}_f}{1 - \bar{Fr}^2} = \frac{0.001 - 0.00274}{1 - 0.80} = -0.0087$$

and the estimated depth 100 m upstream of the gaging station is given by Equation 4.129 as

$$y_1 = y_2 - F(\bar{x}, \bar{y})(x_2 - x_1) = 0.80 - (-0.0087)(100 - 0) = 1.67 \text{ m}$$

Since this calculated flow depth at Section 1 is significantly different from the assumed flow depth ($= 0.80$ m), the calculations must be repeated, starting with the assumption that $y_1 = 1.67$ m. These calculations are summarized in the following table, where the assumed values of y_1 are given in Column 1 and the calculated values of y_1 (using Equation 4.129) in Column 4:

(1) y_1 (m)	(2) $\bar{S}_f$	(3) $\bar{Fr}^2$	(4) y_1 (m)
1.67	0.000761	0.216	0.77
1.5	0.000936	0.268	0.79
1.3	0.00122	0.352	0.83
1.1	0.00164	0.476	0.92
.	.	.	.
.	.	.	.
.	.	.	.
1.01	0.00193	0.559	1.01

These results indicate that the depth 100 m upstream from the gaging station is approximately equal to 1.01 m. ■

4.3.4.2 Direct-Step Method

In applying the direct-step method, the flow conditions at one section are known, the flow conditions at a second section are specified, and the objective is to find the distance between these two sections. With the flow conditions at two channel sections known, the terms on the righthand side of Equation 4.125 are evaluated to determine the distance ΔL between these sections. Computations then proceed to find the distance to another section with specified flow conditions, using the previously specified conditions as given and newly specified conditions to calculate the new interval ΔL.

The main drawbacks of the direct-step method are: (1) the water-surface profile is not computed at predetermined locations, and (2) the method is only suitable for prismatic channels, where the shape of the channel cross-section is independent of the interval, ΔL. In cases where the flow conditions at specific locations in a prismatic or nonprismatic channel are required, the standard-step method should be used.

■ **Example 4.9**

Water flows at 12 m³/s in a trapezoidal concrete channel ($n = 0.015$) of bottom width 4 m, side slopes 2:1 (H:V), and longitudinal slope 0.0009. If depth of flow at a gaging station is measured as 0.80 m, use the direct-step method to estimate the location where the depth is 1.00 m.

Solution

At the location where the depth is 1.00 m:

$$y_1 = 1.00 \text{ m}$$

$$A_1 = [4 + 2y_1]y_1 = [4 + 2(1.00)](1.00) = 6.00 \text{ m}^2$$

$$P_1 = 4 + 2\sqrt{5}y_1 = 4 + 2\sqrt{5}(1.00) = 8.47 \text{ m}$$

$$R_1 = \frac{A_1}{P_1} = \frac{6.00}{8.47} = 0.708 \text{ m}$$

$$V_1 = \frac{Q}{A_1} = \frac{12}{6.00} = 2.00 \text{ m/s}$$

$$S_{f1} = \left[\frac{nQ}{A_1 R_1^{2/3}}\right]^2 = \left[\frac{(0.015)(12)}{(6.00)(0.708)^{\frac{2}{3}}}\right]^2 = 0.00143$$

and where the depth is 0.80 m:

$$y_2 = 0.80 \text{ m}$$

$$A_2 = [4 + 2y_2]y_2 = [4 + 2(0.80)](0.80) = 4.48 \text{ m}^2$$

$$P_2 = 4 + 2\sqrt{5}\,y_2 = 4 + 2\sqrt{5}\,(0.80) = 7.58 \text{ m}$$

$$R_2 = \frac{A_2}{P_2} = \frac{4.48}{7.58} = 0.591 \text{ m}$$

$$V_2 = \frac{Q}{A_2} = \frac{12}{4.48} = 2.679 \text{ m/s}$$

$$S_{f2} = \left[\frac{nQ}{A_2 R_2^{2/3}}\right]^2 = \left[\frac{(0.015)(12)}{(4.48)(0.591)^{\frac{2}{3}}}\right]^2 = 0.00325$$

Substituting the hydraulic parameters at Sections 1 and 2 into Equation 4.125 gives

$$\Delta L = \frac{\left[y + \alpha \frac{V^2}{2g}\right]_2^1}{\bar{S}_f - S_o}$$

$$= \frac{\left(1.00 + \frac{2.000^2}{2 \times 9.81}\right) - \left(0.80 + \frac{2.679^2}{2 \times 9.81}\right)}{\left(\frac{0.00143 + 0.00325}{2}\right) - 0.0009}$$

$$= 26.4 \text{ m}$$

Hence the depth in the channel increases to 1.00 m at a location that is approximately 26.4 m upstream of the section where the depth is 0.80 m. ■

4.3.4.3 Standard-Step Method

In applying the standard-step method, the flow depth is known at one section and the objective is to find the flow depth at a second section a given distance away. The standard-step method is similar to the direct-step method in being based on the solution of Equation 4.125; in the standard-step method, however, ΔL is given and the flow conditions at the second section are unknown.

■ Example 4.10

Water flows in an open channel whose slope is 0.04%. The Manning roughness coefficient of the channel lining is estimated to be 0.035, and the flowrate is 200 m^3/s. At a given section of the channel, the cross-section is trapezoidal, with a bottom width of 10 m, side slopes of 2:1 (H:V) and a depth of flow of 7 m. Use the standard-step method to calculate the depth of flow 100 m upstream from this section, where the cross-section is trapezoidal, with a bottom width of 15 m and side slopes of 3:1 (H:V).

Solution

The channel along with the upstream and downstream channel sections is illustrated in Figure 4.12. In this case, $Q = 200$ m^3/s, $n = 0.035$, and the flow conditions at Section 2 (downstream section) are given as $y_2 = 7$ m, $b_2 = 10$ m, $m_2 = 2$ (side slope is m_2:1). Therefore

$$A_2 = [b_2 + m_2 y_2]y_2 = [10 + (2)(7)](7) = 168 \text{ m}^2$$

$$P_2 = b_2 + 2y_2\sqrt{1 + m_2^2} = 10 + 2(7)\sqrt{1 + 2^2} = 41.3 \text{ m}$$

$$R_2 = \frac{A_2}{P_2} = \frac{168}{41.3} = 4.07 \text{ m}$$

$$V_2 = \frac{Q}{A_2} = \frac{200}{168} = 1.19 \text{ m/s}$$

$$S_{f2} = \left[\frac{nQ}{A_2 R_2^{\frac{2}{3}}}\right]^2 = \left[\frac{(0.035)(200)}{(168)(4.07)^{\frac{2}{3}}}\right]^2 = 0.000267$$

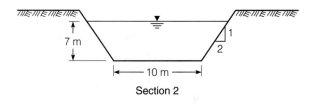

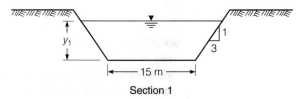

Figure 4.12 ■ Calculation of Water-Surface Profile

Assume $y_1 = 6.90$ m, then

$$A_1 = [b_1 + m_1 y_1] y_1 = [(15) + (3)(6.90)](6.90) = 246 \text{ m}^2$$

$$P_1 = b_1 + 2 y_1 \sqrt{1 + m_1^2} = 15 + 2(6.90)\sqrt{1 + 3^2} = 58.6 \text{ m}$$

$$R_1 = \frac{A_1}{P_1} = \frac{246}{58.6} = 4.20 \text{ m}$$

$$V_1 = \frac{Q}{A_1} = \frac{200}{246} = 0.812 \text{ m/s}$$

$$S_{f1} = \left[\frac{nQ}{A_1 R_1^{\frac{2}{3}}} \right]^2 = \left[\frac{(0.035)(200)}{(246)(4.20)^{\frac{2}{3}}} \right]^2 = 0.000119$$

The average slope between Sections 1 and 2, $\bar{S}_f$, is therefore given by

$$\bar{S}_f = \frac{S_1 + S_2}{2} = \frac{0.000119 + 0.000267}{2} = 0.000193$$

Substituting the flow parameters into the energy equation, Equation 4.125, and taking the velocity coefficients, α_1 and α_2, to be equal to unity yields

$$\Delta L = \frac{\left[y + \alpha \frac{V^2}{2g} \right]_2^1}{\bar{S}_f - S_o}$$

$$= \frac{\left[6.90 + \frac{0.812^2}{2(9.81)}\right] - \left[7.00 + \frac{1.19^2}{2(9.81)}\right]}{0.000193 - 0.0004}$$

$$= 670 \text{ m}$$

Since we are interested in finding the depth 100 m upstream of Section 2, we want to find the value of y_1 that yields $\Delta L = 100$ m, and repeated trials with the energy equation are necessary. These trials are easily implemented and lead to

$$y_1 = 7.02 \text{ m}$$

Therefore, the depth of the flow 100 m upstream is 7.02 m. ■

The computation of a water-surface profile generally begins at a section where the depth of flow is known. In most cases, the depth of flow is known at a section where there is a unique relationship between the depth and the flowrate. Such sections are called *control sections*. A typical control section requires that critical flow conditions occur at that section, in which case the depth of flow and the flowrate are related by Equation 4.85. Examples of control sections include free overfalls on mild channels, where the critical depth of flow occurs at a distance of three to four times the critical depth upstream of the brink of the overfall (Henderson, 1966), and channel constrictions that choke the flow to create critical conditions at the control section.

4.4 Hydraulic Structures

Hydraulic structures are used to regulate, measure, and/or transport water in open channels. These structures are called *control structures* when there is a fixed relationship between the water-surface elevation upstream or downstream of the structure and the flowrate through the structure. Hydraulic structures can be grouped into three categories: (1) flow measuring structures, such as weirs; (2) regulation structures, such as gates; and (3) discharge structures, such as culverts. The performance of typical structures in each of these categories are discussed in the following sections.

4.4.1 ■ Weirs

Weirs are elevated structures in open channels that are used to measure flow and/or control outflow elevations from basins and channels. There are two types of weirs in common use: (1) sharp-crested weirs, and (2) broad-crested weirs.

4.4.1.1 Sharp-Crested Weirs

Sharp-crested, or *thin-plate*, weirs consist of a plastic or metal plate that is set vertically and across the width of the channel. The main types of sharp-crested weirs are rectangular and V-notch weirs. In *suppressed* (uncontracted) rectangular weirs, the rectangular opening spans the entire width of the channel; in *unsuppressed* (contracted) weirs, the rectangular opening spans only a portion of the channel. Both suppressed and unsuppressed rectangular weirs are illustrated in Figure 4.13. An elevation view of the flow over a sharp-crested weir is illustrated in Figure 4.14, where at Section 1, just upstream of the weir, the flow is approximately horizontal, the pressure distribution is approximately hydrostatic, and the head (or energy per unit weight) of the fluid, E_1, is given by

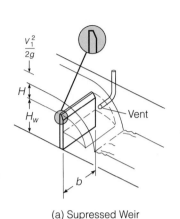
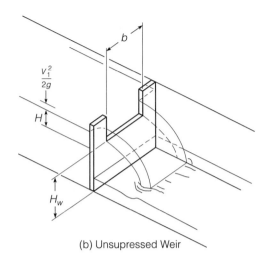

Figure 4.13 ■ Rectangular Weirs

Source: Adapted from Chadwick and Morfett, 1993.

(a) Supressed Weir   (b) Unsupressed Weir

$$E_1 = H + \frac{V_1^2}{2g} \tag{4.130}$$

where H is the elevation of the water surface and V_1 is the average velocity at Section 1. At Section 2, over the crest of the weir, the flow can be approximated as horizontal, with a head, E_2, at any elevation, z_2, given by

$$E_2 = \frac{p_2}{\gamma} + \frac{v_2^2}{2g} + z_2 \tag{4.131}$$

where p_2/γ and v_2 are the pressure head and velocity respectively at elevation z_2. The fluid pressure in Section 2 is equal to atmospheric pressure both at the top and bottom water surfaces, increasing above atmospheric pressure in between the two water surfaces. The bottom water surface is where the water springs clear of the crest at Section 2. If the flow depth at Section 2 is not too deep, then the pressure may be assumed equal to atmospheric pressure throughout the depth and E_2 becomes

$$E_2 = \frac{v_2^2}{2g} + z_2 \tag{4.132}$$

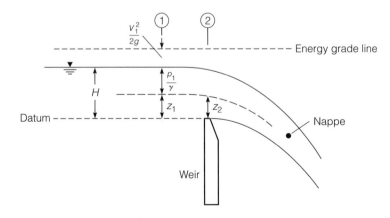

Figure 4.14 ■ Flow over a Sharp-Crested Weir

Assuming that the head loss is negligible along a streamline crossing Section 1, at elevaton z_1, and leaving Section 2 at elevation z_2, then

$$E_1 = E_2 \qquad (4.133)$$

Combining Equations 4.130, 4.132, and 4.133 leads to

$$v_2 = \left[2g \left(H - z_2 + \frac{V_1^2}{2g} \right) \right]^{\frac{1}{2}} \qquad (4.134)$$

The estimated flowrate, $\hat{Q}$, across Section 2 can be calculated by integrating the flowrates across elements of area $b\, dz_2$, where b is the width of the rectangular weir. Therefore

$$
\begin{aligned}
\hat{Q} &= \int_0^H v_2 b\, dz_2 \\
&= \int_0^H \left[2g \left(H - z_2 + \frac{V_1^2}{2g} \right) \right]^{\frac{1}{2}} b\, dz_2 \\
&= \frac{2}{3} b \sqrt{2g} \left[\left(H + \frac{V_1^2}{2g} \right)^{\frac{3}{2}} - \left(\frac{V_1^2}{2g} \right)^{\frac{3}{2}} \right]
\end{aligned}
\qquad (4.135)
$$

This equation assumes that the elevation of the water surface at Section 2 is equal to the elevation of the water surface at Section 1. This condition is physically impossible but does not necessarily lead to a significant error in the estimated flowrate at Section 2. If $V_1^2/2g$ is negligible compared with H, then Equation 4.135 reduces to

$$\hat{Q} = \frac{2}{3} \sqrt{2g} b H^{\frac{3}{2}} \qquad (4.136)$$

This expression gives the flowrate, $\hat{Q}$, over the weir in terms of measurable quantities, b and H, and illustrates why the weir is a hydraulic structure that is used for measuring flowrates in open channels based on the upstream water stage, H.

The weir equation given by Equation 4.135, and approximated by Equation 4.136, was derived with the following theoretical discrepancies: (1) the pressure distribution in the water over the crest of the weir is not uniformly atmospheric; (2) the water surface does not remain horizontal as the water approaches the weir; and (3) viscous effects that cause a nonuniform velocity and a loss of energy between Sections 1 and 2 have been neglected. The error in the flowrate resulting from these theoretical discrepancies are handled by a *discharge coefficient*, C_d, defined by the relation

$$C_d = \frac{Q}{\hat{Q}} \qquad (4.137)$$

where Q is the actual flowrate over the weir. Combining Equations 4.136 and 4.137 leads to the following expression for the flowrate over a weir in terms of the discharge coefficient,

$$\boxed{Q = \frac{2}{3} C_d \sqrt{2g} b H^{\frac{3}{2}}} \qquad (4.138)$$

It can be shown by dimensional analysis that (Franzini and Finnemore, 1997)

$$C_d = f\left(\text{Re, We, } \frac{H}{H_w}\right) \tag{4.139}$$

where Re is a Reynolds number, We is a Weber number and H_w is the height of the crest of the weir above the bottom of the channel. Experiments have shown that H/H_w is the most important variable affecting C_d, with We only important at low heads; Re is usually sufficiently high that viscous effects can be neglected. An empirical formula for C_d is (Rouse, 1946; Blevins, 1984)

$$C_d = 0.611 + 0.075\frac{H}{H_w} \tag{4.140}$$

which is valid for $H/H_w < 5$, and is approximate up to $H/H_w = 10$. For $H/H_w > 15$, the discharge can be computed from the critical flow equation by assuming $y_c = H$ (Chaudhry, 1993). It is convenient to express the discharge formula, Equation 4.138, as

$$Q = C_w b H^{\frac{3}{2}} \tag{4.141}$$

where C_w is called the *weir coefficient* and is related to the discharge coefficient by

$$C_w = \frac{2}{3}C_d\sqrt{2g} \tag{4.142}$$

Taking $C_d = 0.62$ in Equation 4.142 yields $C_w = 1.83$, and Equation 4.141 becomes

$$Q = 1.83bH^{\frac{3}{2}} \tag{4.143}$$

which gives good results if $H/H_w < 0.4$, which is within the usual operating range (Franzini and Finnemore, 1997).

Equation 4.143 is applicable in SI units, where Q is in m³/s, and b and H are in meters. According to the Ackers and colleagues (1978), the accuracy of a discharge formula depends significantly on the location of the gaging station for measuring the upstream head, H, and it is recommended that measurements of H be taken between $4H$ and $5H$ upstream of the weir. The jet of water that flows over the weir is commonly referred to as the *nappe*. The behavior of uncontracted weirs is complicated by the fact that air is trapped beneath the nappe, which tends to be entrained into the jet, thereby reducing the air pressure beneath the nappe and drawing the nappe toward the face of the weir. To avoid this effect, a vent is sometimes placed beneath the weir to maintain atmospheric pressure (see Figure 4.13a).

In the case of unsuppressed (contracted) weirs, the air beneath the nappe is in contact with the atmosphere and venting is not necessary. Experiments have shown that the effect of side contractions is to reduce the effective width of the nappe by $0.1H$ and that flowrate over the weir, Q, can be estimated by

$$Q = C_w(b - 0.1nH)H^{\frac{3}{2}} \tag{4.144}$$

where C_w is the weir coefficient calculated using Equations 4.140 and 4.142, b is the width of the contracted weir, and n is the number of sides of the weir that are contracted, usually equal to 2. Equation 4.144 gives acceptable results as long as $b > 3H$. A type of contracted weir that is related to the rectangular sharp-crested weir is the *Cipolletti* weir, which has a trapezoidal cross-section with side slopes 1:4 (H:V) and is illustrated in Figure 4.15. The advantage of using a Cipolletti weir is that corrections for end contractions are not necessary. The discharge formula can be written simply as

$$Q = C_w b H^{\frac{3}{2}} \qquad (4.145)$$

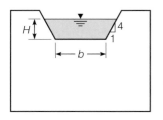

Figure 4.15 ■
Cipolletti Weir

where b is the bottom-width of the Cipolletti weir. The minimum head on standard rectangular and Cipolletti weirs is 6 mm (0.2 ft), and at heads less than 6 mm (0.2 ft) the nappe does not spring free of the crest (Aisenbrey et al., 1974).

The sharp-crested weir is also a *control structure*, since the flowrate is determined by the weir geometry and the stage just upstream of the weir. The control relationship assumes that the water downstream of the weir, called the *tailwater*, does not interfere with the operation of the weir. If the tailwater depth rises above the crest of the weir, then the flowrate becomes influenced by the downstream flow conditions and the weir no longer operates as a control structure. Under these conditions, the weir is *submerged*. The discharge over a submerged weir, Q_s, can be estimated in terms of the upstream and downstream heads on the weir using Villemonte's formula (Villemonte, 1947),

$$\frac{Q_s}{Q} = \left[1 - \left(\frac{y_d}{H} \right)^{\frac{3}{2}} \right]^{0.385} \qquad (4.146)$$

where Q is the calculated flowrate assuming the weir is not submerged, y_d is the head downstream of the weir, and H is the head upstream of the weir. The head downstream of the weir, y_d, is approximately equal to the difference between the downstream water surface elevation and the crest of the weir. Consistent with these definitions, when $y_d = 0$ Equation 4.146 gives $Q_s = Q$. In using Equation 4.146 it is recommended that H be measured at least $2.5H$ upstream of the weir and that y_d be measured beyond the turbulence caused by the nappe (Brater et al., 1996). Weirs should be designed to discharge freely rather than submerged because of greater measurement accuracy.

■ Example 4.11

A weir is to be installed to measure flows in the range of 0.5–1.0 m³/s. If the maximum (total) depth of water that can be accommodated at the weir is 1 m and the width of the channel is 4 m, determine the height of a suppressed weir that should be used to measure the flowrate.

Solution
The flow over the weir is illustrated in Figure 4.16, where the height of the weir is H_w and the flowrate is Q. The height of the water over the crest of the weir, H, is given by

$$H = 1 - H_w$$

Assuming that $H/H_w < 0.4$, then Q is related to H by Equation 4.143, where

$$Q = 1.83 b H^{\frac{3}{2}}$$

Taking $b = 4$ m, and $Q = 1$ m³/s (the maximum flowrate will give the maximum head, H), then

$$H = \left[\frac{Q}{1.83b} \right]^{\frac{2}{3}} = \left[\frac{1}{1.83(4)} \right]^{\frac{2}{3}} = 0.265 \text{ m}$$

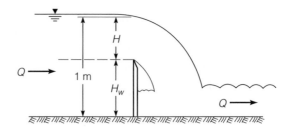

Figure 4.16 ▪
Weir Flow

The height of the weir, H_w, is therefore given by

$$H_w = 1 - 0.265 = 0.735 \text{ m}$$

and

$$\frac{H}{H_w} = \frac{0.265}{0.735} = 0.36$$

The initial assumption that $H/H_w < 0.4$ is therefore validated, and the height of the weir should be 0.735 m.   ▪

V-Notch Weirs.   A *V-notch weir* is a sharp-crested weir that has a V-shaped opening instead of a rectangular-shaped opening. These weirs, also called *triangular weirs*, are typically used instead of rectangular weirs under low-flow conditions, where rectangular weirs tend to be less accurate. V-notch weirs are usually limited to flows of 0.28 m³/s (10 cfs) or less. The basic theory of V-notch weirs is the same as for rectangular weirs, where the theoretical flowrate over the weir, $\hat{Q}$, is given by

$$\hat{Q} = \int_0^H v_2 b \, dz_2$$

$$= \int_0^H \left[2g \left(H - z_2 + \frac{V_1^2}{2g} \right) \right]^{\frac{1}{2}} b \, dz_2 \tag{4.147}$$

where b is the width of the V-notch weir at elevation z_2 and is given by

$$b = 2z_2 \tan \left(\frac{\theta}{2} \right) \tag{4.148}$$

where θ is the angle at the apex of the V-notch, as illustrated in Figure 4.17. Combining Equations 4.147 and 4.148 leads to

$$\hat{Q} = \int_0^H \left[2g \left(H - z_2 + \frac{V_1^2}{2g} \right) \right]^{\frac{1}{2}} 2z_2 \tan \left(\frac{\theta}{2} \right) dz_2 \tag{4.149}$$

The approach velocity, V_1, is usually negligible for the low velocities that are typically handled by V-notch weirs, and therefore Equation 4.149 can be approximated by

$$\hat{Q} = \int_0^H \left[2g(H - z_2) \right]^{\frac{1}{2}} 2z_2 \tan \left(\frac{\theta}{2} \right) dz_2$$

$$= \frac{8}{15} \sqrt{2g} \tan \left(\frac{\theta}{2} \right) H^{\frac{5}{2}} \tag{4.150}$$

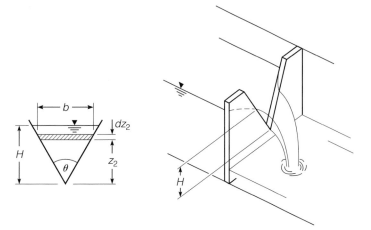

Figure 4.17 ■
V-Notch Weir

Source: Chadwick, Andrew and Morfett, John. *Hydraulics in Civil and Environmental Engineering*, 2e, Figure 13.4, p. 405. Copyright © 1993 by Kluwer Academic Publishers. Reprinted by permission.

As in the case of a rectangular weir, the theoretical discharge given by Equation 4.150 is corrected by a discharge coefficient, C_d, to account for discrepancies in the assumptions leading to Equation 4.150. The actual flowrate, Q, over a V-notch weir is therefore given by

$$Q = \frac{8}{15} C_d \sqrt{2g} \tan\left(\frac{\theta}{2}\right) H^{\frac{5}{2}} \qquad (4.151)$$

where C_d generally depends on Re, We, θ, and H.

The vertex angles used in V-notch weirs are usually between 10° and 90°. Values of C_d for a variety of notch angles, θ, and heads, H, are plotted in Figure 4.18. The

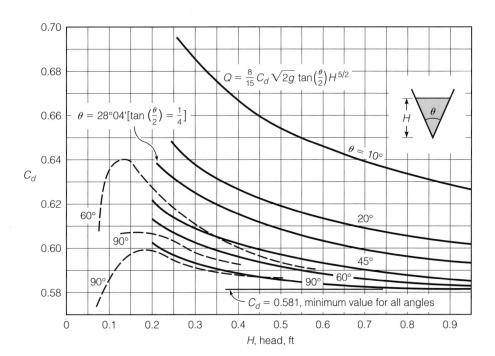

Figure 4.18 ■ Discharge Coefficient in V-Notch Weirs

Source: Franzini, Joseph and Finnemore, E. John. *Fluid Mechanics*, 9e. Copyright © 1997 by The McGraw-Hill Companies. Reprinted by permission.

minimum discharge coefficient corresponds to a notch angle of 90°, and the minimum value of C_d for all angles is 0.581. According to Potter and Wiggert (1991) and White (1994), using $C_d = 0.58$ for engineering calculations is usually acceptable, provided that $20° < \theta < 100°$ and $H > 50$ mm (2 in.). For $H < 50$ mm, both viscous and surface-tension effects may be important and a recommended value of C_d is given by (White, 1994)

$$C_d = 0.583 + \frac{1.19}{(\text{ReWe})^{\frac{1}{6}}} \tag{4.152}$$

where Re is the Reynolds number defined by

$$\text{Re} = \frac{g^{\frac{1}{2}}H^{\frac{1}{2}}}{\nu} \tag{4.153}$$

and We is the Weber number defined by

$$\text{We} = \frac{\rho g H^2}{\sigma} \tag{4.154}$$

where ν is the kinematic viscosity and σ is the surface tension of water. The minimum head on a V-notch weir should be greater than 6 mm (0.2 in.).

■ Example 4.12

A V-notch weir is to be used to measure channel flows in the range 0.1 to 0.2 m³/s. What is the maximum head of water on the weir for a vertex angle of 45°?

Solution

The maximum head of water results from the maximum flow, so $Q = 0.2$ m³/s will be used to calculate the maximum head. The relationship between the head and flowrate is given by Equation 4.151, which can be put in the form

$$H = \left[\frac{15Q}{8C_d\sqrt{2g}\tan(\theta/2)} \right]^{\frac{2}{5}} = \left[\frac{15(0.2)}{8C_d\sqrt{2(9.81)}\tan(45°/2)} \right]^{\frac{2}{5}} = \frac{0.530}{C_d^{2/5}} \text{ m}$$

or

$$H = \frac{1.74}{C_d^{2/5}} \text{ ft}$$

The discharge coefficient as a function of H for $\theta = 45°$ is given in Figure 4.18, and some iteration is necessary to find H. These iterations are summarized in the following table:

Assumed H (ft)	C_d (Fig. 4.18)	Calculated H (ft)
0.40	0.6	2.13
2.13	0.581	2.16
2.16	0.581	2.16

Therefore, the maximum depth expected at the V-notch weir is 2.16 ft = 0.66 m.   ■

In cases where the tailwater depth rises above the crest of the weir, the flowrate is influenced by downstream flow conditions. Under this submerged condition, the discharge over the weir, Q_s, can be estimated in terms of the upstream and downstream heads on the weir using Villemonte's formula (Equation 4.146), with the exceptions that the exponent of y_d/H is taken as 5/2 instead of 3/2, and the unsubmerged discharge, Q, is calculated using Equation 4.151 (Brater et al., 1996).

4.4.1.2   Broad-Crested Weirs

Broad-crested weirs, also called *long-based weirs*, have crest lengths that are significantly longer than sharp-crested weirs. These weirs are usually constructed of concrete, have rounded edges, and are capable of handling much larger discharges than sharp-crested weirs. There are several different designs of broad-crested weirs, of which the rectangular (broad-crested) weir can be considered representative.

Rectangular (Broad-Crested) Weirs.   A typical rectangular weir is illustrated in Figure 4.19. These weirs operate on the theory that the elevation of the weir above the channel bottom is sufficient to create critical flow conditions over the weir. Under these circumstances, the estimated flowrate over the weir, $\hat{Q}$, is given by

$$\hat{Q} = y_c b V_c \tag{4.155}$$

where y_c is the critical depth of flow over the weir, b is the width of the weir, and V_c is the velocity at critical flow. If E_1 is the specific energy of the flow at Section 1 just upstream of the weir, and energy losses between this upstream section and the critical flow section over the weir are negligible, then the energy equation requires that

$$E_1 = H_w + y_c + \frac{V_c^2}{2g} \tag{4.156}$$

where H_w is the height of the weir above the upstream channel. Under critical flow conditions, the Froude number is equal to 1, hence

$$\frac{V_c}{\sqrt{gy_c}} = 1 \tag{4.157}$$

Combining Equations 4.156 and 4.157 yields the following expression for y_c

$$y_c = \frac{2}{3}(E_1 - H_w) = \frac{2}{3}H \tag{4.158}$$

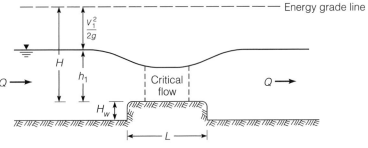

Figure 4.19 ■ Flow over a Rectangular Broad-Crested Weir

Longitudinal section view

where H is the energy of the upstream flow measured relative to the weir elevation. The upstream energy, H, can be written as

$$H = h_1 + \frac{V_1^2}{2g} \tag{4.159}$$

where h_1 is the elevation of the upstream water surface above the weir and V_1 is the average velocity of flow upstream of the weir. The depth h_1 should be measured at least $2.5h_1$ upstream of the weir (Brater et al., 1996). Combining Equations 4.155, 4.157, and 4.158 leads to the following estimate of the flowrate over a rectangular weir

$$\hat{Q} = \sqrt{g}b\left(\frac{2}{3}H\right)^{\frac{3}{2}} \tag{4.160}$$

In reality, energy losses over the weir are not negligible and the estimated flowrate, $\hat{Q}$, must be corrected to account for these losses. The correction factor is the discharge coefficient, C_d, and the actual flowrate, Q, over the weir is given by

$$\boxed{Q = C_d\sqrt{g}b\left(\frac{2}{3}H\right)^{\frac{3}{2}}} \tag{4.161}$$

where values of C_d can be estimated using the relation (Chow, 1959)

$$\boxed{C_d = \frac{0.65}{(1 + H/H_w)^{\frac{1}{2}}}} \tag{4.162}$$

To ensure proper operation of a broad-crested weir, flow conditions are restricted to the operating range $0.08 < h_1/L < 0.50$. For $h_1/L < 0.08$ head losses across the weir cannot be neglected, while for $h_1/L > 0.50$ the streamlines over the weir block are not horizontal (Munson et al., 1994). A broad-crested weir can be assumed to discharge freely if the tailwater level is lower than $0.8H$ above the crest of the weir (Henderson, 1966).

■ Example 4.13

A 20-cm high broad-crested weir is placed in a 2-m wide channel. Estimate the flowrate in the channel if the depth of water upstream of the weir is 50 cm.

Solution

Upstream of the weir, $h_1 = 0.5\text{m} - 0.2\text{m} = 0.30$ m, and

$$H = h_1 + \frac{V_1^2}{2g} = h_1 + \frac{Q^2}{2gA_1^2} = 0.30 + \frac{Q^2}{2(9.81)(0.5 \times 2)^2} = 0.30 + 0.0510Q^2$$

The discharge coefficient, C_d, is given by

$$C_d = \frac{0.65}{(1 + H/H_w)^{\frac{1}{2}}} = \frac{0.65}{[1 + (0.30 + 0.0510Q^2)/0.2]^{\frac{1}{2}}} = \frac{0.65}{[2.5 + 0.255Q^2]^{\frac{1}{2}}}$$

where H_w has been taken as 0.2 m. The discharge over the weir is therefore given by

$$Q = C_d \sqrt{g} b \left(\frac{2}{3}H\right)^{\frac{3}{2}} = \frac{0.65}{[2.5 + 0.255Q^2]^{\frac{1}{2}}} \sqrt{9.81}(2) \left[\frac{2}{3}(0.30 + 0.0510Q^2)\right]^{\frac{3}{2}}$$

$$= 2.22 \left[\frac{(0.30 + 0.0510Q^2)^3}{2.5 + 0.255Q^2}\right]^{\frac{1}{2}}$$

Solving iteratively gives

$$Q = 0.23 \text{ m}^3/\text{s}$$

This solution assumes that the length of the weir is such that $0.08 < h_1/L < 0.5$. ■

4.4.2 ■ Parshall Flume

Although weirs are the simplest structures for measuring the discharge in open channels, the high head losses caused by weirs and the tendency for suspended particles to accumulate behind weirs may be important limitations. The Parshall flume, named after Ralph L. Parshall, provides a convenient alternative to the weir for measuring flowrates in open channels where high head losses and sediment accumulation are of concern. Such cases include flow measurement in wastewater treatment plants and irrigation channels. The Parshall flume, illustrated in Figure 4.20, consists of a converging section that causes critical flow conditions, followed by a steep throat section that provides for a transition to supercritical flow. The unique relationship between the depth of flow and the flowrate under critical flow conditions is the basic principle on which the Parshall flume operates. The transition from supercritical flow to subcritical flow at the exit of the flume usually occurs via a hydraulic jump, but under high tailwater conditions the jump is sometimes submerged.

Within the flume structure, water depths are measured at two locations, one in the converging section, H_a, and the other in the throat section, H_b. The flow depth in the throat section is measured relative to the bottom of the converging section, as illustrated in Figure 4.21. If the hydraulic jump at the exit of the Parshall flume is not submerged, then the discharge through the flume is related to the measured flow depth in the converging section, H_a, by the empirical discharge relations given in Table 4.6, where Q is the discharge in ft^3/s (cfs), W is the width of the throat in ft, and H_a is measured in ft.

Submergence of the hydraulic jump is determined by the ratio of the flow depth in the throat, H_b, to the flow depth in the converging section, H_a, and critical values for the ratio H_b/H_a are given in Table 4.7. Whenever H_b/H_a exceeds the critical values given in Table 4.7, the hydraulic jump is submerged and the discharge is reduced from the values given by the equations in Table 4.6. Corrections to the theoretical flowrates as a function of H_a and the percentage of submergence, H_b/H_a, are given in Figure 4.22 for a throat width of 1 ft and in Figure 4.23 for a throat width of 10 ft. Flow corrections for the 1-ft flume are applied to larger flumes by multiplying the correction for the 1-ft flume by a factor corresponding to the flume size given in Table 4.8. Similarly, flow corrections for the 10-ft flume are applied to larger flumes by multiplying the correction for the 10-ft flume by a factor corresponding to the flume size given in Table 4.9. According to Aisenbrey and colleagues (1974), correct zeroing and reading of the gages is necessary for accurate results, which are usually within 2% for free flows and 5% for submerged flows. Parshall flumes do not reliably measure flowrates when the submergence ratio, H_b/H_a, exceeds 0.95.

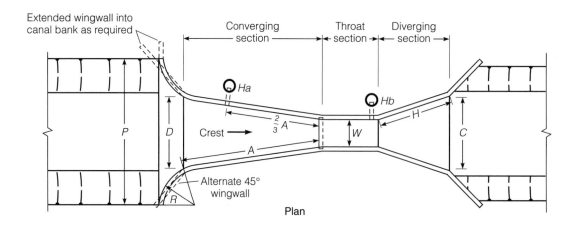

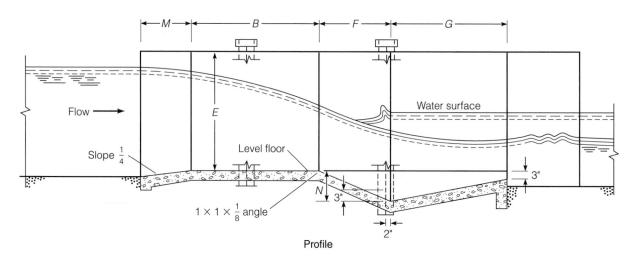

Figure 4.20 ■ Parshall Flume

Parshall flumes can be constructed of concrete, wood, fiberglass, galvanized metal, or any other construction material that can be built to specified dimensions in the field or prefabricated in a shop. As a general rule, the width of the Parshall flume should be about one-third to one-half of the top width of the water surface in the upstream channel at the design discharge and at normal depth (Aisenbrey et al., 1974).

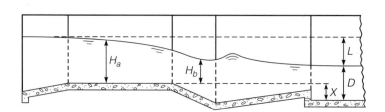

Figure 4.21 ■ Measured Water Depths in Parshall Flume

■ Table 4.6
Parshall Flume
Discharge Equations

Throat width	Equation	Free-flow capacity (cfs)
3 in.	$Q = 0.992H_a^{1.547}$	0.03–1.9
6 in.	$Q = 2.06H_a^{1.58}$	0.05–3.9
9 in.	$Q = 3.07H_a^{1.53}$	0.09–8.9
1 ft to 8 ft	$Q = 4WH_a^{1.522W^{0.026}}$	up to 140
10 ft to 50 ft	$Q = (3.6875W + 2.5)H_a^{1.6}$	up to 2,000

■ Table 4.7
Submergence Criteria in
Parshall Flumes

Throat width	$(H_b/H_a)_{\text{crit}}$
1 in., 2 in., 3 in.	0.5
6 in., 9 in.	0.6
1 ft to 8 ft	0.7
8 ft to 50 ft	0.8

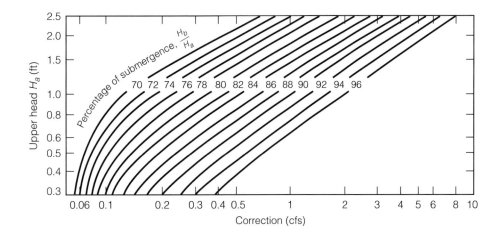

Figure 4.22 ■ Parshall Flume Correction for Submerged Flow (throat width = 1 ft)

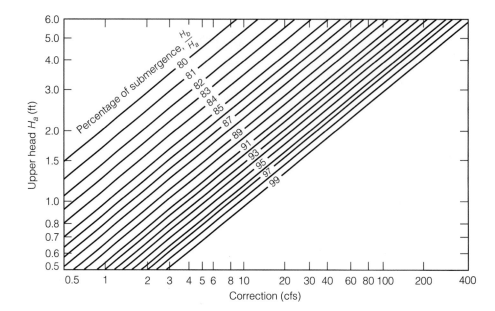

Figure 4.23 ▪ Parshall Flume Correction for Submerged Flow (throat width = 10 ft)

■ **Table 4.8**
Correction Factors for 1-ft Parshall Flumes

Throat width (ft)	Correction factor
1	1.0
1.5	1.4
2	1.8
3	2.4
4	3.1
6	4.3
8	5.4

■ **Table 4.9**
Correction Factors for 10-ft Parshall Flumes

Throat width (ft)	Correction factor
10	1.0
12	1.2
15	1.5
20	2.0
25	2.5
30	3.0
40	4.0
50	5.0

■ **Example 4.14**

Flow is being measured by a Parshall flume that has a throat width of 2 ft. Determine the flowrate through the flume when the water depth in the converging section is 2.00 ft and the depth in the throat section is 1.70 ft.

Solution

From the given data: $W = 2$ ft, $H_a = 2$ ft, and $H_b = 1.7$ ft. According to Table 4.6, the flowrate, Q, is given by

$$Q = 4WH_a^{1.522W^{0.026}} = 4(2)(2)^{1.522 \times 2^{0.026}} = 23.4 \text{ cfs}$$

In this case,

$$\frac{H_b}{H_a} = \frac{1.7}{2} = 0.85$$

Therefore, according to Table 4.7, the flow is submerged. Figure 4.22 gives the flowrate correction for a 1-ft flume as 2 cfs, and Table 4.8 gives the correction factor for a 2-ft flume as 1.8. The flowrate correction, ΔQ, for a 2-ft flume is therefore given by

$$\Delta Q = 2 \times 1.8 = 3.6 \text{ cfs}$$

and the flowrate through the Parshall flume is $Q - \Delta Q$, where

$$Q - \Delta Q = 23.4 - 3.6 = 19.8 \text{ cfs}$$

The flowrate is 19.8 cfs.   ■

4.4.3 ■ Gates

Gates are used to regulate the flow in open channels. They are designed for either overflow or underflow operation, with overflow operation appropriate for channels in which there is a significant amount of floating debris. Two common types of gates are vertical gates and radial (Tainter) gates, which are illustrated in Figure 4.24. Vertical gates are supported by vertical guides with roller wheels, and large hydrostatic forces usually induce significant frictional resistance to raising and lowering the gates. A radial (Tainter) gate consists of an arc-shaped face plate supported by radial arms that are attached to a central horizontal shaft that transmits the hydrostatic force to the supporting structure. Since the vector of the resultant hydrostatic force passes through the axis of the horizontal shaft, only the weight of the gate needs to be lifted to open the gate. Tainter gates are economical to install and are widely used in both underflow and overflow applications. Structural design guidelines for several types of gates can be found in Sehgal (1996).

Figure 4.24 ■
Types of Gates

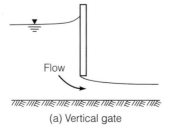

(a) Vertical gate

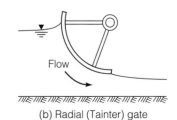

(b) Radial (Tainter) gate

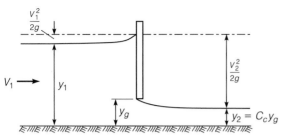

(a) Sectional Elevation Through Vertical Sluice Gate

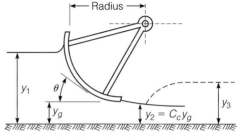

(b) Sectional Elevation Through Radial Gate

Figure 4.25 ■ Flow Through Gates
Source: Chadwick, Andrew and Morfett, John. *Hydraulics in Civil and Environmental Engineering*, 2e, Figure 13.18, p. 433. Copyright © 1993 by Kluwer Academic Publishers. Reprinted by permission.

Applying the energy equation to both vertical and Tainter gates, Figure 4.25, yields

$$y_1 + \frac{V_1^2}{2g} = y_2 + \frac{V_2^2}{2g} \tag{4.163}$$

where Sections 1 and 2 are upstream and downstream of the gate, respectively, and energy losses are neglected. Writing Equation 4.163 in terms of the flowrate, Q, leads to

$$y_1 + \frac{Q^2}{2gb^2 y_1^2} = y_2 + \frac{Q^2}{2gb^2 y_2^2}$$

and solving for Q gives

$$Q = by_1 y_2 \sqrt{\frac{2g}{y_1 + y_2}} \tag{4.164}$$

The depth of flow downstream of the gate, y_2, is less than the gate opening, y_g, since the streamlines of the flow contract as they move past the gate (see Figure 4.25). Denoting the ratio of the downstream depth, y_2, to the gate opening, y_g, by the *coefficient of contraction*, C_c, where

$$C_c = \frac{y_2}{y_g} \tag{4.165}$$

then Equations 4.164 and 4.165 can be combined to yield the following expression for the discharge through a gate

$$\boxed{Q = C_d b y_g \sqrt{2gy_1}} \tag{4.166}$$

where C_d is the *discharge coefficient* or *sluice coefficient* given by

$$\boxed{C_d = \frac{C_c}{\sqrt{1 + C_c \frac{y_g}{y_1}}}} \tag{4.167}$$

The form of the discharge equation given by Equation 4.166 expresses the discharge in terms of an "orifice-flow" velocity, $\sqrt{2gy_1}$, times the flow area through the gate, by_g, times a discharge coefficient, C_d, to account for deviations from the orifice-flow

Figure 4.26 ■ Submerged Flow Through Gates

Source: Chadwick, Andrew and Morfett, John. *Hydraulics in Civil and Environmental Engineering*, 2e, Figure 13.19, p. 434. Copyright © 1993 by Kluwer Academic Publishers. Reprinted by permission.

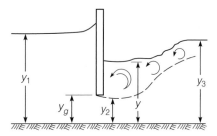

assumption. On the basis of Equation 4.167, the discharge coefficient depends on the amount of flow contraction as measured by C_c and y_g/y_1. In the case of a vertical sluice gate, it has been found that (Chadwick and Morfett, 1993)

$$C_c = 0.61 \tag{4.168}$$

whenever $0 < y_g/E_1 < 0.5$, where E_1 is the specific energy of the flow upstream of the gate, defined by $E_1 = y_1 + \frac{V_1^2}{2g}$. In the case of Tainter gates, the contraction coefficient, C_c, is generally greater than 0.61 and is commonly expressed as a function of the angle θ shown in Figure 4.25. It can be estimated using the relation

$$C_c = 1 - 0.75\left(\frac{\theta}{90}\right) + 0.36\left(\frac{\theta}{90}\right)^2 \tag{4.169}$$

where θ is measured in degrees. Equation 4.169 gives results that are accurate to within ± 5% provided that $\theta < 90°$ (Henderson, 1966).

In cases where the discharge through the gate opening is supercritical and the depth of flow downstream of the gate exceeds the conjugate depth of the gate opening, there is the possibility that the outflow will be submerged and the discharge equation given by Equation 4.166 will not be applicable. This condition is illustrated in Figure 4.26. An approximate analysis of the submerged flow condition assumes that all head losses occur in the flow downstream of the gate, between Sections 2 and 3, in which case the energy equation can be written as

$$y_1 + \frac{Q^2}{2gb^2y_1^2} = y + \frac{Q^2}{2gb^2y_2^2} \tag{4.170}$$

where y is the depth of flow immediately downstream of the gate. Between Sections 2 and 3, the momentum equation (Equation 4.113) can be written as

$$\frac{y^2}{2} + \frac{Q^2}{gb^2y_2} = \frac{y_3^2}{2} + \frac{Q^2}{gb^2y_3} \tag{4.171}$$

and flowrate, Q, can be estimated by simultaneous solution of Equations 4.170 and 4.171, where y_1 and y_3 are usually known and y_2 is estimated by $C_c y_g$.

■ Example 4.15

Water is ponded behind a vertical gate to a height of 4 m in a rectangular channel of width 7 m. Calculate the gate opening that will release 40 m³/s through the gate. How would this discharge be affected by a downstream flow depth of 3.5 m?

Solution

The flowrate, Q, through the gate is given by

$$Q = C_d b y_g \sqrt{2 g y_1}$$

where C_d is the discharge coefficient, b is the channel width ($= 7$ m), y_g is the gate opening, and y_1 is the depth of water behind the gate ($= 4$ m). Substituting the known parameters into the flow equation yields

$$40 = C_d (7) y_g \sqrt{(2)(9.81)(4)}$$

$$= 62.0 C_d y_g \tag{4.172}$$

The discharge coefficient, C_d, is given by

$$C_d = \frac{C_c}{\sqrt{1 + C_c \frac{y_g}{y_1}}}$$

where C_c is the contraction coefficient that can be taken as 0.61, therefore

$$C_d = \frac{0.61}{\sqrt{1 + 0.61 \frac{y_g}{4}}} = \frac{0.61}{(1 + 0.153 y_g)^{\frac{1}{2}}} \tag{4.173}$$

Combining Equations 4.172 and 4.173 yields

$$\frac{y_g}{(1 + 0.153 y_g)^{\frac{1}{2}}} = 1.06$$

which can be reduced to a quadratic equation and solved to give the following (positive) value of y_g,

$$y_g = 1.15 \text{ m}$$

Therefore, when the gate is opened 1.15 m the flowrate through the gate is expected to be 40 m³/s. This flowrate assumes that the flow through the gate is not affected by downstream conditions and that $C_c = 0.61$, which requires that $0 < y_g/E_1 < 0.5$. Since

$$E_1 = y_1 + \frac{Q^2}{2 g b^2 y_1^2} = 4 + \frac{40^2}{2(9.81)(7)^2(4)^2} = 4.10 \text{ m}$$

therefore $y_g/E_1 = 1.15/4.10 = 0.28$, which supports the assumption that $C_c = 0.61$.

Consider now the case where the downstream depth is equal to 3.5 m. In this case, there are two flow possibilities: (1) a hydraulic jump forms downstream of the gate (in which case the gate discharge is unaffected by the tailwater depth); and (2) there is a drowned hydraulic jump at the gate. For a hydraulic jump to occur downstream of the gate, the conjugate depth corresponding to a (downstream) depth of 3.5 m must be greater than the depth of flow emerging from the gate opening. The hydraulic jump equation is given by (Equation 4.115)

$$y_2 = \frac{y_1}{2} \left[-1 + \sqrt{1 + \frac{8 q^2}{g y_1^3}} \right]$$

where $y_2 = 3.5$ m and $q = 40/7 = 5.71$ m²/s. Substituting into the hydraulic jump equation gives

$$3.5 = \frac{y_1}{2}\left[-1 + \sqrt{1 + \frac{8(5.71)^2}{(9.81)y_1^3}}\right]$$

or

$$y_1\left[-1 + \sqrt{1 + 26.6y_1^{-3}}\right] = 7$$

Solving by trial and error gives

$$y_1 = 0.48 \text{ m}$$

Since the depth of flow downstream of the gate is $C_c y_g = 0.61(1.15) = 0.702$ m (which exceeds the conjugate depth of 0.48 m), then the hydraulic jump will be *drowned*.

For the case of a drowned hydraulic jump, the energy equation between the ponded water behind the gate and immediately downstream of the gate is given by

$$y_1 + \frac{Q^2}{2gb^2y_1^2} = y + \frac{Q^2}{2gb^2y_2^2}$$

where y is the depth of water immediately downstream of the gate opening, and y_2 is the height of the water jet emerging from under the gate which can be estimated by

$$y_2 = C_c y_g = (0.61)(1.15) = 0.702 \text{ m}$$

Substituting for y_2 and other known parameters in the energy equation yields

$$4 + \frac{Q^2}{2(9.81)(7)^2(4)^2} = y + \frac{Q^2}{2(9.81)(7)^2(0.702)^2}$$

which simplifies to

$$Q^2 = 488(4 - y) \tag{4.174}$$

The momentum equation between the section just downstream of the gate and the section where the water depth is 3.5 m is given by

$$\frac{y^2}{2} + \frac{Q^2}{gb^2y_2} = \frac{y_3^2}{2} + \frac{Q^2}{gb^2y_3}$$

where y_3 is the water depth of 3.5 m. Substituting the values of known parameters into the momentum equation yields

$$\frac{y^2}{2} + \frac{Q^2}{(9.81)(7)^2(0.702)} = \frac{(3.5)^2}{2} + \frac{Q^2}{(9.81)(7)^2(3.5)}$$

which simplifies to

$$Q^2 = 212(12.3 - y^2) \tag{4.175}$$

Combining Equations 4.174 and 4.175 (eliminating Q) leads to

$$y = 3.23 \text{ m}$$

Substituting this result into either Equation 4.174 or 4.175 yields

$$Q = 19.4 \text{ m}^3/\text{s}$$

Therefore, a downstream depth of 3.5 m reduces the discharge under the gate from 40 m³/s to 19.4 m³/s. ■

4.4.4 ■ Culverts

Culverts are short conduits that are used to pass water under roads and highways. Culverts perform a similar function to that of bridges, but unlike bridges they have spans less than 6 m (20 ft) and can be designed to have a submerged inlet. Typical cross-sections of culverts include circular, arched, rectangular, and oval shapes.

Culvert design typically requires the selection of a barrel cross-section that passes a given flowrate when the water is ponded to a given height at the culvert entrance. The hydraulic analysis of culverts is complicated by the fact that there are several possible flow regimes, with the governing flow equation being determined by the flow regime. The flow regimes can be broadly grouped into *submerged-entrance conditions* and *free-entrance conditions*, which are illustrated in Figures 4.27 and 4.28, respectively. The

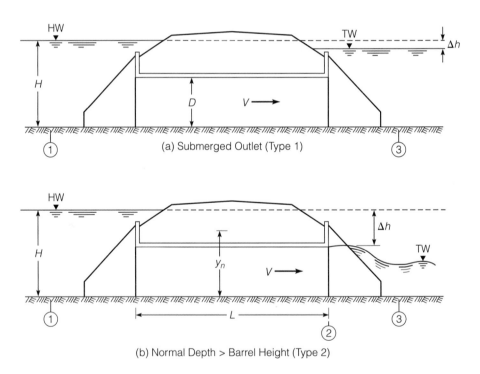

(a) Submerged Outlet (Type 1)

(b) Normal Depth > Barrel Height (Type 2)

Figure 4.27 ■ Flow Through Culvert with Submerged Entrance

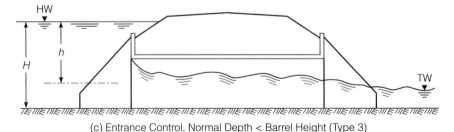

(c) Entrance Control, Normal Depth < Barrel Height (Type 3)

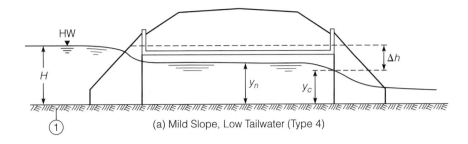

(a) Mild Slope, Low Tailwater (Type 4)

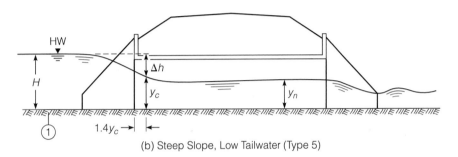

(b) Steep Slope, Low Tailwater (Type 5)

Figure 4.28 ■ Flow Through Culvert with Free Entrance

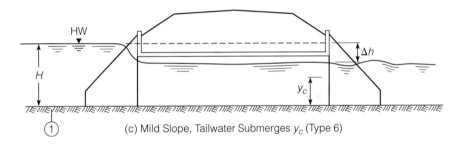

(c) Mild Slope, Tailwater Submerges y_c (Type 6)

entrance to a culvert is regarded as submerged when the depth, H, of water upstream of the culvert exceeds $1.2D$, where D is the diameter of the culvert. In some references this limit is taken as $1.5D$ (e.g., French, 1985). In the case of a submerged entrance, Figure 4.27 indicates that there are three possible flow regimes: (a) the outlet is submerged (Type 1 flow); (b) the outlet is not submerged and the normal depth of flow in the culvert is larger than the culvert diameter, D (Type 2 flow); and (c) the outlet is not submerged and the normal depth of flow in the culvert is less than the culvert diameter (Type 3 flow).

In Type 1 flow, applying the energy equation between Sections 1 (headwater) and 3 (tailwater) leads to

$$\Delta h = h_i + h_f + h_o \tag{4.176}$$

where Δh is the difference between the headwater and tailwater elevations, h_i is the entrance loss, h_f is the head loss due to friction in the culvert, and h_o is the exit loss. Equation 4.176 neglects the velocity head at Sections 1 and 3, which is usually small compared with the other terms. Using the Manning equation to calculate h_f within the culvert, then

$$h_f = \frac{n^2 V^2 L}{R^{\frac{4}{3}}} \tag{4.177}$$

where n is the roughness coefficient, V is the velocity of flow, L is the length, and R is the hydraulic radius of the culvert. The entrance and exit losses, h_i and h_o, are given by

$$h_i = k_e \frac{V^2}{2g} \tag{4.178}$$

$$h_o = \frac{V^2}{2g} \tag{4.179}$$

where k_e is the entrance loss coefficient. Combining Equations 4.176 to 4.179 yields the following form of the energy equation between Sections 1 and 3

$$\Delta h = \frac{n^2 V^2 L}{R^{\frac{4}{3}}} + k_e \frac{V^2}{2g} + \frac{V^2}{2g} \tag{4.180}$$

This equation can also be applied between Section 1 (headwater) and Section 2 (culvert exit) in Type 2 flow, illustrated in Figure 4.27(b), where the velocity head at the exit, $V^2/2g$, is equal to the exit loss in Type 1 flow. Equation 4.180 reduces to the following relationship between the difference in the water-surface elevations on both sides of the culvert, Δh, and the discharge through the culvert, Q,

$$\boxed{Q = A \sqrt{\frac{2g\Delta h}{19.62 n^2 L / R^{\frac{4}{3}} + k_e + 1}}} \tag{4.181}$$

where A is the cross-sectional area of the culvert. It should be noted that Δh is equal to the difference between the headwater and tailwater elevations for a submerged outlet (Type 1 flow), and Δh is equal to the difference between the headwater and the crown of the culvert exit when the normal depth of flow in the culvert exceeds the height of the culvert (Type 2 flow).

Under both Type 1 and Type 2 conditions, the flow is said to be under *outlet control* since the water depth at the outlet influences the discharge through the culvert. In cases where the inlet is submerged and the culvert entrance will not admit water fast enough to fill the culvert (Type 3 flow in Figure 4.27), the culvert inlet behaves like an orifice and the discharge through the culvert, Q, is related to the head on the center of the orifice, h, by the relation

$$\boxed{Q = C_d A \sqrt{2gh}} \tag{4.182}$$

where C_d is the coefficient of discharge and h is equal to the vertical distance from the center of the culvert entrance to the water surface at the entrance. The coefficient of discharge, C_d, is equal to 0.62 for a square-edged entrance and approaches 1 for a well-rounded entrance (Franzini and Finnemore, 1997). In cases where the culvert entrance acts like an orifice, the downstream water level does not influence the flow through the culvert and the flow is said to be under *inlet control*. According to ASCE (1992), Equation 4.182 is applicable only when $H/D \geq 2$, but Franzini and Finnemore (1997) state that the error in Equation 4.182 is less than 2% when $H/D \geq 1.2$.

Submerged entrances usually lead to greater flows through the culvert than unsubmerged entrances. In some cases, however, culverts must be designed so that the entrances are not submerged. Such cases include those in which the top of the culvert forms the base of a roadway. In the case of an unsubmerged entrance, Figure 4.28 indicates that there are three possible flow regimes: (a) the culvert has a mild slope and a low tailwater, in which case the critical depth occurs somewhere near the exit of the culvert (Type 4 flow); (b) the culvert has a steep slope and a low tailwater, in which case the critical depth occurs somewhere near the entrance of the culvert, at approximately $1.4y_c$ downstream from the entrance (Type 5 flow); and (c) the culvert has a mild slope and the tailwater submerges y_c (Type 6 flow).

Applying the energy equation between the headwater and the culvert exit in Type 4 flow gives

$$\Delta h + \frac{V_1^2}{2g} - \frac{V^2}{2g} = h_i + h_f \tag{4.183}$$

where Δh is the difference between the headwater elevation and the elevation of the (critical) water surface at the exit of the culvert, V_1 is the headwater velocity, h_i is the entrance loss given by Equation 4.178, and h_f is the friction loss in the culvert given by Equation 4.177. The velocity of the headwater is *not* neglected, as in the case of a ponded headwater where $H/D > 1.2$. Equation 4.183 yields the following expression for the discharge, Q, through the culvert,

$$\boxed{Q = A_c\sqrt{2g(\Delta h + V_1^2/2g - h_i - h_f)}} \tag{4.184}$$

where A_c is the flow area at the critical flow section at the exit of the culvert. Equation 4.184 is actually an implicit expression for the discharge, since the entrance loss, h_i, and the friction loss, h_f, depend on the discharge, Q.

In Type 5 flow (steep slope, low tailwater), the critical flow depth occurs at the entrance of the culvert. Applying the energy equation between the headwater and the culvert entrance gives

$$\Delta h + \frac{V_1^2}{2g} - \frac{V^2}{2g} = h_i \tag{4.185}$$

where Δh is the difference between the headwater elevation and the elevation of the (critical) water surface at the entrance of the culvert. Equation 4.185 leads to the following expression for the discharge, Q, through the culvert

$$\boxed{Q = A_c\sqrt{2g(\Delta h + V_1^2/2g - h_i)}} \tag{4.186}$$

Finally, in Type 6 flow (mild slope, tailwater submerges y_c), the water surface at the culvert exit is approximately equal to the tailwater elevation. Applying the energy equation between the headwater and the culvert exit gives

$$\Delta h + \frac{V_1^2}{2g} - \frac{V^2}{2g} = h_i + h_f \tag{4.187}$$

where Δh is the difference between the headwater elevation and the tailwater elevation at the exit of the culvert. Equation 4.187 leads to the following expression for the discharge, Q, through the culvert,

$$\boxed{Q = A\sqrt{2g(\Delta h + V_1^2/2g - h_i - h_f)}} \tag{4.188}$$

where A is the flow area at the exit of the culvert.

Determination of Flow Type.   The flow type is determined by the headwater, tailwater, and culvert dimensions. Based on the headwater elevation and culvert dimensions, it is determined whether the culvert entrance is submerged. If $H/D > 1.2$, the culvert entrance is submerged; if not, the entrance is not submerged.

Submerged Entrance.   When the culvert entrance is submerged, the flow is either Type 1, 2, or 3. The flow type and the associated flowrate are determined by the following procedure:

1. If the culvert exit is submerged, then the flow is Type 1 and the discharge is given by Equation 4.181.
2. If the culvert exit is not submerged, the flow is either Type 2 or 3. Assume that the flow is Type 2 and calculate the discharge using Equation 4.181, with the appropriate definition of Δh.
3. Use the flowrate calculated in step 2 to determine the normal depth of flow in the culvert.
4. If the normal depth of flow calculated in step 3 is greater than the height of the culvert, then the flow is Type 2 and the discharge calculated in step 2 is correct.
5. If the normal depth of flow calculated in step 3 is less than the height of the culvert, then the flow is probably Type 3. Calculate the discharge using Equation 4.182 and verify that the normal depth of flow in the culvert is less than the culvert height. If the normal depth is less than the culvert height, then Type 3 flow is confirmed.
6. If neither Type 2 nor Type 3 flow can be confirmed, take the capacity of the culvert to be the lesser of the two calculated discharges.

It is useful to note that circular culverts flow full when the discharge rate exceeds $1.07\, Q_{full}$, where Q_{full} is the full-flow discharge calculated using the Manning equation (Brater et al., 1996).

Unsubmerged Entrance.   When the culvert entrance is unsubmerged, the flow is either Type 4, 5, or 6. The flow type and the associated flowrate are determined by the following procedure:

1. Assume that the flow is Type 4, use Equation 4.184 to calculate the discharge, Q, and use Q to calculate the normal depth, y_n, and critical depth, y_c. If $y_n > y_c$ and the tailwater depth is less than y_c, then Type 4 flow is verified and the calculated discharge is correct.
2. Assume that the flow is Type 5, use Equation 4.186 to calculate the discharge, Q, and use Q to calculate the normal depth, y_n, and critical depth, y_c. If $y_n < y_c$ and the tailwater depth is less than y_c, then Type 5 flow is verified and the calculated discharge is correct.
3. Assume that the flow is Type 6, use Equation 4.188 to calculate the discharge, Q, and use Q to calculate the normal depth, y_n, and critical depth, y_c. If $y_n > y_c$ and the tailwater depth is greater than y_c, then Type 6 flow is verified and the calculated discharge is correct.

Design Considerations.   The geometry of a culvert entrance is an important aspect of culvert design since the culvert entrance exerts a significant influence on the hydraulic characteristics of a culvert. The four standard inlet types are: (1) flush setting in a vertical headwall, (2) wingwall entrance, (3) projecting entrance, and (4) mitered entrance set flush with a sloping embankment. Structural stability, aesthetics, and erosion control are among the factors that influence the selection of the inlet configuration. The entrance loss coefficient, k_e, used to describe the entrance losses in most discharge formulae depend on the pipe material, shape, and entrance type, and can be estimated using the guidelines in Table 4.10.

Local drainage regulations often state the minimum culvert size (usually in the 30 to 60 cm range), with debris potential being an important consideration in determining the minimum acceptable size of the culvert. Some localities require that the engineer assume 25% debris blockage in computing the required size of the culvert. Both minimum and maximum velocities must be considered in designing a culvert. A minimum velocity in a culvert of 0.6 to 0.9 m/s at the design flow is required to assure self-cleansing. The maximum allowable velocity for corrugated metal pipe is 3 m/s (10 ft/s), and there is no specified maximum allowable velocity for reinforced concrete pipe (Debo and Reese, 1995), although velocities greater than 4–5 m/s are rarely used because of potential problems with scour. Outlet protection should be provided where discharge velocities will cause erosion problems. The most common culvert materials are concrete (reinforced and nonreinforced), corrugated aluminum, and corrugated steel. The selection of a

■ **Table 4.10** Culvert Entrance Loss Coefficients	**Culvert type and entrance conditions**	k_e
	Pipe, concrete:	
	Projecting from fill, socket end (groove end)	0.2
	Projecting from fill, square-cut end	0.5
	Headwall or headwall and wingwalls	
	Socket end of pipe (groove end)	0.2
	Square edge	0.5
	Rounded (radius = $1/12\,D$)	0.2
	Mitered to conform to fill slope	0.7
	End section conforming to fill slope	0.5
	Beveled edges, 33.7° or 45° bevels	0.2
	Side- or slope-tapered inlet	0.2
	Pipe, or pipe arched, corrugated metal:	
	Projecting from fill (no headwall)	0.9
	Headwall or headwall and wingwalls square edge	0.5
	Mitered to conform to fill slope, paved or unpaved slope	0.7
	End section conforming to fill slope	0.5
	Beveled edges, 33.7° or 45° bevels	0.2
	Side- or slope-tapered inlet	0.2
	Box, reinforced concrete:	
	Headwall parallel to embankment (no wingwalls)	
	Square edged on 3 edges	0.5
	Rounded on 3 edges	0.2
	Wingwalls at 30° to 75° to barrel	
	Square edged at crown	0.4
	Crown edge rounded	0.2
	Wingwalls at 10° to 25° to barrel	
	Square edged at crown	0.5
	Wingwalls parallel	
	Square edged at crown	0.7
	Side or slope-tapered inlet	0.2

Source: USFHWA (1984a).

■ **Table 4.11** Manning n in Culverts	**Type of conduit**	**Wall and joint description**	**n**
	Concrete pipe	good joints, smooth walls	0.011–0.013
		good joints, rough walls	0.014–0.016
		poor joints, rough walls	0.016–0.017
		badly spalled	0.015–0.020
	Concrete box	good joints, smooth finished walls	0.014–0.018
		poor joints, rough, unfinished walls	0.014–0.018
	Corrugated metal pipes and boxes, annular corrugations	$2\frac{2}{3} \times \frac{1}{2}$ inch corrugations	0.022–0.027
		6×1 inch corrugations	0.022–0.025
		5×1 inch corrugations	0.025–0.026
		3×1 inch corrugations	0.027–0.028
		6×2 inch structural plate	0.033–0.035
		$9 \times 2\frac{1}{2}$ inch structural plate	0.033–0.037

Source: USFHWA (1984a).

culvert material depends on the required structural strength, hydraulic roughness, durability, and corrosion and abrasion resistance. In general, corrugated culverts have significantly higher frictional resistance than concrete culverts, and most cities require the use of concrete pipe for culverts placed in critical areas or within the public right-of-way. Recommended Manning n values for culvert design are given in Table 4.11.

■ Example 4.16

What is the capacity of a 1.22 m by 1.22 m concrete box culvert ($n = 0.013$) with a rounded entrance ($k_e = 0.05$, $C_d = 0.95$) if the culvert slope is 0.5%, the length is 36.6 m, and the headwater level is 1.83 m above the culvert invert? Consider the following cases: (a) free-outlet conditions, and (b) tailwater elevation 0.304 m above the top of the box at the outlet. What must the headwater elevation be in case (b) for the culvert to pass the flow that exists in case (a)?

Solution

Since the headwater depth exceeds 1.2 times the height of the culvert opening, then the culvert entrance is *submerged*. An elevation view of the culvert is shown in Figure 4.29.

(a) For free outlet conditions, two types of flow are possible: the normal depth of flow is greater than the culvert depth (Type 2), or the normal depth of flow is less than the culvert depth (Type 3). To determine the flow type, assume a certain type of flow, calculate the discharge and depth of flow, and see if the assumption is confirmed. If

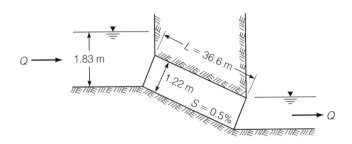

Figure 4.29 ■ Elevation View of Culvert

the assumption is not confirmed, then the initial assumption is incorrect. Assuming Type 2 flow, then the flowrate equation, Equation 4.181, is given by

$$Q = A\sqrt{\frac{2g\Delta h}{19.62n^2L/R^{\frac{4}{3}} + k_e + 1}} \qquad (4.189)$$

where Δh is the difference in water levels between the entrance and exit of the culvert [= $0.183 + 0.005(36.6) - 1.22 = 0.793$ m], n is the roughness coefficient (= 0.013), L is the length of the culvert (= 36.6 m), and R is the hydraulic radius given by

$$R = \frac{A}{P}$$

where A is the cross-sectional area of the culvert, and P is the wetted perimeter of the culvert where

$$A = (1.22)(1.22) = 1.49 \text{ m}^2$$
$$P = 4(1.22) = 4.88 \text{ m}$$

and therefore

$$R = \frac{1.49}{4.88} = 0.305 \text{ m}$$

Substituting known values of the culvert parameters into Equation 4.189 gives

$$Q = (1.49)\sqrt{\frac{2(9.81)(0.793)}{19.62(0.013)^2(36.6)/(0.305)^{\frac{4}{3}} + 0.05 + 1}}$$

which simplifies to

$$Q = 4.59 \text{ m}^3/\text{s}$$

The next step is to calculate the normal depth at a discharge of 4.59 m^3/s using the Manning equation, where

$$Q = \frac{1}{n}AR^{\frac{2}{3}}S_0^{\frac{1}{2}}$$

and S_o is the slope of the culvert (= 0.005). If the normal depth of flow is y_n, then the area, A, wetted perimeter, P, and hydraulic radius, R, are given by

$$A = by_n = 1.22y_n$$
$$P = b + 2y_n = 1.22 + 2y_n$$
$$R = \frac{A}{P} = \frac{1.22y_n}{1.22 + 2y_n}$$

The Manning equation gives

$$4.59 = \frac{1}{0.013}\frac{(1.22y_n)^{\frac{5}{3}}}{(1.22 + 2y_n)^{\frac{2}{3}}}(0.005)^{\frac{1}{2}}$$

which simplifies to

$$\frac{(1.22 y_n)^{\frac{5}{3}}}{(1.22 + 2y_n)^{\frac{2}{3}}} = 0.844$$

and solving iteratively for y_n yields

$$y_n = 1.25 \text{ m}$$

Therefore, the initial assumption that the normal depth is greater than the height of the culvert ($= 1.22$ m) is verified, Type 2 flow is confirmed, and the flow through the culvert is equal to 4.59 m³/s.

(b) In this case, the tailwater is 0.304 m above the culvert outlet, and therefore the difference in water levels between the inlet and outlet, Δh, decreases by 0.304 m to $0.793 - 0.304 = 0.489$ m. The flow equation in this case, Type 1 flow, is the same as Equation 4.189, with $\Delta h = 0.489$ m, which gives

$$Q = (1.49) \sqrt{\frac{2(9.81)(0.489)}{19.62(0.013)^2(36.6)/(0.305)^{\frac{4}{3}} + 0.05 + 1}}$$

which simplifies to

$$Q = 3.60 \text{ m}^3/\text{s}$$

Therefore, when the tailwater depth rises to 0.305 m above the culvert exit, the discharge decreases from 4.59 m³/s to 3.60 m³/s.

When the headwater is at a height x above the culvert inlet and the tailwater is 0.305 m above the outlet, then the flow through the culvert is 4.59 m³/s. The difference between the headwater and tailwater elevations, Δh, is given by

$$\Delta h = [x + 1.22 + 0.005(36.6)] - [1.22 + 0.305] = x - 0.122$$

The flow equation for Type 1 flow (Equation 4.189) requires that

$$4.59 = (1.49) \sqrt{\frac{2(9.81)(x - 0.122)}{19.62(0.013)^2(36.6)/(0.305)^{\frac{4}{3}} + 0.05 + 1}}$$

which leads to

$$x = 0.915 \text{ m}$$

Therefore, the headwater depth at the entrance of the culvert for a flow of 4.59 m³/s is $1.22 + 0.915 = 2.14$ m. ∎

4.5   Design of Open Channels

Constructed open channels frequently serve as major drainageways in urban stormwater management systems, and are used as water-delivery systems for water-supply and irrigation projects. There are three types of constructed open channels: (1) lined (nonerodible) channels, (2) unlined (erodible) channels, and (3) grass-lined channels.

4.5.1 ■ Basic Principles

The objective in designing an open channel is to determine the shape of the channel that will safely accommodate the design flow at a reasonable cost and limit the erosion and deposition of materials in the channel.

Best Hydraulic Section.   The *best hydraulic section* can be determined by requiring that the flow area of the channel be minimized while maintaining the hydraulic capacity. Consider the Manning equation given by

$$Q = \frac{1}{n}AR^{\frac{2}{3}}S_o^{1/2} \tag{4.190}$$

where Q is the flow in the channel, A is the cross-sectional area, R is the hydraulic radius ($= A/P$), P is the wetted perimeter, and S_o is the slope of the channel. Equation 4.190 can be rearranged into the form

$$A = \left(\frac{Qn}{S_o^{1/2}}\right)^{\frac{3}{5}} P^{\frac{2}{5}} \tag{4.191}$$

which demonstrates that for given values of Q, n, and S_o, the area, A, is proportional to the wetted perimeter, P, and minimizing A is tantamount to minimizing P. The best hydraulic section is defined as the section that minimizes the flow area for given values of Q, n, and S_o.

To illustrate the process of determining the best hydraulic section, consider the trapezoidal section shown in Figure 4.30, where the shape parameters are the bottom width, b, and the side slope m:1 (H:V). The flow area, A, and the wetted perimeter, P, are given by

$$A = by + my^2 \tag{4.192}$$

$$P = b + 2y\sqrt{m^2 + 1} \tag{4.193}$$

Eliminating b from Equations 4.192 and 4.193 leads to

$$A = (P - 2y\sqrt{m^2 + 1})y + my^2 \tag{4.194}$$

Eliminating A from Equations 4.191 and 4.194 yields

$$(P - 2y\sqrt{m^2 + 1})y + my^2 = cP^{\frac{2}{5}} \tag{4.195}$$

where c is a constant given by

$$c = \left(\frac{Qn}{S_o^{1/2}}\right)^{\frac{3}{5}} \tag{4.196}$$

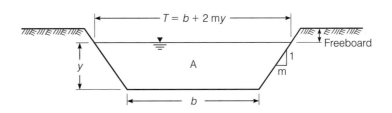

Figure 4.30 ■
Trapezoidal Section

Holding m constant in Equation 4.195, differentiating with respect to y, and setting $\partial P / \partial y$ equal to zero leads to

$$P = 4y\sqrt{1 + m^2} - 2my \qquad (4.197)$$

Similarly, holding y constant in Equation 4.195, differentiating with respect to m, and setting $\partial P / \partial m$ equal to zero leads to

$$m = \frac{\sqrt{3}}{3} \approx 0.577 \qquad (4.198)$$

On the basis of Equations 4.197 and 4.198, the best hydraulic section for a trapezoid is one having the following geometric characteristics

$$P = 2\sqrt{3}y, \qquad b = 2\frac{\sqrt{3}}{3}y, \qquad A = \sqrt{3}y^2 \qquad (4.199)$$

where $P = 3b$, indicating that the sides of the channel have the same length as the bottom of the channel. The best side slope, indicated by Equation 4.198, makes an angle of 60° with the horizontal. In cases where the side slopes are controlled by the angle of repose of the soil surrounding the channel, Equation 4.197 is used with a specified side slope, m, to find the best bottom width to depth ratio. This procedure is usually necessary since the recommended side slopes for lined channels are usually less than 1.5:1 (H:V) or 33.7°.

The area, A, wetted perimeter, P, and top width, T, of the best hydraulic sections for a variety of channel shapes are given in Table 4.12. Although the best hydraulic section appears to be the most economical in terms of excavation and channel lining, it is important to note that this section may not always be the most economical section since (a) the flow area does not include freeboard, and therefore is not the total area to be excavated; (b) it may not be possible to excavate a stable best hydraulic section in the available natural material; (c) for lined channels, the cost of lining may be comparable to the excavation costs; and (d) other factors such as the ease of access to the site and the cost of disposing of removed material may affect the economics of the channel design.

Minimum Permissible Velocity.   The minimum permissible velocity is the lowest velocity that will prevent both sedimentation and vegetative growth in the channel. According to French (1985), an average velocity of 0.6 to 0.9 m/s (2 to 3 ft/s) will prevent sedimentation when the silt load is low and a velocity of 0.75 m/s (2.5 ft/s) is usually sufficient to prevent vegetative growth. These estimates of minimum permissible velocities are only approximate values and can vary significantly in practice. Fortier and Scobey (1926) observed that canals carrying turbid waters are seldom bothered by plant growth, while in channels transporting clear water some plant species flourish at velocities that are significantly in excess of the velocity that will cause erosion in the channel.

■ Table 4.12 Geometric Characteristics of Best Hydraulic Sections	Shape	Best geometry	A	P	T
	Trapezoidal	half of a hexagon	$1.73y^2$	$3.46y$	$2.31y$
	Rectangle	half of a square	$2y^2$	$4y$	$2y$
	Triangle	half of a square	y^2	$2.83y$	$2y$
	Semicircle	—	$0.500\pi y^2$	πy	$2y$
	Parabola	—	$1.89y^2$	$3.77y$	$2.83y$

A velocity of 0.6 m/s (2 ft/s) is sufficient to move a 15-mm diameter organic or a 2-mm sand particle (ASCE, 1982).

Channel Slopes. The relevant slopes in channel design are the longitudinal slope and the side slope of the channel. Longitudinal slopes are determined by land topography and the minimum slope necessary to maintain the required flow velocities. Excavation is usually minimized by laying the channel on a slope equal to the slope of the ground surface. However, if the resulting flow velocity is less than the minimum permissible velocity, then a steeper slope that produces a higher velocity must be used. The side slopes of excavated channels are influenced by the material in which the channel is excavated. Suitable side slopes for channels excavated in various types of materials are shown in Table 4.13. These values are recommended for preliminary design. In deep cuts, side slopes are often steeper above the water surface than below the water surface, and in small drainage ditches, the side slopes are often steeper than they would be in an irrigation channel excavated in the same material (French, 1985). If concrete is the lining material, then side slopes greater than 1:1 usually require the use of forms, and for side slopes greater than 0.75:1 (H:V) the linings must be designed to withstand earth pressures. The U.S. Bureau of Reclamation prefers a 1.5:1 (H:V) slope for usual sizes of lined canals (USBR, 1978).

Freeboard. The *freeboard* is defined as the vertical distance between the water surface and the top of the channel when the channel is carrying the design flow at normal depth. Freeboard is provided to afford a degree of protection associated with the uncertainty in the design parameters. An estimate of the required freeboard for unlined channels is given by the empirical formula (Chow, 1959)

$$F = 0.55\sqrt{Cy} \qquad (4.200)$$

where F is the required freeboard in meters, y is the design flow depth in meters, and C is a coefficient that varies from 1.5 at a flow of 0.57 m³/s to 2.5 for a flow of 85 m³/s or more. The minimum freeboard is usually 30 cm (ASCE, 1992), and additional freeboard must be provided to accommodate the superelevation of the water surface that occurs at channel bends. The superelevation, h_s, of the water surface can be estimated by

$$h_s = \frac{V^2 T}{g r_c} \qquad (4.201)$$

■ Table 4.13	Material	Side slope (H:V)
Recommended Side Slopes in Various Types of Material	Rock	Nearly vertical
	Muck and peat soils	$\frac{1}{4}$:1
	Stiff clay or earth with concrete lining	$\frac{1}{2}$:1 to 1:1
	Earth with stone lining or earth for large channels	1:1
	Firm clay or earth for small ditches	$1\frac{1}{2}$:1
	Loose, sandy earth	2:1
	Sandy loam or porous clay	3:1

Source: Chow (1959).

where V is the average velocity in the channel, T is the top-width of the channel, and r_c is the radius of curvature of the centerline of the channel. The additional freeboard to accommodate the superelevation of the water surface around bends need only be provided in the vicinity of the bend. In order to minimize flow disturbances around bends, it is also recommended that the radius of curvature should be at least three times the channel width (USACE, 1995).

4.5.2 ■ Lined Channels

Lined or *rigid-boundary channels* are used for a variety of purposes, such as to (1) transport water at high velocities to reduce construction and excavation costs, (2) decrease seepage losses, (3) decrease operation and maintenance costs, and (4) ensure the stability of the channel section. According to ASCE (1992), all channels carrying supercritical flow should be lined with concrete, which is continuously reinforced both longitudinally and laterally. A recommended design procedure for lined channels is as follows (French, 1985):

1. Estimate the roughness coefficient, n, and freeboard coefficient, C, for the specified lining material and design flowrate, Q. Manning roughness coefficients for lined channels are listed in Table 4.14. ASCE (1992) has recommended that open-channel designs should not use a roughness coefficient lower than 0.013 for well-troweled concrete, and other finishes should have proportionally higher n values assigned to them.
2. Compute the normal depth of flow, y_n, using the Manning equation

$$Q = \frac{1}{n}AR^{\frac{2}{3}}S_o^{1/2} \tag{4.202}$$

where A is the flow area, R is the hydraulic radius, S_o is the (given) longitudinal slope of the channel, and the shape of the cross-section is specified by the designer. If appropriate, the relative dimensions of the best hydraulic section may be specified.
3. Check the minimum permissible velocity and the Froude number. Repeat steps 2 and 3 if necessary to meet the minimum velocity and subcritical flow requirements.
4. Calculate the required freeboard using Equation 4.200, and increase the freeboard on channel bends by the superelevation height given by Equation 4.201.

As an additional constraint in designing concrete-lined channels, ASCE (1992) has stated that flow velocities should not exceed 2.1 m/s or result in a Froude number greater than 0.8 for nonreinforced linings and that flow velocities should not exceed 5.5 m/s for reinforced linings.

■ Example 4.17

Design a lined trapezoidal channel to carry 20 m³/s on a longitudinal slope of 0.0015. The lining of the channel is to be float-finished concrete. Consider: (a) the best hydraulic section, and (b) a section with side slopes of 1.5:1 (H:V).

Solution

(a) According to Table 4.14, $n = 0.015$. Using the best (trapezoidal) hydraulic section, the bottom width, b, and side slope, m, are related to the depth, y, by the relations

■ Table 4.14	Type	Characteristics	Minimum n	Normal n	Maximum n
Roughness Coefficients in Lined Open Channels	Cement	neat surface	0.010	0.011	0.013
		mortar	0.011	0.013	0.015
	Concrete	trowel finish	0.011	0.013*	0.015
		float finish	0.013	0.015	0.016
		finished, with gravel on bottom	0.015	0.017	0.020
		unfinished	0.014	0.017	0.020
		gunite, good section	0.016	0.019	0.023
		gunite, wavy section	0.018	0.022	0.025
		on good excavated rock	0.017	0.020	—
		on irregular excavated rock	0.022	0.027	—
	Concrete bottom float finished with sides of:	dressed stone in mortar	0.015	0.017	0.020
		random stone in mortar	0.017	0.020	0.024
		cement rubble masonry, plastered	0.016	0.020	0.024
		cement rubble masonry	0.020	0.025	0.030
		dry rubble or riprap	0.020	0.030	0.035
	Gravel bottom with sides of:	formed concrete	0.017	0.020	0.025
		random stone in mortar	0.020	0.023	0.026
		dry rubble or riprap	0.023	0.033	0.036
	Brick	glazed	0.011	0.013*	0.015
		in cement mortar	0.012	0.015*	0.018
	Masonry	cemented rubble	0.017	0.025	0.030
		dry rubble	0.023	0.032	0.035
	Dressed ashlar	—	0.013	0.015	0.017
	Asphalt	smooth	0.013	0.013	—
	Vegetal lining	—	0.030	—	0.500

Source: Chow (1959).
*Chow (1959) recommended this value for use in design.

given by Equations 4.198 and 4.199 as

$$b = 1.15y, \qquad m = 0.58 \quad (= 60° \text{ angle})$$

and, according to Table 4.12,

$$A = 1.73y^2, \qquad P = 3.46y, \qquad T = 2.31y, \qquad R = \frac{A}{P} = 0.5y$$

Substituting the geometric characteristics of the channel into the Manning equation yields

$$20 = \frac{1}{0.015}(1.73y_n^2)(0.5y_n)^{\frac{2}{3}}(0.0015)^{\frac{1}{2}}$$

or

$$y_n^{8/3} = 7.12$$

which leads to

$$y_n = 2.09 \text{ m}$$

and hence the bottom width, b, of the channel is given by

$$b = 1.15 y_n = 1.15(2.09) = 2.40 \text{ m}$$

The flow area is therefore given by

$$A = 1.73 y_n^2 = 1.73(2.09)^2 = 7.6 \text{ m}^2$$

and the average velocity, V, is

$$V = \frac{Q}{A} = \frac{20}{7.6} = 2.6 \text{ m/s}$$

This velocity should be sufficient to prevent sedimentation and vegetative growth. The hydraulic depth, D, is given by

$$D = \frac{A}{T} = \frac{1.73 y_n^2}{2.31 y_n} = 0.749 y_n = 0.749(2.09) = 1.6 \text{ m}$$

and the Froude number, Fr, is

$$\text{Fr} = \frac{V}{\sqrt{gD}} = \frac{2.6}{\sqrt{(9.81)(1.6)}} = 0.66$$

The flow is therefore subcritical (Fr < 1).

The required freeboard, F, can be estimated using Equation 4.200,

$$F = 0.55\sqrt{Cy}$$

where F is the freeboard in meters, y is the design flow depth in meters, and C is a coefficient that varies from 1.5 at a flow of 0.57 m^3/s to 2.5 for a flow of 85 m^3/s. Based on a flow of 20 m^3/s, C can be interpolated as 1.7. The required freeboard is

$$F = 0.55\sqrt{(1.7)(2.09)} = 1.04 \text{ m}$$

The total depth of the channel to be excavated and lined is therefore equal to the normal depth plus the freeboard, 2.09 m + 1.04 m = 3.13 m. The channel is to have a bottom width of 2.40 m, and side slopes of 0.58:1 (H:V).

(b) If the channel side slope is 1.5:1, then $m = 1.5$ and Equation 4.197 gives the perimeter, P, as

$$P = 4y\sqrt{1 + m^2} - 2my = 4y\sqrt{1 + 1.5^2} - 2(1.5)y = 4.21y$$

Also, since

$$P = b + 2y\sqrt{1 + m^2}$$

then

$$b = P - 2y\sqrt{1 + m^2} = 4.21y - 2y\sqrt{1 + 1.5^2} = 0.60y$$

Also,

$$T = b + 2my = 0.6y + 2(1.5)y = 3.6y$$

$$A = (b + my)y = (0.60y + 1.5y)y = 2.10y^2$$

$$R = \frac{A}{P} = \frac{2.10y^2}{4.21y} = 0.499y$$

Substituting the geometric characteristics of the channel into the Manning equation gives

$$20 = \frac{1}{0.015}(2.10y_n^2)(0.499y_n)^{\frac{2}{3}}(0.0015)^{\frac{1}{2}}$$

or

$$y_n^{8/3} = 5.86$$

which leads to

$$y_n = 1.94 \text{ m}$$

and hence the bottom width, b, of the channel is given by

$$b = 0.60y_n = 0.60(1.94) = 1.16 \text{ m}$$

The flow area is therefore given by

$$A = 2.10y_n^2 = 2.10(1.94)^2 = 7.90 \text{ m}^2$$

and the average velocity, V, is

$$V = \frac{Q}{A} = \frac{20}{7.90} = 2.53 \text{ m/s}$$

This velocity should be sufficient to prevent sedimentation and vegetative growth. The hydraulic depth, D, is given by

$$D = \frac{A}{T} = \frac{2.10y_n^2}{3.6y_n} = 0.58y_n = 0.58(1.94) = 1.13 \text{ m}$$

and the Froude number, Fr, is

$$\text{Fr} = \frac{V}{\sqrt{gD}} = \frac{2.53}{\sqrt{(9.81)(1.13)}} = 0.76$$

The flow is therefore subcritical (Fr < 1).

The required freeboard, F, can be estimated using Equation 4.200

$$F = 0.55\sqrt{Cy}$$

Based on a flow of 20 m^3/s, C can be interpolated as 1.7 and the required freeboard is

$$F = 0.55\sqrt{(1.7)(1.94)} = 1.00 \text{ m}$$

The total depth of the channel to be excavated and lined is therefore equal to the normal depth plus the freeboard, 1.94 m + 1.00 m = 2.94 m. The channel is to have a bottom width of 1.16 m and side slopes of 1.5:1 (H:V).   ■

4.5.3 ■ Unlined Channels

Primary objectives in the design of unlined channels are to prevent deposition of suspended sediment, and prevent scour of the perimeter material. Recall that the average shear stress, τ_o, around the perimeter of a channel is given by (see Section 4.2.2)

$$\tau_o = \gamma R S_f \tag{4.203}$$

where γ is the specific weight of the fluid, R is the hydraulic radius of the channel, and S_f is the slope of the energy grade line. This average shear stress, τ_o, is also called the *unit tractive force*. In reality, the boundary shear stress is not uniformly distributed around the perimeter of the channel, and in trapezoidal sections, which is the shape most commonly used for unlined channels, studies have shown that the maximum shear stress on the bottom of the channel, τ_b can be approximated by (Lane, 1955)

$$\boxed{\tau_b = \gamma y S_f} \tag{4.204}$$

and the maximum shear stress on the sides of the channel, τ_s, can be approximated by

$$\boxed{\tau_s = 0.76 \gamma y S_f} \tag{4.205}$$

where y is the depth of flow in the channel.

A particle of (submerged) weight w_p on the bottom of the channel resists the shear force of the flowing fluid by the friction force between the particle and surrounding particles on the bottom of the channel, which is given by $\mu_p w_p$ where μ_p is the coefficient of friction between particles on the bottom of the channel. Therefore, when particle motion on the bottom of the channel is incipient, the shear stress on the bottom of the channel is τ_b', and

$$A_p \tau_b' = \mu_p w_p \tag{4.206}$$

where A_p is the effective surface area of a particle on the bottom of the channel. The coefficient of friction between particles on the bottom of the channel can be related to the *angle of repose*, α, of the particle material by the relation

$$\mu_p = \tan \alpha \tag{4.207}$$

Combining Equations 4.206 and 4.207 leads to the following expression for the critical bottom shear stress under incipient motion

$$\tau_b' = \frac{w_p}{A_p} \tan \alpha \tag{4.208}$$

For particles on the side of the channel, the force on each particle consists of the shear force exerted by the fluid, which acts in the direction of flow, plus the component of the particle weight that acts down the side of the channel. Therefore the total force, F_p, tending to move a particle on the side of the channel is given by

$$F_p = \sqrt{(\tau_s A_p)^2 + (w_p \sin \theta)^2} \tag{4.209}$$

where θ is the angle that the side slope makes with the horizontal. When motion is incipient, the force tending to move the particle is equal to the frictional force, F_f, which is equal to the component of the particle weight normal to the side of the channel, multiplied by the coefficient of friction $(\tan \alpha)$, in which case

$$F_f = w_p \cos \theta \tan \alpha \tag{4.210}$$

When motion is incipient, $F_p = F_f$, and the shear stress on the side of the channel is τ_s'. Equations 4.209 and 4.210 combine to yield

$$w_p \cos \theta \tan \alpha = \sqrt{(\tau_s' A_p)^2 + (w_p \sin \theta)^2}$$

which can be put in the form

$$\tau_s' = \frac{w_p}{A_p} \cos \theta \tan \alpha \sqrt{1 - \frac{\tan^2 \theta}{\tan^2 \alpha}} \tag{4.211}$$

The ratio of the critical shear stress on the side of the channel to the critical shear stress on the bottom of the channel is given by the *tractive force ratio*, K, where

$$\boxed{K = \frac{\tau_s'}{\tau_b'} = \sqrt{1 - \frac{\sin^2 \theta}{\sin^2 \alpha}}} \tag{4.212}$$

On the basis of Equations 4.208 and 4.211, it is clear that the maximum permissible shear stress on the bottom and sides of trapezoidal channels depend on the particle characteristics, as parameterized by the particle weight, w_p; effective area, A_p; and angle of repose, α. Lane (1955) found that the angle of repose for noncohesive material is directly proportional to the size and angularity of the particles and suggested the relationship illustrated in Figure 4.31. The particle sizes to be used with Figure 4.31 are the 75-percentile diameters by weight. Relations between the maximum permissible unit tractive force $(= \gamma R S_f)$, and particle diameters for noncohesive and cohesive materials are illustrated in Figure 4.32 (after Lane, 1955; Chow, 1959). The particle diameters used in Figure 4.32 are 50-percentile diameters by weight, and the permissible tractive forces in these curves can be taken to be conservative estimates (French, 1985). It should be noted that since the unit tractive force $(\gamma R S_f)$ is approximately equal to the shear stress on the bottom of a trapezoidal channel $(\gamma y S_f)$, then the permissible shear stress on the side of the channel can be estimated by multiplying the permissible unit tractive force by the tractive force ratio given in Equation 4.212.

It has been observed that sinuous canals scour more easily than straight channels, and therefore the permissible shear stresses must be adjusted based on the degree of sinuousness of the channel. Correction factors for the permissible unit tractive force in sinuous channels have been suggested by Lane (1955) and are given in Table 4.15.

The following procedure is suggested for designing unlined earthen channels:

1. Estimate the roughness coefficient, n, based on the perimeter characteristics of the channel, and select the freeboard coefficient, C, based on the design flowrate in the channel. The roughness coefficient in the channel can be estimated using Table 4.16.

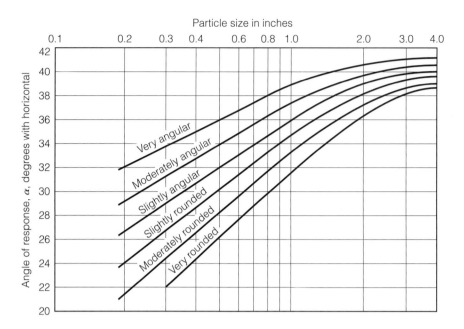

Figure 4.31 ■ Angles of Repose of Noncohesive Material

2. Estimate the angle of repose of the channel material from Figure 4.31.
3. Estimate the channel sinuousness and the tractive force correction factor from Table 4.15.
4. Specify a side-slope angle based on the guidelines in Table 4.13.
5. Estimate the tractive force ratio from Equation 4.212.
6. Estimate the permissible tractive force on the bottom and sides of the channel from Figure 4.32. Correct for sinuousness.
7. Assume that the permissible shear stress on the side of the channel is the limiting factor in the channel design, and determine the normal depth of flow, y_n.
8. Calculate the required bottom width, b, of the channel using the Manning equation and y_n from step 7.
9. Compare the permissible tractive force on the bottom (step 6) with the actual tractive force given by $\gamma y_n S_f$.
10. Compare the design velocity with the minimum permissible velocity and calculate the Froude number of the design flow. Verify that the flow is subcritical.
11. Estimate the required freeboard in the channel using Equation 4.200.

This design procedure is illustrated in the following example.

■ Example 4.18

Design a trapezoidal channel to carry 20 m³/s through a slightly sinuous channel on a slope of 0.0015. The channel is to be excavated in coarse alluvium with a 75-percentile diameter of 2 cm, and with the particles on the perimeter of the channel moderately rounded.

Solution

Step 1: The first step is to estimate the Manning roughness coefficient, n. Based on the data given in Table 4.16, an estimated n value for a uniform section lined with gravel is 0.025. This can be compared with the estimate given by Equation 4.41,

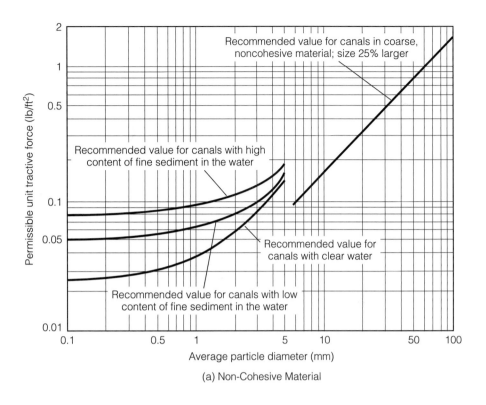

(a) Non-Cohesive Material

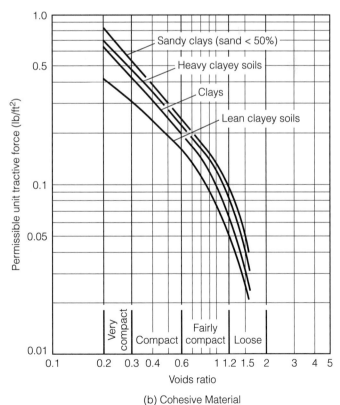

(b) Cohesive Material

Figure 4.32 ■ Permissible Unit Tractive Force for Channels in (a) Noncohesive Material, and (b) Cohesive Material

■ **Table 4.15**
Correction Factors for
Degree of Sinuousness

Degree of sinuousness	Correction factor
Straight channels	1.00
Slightly sinuous channels	0.90
Moderately sinuous channels	0.75
Very sinuous channels	0.60

Source: Lane, E. W. "Design of Stable Channels," Transactions of the American Society of Civil Engineers, v. 120, p. 1234–79. Copyright © 1955 by ASCE. Reprinted by permission.

which states that

$$n = 0.031 d^{\frac{1}{6}}$$

where d is the characteristic particle diameter on the boundary (in feet). Using the 75-percentile diameter, d_{75}, for d yields

$$n = 0.031(d_{75})^{\frac{1}{6}} = 0.031(0.02 \times 3.281)^{\frac{1}{6}} = 0.020$$

This compares favorably with $n = 0.025$ estimated from Table 4.16, and the more conservative value of 0.025 will be used in the design.

Step 2: The angle of repose of the channel material is estimated from Figure 4.31, where

$$d_{75} = 2 \text{ cm} = 0.8 \text{ in.}$$

Therefore, the angle of repose, α, is equal to 32° (based on moderately rounded material).

■ **Table 4.16**
Roughness Coefficients in
Excavated Open Channels

Type	Characteristics	Minimum n	Normal n	Maximum n
Earth, straight, and uniform	clean, recently completed	0.016	0.018	0.020
	clean, after weathering	0.018	0.022*	0.025
	gravel, uniform section, clean	0.022	0.025	0.030
	with short grass, few weeds	0.022	0.027	0.033
Earth, winding, and sluggish	no vegetation	0.023	0.025	0.030
	grass, some weeds	0.025	0.030	0.033
	dense weeds or aquatic plants in deep channels	0.030	0.035	0.040
	earth bottom and rubble sides	0.028	0.030	0.035
	stony bottom and weedy banks	0.025	0.035	0.040
	cobble bottom and clean sides	0.030	0.040	0.050
Dragline-excavated or dredged	no vegetation	0.025	0.028	0.033
	light brush on banks	0.035	0.050	0.060
Rock cuts	smooth and uniform	0.025	0.035	0.040
	jagged and irregular	0.035	0.040	0.050
Channels not maintained, weeds and brush uncut	dense weeds high as flow depth	0.050	0.080	0.120
	clean bottom, brush on sides	0.040	0.050	0.080
	same, highest stage of flow	0.045	0.070	0.110
	dense brush, high stage	0.080	0.100	0.140

Source: Chow (1959).
*Chow (1959) recommended this value for use in design.

Step 3: Since the channel is slightly sinuous, the correction factor, C_s, for the maximum tractive force is given by Table 4.15 as $C_s = 0.90$.

Step 4: Specify a channel side slope of 2:1 (H:V). The angle that the side slope makes with the horizontal, θ, is given by

$$\theta = \tan^{-1}\left(\frac{1}{2}\right) = 26.6°$$

Step 5: The tractive force ratio, K, is given by Equation 4.212 as

$$K = \frac{\tau_s'}{\tau_b'} = \sqrt{1 - \frac{\sin^2\theta}{\sin^2\alpha}} = \sqrt{1 - \frac{\sin^2 26.6°}{\sin^2 32°}} = 0.53$$

Step 6: The permissible tractive force on the bottom of the channel is estimated from Figure 4.32 as 0.33 lb/ft^2 or 15.9 N/m^2 for a median particle size of 20 mm. Correcting this permissible tractive force for sinuousness leads to an allowable shear stress on the bottom of the channel, τ_b', equal to

$$\tau_b' = C_s(15.9) = 0.9(15.9) = 14.3 \text{ N/m}^2$$

and the permissible tractive force on the side of the channel, τ_s', is therefore given by

$$\tau_s' = K\tau_b' = 0.53(14.3) = 7.6 \text{ N/m}^2$$

Step 7: The normal depth of flow, y_n, can now be estimated by assuming that particle motion is incipient on the side of the channel, hence

$$0.76\gamma y_n S_o = 7.6 \text{ N/m}^2$$

which leads to

$$y_n = \frac{7.6}{0.76\gamma S_o} = \frac{7.6}{0.76(9790)(0.0015)} = 0.68 \text{ m}$$

Step 8: The Manning equation requires that

$$Q = \frac{1}{n}AR^{\frac{2}{3}}S_o^{1/2} = \frac{1}{n}\frac{A^{\frac{5}{3}}}{P^{\frac{2}{3}}}S_o^{1/2}$$

Substituting known quantities gives

$$20 = \frac{1}{0.025}\frac{A^{\frac{5}{3}}}{P^{\frac{2}{3}}}(0.0015)^{\frac{1}{2}}$$

which simplifies to

$$\frac{A^{\frac{5}{3}}}{P^{\frac{2}{3}}} = 12.9$$

Since

$$A = [b + my_n]y_n = [b + 2(0.68)](0.68) = 0.68[b + 1.36]$$
$$P = b + 2y_n\sqrt{1 + m^2} = b + 2(0.68)\sqrt{1 + 2^2} = b + 3.04$$

then the Manning equation can be written as

$$\frac{(b+1.36)^{\frac{5}{3}}}{(b+3.04)^{\frac{2}{3}}} = 24.5$$

Solving for b gives

$$b = 24.2 \text{ m}$$

Several values of b actually satisfy the Manning equation, but $b = 24.2$ m is the smallest real solution. Based on this result, the channel should have a width of 24.2 m, side slopes of 2:1, and a normal depth of flow at 20 m³/s of 0.68 m.

Step 9: The tractive force on the bottom of the channel, τ_b, is given by

$$\tau_b = \gamma y_n S_o = (9790)(0.68)(0.0015) = 10 \text{ N/m}^2$$

Since this is less than the maximum permissible tractive force on the bottom (14.2 N/m²), then the channel design is acceptable from the viewpoint of tractive force.

Step 10: The flow area, A, is given by

$$A = [b + my]y = [24.2 + (2)(0.68)](0.68) = 17.4 \text{ m}^2$$

and the average velocity in the channel, V, is given by

$$V = \frac{Q}{A} = \frac{20}{17.4} = 1.1 \text{ m/s}$$

This velocity should be sufficient to prevent sedimentation and vegetative growth. The Froude number, Fr, can be estimated using the relation

$$\text{Fr} = \frac{V}{\sqrt{gD}}$$

where D is the hydraulic depth given by

$$D = \frac{A}{T} = \frac{A}{b+2my} = \frac{17.4}{24.2 + 2(2)(0.68)} = 0.65 \text{ m}$$

Therefore, the Froude number is

$$\text{Fr} = \frac{1.1}{\sqrt{(9.81)(0.65)}} = 0.44$$

The flow is subcritical and acceptable (Fr < 1).

Step 11: The required freeboard, F, can be estimated using Equation 4.200,

$$F = 0.55\sqrt{Cy}$$

where F is the freeboard in meters, y is the design flow depth in meters, and C is a coefficient that varies from 1.5 at a flow of 0.57 m³/s to 2.5 for a flow of 85 m³/s. Based on a flow of 20 m³/s, C can be interpolated as 1.7 and therefore the required freeboard is

$$F = 0.55\sqrt{(1.7)(0.68)} = 0.59 \text{ m}$$

The total depth of the channel to be excavated is therefore equal to the normal depth plus the freeboard, 0.68 m + 0.59 m = 1.27 m. The channel is to have a bottom width of 24.2 m, and side slopes of 2:1 (H:V).   ■

4.5.4 ■ Grass-Lined Channels

Grass-lined channels are commonly used to transmit intermittent irrigation and storm-water flows and to control erosion in agricultural areas. In major stormwater management systems, grass-lined channels are usually preferable to concrete-lined channels because of their increased channel storage, low velocities, and aesthetic benefits. Grass-lined channels cannot withstand prolonged innundation and wetness, and the design criteria for grass-lined channels differs in many ways from the criteria for designing lined channels.

Channel Roughness.   The Manning roughness coefficient, n, in grass-lined channels is a function of the average velocity, V, hydraulic radius, R, and the vegetal type (Coyle, 1975). The *retardance* for various types and conditions of grass cover is given in Table 4.17, and the Manning roughness coefficient, n, as a function of the product VR and the retardance is given in Figure 4.33 (after Coyle, 1975). The selection of the type of grass for a channel is determined primarily by the climate and soil conditions within the channel.

Permissible Velocities.   Permissible velocities for various grasses have been suggested by the U.S. Soil Conservation Service based on the channel slope and erodibility of the soil containing the channel. These guidelines are shown in Table 4.18. In addition, the following restrictions were suggested by French (1985):

1. Where only sparse vegetal cover can be established or maintained, velocities should not exceed 0.9 m/s.
2. Where the vegetation must be established by seeding, velocities in the range of 0.9 to 1.2 m/s are permitted.
3. Where dense sod can be developed quickly or where the normal flow in the channel can be diverted until a vegetal cover is established, velocities of 1.2 to 1.5 m/s can be allowed.
4. On well-established sod of good quality, velocities in the range of 1.5 to 1.8 m/s are permitted.
5. Under very special conditions, velocities as high as 1.8 to 2.1 m/s are permitted.

Channel Cross-Sections.   Channel shapes commonly used for grass-lined channels are trapezoidal, triangular, and parabolic, with the latter two shapes being the most popular. In designing grass-lined channels, the ability of tractors, or other farm-type machinery, to cross the channels during periods of no flow is an important consideration. This may require the side slopes of a channel to be designed to allow tractors to cross rather than for hydraulic efficiency or stability.

Freeboard.   The freeboard requirement, F (in meters), for grass-lined channels can be calculated using the relation (ASCE, 1992)

$$F = 0.152 + \frac{V^2}{2g}$$

(4.213)

Retardance	Cover	Condition
A	reed canary grass	excellent stand, tall (average 90 cm)
	yellow bluestem	excellent stand, tall (average 90 cm)
B	smooth bromegrass	good stand, mowed (average 30–40 cm)
	Bermuda grass	good stand, tall (average 30 cm)
	native grass mixture (little bluestem, blue grama, and other long and short Midwest grasses)	good stand, unmowed
	tall fescue	good stand, unmowed (average 45 cm)
	Lespedeza sericea	good stand, not woody, tall (average 50 cm)
	grass-legume mixture—timothy smooth	good stand, uncut (average 50 cm)
	tall fescue, with bird's foot trefoil or lodino	good stand, uncut (average 45 cm)
	blue grama	good stand, uncut (average 35 cm)
C	bahia	good stand, uncut (15–18 cm)
	Bermuda grass	good stand, mowed (average 15 cm)
	redtop	good stand, headed (40–60 cm)
	grass-legume mixture—summer	good stand, uncut (15–20 cm)
	centipede grass	very dense cover (average 15 cm)
	Kentucky bluegrass	good stand, headed (15–30 cm)
D	Bermuda grass	good stand, cut to 6-cm height
	red fescue	good stand, headed (30–45 cm)
	buffalo grass	good stand, uncut (8–15 cm)
	grass-legume mixture—fall, spring	good stand, uncut (10–13 cm)
	Lespedeza sericea	after cutting to 5-cm height; very good stand before cutting
E	Bermuda grass	good stand, cut to 4-cm height
	Bermuda grass	burned stubble

■ **Table 4.17**
Retardance in
Grass-Lined Channels

Source: Coyle (1975).

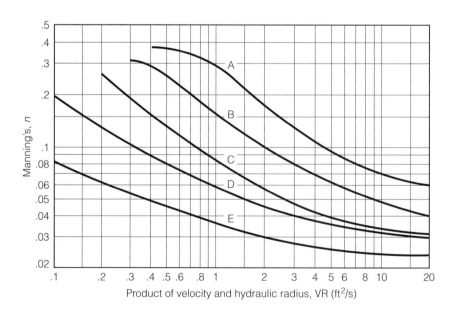

Figure 4.33 ■ Manning
Roughness Coefficient in
Grass-Lined Channels

■ Table 4.18
Permissible Velocities in
Grass-Lined Channels

Cover	Slope range (%)	Erosion-resistant soils (m/s)	Easily eroded soils (m/s)
Bermuda grass	0–5	2.4	1.8
	5–10	2.1	1.5
	> 10	1.8	1.2
Bahia, buffalo grass,	0–5	2.1	1.5
Kentucky bluegrass,	5–10	1.8	1.2
smooth brome, blue	> 10	1.5	0.9
grama, tall fescue			
Grass mixtures, reed	0–5*	1.5	1.2
canary grass	5–10	1.2	0.9
Lespedeza sericea,	0–5[†]	1.1	0.8
weeping lovegrass,			
yellow			
bluestem, redtop,			
alfalfa, red fescue			
Common *Lespedeza*,[‡£]	0–5	1.1	0.8
Sudan grass[¶]			

Source: Coyle (1975).

*Do not use grass channels on slopes greater than 10% except for vegetated side slopes in combination with a stone, concrete, or highly resistant vegetative center section.

[†]Do not use on slopes steeper than 5% except for vegetated side slopes in combination with a stone, concrete, or highly resistant vegetative center section.

[‡]Annuals: Use on mild slopes or as temporary protection until permanent covers are established.

[£]Use on slopes steeper than 5% is not recommended.

[¶]Annual: Use on mild slopes or as temporary protection until permanent cover is established.

where V is the average velocity in the channel, under design conditions. The minimum freeboard should normally be at least 30 cm at the maximum design water-surface elevation and an additional freeboard equal to the superelevation of the water surface should be provided around bends. The superelevation can be estimated using Equation 4.201.

Design Procedure. A two-stage design process for grass-lined channels is recommended. The channel is first designed for a lower bound of the expected retardance, and then the design is checked for the upper bound of the retardance. The rationale for this approach is that lower retardances result in higher velocities, with possible erosion damage, while higher retardances result in lower velocities, with more area required to transmit the flow. The design procedure has two stages (French, 1985).

Stage I 1. Assume a value of the roughness coefficient, n, and determine the corresponding value of VR from Figure 4.33. The (lower-bound) retardance of the channel is specified based on the type of grass to be planted in the channel.

  2. Select the maximum permissible velocity from Table 4.18 corresponding to the specified channel slope, lining material, and soil. Compute the value of R by using the results of step 1.

  3. Using the Manning equation and the assumed value of n, compute

$$VR = \frac{1}{n}R^{\frac{5}{3}}S_o^{1/2} \qquad (4.214)$$

where S_o is the channel slope. The value of R found in step 2 is used in the righthand side of this equation.

4. Repeat Steps 1 to 3 until the values of VR determined in steps 1 and 3 agree.
5. Determine A from the design flow and the maximum permissible velocity.
6. Determine the channel proportions for the calculated values of R and A.

Stage II 1. Assume a depth of flow for the channel identified in Stage I of the design and compute A and R.
2. Compute the average velocity, $V = Q/A$, for the A found in step 1.
3. Compute VR using the results of steps 1 and 2.
4. Use the results of step 3 to determine n from Figure 4.33, using the upper-bound retardance.
5. Use the n from step 4, R from step 1, and the Manning equation to compute V.
6. Repeat steps 1 to 5 until the values of V computed in steps 2 and 5 are approximately equal.
7. Add the appropriate freeboard and check the Froude number.

■ Example 4.19

Design a triangular grass-lined channel to handle an intermittent flow of 0.7 m³/s. The channel is to be excavated in an easily erodible soil, on a longitudinal slope of 2%, and lined with Bermuda grass. During the early stages of channel development, the height of the grass will be maintained at about 4 cm (a retardance of E); during the latter stages of development, the Bermuda grass is expected to be at a height of about 30 cm (a retardance of B).

Solution

Stage I: Easily eroded soil with Bermuda grass on a 2% slope has a maximum permissible velocity of 1.8 m/s (Table 4.18). The following are the results of the trial and error procedure to determine the roughness coefficient, n, and the channel proportions:

Trial no.	Assumed n	VR from Figure 4.33* (m^2/s)	R (m)	VR from Manning eq. (m^2/s)
1	0.050	0.0372	0.0206	0.00434
2	0.040	0.0697	0.0387	0.0155
3	0.030	0.186	0.103	0.106
4	0.025	0.557	0.310	0.799
5	0.027	0.372	0.206	0.375

*VR in Figure 4.33 is in ft²/s, the appropriate unit conversion has been made.

The roughness coefficient, n, is therefore approximately equal to 0.027 and the channel has a hydraulic radius, R, equal to 0.206 m. The flow area, A, is therefore given by

$$A = \frac{Q}{V} = \frac{0.7}{1.8} = 0.389 \text{ m}^2$$

A and R for a triangular channel can be expressed in terms of the depth, y, and size slope, m, by the relations

$$A = my^2 = 0.389 \text{ m}^2$$

$$R = \frac{A}{2y\sqrt{1 + m^2}} = \frac{0.389}{2y\sqrt{1 + m^2}} = 0.206 \text{ m}$$

Solving these equations simultaneously yields two solutions:

$$m = 1.71, \quad y = 0.48 \text{ m} \qquad \text{and} \qquad m = 0.58, \quad y = 0.82 \text{ m}$$

For stable side slopes, use $m = 1.71$.

Stage II: The performance of this design must now be reviewed for the upper bound retardance, B. Therefore, Stage II of the design is as follows:

Trial no.	Assumed y (m)	A (m²)	R (m)	$V = Q/A$ (m/s)	VR (m²/s)	n from Figure 4.33	V from Manning eq. (m/s)
1	0.6	0.62	0.26	1.13	0.29	0.080	0.72
2	0.7	0.84	0.30	0.83	0.25	0.085	0.75
3	0.8	1.1	0.35	0.64	0.22	0.09	0.78
4	0.75	0.96	0.32	0.73	0.23	0.09	0.74

Therefore, the depth of flow and velocity at Stage II are 0.75 m and 0.73 m/s, respectively. The freeboard to be added to the channel can be estimated from Equation 4.213. The required freeboard is given by

$$F = 0.152 + \frac{V^2}{2g} = 0.152 + \frac{0.73^2}{(2)(9.81)} = 0.18 \text{ m}$$

Since this freeboard is less than the minimum freeboard of 0.30 m, use a freeboard of 0.30 m. This freeboard is then added to the flow depth of 0.75 m to yield a total channel depth of 1.05 m. The (triangular) channel is to have side slopes of 1.7:1 (H:V).

In concluding the design, it is important to check the Froude number to verify subcritical flow under Stage II conditions. The hydraulic depth, D, is given by

$$D = \frac{A}{T} = \frac{my^2}{2my} = \frac{y}{2} = \frac{0.75}{2} = 0.38 \text{ m}$$

and the Froude number, Fr, is

$$\text{Fr} = \frac{V}{\sqrt{gD}} = \frac{0.73}{\sqrt{(9.81)(0.38)}} = 0.38$$

The flow is subcritical and the design is satisfactory. ■

4.6   Design of Sanitary-Sewer Systems

Sanitary sewers transport the wastewater of a community to treatment plants or locations of ultimate disposal. Sanitary-sewer systems consist mostly of buried pipe in which wastewater is conveyed under open-channel flow conditions. Key components of sanitary-sewer systems are service connections, manholes, and pump stations. *Service connections* join the sewer system to individual service locations, *manholes* are access chambers placed at fixed intervals along the sewer system with the primary function of facilitating cleaning and repairs, and *pump stations* are used to lift the flow to higher elevations where necessary. Detailed guidance on the design and construction of sanitary sewers is provided by ASCE (1982).

4.6.1   ■   Design Flows

Sewer capacities are based on the present and probable future quantities of domestic, commercial, and industrial wastewater to be conveyed, ground-water infiltration from surrounding soils, and extraneous inflows from sources such as roof leaders, basement drains, and submerged manhole covers. The duration for which the sewer system will be adequate, called the *design period*, is usually on the order of 50 years (ASCE, 1982). In cases where there may be significant additional development beyond the design period, it is generally prudent to secure easements and rights-of-way for future expansions. Pump stations in sewer systems can be designed with shorter design periods since they are relatively short lived and easier to update. Sewage treatment plants, which are also relatively easy to update, are generally designed for periods of about 20 years.

Sewer flows consist of two components: flows contributed through service connections, and flows contributed by infiltration and inflow (I/I). Flows contributed through service connections are usually related to present and projected populations in the service area multiplied by per capita wastewater production rates. Methods of population projection are described in Section 3.5.2, and the uncertainty in these projections resulting from economic and social variables should always be noted. Design population densities in residential areas are typically taken as the saturation densities, and typical values are listed in Table 4.19. The average daily per capita domestic wastewater flows vary considerably within the United States, and several design per capita flowrates are listed in Table 4.20. Local regulatory agencies usually specify the per capita domestic wastewater flowrates to be used in their jurisdiction. Wastewater flows from commercial and industrial areas must be added to the domestic flowrates shown in Table 4.20. Typical flowrates associated with commercial and industrial areas are shown in Table 4.21.

It is fairly common to assume that the average rate of sewage flow, including domestic, commercial, and industrial flows, is equal to the average rate of water consumption,

■ **Table 4.19**
Typical Saturation Densities

Type of area	Density (persons/ha)
Large lots	5–7
Small lots, single-family	75
Small lots, two-family	125
Multistory apartments	2,500

Source: American Concrete Pipe Association (1981).

City	Flowrate (L/d/capita)	Comments
Berkeley, CA	350	
Boston, MA	380	includes infiltration; multiply by 3 when sewer is flowing full
Des Moines, IA	380× factor	factor $= \frac{18+\sqrt{P}}{4+\sqrt{P}}$ where P is the population in thousands
Detroit, MI	980	
Las Vegas, NV	950	
Little Rock, AR	380	
Milwaukee, WI	1,000	plus additional flow for inflow/infiltration
Orlando, FL	950	
Shreveport, LA	570	plus 5,600 L/d/ha infiltration

■ Table 4.20
Average Per Capita Wastewater Domestic Flowrates

Source: ASCE. *Gravity Sanitary Sewer Design and Construction.* Copyright © 1982 by ASCE. Reprinted by permission.

but this assumption should only be made after careful consideration of the nature of the community. For example, sewage flows may only be approximately equal to water consumption during dry weather and, in arid regions, evaporation and other losses may lead to significantly less sewage flows than water-use rates.

Flows contributed by infiltration must generally be added to the peak rate of flow of wastewater contributed by domestic, commercial, and industrial sources. Regulatory agencies in most states specify maximum allowances for infiltration. For sanitary sewers up to 600 mm in diameter, it is common to allow 71 m^3/d/km for the total length of main sewers, laterals, and house connections. However, it should be noted that infiltration rates as high as 140 m^3/d/km have been recorded for sewers submerged in ground water, with rates up to 2350 m^3/d/km in isolated segments. Sewer size apparently has little effect on the infiltration inflow (McGhee, 1991), since larger sewers tend to have better joint workmanship, negating the increased size of the potential infiltration area. The actual amount of infiltration depends on the quality of sewer installation, the height of the ground water table, and the properties of the soil. Expansive soils tend to pull joints apart, while granular soils permit water to move easily through joints and breaks. Inflows from unregulated sources such as roof leaders and foundation drains are difficult to predict and are usually lumped with the infiltration, which is then called the infiltration and inflow (I/I) contribution.

Wastewater flows, exclusive of I/I, vary continuously, with very low flows during the early morning hours and peak flows occurring at various times during the daylight hours. The I/I component remains fairly constant, except during and immediately

City	Commercial	Industrial
Grand Rapids, MI	150–190 L/person/day, office buildings 1,500–1,900 L/day/room, hotels 750 L/d/bed, hospitals 750–1,150 L/d/room, schools	2.3 × 10^6 L/d/ha
Kansas City, MO	47,000 L/d/ha	94,000 L/d/ha
Memphis, TN	19,000 L/d/ha	19,000 L/d/ha
Santa Monica, CA	91,000 L/d/ha	127,000 L/d/ha

■ Table 4.21
Commercial and Industrial Wastewater Flowrates

Source: American Concrete Pipe Association, 1981. Adapted by permission.

following periods of heavy rainfall, when significant increases usually occur. Based on data reported in ASCE (1982), the ratio of the peak flow, Q_{peak}, to the average daily flow, Q_{ave}, can be estimated by the relation

$$\frac{Q_{peak}}{Q_{ave}} = \frac{5.5}{P^{0.18}}$$

(4.215)

where P is the population of the service area in thousands and the peak flow is defined as the maximum flow occurring during a 15-minute period for any 12-month period (American Concrete Pipe Association, 1985). The ratio of the minimum flow, Q_{min}, to the average daily flow can be estimated by the relation

$$\frac{Q_{min}}{Q_{ave}} = 0.2P^{0.16}$$

(4.216)

Equations 4.215 and 4.216 indicate that the ratio of peak flow to minimum flow varies from less than 3:1 for sewers serving large populations to more than 20:1 for sewers serving smaller populations.

The flowrates commonly used in sewer design are the peak and minimum flowrates generated by the projected population of the service area. To estimate these flows, the average daily flow is first estimated by the product of the population projection and the average domestic per capita flowrate. This flow is added to the commercial and industrial contributions to estimate the average wastewater flowrate. The peak and minimum flowrates are then calculated from the average flowrate using the factors given by Equations 4.215 and 4.216, and the (constant) I/I is added to the calculated flowrates to yield the overall peak and minimum flowrates to be accommodated by the sewer system.

■ Example 4.20

A trunk sewer is to be sized for a 25-km^2 city in which the master plan calls for 60% residential, 30% commercial, and 10% industrial development. The residential area is to be 40% large lots, 55% small single-family lots, and 5% multistory apartments. The average domestic wastewater flowrate is 800 L/d/capita, the average commercial flowrate is 25,000 L/d/ha, and the average industrial flowrate is 40,000 L/d/ha. Infiltration and inflow is 1,000 L/d/ha for the entire area. Estimate the peak and minimum flows to be handled by the trunk sewer.

Solution

From the given data, the total area of the city is 25 km^2 = 2,500 ha and the residential area is 60% of 2,500 ha = 1,500 ha. The residential saturation densities can be taken as those in Table 4.19. Taking the per capita flowrate as 800 L/d/person (= 9.26×10^{-6} m^3/s/person) gives the wastewater flows in the following table

Type	Area (ha)	Density (persons/ha)	Population	Flow (m^3/s)
Large lots	0.40(1500) = 600	6	3,600	0.03
Small single-family lots	0.55(1500) = 825	75	61,875	0.57
Multistory apartments	0.05(1500) = 75	2,500	187,500	1.74
Total			252,975	2.34

The commercial sector of the city covers 30% of 2500 ha = 750 ha, with a flowrate per unit area of 25,000 L/d/ha = 2.89×10^{-4} m³/s/ha. Hence the average flow from the commercial sector is $(2.89 \times 10^{-4})(750) = 0.22$ m³/s.

The industrial sector of the city covers 10% of 2500 ha = 250 ha, with a flowrate per unit area of 40,000 L/d/ha = 4.63×10^{-4} m³/s/ha. Hence the average flow from the commercial sector is $(4.63 \times 10^{-4})(250) = 0.12$ m³/s.

The infiltration and inflow from the entire area is 1000 L/ha × 2500 ha = 2.5×10^6 L/d = 0.03 m³/s.

On the basis of these calculations, the average daily wastewater flow (excluding I/I) is $2.34 + 0.22 + 0.12 = 2.68$ m³/s. Assume that the total population, P, of the city is equal to the residential population of 252,975, then the peak and minimum flow ratios can be estimated by Equations 4.215 and 4.216 as

$$\frac{Q_{\text{peak}}}{Q_{\text{ave}}} = \frac{5.5}{P^{0.18}} = \frac{5.5}{(252.975)^{0.18}} = 2.0$$

$$\frac{Q_{\text{min}}}{Q_{\text{ave}}} = 0.2P^{0.16} = 0.2(252.975)^{0.16} = 0.48$$

The peak and minimum flows are estimated by multiplying the average wastewater flows by these factors and adding the I/I. Thus

$$\text{Peak flow} = 2.0(2.68) + 0.03 = 5.39 \text{ m}^3/\text{s}$$

$$\text{Minimum flow} = 0.48(2.68) + 0.03 = 1.32 \text{ m}^3/\text{s} \qquad ■$$

4.6.2 ■ Hydraulics of Sewers

Design guidelines published by the American Society of Civil Engineers (ASCE, 1982) and the Water Environment Federation (formerly the Water Pollution Control Federation; WPCF, 1982) state that sanitary sewers through 375 mm (15 in.) in diameter be designed to flow half full at the design flowrate, with larger sewers designed to flow three-fourths full. These guidelines reflect the fact that smaller wastewater flows are much more uncertain than larger flows. Since sanitary sewers transport a significant amount of suspended solids, the prevention of solids deposition by specifying minimum permissible velocities is an important aspect of the hydraulic design of sanitary sewers. Minimum permissible velocities are sometimes called *self-cleansing* velocities. The specification of maximum permissible velocities is also important to prevent excessive scouring of the sewer pipe. ASCE (1982) recommends that flow velocities in sanitary sewers should not be less than 0.60 m/s (2 ft/s) or greater than 3.5 m/s (10 ft/s).

In accordance with these requirements, it is usually necessary to calculate the depth of flow and average velocity in circular pipes for given values of the pipe diameter, D, flowrate, Q, and pipe slope, S_o, where, in uniform flow, the slope of the channel, S_o, is equal to the slope of the energy grade line, S_f. Consider the circular-pipe cross-section in Figure 4.34, where h is the depth of flow and θ is the water-surface angle. The depth of flow, h, cross-sectional area, A, and wetted perimeter, P, can be expressed in terms of θ by the geometric relations

$$h = \frac{D}{2}\left[1 - \cos\left(\frac{\theta}{2}\right)\right] \qquad (4.217)$$

$$A = \left(\frac{\theta - \sin\theta}{8}\right)D^2 \qquad (4.218)$$

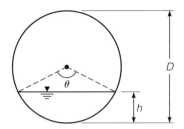

Figure 4.34 ■ Flow in Partially Filled Pipe

$$P = \frac{D\theta}{2} \tag{4.219}$$

The Manning equation is commonly applied to describe the flow in sanitary sewers and can be written in the form

$$Q = \frac{A}{n}\left(\frac{A}{P}\right)^{\frac{2}{3}} S_o^{1/2} \tag{4.220}$$

where n is the roughness coefficient. Combining Equations 4.218 to 4.220 leads to the following form of the Manning equation

$$\boxed{\theta^{-\frac{2}{3}}(\theta - \sin\theta)^{\frac{5}{3}} - 20.16 n Q D^{-\frac{8}{3}} S_o^{-1/2} = 0} \tag{4.221}$$

where the only unknown is θ and the quantities n, Q, D, and S_o are regarded as given for this analysis. The solution of Equation 4.221 (θ) is substituted into Equation 4.217 to obtain the depth of flow in the sewer pipe. The average flow velocity is obtained by first calculating the flow area, A, from Equation 4.218, and then calculating the velocity, V, by

$$V = \frac{Q}{A} \tag{4.222}$$

The only obstacle to solving Equation 4.221 for θ, and hence obtaining h and V, is that Equation 4.221 must be solved iteratively. A variety of techniques, including Newton's method, are available for this purpose. Combining Equations 4.221 and 4.217 gives a nonlinear relationship between the flowrate, Q, and the depth of flow, h, and it can be shown (see Problem 4.89) that the maximum flowrate occurs when $h/D \approx 0.94$. This condition manifests itself as a flow instability when the pipe is flowing almost full, and there is a tendency for the pipe to run temporarily full at irregular intervals (Henderson, 1966). This instability is avoided in practice by designing pipes such that $h/D \leq 0.75$. Another interesting feature of open-channel flow in circular pipes is that the velocity is the same whether the pipe flows half full or completely full (see Problem 4.90). According to the Manning equation (Equation 4.221), the velocity increases with depth of flow until it reaches a maximum at $h/D \approx 0.94$; the velocity then decreases with increasing depth and becomes equal to the half-full velocity when the pipe flows full (Gribbin, 1997).

■ Example 4.21

Water flows at a rate of 4 m³/s in a circular concrete sewer of diameter 1,500 mm and a Manning n of 0.015. If the slope of the sewer is 1%, calculate the depth of flow and velocity in the sewer.

Solution

According to Equation 4.221,

$$\theta^{-\frac{2}{3}}(\theta - \sin\theta)^{\frac{5}{3}} - 20.16nQD^{-\frac{8}{3}}S_o^{-1/2} = 0$$

where in this case $n = 0.015$, $Q = 4 \text{ m}^3/\text{s}$, $S_o = 0.01$, and $D = 1.5$ m. Therefore

$$\theta^{-\frac{2}{3}}(\theta - \sin\theta)^{\frac{5}{3}} - 20.16(0.015)(4)(1.5)^{-\frac{8}{3}}(0.01)^{-\frac{1}{2}} = 0$$

which leads to

$$\theta^{-\frac{2}{3}}(\theta - \sin\theta)^{\frac{5}{3}} - 4.10 = 0$$

Solving this equation iteratively for θ yields

$$\theta = 3.50 \text{ radians}$$

Therefore, the normal flow depth, h, and area, A, are given by Equations 4.217 and 4.218 as

$$h = \frac{D}{2}\left[1 - \cos\left(\frac{\theta}{2}\right)\right] = \frac{1.5}{2}\left[1 - \cos\left(\frac{3.50}{2}\right)\right] = 0.88 \text{ m}$$

$$A = \left(\frac{\theta - \sin\theta}{8}\right)D^2 = \left(\frac{3.50 - \sin 3.50}{8}\right)(1.5)^2 = 1.08 \text{ m}^2$$

The average flow velocity, V, in the sewer is given by

$$V = \frac{Q}{A} = \frac{4}{1.08} = 3.70 \text{ m/s} \qquad\blacksquare$$

The value of the Manning n in sanitary sewers will generally approach a constant which is not a function of the pipe material, but is determined by the grit accumulation and slime build-up on the pipe walls. According to ASCE (1982), this n value is usually on the order of 0.013. A higher n value should be used in cases where additional roughness sources are known or anticipated. Some experiments have shown that the value of n for concrete and vitrified clay pipes flowing partly full is greater than for the same pipe flowing full (Yarnell and Woodward, 1920; Wilcox, 1924). These results have been summarized by Camp (1946). The values of n/n_{full} as a function of the relative flow depth, h/D, are given in Table 4.22, where n_{full} is the n value when the pipe is flowing full. Table 4.22 indicates that the n value for pipes flowing partially full can be as much as 29% higher than the full-flow n value. Similar variations in n value with depth have not been found for PVC pipes (Neale and Price, 1964). ASCE (1982) recommends that until more information and better analyses are available, the decision to use a constant or variable n must be left to the engineer. However, it seems prudent to either incorporate the variation of n with flow depth or apply a factor of safety on the order of 1.3 to the selected constant n value to accommodate the uncertainty of n with depth. Head losses in bends are primarily associated with large-diameter pipes (> 915 mm), are difficult to quantify, and are usually accounted for by increasing the Manning n in the bend by 25% to 40% (McGhee, 1991).

	h/D	n/n_{full}
▪ **Table 4.22**	0	1.00
Variation of the Manning n	0.1	1.22
with Depth	0.2	1.28
	0.3	1.29
	0.4	1.28
	0.5	1.25
	0.6	1.22
	0.7	1.18
	0.8	1.14
	0.9	1.08
	1.0	1.00

Source: Adapted from Camp (1946).

4.6.3 ▪ Sewer-Pipe Material

A variety of pipe materials are used in practice, including concrete, vitrified clay, cast iron, ductile iron, and various thermoplastic materials including PVC. Pipes are broadly classified as either *rigid pipes* or *flexible pipes*. Rigid pipes derive a substantial part of their load-carrying capacity from the structural strength inherent in the pipe wall, while flexible pipes derive their load-carrying capacity from the interaction of the pipe and the embedment soils affected by the deflection of the pipe to the equilibrium point under load. Pipe materials classified as rigid and flexible are listed in Table 4.23, and a summary of the advantages and disadvantages to using various pipe materials, along with typical n values, are given in Table 4.24.

The type of pipe material to be used in any particular case is dictated by several factors, including the type of wastewater to be transported (residential, industrial, or combination), scour and abrasion conditions, installation requirements, type of soil, trench-load conditions, bedding and initial backfill material available, infiltration/exfiltration requirements, and cost effectiveness. Nonreinforced and reinforced concrete sewer pipe is used frequently in sanitary-sewer systems, and the commercially available pipe diameters are given in Table 4.25 (American Concrete Pipe Association, 1985). Numerous test programs have established values for the roughness coefficient of concrete pipe from 0.009 to 0.011, however, higher values of 0.012 and 0.013 are typically used in design to account for the possibility of slime or grease buildup in sanitary sewers (American Concrete Pipe Association, 1985).

4.6.4 ▪ System Layout

The layout of a sanitary-sewer system begins with the selection of an outlet, the determination of the tributary area, the location of the *trunk* and *main* sewers, the determination of whether there is a need for, and the location of, pumping stations and force

	Rigid pipe	Flexible pipe
▪ **Table 4.23**	Concrete	Ductile iron
Rigid and Flexible Pipe	Cast iron	Steel
Materials	Vitrified clay	Thermoplastic (e.g., PVC)

Material	Advantages	Disadvantages	Manning n	Available diameters
Concrete	wide range of structural and pressure strengths wide range of nominal diameters wide range of laying lengths	high weight subject to corrosion when acids are present	0.011–0.015*	300 mm–3,600 mm‡
Vitrified clay	high resistance to chemical corrosion high resistance to abrasion wide range of fittings available	limited range of sizes available high weight subject to shear and beam breakage when improperly bedded	0.010–0.015	100 mm–1,070 mm
Cast iron	long laying lengths high pressure and load-bearing capacity	subject to corrosion where acids are present subject to chemical attack in corrosive soils subject to shear and beam breakage when improperly bedded high weight	0.011–0.015†	50 mm–1,220 mm
Ductile iron	long laying lengths high pressure and load-bearing capacity high impact strength high beam strength	subject to corrosion where acids are present subject to chemical attack in corrosive soils high weight	0.011–0.015†	50 mm–1,220 mm
PVC	Light weight long laying lengths high impact strength ease in field cutting and tapping	Subject to attack by certain organic chemicals subject to excessive deflection when improperly bedded and haunched limited range of sizes available subject to surface changes effected by long-term ultraviolet exposure	0.010–0.015	100 mm–375 mm

■ **Table 4.24** Sewer-Pipe Materials

Source: Adapted from ASCE (1982).
*The n values shown are for new pipes.
† Cement-lined and seal coated.
‡ Diameters are for reinforced-concrete pipe. Nonreinforced concrete pipe is available in nominal diameters from 100 mm to 900 mm.

mains; the location of underground rock formations; the location of water and gas lines, electrical, telephone, and television wires, and other underground utilities. The trunk sewer refers to the sewer pipeline that receives many tributary branches and serves a large territory, and a main sewer is a principal sewer to which branch sewers are tributary. In larger systems, the main sewer is also called the trunk sewer. The selected system outlet depends on the scope and objectives of the particular project and may consist of a pumping station, an existing trunk or main sewer, or a treatment plant.

Since the flows in sewer systems are driven by gravity, preliminary layouts are generally made using topographic maps. Trunk and main sewers are located at the lower elevations of the service area, although existing roadways and the availability of rights-of-way

■ **Table 4.25**
Available Sizes of
Concrete Pipe

	Nonreinforced pipe		Reinforced Pipe	
	Diameter (mm)	Diameter (in.)	Diameter (mm)	Diameter (in.)
	100	4	—	—
	150	6	—	—
	205	8	—	—
	255	10	—	—
	305	12	305	12
	380	15	380	15
	455	18	455	18
	535	21	535	21
	610	24	610	24
	685	27	685	27
	760	30	760	30
	840	33	840	33
	915	36	915	36
	—	—	1,065	42
	—	—	1,220	48
	—	—	1,370	54
	—	—	1,525	60
	—	—	1,675	66
	—	—	1,830	72
	—	—	1,980	78
	—	—	2,135	84
	—	—	2,285	90
	—	—	2,440	96
	—	—	2,590	102
	—	—	2,745	108

may affect the exact locations. In developed areas, sanitary sewers are commonly located at or near the centers of roadways and alleys. In very wide streets, however, it may be more economical to install sanitary sewers on both sides of the street. When sanitary sewers are in close proximity to public water supplies, it is common practice to use pressure-type sewer pipe, concrete encasement of the sewer pipe, or sewer pipe with joints that meet stringent infiltration/exfiltration requirements, or at least to put water pipes and sewer pipes on opposite sides of the street. Most building codes prohibit sanitary-sewer installation in the same trench as water mains.

Manholes provide access to the sewer system for preventitive maintenance and emergency service and are generally located at the junctions of sanitary sewers, at changes of grade or alignment, and at the beginning of the sewer system. A typical manhole is illustrated in Figure 4.35(a), indicating that manhole covers are typically 0.61 m (2 ft) in diameter, with a working space in the manhole typically 1.2 m (4 ft) in diameter. The invert of the manhole should conform to the shape and slope of the incoming and outgoing sewer lines. In some cases, a section of pipe is laid through the manhole and the upper portion is sawed or broken off. More commonly, U-shaped channels are specified. The spacing between manholes varies with local conditions and methods of sewer maintenance and is often in the range of 90 to 150 m, with spacings of 150 to 300 m used for larger sewers that a person can walk through. In cases where a sewer pipe enters a manhole at an elevation considerably higher than the outgoing pipe, it is generally not acceptable to let the incoming wastewater simply pour into the manhole, since this does not provide an acceptable workspace for maintenance and repair. Under

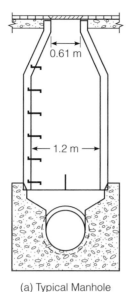

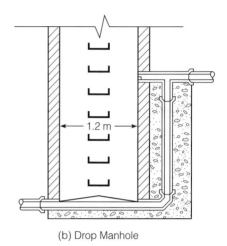

Figure 4.35 ■ Typical Manholes

ASCE. *Gravity Sewer Design and Construction.* Copyright © 1982 by ASCE. Reprinted by permission.

(a) Typical Manhole   (b) Drop Manhole

these conditions, a *drop manhole*, illustrated in Figure 4.35(b), is used. Drop manholes are typically specified when the invert of the inflow pipe is more than 0.6 m above the elevation that would be obtained by matching the crowns of the inflow and outflow pipes.

Sanitary sewers should be buried to a sufficient depth that they can receive the contributed flow from the tributary area by gravity flow. Deep basements and buildings on land substantially below street level may require individual pumping facilities. In northern states, a cover of 3 m is typically required to prevent freezing, while in southern states the minimum cover is dictated by traffic loads, and ranges upward from 0.75 m, depending on the pipe size and anticipated loads (McGhee, 1991). The depth of sanitary sewers is such that they pass under all other utilities, with the possible exception of storm sewers. Sanitary sewers typically have a minimum diameter of 205 mm (8 in.), and it is common practice to lay service connections at a slope of 2%, with a minimum slope of 1%. In some developments where the houses are set well back from the street, the length and slope of the house connections may determine the minimum sewer depths. Service connections typically use 150- or 205-mm diameter pipe (Corbitt, 1990). The diameters of sanitary sewers should never be allowed to decrease in the downstream direction, since this can cause sediment accumulation and blockage where this reduction occurs.

Pump stations, also known as *lift stations*, are frequently necessary in flat terrain to raise the wastewater to a higher elevation so that gravity flow can continue at reasonable slopes and depths. Pump stations typically have a wet well–dry well configuration, where the wet well receives the wastewater flow and is sized for a 10- to 30-min detention time with a bottom slope of 2:1, and the dry well is the pump and motor area. Ventilation, humidity control (dry well), and standby-power supply are important design considerations. For smaller stations ($< 3,800$ m³/d), at least two pumps should be provided; for larger pump stations, three or more pumps. In both cases, the pump capacities should be sufficient to pump at the maximum wastewater flowrate if any one pump is out of service (Corbitt, 1990). To avoid clogging, the associated suction and discharge piping should be at least 100 mm in diameter and the pump should be capable of passing 75-mm diameter solids.

4.6.5 ■ Sulfide Generation

Hydrogen-sulfide (H_2S) generation is a common problem in sanitary sewers. Among the problems associated with H_2S generation are odor, health hazard to maintenance crews, and corrosion of unprotected sewer pipes produced from cementitious materials and metals. The design of sanitary sewers seeks to preclude septic conditions, and to provide an environment that is relatively free of H_2S. Generation of H_2S in sanitary sewers results primarily from the bacterial reduction of sulfate under anaerobic conditions. Corrosive conditions occur when sulfuric acid (H_2SO_4) is derived through the oxidation of hydrogen sulfide by bacterial action on the exposed sewer pipe wall. Concrete pipes, asbestos-cement pipes, and mortar linings on ferrous pipes will experience surface reactions in which the surface material is converted to an expanding, pasty mass which may fall away and expose new surfaces to corrosive attack. Hydrogen sulfide gas is extremely toxic and can cause death at concentrations as low as 300 ppm (0.03%) in air. A person who ignores the first odor of the gas quickly loses the ability to smell the gas, eliminating further warning and leading to deadly consequences.

Significant factors that contribute to H_2S generation are high wastewater temperatures and low flow velocities. The potential for sulfide generation can be assessed using the formula (Pomeroy and Parkhurst, 1977)

$$Z = 0.308 \frac{\text{EBOD}}{S_o^{0.50} Q^{0.33}} \times \frac{P}{B} \tag{4.223}$$

where EBOD is the *effective biochemical oxygen demand* in mg/L, defined by the relation

$$\text{EBOD} = \text{BOD} \times 1.07^{T-20} \tag{4.224}$$

where BOD is the average five-day *biochemical oxygen demand* (mg/L at 20°C) during the highest 6-hour flow period of the day, T is the temperature of the wastewater in the sewer (°C), S_o is the slope of the sewer, Q is the flowrate in the sewer (m^3/s), P is the wetted perimeter (m), and B is the top width (m) of the sewer flow. The relationship given by Equation 4.223 is commonly referred to as the *Z formula*, and the relationship between the calculated Z value and the potential for sulfide generation is given in Table 4.26.

It is usually impractical or impossible to design a sulfide-free sewer system, and engineers endeavor to minimize sulfide generation and use corrosion-resistant materials to the extent possible. Wastewater treatment plants are usually located at the terminus of sewer systems, and chlorination with either elemental chlorine or hypochlorite quickly destroys sulfide and odorous organic sulfur compounds. Chlorination in sanitary sewers is generally considered impractical. The dissolution of air or oxygen in the wastewater as it moves through the sewer system, however, is an effective sulfide control measure.

■ Table 4.26 Sulfide Generation Based on Z Values	Z Values	Sulfide condition
	$Z < 5,000$	sulfide rarely generated
	$5,000 \leq Z \leq 10,000$	marginal condition for sulfide generation
	$Z > 10,000$	sulfide generation common

Source: ASCE. *Gravity Sanitary Sewer Design and Construction.* Copyright © 1982 by ASCE. Reprinted by permission.

■ Example 4.22

A 915-mm diameter concrete sewer is laid on a slope of 0.9% and is to carry 1.7 m³/s of domestic wastewater. If the five-day BOD of the wastewater at 20°C is expected to be 300 mg/L, determine the potential for sulfide generation when the wastewater temperature is 25°C.

Solution

From the given data, $Q = 1.7$ m³/s, $D = 915$ mm $= 0.915$ m, $S_o = 0.009$, and Equation 4.221 gives the flow angle θ by the relation

$$\theta^{-\frac{2}{3}}(\theta - \sin\theta)^{\frac{5}{3}} - 20.16nQD^{-\frac{8}{3}}S_o^{-1/2} = 0$$

Taking the Manning n as 0.013, Equation 4.221 gives

$$\theta^{-\frac{2}{3}}(\theta - \sin\theta)^{\frac{5}{3}} - 20.16(0.013)(1.7)(0.915)^{-\frac{8}{3}}(0.009)^{-\frac{1}{2}} = 0$$

which simplifies to

$$\theta^{-\frac{2}{3}}(\theta - \sin\theta)^{\frac{5}{3}} = 5.95$$

and yields

$$\theta = 4.3 \text{ radians}$$

The flow perimeter, P, is given by Equation 4.219 as

$$P = \frac{D\theta}{2}$$

and the top width, B, can be inferred from Figure 4.34 as

$$B = 2\left(\frac{D}{2}\right)\sin\left(\frac{\theta}{2}\right) = D\sin\left(\frac{\theta}{2}\right)$$

Hence

$$\frac{P}{B} = \frac{D\theta/2}{D\sin(\theta/2)} = \frac{\theta}{2\sin(\theta/2)}$$

and in this case

$$\frac{P}{B} = \frac{4.3}{2\sin(4.3/2)} = 2.57$$

The effective BOD, EBOD at $T = 25°C$, is given by Equation 4.224 as

$$\text{EBOD} = \text{BOD} \times 1.07^{T-20} = 300 \times 1.07^{25-20} = 421 \text{ mg/L}$$

According to the Z formula, Equation 4.223,

$$Z = 0.308\frac{\text{EBOD}}{S_o^{0.50}Q^{0.33}} \times \frac{P}{B} = 0.308\frac{421}{(0.009)^{0.50}(1.7)^{0.33}} \times 2.57 = 2,948$$

Therefore, according to Table 4.26, hydrogen sulfide will be rarely generated. ■

4.6.6 ■ Design Computations

Basic information required prior to computing the sizes and slopes of the sewer pipes includes (1) a topographic map showing the proposed locations of the sewer lines, (2) the tributary areas to each line, (3) the (final) ground-surface elevations along each line, (4) elevations of the basements of low-lying houses and other buildings, and (5) the elevations of existing sanitary sewers which must be intercepted.

After the sewer layout has been developed, a typical computation form useful in sizing sanitary sewers is shown in Figure 4.36 (ASCE, 1982). Design computations begin with the characteristics of the sewer-pipe configuration in columns 1 to 5 and lead to the computation of the sewer slopes in column 14, diameters in column 15, and sewer invert elevations in columns 21 and 22. The steps to be followed in using Figure 4.36 in the design of sanitary sewers are as follows:

1. Computations begin with the uppermost pipe in the sewer system.
2. List the pipeline number in column 1 (usually starting from the number 1), list the street location of the pipe in column 2, list the beginning and ending manhole numbers in columns 3 and 4 (the manhole numbers usually start from 1 at the uppermost manhole), and list the length of the sewer pipe in column 5. The ground-surface elevations at the upstream and downstream manhole locations are listed in columns 23 and 24. These figures are used as reference elevations to ensure that the cover depth is acceptable and to compare the pipe slope with the ground slope.
3. In column 6, list the land area that will contribute wastewater flow to the sewer line. The contributing land area can be estimated using a topographic map and the proposed development plan.
4. In column 7, list the total area contributing wastewater flow to the sewer pipe. This total contributing area is the sum of the area that contributes directly to the pipe (listed in column 6) and the area that contributes flow to the upstream pipes that feed the sewer pipe.
5. In column 8, the contribution of infiltration and inflow (I/I) to the pipe flow is listed. I/I is usually calculated by multiplying the length of the pipe by a design inflow rate in $m^3/d/km$ and adding this inflow to the I/I contribution from all upstream connected pipes.
6. In column 9, the maximum sewage flow is calculated by multiplying the contributing area listed in column 7 by the average wastewater flowrate (usually given in $m^3/d/ha$) and the peaking factor (derived using a relation similar to Equation 4.215).
7. In column 10, the peak design flow is calculated as the sum of I/I (column 8) and the peak wastewater flow (column 9).
8. In columns 11, 12, and 13, the minimum design flow is calculated using a similar procedure to that used in calculating the peak design flow. The I/I contribution (column 11) is the same as calculated in step 5 (column 8); the mimimum wastewater flow (column 12) is the average wastewater flow multiplied by the minimum-flow factor (derived using a relation similar to Equation 4.216); and the minimum design flow (column 13) is the sum of I/I (column 11) and the minimum wastewater flow (column 12).
9. For a given pipe diameter, the slope of the sewer (column 14) is equal to the steeper of the ground slope and the slope that yields the minimum permissible velocity (0.60 m/s) at the minimum design flow. The relationship between the pipe slope and the flow velocity is given by Equations 4.218 to 4.222. Once the slope is determined (using the minimum flowrate), the flow depth and velocity at the maximum flowrate (column 10) is calculated. If the flow depth under maximum-flow

Line No. (1)	Location (2)	Manhole No.		Length (m) (5)	Area			Maximum Flow				Minimum Flow				Slope of Sewer (14)	Diam (mm) (15)	Min Velocity (m/s) (16)	Max Velocity (m/s) (17)	Max Depth (mm) (18)	Manhole Invert Drop (m) (19)	Fall in Sewer (m) (20)	Sewer Invert Elevation		Ground Surface Elevation	
		From (3)	To (4)		Increment (ha) (6)	Total (ha) (7)		I/I (m³/s) (8)	Sewage (m³/s) (9)	Total (m³/s) (10)		I/I (m³/s) (11)	Sewage (m³/s) (12)	Total (m³/s) (13)									Upper End (m) (21)	Lower End (m) (22)	Upper End (m) (23)	Lower End (m) (24)

Figure 4.36 ■ Typical Computation Form for Design of Sanitary Sewers

conditions exceeds the acceptable limit (0.5D or 0.75D), or the maximum-flow velocity exceeds an acceptable limit (3.5 m/s), then the pipe diameter is increased to the next commercial size (e.g., see Table 4.25), and the slope (column 14) is recalculated. This iteration may need to be repeated until the minimum permissible velocity, acceptable flow depth, and maximum-velocity criteria are all met. After this iteration, the diameter of the sewer pipe is listed in column 15, the minimum velocity in column 16, the maximum velocity in column 17, and the maximum-flow depth in column 18. In cul-de-sacs and other dead-end street sections, it is often impossible to satisfy the minimum velocity requirements. Such pipes will usually require periodic flushing to scour accumulated sediments.

10. A drop in the sewer invert at the manhole upstream of a sewer pipe is necessary when either the sewer line changes direction or the diameter of the sewer is increased (over the upstream pipe). If a change in direction occurs, it is advisable to drop the pipe invert by 30 mm to compensate for the energy losses. If the diameter of the sewer pipe leaving a manhole is larger than the diameter of the sewer pipe entering the manhole, head losses are accounted for either by matching the crown elevations of the entering and leaving pipes or by aligning points that are 80% of the diameter from the pipe inverts. It is usually preferable to align the crowns of the pipes because the associated drop in the invert will always exceed 30 mm, losses associated with a change in direction can be neglected, and aligning the crowns of the pipes ensures that the smaller sewer will not flow full as a result of the backwater from the larger pipe unless the larger pipe also flows full. The drop in the sewer invert resulting from either a change in pipe direction or a change in pipe diameter is listed in column 19.

11. The fall in the sewer line is equal to the product of the slope (column 14) and the length of the sewer (column 5) and is listed in column 20.

12. The invert elevations of the upper and lower ends of the sewer line are listed in columns 21 and 22. The difference in these elevations is equal to the fall in the sewer calculated in column 20, and the invert elevation at the upper end of the sewer differs from the invert elevation at the lower end of the upstream pipe by the manhole invert drop listed in column 19.

13. Repeat steps 3 to 12 for all connected pipes, proceeding downstream until the sewer main is reached. Then repeat steps 3 to 12 for the sewer main until the connection to the next lateral is reached.

14. Repeat steps 3 to 13, beginning with the outermost pipe in the next sewer lateral. Continue designing the sewer main until the outlet point of the sewer system is reached.

The design procedure described here is usually automated to some degree and is most easily implemented using a spreadsheet program. The design procedure is illustrated by the following example.

■ Example 4.23

A sewer system is to be designed to service the residential area shown in Figure 4.37. The average per capita wastewater flowrate is estimated to be 800 L/d/capita, and the infiltration and inflow (I/I) is estimated to be 70 m^3/d/km. The sewer system is to join an existing main sewer at manhole (MH) 5, where the average wastewater flow is 0.37 m^3/s, representing the contribution of approximately 100,000 people. The I/I contribution to the flow in the main sewer at MH 5 is negligible. The existing sewer main at MH 5 is 1,065 mm in diameter, has an invert elevation of 55.35 m, and is laid on a slope of

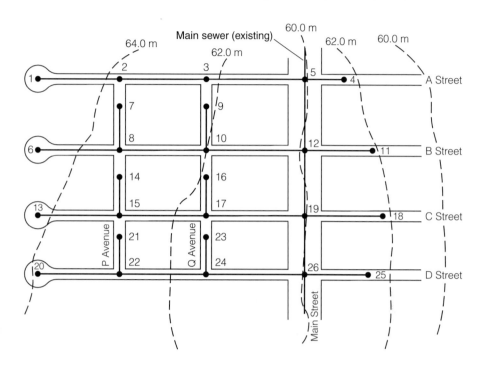

Figure 4.37 ■ Residential Sewer Project

0.9%. The layout of the proposed sewer system shown in Figure 4.37 is based on the topography of the area. Pipe lengths, contributing areas, and ground-surface elevations are given in the following table:

Line no. (1)	Location (2)	Manhole no.		Length (m) (5)	Contributing area (ha) (6)	Ground surface elevation	
		From (3)	To (4)			Upper end (m) (28)	Lower end (m) (29)
0	Main Street	—	5	—	—	—	60.04
1	A Street	1	2	53	0.47	65.00	63.80
2	A Street	2	3	91	0.50	63.80	62.40
3	A Street	3	5	100	0.44	62.40	60.04
4	A Street	4	5	89	0.90	61.88	60.04
5	Main Street	5	12	69	0.17	60.04	60.04
6	B Street	6	8	58	0.43	65.08	63.20
7	P Avenue	7	8	50	0.48	63.60	63.20
8	B Street	8	10	91	0.39	63.20	62.04
9	Q Avenue	9	10	56	0.88	62.72	62.04
10	B Street	10	12	97	0.45	62.04	60.04
11	B Street	11	12	125	0.90	61.88	60.04
12	Main Street	12	19	75	0.28	60.04	60.20
13	C Street	13	15	57	0.60	64.40	62.84
14	P Avenue	14	15	53	0.76	63.24	62.84
15	C Street	15	17	97	0.51	62.84	61.60
16	Q Avenue	16	17	63	0.94	62.12	61.60
17	C Street	17	19	100	0.46	61.60	60.20
18	C Street	18	19	138	1.41	61.92	60.20
19	Main Street	19	26	78	0.30	60.20	60.08

Design the sewer system between A Street and C Street for a saturation density of 130 persons/ha. Local municipal guidelines require that the sewer pipes have a minimum cover of 2 m, a minimum slope of 0.08%, a peak flow factor of 3.0, a minimum flow factor of 0.5, and a minimum allowable pipe diameter of 150 mm.

Solution

From the given data, the average wastewater flow is 800 L/d/person × 130 persons/ha = 104,000 L/d/ha = 0.0722 m³/min/ha. The infiltration and inflow (I/I) is 70 m³/d/km = 4.86×10^{-5} m³/min/m.

The results of the design computations are shown in Table 4.27. The computations begin with Line 0, which is the existing sewer main that must be extended to accommodate the sewer lines in the proposed residential development. The average flow in the sewer main is 0.37 m³/s = 22.2 m³/min, hence the maximum flow is 3.0 × 22.2 = 66.6 m³/min (column 10) and the minimum flow is 0.5 × 22.2 = 11.1 m³/min (column 13). With a slope of 0.009 (column 14) and a diameter of 1,065 mm (column 15), the velocity at the minimum flowrate is calculated using Equations 4.218 to 4.222 as 1.75 m/s (column 16). The velocity at the maximum flowrate is 2.88 m/s (column 17) with a maximum depth of flow, calculated using Equation 4.217 as 476 mm, which is 45% of the pipe diameter. The invert elevation of the main sewer at MH 5 is 55.35 m (column 22) and the ground-surface elevation at MH 5 is 60.04 m (column 24).

The design of the sewer system begins with Line 1 on A Street, which goes from MH 1 to MH 2 and is 53 m long. The area contributing wastewater flow is 0.47 ha (column 7), hence the average wastewater flow in Line 1 is 0.0722 m³/min/ha × 0.47 ha = 0.0339 m³/min. The I/I contribution is 4.86×10^{-5} m³/min/m × 53 m = 0.0026 m³/min (columns 8 and 11), the peak wastewater flow is 3.0 × 0.0339 = 0.102 m³/min (column 9), giving a total peak flow of 0.102 + 0.0026 = 0.105 m³/min (column 10). Similarly, the minimum wastewater flow is 0.5 × 0.0339 = 0.0170 m³/min (column 12), and the minimum total flow is 0.0170 + 0.0026 = 0.0196 m³/min (column 13). Using the minimum allowable pipe diameter of 150 mm, and a slope of 0.047 (column 14) yields a velocity of 0.60 m/s (column 16), which is the minimum permissible velocity. At the peak flow of 0.105 m³/s, the velocity is 0.99 m/s (column 17) and the depth of flow is 23 mm (column 18). The peak velocity is less than the maximum allowable velocity of 3.5 m/s, and the depth of flow is (much) less than the maximum desirable depth of flow of 75 mm (half full), and so the sewer line is acceptable. With a slope of 0.047 and a length of 53 m, the fall in the sewer is 0.047 × 53 m = 2.49 m (column 20). The ground elevation at the upstream end of the sewer is 65.00 m (column 23). With a cover of 2.11 m (slightly above the minimum cover of 2 m), the invert elevation of the upstream end of the pipe is taken as 65.00 − 2.11 − 0.15 = 62.74 m (column 21), and therefore the invert elevation of the downstream end of the pipe is equal to the upstream invert minus the fall in the sewer, which is 62.74 − 2.49 = 60.25 m (column 22).

The design of Lines 2 and 3 follow the same sequence as for Line 1, with the exception that the wastewater flows in each pipe are derived from the sum of the contributing areas of all upstream pipes plus the pipe being designed and that the I/I flow in each pipe is the sum of all upstream I/I flows plus the I/I contribution to the pipe being designed. Using this approach, the invert elevation at the end of Line 3 is 56.27 m (column 22), where the sewer lateral joins the main sewer. The crown elevation of the 150 mm lateral (Line 3) is 56.27 + 0.15 = 56.42 m, which matches the crown elevation of the sewer main of 55.35 + 1.065 = 56.42 m. Line 4 is designed from the other side of Main Street and joins the main sewer at MH 5 with an invert elevation of 56.27 m, which also aligns the crown of the sewer with that of the main.

■ **Table 4.27** Sewer Design Calculations

Line No. (1)	Location (2)	Manhole No. From (3)	Manhole No. To (4)	Length (m) (5)	Area Increment (ha) (6)	Area Total (ha) (7)	Max Flow I/I (m³/min) (8)	Max Flow Sewage (m³/min) (9)	Max Flow Total (m³/min) (10)	Min Flow I/I (m³/min) (11)	Min Flow Sewage (m³/min) (12)	Min Flow Total (m³/min) (13)	Slope of Sewer (14)	Diam (mm) (15)	Min Velocity (m/s) (16)	Max Velocity (m/s) (17)	Max Depth (mm) (18)	Invert Drop (m) (19)	Fall in Sewer (m) (20)	Sewer Invert Upper End (m) (21)	Sewer Invert Lower End (m) (22)	Ground Surface Upper End (m) (23)	Ground Surface Lower End (m) (24)
0	Main Street	–	5	–	–	–	–	–	66.6	–	–	11.1	0.009	1065	1.75	2.88	476	–	–	–	55.35	–	60.04
1	A Street	1	2	53	0.47	0.47	0.0026	0.102	0.105	0.0026	0.0170	0.0196	0.047	150	0.60	0.99	23	–	2.49	62.74	60.25	65.00	63.80
2	A Street	2	3	91	0.50	0.97	0.0070	0.210	0.217	0.0070	0.0350	0.0420	0.024	150	0.60	0.97	40	–	2.18	60.25	58.07	63.80	62.40
3	A Street	3	5	100	0.44	1.41	0.0120	0.305	0.317	0.0120	0.0509	0.0629	0.018	150	0.61	0.97	52	–	1.80	58.07	56.27	62.40	60.04
4	A Street	4	5	89	0.90	0.90	0.0043	0.195	0.199	0.0043	0.0325	0.0368	0.027	150	0.60	0.98	37	–	2.40	58.67	56.27	61.88	60.04
5	Main Street	5	12	69	0.17	309.96	0.0197	67.14	67.16	0.0197	11.19	11.21	0.001	1220	0.78	1.24	879	0.155	0.07	55.20	55.13	60.04	60.04
6	B Street	6	8	58	0.43	0.43	0.0028	0.0932	0.0960	0.0028	0.0155	0.0183	0.050	150	0.60	0.99	22	–	2.90	62.90	60.00	65.08	63.20
7	P Avenue	7	8	50	0.48	0.48	0.0024	0.104	0.106	0.0024	0.0173	0.0197	0.048	150	0.60	1.00	23	–	2.40	61.34	58.99	63.60	63.20
8	B Street	8	10	91	0.39	1.30	0.0097	0.282	0.292	0.0097	0.0469	0.0566	0.019	150	0.60	0.97	49	–	1.73	58.99	57.26	63.20	62.04
9	Q Avenue	9	10	56	0.88	0.88	0.0027	0.191	0.194	0.0027	0.0318	0.0345	0.029	150	0.60	1.00	36	–	1.62	60.44	58.82	62.72	62.04
10	B Street	10	12	97	0.45	2.67	0.0171	0.578	0.595	0.0171	0.0964	0.114	0.011	205	0.61	0.95	86	0.055	1.07	57.21	56.14	62.04	60.04
11	B Street	11	12	125	0.90	0.90	0.0061	0.195	0.201	0.0061	0.0325	0.0386	0.026	150	0.60	0.97	37	–	3.25	59.45	56.20	61.88	60.04
12	Main Street	12	19	75	0.28	313.81	0.0465	67.97	68.02	0.0465	11.33	11.38	0.001	1220	0.79	1.24	887	–	0.08	55.13	55.06	60.04	60.20
13	C Street	13	15	57	0.60	0.60	0.0028	0.130	0.133	0.0028	0.0217	0.0245	0.040	150	0.60	1.00	27	–	2.28	62.20	59.92	64.40	62.84
14	P Avenue	14	15	53	0.76	0.76	0.0026	0.165	0.168	0.0026	0.0274	0.0300	0.034	150	0.61	1.02	32	–	1.80	60.38	58.58	63.24	62.84
15	C Street	15	17	97	0.51	1.87	0.0101	0.405	0.415	0.0101	0.0675	0.0776	0.015	150	0.61	0.98	63	–	1.46	58.58	57.12	62.84	61.60
16	Q Avenue	16	17	63	0.94	0.94	0.0031	0.204	0.207	0.0031	0.0339	0.0370	0.028	150	0.60	1.01	37	–	1.76	59.90	58.21	62.12	61.60
17	C Street	17	19	100	0.48	3.27	0.0180	0.708	0.726	0.0180	0.1180	0.1360	0.010	205	0.60	0.96	83	0.055	1.00	57.07	56.07	61.60	60.20
18	C Street	18	19	138	1.41	1.41	0.0067	0.305	0.312	0.0067	0.0509	0.0576	0.019	150	0.60	0.99	51	–	2.62	58.75	56.13	61.92	60.20
19	Main Street	19	26	78	0.30	318.79	0.0750	69.05	69.13	0.0750	11.51	11.59	0.001	1220	0.79	1.25	900	–	0.08	55.06	54.98	60.20	60.08

The main sewer leaving MH 5 (Line 5) is designed next. The tributary area to Line 5 is the sum of the contributing area of all contributing sewer laterals (Lines 1 to 4, $1.41 + 0.9 = 2.31$ ha) plus the equivalent area of the average flow in the main sewer upstream of MH 5 (0.37 m^3/s ÷ 0.0722 m^3/min/ha $= 307.48$ ha) plus the area that contributes directly to Line 5 (0.17 ha). Hence the total contributing area is $2.31 + 307.48 + 0.17 = 309.96$ ha (column 7). The I/I contribution to Line 5 is the sum of the I/I contributions to all upstream laterals ($0.012 + 0.0043 = 0.0163$ m^3/min) plus the I/I contribution directly to Line 5 (69 m $\times 4.86 \times 10^{-5}$ m^3/min/m $= 0.0034$ m^3/min) for a total I/I contribution of $0.0163 + 0.0034 = 0.0197$ m^3/min (column 8). The peak flow (67.16 m^3/min) and the minimum flow (11.21 m^3/min) are calculated using the flow factors as previously described. Using a pipe diameter of 1,065 mm (the diameter of the upstream sewer section) on a slope that yields the minimum permissible velocity (0.60 m/s) is not acceptable since the capacity of the sewer pipe will be exceeded under peak-flow conditions. Using the next larger commercial size of 1,220 mm (column 15) on a slope of 0.001 (column 14) yields flow conditions where the pipe flows slightly less than three-fourths full (879 mm), and the minimum and maximum velocity criteria are met. Using the smallest slope (0.001) that meets the depth-of-flow and velocity criteria is a good choice since the ground surface is flat, minimizing excavation. Aligning the crowns of the incoming and outgoing sewer pipes at MH 5 requires a drop of $1.220 - 1.065 = 0.155$ m (column 19), hence the sewer invert leaving MH 5 is at elevation $55.35 - 0.155 = 55.20$ m (column 21). The drop in the sewer line is 0.001×69 m $= 0.074$ m (column 20), and the invert elevation at the end of Line 5 is $55.20 - 0.07 = 55.13$ m.

The design of the other sewer laterals and the main sewer line follows the same procedure. In cases where laterals intersect with crown elevations more than 0.6 m apart (vertically), a drop manhole must be used (see Figure 4.35). Drop manholes are required at MH 8, MH 10, MH 15, and MH 17. All laterals intersecting the main sewer have crown elevations that match those of the main sewer. All pipes in the sewer system are made of concrete; the 150 mm and 205 mm pipes are nonreinforced, and the sewer main consists entirely of reinforced concrete pipe (RCP). ■

4.7 Computer Models

Several good computer models are available for simulating flow in open channels, and many such models are parts of more comprehensive models that simulate surface runoff from rainfall (see Chapter 6). In engineering practice, the use of computer models to apply the fundamental principles covered in this chapter is sometimes essential. There are usually a variety of models to choose from for a particular application, but in doing work to be reviewed by regulatory bodies, models developed and maintained by agencies of the U.S. government have the greatest credibility and, perhaps more important, are almost universally acceptable in supporting permit applications and defending design protocols. A secondary guideline in choosing a model is that the simplest model that will accomplish the design objectives should be given the most serious consideration. Two of the more widely used models developed and endorsed by agencies of the U.S. government are described briefly here.

■ HEC-2/HEC-RAS. These models were developed and are maintained by the U.S. Army Corps of Engineers Hydrologic Engineering Center. HEC-2, by far the most widely used program for calculating water-surface profiles in natural channels, computes water-surface profiles for one-dimensional, gradually varied steady flows in natural or man-made channels with small slopes. Computations are based on the standard-step method. Flow through bridges, culverts, and weirs can be modeled as well as the effects of buildings in flood plains. HEC-RAS (River Analysis System), in the process of replacing HEC-2, has many of the same capabilities as HEC-2, but with improved user graphics and reporting facilities.

■ WSP2. This model was developed and is maintained by National Resources Conservation Service (NRCS) in the U.S. Department of Agriculture. WSP2 calculates the water-surface profile in one-dimensional, gradually varied steady flow. Calculations are made using the standard-step method, and the capability to handle flow through hydraulic structures is quite limited.

Besides these two models, there are certainly many other good models capable of performing the same tasks.

Summary

The fundamental equations governing flow in open channels are the continuity, momentum, and energy equations. In the case of uniform flow, the momentum equation can be approximated by either the Darcy-Weisbach or Manning equation. In cases of steady nonuniform flow, the energy equation is used to calculate the water-surface profile using either the direct-integration, direct-step, or standard-step method. Abrupt transitions from supercritical to subcritical flow are accomplished via a hydraulic jump. Hydraulic structures commonly used in engineering practice include weirs and Parshall flumes for measuring flowrates, gates for regulating flows, and culverts for passing water under roadways. In weirs and Parshall flumes, the flowrate and depth have a unique relationship, allowing the flowrate to be estimated from a depth measurement. The flowrate through gates is expressed in terms of the water depths upstream and downstream of the gate, with such relationships commonly referred to as *rating curves*. The flowrate through a culvert can depend on either the headwater depth or both the headwater and tailwater depths, and some iteration is usually necessary to determine the flow conditions in the culvert. Applying the theory of open-channel flow in design, three types of channels are considered: lined channels, unlined channels, and grass-lined channels. For each type of channel, design considerations include the shape of the channel section that minimizes flow area, minimum permissible velocity to prevent deposition of suspended material, selection of a stable side slope, and an allowance for adequate freeboard under design flow conditions. Special considerations for unlined channels include limitations on shear stresses exerted by flowing water on the (erodible) channel perimeter; a special consideration in grass-lined channels includes accounting for the dependence of the Manning roughness coefficient on the flow velocity and hydraulic radius. The principles of channel design are particularly useful in sizing major stormwater drainage channels. Sanitary sewers convey wastewater to treatment plants or disposal locations. Because the sewer pipes are designed to flow partially full, the principles of open-channel flow are applicable. Design criteria generally include maximum and minimum allowable velocities and maximum allowable depths of flow.

Problems

4.1. An open channel has a trapezoidal cross-section with a bottom width of 5 m and side slopes of 2:1 (H:V). If the depth of flow is 2 m and the average velocity in the channel is 1 m/s, calculate the discharge in the channel.

4.2. Water flows at 8 m³/s through a rectangular channel 4 m wide and 3 m deep. If the flow velocity is 1 m/s, calculate the depth of flow in the channel. If this channel expands (downstream) to a width of 5 m and the depth decreases by 0.5 m from the upstream depth, then what is the flow velocity in the expanded section?

4.3. Show that for circular pipes of diameter D the hydraulic radius, R, is related to the pipe diameter by $4R = D$.

4.4. A trapezoidal channel is to be excavated at a site where permit restrictions require that the channel have a bottom width of 5 m, side slopes of 1.5:1 (H:V) and a depth of flow of 1.8 m. If the soil material erodes when the shear stress on the perimeter of the channel exceeds 3.5 N/m², determine the appropriate slope and flow capacity of the channel. Use the Darcy-Weisbach equation and assume that the excavated channel has an equivalent sand roughness of 3 mm.

4.5. Water flows in a 8-m wide rectangular channel that has a longitudinal slope of 0.0001. The channel has an equivalent sand roughness of 2 mm. Calculate the uniform flow depth in the channel when the flowrate is 15 m³/s. Use the Darcy-Weisbach equation.

4.6. Water flows at a depth of 2.20 m in a trapezoidal, concrete-lined section ($k_s = 2$ mm) with a bottom width of 3.6 m and side slopes of 2:1 (H:V). The slope of the channel is 0.0006 and the water temperature is 20°C. Assuming uniform-flow conditions, estimate the average velocity and flowrate in the channel. Use both the Darcy-Weisbach and Manning equations and compare your answers.

4.7. Water flows in a trapezoidal channel that has a bottom width of 5 m, side slopes of 2:1 (H:V), and a longitudinal slope of 0.0001. The channel has an equivalent sand roughness of 1 mm. Calculate the uniform flow depth in the channel when the flowrate is 18 m³/s. Is the flow hydraulically rough, smooth, or in transition? Would the Manning equation be valid in this case?

4.8. Show that the Manning n, can be expressed in terms of the Darcy friction factor, f, by the following relation

$$n = \frac{f^{\frac{1}{2}} R^{\frac{1}{6}}}{8.86}$$

where R is the hydraulic radius of the flow. Does this relationship conclusively show that n is a function of the flow depth?

4.9. Water flows at 20 m³/s in a trapezoidal channel that has a bottom width of 2.8 m, side slope of 2:1 (H:V), longitudinal slope of 0.01, and a Manning n of 0.015. (a) Use the Manning equation to find the normal depth of flow, and (b) determine the equivalent sand roughness of the channel. Assume the flow is fully turbulent.

4.10. It has been shown that in fully turbulent flow the Manning n can be related to the height, d, of the roughness projections by the relation

$$n = 0.031 d^{\frac{1}{6}}$$

If the estimated roughness height, d, in a channel is 30 mm, then determine the percentage error in n resulting from a 70% error in estimating d.

4.11. The sections in the floodplain shown in Figure 4.4 have the following values of the Manning n:

Section	n
1	0.040
2	0.030
3	0.015
4	0.013
5	0.017
6	0.035
7	0.060

If the depth of flow in the floodplain is 5 m, use the formulae in Table 4.2 to estimate the composite roughness.

4.12. Use Equation 4.44 to show that the velocity in an open channel is equal to the depth-averaged velocity at a distance of $0.368d$ from the bottom of an open channel, where d is the depth of flow.

4.13. Use Equation 4.44 to show that the depth-averaged velocity in an open channel of depth d can be estimated by averaging the velocities at $0.2d$ and $0.8d$ from the bottom of the channel.

4.14. Water flows at 8.4 m³/s in a trapezoidal channel with a bottom width of 2 m and side slopes of 2:1 (H:V). Over a distance of 100 m, the bottom width expands to 2.5 m, with the side slopes remaining constant at 2:1. If the depth of flow at both of these sections is 1 m and the channel slope is 0.001, calculate the head loss between the sections. What is the power in kilowatts that is dissipated?

4.15. Use the Darcy-Weisbach equation to show that the head loss per unit length, S, between any two sections in an open channel can be estimated by the relation

$$S = \frac{\bar{f}}{4\bar{R}} \frac{\bar{V}^2}{2g}$$

where $\bar{f}$, $\bar{R}$, and $\bar{V}$ are the average friction factor, hydraulic radius, and flow velocity, respectively, between the upstream and downstream sections.

4.16. Determine the critical depth for 30 m³/s flowing in a rectangular channel with width 5 m. If the depth of flow is equal to 3 m, is the flow supercritical or subcritical?

4.17. Determine the critical depth for 50 m³/s flowing in a trapezoidal channel with bottom width 4 m and side slopes of 1.5:1 (H:V). If the depth of flow is 3 m, calculate the Froude number and state whether the flow is subcritical or supercritical.

4.18. A rectangular channel 2 m wide carries 3 m³/s of water at a depth of 1.2 m. If an obstruction 40 cm wide is placed in the middle of this channel, find the elevation of the water surface at the constriction. What is the minimum width of the constriction that will not cause a rise in the water surface upstream?

4.19. Water flows at 1 m³/s in a rectangular channel of width 1 m and depth 1 m. What is the maximum contraction of the channel that will not choke the flow?

4.20. A rectangular channel 3 m wide carries 4 m³/s of water at a depth of 1.5 m. If an obstruction 15 cm high is placed across the channel, calculate the elevation of the water surface over the obstruction. What is the maximum height of the obstruction that will not cause a rise in the water surface upstream?

4.21. Water flows at 4.3 m³/s in a rectangular channel of width 3 m and depth of flow 1 m. If the channel width is decreased by 0.75 m and the bottom of the channel is raised by 0.25 m, what is the depth of flow in the constriction?

4.22. Water flows at 18 m³/s in a trapezoidal channel with a bottom width of 5 m and side slopes of 2:1 (H:V). The depth of flow in the channel is 2 m. If a bridge pier of width 50 cm is placed in the middle of the channel, what is the depth of flow adjacent to the pier? What is the maximum width of a pier that will not cause a rise in the water surface upstream of the pier?

4.23. Water flows at 15 m³/s in a trapezoidal channel with a bottom width of 4.5 m and side slopes of 1.5:1 (H:V). The depth of flow in the channel is 1.9 m. If a step of height 15 cm is placed in the channel, what is the depth of flow over the step? What is the maximum height of the step that will not cause a rise in the water surface upstream of the step?

4.24. Water flows at 36 m³/s in a rectangular channel of width 10 m and a Manning n of 0.030. If the depth of flow at a channel section is 3 m and the slope of the channel is 0.001, classify the water-surface profile. What is the slope of the water surface at the observed section? Would the shape of the water-surface profile be much different if the depth of flow were equal to 2 m?

4.25. Water flows at 30 m³/s in a rectangular channel of width 8 m. The Manning n of the channel is 0.035. Determine the range of channel slopes that would be classified as steep and the range of channel slopes that would be classified as mild.

4.26. Water flows in a trapezoidal channel where the bottom width is 6 m and has side slopes of 2:1 (H:V). The channel lining has an estimated Manning n of 0.045, and the slope of the channel is 1.5%. When the flowrate is 80 m³/s, the depth of flow at a gaging station is 5 m. Classify the water-surface profile, state whether the depth increases or decreases in the downstream direction, and calculate the slope of the water surface at the gaging station. On the basis of this water-surface slope, estimate the depth of flow 100 m downstream and upstream of the gaging station.

4.27. Derive an expression relating the conjugate depths in a hydraulic jump where the slope of the channel is equal to S_o. [*Hint*: Assume that the length of the jump is equal to $5y_2$ and that the shape of the jump between the upstream and downstream depths can be approximated by a trapezoid.]

4.28. If 100 m³/s of water flows in a channel 8 m wide at a depth of 0.9 m, calculate the downstream depth required to form a hydraulic jump and the fraction of the initial energy lost in the jump.

4.29. The head loss, h_L, across a hydraulic jump is described by the equation:

$$y_1 + \frac{V_1^2}{2g} = y_2 + \frac{V_2^2}{2g} + h_L$$

where the subscripts 1 and 2 refer to the upstream and downstream locations, respectively. Show that the dimensionless head loss, h_L/y_1, is given by

$$\frac{h_L}{y_1} = 1 - \frac{y_2}{y_1} + \frac{Fr_1^2}{2}\left[1 - \left(\frac{y_1}{y_2}\right)^2\right]$$

where Fr_1 is the upstream Froude number.

4.30. Water flows in a horizontal trapezoidal channel at 21 m³/s, where the bottom width of the channel is 2 m, side slopes are 1:1, and depth of flow is 1 m. Calculate the downstream depth required for a hydraulic jump to form at this location. What would be the energy loss in the jump?

4.31. Water flows at 10 m³/s in a rectangular channel of width 5.5 m. The slope of the channel is 0.15% and the Manning roughness coefficient is 0.038. Estimate the depth 100 m upstream of a section where the flow depth is 2.2 m using: (a) the direct-integration method, and (b) the direct-step method. Approximately how far upstream of this section would you expect to find uniform flow?

4.32. Water flows at 5 m³/s in a 4-m wide rectangular channel that is laid on a slope of 4%. If the channel has a Manning n of 0.05 and the depth at a given section is 1.5 m, how far upstream or downstream is the depth equal to 1 m?

4.33. If the depth of flow in the channel described in Problem 4.9 is measured as 1.4 m, find the location where the depth is 1.6 m.

4.34. Water flows at 11 m³/s in a rectangular channel of width 5 m. The slope of the channel is 0.1% and the Manning roughness coefficient is equal to 0.035. If the depth of flow at a selected section is 2 m, then calculate the upstream depths at 20-m intervals along the channel until the depth of flow is within 5% of the uniform depth.

4.35. Water flows in an open channel whose slope is 0.05%. The Manning roughness coefficient of the channel lining is estimated to be 0.040 when the flowrate is 250 m^3/s. At a given section of the channel, the cross-section is trapezoidal with a bottom width of 12 m, side slopes of 2:1 (H:V), and a depth of flow of 8 m. Use the standard-step method to calculate the depth of flow 100 m upstream from this section where the cross-section is trapezoidal with a bottom width of 16 m and side slopes of 3:1 (H:V).

4.36. Water flows at 10 m^3/s in a 5-m wide channel. What is the height of a suppressed rectangular (sharp-crested) weir that will cause the depth of flow in the channel to be 2 m?

4.37. A sharp-crested weir is to be installed to measure flows that are expected to be on the order of 1 to 3 m^3/s. If the depth of the channel is such that the maximum depth of water that can be accommodated at the weir is 3.0 m and the width of the channel is 5 m, determine the height of a suppressed weir that can be used to measure the flowrate.

4.38. The stage in a river is to be maintained at an elevation of 20.00 m behind a weir. If the river is 10 m wide and 1 m deep and the flowrate in the river is 5 m^3/s, what is the crest elevation of the suppressed weir that must be used?

4.39. A contracted weir is to be used to measure the flowrate in an open channel that is 5 m wide and 2 m deep. It is desired that the depth of water over the crest of the weir be no less than 0.5 m when the flowrate is 1 m^3/s and that the water level be at least 30 cm below the top of the channel. What should the crest length of the weir be, and how far below the top of the channel should the crest of the weir be located? Sketch the weir.

4.40. Water flows at a rate of 5 m^3/s in a rectangular channel of width 10 m. The normal depth of flow in the channel is 5 m. If a sharp-crested (suppressed) weir of height 5 m is installed across the channel, determine the new flow depth in the channel for the same flowrate. What would the flow depth be if the rectangular weir were contracted (on both sides) to a width of 7 m? What would the flow depth be if a 7-m wide Cipolleti weir is used?

4.41. A Cipolletti weir is to be designed to measure the flow in a channel. Under design conditions, the normal depth of flow is 2 m and the flow rate is 5 m^3/s. Design the weir with a top width of 10 m.

4.42. A 0.8-m high suppressed rectangular weir is placed in a 1-m wide rectangular channel. If the head over the weir is observed to be 20 cm, estimate the flowrate in the channel. If the downstream water depth rises to 5 cm above the crest of the weir and the upstream water depth remains at 20 cm above the crest, estimate the new flowrate in the channel.

4.43. A V-notch weir is to be used to measure a maximum flow of 12 L/s. If a notch angle of 30° is selected, what should the depth of the notch be?

4.44. If a V-notch weir has a notch angle of 50° and a height of 50 cm, estimate the capacity of the weir.

4.45. A 90° V-notch weir is to be used to measure flows as low as 1 L/min. Estimate the depth of flow over the weir, accounting for viscous and surface tension effects if necessary.

4.46. A 70° V-notch weir is observed to have a 40 mm head. Estimate the flowrate over the weir.

4.47. A V-notch weir is to be designed to measure the flowrate in an irrigation channel. Under design conditions, the flowrate in the channel is 0.30 m^3/s and the depth of flow is 1 m. Design the weir with a top width of 1 m.

4.48. A 25-cm high broad-crested weir is placed in a 1.5-m wide channel. If the maximum depth of water that can be measured upstream of the weir is equal to 75 cm, what is the maximum flowrate that can be measured by the weir?

4.49. A broad-crested rectangular weir of length 1 m, width 1 m, and height 30 cm is being considered to measure the flow in a canal. For what range of flows would this weir length be adequate?

4.50. A broad-crested weir is to be used to measure the flow in an irrigation channel. The design section upstream of the weir is rectangular with a width of 3 m, and the depth of flow is 4 m at a flowrate of 5 m^3/s. Design the height and length of the weir.

4.51. A broad-crested weir is to be constructed to measure the flowrate in a 3-m wide channel. If the flowrate in the channel is expected to be in the range of 0.8–1.5 m^3/s and the depth of the channel is 2.5 m, select a height and length of weir such that the water level in the channel does not rise any higher than 2 m above the bottom of the channel.

4.52. Flow is being measured by a Parshall flume that has a throat width of 3 ft. Determine the flowrate through the flume when the water depth in the converging section is 1.50 ft and the depth in the throat section is 1.05 ft.

4.53. A Parshall flume has a throat width of 20 ft, and the water depth in the converging section is 4.00 ft and 3.60 ft in the throat section. Estimate the discharge through the flume. What would the discharge through the flume be if the water depth in the throat section were equal to 2.00 ft?

4.54. A Parshall flume is to be designed to measure the discharge in a channel. If the design flowrate is 1 m^3/s, determine the width of the flume to be used such that the depth of flow immediately upstream of the throat under design conditions is 1 m. By what percentage is the capacity of the flume reduced if the downstream depth is 0.85 m?

4.55. A vertical gate is installed at the end of a canal that is 7 m wide; the depth of flow is 3.5 m. If the gate has a width of 3 m, estimate the flow through the gate when the gate is raised 0.5 m. Neglect the effect of the downstream water depth.

4.56. If a vertical gate has an opening of 70 cm and discharges 16 m^3/s into a 5-m wide channel, estimate the maximum depth of water downstream of the gate for which the discharge under the gate will not be "drowned." (*Hint*: As long as a hydraulic jump can form, the discharge will not be drowned.)

4.57. Water flows under a sluice gate and into a 3-m wide rectangular channel where the flowrate and depth of flow are 4 m^3/s and 1.1 m, respectively. Estimate the minimum gate opening to prevent the formation of a hydraulic jump. What energy loss (in kilowatts) can be expected in a hydraulic jump?

4.58. Water is ponded behind a vertical gate to a height of 5 m in a rectangular channel of width 8 m. Calculate the gate opening that will release 50 m^3/s through the gate. How would this discharge be affected by a downstream flow depth of 4 m?

4.59. The contraction coefficient for a radial gate is given by

$$C_c = 1 - 0.75\left(\frac{\theta}{90}\right) + 0.36\left(\frac{\theta}{90}\right)^2$$

Find the value of θ that gives the minimum value of C_c.

4.60. A 500-mm diameter concrete drainage culvert ($n = 0.013$) is to be placed under a roadway. During the design storm, it is expected that water will pond behind the culvert to a height 20 cm above the top of the culvert. If the culvert entrance is to be well rounded ($k_e = 0.05$, $C_d = 0.95$), the slope of the culvert is 2%, the length of the culvert is 20 m, and the exit is not submerged, estimate the discharge through the culvert.

4.61. What is the capacity of a 1.5 m by 1.5 m concrete box culvert ($n = 0.013$) with a rounded entrance ($k_e = 0.05$, $C_d = 0.95$) if the culvert slope is 0.7%, the length is 40 m, and the headwater level is 2 m above the culvert invert? Consider the following cases: (a) free-outlet conditions, and (b) tailwater elevation 0.5 m above the top of the box at the outlet. What must the headwater elevation be in case (b) for the culvert to pass the flow that exists in case (a)?

4.62. A concrete pipe culvert ($n = 0.013$) is to be sized to pass 0.80 m^3/s when the headwater depth is 1 m. Site conditions require that the longitudinal slope of the culvert be 2% and the length of the culvert be 10 m. If the culvert entrance conditions are such that $k_e = 0.5$, $C_d \approx 1$, and the discharge is free, determine the required diameter of the culvert.

4.63. A circular concrete culvert is designed to pass water under a roadway. Under design conditions, the headwater depth is 2 m, the tailwater depth is 1 m, the flow through the culvert is 1 m^3/s, the length of the culvert is 15 m, and the slope is 1.5%. Determine the required diameter of the culvert.

4.64. A 2 m × 2 m concrete box culvert is to handle a design flow of 4 m^3/s. The culvert is to be laid on a slope of 0.1, be 25 m long, and have a low tailwater elevation. Assuming an entrance-loss coefficient of 0.1, determine the headwater depth under design conditions.

4.65. Show that the best hydraulic section for a rectangular shaped section is one in which the bottom width is equal to twice the flow depth.

4.66. Show that the best hydraulic section for a triangular shaped section is one in which the top width is equal to twice the flow depth.

4.67. How do the side slopes in the best trapezoidal section compare with the side slopes in the best triangular section?

4.68. Water flows in a concrete-lined trapezoidal channel on a slope of 0.5%. The channel has a bottom width of 10 m, side slopes of 2:1 (H:V), and a flowrate of 150 m^3/s. If the temperature of the water is 20°C: (a) Calculate the critical depth. (b) Calculate the normal depth. (c) Do you expect the velocity in the channel to prevent sedimentation and the growth of vegetation? (d) What freeboard would be appropriate for this channel? (e) If the actual depth of flow is 3 m, classify the water-surface profile.

4.69. If the best trapezoidal section is to be excavated in loose sandy earth, what side slope would you use? How would this compare with a trapezoidal channel excavated in stiff clay?

4.70. A trapezoidal channel has been designed with a bottom width of 10 m, side slopes of 2:1 (H:V), and a longitudinal slope of 0.053%. The Manning n of the channel material is estimated to be 0.030. If the design flowrate is 28 m^3/s and the radius of curvature of the channel is 100 m, determine the design depth of flow in the channel and the minimum freeboard that must be provided. Is this channel design acceptable from the viewpoint of minimum permissible velocity?

4.71. Design a (straight) lined trapezoidal channel to carry 30 m^3/s on a longitudinal slope of 0.002. The lining of the channel is to be float-finished concrete.

4.72. A lined rectangular channel is to be constructed on a slope of 0.042% to handle a design flowrate of 4.5 m^3/s. The lining of the channel is to be unfinished concrete, and the maximum radius of curvature is 150 m. Determine the depth and width of the channel to be excavated.

4.73. A lined (straight) triangular channel is to be constructed on a slope of 0.032% to handle a design flowrate of 4.4 m^3/s. The lining of the channel is to be smooth asphalt. Determine the dimensions of the channel to be excavated.

4.74. A parabolic-shaped channel is to be excavated and lined with mortar. If the channel is to carry 6 m^3/s on a slope of 0.05%, determine the depth of channel to be excavated. Give the equation of the channel side.

4.75. Design a trapezoidal channel to carry 30 m^3/s through a slightly sinuous channel on a slope of 0.002. The channel is to be excavated in coarse alluvium with a 75-percentile diameter of 2.5 cm, with the particles on the perimeter of the channel moderately rounded.

4.76. Design a trapezoidal channel to carry 10 m^3/s through a straight channel on a slope of 0.2%. The channel will be excavated in coarse alluvium with a 75-percentile diameter of

1 cm, with the particles on the channel perimeter moderately rounded.

4.77. A straight trapezoidal channel is to be excavated on a slope of 0.1% and have a design discharge of 27 m^3/s. Right-of-way constraints require the top width of the channel to be 17 m, and the channel is expected to consist of clean gravel (moderately angular) with a 75-percentile diameter of 3 cm. If freeboard is taken to be zero, determine the required bottom width and side slopes of the channel.

4.78. The theoretical tractive-force ratio, K, in trapezoidal channels is given by Equation 4.212, and the bottom and side shear stresses are given by Equations 4.204 and 4.205. Use these relations to identify the side slopes under which the critical shear stress will be on the side of an unlined (trapezoidal) channel. If the side slopes of a particular channel are 25°, and the angle of repose is 30°, will the critical shear stress on the sides be reached before the critical shear stress on the bottom is exceeded?

4.79. Design a triangular grass-lined channel to handle an intermittent flow of 5 m^3/s. The channel is to be excavated in an erosion-resistant soil, on a longitudinal slope of 0.025, and lined with Bermuda grass. The retardance is expected to vary between E and B.

4.80. Design a triangular grass-lined channel to carry 2 m^3/s on a slope of 1%. The retardance is expected to vary between A and C, and the grass can be taken as similar to Kentucky bluegrass. Make a conservative assumption on the soil type.

4.81. Design a parabolic grass-lined channel to handle an intermittent flow of 8 m^3/s. The channel is to be excavated in an erosion-resistant soil, on a longitudinal slope of 0.020, and lined with Kentucky bluegrass. The retardance is expected to vary between E and C.

4.82. A trapezoidal drainage channel on a golf course is lined with Kentucky bluegrass and is expected to convey 10 m^3/s under heavy-rainfall conditions. The slope of the drainage channel is 0.001, and the soil is easily eroded. If the channel has a bottom width of 3 m, side slopes of 2:1 (H:V), and a total depth of 2 m, assess the adequacy of the existing channel.

4.83. Estimate the maximum and minimum wastewater flowrates from a 65-ha residential development that will consist of 10% large lots, 75% small single-family lots, and 15% small-two family lots. Assume an average per-capita flowrate of 350 L/d/person.

4.84. The master plan of a 45-km^2 city calls for land usages that are 65% residential, 25% commercial, and 10% industrial. The residential development is to be 15% large lots, 75% small single-family lots, and 10% multistory apartments. The average domestic wastewater flowrate is 500 L/d/person, the average commercial flowrate is 50,000 L/d/ha, and the average industrial flowrate is 90,000 L/d/ha. Infiltration and inflow is 1,500 L/d/ha for the entire area. Estimate the peak and minimum flows to be handled by the trunk sewer.

4.85. Show that the Manning equation can be written in the form given by Equation 4.221.

4.86. Water flows at a rate of 7 m^3/s in a circular concrete sewer of diameter 1,600 mm. If the slope of the sewer is 0.01, calculate the depth of flow and velocity in the sewer. What diameter of pipe would be required for the pipe to flow three-quarters full?

4.87. Water flows at 3 m^3/s in a circular concrete sewer laid on a slope of 0.005. If a velocity of 2 m/s is desired in the sewer, calculate the diameter of sewer pipe that should be used. What would the depth of flow be in the sewer pipe?

4.88. Water flows at 3.5 m^3/s in a 1,400-mm diameter concreter sewer ($n = 0.015$). Find the slope at which the sewer flows half full.

4.89. Show that the maximum flowrate in a circular pipe occurs when $h/D \approx 0.94$, where h is the depth of flow and D is the diameter of the pipe. Assume that the Manning n is constant with depth.

4.90. Show that the flow velocity in a circular pipe is the same whether the pipe is flowing half full or completely full.

4.91. A minimum wastewater flow of 0.15 m^3/s is to be transported in a 915-mm diameter concrete pipe. Find the slope of the pipe that will give a minimum velocity of 0.6 m/s. If the maximum flowrate is 0.29 m^3/s, assess the adequacy of the pipe in terms of maximum flow depth and maximum allowable velocity. Assuming the pipe is to carry an average flow of 0.25 m^3/s, assess the potential for hydrogen sulfide generation if the temperature of the wastewater is 25°C and the five-day BOD is 250 mg/L at 20°C.

4.92. A 1,220-mm diameter concrete sewer is to carry 0.50 m^3/s of domestic wastewater at a temperature of 23°C on a slope of 0.009. Estimate the maximum five-day BOD for hydrogen-sulfide generation not to be a problem.

4.93. A sanitary sewer is to transport wastewater with a five-day BOD of 300 mg/L at 20°C. The flowrate of the wastewater is 0.01 m^3/s, the diameter of the pipe is 205 mm, and the slope is 0.1%. Assess the potential for sulfide generation.

4.94. A previously designed sewer line is made of 535-mm diameter reinforced concrete pipe on a slope of 0.001 and enters a manhole with a crown elevation of 2.50 m. The (average) design flowrate of the sewer line is 0.130 m^3/s. It is expected that an additional flow of 0.02 m^3/s will enter at the manhole, and the downstream sewer line leaving the manhole will carry an average flow of 0.15 m^3/s. The downstream sewer line is to be designed for a peak-flow factor of 2.0, a minimum-flow factor of 0.5, and a maximum allowable flow depth equal to 75% of the pipe diameter. If the ground-surface elevation upstream of the new sewer line is 5.00 m, the ground-surface elevation at the manhole downstream of the new sewer line is 4.50 m, and the new sewer line is 150 m long, determine: (a) the diameter and slope of the new sewer line, and (b) the crown and invert elevations at the upstream and downstream ends of the new sewer line.

4.95. A sewer system is to be designed to service the area shown in Figure 4.37. The average per capita wastewater flowrate is estimated to be 1,000 L/d/person, and the infiltration and inflow (I/I) is estimated to be 100 m³/d/km. The sewer system is to join an existing main sewer at manhole (MH) 5, where the average wastewater flow is 0.40 m³/s, representing the contribution of approximately 120,000 people.

The I/I contribution to the flow in the main sewer at MH 5 is negligible, and the main sewer at MH 5 is 1,220 mm in diameter, has an invert elevation of 55.00 m, and is laid on a slope of 0.8%. The layout of the sewer system shown in Figure 4.37 is based on the topography of the area, and the pipe lengths, contributing areas, and ground-surface elevations are shown in the following table:

Line no. (1)	Location (2)	Manhole no. From (3)	To (4)	Length (m) (5)	Contributing area (ha) (6)	Upper end (m) (28)	Lower end (m) (29)
0	Main Street	—	5	—	—	—	60.04
1	A Street	1	2	55	0.47	65.00	63.80
2	A Street	2	3	90	0.50	63.80	62.40
3	A Street	3	5	100	0.44	62.40	60.04
4	A Street	4	5	90	0.90	61.88	60.04
5	Main Street	5	12	70	0.17	60.04	60.04
6	B Street	6	8	60	0.43	65.08	63.20
7	P Avenue	7	8	50	0.48	63.60	63.20
8	B Street	8	10	90	0.39	63.20	62.04
9	Q Avenue	9	10	55	0.88	62.72	62.04
10	B Street	10	12	95	0.45	62.04	60.04
11	B Street	11	12	125	0.90	61.88	60.04
12	Main Street	12	19	75	0.28	60.04	60.20
13	C Street	13	15	55	0.60	64.40	62.84
14	P Avenue	14	15	55	0.76	63.24	62.84
15	C Street	15	17	95	0.51	62.84	61.60
16	Q Avenue	16	17	65	0.94	62.12	61.60
17	C Street	17	19	100	0.46	61.60	60.20
18	C Street	18	19	140	1.41	61.92	60.20
19	Main Street	19	26	80	0.30	60.20	60.08

Design the sewer system between A Street and C Street for a saturation density of 150 persons/ha. Municipal guidelines require that the sewer pipes have a minimum cover of 2 m, a minimum slope of 0.08%, a peak flow factor of 2.5, a minimum flow factor of 0.7, and a minimum allowable pipe diameter of 150 mm.

5 Probability and Statistics in Water-Resources Engineering

5.1 Introduction

All hydrologic phenomena obey the laws of nature and can, in theory, be described by the solution of the relevant fundamental equations. However, in most cases, the uncertainty and amount of detail that needs to be known about the physical system, combined with the limitations of analytical and numerical techniques for solving systems of equations, preclude an exact description of most natural phenomena. Processes that can be specified with certainty are called *deterministic*, while processes that cannot be specified with certainty are called *stochastic*. The two types of stochastic processes commonly encountered in engineering practice are: (1) processes where the outcome is uncertain because of the lack of information on the process or parameters affecting the process, and (2) processes where the outcome is uncertain because the process generating the outcome is fundamentally random. In the first case, the uncertainty can be reduced, or even eliminated, by additional measurements of the variables affecting the process or by the use of more accurate models. In the second case, randomness is intrinsic to the process and uncertainty in the outcome may not be reduced by additional measurements. Whether uncertainty in an outcome is due to a lack of information or the intrinsic randomness of a process, an uncertain outcome is generally defined by a probability distribution that describes the relative likelihood of all possible outcomes. In practice, uncertain variables are collectively referred to as *random variables*.

A common application of probability theory in water-resources engineering involves the assignment of an exceedance probability, P_e, to hydrologic events. A system designed to handle a particular hydrologic event can be expected to fail with a probability equal to the probability of exceedance, P_e, of the design event. This probability of system failure is called the *risk* of failure, and the probability that the system will not fail, $1 - P_e$, is called the *reliability* of the system. Exceedance probabilities of hydrologic events are usually stated in terms of the probability that a given event is exceeded in any given year. The average number of years between exceedances is called the *return period*, T, which is related to the exceedance probability by

$$T = \frac{1}{P_e} \qquad (5.1)$$

The return period, T, is sometimes called the *recurrence interval* and is usually a criterion for selecting the design events in water-resource systems.

5.2 Probability Distributions

A *probability function* defines the relationship between the outcome of a random process and the probability of occurrence of that outcome. If the sample space, S, contains discrete elements, then the sample space is called a *discrete sample space*; if S is a continuum, the sample space is called a *continuous sample space*. The sample space of a random variable is commonly denoted by an upper-case letter (e.g., X), and the corresponding lower case letter denotes an element of the sample space (e.g., x). *Discrete probability distributions* describe the probability of outcomes in discrete sample spaces, while *continuous probability distributions* describe the probability of outcomes in continuous sample spaces.

5.2.1 ■ Discrete Probability Distributions

If X is the sample space of a discrete random variable, with outcomes $x_1, x_2, \ldots, x_N$, then the probability of occurrence of each element, x_n, can be written as a *probability function*, $f(x_n)$, which is sometimes referred to as the *probability distribution function*. If $f(x_n)$ is a valid probability function, then it must necessarily have the following properties:

$$f(x_n) \geq 0 \qquad \forall x_n \tag{5.2}$$

$$\sum_{n=1}^{N} f(x_n) = 1 \tag{5.3}$$

where N is the number of elements in X. In many engineering applications, the quantity of interest is the probability that a random process will generate an outcome that is greater than or equal to some outcome x_n, where x_n is a design variable and exceedance of x_n will result in system failure.

Based on the definition of the probability distribution function, the probability that an outcome is less than or equal to x_n is given by

$$P(x_i \leq x_n) = \sum_{x_i \leq x_n} f(x_i) \tag{5.4}$$

and the probability that the outcome is greater than x_n is given by

$$P(x_i > x_n) = 1 - P(x_i \leq x_n) \tag{5.5}$$

The function $P(x_i \leq x_n)$ is called the *cumulative distribution function* and is commonly written as $F(x_n)$, where

$$F(x_n) = P(x_i \leq x_n) \tag{5.6}$$

These definitions of the probability distribution function, $f(x_n)$, and cumulative distribution function, $F(x_n)$, can be expanded to describe many random variables simultaneously. As an illustration, consider the case of two discrete random variables, X and Y, with elements x_p and y_q, where $p \in [1, N]$ and $q \in [1, M]$. The *joint probability distribution function* of X and Y, $f(x_p, y_q)$, is the probability of x_p and y_q occurring simultaneously. Such distributions are called *bivariate probability distributions*. If more than two variables are involved, the joint probability distribution is referred to as a *multivariate probability distribution*. Based on the definition of the bivariate probability distribution,

$f(x_p, y_q)$, the following conditions must necessarily be satisfied

$$f(x_p, y_q) \geq 0 \qquad \forall x_p, y_q \tag{5.7}$$

$$\sum_{p=1}^{N} \sum_{q=1}^{M} f(x_p, y_q) = 1 \tag{5.8}$$

As in the case of a single random variable, the cumulative distribution function, $F(x_p, y_q)$, is defined as the probability that the outcomes are simultaneously less than or equal to x_p and y_q, respectively. Hence $F(x_p, y_q)$ is related to the joint probability distribution by the following relation

$$F(x_p, y_q) = \sum_{x_i \leq x_p} \sum_{y_j \leq y_q} f(x_i, y_j) \tag{5.9}$$

In multivariate analysis, the probability distribution of any single random variate is called the *marginal probability distribution function*. In the case of a bivariate probability function, $f(x_p, y_q)$, the marginal probability of x_p, $g(x_p)$, is given by the relation

$$g(x_p) = \sum_{q=1}^{M} f(x_p, y_q) \tag{5.10}$$

Clearly, there are several probability functions that can be derived from the basic function describing the probability of occurrence of discrete events. The particular function of interest in any given case (e.g., cumulative or joint) is dependent on the question being asked.

5.2.2 ■ Continuous Probability Distributions

If X is a random variable with a continuous sample space, then there are an infinite number of possible outcomes and the probability of occurrence of any single value of X is actually zero. This problem is addressed by defining the probability of an outcome being in the range $[x, x + \Delta x]$, as $f(x)\Delta x$, where $f(x)$ is the *probability density function*. Based on this definition, any valid probability density function must satisfy the following conditions,

$$f(x) \geq 0 \qquad \forall x \tag{5.11}$$

$$\int_{-\infty}^{+\infty} f(x')dx' = 1 \tag{5.12}$$

The cumulative distribution function, $F(x)$, describes the probability that the outcome of a random experiment will be less than or equal to x and is related to the probability density function by the equation

$$F(x) = \int_{-\infty}^{x} f(x')dx' \tag{5.13}$$

which can also be written as

$$f(x) = \frac{dF(x)}{dx} \tag{5.14}$$

In describing the probability distribution of more than one random variable, the *joint probability density function* is used. In the case of two variables, X and Y, the probability that x will be in the range $[x, x + \Delta x]$ and y will be in the range $[y, y + \Delta y]$ is approximated by $f(x, y)\Delta x\Delta y$, where $f(x, y)$ is the joint probability density function of x and y. As in the case of discrete random variables, the bivariate cumulative distribution function, $F(x, y)$, and marginal probability distributions, $g(x)$ and $h(y)$, are defined as

$$F(x, y) = \int_{-\infty}^{x} \int_{-\infty}^{y} f(x', y')\, dx'\, dy' \tag{5.15}$$

$$g(x) = \int_{-\infty}^{\infty} f(x, y')\, dy' \tag{5.16}$$

$$h(y) = \int_{-\infty}^{\infty} f(x', y)\, dx' \tag{5.17}$$

5.2.3 ■ Mathematical Expectation and Moments

Assuming that x_i is an outcome of the discrete random variable X, $f(x_i)$ is the probability distribution function of X, and $g(x_i)$ is an arbitrary function of x_i, then the expected value of the function g, represented by $\langle g \rangle$, is defined by the equation

$$\langle g \rangle = \sum_{i=1}^{N} g(x_i) f(x_i) \tag{5.18}$$

where N is the number of possible outcomes in the sample space, X. The expected value of a random function is the *arithmetic average* of the function calculated from an infinite number of random experiments, and the analogous result for a function of a continuous random variable is given by

$$\langle g \rangle = \int_{-\infty}^{+\infty} g(x') f(x')\, dx' \tag{5.19}$$

where $g(x)$ is a continuous random function and $f(x)$ is the probability density function of x.

Several random functions are particularly useful in characterizing the distribution of random variables. The first is simply

$$g(x) = x \tag{5.20}$$

In this case $\langle g \rangle = \langle x \rangle$ corresponds to the arithmetic average of the outcomes over an infinite number of realizations. The quantity $\langle x \rangle$ is called the *mean* of the random variable and is usually denoted by μ_x. According to Equations 5.18 and 5.19, μ_x is defined for both discrete and continuous random variables by

$$\mu_x = \begin{cases} \sum_{i=1}^{N} x_i f(x_i) & \text{(discrete)} \\[2mm] \int_{-\infty}^{+\infty} x' f(x')\, dx' & \text{(continuous)} \end{cases} \tag{5.21}$$

A second random function that is frequently used is

$$g(x) = (x - \mu x)^2 \tag{5.22}$$

which equals the square of the deviation of a random outcome from its mean. The expected value of this quantity is referred to as the *variance* of the random variable and is usually denoted by σ_x^2. According to Equations 5.18 and 5.19, the variance of discrete and continuous random variables are given by the relations

$$\sigma_x^2 = \begin{cases} \sum_{i=1}^{N} (x_i - \mu_x)^2 f(x_i) & \text{(discrete)} \\ \int_{-\infty}^{+\infty} (x' - \mu_x)^2 f(x') dx' & \text{(continuous)} \end{cases} \tag{5.23}$$

The square root of the variance, σ_x, is called the *standard deviation* of x and measures the average magnitude of the deviation of the random variable from its mean. Random outcomes occur that are either less than or greater than the mean, μ_x, and the symmetry of these outcomes about μ_x is measured by the *skewness* or *skewness coefficient*, which is the expected value of the function

$$g(x) = \frac{(x - \mu_x)^3}{\sigma_x^3} \tag{5.24}$$

If the random outcomes are symmetrical about the mean, the skewness is equal to zero. There is no universal symbol that is used to represent the skewness, but in this text skewness will be represented by g_x. For discrete and continuous random variables, we have

$$g_x = \begin{cases} \frac{1}{\sigma_x^3} \sum_{i=1}^{N} (x_i - \mu_x)^3 f(x_i) & \text{(discrete)} \\ \frac{1}{\sigma_x^3} \int_{-\infty}^{+\infty} (x' - \mu_x)^3 f(x') dx' & \text{(continuous)} \end{cases} \tag{5.25}$$

The variables μ_x, σ_x, and g_x are measures of the average, variability about the average, and the symmetry about the average, respectively.

■ Example 5.1

A water-resource system is designed such that the probability, $f(x_i)$, that the system capacity is exceeded x_i times during the 50-year design life is given by the following discrete probability distribution:

x_i	$f(x_i)$
0	0.13
1	0.27
2	0.28
3	0.18
4	0.09
5	0.03
6	0.02
>6	0.00

What is the mean number of system failures expected in 50 years? What is the variance and skewness of the number of failures?

Solution

The mean number of failures, μ_x, is defined by Equation 5.21, where

$$\mu_x = \sum_{i=1}^{N} x_i f(x_i)$$

$$= (0)(0.13)+(1)(0.27)+(2)(0.28)+(3)(0.18)+(4)(0.09)+(5)(0.03)+(6)(0.02)$$

$$= 2$$

The variance of the number of failures, σ_x^2, is defined by Equation 5.23 as

$$\sigma_x^2 = \sum_{i=1}^{N} (x_i - \mu_x)^2 f(x_i)$$

$$= (0-2)^2(0.13) + (1-2)^2(0.27) + (2-2)^2(0.28) + (3-2)^2(0.18)$$

$$\quad + (4-2)^2(0.09) + (5-2)^2(0.03) + (6-2)^2(0.02)$$

$$= 1.92$$

The skewness coefficient, g_x, is defined by Equation 5.25 as

$$g_x = \frac{1}{\sigma_x^3} \sum_{i=1}^{N} (x_i - \mu_x)^3 f(x_i)$$

$$= \frac{1}{(1.92)^{\frac{3}{2}}} \left[(0-2)^3(0.13) + (1-2)^3(0.27) + (2-2)^3(0.28) + (3-2)^3(0.18) \right.$$

$$\left. + (4-2)^3(0.09) + (5-2)^3(0.03) + (6-2)^3(0.02) \right]$$

$$= 0.631$$ ■

■ **Example 5.2**

The probability density function, $f(t)$, of the time between storms during the summer in Miami is estimated as

$$f(t) = \begin{cases} 0.014e^{-0.014t} & : \quad t > 0 \\ 0 & : \quad t \leq 0 \end{cases}$$

where t is the time interval between storms in hours. Estimate the mean, standard deviation, and skewness of t.

Solution

The mean interstorm time, μ_t, is given by Equation 5.21 as

$$\mu_t = \int_0^\infty t' f(t')\, dt'$$

$$= \int_0^\infty t'(0.014e^{-0.014t'})\, dt'$$

$$= 0.014 \left(\int_0^\infty t' e^{-0.014t'}\, dt' \right)$$

The quantity in parentheses can be conveniently integrated by using the result (Dwight, 1961)

$$\int_0^\infty x e^{-ax}\, dx = \frac{1}{a^2}$$

Hence, the expression for μ_t can be written as

$$\mu_t = 0.014\left(\frac{1}{0.014^2}\right) = 71 \text{ hours}$$

The variance of t, σ_t^2, is given by Equation 5.23 as

$$\sigma_t^2 = \int_0^\infty (t' - \mu_t)^2 f(t')\, dt'$$

$$= \int_0^\infty (t' - 71)^2 0.014 e^{-0.014t'}\, dt'$$

$$= 0.014 \int_0^\infty (t'^2 - 142t' + 5041) e^{-0.014t'}\, dt'$$

Using the integration results (Dwight, 1961)

$$\int_0^\infty x^2 e^{-ax}\, dx = \frac{2}{a^3}; \qquad \int_0^\infty x e^{-ax}\, dx = \frac{1}{a^2}; \qquad \int_0^\infty e^{-ax}\, dx = \frac{1}{a}$$

the expression for σ_t^2 can be written as

$$\sigma_t^2 = 0.014\left[\frac{2}{0.014^3} - 142\left(\frac{1}{0.014^2}\right) + 5041\left(\frac{1}{0.014}\right)\right] = 5{,}102 \text{ h}^2$$

The standard deviation, σ_t, of the storm interevent time is therefore equal to $\sqrt{5102} = 71$ h.

The skewness of t, g_t, is given by Equation 5.25 as

$$g_t = \frac{1}{\sigma_t^3} \int_0^\infty (t' - \mu_t)^3 f(t')\, dt'$$

$$= \frac{1}{(71)^3} \int_0^\infty (t' - 71)^3 0.014 e^{-0.014t'}\, dt'$$

$$= \frac{0.014}{(71)^3} \int_0^\infty (t'^3 - 213t'^2 + 15123t' - 357911) e^{-0.014t'}\, dt'$$

Using the previously cited integration formulae plus (Dwight, 1961)

$$\int_0^\infty x^3 e^{-ax}\, dx = \frac{6}{a^4}$$

gives

$$g_t = 3.91 \times 10^{-8}\left(\frac{6}{0.014^4} - 213\frac{2}{0.014^3} + 15123\frac{1}{0.014^2} - 357911\frac{1}{0.014}\right) = 2.1$$

The positive skewness ($= 2.1$) indicates that the probability distribution of t has a long tail to the right of μ_t (toward higher values of t). ■

5.2.4 ■ Return Period

Consider a random variable X, with the outcome having a return period T given by x_T. Let p be the probability that $X \geq x_T$ in any observation, or $p = P(X \geq x_T)$. For each observation, there are two possible outcomes, either $X \geq x_T$ with probability p, or $X < x_T$ with probability $1 - p$. Assuming that all observations are independent, then the probability of a return period τ is the probability of $\tau - 1$ observations where $X < x_T$ followed by an observation where $X \geq x_T$. The expected value of τ, $\langle \tau \rangle$, is therefore given by

$$\langle \tau \rangle = \sum_{\tau=1}^{\infty} \tau (1 - p)^{\tau-1} p$$

$$= p + 2(1 - p)p + 3(1 - p)^2 p + 4(1 - p)^3 p + \cdots$$

$$= p[1 + 2(1 - p) + 3(1 - p)^2 + 4(1 - p)^3 + \cdots] \tag{5.26}$$

The expression within the brackets has the form of the power series expansion

$$(1 + x)^n = 1 + nx + \frac{n(n - 1)}{2}x^2 + \frac{n(n - 1)(n - 2)}{6}x^3 + \cdots \tag{5.27}$$

with $x = -(1 - p)$ and $n = -2$. Equation 5.26 can therefore be written in the form

$$\langle \tau \rangle = \frac{p}{[1 - (1 - p)]^2} = \frac{1}{p} \tag{5.28}$$

Since $T = \langle \tau \rangle$, Equation 5.28 can be expressed as

$$T = \frac{1}{p} \tag{5.29}$$

or

$$\boxed{T = \frac{1}{P(X \geq x_T)}} \tag{5.30}$$

which can also be written as

$$\boxed{P(X \geq x_T) = \frac{1}{T}} \tag{5.31}$$

In engineering practice it is more common to describe an event by its return period than its exceedance probability. For example, floodplains are usually delineated for the "100-year flood," which has an exceedance probability of 1% in any given year. However, because of the misconception that the 100-year flood occurs once every 100 years, ASCE (1996a) recommends that the reporting of return periods should be avoided, and annual exceedance probabilities reported instead.

■ Example 5.3

Analyses of the maximum annual floods over the past 150 years in a small river indicates the following cumulative distribution:

n	Flow, x_n (m^3/s)	$P(X < x_n)$
1	0	0
2	25	0.19
3	50	0.35
4	75	0.52
5	100	0.62
6	125	0.69
7	150	0.88
8	175	0.92
9	200	0.95
10	225	0.98
11	250	1.00

Estimate the magnitudes of the floods with return periods of 10, 50, and 100 years.

Solution

According to Equation 5.31, floods with return periods, T, of 10, 50, and 100 years have exceedance probabilities of $1/10 = 0.10$, $1/50 = 0.02$, and $1/100 = 0.01$. These exceedance probabilities correspond to cumulative probabilities of $1 - 0.1 = 0.9$, $1 - 0.02 = 0.98$, and $1 - 0.01 = 0.99$, respectively. Interpolating from the given cumulative probability distribution gives the following results

Return period (years)	Flow (m^3/s)
10	163
50	225
100	238

■

5.2.5 ■ Common Probability Functions

There are an infinite number of valid probability distributions. Engineering analyses, however, are typically confined to well-defined probability distributions that are relatively simple, associated with identifiable processes, and can be fully characterized by a relatively small number of parameters. Probability functions that can be expressed analytically are the functions of choice in engineering applications. In utilizing theoretical probability distribution functions to describe observed phenomena, it is important to understand the fundamental processes that lead to the probability distributions, since there is an implicit assumption that the theoretical and observed processes are the same. Probability distribution functions are either discrete or continuous. Commonly encountered discrete and continuous probability distribution functions, and their associated processes, are described in the following sections.

Binomial Distribution.   The *binomial distribution*, also called the *Bernoulli distribution*, is a discrete probability distribution that describes the probability of outcomes as either successes or failures. For example, in studying the probability of the annual-maximum flow in a river exceeding a certain value, the maximum flow in any year may be deemed either a success (the flow exceeds a specified value) or a failure (the flow does not exceed the specified given value). Since many engineered systems are designed to operate below certain threshold (design) conditions, the analysis in terms of success and failure is fundamental to the assessment of system reliability. Specifically, the binomial distribution describes the probability of n successes in N trials, given that the outcome in any one trial is independent of the outcome in any other trial and the probability of a success in any one trial is p. Such trials are called *Bernoulli trials*, and the process generating Bernoulli trials is called a *Bernoulli process*. The binomial (Bernoulli) probability distribution, $f(n)$, is given by

$$f(n) = \frac{N!}{n!(N-n)!} p^n (1-p)^{N-n} \qquad (5.32)$$

This discrete probability distribution follows directly from permutation and combination analysis and can also be written in the form

$$f(n) = \binom{N}{n} p^n (1-p)^{N-n} \qquad (5.33)$$

where

$$\binom{N}{n} = \frac{N!}{n!(N-n)!} \qquad (5.34)$$

is commonly referred to as the *binomial coefficient*. The parameters of the binomial distribution are the total number of outcomes, N, and the probability of success, p, in each outcome, and the binomial probability distribution is illustrated in Figure 5.1. The mean, variance, and skewness coefficient of the binomial distribution are given by

$$\mu_n = Np, \quad \sigma_n^2 = Np(1-p), \quad g = \frac{1-2p}{\sqrt{Np(1-p)}} \qquad (5.35)$$

The Bernoulli probability distribution, $f(n)$, is symmetric for $p = 0.5$, skewed to the right for $p < 0.5$, and skewed to the left for $p > 0.5$.

■ Example 5.4

The capacity of a stormwater-management system is designed to accommodate a storm with a return period of 10 years. What is the probability that the stormwater system will fail once in 20 years? What is the probability that the system fails *at least* once in 20 years?

Figure 5.1 ■ Binomial Probability Distribution

Source: Benjamin, Jack and Cornell, C. Allin. *Probability, Statistics, and Decision for Civil Engineers.* Copyright © 1970 by The McGraw-Hill Companies. Reprinted by permission of The McGraw-Hill Companies.

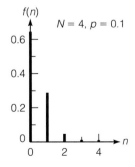
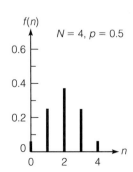

Solution

The stormwater-management system fails when the magnitude of the design storm is exceeded. The probability, p, of the 10-year storm being exceeded in any one year is $1/10 = 0.1$. The probability of the 10-year storm being exceeded once in 20 years is given by the binomial distribution, Equation 5.32, as

$$f(n) = \frac{N!}{n!(N-n)!} p^n (1-p)^{N-n}$$

where $n = 1$, $N = 20$, and $p = 0.1$. Substituting these values gives

$$f(1) = \frac{20!}{1!(20-1)!}(0.1)^1(1-0.1)^{20-1} = 0.27$$

Therefore, the probability that the stormwater-management system fails once in 20 years is 27%.

The probability, P, of the 10-year storm being equaled or exceeded *at least* once in 20 years is given by

$$P = \sum_{i=1}^{20} f(i) = 1 - f(0) = 1 - \frac{20!}{0!(20-0)!}(0.1)^0(1-0.1)^{20-0} = 1 - 0.12 = 0.88$$

The probability that the design event will be exceeded (at least once) is referred to as the *risk of failure*, and the probability that the design event will not be exceeded is called the *reliability* of the system. In this example, the risk of failure is 88% over 20 years and the reliability of the system over the same period is 12%. ■

Geometric Distribution.   The *geometric distribution* is a discrete probability distribution function that describes the number of Bernoulli trials, n, up to and including the one in which the first success occurs. The probability that the first success in a Bernoulli trial occurs on the nth trial is found by noting that for the first exceedance to be on the nth trial, there must be $n-1$ preceding trials without success followed by a success. The probability of this sequence of events is $(1-p)^{n-1}p$, and therefore the (geometric) probability distribution of the first success being in the nth trial, $f(n)$, is given by

$$f(n) = p(1-p)^{n-1} \qquad (5.36)$$

The parameter of the geometric distribution is the probability of success, p, and the geometric distribution is illustrated in Figure 5.2. The mean and variance are given by

$$\mu_n = \frac{1}{p}, \quad \sigma_n^2 = \frac{1-p}{p^2} \qquad (5.37)$$

The probability distribution of the number of Bernoulli trials between successes can be found from the geometric distribution by noting that the probability that n trials elapse between successes is the same as the probability that the first success is in the $n+1$st trial. Hence the probability n trials between successes is equal to $p(1-p)^n$.

■ **Example 5.5**

A stormwater-management system is designed for a 10-year storm. Assuming that exceedance of the 10-year storm is a Bernoulli process, what is the probability that the capacity of the stormwater-management system will be exceeded in the 5th year

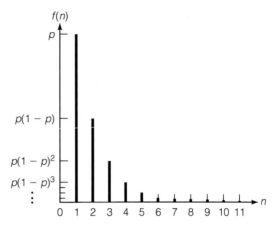

Figure 5.2 ■ Geometric Probability Distribution
Source: Benjamin, Jack and Cornell, C. Allin. *Probability, Statistics, and Decision for Civil Engineers.* Copyright © 1970 by The McGraw-Hill Companies. Reprinted by permission of The McGraw-Hill Companies.

after the system is constructed? What is the probability that the system capacity will be exceeded within the first 5 years?

Solution

From the given data, $n = 5$, $p = 1/10 = 0.1$, and the probability that the 10-year storm will be exceeded (for the first time) in the 5th year is given by the geometric distribution (Equation 5.36) as

$$f(5) = (0.1)(1 - 0.1)^4 = 0.066 = 6.6\%$$

The probability, P, that the 10-year storm will be exceeded within the first 5 years is given by

$$P = \sum_{n=1}^{5} p(1 - p)^{n-1} = \sum_{n=1}^{5} (0.1)(1 - 0.1)^{n-1} = \sum_{n=1}^{5} (0.1)(0.9)^{n-1}$$

$$= 0.1 + 0.09 + 0.081 + 0.073 + 0.066$$

$$= 0.41 = 41\%$$ ■

Poisson Distribution. The *Poisson process* is a limiting case of an infinite number of Bernoulli trials, where the number of successes, n, becomes large and the probability of success in each trial, p, becomes small in such a way that the expected number of successes, Np, remains constant. Under these circumstances it can be shown that (Thiébaux, 1994; Benjamin and Cornell, 1970)

$$\lim_{N \to \infty,\, p \to 0} \frac{N!}{n!(N - n)!} p^n (1 - p)^{N-n} = \frac{(Np)^n e^{-Np}}{n!} \tag{5.38}$$

Denoting the expected number of events, Np, by λ, the probability distribution of the number of events, n, in an interval in which the expected number of events is given by λ is described by the *Poisson distribution*

$$\boxed{f(n) = \frac{\lambda^n e^{-\lambda}}{n!}} \tag{5.39}$$

Haan (1977) describes the Poisson process as corresponding to the case where, for a given interval of time, the measurement increments decrease, resulting in a corresponding decrease in the event probability within the measurement increment, in such a way that the total number of events expected in the overall time interval remains constant and equal to λ. The only parameter in the Poisson distribution is λ. The mean, variance, and skewness of the distribution are given by

$$\mu_n = \lambda, \quad \sigma_n^2 = \lambda, \quad g_n = \lambda^{-\frac{1}{2}} \tag{5.40}$$

Poisson distributions for several values of λ are illustrated in Figure 5.3. The Poisson process for a continuous time is defined in a similar way to the Bernoulli process on a discrete time scale. The assumptions underlying the Poisson process on a continuous time are: (1) the probability of an event in a short interval between t and $t + \Delta t$ is $\lambda \Delta t$ (proportional to the length of the interval) for all values of t; (2) the probability of more than one event in any short interval t to $t + \Delta t$ is negligible in comparison to $\lambda \Delta t$; and (3) the number of events in any interval of time is independent of the number of events in any other nonoverlapping interval of time.

Extending the result for the discrete Poisson process given by Equation 5.39 to the continuous Poisson process, the probability distribution of the number of events, n, in time t for a Poisson process is given by

$$f(n) = \frac{(\lambda t)^n e^{-\lambda t}}{n!} \tag{5.41}$$

The parameter of the Poisson distribution for a continuous process is λt, where λ is the average rate of occurrence of the event. The mean, variance, and skewness of the Poisson

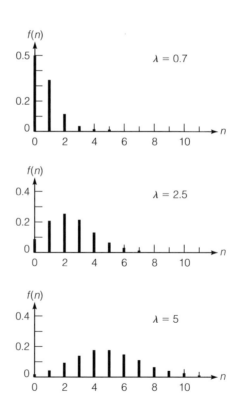

Figure 5.3 ■ Poisson Probability Distribution

Source: Benjamin, Jack and Cornell, C. Allin. *Probability, Statistics, and Decision for Civil Engineers.* Copyright © 1970 by The McGraw-Hill Companies. Reprinted by permission of The McGraw-Hill Companies.

distribution are given by

$$\mu_n = \lambda t, \quad \sigma_n^2 = \lambda t, \quad g_n = (\lambda t)^{-\frac{1}{2}} \tag{5.42}$$

The occurrence of storms and major floods have both been successfully described as Poisson processes (Borgman, 1963; Shane and Lynn, 1964).

■ Example 5.6

A flood-control system is designed for a runoff event with a 50-year return period. Assuming that exceedance of the 50-year runoff event is a Poisson process, what is the probability that the design event will be exceeded twice in the first 10 years of the system? What is the probability that the design event will be exceeded more than than twice in the first 10 years?

Solution

The probability distribution of the number of exceedances is given by Equation 5.39 (since the events are discrete). Over a period of 10 years, the expected number of exceedances, λ, of the 50-year event is given by

$$\lambda = Np = (10)\left(\frac{1}{50}\right) = 0.2$$

and the probability of the design event being exceeded twice is given by Equation 5.39 as

$$f(2) = \frac{0.2^2 e^{-0.2}}{2!} = 0.016 = 1.6\%$$

The probability, P, that the design event is exceeded more than twice in 10 years can be calculated using the relation

$$P = 1 - [f(0) + f(1) + f(2)]$$

$$= 1 - \left[\frac{0.2^0 e^{-0.2}}{0!} + \frac{0.2^1 e^{-0.2}}{1!} + \frac{0.2^2 e^{-0.2}}{2!}\right]$$

$$= 1 - [0.819 + 0.164 + 0.016] = 0.001$$

Therefore, there is only a 0.1% chance that the 50-year design event will be exceeded more than twice in any 10-year interval. ■

Exponential Distribution. The *exponential distribution* describes the probability distribution of the time between occurrences of an event in a continuous Poisson process. For any time duration, t, the probability of zero events during time t is given by Equation 5.41 as $e^{-\lambda t}$. Therefore, the probability that the time between events is less than t is equal to $1 - e^{-\lambda t}$. This is equal to the cumulative probability distribution, $F(t)$, of the time, t, between occurrences, which is equal to the integral of the probability density function, $f(t)$. The probability density function can therefore be derived by differentiating the cumulative distribution function, hence

$$f(t) = \frac{d}{dt} F(t)$$

$$= \frac{d}{dt}(1 - e^{-\lambda t}) \tag{5.43}$$

which simplifies to

$$f(t) = \lambda e^{-\lambda t} \tag{5.44}$$

This distribution is called the exponential probability distribution. The mean, μ_t, variance, σ_t^2, and skewness coefficient, g_t, of the exponential distribution are given by

$$\mu_t = \frac{1}{\lambda}, \quad \sigma_t^2 = \frac{1}{\lambda^2}, \quad g_t = 2 \tag{5.45}$$

The positive and constant value of the skewness indicates that the distribution is skewed to the right for all values of λ. The exponential probability distribution is illustrated in Figure 5.4. The cumulative distribution function, $F(t)$, of the exponential distribution function is given by

$$F(t) = \int_0^t \lambda e^{-\lambda \tau} d\tau \tag{5.46}$$

which leads to

$$F(t) = 1 - e^{-\lambda t} \tag{5.47}$$

In hydrologic applications, the exponential distribution is sometimes used to describe the interarrival times of random shocks to hydrologic systems, such as slugs of polluted runoff entering streams (Chow et al., 1988) and the arrival of storm events (Bedient and Huber, 1992).

■ Example 5.7

In the 86-year period between 1903 and 1988, Florida has been hit by 56 hurricanes (Winsberg, 1990). Assuming that the time interval between hurricanes is given by the exponential distribution, what is the probability that there will be less than 1 year between hurricanes? What is the probability that there will be less than 2 years between hurricanes?

Solution

The mean time between hurricanes, μ_t, is estimated as

$$\mu_t \approx \frac{86}{56} = 1.54 \text{ y}$$

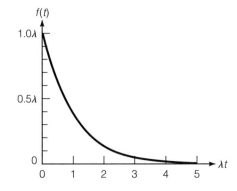

The mean, μ_t, is related to the parameter, λ, of the exponential distribution by Equation 5.45, which gives

$$\lambda = \frac{1}{\mu_t} = \frac{1}{1.54} = 0.649 \text{ y}^{-1}$$

The cumulative probability distribution, $F(t)$, of the time between hurricanes is given by Equation 5.47 as

$$F(t) = 1 - e^{-\lambda t} = 1 - e^{-0.649t}$$

The probability that there will be less than one year between hurricanes is given by

$$F(1) = 1 - e^{-0.649(1)} = 0.48 = 48\%$$

and the probability that there will be less than two years between hurricanes is

$$F(2) = 1 - e^{-0.649(2)} = 0.73 = 73\% \qquad ■$$

Gamma/Pearson Type III Distribution.   The *gamma distribution* describes the probability of the time to the nth occurrence of an event in a Poisson process. This probability distribution is derived by finding the probability distribution of $t = t_1 + t_2 + \cdots + t_n$, where t_i is the time from the $i - 1$th to the ith event. This is simply equal to the sum of time intervals between events. The probability distribution of the time, t, to the nth event can be derived from the exponential distribution to yield

$$f(t) = \frac{\lambda^n t^{n-1} e^{-\lambda t}}{(n-1)!} \qquad (5.48)$$

where $f(t)$ is the gamma probability distribution, illustrated in Figure 5.5 for several values of n. The mean, μ_t, variance, σ_t^2, and skewness coefficient, g_t, of the gamma

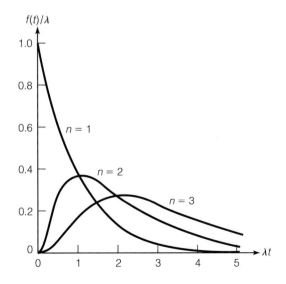

Figure 5.5 ■ Gamma Probability Distribution
Source: Benjamin, Jack and Cornell, C. Allin. *Probability, Statistics, and Decision for Civil Engineers*. Copyright © 1970 by The McGraw-Hill Companies. Reprinted by permission of The McGraw-Hill Companies.

distribution are given by

$$\mu_t = \frac{n}{\lambda}, \quad \sigma_t^2 = \frac{n}{\lambda^2}, \quad g_t = \frac{2}{n} \tag{5.49}$$

In cases where n is not an integer, the gamma distribution, Equation 5.48, can be written in the more general form

$$f(t) = \frac{\lambda^n t^{n-1} e^{-\lambda t}}{\Gamma(n)} \tag{5.50}$$

where $\Gamma(n)$ is the gamma function defined by the equation

$$\Gamma(n) = \int_0^{\infty} t^{n-1} e^{-t} dt \tag{5.51}$$

which has the property that

$$\Gamma(n) = (n-1)!, \quad n = 1, 2, \ldots \tag{5.52}$$

In hydrologic applications, the gamma distribution is useful for describing skewed variables without the need for transformation and has been used to describe the distribution of the depth of precipitation in individual storms. On the basis of the fundamental process leading to the gamma distribution, it should be clear why the gamma distribution is equal to the exponential distribution when $n = 1$. The gamma distribution given by Equation 5.50 can also be written in the form

$$f(t) = \frac{\lambda(\lambda t)^{n-1} e^{-\lambda t}}{\Gamma(n)} \tag{5.53}$$

Using the transformation

$$x = \lambda t \tag{5.54}$$

the probability density function of x, $p(x)$, is related to $f(t)$ by the relation

$$p(x) = f(t) \left| \frac{dt}{dx} \right| = f(t) \frac{1}{\lambda} \tag{5.55}$$

Combining Equations 5.53 to 5.55 and replacing n by α yields

$$p(x) = \frac{1}{\Gamma(\alpha)} x^{\alpha-1} e^{-x}, \quad x \geq 0 \tag{5.56}$$

This distribution is commonly referred to as the *one-parameter gamma distribution* (Yevjevich, 1972). The mean, variance, and skewness coefficient of the one-parameter gamma distribution are given by

$$\mu_x = \alpha, \quad \sigma_x^2 = \alpha, \quad g_x = \frac{2}{\sqrt{\alpha}} \tag{5.57}$$

Application of the one-parameter gamma distribution in hydrology is quite limited, primarily because of the difficulty in fitting a one-parameter distribution to observed hydrologic events (Yevjevich, 1972). A more general gamma distribution that does not have a lower bound of zero can be obtained by replacing x in Equation 5.56 by $(x-\gamma)/\beta$

to yield the following probability density function

$$p(x) = \frac{1}{\beta^\alpha \Gamma(\alpha)} (x - \gamma)^{\alpha-1} e^{-(x-\gamma)/\beta}, \quad x \geq \gamma \tag{5.58}$$

where the parameter γ is the lower bound of the distribution, β is a scale parameter, and α is a shape parameter. The distribution function given by Equation 5.58 is called the *three-parameter gamma distribution*. The mean, variance, and skewness coefficient of the three-parameter (α, β, and γ) gamma distribution are given by

$$\mu_x = \gamma + \alpha\beta, \quad \sigma_x^2 = \alpha, \quad g_x = \frac{2}{\sqrt{\alpha}} \tag{5.59}$$

The three-parameter gamma distribution has the desirable properties (from a hydrologic viewpoint) of being bounded on the left and a positive skewness. The three-parameter gamma distribution is commonly called the *Pearson Type III distribution*, which was first applied in hydrology to describe the probability distribution of annual-maximum flood peaks (Foster, 1924). When the measured data are very positively skewed, the data are usually log-transformed (i.e., x is the logarithm of a variable), and the distribution is called the *log–Pearson Type III distribution*. This distribution is widely used in hydrology, primarily because it has been (officially) recommended for application to flood flows by the U.S. Interagency Advisory Committee on Water Data (1982). An example of frequency analysis using the log–Pearson Type III distribution is given in Section 5.3.3.

■ Example 5.8

A commercial area experiences significant flooding whenever a storm yields more than five inches of rainfall in 24 hours. In a typical year, there are four such storms. Assuming that these rainfall events are a Poisson process, what is the probability that it will take more than two years to have four flood events?

Solution

The probability density function of the time for four flood events is given by the gamma distribution, Equation 5.48. From the given data: $n = 4$, $\mu_t = 365$ d ($= 1$ year), and

$$\lambda = \frac{n}{\mu_t} = \frac{4}{365} = 0.01096 \text{ d}^{-1}$$

The probability of up to t_o days elapsing before the fourth flood event is given by the cumulative distribution function

$$F(t_o) = \int_0^{t_o} f(t)\, dt = \int_0^{t_o} \frac{\lambda^n t^{n-1} e^{-\lambda t}}{(n-1)!}\, dt$$

$$= \frac{(0.01096)^4}{(4-1)!} \int_0^{t_o} t^{4-1} e^{-0.01096t}\, dt = 2.405 \times 10^{-9} \int_o^{t_o} t^3 e^{-0.01096t}\, dt$$

According to Dwight (1961)

$$\int t^3 e^{at}\, dt = e^{at} \left(\frac{t^3}{a} - \frac{3t^2}{a^2} + \frac{6t}{a^3} - \frac{6}{a^4} \right)$$

Using this relation to evaluate $F(t_o)$ gives

$$F(t_o) = -2.405 \times 10^{-9}\left[e^{-0.01096t}\left(\frac{t^3}{0.01096} + \frac{3t^2}{0.01096^2} + \frac{6t}{0.01096^3} + \frac{6}{0.01096^4}\right)\right]_0^{t_o}$$

Taking $t_o = 730$ d ($= 2$ years) gives

$F(730)$

$$= -2.405 \times 10^{-9}\left[e^{-0.01096t}\left(\frac{t^3}{0.01096} + \frac{3t^2}{0.01096^2} + \frac{6t}{0.01096^3} + \frac{6}{0.01096^4}\right)\right]_0^{730}$$

$$= -2.405 \times 10^{-9}\{[e^{-8}(3.549 \times 10^{10} + 1.331 \times 10^{10} + 0.333 \times 10^{10} + 0.042 \times 10^{10})]$$

$$- [1(4.158 \times 10^{8})]\}$$

$$= 0.958$$

Hence, the probability that four flood events will occur in less than two years is 0.958, and the probability that it will take more than two years to have four flood events is $1 - 0.958 = 0.042$ or 4.2%. ∎

Normal Distribution.   The *normal distribution*, also called the *Gaussian distribution*, is a symmetrical bell-shaped curve describing the probability density of a continuous random variable. The functional form of the normal distribution is given by

$$f(x) = \frac{1}{\sigma_x\sqrt{2\pi}} \exp\left[-\frac{1}{2}\left(\frac{x - \mu_x}{\sigma_x}\right)^2\right] \tag{5.60}$$

where the parameters μ_x and σ_x are equal to the mean and standard deviation of x, respectively. Normally-distributed random variables are commonly described by the shorthand notation $N(\mu, \sigma^2)$, and the shape of the normal distribution is illustrated in Figure 5.6. Most hydrologic variables cannot (theoretically) be normally distributed, since the range of normally distributed random variables is from $-\infty$ to ∞, and negative values of many hydrologic variables do not make sense. However, if the mean of a random variable is more than three or four times greater than its standard deviation, errors in the normal-distribution assumption can, in many cases, be neglected (Haan, 1977).

It is sometimes convenient to work with the *standard normal deviate, z*, which is defined by

$$z = \frac{x - \mu_x}{\sigma_x} \tag{5.61}$$

Figure 5.6 ■ Normal Probability Distribution
Source: Benjamin, Jack and Cornell, C. Allin. *Probability, Statistics, and Decision for Civil Engineers.* Copyright © 1970 by The McGraw-Hill Companies. Reprinted by permission of The McGraw-Hill Companies.

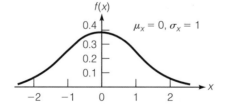

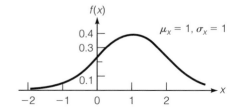

where x is normally distributed. The probability density function of z is therefore given by

$$f(z) = \frac{1}{\sqrt{2\pi}} e^{-z^2/2} \tag{5.62}$$

Equation 5.61 guarantees that z is normally distributed with a mean of zero and a variance of unity, and is therefore a $N(0,1)$ variate. The cumulative distribution, $F(z)$, of the standard normal deviate is given by

$$F(z) = \int_{-\infty}^{z} f(z')\, dz' = \frac{1}{\sqrt{2\pi}} \int_{-\infty}^{z} e^{-z'^2/2}\, dz' \tag{5.63}$$

where $F(z)$ is sometimes referred to as the area under the standard normal curve. These values are tabulated in Appendix D.1. Values of $F(z)$ can be approximated by the analytic relation (Abramowitz and Stegun, 1965)

$$F(z) \approx \begin{cases} B & : \quad z < 0 \\ 1 - B & : \quad z > 0 \end{cases} \tag{5.64}$$

where

$$B = \frac{1}{2}\left[1 + 0.196854|z| + 0.115194|z|^2 + 0.000344|z|^3 + 0.019527|z|^4\right]^{-4} \tag{5.65}$$

and the error in $F(z)$ using Equation 5.65 is less than 0.00025. Conditions under which any random variable can be expected to follow a normal distribution are specified by the *central limit theorem*:

■ **Theorem 5.1 (Central Limit Theorem)**   *If* S_n *is the sum of* n *independently and identically distributed random variables* X_i, *each having a mean* μ *and variance* σ^2, *then in the limit as* n *approaches infinity, the distribution of* S_n *approaches a normal distribution with mean* $n\mu$ *and variance* $n\sigma^2$.

Unfortunately, not many hydrologic variables are the sum of independent and identically distributed random variables. Under some very general conditions, however, it can be shown that if $X_1, X_2, \ldots, X_n$ are n random variables with $\mu_{X_i} = \mu_i$ and $\sigma_{X_i}^2 = \sigma_i^2$, then the sum, S_n, defined by

$$S_n = X_1 + X_2 + \cdots + X_n \tag{5.66}$$

is a random variable whose probability distribution approaches a normal distribution with mean, μ, and variance, σ^2 given by

$$\mu = \sum_{i=1}^{n} \mu_i \tag{5.67}$$

$$\sigma^2 = \sum_{i=1}^{n} \sigma_i^2 \tag{5.68}$$

The application of this result generally requires a large number of independent variables to be included in the sum S_n, and the probability distribution of each X_i has negligible influence on the distribution of S_n. In hydrologic applications, annual averages of variables such as precipitation and evaporation are assumed to be calculated from the sum of independent identically distributed measurements, and they tend to follow normal distributions (Chow et al., 1988).

■ Example 5.9

The annual rainfall in the Upper Kissimmee River basin has been estimated to have a mean of 51.2 in. and a standard deviation of 6.14 in. (Chin, 1993b). Assuming that the annual rainfall is normally distributed, what is the probability of having an annual rainfall of less than 40 in.?

Solution

From the given data: $\mu_x = 51.2$ in., and $\sigma_x = 6.14$ in. For $x = 40$ in., the standard normal deviate, z, is given by Equation 5.61 as

$$z = \frac{x - \mu_x}{\sigma_x} = \frac{40 - 51.2}{6.14} = -1.82$$

The probability that $z \leq -1.82$, $F(-1.82)$, is given by Equation 5.64, where

$$B = \frac{1}{2}\left[1 + 0.196854|z| + 0.115194|z|^2 + 0.000344|z|^3 + 0.019527|z|^4\right]^{-4}$$

$$= \frac{1}{2}\left[1 + 0.196854|-1.82| + 0.115194|-1.82|^2 + 0.000344|-1.82|^3 \right.$$

$$\left. + 0.019527|-1.82|^4\right]^{-4}$$

$$= 0.034$$

and therefore

$$F(-1.82) = 0.034 = 3.4\%$$

There is a 3.4% probability of having an annual rainfall less than 40 in.   ■

Log-normal Distribution.   In cases where the random variable, X, is equal to the product of many random variables $X_1, X_2, \ldots, X_n$, the logarithm of X is equal to the sum of many random variables where

$$\ln X = \ln X_1 + \ln X_2 + \cdots + \ln X_n \tag{5.69}$$

Therefore, according to the central limit theorem, $\ln X$ will be asymptotically normally distributed and X is said to have a *log-normal distribution*. Defining the random variable, Y, by the relation

$$Y = \ln X \tag{5.70}$$

then if Y is normally distributed, the theory of random functions can be used to show that the probability distribution of X, the log-normal distribution, is given by

$$f(x) = \frac{1}{x\sigma_y \sqrt{2\pi}} \exp\left[-\frac{(\ln x - \mu_y)^2}{2\sigma_y^2}\right], \quad x > 0 \tag{5.71}$$

where μ_y and σ_y^2 are the mean and variance of Y, respectively. The mean, variance, and skewness of a log-normally distributed variable, X, are given by

$$\mu_x = \exp(\mu_y + \sigma_y^2/2), \quad \sigma_x^2 = \mu_x^2[\exp(\sigma_y^2) - 1], \quad g_x = 3C_v + C_v^3 \qquad (5.72)$$

where C_v is the *coefficient of variation* defined as

$$C_v = \frac{\sigma_x}{\mu_x} \qquad (5.73)$$

If Y is defined by the relation

$$Y = \log_{10} X \qquad (5.74)$$

then Equation 5.71 still describes the probability density of X, with $\ln x$ replaced by $\log x$, and the moments of X are related to the moments of Y by

$$\mu_x = 10^{(\mu_y + \sigma_y^2/2)}, \quad \sigma_x^2 = \mu_x^2\left[10^{(\sigma_y^2)} - 1\right], \quad g_x = 3C_v + C_v^3 \qquad (5.75)$$

In hydrologic applications, the log-normal distribution has been found to reasonably describe such variables as daily precipitation depths, daily peak discharge rates, and the distribution of hydraulic conductivity in porous media (Viessman and Lewis, 1996).

■ Example 5.10

Annual-maximum discharges in the Guadalupe River near Victoria, Texas, show a mean of 801 m³/s and a standard deviation of 851 m³/s. If the capacity of the river channel is 900 m³/s, and the flow is assumed to follow a log-normal distribution, what is the probability that the maximum discharge will exceed the channel capacity?

Solution

From the given data: $\mu_x = 801$ m³/s; and $\sigma_x = 851$ m³/s. Equation 5.72 gives

$$\exp(\mu_y + \sigma_y^2/2) = 801 \qquad (= \mu_x)$$

and

$$(801)^2\left[\exp(\sigma_y^2) - 1\right] = (851)^2 \qquad (= \sigma_x^2)$$

which are solved to yield

$$\mu_y = 6.31 \quad \text{and} \quad \sigma_y = 0.870$$

When $x = 900$ m³/s, the log-transformed variable, y, is given by $y = \ln 900 = 6.80$ and the normalized random variate, z, is given by

$$z = \frac{y - \mu_y}{\sigma_y} = \frac{6.80 - 6.31}{0.870} = 0.563$$

The probability that $z \leq 0.563$, $F(0.563)$, in a standard normal distribution is given by Equation 5.64, where

$$B = \frac{1}{2}\left[1 + 0.196854|z| + 0.115194|z|^2 + 0.000344|z|^3 + 0.019527|z|^4\right]^{-4}$$

$$= \frac{1}{2}\left[1 + 0.196854|0.563| + 0.115194|0.563|^2 + 0.000344|0.563|^3\right.$$

$$\left. + 0.019527|0.563|^4\right]^{-4}$$

$$= 0.286$$

which yields

$$F(0.563) = 1 - 0.286 = 0.714 = 71.4\%$$

The probability that the maximum discharge in the river is greater than channel capacity of 900 m³/s is therefore equal to $1 - 0.714 = 0.286 = 28.6\%$. This is the probability of flooding in any given year. ■

Uniform Distribution.   The *uniform distribution* describes the behavior of a random variable in which all possible outcomes are equally likely within the range $[a, b]$. The uniform distribution can be applied to either discrete or continuous random variables. For a continuous random variable, x, the uniform probability density function, $f(x)$, is given by

$$f(x) = \frac{1}{b-a}, \qquad a \le x \le b \tag{5.76}$$

where the parameters a and b define the range of the random variable. The mean, μ_x, and variance, σ_x^2, of a uniformly distributed random variable are given by

$$\mu_x = \frac{1}{2}(a+b), \quad \sigma_x^2 = \frac{1}{12}(b-a)^2 \tag{5.77}$$

■ Example 5.11

Conflicting data from several remote rain gages in a region of the Amazon basin indicate that the annual rainfall in 1997 was between 810 mm and 1,080 mm. Assuming that the uncertainty can be described by a uniform probability distribution, what is the probability that the rainfall is greater than 1,000 mm?

Solution

In this case, $a = 810$ mm, $b = 1,080$ mm, and the uniform probability distribution of the 1997 rainfall is given by Equation 5.76 as

$$f(x) = \frac{1}{1080 - 810} = 0.00370 \quad \text{for} \quad 810 \text{ mm} \le x \le 1,080 \text{ mm}$$

The probability, P, that $x \ge 1,000$ mm is therefore given by

$$P = 0.00370 \times (1080 - 1000) = 0.296 = 29.6\%$$
■

Extreme-Value Distributions.   Extreme values are either maxima or minima of random variables. Consider the sample set $X_1, X_2, \dots, X_n$, and let Y be the largest of the sample values. If $F(y)$ is the probability that $Y < y$ and $P_{X_i}(x_i)$ is the probability that

$X_i < x_i$, then the probability that $Y < y$ is equal to the probability that all the $x_i s$ are less than y, which means that

$$F(y) = P_{X_1}(y)P_{X_2}(y) \ldots P_{X_n}(y) = [P_X(y)]^n \tag{5.78}$$

This equation gives the cumulative probability distribution of the extreme value, Y, in terms of the cumulative probability distribution of each outcome, which is also called the *parent distribution*. The probability density function, $f(y)$ of the extreme value, Y, is therefore given by

$$\boxed{f(y) = \frac{d}{dy}F(y) = n[P_X(y)]^{n-1}p_X(y)} \tag{5.79}$$

where $p_X(y)$ is the probability density function of each outcome. Equation 5.79 indicates that the probability distribution of extreme values, $f(y)$, depends on both the sample size and the parent distribution. Equation 5.79 was derived for maximum values, and a similar result can be derived for minimum values (see Problem 5.19).

Distributions of extreme values selected from large samples of many probability distributions have been shown to converge to one of three types of extreme-value distribution (Fisher and Tippett, 1928): Type I, Type II, or Type III. In Type I distributions, the parent distribution is unbounded in the direction of the desired extreme, and all moments of the distribution exist. In Type II distributions, the parent distribution is unbounded in the direction of the desired extreme and all moments of the distribution do not exist. In Type III distributions, the parent distribution is bounded in the direction of the desired extreme. The distributions of maxima of hydrologic variables are typically of Type I, since most hydrologic variables are (quasi-) unbounded to the right; the distributions of minima are typically of Type III, since many hydrologic variables are bounded on the left by zero.

Extreme-Value Type I (Gumbel) Distribution. The *extreme-value Type I distribution* requires that the parent distribution be unbounded in the direction of the extreme value. Specifically, Type I distributions require that the parent distribution falls off in an exponential manner, such that the upper tail of the cumulative distribution function, $P_X(x)$, can be expressed in the form

$$P_X(x) = 1 - e^{-g(x)}$$

with $g(x)$ an increasing function of x (Benjamin and Cornell, 1970). In using the Type I distribution to estimate maxima, parent distributions with the unbounded property include the normal, log-normal, exponential, and gamma distributions. Using the normal distribution as the parent distribution, the probability density function for the Type I extreme-value distribution is given by (Gumbel, 1958)

$$\boxed{f(x) = \frac{1}{a}\exp\left\{\pm\frac{x-b}{a} - \exp\left[\pm\frac{x-b}{a}\right]\right\}, \quad -\infty < x < \infty, \quad -\infty < b < \infty, \quad a > 0}$$

$$\tag{5.80}$$

where a and b are scale and location parameters, b is the mode of the distribution and the minus of the $\pm$ used for maximum values. The plus of the $\pm$ is used for minimum values. The Type I distribution for maximum values is illustrated in Figure 5.7. The Type I extreme-value distribution is sometimes referred to as the *Gumbel extreme-value* distribution, the *Fisher-Tippett Type I* distribution, or the *double-exponential* distribution.

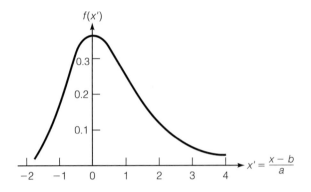

Figure 5.7 ■ Type I Extreme-Value Probability Distribution

Source: Benjamin, Jack and Cornell, C. Allin. *Probability, Statistics, and Decision for Civil Engineers.* Copyright © 1970 by The McGraw-Hill Companies. Reprinted by permission of The McGraw-Hill Companies.

The mean, variance, and skewness coefficient for the extreme-value Type I distribution applied to maximum values are

$$\mu_x = b + 0.577a, \quad \sigma_x^2 = 1.645a^2, \quad g_x = 1.1396 \tag{5.81}$$

and for the Type I distribution applied to minimum values

$$\mu_x = b - 0.577a, \quad \sigma_x^2 = 1.645a^2, \quad g_x = -1.1396 \tag{5.82}$$

By using the transformation

$$y = \frac{x - b}{a} \tag{5.83}$$

the extreme value Type I distribution can be written in the form

$$f(y) = \exp[\pm y - \exp(\pm y)] \tag{5.84}$$

which yields the following cumulative distribution functions

$$\boxed{F(y) = \exp[-\exp(-y)] \qquad \text{(maxima)}} \tag{5.85}$$

$$\boxed{F(y) = 1 - \exp[-\exp(y)] \qquad \text{(minima)}} \tag{5.86}$$

These cumulative distributions are most useful in determining the return periods of extreme events, such as flood flows (annual-maximum flows), maximum rainfall, and maximum wind speed (Gumbel, 1954). The extreme-value Type I (Gumbel) distribution is used extensively in flood studies in the United Kingdom and in many other parts of the world (Cunnane, 1988).

■ Example 5.12

The annual-maximum discharges in the Guadalupe River near Victoria, Texas, between 1935 and 1978 show a mean of 811 m³/s and a standard deviation of 851 m³/s. Assuming that the annual-maximum flows are described by an extreme-value Type I (Gumbel) distribution, estimate the annual-maximum flowrate with a return period of 100 years.

Solution

From the given data: $\mu_x = 811$ m^3/s, and $\sigma_x = 851$ m^3/s. According to Equation 5.81, the scale and location parameters, a and b, are derived from μ_x and σ_x by

$$a = \frac{\sigma_x}{\sqrt{1.645}} = \frac{851}{\sqrt{1.645}} = 664 \text{ m}^3/\text{s}$$

$$b = \mu_x - 0.577a = 811 - 0.577(664) = 428 \text{ m}^3/\text{s}$$

These parameters are used to transform the annual maxima, X, to the normalized variable, Y, where

$$Y = \frac{X - b}{a} = \frac{X - 428}{664}$$

For a return period of 100 years, the exceedance probability is $1/100 = 0.01$ and the cumulative probability, $F(y)$, is $1 - 0.01 = 0.99$. The extreme-value Type I probability distribution given by Equation 5.85 yields

$$0.99 = \exp[-\exp(-y_{100})]$$

where y_{100} is the event with a return period of 100 years. Solving for y_{100} gives

$$y_{100} = -\ln(-\ln 0.99) = 4.60$$

The annual-maximum flow with a return period of 100 years, x_{100}, is therefore given by

$$y_{100} = \frac{x_{100} - 428}{664}$$

which leads to

$$x_{100} = 428 + 664y_{100} = 428 + 664(4.60) = 3{,}482 \text{ m}^3/\text{s} \qquad ■$$

Extreme-Value Type III (Weibull) Distribution.   The extreme-value Type III distribution requires that the parent distribution be bounded in the direction of the extreme value. Specifically, Type III distributions (for minima) requires that the lefthand tail of the cumulative distribution function, $P_X(x)$, rises from zero for values of $x \geq c$ such that

$$P_X(x) = \alpha(x - c)^a, \qquad x \geq c$$

where c is the lower limit of x (Benjamin and Cornell, 1970). The extreme-value Type III distribution has mostly been used in hydrology to estimate low streamflows, which are bounded on the left by zero. The extreme-value Type III distribution for minimum values is commonly called the *Weibull distribution* and is given by

$$f(x) = ax^{a-1}b^{-a} \exp\left[-\left(\frac{x}{b}\right)^a\right], \qquad x \geq 0, \quad a, b > 0 \qquad (5.87)$$

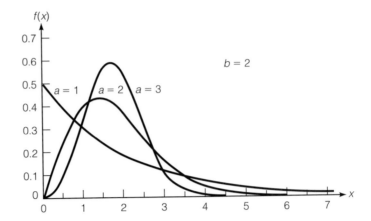

Figure 5.8 ■ Type III Extreme-Value (Weibull) Probability Distribution

The Weibull distribution is illustrated in Figure 5.8. The mean and variance of the distribution are given by

$$\mu_x = b\Gamma\left(1 + \frac{1}{a}\right), \quad \sigma_x^2 = b^2\left[\Gamma\left(1 + \frac{2}{a}\right) - \Gamma^2\left(1 + \frac{1}{a}\right)\right] \tag{5.88}$$

and the skewness coefficient is

$$g_x = \frac{\Gamma(1 + 3/a) - 3\Gamma(1 + 2/a)\Gamma(1 + 1/a) + 2\Gamma^3(1 + 1/a)}{[\Gamma(1 + 2/a) - \Gamma^2(1 + 1/a)]^{\frac{3}{2}}} \tag{5.89}$$

Where $\Gamma(n)$ is the gamma function defined by Equation 5.51 the cumulative Weibull distribution is given by

$$F(x) = 1 - \exp\left[-\left(\frac{x}{b}\right)^a\right] \tag{5.90}$$

which is useful in determining the return period of minimum events. If the lower bound of the parent distribution is nonzero, then a displacement parameter must be added to the Type III extreme-value distributions for minimums, and the probability density function becomes (Haan, 1977)

$$f(x) = a(x - c)^{a-1}(b - c)^{-a}\exp\left[-\left(\frac{x - c}{b - c}\right)^a\right] \tag{5.91}$$

and the cumulative distribution then becomes

$$F(x) = 1 - \exp\left[-\left(\frac{x - c}{b - c}\right)^a\right] \tag{5.92}$$

Equation 5.91 is sometimes called the *three-parameter Weibull distribution* or the *bounded exponential distribution*. The mean and variance of the three-parameter Weibull distribution are given by

$$\mu_x = c + (b - c)\Gamma\left(1 + \frac{1}{a}\right), \quad \sigma_x^2 = (b - c)^2\left[\Gamma\left(1 + \frac{2}{a}\right) - \Gamma^2\left(1 + \frac{1}{a}\right)\right] \tag{5.93}$$

and the skewness coefficient of the three-parameter Weibull distribution is given by Equation 5.89.

■ Example 5.13

The annual-minimum flows in a river at the location of a water-supply intake have a mean of 123 m³/s and a standard deviation of 37 m³/s. Assuming that the annual low flows have a lower bound of 0 m³/s and are described by an extreme-value Type III distribution, what is the probability that an annual-low flow will be less than 80 m³/s?

Solution

From the given data: $\mu_x = 123$ m³/s, and $\sigma_x = 37$ m³/s. According to Equation 5.88, the parameters of the probability distribution, a and b, must satisfy the relations

$$b\Gamma\left(1 + \frac{1}{a}\right) = 123 \qquad (= \mu_x) \tag{5.94}$$

and

$$b^2\left[\Gamma\left(1 + \frac{2}{a}\right) - \Gamma^2\left(1 + \frac{1}{a}\right)\right] = (37)^2 \qquad (= \sigma_x^2)$$

Combining these equations to eliminate b yields

$$\frac{123^2}{\Gamma^2\left(1 + \frac{1}{a}\right)}\left[\Gamma\left(1 + \frac{2}{a}\right) - \Gamma^2\left(1 + \frac{1}{a}\right)\right] = 37^2$$

or

$$\frac{\Gamma\left(1 + \frac{2}{a}\right) - \Gamma^2\left(1 + \frac{1}{a}\right)}{\Gamma^2\left(1 + \frac{1}{a}\right)} = 0.0905$$

This equation can be solved iteratively for a by defining $f(a)$ as

$$f(a) = \frac{\Gamma\left(1 + \frac{2}{a}\right) - \Gamma^2\left(1 + \frac{1}{a}\right)}{\Gamma^2\left(1 + \frac{1}{a}\right)} = \frac{\Gamma\left(1 + \frac{2}{a}\right)}{\Gamma^2\left(1 + \frac{1}{a}\right)} - 1$$

and selecting different values for a until $f(a) = 0.0905$. Using the tabulated values of the gamma function in Appendix E.2:

a	$\Gamma(1 + \frac{2}{a})$	$\Gamma(1 + \frac{1}{a})$	$f(a)$
2.00	1.000	0.886	0.273
3.00	0.903	0.893	0.132
4.00	0.886	0.906	0.079
3.79	0.888	0.904	0.0866
3.69	0.888	0.903	0.0890
3.65	0.889	0.903	0.0903

Therefore, $a = 3.65$ and Equation 5.94 gives

$$b = \frac{123}{\Gamma(1 + \frac{1}{a})} = \frac{123}{\Gamma(1 + \frac{1}{3.65})} = \frac{123}{0.903} = 136$$

The cumulative (Weibull) distribution of the annual-low flows, X, is given by Equation 5.90 as

$$F(x) = 1 - \exp\left[-\left(\frac{x}{b}\right)^a\right] = 1 - \exp\left[-\left(\frac{x}{136}\right)^{3.65}\right]$$

and the probability that $X \leq 80$ m^3/s is given by

$$F(80) = 1 - \exp\left[-\left(\frac{80}{136}\right)^{3.65}\right] = 0.134 = 13.4\%$$

There is a 13.4% probability that the annual-low flow is less than 80 m^3/s.   ■

Chi-Square Distribution.   Unlike the previously described probability distributions, the chi-square distribution is not used to describe a hydrologic process but is more commonly used in testing how well observed outcomes of hydrologic processes are described by theoretical probability distributions. The *chi-square distribution* is the probability distribution of a variable obtained by adding the squares of v normally distributed random variables, all of which have a mean of zero and a variance of 1. That is, if $X_1, X_2, \ldots, X_v$ are normally distributed random variables with mean of zero and variance of 1 and a new random variable χ^2 is defined such that

$$\chi^2 = \sum_{i=1}^{v} X_i^2 \tag{5.95}$$

then the probability density function of χ^2 is defined as the chi-square distribution and is given by

$$\boxed{f(x) = \frac{x^{-(1-v/2)}e^{x/2}}{2^{v/2}\Gamma(v/2)}, \qquad x, v > 0} \tag{5.96}$$

where $x = \chi^2$ and v is called the number of *degrees of freedom*. The chi-square distribution is illustrated in Figure 5.9. The mean and variance of the distribution are given by

$$\mu_{\chi^2} = v, \qquad \sigma_{\chi^2}^2 = 2v \tag{5.97}$$

The chi-square distribution is commonly used in determining the confidence intervals of various sample statistics. The cumulative chi-square distribution is given in Appendix D.2.

Figure 5.9 ■ Chi-Square Probability Distribution

Source: Benjamin, Jack and Cornell, C. Allin. *Probability, Statistics, and Decision for Civil Engineers.* Copyright © 1970 by The McGraw-Hill Companies. Reprinted by permission of The McGraw-Hill Companies.

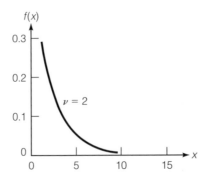
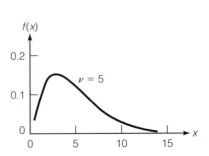

■ **Example 5.14**

A random variable is calculated as the sum of the squares of 10 normally distributed variables with a mean of zero and a standard deviation of 1. What is the probability that the sum is greater than 20? What is the expected value of the sum?

Solution

The random variable cited here is a χ^2 variate with 10 degrees of freedom. The cumulative distribution of the χ^2 variate is given in Appendix D.2 as a function of the number of degrees of freedom, ν, and the exceedance probability, α. In this case, for $\nu = 10$, and $\chi^2 = 20$, Appendix D.2 gives (by interpolation)

$$\alpha = 0.031$$

Therefore, the probability that the sum of the squares of 10 N(0,1) variables exceeds 20 is 0.031 or 3.1%.

The expected value of the sum is, by definition, equal to the mean, μ_{χ^2}, which is given by Equation 5.97 as

$$\mu_{\chi^2} = \nu = 10$$ ■

5.3 Analysis of Hydrologic Data

5.3.1 ■ Estimation of Population Distribution

Utilization of hydrologic data in engineering practice is typically a two-step procedure. In the first step, the cumulative probability distribution of the measured data is compared with a variety of theoretical distribution functions and the theoretical distribution function that best fits the probability distribution of the measured data is taken as the population distribution. In the second step, the exceedance probabilities of selected events are estimated using the theoretical population distribution. Some federal agencies require the use of specific probability distributions and, in some cases, also specify the minimum size of the database to be used in estimating the parameters of the distribution.

In estimating probability distributions from measured data, the measurements are usually assumed to be independent and drawn from identical probability distributions. In cases where the observations are correlated, then alternative methods such as *time-series analysis* are used in lieu of probability theory. The most common methods of estimating population distributions from measured data are: (1) visual comparison of the probability distribution of the measured data with various theoretical distributions, and picking the closest distribution that is consistent with the underlying process generating the sample data, and (2) using hypothesis-testing methods to assess whether various probability distributions are consistent with the probability distribution of the measured data.

5.3.1.1 Probability Distribution of Observed Data

The first step in plotting the probability distribution of observed data is to rank the data, such that for N observations a rank of 1 is assigned to the observation with the largest magnitude and a rank of N is assigned to the observation with the lowest magnitude.

The exceedance probability of the m-ranked observation, x_m, denoted by $P_X(X > x_m)$, is commonly estimated by the relation

$$P_X(X > x_m) = \frac{m}{N+1}, \qquad m = 1, \ldots, N \qquad (5.98)$$

or as a cumulative distribution function

$$P_X(X < x_m) = 1 - \frac{m}{N+1}, \qquad m = 1, \ldots, N \qquad (5.99)$$

Equation 5.98 is called the *Weibull formula* (Weibull, 1939) and is widely used in practice. The main drawback of the Weibull formula for estimating the cumulative probability distribution from measured data is that it is asymptotically exact (as the number of observations approaches infinity) only for a population with an underlying uniform distribution, which is relatively rare in nature. To address this shortcoming, Gringorten (1963) proposed that the exceedance probability of observed data be estimated using the relation

$$P_X(X > x_m) = \frac{m-a}{N+1-2a}, \qquad m = 1, \ldots, N \qquad (5.100)$$

where a is a parameter that depends on the population distribution. For a normal (or log-normal) distribution, $a = 0.375$; for a Gumbel distribution, $a = 0.44$. Bedient and Huber (1992) suggest that $a = 0.40$ is a good compromise for the usual situation in which the exact distribution is unknown.

■ Example 5.15

The annual peak flows in the Guadalupe River near Victoria, Texas, between 1965 and 1978 are as follows:

Year	Peak flow (ft^3/s)
1965	15,000
1966	9,790
1967	70,000
1968	44,300
1969	15,200
1970	9,190
1971	9,740
1972	58,500
1973	33,100
1974	25,200
1975	30,200
1976	14,100
1977	54,500
1978	12,700

Use the Weibull and Gringorten formulae to estimate the cumulative probability distribution of annual peak flow and compare the results.

Solution

The ranking of flows between 1965 and 1978 are shown in columns 1 and 2 of the following table:

(1) Rank, m	(2) Flow, q_m (ft³/s)	(3) Weibull $P(Q > q_m)$	(4) Gringorten $P(Q > q_m)$
1	70,000	0.067	0.042
2	58,500	0.133	0.113
3	54,500	0.200	0.183
4	44,300	0.267	0.253
5	33,100	0.333	0.324
6	30,200	0.400	0.394
7	25,200	0.467	0.465
8	15,200	0.533	0.535
9	15,000	0.600	0.606
10	14,100	0.667	0.676
11	12,700	0.733	0.746
12	9,790	0.800	0.817
13	9,740	0.867	0.887
14	9,190	0.933	0.958

In this case, $N = 14$ and the Weibull exceedance probabilities are given by Equation 5.98 as

$$P(Q > q_m) = \frac{m}{14 + 1} = \frac{m}{15}$$

These probabilities are tabulated in column 3. The Gringorten exceedance probabilities are given by (assuming $a = 0.40$)

$$P(Q > q_m) = \frac{m - a}{N + 1 - 2a} = \frac{m - 0.40}{14 + 1 - 2(0.40)} = \frac{m - 0.40}{14.2}$$

and these probabilities are tabulated in column 4. The probability functions estimated using the Weibull and Gringorten formulae are compared in Figure 5.10,

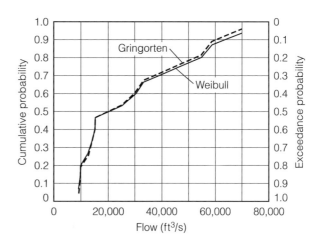

Figure 5.10 ■ Comparison Between Weibull and Gringorten Probability Distributions

where the distributions appear very similar. At high flows, the Gringorten formula assigns a lower exceedance probability and a correspondingly higher return period. ■

Visual comparison of the probability distribution calculated from observed data with various theoretical population distributions is facilitated by using special probability graph paper for each theoretical distribution. For example, on normal probability paper, the cumulative normal distribution plots as a straight line and the normality of the distribution of the sample data is assessed on the basis of how well this observed distribution fits a straight line. The construction of probability graph paper for various probability distributions is described by Haan (1977).

5.3.1.2 Hypothesis Tests

As an alternative to the visual comparison of observed probability distributions with various theoretical probability distributions, a quantitative comparison can be made using hypothesis tests. The two most common hypothesis tests for assessing whether an observed probability distribution can be approximated by a given (theoretical) population distribution are the *chi-square test* and the *Kolmogorov-Smirnov test*.

The Chi-Square Test.   Based on sampling theory, it is known that if the N outcomes are divided into M classes, with X_m being the number of outcomes in class m, and p_m being the theoretical probability of an outcome being in class m, then the random variable,

$$\chi^2 = \sum_{m=1}^{M} \frac{(X_m - Np_m)^2}{Np_m}$$

(5.101)

has a chi-square distribution. The number of degrees of freedom is $M - 1$ if the expected frequencies can be computed without having to estimate the population parameters from the sample statistics, while the number of degrees of freedom is $M - 1 - n$ if the expected frequencies are computed by estimating n population parameters from sample statistics. In applying the chi-square goodness of fit test, the null hypothesis is taken as H_o: The samples are drawn from the proposed probability distribution. The null hypothesis is accepted at the α significance level if $\chi^2 \in [0, \chi_\alpha^2]$ and rejected otherwise.

■ Example 5.16

Analysis of a 47-year record of annual rainfall indicates the following frequency distribution:

Range (mm)	Number of outcomes	Range (mm)	Number of outcomes
< 1,000	2	1,250–1,300	7
1,000–1,050	3	1,300–1,350	5
1,050–1,100	4	1,350–1,400	3
1,100–1,150	5	1,400–1,450	2
1,150–1,200	6	1,450–1,500	2
1,200–1,250	7	> 1,500	1

The measured data also indicate a mean of 1,225 mm and a standard deviation of 151 mm. Using a 5% significance level, assess the hypothesis that the annual rainfall is drawn from a normal distribution.

Solution

The first step in the analysis is to derive the theoretical frequency distribution. Appendix D.1 gives the cumulative probability distribution of the standard normal deviate, z, which is defined as

$$z = \frac{x - \mu_x}{\sigma_x} = \frac{x - 1225}{151}$$

where x is the annual rainfall. Converting the annual rainfall amounts into standard normal deviates, z, yields:

Rainfall (mm)	z	$P(Z < z)$
1,000	−1.49	0.07
1,050	−1.16	0.12
1,100	−0.83	0.20
1,150	−0.50	0.31
1,200	−0.17	0.43
1,250	0.17	0.57
1,300	0.50	0.69
1,350	0.83	0.80
1,400	1.16	0.88
1,450	1.49	0.93
1,500	1.82	0.97

and therefore the theoretical frequencies are given by

Range (mm)	Theoretical probability, p_m	Theoretical outcomes, $N p_m$
< 1,000	0.07	3.29
1,000–1,050	0.05	2.35
1,050–1,100	0.08	3.76
1,100–1,150	0.11	5.17
1,150–1,200	0.12	5.64
1,200–1,250	0.14	6.58
1,250–1,300	0.12	5.64
1,300–1,350	0.11	5.17
1,350–1,400	0.08	3.76
1,400–1,450	0.05	2.35
1,450–1,500	0.04	1.88
> 1,500	0.03	1.41

where the total number of observations, N, is equal to 47. Based on the observed and theoretical frequency distributions, the chi-square statistic is given by Equation 5.101 as

$$\chi^2 = \frac{(2-3.29)^2}{3.29} + \frac{(3-2.35)^2}{2.35} + \frac{(4-3.76)^2}{3.76} + \frac{(5-5.17)^2}{5.17} + \frac{(6-5.64)^2}{5.64}$$

$$+ \frac{(7-6.58)^2}{6.58} + \frac{(7-5.64)^2}{5.64} + \frac{(5-5.17)^2}{5.17} + \frac{(3-3.76)^2}{3.76} + \frac{(2-2.35)^2}{2.35}$$

$$+ \frac{(2-1.88)^2}{1.88} + \frac{(1-1.41)^2}{1.41}$$

$$= 1.42$$

Since both the mean and standard deviation were estimated from the measured data, the χ^2 statistic has $M - 1 - n$ degrees of freedom, where $M = 12$ (= number of intervals), and $n = 2$ (= number of population parameters estimated from measured data), hence $12 - 1 - 2 = 9$ degrees of freedom. Using a 5% significance level, the hypothesis that the observations are drawn from a normal distribution is accepted if

$$0 \le 1.42 \le \chi^2_{0.05}$$

Appendix D.2 gives that for 9 degrees of freedom, $\chi^2_{0.05} = 16.919$. Since $0 \le 1.42 \le 16.919$, the hypothesis that the annual rainfall is drawn from a normal distribution is *accepted*.   ■

Kolmogorov-Smirnov Test.   As an alternative to the chi-square goodness of fit test (to assess the hypothesis that observations are drawn from a population with a given theoretical probability distribution), the Kolmogorov-Smirnov test can also be used. This test differs from the chi-square test in that no parameters from the theoretical probability distribution need to be estimated from the observed data. In this sense, the Kolmogorov-Smirnov test is called a *nonparametric* test. The procedure for implementing the Kolmogorov-Smirnov test is as follows (Haan, 1977):

1. Let $P_X(x)$ be the specified theoretical cumulative distribution function under the null hypothesis.
2. Let $S_N(x)$ be the sample cumulative distribution function based on N observations. For any observed x, $S_N(x) = k/N$, where k is the number of observations less than or equal to x.
3. Determine the maximum deviation, D, defined by

$$\boxed{D = \max|P_X(x) - S_N(x)|} \qquad (5.102)$$

4. If, for the chosen significance level, the observed value of D is greater than or equal to the critical value of the Kolmogorov-Smirnov statistic tabulated in Appendix D.3, the hypothesis is rejected.

■ **Example 5.17**

Use the Kolmogorov-Smirnov test at the 10% significance level to assess the hypothesis that the data in Example 5.16 are drawn from a normal distribution.

Solution

Based on the given data, the measured and theoretical probability distributions of the annual rainfall are as follows:

(1) Rainfall, x x (mm)	(2) Normalized rainfall $(x - \mu_x)/\sigma_x$	(3) $S_N(x)$	(4) $P_X(x)$	(5) $|P_X(x) - S_N(x)|$
1,000	−1.490	2/47 = 0.043	0.068	0.025
1,050	−1.159	5/47 = 0.106	0.123	0.017
1,100	−0.828	9/47 = 0.191	0.204	0.013
1,150	−0.497	14/47 = 0.298	0.310	0.012
1,200	−0.166	20/47 = 0.426	0.434	0.008
1,250	0.166	27/47 = 0.574	0.566	0.008
1,300	0.497	34/47 = 0.723	0.690	0.033
1,350	0.828	39/47 = 0.830	0.796	0.034
1,400	1.159	42/47 = 0.894	0.877	0.017
1,450	1.490	44/47 = 0.936	0.932	0.004
1,500	1.821	46/47 = 0.979	0.966	0.013

The normalized rainfall amounts in column 2 are calculated using the rainfall amount, x, in column 1; mean, μ_x, of 1,225 mm; and standard deviation, σ_x, of 151 mm. The sample cumulative distribution function, $S_N(x)$, in column 3 is calculated using the relation

$$S_N(x) = \frac{k}{N}$$

where k is the number of measurements less than or equal to the given rainfall amount, x, and $N = 47$ is the total number of measurements. The theoretical cumulative distribution function, $P_X(x)$, in column 4 is calculated using the normalized rainfall (column 2) and the standard normal distribution function given in Appendix D.1. The absolute value of the difference between the theoretical and sample distribution functions are given in column 5. The maximum difference, D, between the theoretical and sample distribution functions is equal to 0.034 and occurs at an annual rainfall of 1,350 mm. For a sample size of 11 and a significance level of 10%, Appendix D.3 gives the critical value of the Kolmogorov-Smirnov statistic as 0.352. Since $0.034 < 0.352$, the hypothesis that the measured data are from a normal distribution is accepted at the 10% significance level.

■

According to Haan (1977), neither the chi-square test nor the Kolmogorov-Smirnov test are very powerful in the sense that the probability of accepting a false hypothesis is quite high when these tests are used. This is especially true for small samples. Nevertheless, these methods are widely used in engineering applications.

5.3.2 ■ Estimation of Population Parameters

The probability distribution of a random variable X is typically written in the form $f_X(x)$. However, this distribution could also be written in the form $f_X(x; \theta_1, \theta_2, \ldots, \theta_m)$ to indicate that the probability distribution of the random variable also depends on the values of the m parameters $\theta_1, \theta_2, \ldots, \theta_m$. This explicit expression for the probability distribution is particularly relevant when the population parameters are estimated using sample statistics, which are themselves random variables. The most common methods for estimating the parameters from measured data are: the method of moments, and the maximum likelihood method.

5.3.2.1 Method of Moments

The *method of moments* is based on the observation that the parameters of a probability distribution can usually be expressed in terms of the first few moments of the distribution. These moments can be estimated using sample statistics, and then the parameters of the distribution can be calculated using the relationship between the population parameters and the moments.

■ **Example 5.18**

The number of hurricanes hitting Florida each year between 1903 and 1988 is given in the following table (Winsberg, 1990):

	00	01	02	03	04	05	06	07	08	09
1900				1	1	0	2	0	0	1
1910	1	1	0	0	0	1	2	1	0	1
1920	0	1	0	0	2	1	2	0	2	1
1930	0	0	1	2	0	2	1	0	0	1
1940	0	1	0	0	1	2	1	2	2	1
1950	2	0	0	1	0	0	1	0	0	0
1960	1	0	0	3	0	1	2	0	1	0
1970	0	0	1	0	0	1	0	0	0	1
1980	0	0	0	0	0	2	0	1	0	

If the number of hurricanes per year is described by a Poisson distribution, estimate the parameters of the probability distribution. What is the probability of three hurricanes hitting Florida in one year?

Solution

The Poisson probability distribution, $f(n)$, is given by

$$f(n) = \frac{(\lambda t)^n e^{-\lambda t}}{n!}$$

where n is the number of occurrences per year and the parameter λt is related to the mean, μ_n, and standard deviation, σ_n, of n by

$$\mu_n = \lambda t, \qquad \sigma_n = \sqrt{\lambda t}$$

Based on the 86 years of data (1903–1988), the mean number of occurrences per year, μ_n, is estimated by the first-order moment, $\bar{N}$, as

$$\mu_n \approx \bar{N} = \frac{1}{86} \sum n_i = \frac{1}{86}(56) = 0.65$$

Hence $\lambda t \approx 0.65$ and the probability distribution of the number of hurricanes per year in Florida is given by

$$f(n) = \frac{0.65^n e^{-0.65}}{n!}$$

Putting $n = 3$ gives the probability of three hurricanes in one year as

$$f(3) = \frac{0.65^3 e^{-0.65}}{3!} = 0.02 = 2\%$$

It is interesting to note that the probability of at least one hurricane hitting Florida in any year is given by $1 - f(0) = 1 - e^{-0.65} = 0.48 = 48\%$. ■

5.3.2.2 Maximum-Likelihood Method

The *maximum-likelihood method* selects the population parameters that maximize the likelihood of the observed outcomes. Consider the case of n independent outcomes $x_1, x_2, \ldots, x_n$, where the probability of any outcome, x_i, is given by $p_X(x_i; \theta_1, \theta_2, \ldots, \theta_m)$, where $\theta_1, \theta_2, \ldots, \theta_m$ are the population parameters. The probability of the n observed (independent) outcomes is then given by the product of the probabilities of each of the outcomes. This product is called the *likelihood function*, $L(\theta_1, \theta_2, \ldots, \theta_m)$, where

$$L(\theta_1, \theta_2, \ldots, \theta_m) = \prod_{i=1}^{n} p_X(x_i; \theta_1, \theta_2, \ldots, \theta_m) \tag{5.103}$$

The values of the parameters that maximize the value of L are called the *maximum-likelihood estimates* of the parameters. Since the form of the probability function, $p_X(x; \theta_1, \theta_2, \ldots, \theta_m)$, is assumed to be known, the maximum-likelihood estimates can be derived from Equation 5.103 equating the partial derivatives of L with respect to each of the parameters, θ_i, equal to zero. This leads to the following m equations

$$\frac{\partial L}{\partial \theta_i} = 0, \qquad i = 1, m \tag{5.104}$$

This set of m equations can then be solved simultaneously to yield the m maximum-likelihood parameters $\hat{\theta}_1, \hat{\theta}_2, \ldots, \hat{\theta}_m$. In some cases, it is more convenient to maximize the natural logarithm of the likelihood function than the likelihood function itself. This approach is particularly convenient when the probability distribution function involves an exponential term. It should be noted that since the logarithmic function is monotonic, values of the estimated parameters that maximize the logarithm of the likelihood function also maximize the likelihood function.

■ Example 5.19

Use the maximum-likelihood method to estimate the parameter in the Poisson distribution of hurricane hits described in Example 5.18. Compare this result with the parameter estimate obtained using the method of moments.

Solution

The Poisson probability distribution can be written as

$$f(n) = \frac{\theta^n e^{-\theta}}{n!}$$

where $\theta = \lambda t$ is the population parameter of the Poisson distribution. Define the likelihood function, L', for the 86 years of data as

$$L' = \prod_{i=1}^{86} f(n_i)$$

where n_i is the number of occurrences in year i. It is more convenient to work with the log-likelihood function, L, which is defined as

$$L = \ln L' = \sum_{i=1}^{86} \ln f(n_i)$$

$$= \sum_{i=1}^{86} \left[\ln \left(\frac{1}{n_i!} \right) + n_i \ln \theta - \theta \right]$$

$$= \sum_{i=1}^{86} \ln \left(\frac{1}{n_i!} \right) + \ln \theta \sum_{i=1}^{86} n_i - 86\theta$$

Taking the derivative with respect to θ and putting $\partial L / \partial \theta = 0$ gives

$$\frac{\partial L}{\partial \theta} = \frac{1}{\theta} \sum_{i=1}^{86} n_i - 86 = 0$$

which leads to

$$\theta = \frac{1}{86} \sum_{i=1}^{86} n_i$$

This is the same estimate of θ that was derived in the previous example using the method of moments.   ■

The method of moments and the maximum-likelihood method do not always yield the same estimates of the population parameters. The maximum-likelihood method is generally preferred over the method of moments, particularly for large samples (Haan, 1977). The method of moments is severely affected if the data contain errors in the tails of the distribution, where the moment arms are long (Chow, 1954), and is particularly severe in highly skewed distributions (Haan, 1977).

5.3.3 ■ Frequency Analysis

Frequency analysis is concerned with estimating the relationship between an event, x, and the return period, T, of that event. Recalling that T is related to the exceedance probability of x, $P_X(X \geq x)$, by the relation

$$T = \frac{1}{P_X(X \geq x)} \tag{5.105}$$

then T is related to the cumulative distribution function of x, $P_X(X < x)$, by the relation

$$T = \frac{1}{1 - P_X(X < x)} \tag{5.106}$$

Many cumulative distribution functions cannot be expressed analytically and are tabulated as functions of normalized variables, X', where

$$X' = \frac{X - \mu_x}{\sigma_x} \tag{5.107}$$

and μ_x and σ_x are the mean and standard deviation of the population respectively. The cumulative probability distribution of X', $P_{X'}(X' < x')$, for many distributions depends only on x' and the skewness coefficient, g_x, of the population and can be readily tabulated in statistical tables. Denoting the value of X' with return period T by x'_T, for many distributions

$$x'_T = K_T(T, g_x) \tag{5.108}$$

where $K(T, g_x)$ is derived from the cumulative distribution function of X' and is commonly called the *frequency factor*. Combining Equations 5.107 and 5.108 leads to

$$\boxed{x_T = \mu_x + K_T\sigma_x} \tag{5.109}$$

where x_T is the realization of X with return period T. The frequency factor is applicable to many, but not all, probability distributions. The frequency factors for a few probability distributions that are commonly used in practice are described here.

Normal Distribution. In the case of a normal distribution, the variable X' is equal to the standard normal deviate, and has a N(0,1) distribution. The cumulative distribution function of the standard normal deviate is given in Appendix D.1, and the frequency factor is equal to the standard normal deviate corresponding to a given exceedance probability or return period. The frequency factor (= standard normal deviate), K_T, can also be approximated by the empirical relation (Abramowitz and Stegun, 1965)

$$\boxed{K_T = w - \frac{2.515517 + 0.802853w + 0.010328w^2}{1 + 1.432788w + 0.189269w^2 + 0.001308w^3}} \tag{5.110}$$

where

$$w = \left[\ln\left(\frac{1}{p^2}\right)\right]^{\frac{1}{2}}, \qquad 0 < p \le 0.5 \tag{5.111}$$

and p is the exceedance probability (= $1/T$). When $p > 0.5$, $1 - p$ is substituted for p in Equation 5.111 and the value of K_T is computed using Equation 5.110 is given a negative sign. According to Abramowitz and Stegun (1965), the error in using Equation 5.110 to estimate the frequency factor is less than 0.00045.

▪ Example 5.20

Annual rainfall at a given location is normally distributed with a mean of 127 cm and a standard deviation of 19 cm. Estimate the magnitude of the 50-year annual rainfall.

Solution

The 50-year rainfall, x_{50}, can be written in terms of the frequency factor, K_{50}, as

$$x_{50} = \mu_x + K_{50}\sigma_x$$

From the given data, $\mu_x = 127$ cm, $\sigma_x = 19$ cm, and the exceedance probability, p, is given by

$$p = \frac{1}{T} = \frac{1}{50} = 0.02$$

The intermediate variable, w, is given by Equation 5.111 as

$$w = \left[\ln\left(\frac{1}{p^2}\right)\right]^{\frac{1}{2}} = \left[\ln\left(\frac{1}{0.02^2}\right)\right]^{\frac{1}{2}} = 2.797$$

Equation 5.110 gives the frequency factor, K_{50}, as

$$K_{50} = w - \frac{2.515517 + 0.802853w + 0.010328w^2}{1 + 1.432788w + 0.189269w^2 + 0.001308w^3}$$

$$= 2.797 - \frac{2.515517 + 0.802853(2.797) + 0.010328(2.797)^2}{1 + 1.432788(2.797) + 0.189269(2.797)^2 + 0.001308(2.797)^3}$$

$$= 2.054$$

The 50-year rainfall is therefore given by

$$x_{50} = \mu_x + K_{50}\sigma_x = 127 + 2.054(19) = 166 \text{ cm} \qquad \blacksquare$$

Log-normal Distribution.   In the case of a log-normal distribution, the random variable is first transformed using the relation

$$Y = \ln X \qquad (5.112)$$

and the value of Y with return period T, y_T, is given by

$$y_T = \mu_y + K_T\sigma_y \qquad (5.113)$$

where μ_y and σ_y are the mean and standard deviation of Y and K_T is the frequency factor of the standard normal deviate with return period T. The value of the original variable, X, with return period T, x_T, is then given by

$$x_T = \ln^{-1} y_T = e^{y_T} \qquad (5.114)$$

■ Example 5.21

The annual rainfall at a given location has a log-normal distribution with a mean of 127 cm and a standard deviation of 19 cm. Estimate the magnitude of the 50-year annual rainfall.

Solution

From the given data, $\mu_x = 127$ cm, $\sigma_x = 19$ cm, and the mean, μ_y, and standard deviation, σ_y, of the log-transformed variable, $Y = \ln X$, are related to μ_x and σ_x by Equation 5.72, where

$$\mu_x = \exp\left(\mu_y + \frac{\sigma_y^2}{2}\right), \quad \sigma_x^2 = \mu_x^2[\exp(\sigma_y^2) - 1]$$

These equations can be put in the form

$$\mu_y = \frac{1}{2} \ln\left(\frac{\mu_x^4}{\mu_x^2 + \sigma_x^2}\right), \quad \sigma_y = \sqrt{\ln \frac{\sigma_x^2 + \mu_x^2}{\mu_x^2}}$$

which lead to

$$\mu_y = \frac{1}{2} \ln\left(\frac{127^4}{127^2 + 19^2}\right) = 4.83$$

and

$$\sigma_y = \sqrt{\ln \frac{19^2 + 127^2}{127^2}} = 0.149$$

Since the rainfall, X, is log-normally distributed, $Y \, (= \ln X)$ is normally distributed and the value of Y corresponding to a 50-year return period, y_{50}, is given by

$$y_{50} = \mu_y + K_{50}\sigma_y$$

where K_{50} is the frequency factor for a normal distribution and a return period of 50 years. In the previous example it was shown that $K_{50} = 2.054$, therefore

$$y_{50} = 4.83 + 2.054(0.149) = 5.14$$

and the corresponding 50-year rainfall, x_{50}, is given by

$$x_{50} = e^{y_{50}} = e^{5.14} = 171 \text{ cm} \qquad\qquad ■$$

Gamma/Pearson Type III Distribution.   In the case of the Pearson Type III distribution, also called the three-parameter gamma distribution, the frequency factor depends on both the return period, T, and the skewness coefficient, g_x. The frequency factor for the gamma/Pearson Type III distribution, K_T, can be estimated using the relation (Kite, 1977)

$$K_T = x_T' + (x_T'^2 - 1)k + \frac{1}{3}(x_T'^3 - 6x_T')k^2 - (x_T'^2 - 1)k^3 + x_T'k^4 + \frac{1}{3}k^5 \qquad (5.115)$$

where x_T' is the standard normal deviate corresponding to the return period T and k is related to the skewness coefficient by

$$k = \frac{g_x}{6} \qquad\qquad (5.116)$$

Clearly, when the skewness, g_x, is equal to zero, then Equation 5.115 indicates that $K_T = x_T'$, and the gamma/Pearson Type III distribution is identical to the normal distribution.

■ Example 5.22

The annual-maximum discharges in a river show a mean of 811 m³/s, a standard deviation of 851 m³/s, and a skewness of 2.94. Assuming a Pearson Type III distribution, estimate the 100-year discharge.

Solution

From the given data: $\mu_x = 811$ m³/s, $\sigma_x = 851$ m³/s, and $g_x = 2.94$. The standard normal deviate corresponding to a 100-year return period, x'_{100}, is 2.33 (Appendix D), $k = g_x/6 = 2.94/6 = 0.490$, and the frequency factor K_{100} is given by Equation 5.115 as

$$K_{100} = x'_{100} + (x'^2_{100} - 1)k + \frac{1}{3}(x'^3_{100} - 6x'_{100})k^2 - (x'^2_{100} - 1)k^3 + x'_{100}k^4 + \frac{1}{3}k^5$$

$$= 2.33 + (2.33^2 - 1)(0.490) + \frac{1}{3}[2.33^3 - 6(2.33)](0.490)^2$$

$$- (2.33^2 - 1)(0.490)^3 + (2.33)(0.490)^4 + \frac{1}{3}(0.490)^5$$

$$= 4.02$$

The 100-year discharge, x_{100}, is therefore given by

$$x_{100} = \mu_x + K_{100}\sigma_x = 811 + (4.02)(851) = 4{,}230 \text{ m}^3/\text{s} \qquad ■$$

Log–Pearson Type III Distribution.   In the case of a log–Pearson Type III distribution, the random variable is first transformed using the relation

$$Y = \ln X \qquad (5.117)$$

The value of Y with return period T, y_T, is given by

$$y_T = \mu_y + K_T\sigma_y \qquad (5.118)$$

where μ_y and σ_y are the mean and standard deviation of Y and K_T is the frequency factor of the Pearson Type III distribution with return period T and skewness coefficient g_y. The value of the original variable, X, with return period T, x_T, is then given by

$$x_T = \ln^{-1} y_T = e^{y_T} \qquad (5.119)$$

It should be clear that whenever the skewness of Y ($= \ln X$) is equal to zero, the log–Pearson Type III distribution is identical to the log-normal distribution.

■ **Example 5.23**

The logarithms of annual-maximum discharges (m³/s) in a river show a mean of 6.33, a standard deviation of 0.862, and a skewness of -0.833. Assuming a log–Pearson Type III distribution, estimate the 100-year discharge.

Solution

From the given data, $\mu_y = 6.33$, $\sigma_y = 0.862$, and $g_y = -0.833$, which yields $k = g_y/6 = -0.833/6 = -0.139$. The standard normal deviate corresponding to a 100-year return period, x'_{100}, is 2.33 (Appendix D), and the frequency factor for a 100-year return period, K_{100}, is given by Equation 5.115 as

$$K_{100} = x'_{100} + (x'^2_{100} - 1)k + \frac{1}{3}(x'^3_{100} - 6x'_{100})k^2 - (x'^2_{100} - 1)k^3 + x'_{100}k^4 + \frac{1}{3}k^5$$

$$= 2.33 + (2.33^2 - 1)(-0.139) + \frac{1}{3}[2.33^3 - 6(2.33)](-0.139)^2$$

$$-(2.33^2 - 1)(-0.139)^3 + (2.33)(-0.139)^4 + \frac{1}{3}(-0.139)^5$$

$$= 1.72$$

The 100-year log-discharge, y_{100}, is therefore given by

$$y_{100} = \mu_y + K_{100}\sigma_y = 6.33 + (1.72)(0.862) = 7.81$$

and the 100-year discharge, x_{100}, is

$$x_{100} = e^{y_{100}} = e^{7.81} = 2{,}465 \text{ m}^3/\text{s} \qquad ■$$

Extreme-Value Type I Distribution. In the case of an extreme-value Type I distribution, the frequency factor, K_T, can be written as (Chow, 1953)

$$\boxed{K_T = -\frac{\sqrt{6}}{\pi}\left\{0.5772 + \ln\left[\ln\left(\frac{T}{T-1}\right)\right]\right\}} \qquad (5.120)$$

■ **Example 5.24**

The annual-maximum discharges in a river have a mean of 811 m^3/s and a standard deviation of 851 m^3/s. Assuming that the annual maxima are described by an extreme-value Type I distribution, estimate the 100-year discharge.

Solution

From the given data: $\mu_x = 811$ m^3/s, $\sigma_x = 851$ m^3/s, and the 100-year discharge, x_{100}, is given by

$$x_{100} = \mu_x + K_{100}\sigma_x = 811 + 851 K_{100}$$

The 100-year frequency factor, K_{100}, is given by Equation 5.120 as

$$K_T = -\frac{\sqrt{6}}{\pi}\left\{0.5772 + \ln\left[\ln\left(\frac{T}{T-1}\right)\right]\right\}$$

$$= -\frac{\sqrt{6}}{\pi}\left\{0.5772 + \ln\left[\ln\left(\frac{100}{100-1}\right)\right]\right\}$$

$$= 3.14$$

which gives

$$x_{100} = 811 + 851(3.14) = 3{,}480 \text{ m}^3/\text{s} \qquad ■$$

Frequency analyses are generally based on assumed (population) probability distributions and sample estimates of the population parameters. The uncertainty of sample estimates of parameters requires caution when using data records shorter than 10 years and when estimating variables with recurrence intervals longer than twice the record length (Viessman and Lewis, 1966).

5.4   Floods

A flood is defined as a high flow that exceeds the capacity of a stream or drainage channel, and the elevation at which the flood overflows the embankments is called the *flood stage*. A *flood plain* is the normally dry land adjoining rivers, streams, lakes, bays, or oceans that is inundated during flood events. In analyzing floods, engineers and hydrologists commonly utilize the so-called *annual flood series*, which consists of the maximum flowrate for each year of record. Typically, annual peak flows with return periods from 1.5 to 3 years represent bankfull conditions, with larger flows causing inundation of the floodplain (McCuen, 1989).

Flood plains are usually relatively flat lands where roads and buildings are easily constructed. Regulatory mechanisms used to control development in flood plains include flood insurance requirements, open space reservations, and building restrictions. *Regulatory floods* are the bases for land-use controls. In the United States, regulatory floods for flood plain management are specified by the National Flood Insurance Program (NFIP), managed by the Federal Emergency Management Agency (FEMA). The 100-year flood (1% annual exceedance probability) has been adopted by FEMA as the base flood for delineating flood plains, and the area covered by the base flood is identified on a Flood Insurance Rate Map (FIRM). Communities participating in the NFIP are required to regulate development in the flood plain. Floods larger than and smaller than the regulatory flood are also identified on the FIRM for informational purposes. Updates to FIRMs are frequently necessary, since encroachment onto floodplains reduces the capacity of the watercourse and increases the spatial extent of the flood plain.

To appreciate the importance of flood plains, it should be noted that approximately 7%–10% of the land in the United States is in a flood plain; in the 1970s flood-related deaths were 200 per year, with another 80,000 per year forced from their homes. The largest flood plain areas in the United States are in the south, and the most populated flood plains are along the north Atlantic coast, the Great Lakes region, and in California. The cross-section of a typical flood plain is shown in Figure 5.11. For specified flows, flood plains are delineated using open-channel backwater computations facilitated by computer models such as HEC-2/HEC-RAS and WSP2, described in Chapter 4.

In addition to flood plain delineation, flood flows are used for many other important purposes, such as the sizing of bridges and culverts, as well as determining the design

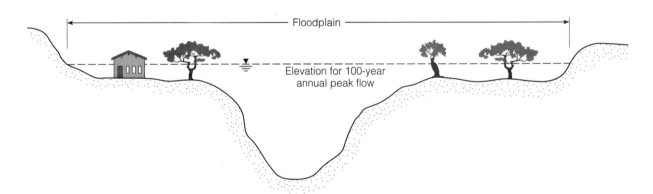

Figure 5.11 ■ Cross-Section of Floodplain

capacities of levees, spillways, and other control structures. Since available flood records are usually too short to define the frequency of flows with long return periods, such as 100 years, the methods of frequency analysis described in this chapter are commonly used.

Summary

Many design variables in water-resources engineering are either intrinsically random or uncertain, and such variables can only be defined using probability distributions. Useful parameters of probability distributions include the mean, variance, and skewness, which are defined in terms of mathematical expectation and moments. A large number of natural phenomena can be approximated by theoretical random processes, for which the outcomes are described by known probability functions. Discrete random processes in which the outcomes fall into only two categories can be approximated as Bernoulli processes, in which cases the binomial and geometric distributions are used in various analyses. Continuous random processes can be approximated as Poisson processes, where the Poisson, exponential, and gamma distributions are used in analyses.

Outcomes derived from the sum of a large number of independent and identically distributed random variables can be approximated by the normal distribution, and outcomes derived from the product of these variables can be approximated by the lognormal distribution. In cases where only extreme values are considered, the Gumbel (for maxima) and Weibull (for minima) are appropriate. Frequency analyses use various techniques for estimating the population probability distribution from measured data. The relationship between the outcome of a hydrologic process and its return period is usually the basis on which design events are specified and the reliability of engineered systems are assessed.

Problems

5.1. A flood-control system is designed such that the probability that the system capacity is exceeded X times in 30 years is given by the following discrete probability distribution:

x_i	$f(x_i)$
0	0.04
1	0.14
2	0.23
3	0.24
4	0.18
5	0.10
6	0.05
7	0.02
8	0.01
>9	0.00

What is the mean number of system failures expected in 30 years? What is the variance and skewness of the number of failures?

5.2. The probability distribution of the time between flood-

ing in a residential area is given by

$$f(t) = \begin{cases} 0.143e^{-0.143t} & : \quad t > 0 \\ 0 & : \quad t \le 0 \end{cases}$$

where t is the time interval between flood events in days. Estimate the mean, standard deviation, and skewness of t.

5.3. The maximum annual flows in a river are known to vary between 4 m^3/s and 10 m^3/s and have a probability distribution of the form

$$f(x) = \frac{\alpha}{x^2}$$

where x is the maximum annual flow. Determine (a) the value of α, (b) the return period of a maximum flow of 7 m^3/s, and (c) the mean and standard deviation of the maximum annual flows.

5.4. The flows during the summer in a regulated channel vary between 1 m^3/s and 2 m^3/s, and all flows between these values are equally likely. Derive an expression for the probability density function of the river flow, and use this distribution to calculate the mean, variance, and skewness of the steamflows. Identify the flowrate that has a return period of 50 years. (*Hint*: Your derived expression must satisfy the relations: $f(Q) \geq 0$ and $\int_{-\infty}^{\infty} f(Q)\, dQ = 1$.)

5.5. The average rainfall over a certain area is estimated to be 1,524 mm, and the rainfall is as likely to be above average in any year (wet year) as it is to be below average. For any 20-year interval, what is the probability of experiencing the following events: (a) one wet year with 19 dry years, and (b) 10 wet years with 10 dry years?

5.6. Use the theory of combinations to show that if the probability of an event occurring in any trial is p, then if all trials are independent, the probability of the event occurring n times in N trials, $f(n)$, is given by

$$f(n) = \frac{N!}{n!(N-n)!} p^n (1-p)^{N-n}$$

5.7. A flood-control system is designed for a rainfall event with a return period of 25 years. What is the probability that the design storm will be exceeded five times in 10 years? What is the probability that the design rainfall will be equaled or exceeded at least once in a 10-year period?

5.8. A water-resource system is designed for a hydrologic event with a 25-year return period. Assuming that the hydrologic events can be taken as a Bernoulli process, what is the probability that the design event will be exceeded once in the first five years of system operation? What is the probability that the design event will be exceeded more than than once in five years?

5.9. A stormwater management system is being designed for a (design) life of 10 years, and a (design) runoff event is to be selected such that there is only a 1% chance that the system capacity will be exceeded during the design life. Determine the return period of the design runoff event. If the cumulative distribution function of the runoff, Q (in m^3/s), is given by

$$F(Q) = \exp[-\exp(-Q)]$$

what is the design Q?

5.10. A water-resource system is designed for a 50-year storm. Assuming that the occurrence of the 50-year storm is a Bernoulli process, what is the probability that the 50-year storm will be equaled or exceeded one year after the system is constructed? What is the probability that the 50-year storm will occur within the first six years?

5.11. South Florida experiences a six-month wet season between May and October. Within the wet season there is, on average, 35 days with more than 1 inch of rainfall. Assuming that the time between 1-in. rainfall events is exponentially distributed, what is the probability that there is less than one week between 1-in rainfall events during the wet season?

5.12. A drainage system is designed for a storm with a return period of 10 years. What is the average interval between floods? What is the probability that there will be more than six months between floods?

5.13. Explain why a gamma distribution with $n = 1$ is the same as an exponential distribution.

5.14. A residential development in the southern United States floods whenever more than 4 inches of rain falls in 24 hours. In a typical year, there are three such storms. Assuming that these rainfall events are a Poisson process, what is the probability that it will take more than one year to have three flood events?

5.15. The annual rainfall in Everglades National Park has been estimated to have a mean of 55.6 in. and a standard deviation of 11.1 in. (Chin, 1993a). Assuming that the annual rainfall is normally distributed, what is the probability of having (a) an annual rainfall of less than 50 inches, (b) an annual rainfall of less than 60 inches, and (c) an annual rainfall between 50 and 60 inches?

5.16. Annual-maximum discharges in a river show a mean of 620 m^3/s and a standard deviation of 311 m^3/s. If the capacity of the river is 780 m^3/s and the flow can be assumed to follow a log-normal distribution, what is the probability that the maximum discharge will exceed the channel capacity?

5.17. The annual rainfall in a rural town is shown to be log-normally distributed with a mean of 114 cm and a standard deviation of 22 cm. Estimate the rainfall having a return period of 100 years. How would your estimate differ if the rainfall is normally distributed?

5.18. Streamflow data collected from several sources indicate that the peak flow in a drainage channel at a given location is somewhere in the range between 95 and 115 m^3/s. Assuming that the uncertainty can be described by a uniform distribution, what is the probability that the peak flow is greater than 100 m^3/s?

5.19. The probability distribution for the maximum values in a sample set is given by Equation 5.79. Derive a similar expression for the minimum values in a sample set.

5.20. The annual-maximum discharges in a river show a mean of 480 m^3/s and a standard deviation of 320 m^3/s. Assuming that the annual-maximum flows are described by an extreme-value Type I (Gumbel) distribution, use Equation 5.85 to estimate the annual-maximum flowrates corresponding to return periods of 50 and 100 years.

5.21. The annual-minimum rainfall amounts have been tabulated for a region and show a mean of 710 mm and a standard deviation of 112 mm. Assuming that these minima are described by an extreme-value Type I distribution, estimate the average interval between years where the minimum rainfall exceeds 600 mm.

5.22. The annual-low flows in a drainage channel have a mean of 43 m^3/s and a standard deviation of 12 m^3/s. Assuming that the annual-low flows have a lower bound of 0 m^3/s and are described by an extreme-value Type III distribution, what is the probability that an annual-low flow will be less than 10 m^3/s?

5.23. Measurements of annual-minimum flows in a river indicate that the minimum annual-minimum is 9 m^3/s, an annual-minimum flow of 34 m^3/s has a return of 10 years, and an annual-minimum flow of 76 m^3/s has a return period of 50 years. If these data follow an extreme-value Type III distribution, what would the 100-year annual-minimum flow be?

5.24. A random variable, X, is equal to the sum of the squares of 15 normally distributed variables with a mean of zero and a standard deviation of 1. What is the probability that X is greater than 25? What is the expected value of X?

5.25. Exceedance probabilities of measured data can be estimated using the Gringorten formula

$$P_X(X > x_m) = \frac{m - a}{N + 1 - 2a}$$

where x_m is a measurement with rank m, N is the total number of measurements, and a is a constant that depends on the population probability distribution. Values of a are typically in the range of 0.375 to 0.44. Determine the formula for the return period of x_m. What is the maximum error in the return period that can result from uncertainty in a?

5.26. The annual peak flows in a river between 1980 and 1996 are as follows:

Year	Peak flow (ft^3/s)
1980	8,000
1981	5,550
1982	3,390
1983	6,390
1984	5,889
1985	7,182
1986	10,584
1987	11,586
1988	8,293
1989	9,193
1990	5,142
1991	7,884
1992	4,132
1993	12,136
1994	5,129
1995	7,236
1996	6,222

Use the Weibull and Gringorten formulae to estimate the cumulative probability distribution of peak flow and compare the results.

5.27. Analysis of a 50-year record of annual rainfall indicates the following frequency distribution:

Range (in.)	Number of outcomes
< 37.8	4
37.8–41.9	5
41.9–44.8	6
44.8–47.3	6
47.3–49.6	5
49.6–51.9	7
51.9–54.4	4
54.4–57.3	3
57.3–61.4	6
> 61.4	4

The measured data also indicate a mean of 49.6 in. and a standard deviation of 9.2 in. Using the chi-square test with a 5% significance level, assess the hypothesis that the annual rainfall is drawn from a normal distribution.

5.28. Use the Kolmogorov-Smirnov test at the 10% significance level to assess the hypothesis that the data in Problem 5.27 is drawn from a normal distribution.

5.29. A sample of a random variable has a mean of 1.6, and the random variable is assumed to have an exponential distribution. Use the method of moments to determine the parameter (λ) in the exponential distribution.

5.30. A sample of a random variable has a mean of 1.6 and a standard deviation of 1.2. If the random variable is assumed to have a log-normal distribution, apply the method of moments to estimate the parameters in the log-normal distribution.

5.31. A sample of random data is to be fitted to the exponential probability distribution given by Equation 5.44. Derive the maximum likelihood estimate of the distribution parameter λ. Compare your result with the method of moments estimate.

5.32. Derive the maximum-likelihood estimator of the parameters in the normal distribution.

5.33. The annual rainfall at a given location is normally distributed with a mean of 152 cm and a standard deviation of 30 cm. Estimate the magnitude of the 10-year and 100-year annual rainfall amounts.

5.34. The annual rainfall at a given location has a log-normal distribution with a mean of 152 cm and a standard deviation of 30 cm. Estimate the magnitude of the 10-year and 100-year annual rainfall amounts. Compare your results with those obtained in Problem 5.33.

5.35. The annual-maximum discharges in a river show a mean of 480 m^3/s, a standard deviation of 320 m^3/s, and a skewness of 2.13. Assuming a Pearson Type III distribution, use the frequency factor to estimate the 50-year and 100-year annual-maximum flows.

5.36. The annual-maximum discharges in a river show a mean of 480 m^3/s, a standard deviation of 320 m^3/s, and a log-transformed skewness of 0.1. Assuming a log–Pearson Type III distribution, estimate the 100-year annual-maximum flow.

5.37. The annual-maximum discharges in a river have a mean of 480 m^3/s and a standard deviation of 320 m^3/s. Assuming that the annual maxima are described by an extreme/value Type I distribution, use the frequency factor to estimate the 50-year and 100-year maximum flows. Compare your results with those obtained in Problem 5.20.

5.38. Consider the series of annual-maximum streamflow measurements given in the following table: (a) Estimate the return period of a 30 m^3/s (annual-maximum) flow using the Weibull formula; and (b) Estimate the return period of a 30 m^3/s (annual-maximum) flow using the Gringorten formula. It is postulated that the observed data follow the extreme-value Type I distribution. (c) Use the method of moments to estimate the parameters (a and b) of the extreme-value Type I distribution; (d) Use the chi-square test with six classes to assess whether the data fit an extreme-value Type I

distribution; and (e) Assuming that the data are adequately described by an exteme-value Type I distribution, use the frequency factor to determine the annual-maximum streamflow with a 100-year return period.

Year	Flow (m^3/s)
1970	20
1971	23
1972	13
1973	28
1974	35
1975	19
1976	14
1977	10
1978	31
1979	25
1980	9
1981	40
1982	22
1983	19
1984	17

6 Surface-Water Hydrology

6.1 Introduction

Surface-water hydrology is the science dealing with the movement, distribution, and properties of water above the surface of the earth. Applications of surface-water hydrology in engineering practice usually include the modeling of rainfall events, and predicting the quantity and quality of the resulting surface runoff. The temporal distribution of rainfall at a given location is called a *hyetograph*, the temporal distribution of surface runoff at any location is called a *hydrograph*, and the temporal distribution of pollutant concentration in the runoff is called a *pollutograph*. The estimation of hydrographs and pollutographs from hyetographs and the design of systems to control the quantity and quality of surface runoff are the responsibility of a water-resources engineer.

The land area that can contribute to the runoff at any particular location is determined by the shape and topography of the region surrounding the location. The potential contributing area is called the *watershed*; the area within a watershed over which rainfall occurs is called the *catchment* area. In most engineering applications, the watershed and catchment areas are taken to be the same. Classes of pollutants contained in surface-water runoff that are typically of interest include suspended solids, heavy metals, nutrients, organics, oxygen-demanding substances, and bacteria (ASCE, 1992).

6.2 Rainfall

The precipitation of water vapor from the atmosphere occurs in many forms, the most important of which are rain and snow. Hail and sleet are less frequent forms of precipitation. Engineered drainage systems in most urban communities are designed primarily for rainfall. The formation of precipitation usually results from the lifting of moist air masses within the atmosphere, which results in the cooling and condensation of moisture. The four conditions that must be present for the production of precipitation are: (1) cooling of the atmosphere, (2) condensation of water droplets onto nuclei, (3) growth of water droplets, and (4) mechanisms to cause a sufficient density of the droplets.

Cloud droplets generally form on condensation nuclei, which are usually less than 1 micron in diameter and typically consist of sea salts, dust, or combustion by-products. In pure air, condensation of water vapor to form liquid water droplets occurs only when the air is supersaturated. Typically, once condensation begins, water droplets and ice crystals are both formed. However, the greater saturation vapor pressure over water compared to over ice results in a vapor pressure gradient that causes the growth of ice crystals at the expense of water droplets. This process is called the *ice crystal process*. When the condensed moisture is large enough, the precipitation falls to the ground. Moisture droplets larger than about 0.1 mm are large enough to fall, and these drops grow as they collide and coalesce to form larger particles. Rain drops falling to the ground are typically in the size range of 0.5–3 mm, while snow flakes are considerably larger.

The main mechanisms of air-mass lifting are frontal lifting, orographic lifting, and convective lifting. In *frontal lifting*, warm air is lifted over cooler air by frontal passage. The resulting precipitation events are called *cyclonic* or *frontal* storms, and the zone where the warm and cold air masses meet is called a *front*. Frontal precipitation is the dominant type of precipitation in the north-central United States and other continental areas (Elliot, 1995). In a *warm front*, warm air advances over a colder air mass with a relatively slow rate of ascent (0.3%–1% slope) causing a large area of precipitation in advance of the front, typically 300 to 500 km (200 to 300 mi) ahead of the front. In a *cold front*, warm air is pushed upward at a relatively steep slope (0.7%–2%) by advancing cold air, leading to smaller precipitation areas in advance of the cold front. Precipitation rates are generally higher in advance of cold fronts than in advance of warm fronts. In *orographic lifting*, warm air rises as it flows over hills or mountains, and the resulting precipitation events are called *orographic storms*. An example of the orographic effect can be seen in the northwestern United States, where the westerly air flow results in higher precipitation and cooler temperatures to the west of the Cascade mountains (e.g., Seattle, Washington) than to the east of the Cascade mountains (e.g., Boise, Idaho). Orographic precipitation is a major factor in most mountainous areas. In *convective lifting*, air rises as by virtue of being warmer and less dense than the surrounding air, and the resulting precipitation events are called *convective storms* or, more commonly, *thunderstorms*. Convective precipitation is common during the summer months in the central United States and other continental climates with moist summers. Convective storms are typically of short duration, with small areal coverage, and usually occur on hot midsummer days as late afternoon storms.

6.2.1 ■ Local Rainfall

Rainfall is typically measured using rain gages operated by government agencies such as the National Weather Service and local drainage districts. Rainfall amounts are described by the volume of rain falling per unit area and is given as a depth of water. In the United States, much of the rainfall data reported by the National Weather Service (NWS) are collected using a standard rain gage that consists of a 20.3 cm (8 in.) diameter funnel that passes water into a cylindrical measuring tube, and the whole assembly is placed within an overflow can. The measuring tube has a cross-sectional area one-tenth that of the collector funnel; therefore, a 2.5 mm (0.1 in.) rainfall will occupy a depth of 25 mm (1 in.) in the collector tube. The capacity of the collector tube is 50 mm (2 in.), and rainfall in excess of this amount collects in the overflow can. The manual NWS gage is primarily used for collecting daily rainfall amounts; automatic-recording gages are usually used for shorter durations and for collecting data in remote locations. The accuracy of both manual and automatic-recording rain gages are typically on the order of 0.25 mm (0.01 in.). Rain gage measurements are actually point measurements of rainfall and may only be representative of a small area surrounding the rain gage. Areas on the order of 25 km^2 (10 mi^2) have been taken as characteristic of rain gage measurements

(Gupta, 1989; Ponce, 1989), although considerably smaller characteristic areas can be expected in regions where convection storms are common.

Rainfall measurements are seldom used directly in design applications, but rather the statistics of the rainfall measurements are typically used. The most common form of presenting rainfall statistics is in the form of *intensity-duration-frequency* (IDF) curves, which express the relationship between the intensity in a rainstorm and the averaging time (= duration), with each relationship having a given return period. To fully understand the meaning and application of the IDF curve, it is best to review how this curve is calculated from raw rainfall measurements. The data required to calculate the IDF curve is a record of rainfall measurements in the form of the depth of rainfall during fixed intervals of time, Δt, typically on the order of 5 minutes. Local rainfall data are usually in the form of daily totals for nonrecording gages, with smaller time increments used in recording gages. For a rainfall record containing several years of data, the following computations lead to the IDF curve:

1. For a given duration of time (= averaging period), starting with Δt, determine the maximum rainfall amount for this duration in *each* year.
2. The precipitation amounts, one for each year, are rank ordered, and the return period, T, for each precipitation amount is estimated using the Weibull formula

$$T = \frac{n+1}{m} \tag{6.1}$$

where n is the number of years of data and m is the rank of the data corresponding to the event with return period T. As an alternative to using the Weibull formula to estimate the probability distribution of the precipitation amounts, the extreme-value Type I distribution may be assumed. In this case, the probability distribution of the precipitation amounts are derived from the mean and standard deviation of the annual maximum depths (Akan, 1993).
3. Steps 1 and 2 are repeated, with the duration increased by Δt. A maximum duration needs to be specified, and is typically on the order of 1 to 2 h.
4. For each return period, T, the precipitation amount versus duration can be plotted. This relationship is called the depth-duration-frequency curve. Dividing the precipitation amount by the corresponding duration yields the average intensity, which is plotted versus the duration, for each return period, to yield the IDF curve.

This procedure is illustrated in the following example.

■ Example 6.1

A rainfall record contains 32 years of rainfall measurements at 5-minute intervals. The maximum rainfall amounts for intervals of 5 min, 10 min, 15 min, 20 min, 25 min, and 30 min have been calculated and ranked. The top three rainfall amounts, in millimeters, for each time increment are given in the following table.

Rank	Δt in minutes					
	5	10	15	20	25	30
1	12.1	18.5	24.2	28.3	29.5	31.5
2	11.0	17.9	22.1	26.0	28.4	30.2
3	10.7	17.5	21.9	25.2	27.6	29.9

Calculate the IDF curve for a return period of 20 years.

Solution

For each time interval, Δt, there are $n = 32$ ranked rainfall amounts of annual maxima. The relationship between the rank, m, and the return period, T, is given by Equation 6.1 as

$$T = \frac{n+1}{m} = \frac{32+1}{m} = \frac{33}{m}$$

The return period can therefore be used in lieu of the rankings, and the given data can be put in the form:

Return Period, T (years)	Δt in minutes					
	5	10	15	20	25	30
33	12.1	18.5	24.2	28.3	29.5	31.5
16.5	11.0	17.9	22.1	26.0	28.4	30.2
11	10.7	17.5	21.9	25.2	27.6	29.9

The rainfall increments with a return period, T, of 20 years can be linearly interpolated between the rainfall increments corresponding to $T = 33$ years and $T = 16.5$ years to yield:

Duration (min)	5	10	15	20	25	30
Rainfall (mm)	11.2	18.0	22.5	26.5	28.6	30.5

and the average intensities for each duration are obtained by dividing the rainfall amounts by the corresponding duration to yield:

Duration (min)	5	10	15	20	25	30
Intensity (mm/h)	134	108	90	79	69	61

These points define the IDF curve for a return period of 20 years.   ■

■ Example 6.2

How many years of rainfall data are required to derive the IDF curve for a return period of 10 years?

Solution

The return period, T, is related to the number of years of data, n, and the ranking, m, by

$$T = \frac{n+1}{m}$$

For $T = 10$ years and $m = 1$,

$$n = mT - 1 = (1)(10) - 1 = 9 \text{ years}$$

Therefore a minimum of 9 years of data are required to estimate the IDF curve for a 10-year return period.   ■

■ Table 6.1	Return period	
Factors for Converting Partial-Duration Series to Annual Series	(years)	Factor
	2	0.88
	5	0.96
	10	0.99
	25	1.00
	> 25	1.00

Source: Frederick et al. (1977).

The previous example has illustrated the derivation of IDF curves from n annual maxima of rainfall measurements, where the series of annual maxima is simply called the *annual series*. As an alternative to using a n-year annual series to derive IDF curves, a *partial-duration series* is sometimes used in which the largest n rainfall amounts in a n-year record are selected for each duration, regardless of the year in which the rainfall amounts occur. In this case, the return period, T, assigned to each rainfall amount is still calculated using Equation 6.1. The frequency distribution of rainfall amounts derived using the partial duration series differs from that derived using the annual series. However, the return periods calculated from the partial-duration series can be converted to corresponding return periods from the annual series using the empirical factors in Table 6.1 for return periods in excess of 2 years.

■ Example 6.3

A 10-year rainfall record measures the rainfall increments at 5-minute intervals. The top six rainfall increments derived from the partial-duration series are as follows:

Rank	1	2	3	4	5	6
5-min rainfall (mm)	22.1	21.9	21.4	20.7	20.3	19.8

Estimate the frequency distribution of the annual maxima with return periods greater than 2 years.

Solution

The ranked data were derived from a 10-year partial-duration series ($n = 10$), where the return period, T, is given by

$$T = \frac{n+1}{m} = \frac{10+1}{m} = \frac{11}{m}$$

Using this (Weibull) relation, the frequency distribution of the 5-min rainfall increments is given by

Return period (y)	11	5.5	3.7	2.8	2.2	1.8
5-min rainfall (mm)	22.1	21.9	21.4	20.7	20.3	19.8

The calculated return periods for the partial-duration series can be converted to return periods for the annual series of maxima by using the factors in Table 6.1. Applying linear interpolation in Table 6.1 leads to the following conversion factors:

Return period (y)	11	5.5	3.7	2.8	2.2	1.8
Factor	0.99	0.96	0.93	0.90	0.89	—

Applying these factors to the partial-duration frequency distribution leads to the following frequency distribution of annual-maxima 5-min rainfall increments:

Return period (y)	10.9	5.3	3.4	2.5	2.0
5-min rainfall (mm)	22.1	21.9	21.4	20.7	20.3

This analysis can be repeated for different durations in order to determine the IDF curves from a partial duration series.   ■

The frequency distributions of local rainfall in the United States have been published by Hershfield (1961) for storm durations from 30 min to 24 hours, and return periods from 1 to 100 years. Hershfield's paper is commonly referred to as TP-40 (an acronym for the U.S. Weather Bureau* *Technical Paper Number 40*, published by Hershfield in 1961) or simply the *Rainfall Frequency Atlas*. The frequency distributions in TP-40 were derived from data at approximately 4,000 stations by assuming a Gumbel distribution. The data in TP-40 for the 11 western states have been updated by Miller and colleagues (1973). The U.S. Weather Bureau has also published rainfall frequency maps for storm durations from 2 to 10 days (USWB, 1964), and this report is commonly referred to as TP-49 (an acronym for the U.S. Weather Bureau *Technical Paper Number 49*). In designing urban drainage systems, rainfall durations of less than 60 min and sometimes as small as 5 min are commonly used (Wenzel, 1982). The rainfall statistics for durations from 5 to 60 min have been published by Frederick and colleagues (1977) for the eastern and central United States. This paper is commonly referred to as HYDRO-35 (an acronym for the NOAA *Technical Memorandum Number NWS HYDRO-35*), and these results partially supersede those in TP-40.

A typical IDF curve is illustrated in Figure 6.1. Although IDF curves are available for many locations within the United States, usually from a local water management or drainage district, there are some locations for which IDF curves are not available. In these cases, IDF curves can be estimated from the published National Weather Service frequency distributions in TP-40 using a methodology proposed by Chen (1983). This method is based on three rainfall depths derived from TP-40: the 10-year 1-h rainfall (R_1^{10}) shown in Figure 6.2, the 10-year 24-h rainfall (R_{24}^{10}) shown in Figure 6.3, and the 100-year 1-hour rainfall (R_1^{100}) shown in Figure 6.4. The IDF curve can then be expressed

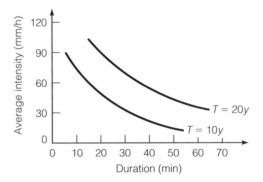

Figure 6.1 ■ Intensity-Duration-Frequency (IDF) Curve

---

*The United States Weather Bureau is now called the National Weather Service.

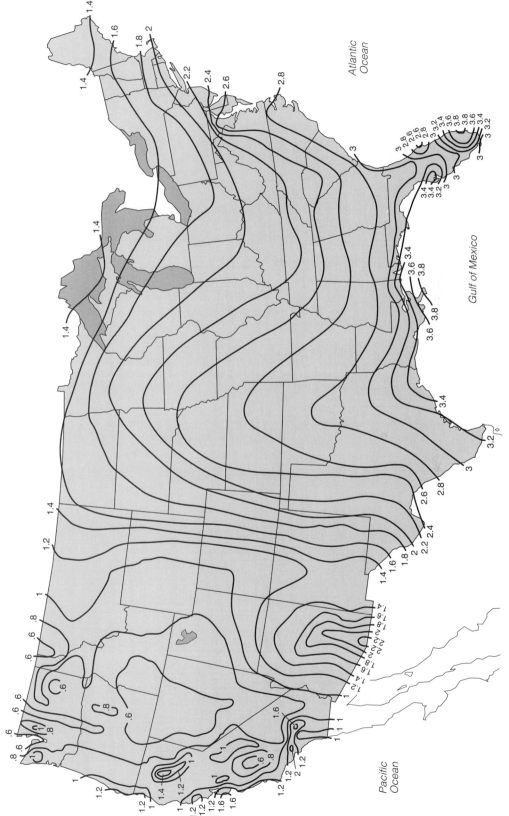

Figure 6.2 ■ 10-year 1-hour Rainfall (inches)

305

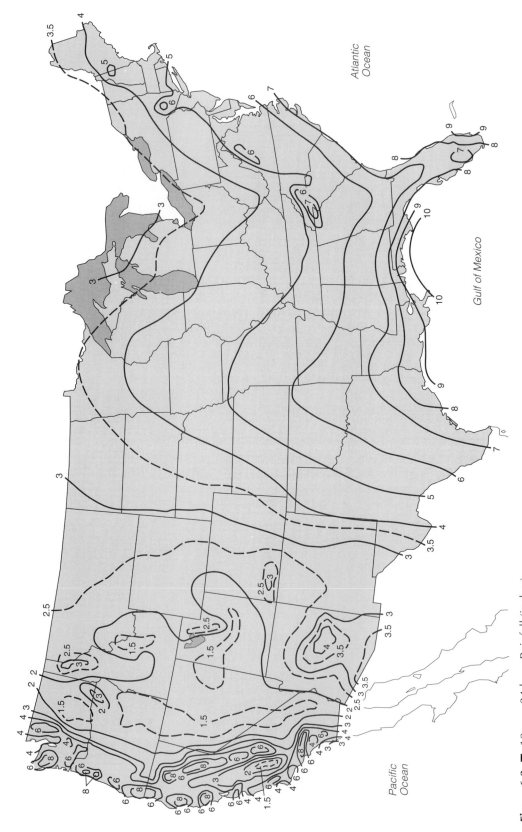

Figure 6.3 ■ 10-year 24-hour Rainfall (inches)

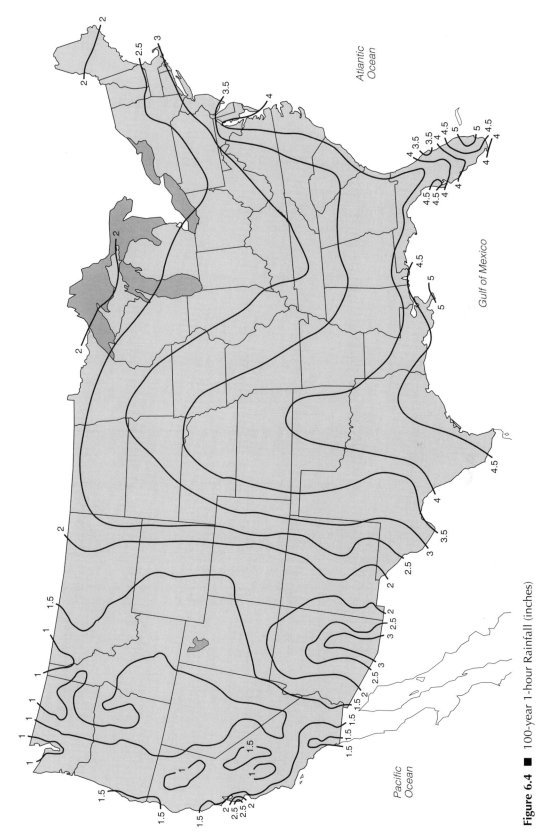

Figure 6.4 ■ 100-year 1-hour Rainfall (inches)
Source: Chen, C. 1983. "Rainfall intensity-duration-frequency formulas," *Journal of Hydraulic Engineering,* Vol. 109, 12. Reprinted by permission of ASCE.

as a relationship between the average intensity, i (in./h), for a rainfall duration t (min) by

$$i = \frac{a}{(t + b_1)^{c_1}}$$ (6.2)

where a is a constant given by

$$a = a_1 R_1^{10}[(x - 1)\log(T_p/10) + 1]$$ (6.3)

a_1, b_1, and c_1 are empirical functions of R_1^{10}/R_{24}^{10} derived from Figure 6.5, x is defined by

$$x = \frac{R_1^{100}}{R_1^{10}}$$ (6.4)

and T_p is the return period for the partial-duration series, which is assumed to be related to return period, T, for the annual-maximum series by the relation

$$T_p = -\frac{1}{\ln(1 - 1/T)}$$ (6.5)

For $T > 10$ years there is not a significant difference between T and T_p.

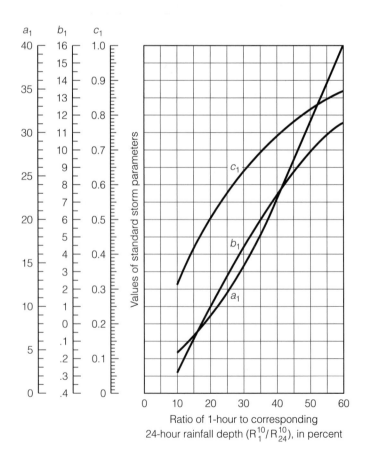

Figure 6.5 ■ Constants in Chen IDF Curve

■ Example 6.4

Estimate the IDF curve for 50-year storms in Miami using the Chen method. What is the average intensity of a 50-year 1-hour storm?

Solution

For Miami, $R_1^{10} = 3.6$ in. (Figure 6.2), $R_{24}^{10} = 9$ in. (Figure 6.3), and $R_1^{100} = 4.7$ in. (Figure 6.4). For a return period, T, equal to 50 years, then

$$T_p = -\frac{1}{\ln(1 - 1/T)} = -\frac{1}{\ln(1 - 1/50)} = 49.5 \text{ years}$$

$$x = \frac{R_1^{100}}{R_1^{10}} = \frac{4.7}{3.6} = 1.31$$

Since $R_1^{10}/R_{24}^{10} = 3.6/9 = 0.4 = 40\%$, then Figure 6.5 gives $a_1 = 22.8$, $b_1 = 7.5$, and $c_1 = 0.74$. Equation 6.3 gives

$$a = a_1 R_1^{10}[(x - 1)\log(T_p/10) + 1] = (22.8)(3.6)[(1.31 - 1)\log(49.5/10) + 1] = 99.8$$

and therefore the IDF curve is given by

$$i = \frac{a}{(t + b_1)^{c_1}} = \frac{99.8}{(t + 7.5)^{0.74}}$$

For a storm of duration, t, equal to 1 hour ($= 60$ min),

$$i = \frac{99.8}{(60 + 7.5)^{0.74}} = 4.4 \text{ in./h}$$

A 50-year 1-hour storm in Miami therefore has an average intensity of 4.4 in./h.   ■

    Wenzel (1982) developed empirical IDF curves for several large cities in the United States using the Chen (1983) form of the IDF curves given by Equation 6.2. Values of a, b_1, and c_1 for the IDF curves corresponding to a 10-year return period are given in Table 6.2. These IDF curves are particularly useful to engineers since most drainage systems are designed for rainfall events with return periods on the order of 10 years.

Table 6.2 Ten-Year IDF Constants for Major U.S. Cities	**City**	**a**	**b_1**	**c_1**
	Atlanta	64.1	8.16	0.76
	Chicago	60.9	9.56	0.81
	Cleveland	47.6	8.86	0.79
	Denver	50.8	10.50	0.84
	Houston	98.3	9.30	0.80
	Los Angeles	10.9	1.15	0.51
	Miami	79.9	7.24	0.73
	New York	51.4	7.85	0.75
	Santa Fe	32.2	8.54	0.76
	St. Louis	61.0	8.96	0.78

Source: Wenzel (1982).

6.2.2 ■ Spatially Averaged Rainfall

The utilization of point rainfall data to calculate surface runoff usually requires the estimation of catchment-averaged rainfall. Rainfall is never uniformly distributed in space, and spatially averaged rainfall tends to be scale dependent and not statistically homogeneous (in space). The scale dependence of spatially averaged rainfall can be derived from rain gage measurements using numerical interpolation schemes such as kriging (Journel, 1989).

Consider the averaging area and distribution of rain gages illustrated in Figure 6.6. The objective is to estimate the average rainfall over the given area based on rainfall measurements at the individual rain gages. The basis of the estimation scheme is an interpolation function that estimates the rainfall at any point in the given area, usually as a weighted average of the rainfall measurements at the individual rain gages. A linear interpolation function has the form

$$\hat{P}(\mathbf{x}) = \sum_{i=1}^{N} \lambda_i P(\mathbf{x}_i) \tag{6.6}$$

where $\hat{P}(\mathbf{x})$ is the (interpolated) rainfall at location $\mathbf{x}$, $P(\mathbf{x}_i)$ is the measured precipitation at rain gage i that is located at $\mathbf{x}_i$, λ_i is the weight given to measurements at station i, and these weights generally satisfy the relation

$$\sum_{i=1}^{N} \lambda_i = 1 \tag{6.7}$$

There are a variety of ways to estimate the weights, λ_i, depending on the underlying assumptions about the spatial distribution of the rainfall. Some of the more common assumptions are as follows:

1. The rainfall is uniformly distributed in space. Under this assumption, equal weight is assigned to each station

$$\lambda_i = \frac{1}{N} \tag{6.8}$$

The estimated rainfall at any point is simply equal to the arithmetic average of the measured data.

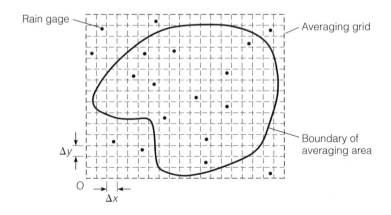

Figure 6.6 ■ Estimation of Spatially Averaged Rainfall

2. The weights are calculated from the kriging weights (Journel, 1989), using either the covariance function or the variogram of the rainfall. This method is equivalent estimating the point rainfall from the contours of equal rainfall (isohyets) derived from the measured data.
3. The rainfall at any point is estimated by the rainfall at the nearest station. Under this assumption, $\lambda_i = 1$ for the nearest station, and $\lambda_i = 0$ for all other stations. This methodology is the discrete equivalent of the graphical *Thiessen polygon method* (Thiessen, 1911) that has been used for many years in hydrology. This approach can reasonably be called the *discrete Thiessen method*.
4. The weight assigned to each measurement station is inversely proportional to the distance to some power from the estimation point to the measurement station. This method is called the *reciprocal-distance method* (Simanton and Osborn, 1980, Wei and McGuinness, 1973).

After specifying the station weights in the rainfall interpolation formula, the next step is to numerically discretize the averaging area by an averaging grid as indicated in Figure 6.6. The definition of the averaging grid requires specification of the origin, O; discretizations in the x and y directions, Δx and Δy; and the number of cells in each of the coordinate directions. The rainfall, $\hat{P}(\mathbf{x}_j)$, at the center, $\mathbf{x}_j$, of each cell is then calculated using the interpolation formula (Equation 6.6) with specified weights, and the average rainfall over the entire area, $\bar{P}$ is given by

$$\bar{P} = \frac{1}{A} \sum_{j=1}^{J} \hat{P}(\mathbf{x}_j) A_j \qquad (6.9)$$

where A is the averaging area, J is the number of cells that contain a portion of the averaging area, and A_j is the amount of the averaging area contained in cell j. This method is well suited to spreadsheet calculation, or to computer programs written specifically for this task.

■ Example 6.5

The spatially averaged rainfall is to be calculated for the catchment area shown in Figure 6.7. There are five rain gages in close proximity to the catchment area, and the Cartesian

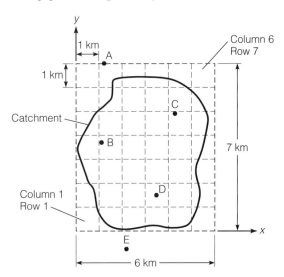

Figure 6.7 ■
Catchment Area

coordinates of these gages are as follows:

Gage	x (km)	y (km)
A	1.3	7.0
B	1.0	3.7
C	4.2	4.9
D	3.5	1.4
E	2.1	−1.0

The rainfall measured at each of the gages during a 1-hour interval are 60 mm at A, 90 mm at B, 65 mm at C, 35 mm at D, and 20 mm at E. Use the discrete Thiessen method with the 1 km × 1 km grid shown in Figure 6.7 to estimate the average rainfall over the catchment area during the 1-hour interval.

Solution

Using the discrete Thiessen method the weights, λ_i, are assigned to each cell in Figure 6.7 using the convention that the nearest station has a weight of 1, and all other stations have weights of zero. Computations are summarized in Table 6.3, where the row numbers increase from the bottom to the top, and the column numbers increase from left to right in Figure 6.7. The station weights assigned to each cell, based on the rain gage closest to the center of the cell, are shown in columns 3 to 7 in Table 6.3, and area, A_i of the catchment contained in each cell, i, is shown in column 8. The rainfall amount, P_i, assigned to each cell is equal to the rainfall at the nearest station (according to the discrete Thiessen method) and is given in column 9. The total area, A, of the catchment is obtained by summing the values in column 8 and is equal to 29.61 km^2. The average rainfall over the catchment, $\overline{P}$, is given by

$$\overline{P} = \frac{1}{A} \sum P_i A_i = \frac{1}{29.61} \sum 1721.2 = 58 \text{ mm}$$

The average rainfall over the entire catchment is therefore equal to 58 mm.   ■

6.2.3   ■   Design Rainfall

A hypothetical rainfall event associated with a specified return period is usually the basis for the design and analysis of stormwater-management systems. However, it should be noted that the return period of a rainfall event is not equal to the return period of the resulting runoff, and therefore the reliability of surface-water management systems will depend on such factors as the antecedent moisture conditions in the catchment. In contrast to the single-event design-storm approach, a continuous-simulation approach is sometimes used where a historical rainfall record is used as input to a rainfall-runoff model; the resulting runoff is analyzed to determine the hydrograph corresponding to a given return period. The design-storm approach is more widely used in engineering practice than the continuous-simulation approach. Design storms can be either synthetic or actual (historic) design storms, with synthetic storms defined from historical rainfall statistics.

Synthetic design storms are characterized by their return period, duration, depth, temporal distribution, and spatial distribution. The selection of these quantities for design purposes are described in the following sections.

■ **Table 6.3**
Calculation of
Catchment-Averaged
Rainfall

(1) Row	(2) Column	(3) A	(4) B	(5) C	(6) D	(7) E	(8) Area, A_i (km^2)	(9) Rainfall, P_i (mm)	(10) $P \times A$ (km$^2 \cdot$mm)
				Weights, λ_i					
1	1					1	0.01	20	0.2
	2					1	0.7	20	14
	3				1		0.95	35	33.25
	4				1		0.9	35	31.5
	5				1		0.8	35	28
	6				1		0.1	35	3.5
2	1					1	0.15	20	17
	2				1		1	35	35
	3				1		1	35	35
	4				1		1	35	35
	5				1		1	35	35
	6				1		0.3	35	10.5
3	1		1				0.7	90	63
	2		1				1	90	90
	3				1		1	35	35
	4				1		1	35	35
	5				1		1	35	35
	6				1		0.2	35	7
4	1		1				0.95	90	85.5
	2		1				1	90	90
	3		1				1	90	90
	4			1			1	65	65
	5			1			1	65	65
	6			1			0.4	65	26
5	1		1				0.6	90	54
	2		1				1	90	90
	3		1				1	90	90
	4			1			1	65	65
	5			1			1	65	65
	6			1			0.75	65	48.75
6	1	1					0	60	0
	2	1					0.7	60	42
	3			1			1	65	65
	4			1			1	65	65
	5			1			1	65	65
	6			1			0.35	65	22.75
7	1	1					0	60	0
	2	1					0.15	60	9
	3	1					0.55	60	33
	4			1			0.45	65	29.25
	5			1			0.2	65	13
	6			1			0	65	0
Total							29.61		1,721.2

Land Use	Design Storm Return Period (years)
Minor drainage systems:	
Residential	2–5
High-value general commercial area	2–10*
Airports (terminals, roads, aprons)	2–10
High-value downtown business areas	5–10
Major drainage-system elements	up to 100 years

■ Table 6.4
Typical Return Periods

Source: ASCE, 1992. *Design and Construction of Urban Stormwater Management Systems.* p. 70. Reprinted by permission of ASCE.
*According to Burton (1996), return periods of 10 to 30 years are commonly used in practice for designing storm sewers in commercial and high-value districts.

6.2.3.1   Return Period

The return period of a design rainfall should be selected on the basis of economic efficiency (ASCE, 1992). In practice, however, the return period is usually selected on the basis of level of protection. Typical return periods are given in Table 6.4, although longer return periods are sometimes used. In selecting the return period for a particular project, local drainage manuals and regulations should also be reviewed.

6.2.3.2   Rainfall Duration

The design duration of a storm is usually selected on the basis of the time-response characteristics of the catchment. The time response of a catchment is measured by the travel time of surface water from the most remote point of the catchment to the catchment outlet and is called the *time of concentration*. Current practice is to select the duration of the design rainfall as equal to the time of concentration. This approach usually leads to the maximum peak runoff for a given return period. For the design of detention basins, however, the duration causing the largest detention volume is most critical, and several different storm durations may need to be tried to identify the most critical design-storm duration (Akan, 1993). For catchments with high infiltration losses, the duration of the critical rainfall associated with the maximum peak runoff may be shorter than the time of concentration (Chen and Wong, 1993). Many drainage districts require that the performance of drainage systems be analyzed for a standard 24-hour storm with a specified return period, typically on the order of 25 years.

6.2.3.3   Rainfall Depth

The design-rainfall depth for a selected return period and duration is obtained directly from the intensity-duration-frequency (IDF) curve of the catchment. The IDF curve can be estimated from rainfall measurements, derived from National Weather Service (NWS) publications such as TP-40 (Hershfield, 1961), or obtained from regulatory manuals that govern local drainage designs.

6.2.3.4   Temporal Distribution

Realistic temporal distributions of rainfall within design storms are best determined from historical rainfall measurements. In many cases, however, either the data are not

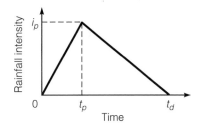

Figure 6.8 ■ Triangular
Rainfall Distribution
Adapted from ASCE (1992).

available or such a detailed analysis cannot be justified. Under these conditions, the designer must resort to empirical distributions. Frequently used methods for estimating the rainfall distribution in storms are the triangular method, alternating-block method, and the NRCS 24-hour hyetograph.

Triangular Method A common approximation for the rainfall distribution is the triangular distribution shown in Figure 6.8. In this distribution, the time to peak, t_p, is related to the (intensity-weighted) mean of the rainfall distribution, $\bar{t}$, by the relation

$$\boxed{\frac{t_p}{t_d} = 3\frac{\bar{t}}{t_d} - 1}$$
(6.10)

where t_d is the duration of the storm. Using this equation, the nondimensional time to peak, t_p/t_d, can be related to the nondimensional mean of the rainfall distribution. Yen and Chow (1980) and El-Jabi and Sarraf (1991) studied a wide variety of storms and found that t_p/t_d was typically in the range of 0.32 to 0.51, while McEnroe (1986) found much shorter time to peak by considering only 1-hour storms with return periods of two years or greater.

■ Example 6.6

The IDF curve for 10-year storms in Houston, Texas, is given by

$$i = \frac{2497}{(t_d + 9.30)^{0.80}}$$

where i is the average intensity in mm/h and t_d is the duration in minutes. Assuming that the mean of the rainfall distribution is equal to 40% of the rainfall duration, estimate the triangular hyetograph for a 1-hour storm.

Solution

For a 1-hour storm, $t_d = 60$ min and the average intensity, i, is given by

$$i = \frac{2497}{(t_d + 9.30)^{0.80}} = \frac{2497}{(60 + 9.30)^{0.80}} = 84.1 \text{ mm/h}$$

The rainfall amount, P, in 1 hour is therefore given by

$$P = 84.1 \text{ mm/h} \times 1 \text{ h} = 84.1 \text{ mm}$$

The peak of the triangular hyetograph occurs at t_p, where Equation 6.10 gives

$$\frac{t_p}{t_d} = 3\frac{\bar{t}}{t_d} - 1$$

From the given data, $\bar{t}/t_d = 0.4$, in which case

$$\frac{t_p}{t_d} = 3(0.4) - 1 = 0.2$$

and

$$t_p = 0.2t_d = 0.2(1) = 0.2 \text{ h}$$

The triangular hyetograph has a peak equal to i_p, a base of t_d and an area under the hyetograph of 84.1 mm. Therefore, using the formula for the area of a triangle

$$\frac{1}{2}i_p t_d = 84.1$$

leads to

$$i_p = \frac{84.1}{\frac{1}{2}t_d} = \frac{84.1}{\frac{1}{2}(1)} = 168 \text{ mm/h}$$

The triangular hyetograph for this 10-year 1-hour storm is illustrated in Figure 6.9.   ■

Alternating-Block Method.   A hyetograph can be derived from the IDF curve for a selected storm duration and return period using the *alternating-block method*. The hyetograph generated by the alternating-block method describes the rainfall in n time intervals of duration Δt, for a total storm duration of $n\Delta t$. The procedure for determining the alternating-block hyetograph from the IDF curve is as follows:

1. Select a return period and duration for the storm.
2. Read the average intensity from the IDF curve for storms of duration Δt, $2\Delta t$, ..., $n\Delta t$. Determine the corresponding precipitation depths by multiplying each intensity by the corresponding duration.
3. Calculate the difference between successive precipitation depth values. These differences are equal to the amount of precipitation during each additional unit of time Δt.
4. Reorder the precipitation increments into a time sequence with the maximum intensity occurring at the center of the storm, and the remaining precipitation increments arranged in descending order alternately to the right and left of the central block.

Figure 6.9 ■ Estimated 10-year 1-hour Rainfall Hyetograph in Houston, Texas

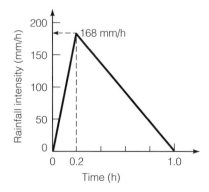

The alternating-block method assumes that the maximum rainfall for any duration less than or equal to the total storm duration has the same return period (ASCE, 1992). Field data indicate that this assumption is very conservative, particularly for longer storms.

■ Example 6.7

The IDF curve for 10-year storms in Houston, Texas, is given by

$$i = \frac{2497}{(t + 9.30)^{0.80}} \text{ mm/h}$$

where t is the duration in minutes. Use the alternating-block method to calculate the hyetograph for a 10-year 1-hour storm using 9 time intervals. Compare this result with the triangular hyetograph determined in Example 6.6.

Solution

For a 60-minute storm with 9 time intervals, the time increment, Δt, is given by

$$\Delta t = \frac{60}{9} = 6.67 \text{ min}$$

The average intensities for storm durations equal to multiples of Δt are derived from the IDF curve using $t = \Delta t, 2\Delta t, \ldots, 9\Delta t$ and the results are given in column 3 of the following table:

(1) Increment	(2) t (min)	(3) i (mm/h)	(4) it (mm)	(5) Rainfall amount (mm)	(6) Intensity (mm/h)
1	6.67	272	30.2	30.2	272
2	13.33	206	45.8	15.6	140
3	20.00	167	55.7	9.90	89.1
4	26.67	142	63.1	7.40	66.6
5	33.33	124	68.9	5.80	52.2
6	40.00	110	73.3	4.40	39.6
7	46.67	99.8	77.6	4.30	38.7
8	53.33	91.2	81.1	3.50	31.5
9	60.00	84.1	84.1	3.00	27.0

The precipitation for each rainfall duration, t, is given in column 4 ($=$ col. 2 × col. 3), the rainfall increments corresponding to the duration increments are given in column 5, and the corresponding intensities are given in column 6. In accordance with the alternating-block method, the maximum intensity ($=$ 272 mm/h) is placed at the center of the storm, and the other intensities are arranged in descending order alternately to the right and left of the center block. The alternating-block hyetograph is therefore given by:

Time (min)	Average intensity (mm/h)
0–6.67	27.0
6.67–13.33	38.7
13.43–20.00	52.2
20.00–26.67	89.1
26.67–33.33	272
33.33–40.00	140
40.00–46.67	66.6
46.67–53.33	39.6
53.33–60.00	31.5

In the triangular hyetograph derived in Example 6.6, the location of the peak intensity can be varied, while in the alternating-block method the peak intensity always occurs at 50% of the storm duration. Also, the triangular hyetograph derived in the previous example resulted in a maximum intensity of 168 mm/h, while the alternating-block method resulted in a much higher maximum intensity of 272 mm/h. This difference is a result of the short duration used to calculate the peak intensity and reflects the conservative nature of the alternating-block method.   ■

NRCS 24-h Hyetographs   The Natural Resources Conservation Service (formerly the Soil Conservation Service) has developed 24-h rainfall distributions for four geographic regions in the United States (SCS, 1986). These rainfall distributions are approximately consistent with local IDF curves, and the geographic boundaries corresponding to these rainfall distributions are shown in Figure 6.10. Types I and IA rainfall distributions are characteristic of a pacific maritime climate with wet winters and dry summers and cover the coastal regions of California, Oregon, Washington, and the entire states of Alaska and Hawaii. Type II rainfall is characteristic of most regions in the United States, with the exceptions of the Gulf coast regions of Texas, Louisiana, Alabama, South Florida, and most of the Atlantic coastline, which are characterized by Type III rainfall, where tropical storms are prevalent and produce large 24-hour rainfall amounts. The 24-h rainfall hyetographs are illustrated in Figure 6.11, where the abcissa is the time in hours, and the ordinate gives the dimensionless precipitation, P/P_T, where P is the cumulative rainfall (a function of time) and P_T is the total 24-hour rainfall amount. The coordinates of the rainfall distributions are given in Table 6.5. The peak rainfall intensity occurs at the time when the slope of the cumulative rainfall distribution is steepest, which for the NRCS 24-h hyetographs are 8.2 h (Type IA), 10.0 h (Type I), 12.0 h (Type II), and 12.2 h (Type III). Comparing the NRCS 24-h rainfall distributions indicates that Type IA yields the least intense storms and Type II the most intense storms.

■ **Example 6.8**

The precipitation resulting from a 10-year 24-hour storm on the Gulf coast of Texas is estimated to be 180 mm. Calculate the NRCS 24-h hyetograph.

Solution

On the Gulf coast of Texas, 24-hour storms are characterized by Type III rainfall. The hyetograph is determined by multiplying the ordinates of the Type III hyetograph in Table 6.5 by $P_T = 180$ mm to yield the following select points (this example is for illustrative purposes only and not all hyetograph points are shown):

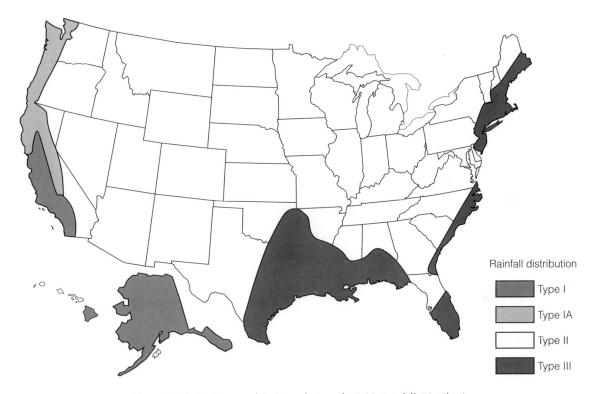

Figure 6.10 ■ Geographic Boundaries of NRCS Rainfall Distributions

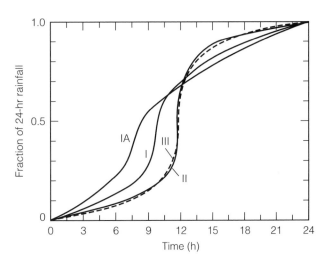

Figure 6.11 ■ NRCS 24-h
Rainfall Distributions

■ **Table 6.5**
NRCS 24-h Rainfall
Distributions

Time (h)	Type I P/P_T	Type IA P/P_T	Type II P/P_T	Type III P/P_T
0.0	0.000	0.000	0.000	0.000
0.5	0.008	0.010	0.005	0.005
1.0	0.017	0.022	0.011	0.010
1.5	0.026	0.036	0.017	0.015
2.0	0.035	0.051	0.023	0.020
2.5	0.045	0.067	0.029	0.026
3.0	0.055	0.083	0.035	0.032
3.5	0.065	0.099	0.041	0.037
4.0	0.076	0.116	0.048	0.043
4.5	0.087	0.135	0.056	0.050
5.0	0.099	0.156	0.064	0.057
5.5	0.112	0.179	0.072	0.065
6.0	0.126	0.204	0.080	0.072
6.5	0.140	0.233	0.090	0.081
7.0	0.156	0.268	0.100	0.089
7.5	0.174	0.310	0.110	0.102
8.0	0.194	0.425	0.120	0.115
8.5	0.219	0.480	0.133	0.130
9.0	0.254	0.520	0.147	0.148
9.5	0.303	0.550	0.163	0.167
10.0	0.515	0.577	0.181	0.189
10.5	0.583	0.601	0.203	0.216
11.0	0.624	0.623	0.236	0.250
11.5	0.655	0.644	0.283	0.298
12.0	0.682	0.664	0.663	0.500
12.5	0.706	0.683	0.735	0.702
13.0	0.728	0.701	0.776	0.751
13.5	0.748	0.719	0.804	0.785
14.0	0.766	0.736	0.825	0.811
14.5	0.783	0.753	0.842	0.830
15.0	0.799	0.769	0.856	0.848
15.5	0.815	0.785	0.869	0.867
16.0	0.830	0.800	0.881	0.886
16.5	0.844	0.815	0.893	0.895
17.0	0.857	0.830	0.903	0.904
17.5	0.870	0.844	0.913	0.913
18.0	0.882	0.858	0.922	0.922
18.5	0.893	0.871	0.930	0.930
19.0	0.905	0.884	0.938	0.939
19.5	0.916	0.896	0.946	0.948
20.0	0.926	0.908	0.953	0.957
20.5	0.936	0.920	0.959	0.962
21.0	0.946	0.932	0.965	0.968
21.5	0.956	0.944	0.971	0.973
22.0	0.965	0.956	0.977	0.979
22.5	0.974	0.967	0.983	0.984
23.0	0.983	0.978	0.989	0.989
23.5	0.992	0.989	0.995	0.995
24.0	1.000	1.000	1.000	1.000

Source: SCS (1986).

Time (h)	P/P_T	Cumulative precipitation, P (mm)
0	0	0
3	0.032	6
6	0.072	13
9	0.148	27
12	0.500	90
15	0.848	153
18	0.922	166
21	0.968	174
24	1.000	180

■

6.2.3.5 Spatial Distribution

The spatial distribution of storms is usually important in calculating the runoff from large catchments. For any given return period and duration, the average rainfall depth over an area is generally less than the point rainfall depth. The *areal-reduction factor* is defined as the ratio of the areal-average rainfall to the point rainfall depth, and can be estimated using Figure 6.12 (Miller et al., 1973). Areal-averaged rainfall is usually assumed to be distributed uniformly over the catchment, and for areas less than 25 km² (10 mi²), areal-reduction factors are not usually required (Jens, 1979). The areal-reduction factor, F, given in Figure 6.12 can be approximated algebraically by the relation (Leclerc and Schaake, 1972)

$$F = 1 - \exp(-1.1t_d^{\frac{1}{4}}) + \exp(-1.1t_d^{\frac{1}{4}} - 0.01A) \tag{6.11}$$

where t_d is the rainfall duration in hours, and A is the catchment area in square miles. Caution should be exercised in using generalized areal-reduction factors, since local and regional effects may lead to significantly different reduction factors.

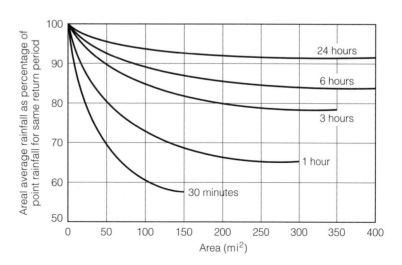

Figure 6.12 ■ Areal-Reduction Factor Versus Catchment Area and Storm Duration
Source: Miller et al. (1973).

■ **Example 6.9**

The local rainfall resulting from a 10-year 24-hour storm on the Gulf coast of Texas is 180 mm. Estimate the average rainfall on a 100 km^2 catchment.

Solution

From the given data: $t_d = 24$ h, $A = 100$ km$^2 = 40$ mi^2, and the areal-reduction factor is given by Equation 6.11 as

$$F = 1 - \exp[-1.1t_d^{\frac{1}{4}}] + \exp[-1.1t_d^{\frac{1}{4}} - 0.01A]$$

$$= 1 - \exp[-1.1(24)^{\frac{1}{4}}] + \exp[-1.1(24)^{\frac{1}{4}} - 0.01(40)]$$

$$= 0.97$$

Therefore, for a local average rainfall of 180 mm, the average precipitation over a 100 km^2 catchment is expected to be 0.97×180 mm $= 175$ mm.   ■

6.2.3.6 Probable Maximum Precipitation

The *probable maximum precipitation* (PMP) for a given storm duration is the maximum amount of precipitation that is physically possible and characteristic of a particular geographic region at a certain time of year. The PMP is usually used in constructing the design storm for water-resource systems where a system failure would result in a significant loss of life. An example of such a case would be in the design of a large dam upstream of a densely populated area. The storm hyetograph corresponding to the PMP is called the *probable maximum storm* (PMS). Precipitation amounts, P, in the world's greatest observed rainfalls can be approximated by (World Meteorological Organization, 1983)

$$P = 422t_d^{0.475} \tag{6.12}$$

where P is the rainfall amount in millimeters and t_d is the storm duration in hours. Several of the world's greatest observed rainfalls are listed in Table 6.6. A procedure that can be used to estimate the PMS in the United States east of the 105th meridian can be found in Hansen and colleagues (1982). For areas west of the 105th meridian, the National Academy of Sciences (1983) has prepared a report that specifies the relevant National Weather Service publications containing PMS estimates.

■ **Table 6.6**
World's Greatest Observed Point Rainfalls

Duration	Precipitation (cm)	Location	Date
1 min	3.8	Barst, Guadeloupe (West Indies)	26 November 1970
20 min	20.6	Curtea-de-Arges, Rumania	7 July 1889
1 day	187	Cilaos, La Réunion (Indian Ocean)	15–16 March 1952
1 week	411	Cilaos, La Réunion	12–19 March 1952
1 month	930	Cherrapunji, India	July 1861
1 year	2,650	Cherrapunji, India	1 August 1860–31 July 1861

Source: Singh (1992).

■ **Example 6.10**

Estimate the maximum amount of rainfall that can be expected in 24 hours.

Solution

According to Equation 6.12, for a 24-hour storm ($t_d = 24$ h) the maximum rainfall amount, P, is given by

$$P = 422t_d^{0.475} = 422(24)^{0.475} = 1910 \text{ mm}$$

It is noteworthy that this 24-hour rainfall is significantly greater than the average annual rainfall in Miami, Florida (1,520 mm). ■

6.3 Rainfall Abstractions

The processes of interception, infiltration, and depression storage are commonly referred to as *rainfall abstractions*. These processes must generally be accounted for in estimating the surface runoff resulting from a given rainfall event.

6.3.1 ■ Interception

Interception is the process by which rainfall is abstracted prior to reaching the ground. The wetting of surface vegetation is typically the primary form of interception, although rainfall is also intercepted by buildings and other above-ground structures. In urban areas, the density of vegetation is usually not sufficient to cause an appreciable amount of interception. However, in areas where there is a significant amount of vegetation, such as wooded areas, interception can significantly reduce the amount of rainfall that reaches the ground. Therefore, in projects that involve the clearing of wooded areas, engineers must be prepared to account for the increased runoff that will occur as a result of reduced interception.

Methods used for estimating interception are mostly empirical, where the amount of interception is expressed either as a fraction of the amount of precipitation or as an empirical function of the rainfall amount. The interception percentages over seasonal and annual time scales for several types of vegetation have been summarized by Woodall (1984) and are given in Table 6.7. These data indicate that tree interception can abstract as much as 48% of rainfall amount (*Picea abies*) and grasses on the order of 13%. Caution should be exercised expressing interception simply as a percentage of rainfall, since the storm characteristics (intensity and duration), local climate, and the age and density of the vegetation have a significant influence of the interception percentages. Where possible, local data should be used to estimate interception losses.

Many interception functions are similar to that suggested by Horton (1919), where the interception, I, for a single storm, is related to the rainfall amount, P, by an equation of the form

$$\boxed{I = a + bP^n} \tag{6.13}$$

where a and b are constants. When I is measured in millimeters, values suggested by Horton (1919) are $n = 1$ (for most vegetative covers); a between 0.02 mm for shrubs and 0.05 mm for pine woods; and b between 0.18 and 0.20 for orchards and woods and 0.40 for shrubs. Surface vegetation generally has a finite interception capacity, which should not be exceeded by the estimated interception amount. The interception storage capacity of surface vegetation can range from less than 0.3 mm to 13 mm, with a typical value for turf grass of 1.3 mm. When rainfall abstractions are given as a depth, this is equal to

■ Table 6.7
Interception Percentages
in Selected Studies

Cover Type	Season	Interception (%)	Reference
Conifers			
Picea abies	year	48	Leyton et al. (1967)
Tsuga canadensis	summer	33	Voigt (1960)
Pseudotsuga	year	36	Aussenac and Boulangeat (1980)
Pseudotsuga	summer	24	Rothacher (1963)
	winter	14	
Pinus radiata	year	26	Feller (1981)
Pinus radiata	year	19	Smith (1974)
Pinus resinosa	summer	19	Voigt (1960)
Pinus strobus	year	16	Helvey (1967)
Pinus taeda	year	14	Swank et al. (1972)
Evergreen hardwoods			
Notofagus sp.	year	33	Aldridge and Jackson (1973)
Notofagus/Podocarpus	summer	30	Rowe (1979)
	winter	21	
Acacia	year	19	Beard (1962)
Eucalyptus regnans	year	19	Feller (1981)
Melaleuca quinquenervia	summer	19	Woodall (1984)
Moist tropical forest	summer	16	Jackson (1971)
Mixed eucalypts	year	11	Smith (1974)
Deciduous hardwoods			
Carpinus sp.	year	36	Leyton et al. (1967)
Fagus grandifolia	summer	25	Voigt (1960)
Fagus silvatica	summer	21	Aussenac and Boulangeat (1980)
	winter	6	
Liriodendron	year	10	Helvey (1964)
Grasses			
Themeda sp.	year	13	Beard (1962)
Cymbopogon sp.	year	13	Beard (1962)
Soil cover			
Hardwood litter	year	3	Helvey (1964)
Pinus strobus litter	year	3	Helvey (1967)
Oak litter	year	2	Blow (1955)
Pinus taeda litter	year	4	Swank et al. (1972)

Source: Woodall (1984).

the volume of abstraction per unit area. More sophisticated interception functions have been suggested to account for the limited storage capacity of surface vegetation and evaporation during the storm (Meriam, 1960; Gray, 1973; Brooks et al., 1991), where the interception, I, is expressed in the form

$$I = S(1 - e^{-\frac{P}{S}}) + KEt \tag{6.14}$$

where S is the available storage, P is the amount of rainfall during the storm, K is the ratio of the surface area of the leaves to the projection of the vegetation on the ground (called the *leaf area index*), E is the evaporation rate during the storm, and t is the duration of the storm. The surface area of only one side of each leaf is counted in calculating the leaf area index.

■ **Example 6.11**

A pine forest is to be cleared for a commercial development in which all the trees on the site will be removed. The IDF curve for a 20-year rainfall is given by the relation

$$i = \frac{2819}{t + 16}$$

where i is the rainfall intensity in mm/h and t is the duration in minutes. The storage capacity of the trees in the forest is estimated as 6 mm, the leaf area index is 7, and the evaporation rate during the storm is estimated as 0.2 mm/h. (a) Determine the increase in precipitation reaching the ground during a 20-min storm that will result from clearing the site; and (b) compare your result with the interception predicted by the Horton-type empirical equation of the form $I = a + bP^n$, where a and b are constants and P is the precipitation amount.

Solution

(a) For a 20-min storm, the average intensity, i, is given by the IDF equation as

$$i = \frac{2819}{t + 16} = \frac{2819}{20 + 16} = 78 \text{ mm/h}$$

and the precipitation amount, P, is given by

$$P = it = (78)\left(\frac{20}{60}\right) = 26 \text{ mm}$$

The interception, I, of the wooded area can be estimated by Equation 6.14, where $S = 6$ mm, $P = 26$ mm, $K = 7$, $E = 0.2$ mm/h, and $t = 20/60$ h $= 0.33$ h. Hence

$$I = S(1 - e^{-\frac{P}{S}}) + KEt = 6(1 - e^{-\frac{26}{6}}) + (7)(0.2)(0.33) = 5.9 + 0.5 = 6.4 \text{ mm}$$

The wooded area intercepts approximately 6.4 mm of the 26 mm that falls on the wooded area. Prior to clearing the wooded area, the rainfall reaching the ground in a 20-min storm is $26 - 6.4 = 19.6$ mm. After clearing the wooded area, the rainfall reaching the ground is expected to be 26 mm, an increase of 33% over the incident rainfall prior to clearing the wooded area. These calculations also show that evaporation contributes only 0.5 mm of the 6.4 mm intercepted, indicating that evaporation during a storm contributes relatively little to interception.

(b) Using the interception formula $I = a + bP^n$ for pine woods, it can be assumed that $n = 1$, $a = 0.05$ mm, and $b = 0.19$. Hence

$$I = a + bP^n = 0.05 + 0.19(26) = 5 \text{ mm}$$

The pine woods are estimated to intercept 5 mm of rainfall, in which case the predevelopment rainfall reaching the ground is 26 mm $- 5$ mm $= 21$ mm and the postdevelopment rainfall reaching the ground is 26 mm, an increase of 24% over predevelopment conditions. ■

6.3.2 ■ Depression Storage

Water that accumulates in surface depressions during a storm is called *depression storage*. This portion of rainfall does not contribute to surface runoff; it either infiltrates or evaporates following the rainfall event. Depression storage is generally expressed as

	Depression storage	
Surface Type	**(mm)**	**Reference**
Pavement:		
Steep	0.5	Pecher (1969), Viessman et al. (1977)
Flat	1.5, 3.5	Pecher (1969), Viessman et al. (1977)
Impervious	1.3–2.5	Tholin and Kiefer (1960)
Lawns	2.5–5.1	Hicks (1944)
Pasture	5.1	ASCE (1992)
Forest litter	7.6	ASCE (1992)

■ **Table 6.8**
Typical Values of
Depression Storage

an average depth over the catchment area, and typical depths of depression storage are given in Table 6.8. These typical values are for moderate slopes; the values would be larger for flat slopes, and smaller for steep slopes. In estimating surface runoff from rainfall, depression storage is usually deducted from the initial rainfall.

■ **Example 6.12**

A 10-min storm produces 12 mm of rainfall on an impervious parking lot. Estimate the fraction of this rainfall that becomes surface runoff.

Solution

On the impervious parking lot, water trapped in depression storage forms puddles and does not contribute to runoff. Depression storage is typically in the range of 1.3–2.5 mm, with an average value of 1.9 mm. The fraction of runoff, C, is therefore estimated by

$$C = \frac{12 - 1.9}{12} = 0.84$$

This indicates an abstraction of 16% for a 12-mm storm, which corresponds to 84% runoff. Clearly, the fraction of runoff will increase for higher precipitation amounts.

■

6.3.3 ■ Infiltration

The process by which water seeps into the ground through the soil surface is called *infiltration* and is usually the dominant rainfall abstraction process. A variety of models are used to describe the infiltration process, with no one model best for all cases. To appreciate the assumptions and limitations of the various models, it is important to look first at the fundamental process of infiltration.

6.3.3.1 The Infiltration Process

Infiltration describes the entry of water into the soil through the soil surface, and *percolation* describes the movement of water within the soil. The infiltration rate is equal to the percolation rate just below the ground surface and can be described by Darcy's law (see Chapter 7):

$$q = -K(\theta)\frac{\partial h}{\partial z} \tag{6.15}$$

where q is the vertical flux of water in the soil (i.e., volumetric flowrate per unit area); $K(\theta)$ is the vertical hydraulic conductivity expressed as a function of the moisture content (= the volume of water per unit volume of the soil), θ; and h is the piezometric head of the pore water defined by

$$h = \frac{p}{\gamma} + z \qquad (6.16)$$

where p is the water pressure in the pores of the soil, γ is the specific weight of water, and z is the vertical coordinate (positive upward). A negative pore pressure indicates that the pressure is below atmospheric pressure. In the unsaturated soil beneath the ground surface, the soil moisture is usually under tension, with negative pore pressures.

Laboratory and field experiments indicate that there is a fairly stable relationship between the pore pressure, p, and moisture content, θ, that is unique to each soil. A typical relationship between $-p/\gamma$ and θ is illustrated in Figure 6.13(a), where the *capillary potential*, $\psi(\theta)$, defined by

$$\psi(\theta) = -\frac{p}{\gamma} \qquad (6.17)$$

is commonly used in lieu of the pressure head, and is closely related to the *matric potential*, which is defined as p/γ. The typical moisture retention curve, Figure 6.13(a), indicates that when the pores are filled with water at atmospheric pressure the moisture content is equal to the saturated moisture content, θ_s, and the pore pressure and capillary potential are both equal to zero. The saturated moisture content, θ_s, is numerically equal to the porosity of the soil. As the moisture content is reduced, the pore pressure decreases and the capillary potential increases in accordance with Equation 6.17. This trend continues until the moisture content is equal to θ_o, at which point the pore water becomes discontinuous and further reductions in the pore pressure do not result in a decreased moisture content.

The moisture retention curve shown in Figure 6.13(a) can be used to describe the equilibrium distribution of water in the soil column between the water table and the ground surface, where the *water table* identifies the top on the saturated region in a soil. Assuming that the pore water forms a continuum above the water table, the pressure distribution must be hydrostatic and described by

$$\frac{p}{\gamma} + z = \text{constant} \qquad (6.18)$$

Figure 6.13 ■ Typical Moisture Retention Curve and $C(\theta)$

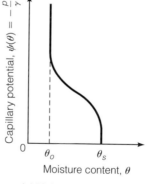

(a) Moisture Retention Curve

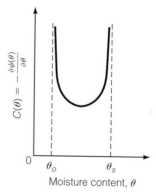

(b) Derived Function, $C(\theta)$

Defining $z = 0$ at the water table, where the pressure is atmospheric ($p = 0$), Equation 6.18 gives the pressure distribution in the soil moisture above the water table as

$$\frac{p}{\gamma} + z = 0 \tag{6.19}$$

or

$$-\frac{p}{\gamma} = z \tag{6.20}$$

Combining Equations 6.20 and 6.17 yields the following relationship between the height, z, above the water table and the capillary potential, $\psi(\theta)$

$$z = \psi(\theta) \tag{6.21}$$

Combining Equation 6.21 with the moisture retention curve, Figure 6.13(a), leads to the moisture distribution as a function of the elevation above the water table, as shown in Figure 6.14. This moisture distribution is typical of what is expected after a surface recharge (such as rainfall) has been fully drained by gravity throughout the soil column.

Combining Darcy's law, Equation 6.15, with the definition of the piezometric head yields the following expression for the vertical seepage velocity in terms of the capillary potential

$$
\begin{aligned}
q &= -K(\theta)\frac{\partial h}{\partial z} \\
&= -K(\theta)\frac{\partial}{\partial z}\left(\frac{p}{\gamma} + z\right) = -K(\theta)\frac{\partial}{\partial z}\left[-\psi(\theta) + z\right] \\
&= K(\theta)\left[\frac{\partial \psi(\theta)}{\partial z} - 1\right]
\end{aligned}
\tag{6.22}
$$

The chain rule of differentiation guarantees that

$$\frac{\partial \psi(\theta)}{\partial z} = \frac{\partial \psi(\theta)}{\partial \theta}\frac{\partial \theta}{\partial z} \tag{6.23}$$

where $\partial\psi/\partial\theta$ is a soil property derived from the moisture retention curve, Figure 6.13(a). Combining Equations 6.23 and 6.22 yields the following relationship between the vertical seepage velocity, q, and the moisture content, θ:

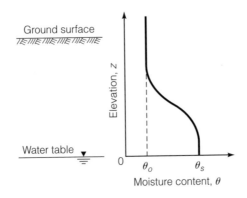

Figure 6.14 ■ Equilibrium Moisture Distribution Between Ground Surface and Water Table

$$q = K(\theta)\left[\frac{\partial\psi(\theta)}{\partial\theta}\frac{\partial\theta}{\partial z} - 1\right]$$

$$= -K(\theta)\left[C(\theta)\frac{\partial\theta}{\partial z} + 1\right] \tag{6.24}$$

where $C(\theta)$ is a soil property defined by

$$C(\theta) = -\frac{\partial\psi(\theta)}{\partial\theta} \tag{6.25}$$

and is derived directly from the moisture retention curve for the soil. The functional form of $C(\theta)$ corresponding to a typical moisture retention curve is illustrated in Figure 6.13(b), where it should be noted that $C(\theta)$ is always positive. Although the vertical seepage flux, q, given by Equation 6.24 is numerically equal to the infiltration rate, f, Equation 6.24 requires that q is positive in the upward direction, while f is conventionally taken as positive in the downward direction. Consequently, the infiltration rate, f, is simply equal to negative q, in which case Equation 6.24 yields the following theoretical expression for the infiltration rate,

$$f = K(\theta)\left[C(\theta)\frac{\partial\theta}{\partial z} + 1\right] \tag{6.26}$$

Practical application of Equation 6.26 requires specification of the hydraulic conductivity function $K(\theta)$. According to Bear (1979), field experiments indicate that $K(\theta)$ can be adequately described by

$$K(\theta) = K_o\left(\frac{\theta - \theta_o}{\theta_s - \theta_o}\right)^3 \tag{6.27}$$

where K_o is a property of the soil that is equal to the hydraulic conductivity at saturation. According to Equation 6.27, the hydraulic conductivity, $K(\theta)$, increases monotonically from zero to K_o as θ increases from θ_o to θ_s. Combining Equations 6.26 and 6.27 yields the infiltration equation

$$\boxed{f = K_o\left(\frac{\theta - \theta_o}{\theta_s - \theta_o}\right)^3\left[C(\theta)\frac{\partial\theta}{\partial z} + 1\right]} \tag{6.28}$$

The fundamental infiltration process is illustrated in Figure 6.15 for the case where water is ponded above the ground surface. The initial moisture distribution between the ground surface and the water table (before ponding) approximates the equilibrium distribution shown in Figure 6.14, where the conditions at the surface are described by $\theta = \theta_o$ and $\partial\theta/\partial z = 0$. Under these conditions, Equation 6.28 indicates that

$$f = 0 \tag{6.29}$$

Immediately after infiltration begins, the soil just below the ground surface becomes saturated, but it is still unsaturated further down in the soil column, leading to a sharp moisture gradient near the surface. Under these circumstances, $\theta = \theta_s$ and $\partial\theta/\partial z > 0$ at the ground surface. The infiltration rate is given by Equation 6.28 as

$$f = K_o\left[C(\theta_s)\frac{\partial\theta}{\partial z} + 1\right] \tag{6.30}$$

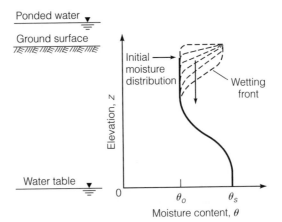

Figure 6.15 ■
Infiltration Process

As infiltration proceeds, K_o and $C(\theta_s)$ remain constant, and the moisture-content gradient, $\partial\theta/\partial z$, gradually decreases to zero as the wetting front penetrates the soil column (see Figure 6.15). Therefore, the infiltration rate gradually decreases from its maximum value given by Equation 6.30 ($\partial\theta/\partial z > 0$), at the beginning of the infiltration process, to the asymptotic minimum infiltration rate when $\partial\theta/\partial z = 0$ and

$$f = K_o \tag{6.31}$$

The minimum infiltration rate is therefore equal to the (vertical) saturated hydraulic conductivity of the soil. As the ponded water continues to infiltrate into the soil, conditions near the ground surface remain approximately constant, with infiltration proceeding at the minimum rate. Eventually, the entire soil column becomes saturated and the recharge rate at the water table equals the infiltration rate.

Several models are commonly used to estimate infiltration, and the validity of each of these models should be viewed relative to their consistency with the theoretical infiltration process described in this section. The models most frequently used in engineering practice are the Horton, Green-Ampt, and Natural Resources Conservation Service (NRCS) models. No single approach works best for all situations, and in most cases the methods are limited by knowledge of their site-specific parameters. Many of these models distinguish between the actual infiltration rate, f, and the *potential infiltration rate*, f_p, which is equal to the infiltration rate when water is ponded at the surface.

6.3.3.2 Horton Model

Horton (1939, 1940) proposed the following empirical equation to describe the decline in the potential infiltration rate, f_p, as a function of time

$$\boxed{f_p = f_c + (f_o - f_c)e^{-kt}} \tag{6.32}$$

where f_o is the initial (maximum) infiltration rate, f_c is the asymptotic (minimum) infiltration rate ($t \to \infty$), and k is a decay constant. It has been shown previously that the asymptotic minimum infiltration rate must be equal to the saturated hydraulic conductivity of the soil. Equation 6.32 describes an infiltration capacity that decreases exponentially with time, ultimately approaching a constant value, and assumes an infiltration

■ **Table 6.9**
Typical Values of Horton
Infiltration Parameters

Soil type	f_o (mm/h)	f_c (mm/h)	k (min^{-1})
Alphalpha loamy sand	483	36	0.64
Carnegie sandy loam	375	45	0.33
Dothan loamy sand	88	67	0.02
Fuquay pebbly loamy sand	158	61	0.08
Leefield loamy sand	288	44	0.13
Tooup sand	584	46	0.55

Source: Rawls et al. (1976).

process described by the relation

$$\frac{df_p}{dt} = -k(f_p - f_c) \tag{6.33}$$

The Horton model fits well with experimental data (Singh, 1989). Typical values of f_o, f_c, and k are given in Table 6.9, and Singh (1992) recommends that f_o/f_c be on the order of 5. The variability of the infiltration parameters in the Horton model reflects the condition that infiltration depends on several factors that are not explicitly accounted for in Table 6.9, such as the initial moisture content and organic content of the soil, vegetative cover, and season (Linsley et al., 1982).

The temporal variation in infiltration rate given by Equation 6.32 is for when the water is ponded above the soil column, and the functional form of this equation is illustrated in Figure 6.16. The ponding of water above the ground surface will occur only when the rainfall rate exceeds the infiltration capacity of the soil. To accommodate this limitation of the Horton model, the potential infiltration, f_p, can be expressed in terms of the cumulative infiltration, F, by an implicit relationship. The cumulative infiltration as a function of time is given by

$$F(t) = \int_0^t f_p(\tau)d\tau$$

$$= \int_0^t \left[f_c + (f_o - f_c)e^{-k\tau} \right]d\tau$$

$$= f_c t + \frac{f_o - f_c}{k}(1 - e^{-kt}) \tag{6.34}$$

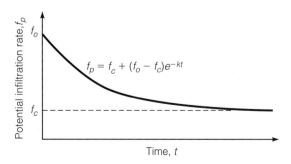

Figure 6.16 ■ Horton
Infiltration Model

Equations 6.32 and 6.34 form an implicit relationship between the cumulative infiltration, F, and the potential infiltration rate, f_p, where t is simply a parameter in the relationship. Hence, at any time during a rainfall event, if the cumulative infiltration is known, it can be substituted into Equation 6.34 to obtain the value of the parameter t, which is then substituted into Equation 6.32 to obtain the corresponding infiltration capacity.

▪ Example 6.13

A catchment soil has Horton infiltration parameters: $f_o = 100$ mm/h, $f_c = 20$ mm/h, and $k = 2$ min^{-1}. What rainfall rate would result in ponding from the beginning of the storm? If this rainfall rate is maintained for 40 minutes, describe the infiltration as a function of time during the storm.

Solution

According to the Horton model of infiltration, the potential infiltration rate varies between a maximum of 100 mm/h ($= f_o$) and an (asymptotic) minimum of 20 mm/h ($= f_c$). Any storm in which the rainfall rate exceeds 100 mm/h during the entire storm will cause ponding from the beginning of the storm. Under these circumstances, the infiltration rate, f, as a function of time is given by Equation 6.32 as

$$f = f_c + (f_o - f_c)e^{-kt} = 20 + (100 - 20)e^{-2t}$$
$$= 20 + 80e^{-2t}, \qquad 0 \le t \le 40 \text{ min}$$

▪

▪ Example 6.14

A catchment soil is found to have the following Horton infiltration parameters: $f_o = 100$ mm/h, $f_c = 20$ mm/h, and $k = 2$ min^{-1}. The design storm is given by the following hyetograph:

Interval (min)	Average rainfall (mm/h)
0–10	10
10–20	20
20–30	80
30–40	100
40–50	80
50–60	10

Estimate the time at which ponding begins.

Solution

According to Equation 6.34, the cumulative infiltration, F, is given by

$$F = f_c t + \frac{f_o - f_c}{k}(1 - e^{-kt})$$
$$= 20\frac{t}{60} + \frac{100 - 20}{2(60)}(1 - e^{-2t})$$
$$= 0.333t + 0.667(1 - e^{-2t}) \qquad (6.35)$$

where t is in minutes and F is in millimeters. The infiltration capacity, f_p, as a function of t is given by Equation 6.32 as

$$
\begin{aligned}
f_p &= f_c + (f_o - f_c)e^{-kt} \\
&= 20 + (100 - 20)e^{-2t} \\
&= 20 + 80e^{-2t}
\end{aligned}
\tag{6.36}
$$

where f_p is in mm/h. Each 10-minute increment of the storm will now be taken sequentially.

$t = 0–10$ min: During this period the rainfall intensity, $i\ (= 10$ mm/h$)$, is less than the minimum infiltration rate, $f_c\ (= 20$ mm/h$)$, and no ponding occurs. The cumulative infiltration, F, after 10 min is given by

$$
F = i\Delta t = 10\left(\frac{10}{60}\right) = 1.67 \text{ mm}
$$

$t = 10–20$ min: During this period, the rainfall intensity, $i\ (= 20$ mm/h$)$, is equal to the minimum infiltration rate, $f_c\ (= 20$ mm/h$)$, and no ponding occurs. The cumulative infiltration, F, after 20 min is given by

$$
F = i\Delta t + 1.67 = 20\left(\frac{10}{60}\right) + 1.67 = 5.01 \text{ mm}
$$

$t = 20–30$ min: During this period, the rainfall intensity, $i\ (= 80$ mm/h$)$, exceeds the minimum infiltration rate, $f_c\ (= 20$ mm/h$)$, and therefore ponding is possible. At $t = 20$ min, $F = 5.01$ mm and Equation 6.35 gives

$$
5.01 = 0.333t + 0.667(1 - e^{-2t})
$$

which can be solved by trial and error to yield $t = 13.0$ min. Equation 6.36 gives the corresponding infiltration capacity, f_p, as

$$
f_p = 20 + 80e^{-2(13.0)} = 20.0 \text{ mm/h}
$$

Since the rainfall rate (80 mm/h) exceeds the infiltration capacity (20 mm/h) from the beginning of the time interval, ponding starts at the beginning of the time interval, at $t = 20$ min.   ■

6.3.3.3   Green-Ampt Model

This physically-based semi-empirical model was first proposed by Green and Ampt (1911) and was put on a firm physical basis by Philip (1954). The Green-Ampt model is sometimes called the *delta function model* (Salvucci and Entekhabi, 1994; Philip, 1993) and is today one of the most realistic models of infiltration available to the engineer. A typical vertical section of soil is shown in Figure 6.17, where it is assumed that water is ponded to a depth H on the ground surface and that there is a sharp interface between the wetted soil and the dry soil. This interface is called the *wetting front*, and as the ponded water infiltrates into the soil the wetting front moves downward. Flow through

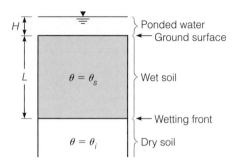

Figure 6.17 ■ Green-Ampt Soil Column

saturated porous media is described by Darcy's law, which can be written in the form

$$q_s = -K_s \frac{dh}{ds} \tag{6.37}$$

where q_s is the flow per unit area or *specific discharge* in the s-direction, K_s is the hydraulic conductivity of the saturated soil, h is the piezometric head, and dh/ds is the gradient of the piezometric head in the s-direction. In reality, K_s is usually less than the hydraulic conductivity at saturation, because of entrapped air which prevents complete saturation. Using a finite-difference approximation to the Darcy equation over the depth of the wet soil, L, leads to

$$f_p = -K_s \frac{(-\Phi_f) - (H + L)}{L} \tag{6.38}$$

where f_p is the potential infiltration rate (equal to the specific discharge when water is ponded above the ground surface), $-\Phi_f$ is the head at the wetting front, and $H + L$ is the head at the top of the wet soil. The suction head Φ_f is defined as $-p/\gamma$, where p is the pressure head in the water at the wetting front and p is generally negative (i.e., below atmospheric pressure). Assuming that the soil was initially dry, then the total volume of water infiltrated, F, is given by

$$F = (n - \theta_i)L \tag{6.39}$$

where n is the porosity of the soil (equal to volumetric water content at saturation) and θ_i is the initial volumetric water content of the dry soil. The volumetric water content of a soil is the volume of water in a soil sample divided by the volume of the sample. In reality, the value of n used in Equation 6.39 should be reduced to account for the entrapment of air. Assuming that $H \ll (L + \Phi_f)$, then Equations 6.38 and 6.39 can be combined to yield

$$f_p = K_s + \frac{K_s(n - \theta_i)\Phi_f}{F} \tag{6.40}$$

Recognizing that the potential infiltration rate, f_p, and the cumulative infiltrated amount, F, are related by

$$f_p = \frac{dF}{dt} \tag{6.41}$$

then Equations 6.40 and 6.41 can be combined to yield the following differential equation for F,

$$\frac{dF}{dt} = K_s + \frac{K_s(n - \theta_i)\Phi_f}{F} \tag{6.42}$$

Separation of variables allows this equation to be written as

$$\int_0^F \frac{F}{K_sF + K_s(n - \theta_i)\Phi_f} dF = \int_0^t dt \tag{6.43}$$

which incorporates the initial condition that $F = 0$ when $t = 0$. Integrating Equation 6.43 leads to the following equation for the cumulative infiltration versus time

$$K_st = F - (n - \theta_i)\Phi_f \ln\left[1 + \frac{F}{(n - \theta_i)\Phi_f}\right] \tag{6.44}$$

This equation assumes that water has been continuously ponded above the soil column from time $t = 0$, which will happen only if the rainfall intensity exceeds the infiltration capacity from the beginning of the storm. If the rainfall intensity, i, is initially less than the infiltration capacity, then the actual infiltration, f, will be equal to the rainfall intensity until the infiltration capacity, f_p (which continually decreases) becomes equal to the rainfall intensity. Assuming that this happens at time, t_p, then

$$f = i, \qquad t < t_p \tag{6.45}$$

$$f = f_p = i, \qquad t = t_p \tag{6.46}$$

Combining Equations 6.46 and 6.40 leads to the following expression for the total infiltrated volume, F_p, at $t = t_p$,

$$F_p = \frac{\Phi_f(n - \theta_i)}{i/K_s - 1} \tag{6.47}$$

Since the infiltration has been equal to i up to this point, then the time, t_p, at which the infiltrated volume becomes equal to F_p is given by

$$t_p = \frac{F_p}{i} \tag{6.48}$$

For $t > t_p$, the rainfall intensity exceeds the potential infiltration, and therefore infiltration continues at the potential rate given by

$$f = f_p = K_s + \frac{K_s(n - \theta_i)\Phi_f}{F}, \qquad t > t_p \tag{6.49}$$

Beyond $t = t_p$, the cumulative infiltration as a function of time can be written in the form (Mein and Larson, 1973)

$$K_s(t - t_p + t_p') = F - (n - \theta_i)\Phi_f \ln\left[1 + \frac{F}{(n - \theta_i)\Phi_f}\right] \tag{6.50}$$

where t_p' is the equivalent time to infiltrate F_p under the condition of surface ponding from $t = 0$.

Typical values of the Green-Ampt parameters for several soil types are given in Table 6.10. These values should not be used for bare soils with crusted surfaces. Methods for

USDA soil-texture class	Hydraulic conductivity K_s (mm/h)	Wetting front suction head Φ_f (mm)	Porosity n	Field capacity θ_o	Wilting point θ_w
Sand	120	49	0.437	0.062	0.024
Loamy sand	30	61	0.437	0.105	0.047
Sandy loam	11	110	0.453	0.190	0.085
Loam	3	89	0.463	0.232	0.116
Silt loam	7	170	0.501	0.284	0.135
Sandy clay loam	2	220	0.398	0.244	0.136
Clay loam	1	210	0.464	0.310	0.187
Silty clay loam	1	270	0.471	0.342	0.210
Sandy clay	1	240	0.430	0.321	0.221
Silty clay	1	290	0.479	0.371	0.251
Clay	0.3	320	0.475	0.378	0.265

Source: Rawls et al. (1983). "Green-Ampt Infiltration Parameters from Soils Data," Journal of Hydraulic Engineering, vol. 109, n. 1. Reprinted by permission of ASCE.

estimating Green-Ampt soil parameters from readily available soil characteristics have been presented by Rawls and Brakensiek (1982) and Rawls and colleagues (1983). The field capacity, θ_o, and wilting point, θ_w, give some indication of the dry-soil moisture content, θ_i, where the field capacity is the residual water content after gravity drainage, and the wilting point is the limiting water content below which plants cannot extract water for transpiration. ASCE (1996b) recommends caution in applying the Green-Ampt method in forested areas, where the basic assumptions of the method may not be valid.

■ Example 6.15

A catchment area consists almost entirely of loamy sand, which typically has a saturated hydraulic conductivity of 30 mm/h, average suction head of 61 mm, porosity of 0.44, field capacity of 0.105, wilting point of 0.047, and depression storage of 5 mm. You are in the process of designing a stormwater-management system to handle the runoff from this area, and have selected the following 1-hour storm as the basis of your design:

Interval (min)	Ave. rainfall (mm/h)
0–10	10
10–20	20
20–30	80
30–40	100
40–50	80
50–60	10

Determine the runoff versus time for average initial moisture conditions, and contrast the amount of rainfall with the amount of runoff. Use the Green-Ampt method for your calculations and assume that the initial moisture conditions are midway between the field capacity and wilting point.

Solution

The basic equations of the Green-Ampt model are Equations 6.49 and 6.50. In the present case: $K_s = 30$ mm/h; $\theta_i = \frac{1}{2}(0.105 + 0.047) = 0.076$; $n = 0.44$; and $\Phi_f =$

61 mm. The infiltration capacity, f_p, as a function of the cumulative infiltration, F, is

$$
\begin{aligned}
f_p &= K_s + \frac{K_s(n - \theta_i)\Phi_f}{F} \\
&= 30 + \frac{30(0.44 - 0.076)(61)}{F} \\
&= 30 + \frac{666.12}{F}
\end{aligned}
\tag{6.51}
$$

The cumulative infiltration as a function of time is given by

$$
K_s(t - t_p + t_p') = F - (n - \theta_i)\Phi_f \ln\left[1 + \frac{F}{(n - \theta_i)\Phi_f}\right]
$$

$$
30(t - t_p + t_p') = F - (0.44 - 0.076)61 \ln\left[1 + \frac{F}{(0.44 - 0.076)61}\right]
$$

$$
= F - 22.2 \ln(1 + 0.0450F)
\tag{6.52}
$$

If ponding occurs from $t = 0$, then

$$
30t = F - 22.2 \ln(1 + 0.0450F)
\tag{6.53}
$$

Each 10-minute increment in the storm will now be taken sequentially, and the computation of the runoff is summarized in Table 6.11.

$t = $ **0–10 min:** During this period, the rainfall intensity, i ($= 10$ mm/h) is less than the saturated hydraulic conductivity, K_s ($= 30$ mm/h), and therefore no ponding occurs. The entire rainfall amount of $10 \times (10/60) = 1.7$ mm is infiltrated.

$t = $ **10–20 min:** During this period, the rainfall rate (20 mm/h) still exceeds the minimum infiltration capacity (30 mm/h) and all the rainfall infiltrates. The rainfall amount during this period is $20 \times (10/60) = 3.3$ mm, and the cumulative infiltration after 20 minutes is 1.7 mm + 3.3 mm = 5.0 mm.

■ **Table 6.11**
Computation of Rainfall Excess Using Green-Ampt Model

Time (min)	F (mm)	Δt (h)	ΔF (mm)	i (mm/h)	iΔt (mm)	Depression storage, ΔS (mm)	Total storage (mm)	Runoff (mm)
0	0.0							
		0.167	1.7	10	1.7	0	0	0
10	1.7							
		0.167	3.3	20	3.3	0	0	0
20	5.0							
		0.167	12.9	80	13.4	0.5	0.5	0
30	17.9							
		0.167	9.9	100	16.7	4.5	5.0	2.3
40	27.8							
		0.167	8.5	80	13.4	0	5.0	4.9
50	36.3							
		0.167	6.7	10	1.7	−5.0	0	0
60	43.0							
Total:					50.2			7.2

$t = $ **20–30 min:** During this period the rainfall rate (= 80 mm/h) exceeds the minimum infiltration capacity (= 30 mm/h), and therefore ponding is possible. Use Equation 6.51 to determine the cumulative infiltration, F_{80}, corresponding to a potential infiltration rate of 80 mm/h:

$$80 = 30 + \frac{666.12}{F_{80}}$$

which leads to

$$F_{80} = 13.3 \text{ mm}$$

After 20 min (0.334 h) the total infiltration was 5.0 mm, therefore if $0.334 + t'$ is the time when the cumulative infiltration is 13.3 mm, then

$$5.0 + 80t' = 13.3$$

which leads to

$$t' = 0.104 \text{ h} \ (= 6.2 \text{ min})$$

Therefore, the time at which ponding occurs, t_p, is

$$t_p = 0.334 + 0.104$$
$$= 0.438 \text{ h} \ (= 26.3 \text{ min})$$

The next step is to find the time, t'_p, that it would take for 13.3 mm to infiltrate, if infiltration occurs at the potential rate from $t = 0$. Infiltration as a function of time is given by Equation 6.53, therefore

$$30t'_p = 13.3 - 22.2 \ln[1 + (0.0450)(13.3)]$$

which leads to

$$t'_p = 0.096 \text{ h} \ (= 5.8 \text{ min})$$

Therefore, the equation for the cumulative infiltration as a function of time after $t = 0.438$ h is given by Equation 6.52, which can be written as

$$30(t - 0.438 + 0.096) = F - 22.2 \ln(1 + 0.0450F)$$

or

$$30(t - 0.342) = F - 22.2 \ln(1 + 0.0450F) \tag{6.54}$$

At the end of the current time period, $t = 3(0.167) = 0.501$ h, and substituting this value into Equation 6.54 leads to

$$F = 17.9 \text{ mm}$$

Since the rainfall during this period is 13.4 mm, and the cumulative infiltration (= cumulative rainfall) up to the beginning of this period is 5 mm, then the amount of rainfall that does not infiltrate is equal to $5 + 13.4 - 17.9 = 0.5$ mm. This excess amount goes toward filling up the depression storage, which has a maximum capacity of 5 mm. The available depression storage at the end of this time period is $5 - 0.5 = 4.5$ mm. Since the depression storage is not filled, there is no runoff.

$t = $ **30–40 min:** Since the rainfall rate is now higher than during the previous time interval, ponding continues to occur. The time at the end of this period is 0.668 h. Substituting this value for t into Equation 6.54 gives the cumulative infiltration at the end of the time period as

$$F = 27.8 \text{ mm}$$

Since the cumulative infiltration up to the beginning of this period is 17.9 mm, then the infiltrated amount during this period is $27.8 - 17.9 = 9.9$ mm. The rainfall during this period is 16.7 mm; therefore, the amount of rain that does not infiltrate is $16.7 - 9.9 = 6.8$ mm. Since there is 4.5 mm in available depression storage, then the amount of runoff is $6.8 - 4.5 = 2.3$ mm. The depression storage is filled at the end of this time period.

$t = $ **40–50 min:** The rainfall rate (80 mm/h) is still higher than the infiltration capacity and ponding continues. The time at the end of this period is 0.835 h, and Equation 6.54 gives the cumulative infiltration at the end of this time period as

$$F = 36.3 \text{ mm}$$

The cumulative infiltration up to the beginning of this period is 27.8 mm; therefore, the infiltration during this time interval is $36.3 - 27.8 = 8.5$ mm. The rainfall during this period is 13.4 mm, therefore the amount of rain that does not infiltrate is $13.4 - 8.5 = 4.9$ mm. Since the depression storage is full, then all 4.9 mm is contributed to runoff.

$t = $ **50–60 min:** The rainfall rate (10 mm/h) is below the minimum infiltration capacity (30 mm/h). Under these circumstances, the ponded water will infiltrate and, if there is sufficient infiltration capacity, some of the rainfall may also infiltrate. The time at the end of this period is 1 h, and if infiltration continues at the potential rate, then the cumulative infiltration at the end of the period is given by Equation 6.54 as

$$F = 44 \text{ mm}$$

Because the cumulative infiltration up to the beginning of this period is 36.3 mm, the potential infiltration during this period is $44 - 36.3 = 7.7$ mm. The rainfall during this period is 1.7 mm and there is 5 mm of depression storage to infiltrate. Since the rainfall plus depression storage is equal to 6.7 mm, then all the rainfall and depression storage is infiltrated during this time interval. The cumulative infiltration at the end of this period is $36.3 \text{ mm} + 6.7 \text{ mm} = 43$ mm.

The total rainfall during this storm is 50.2 mm and the total runoff is 7.2 mm. The runoff is therefore equal to 14% of the rainfall.   ■

According to Ward and Dorsey (1995), the Green-Ampt method best describes the infiltration process in profiles where a *piston-type* infiltration process occurs. In many agricultural soils, however, piston-type flow is not the primary mechanism of infiltration (Thomas and Phillips, 1979; Quisenberry and Phillips, 1976).

6.3.3.4 NRCS Curve-Number Model

The *curve-number model* was developed in 1954 by the Natural Resources Conservation Service* (NRCS), within the U.S. Department of Agriculture. The curve-number model was published in 1954 in the first edition of the National Engineering Handbook, which has subsequently been revised several times (SCS, 1993). This empirical method is the

---

*The Natural Resources Conservation Service (NRCS) was formerly called the Soil Conservation Service (SCS).

most widely used method for estimating *rainfall excess* (= rainfall minus abstractions) in the United States, and the popularity of this method is due to its ease of application, lack of serious competition, and extensive data base of parameters. The curve-number model was originally developed for calculating rainfall excess in small agricultural watersheds; then, because of its overwhelming success, it was subsequently adapted to urban catchments.

The NRCS model separates the rainfall into three components: *rainfall excess, Q*, *initial abstraction, I_a*, and *retention, F*. These components are illustrated graphically in Figure 6.18. The initial abstraction includes the portion of the rainfall that is not available for either infiltration or runoff and includes the portion of the rainfall that is used to wet surfaces prior to reaching the ground (interception). The initial abstraction is generally returned to the atmosphere by evaporation. If the amount of rainfall is less than the initial abstraction, then neither infiltration nor runoff occurs. The retention, F, is the portion of the rainfall reaching the ground that is retained by the catchment and consists primarily of the infiltrated volume. The basic assumption of the NRCS model is that for any rainfall event the precipitation, P, runoff, Q, retention, F, and initial abstraction, I_a, are related by

$$\frac{F}{S} = \frac{Q}{P - I_a} \tag{6.55}$$

where S is the *potential maximum retention* and measures the retention capacity of the soil. The maximum retention, S, does not include I_a. The rationale for Equation 6.55 is that for any rainfall event, the portion of available storage (= S) that is filled, F, is equal to the portion of available water (= $P - I_a$) that appears as runoff, Q. Equation 6.55 is of course only applicable when $P > I_a$. Conservation of mass requires that

$$F = P - Q - I_a \tag{6.56}$$

Eliminating F from Equations 6.55 and 6.56 yields

$$Q = \frac{(P - I_a)^2}{(P - I_a) + S}, \qquad P > I_a \tag{6.57}$$

Empirical data indicate that the initial abstraction, I_a, is directly related to the maximum retention, S, and the following relation is commonly assumed

$$I_a = 0.2S \tag{6.58}$$

Recent research has indicated that the factor of 0.2 is probably adequate for large storms in rural areas, but it is likely an overestimate for small to medium storms and is probably too high for urban areas (Singh, 1992). However, since the storage capacity, S, of many

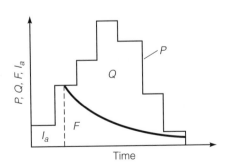

Figure 6.18 ■ Components in the NRCS Curve-Number Model

catchments has been determined based on Equation 6.58, it is recommended that the 0.2 factor be retained when using storage estimates calibrated from field measurements of rainfall and runoff.

Combining Equations 6.57 and 6.58 leads to

$$Q = \frac{(P - 0.2S)^2}{P + 0.8S}, \qquad P > 0.2S \tag{6.59}$$

This equation is the basis for estimating the volume of runoff, Q, from a volume of rainfall, P, given the maximum retention, S. It should be noted, however, that the NRCS method was originally developed as a runoff index for 24-h rainfall amounts and should be used with caution in attempting to analyze incremental runoff amounts during the course of a storm (Kibler, 1982). The total infiltration for a rainfall event is given by the combination of Equations 6.56, 6.58, and 6.59, which yields

$$F = \frac{(P - 0.2S)S}{P + 0.8S}, \qquad P > 0.2S \tag{6.60}$$

The infiltration rate, f, can be derived from Equation 6.60 by differentiation, where

$$f = \frac{dF}{dt} = \frac{S^2 i}{(P + 0.8S)^2} \tag{6.61}$$

and i is the rainfall intensity given by

$$i = \frac{dP}{dt} \tag{6.62}$$

The functional form of the infiltration rate formula given by Equation 6.61 is not physically realistic since it requires that the infiltration rate be dependent on the rainfall intensity (Morel-Seytoux and Verdin, 1981). In spite of this limitation, the NRCS curve-number model is widely used in practice.

Instead of specifying S directly, a *curve number*, CN, is usually specified where CN is related to S by

$$CN = \frac{1000}{10 + 0.0394S} \tag{6.63}$$

where S is given in millimeters. Equation 6.63 is modified from the original formula, which required S in inches. Clearly, in the absence of available storage ($S = 0$, impervious surface) the curve number is equal to 100, and for an infinite amount of storage the curve number is equal to zero. The curve number therefore varies between 0 and 100.

In practical applications, the curve number is considered to be a function of three factors: (1) soil group, (2) soil cover, and (3) antecedent runoff condition. Soils are classified into four groups: A, B, C, and D; descriptions of these groups are given in Table 6.12. Soils are grouped based on profile characteristics that include depth, texture, organic matter content, structure, and degree of swelling when saturated. Minimum infiltration rates associated with the soil groups are given in Table 6.12. The NRCS has classified more than 4,000 soils into these four groups. Local NRCS offices can usually provide information on local soils, but it should be noted that activities such as the operation of heavy equipment can substantially change the local soil characteristics (Haan et al., 1994). Soil covers are classified by land use, and the curve numbers corresponding

■ **Table 6.12**
Description of NRCS
Soil Groups

Group	Description	Minimum infiltration rate (mm/h)
A	Deep sand; deep loess; aggregated silts	7.6–11
B	Shallow loess; sandy loam	3.8–7.6
C	Clay loams; shallow sandy loam; soils low in organic content; soils usually high in clay	1.3–3.8
D	Soils that swell significantly when wet; heavy plastic clays; certain saline soils	0–1.3

to a variety of land uses and soil groups are given in Table 6.13. The *antecedent runoff condition* (ARC) is a measure of the actual available storage relative to the average available storage at the beginning of the rainfall event. The antecedent runoff condition is closely related to the antecedent moisture content of the soil and is grouped into three categories: ARCI, ARCII, and ARCIII. The average curve numbers normally cited for a particular land area correspond to ARCII conditions, and these curve numbers can be adjusted for drier than normal conditions (ARCI) or wetter than normal conditions (ARCIII) by using Table 6.14. According to Chow and colleagues (1988), the curve num-

■ **Table 6.13**
Curve Numbers for Various
Urban Land Uses

Land-use description	Curve numbers for hydrologic soil group			
	A	B	C	D
Lawns, open spaces, parks, golf courses:				
Good condition: grass cover on 75% or more of the area	39	61	74	80
Fair condition: grass cover on 50% to 75% of the area	49	69	79	84
Poor condition: grass cover on 50% or less of the area	68	79	86	89
Paved parking lots, roofs, driveways, etc.:	98	98	98	98
Streets and roads:				
Paved with curbs and storm sewers	98	98	98	98
Gravel	76	85	89	91
Dirt	72	82	87	89
Paved with open ditches	83	89	92	93
Commercial and business areas (85% impervious)	89	92	94	95
Industrial districts (72% impervious)	81	88	91	93
Row houses, town houses, and residential with lot sizes $\frac{1}{8}$ ac or less (65% impervious)	77	85	90	92
Residential average lot size:				
$\frac{1}{4}$ ac (38% impervious)	61	75	83	87
$\frac{1}{3}$ ac (30% impervious)	57	72	81	86
$\frac{1}{2}$ ac (25% impervious)	54	70	80	85
1 ac (20% impervious)	51	68	79	84
2 ac (12% impervious)	46	65	77	82

■ **Table 6.14**
Antecedent Runoff
Condition Adjustments

CN for AMCII	Corresponding CN for condition	
	AMCI	AMCIII
100	100	100
95	87	99
90	78	98
85	70	97
80	63	94
75	57	91
70	51	87
65	45	83
60	40	79
55	35	75
50	31	70
45	27	65
40	23	60
35	19	55
30	15	50
25	12	45
20	9	39
15	7	33
10	4	26
5	2	17
0	0	0

ber adjustments in Table 6.14 can be approximated by the relations

$$CN(I) = \frac{4.2CN(II)}{10 - 0.058CN(II)} \qquad (6.64)$$

and

$$CN(III) = \frac{23CN(II)}{10 + 0.13CN(II)} \qquad (6.65)$$

where CN(I), CN(II), and CN(III) are the curve numbers under AMCI, AMCII, and AMCIII conditions, respectively. The guidelines for selecting curve numbers given in Tables 6.13 and 6.14 are useful in cases where site-specific data on the maximum retention, S, are either not available or cannot be reasonably estimated. Whenever S is available for a catchment, then the curve number should be estimated directly using Equation 6.63, and, if necessary, adjusted using Table 6.14. As a precautionary note, the NRCS does not recommend the use of the curve-number model when CN is less than 40.

■ **Example 6.16**

The drainage facilities of a catchment are to be designed for a rainfall of return period 25 years and duration 2 hours, where the IDF curve for 25-year storms is given by

$$i = \frac{830}{t + 33}$$

where i is the rainfall intensity in cm/h, and t is the storm duration in minutes. A double-ring infiltrometer test on the soil shows that the minimum infiltration rate is on the order of 5 mm/h. The urban area being developed consists of mostly open space with less than 50% grass cover. Use the NRCS method to estimate the total amount of runoff (in cm), assuming the soil is in average condition at the beginning of the design storm. Estimate the percentage increase in runoff that would occur if heavy rainfall occurs within the previous five days and the soil is saturated.

Solution

The amount of rainfall can be estimated using the IDF curve. For a 2-hour 25-year storm, the average intensity is given by

$$i = \frac{830}{120 + 33} = 5.42 \text{ cm/h}$$

and the total amount of rainfall, P, in the storm is equal to $(5.42)(2) = 10.8$ cm.

Since the minimum infiltration rate of the soil is 5 mm/h, then according to Table 6.12 it can be inferred that the soil is in Group B. The description of the area, open space with less than 50% grass and Group B soil, is cited in Table 6.13 to have a curve number, equal to 79. The curve number, CN, and soil storage, S (in mm), are related by Equation 6.63 as

$$\text{CN} = \frac{1000}{10 + 0.0394S}$$

and therefore the storage, S, in this case is given as

$$S = \frac{1}{0.0394}\left(\frac{1000}{\text{CN}} - 10\right) = \frac{1}{0.0394}\left(\frac{1000}{79} - 10\right) = 67.5 \text{ mm}$$

The runoff amount, Q, can be calculated from the rainfall amount, P ($= 10.8$ cm), and the maximum storage, S ($= 6.75$ cm) using Equation 6.59 where

$$Q = \frac{(P - 0.2S)^2}{P + 0.8S}, \qquad P > 0.2S$$

Since P ($= 10.8$ cm) $> 0.2S$ ($= 1.35$ cm), then this equation is valid and

$$Q = \frac{[10.8 - 0.2(6.75)]^2}{10.8 + 0.8(6.75)} = 5.51 \text{ cm}$$

Hence, there is 5.51 cm of runoff from the storm with a rainfall amount of 10.8 cm. When the soil is saturated, Table 6.14 indicates that the curve number, CN, increases to 94 and the maximum available soil storage is given by

$$S = \frac{1}{0.0394}\left(\frac{1000}{\text{CN}} - 10\right) = \frac{1}{0.0394}\left(\frac{1000}{94} - 10\right) = 16.2 \text{ mm}$$

and the runoff amount, Q, is given by

$$Q = \frac{[10.8 - 0.2(1.62)]^2}{10.8 + 0.8(1.62)} = 9.07 \text{ cm}$$

Therefore, under saturated conditions, the runoff amount increases from 5.51 cm to 9.07 cm, which corresponds to a 64.6% increase. ■

6.3.3.5   Comparison of Infiltration Methods

The Horton, Green-Ampt, and curve-number methods are all used in engineering practice and are justified by their inclusion in the ASCE Manual of Practice on the Design and Construction of Urban Stormwater Management Systems (ASCE, 1992). Some comparative studies (Van Mullem, 1991; Hjelmfelt, 1991) have indicated that the Green-Ampt method performs better than the curve-number method in predicting runoff volumes, and hence peak runoff rates. This result is not unexpected, given the relatively poor performance of the curve-number method when applied to individual storms (Willeke, 1997) and the general recognition that the curve-number method best represents a long-term expected-value relationship between rainfall and runoff (Smith, 1997). The measurable nature of the Green-Ampt parameters and the physical basis of the model are desirable features; however, both the Green-Ampt and Horton models suffer from being unbounded (Ponce and Hawkins, 1996), in that they allow an unlimited amount of infiltration and they do not account for the effects of spatial variability in parameters such as the saturated hydraulic conductivity (Woolhiser et al., 1996). Although the Green-Ampt method may be the most accurate method for calculating the rainfall excess from an individual storm, a powerful reason for using the curve-number method is that it is a method that is supported by a U.S. government agency (the Department of Agriculture), which gives its users basic protection in litigation (Smith, 1997). The authoritative origin of the curve-number method apparently qualifies as a defense in legal proceedings that an engineer has performed a hydrologic analysis in accordance with generally accepted standards (Willeke, 1997).

6.3.4 ■ Rainfall Excess on Composite Areas

In many cases, a catchment can be delineated into several subcatchments with different abstraction characteristics. The runoff from such composite catchments depends on how the subcatchments are connected, and this is illustrated in Figure 6.19 for the case of pervious and impervious subcatchments within a larger catchment. The subcatchments

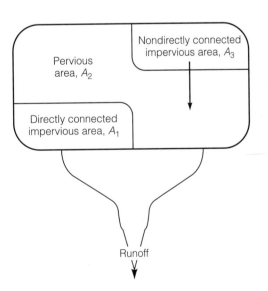

Figure 6.19 ■ Runoff from Composite Catchment

consist of a pervious area and two separate impervious areas, with only one of the impervious areas directly connected to the catchment outlet. The runoff from the composite catchment is equal to the runoff from the impervious area, A_1, that is directly connected to the catchment outlet, plus the runoff from the pervious area, A_2, which assimilates a portion of the runoff from the impervious area, A_3, that is not directly connected to the catchment outlet. To calculate the rainfall excess on the composite catchment, the rainfall excesses on the each of the subcatchments are first calculated using methods such as the NRCS curve-number method, and then these rainfall excesses are routed to the outlet of the composite catchment.

Rainfall abstractions in each subcatchment can be defined by a functional relation such as

$$Q_i = f_i(P), \qquad i = [1, 3] \tag{6.66}$$

where Q_i is the rainfall excess on subcatchment i, and f_i is an abstraction function that relates Q_i to the precipitation, P, on the subcatchment. If the NRCS curve-number method is used to calculate the rainfall excess on subcatchment i, then f_i is given by

$$f_i(P) = \begin{cases} \dfrac{(P - 0.2S_i)^2}{P + 0.8S_i} & \text{if } P > 0.2S_i \\ 0 & \text{if } P \le 0.2S_i \end{cases} \tag{6.67}$$

where S_i is the available storage in the subcatchment i. It can usually be assumed that the precipitation depth, P, on each subcatchment is the same. Since the rainfall excess, Q_3, from the nondirectly connected impervious area, A_3, drains into the pervious area, A_2, the effective precipitation, P_{eff}, on the pervious area is given by

$$P_{\text{eff}} = P + Q_3 \frac{A_3}{A_2} \tag{6.68}$$

The total rainfall excess, Q, from the composite catchment is equal to the sum of the rainfall excess on the directly connected impervious area plus the rainfall excess on the (directly connected) pervious area. Therefore

$$\begin{aligned} Q &= f_1(P)\frac{A_1}{A} + f_2(P_{\text{eff}})\frac{A_2}{A} \\ &= f_1(P)\frac{A_1}{A} + f_2\left(P + Q_3\frac{A_3}{A_2}\right)\frac{A_2}{A} \end{aligned} \tag{6.69}$$

where A is the total area of the composite catchment given by

$$A = \sum_{i=1}^{3} A_i \tag{6.70}$$

This routing methodology for estimating the rainfall excess on composite catchments can be extended to any arrangement of subcatchments, provided that the flow paths to the catchment outlet are clearly defined.

■ Example 6.17

The commercial site illustrated in Figure 6.20 covers 25 ha, of which the parking lot covers 7.5 ha, the building covers 6.3 ha, open grass covers 10 ha, and the site is graded such that all of the runoff is routed to a grass retention area that covers 1.2 ha. Both the

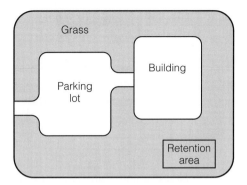

Figure 6.20 ■
Commercial Site

roof of the building and the parking lot are impervious (CN = 98) and drain directly onto the grassed area, which contains Type B soil and is in good condition. For a storm with a precipitation of 200 mm, calculate the surface runoff that enters the retention area.

Solution

In this case, both the parking lot and the roof of the building are not directly connected to the retention area, and therefore the runoff must be routed to the retention area through the grass. For the parking lot and building, CN = 98 and the available storage, S_1, can be derived from Equation 6.63 as

$$S_1 = \frac{1}{0.0394}\left(\frac{1000}{\text{CN}} - 10\right) = \frac{1}{0.0394}\left(\frac{1000}{98} - 10\right) = 5 \text{ mm}$$

For a precipitation, P, of 200 mm, the rainfall excess, Q_1, from the parking lot and building is given by

$$Q_1 = \frac{(P - 0.2S_1)^2}{P + 0.8S_1} = \frac{[200 - (0.2)(5)]^2}{200 + (0.8)(5)} = 194 \text{ mm}$$

The total area of the parking lot and building is equal to 7.5 ha + 6.3 ha = 13.8 ha, and the grassed area (including the retention area) is 10 ha + 1.2 ha = 11.2 ha. The effective rainfall, P_{eff}, on the grassed area is therefore given by

$$P_{\text{eff}} = 200 + 194\frac{13.8}{11.2} = 439 \text{ mm}$$

The grassed area contains Type B soil, is in good condition, and Table 6.13 gives CN = 61. The available storage, S_2, is derived from Equation 6.63 as

$$S_2 = \frac{1}{0.0394}\left(\frac{1000}{\text{CN}} - 10\right) = \frac{1}{0.0394}\left(\frac{1000}{61} - 10\right) = 162 \text{ mm}$$

The rainfall excess from the 11.2-ha grassed area, Q_2, is given by

$$Q_2 = \frac{(P_{\text{eff}} - 0.2S_2)^2}{P_{\text{eff}} + 0.8S_2} = \frac{[439 - (0.2)(162)]^2}{439 + (0.8)(162)} = 291 \text{ mm}$$

and the rainfall excess, Q, from the entire 25-ha composite catchment is

$$Q = 291\frac{11.2}{25} = 130 \text{ mm}$$

Therefore, a rainfall of 200 mm on the composite catchment will result in 130 mm of runoff, which will be disposed of via infiltration in the retention area.   ■

A commonly used approximation in estimating the rainfall excess from composite areas is to apply the NRCS curve-number method with an area-weighted curve number. In a composite catchment with n subcatchments, the area-weighted curve number, CN_{eff}, is estimated by

$$CN_{eff} = \frac{1}{A} \sum_{i=1}^{n} CN_i A_i \qquad (6.71)$$

where CN_i and A_i are the curve number and area of subcatchment i and A is the total area of the composite catchment. If the flow paths in the composite catchment are known, then it is generally preferable that the rainfall excesses be routed through the catchment, as described previously.

■ Example 6.18

The commercial site illustrated in Figure 6.20 covers 25 ha, of which the parking lot covers 7.5 ha, the building covers 6.3 ha, open grass covers 10 ha, and the site is graded such that all of the runoff is routed to a grass retention area that covers 1.2 ha. Both the roof of the building and the parking lot are impervious (CN = 98) and drain directly onto the grassed area (CN = 61). For a storm with a precipitation of 200 mm, use an area-weighted curve number to calculate the rainfall excess that enters the retention area. Contrast your result with that obtained using the routing method in Example 6.17.

Solution

The area-weighted curve number, CN_{eff}, is given by Equation 6.71 as

$$CN_{eff} = \frac{1}{A} \sum_{i=1}^{n} CN_i A_i$$

$$= \frac{1}{25}[(98)(7.5) + (98)(6.3) + (10)(61) + (1.2)(61)]$$

$$= 81$$

The available storage, S, corresponding to CN_{eff}, is

$$S = \frac{1}{0.0394}\left(\frac{1000}{CN_{eff}} - 10\right) = \frac{1}{0.0394}\left(\frac{1000}{81} - 10\right) = 60 \text{ mm}$$

and the rainfall excess, Q, resulting from a rainfall, P, of 200 mm is given by

$$Q = \frac{(P - 0.2S)^2}{P + 0.8S} = \frac{[200 - (0.2)(60)]^2}{200 + (0.8)(60)} = 143 \text{ mm}$$

This estimated runoff of 143 mm is 10% higher than the 130 mm estimated by routing the rainfall excess from the subcatchments.   ■

6.4   Runoff Models

Runoff models predict the temporal distribution of runoff at a catchment outlet based on the temporal distribution of effective rainfall and the catchment characteristics. The *effective rainfall* is defined as the incident rainfall minus the abstractions and is sometimes referred to as the *rainfall excess* or the *runoff*. The most important abstractions are usually infiltration and depression storage, and the catchment characteristics that are usually most important in translating the effective rainfall distribution to a runoff distribution at the catchment outlet are those related to the topography and surface cover of the catchment. Runoff models are classified as either *distributed-parameter* models or *lumped-parameter* models. Distributed-parameter models account for runoff processes on scales smaller than the size of the catchment, such as accounting for the runoff from every roof, over every lawn, and in every street gutter, while lumped-parameter models consider the entire catchment as a single hydrologic element, with the runoff characteristics described by one or more (lumped) parameters.

In cases where the surface runoff flows into an unlined drainage channel that penetrates an aquifer, the flow in the drainage channel originates from both surface-water runoff and ground-water inflow, and these flow components must generally be modeled separately. The flow resulting from surface runoff is called *direct runoff*, and the flow resulting from ground-water inflow is subdivided into *base flow* and *interflow*, which is sometimes referred to as *throughflow*. Base flow is typically (quasi-) independent of the rainfall event, is equal to the flow of ground-water into the drainage channel, and depends on the difference between the ground-water elevation and the water-surface elevation in the drainage channel (Chin, 1991). Interflow is the inflow to the drainage channel that occurs between the ground surface and the water table and is typically caused by a low-permeability subsurface layer that impedes the vertical infiltration of rainwater. The direct runoff resulting from a storm event is added to the base flow and interflow to yield the flow hydrograph in the drainage channel.

A wide variety of models are available for calculating the runoff from rainfall, and the applicability of these models must be assessed in light of the fundamental rainfall-runoff process. The applicability of various runoff models can be broadly associated with the *scale* of the catchment, which can be classified as *small*, *midsize*, or *large* (Ponce, 1989). In small catchments, the response to rainfall events is sufficiently rapid and the catchment is sufficiently small that runoff during a relatively short storm (less than 1 h duration) can be adequately modeled by assuming a constant rainfall in space and time. The *rational method* is the most widely used runoff model in small catchments. In midsize catchments, the slower response to rainfall events requires that the temporal distribution of rainfall be accounted for; however, the catchment is still smaller than the characteristic storm scale, and the rainfall can be assumed to be uniform over the catchment. *Unit hydrograph* models are the most widely used runoff models in midsize catchments. In large catchments, both the spatial and temporal variations in precipitation events must be incorporated in the runoff model, and models that explicitly incorporate routing methodologies are the most appropriate. Runoff regimes within a catchment vary from overland flow at the smallest scales to river flow at the largest scales, and runoff models must necessarily accommodate this scale effect. Small catchments have predominantly overland flow runoff, while large catchments typically have a significant amount of runoff in identifiable river or drainage channels. As a consequence, the channel storage characteristics increase significantly from small catchments to large catchments.

6.4.1 ■ Time of Concentration

The parameter that is most often used to characterize the response of a catchment to a rainfall event is the *time of concentration*. The time of concentration is defined as the time to equilibrium of a catchment under a steady rainfall excess. Alternatively, the time of concentration is sometimes defined as the longest travel time that it takes a particle of water to reach the discharge point of a catchment (Wanielista et al., 1997). Most equations for estimating the time of concentration, t_c, express t_c as function of the rainfall intensity, i, catchment length scale, L, average catchment slope, S_o, and a parameter that describes the catchment surface, C, hence the equations for t_c typically have the functional form

$$t_c = f(i, L, S_o, C) \tag{6.72}$$

The time of concentration of a catchment consists of the time of overland flow and the travel time in drainage channels leading to the catchment outlet.

6.4.1.1   Overland Flow

There are several equations that are commonly used to estimate the time of concentration for overland flow. The most popular equations are described here.

Kinematic Wave Equation.   A fundamental expression for the time of concentration in overland flow can be derived by considering the one-dimensional approximation of the surface-runoff process illustrated in Figure 6.21. The boundary of the catchment area is at $x = 0$, i_e is the rainfall-excess rate, y is the runoff flow depth, and q is the volumetric flow rate per unit width of the catchment area. Within the control volume of length Δx, the law of conservation of mass requires that the net mass inflow is equal to the rate of change of mass within the control volume. This law can be stated mathematically by the relation

$$\left[(\rho q) - \frac{\partial(\rho q)}{\partial x} \frac{\Delta x}{2} \right] + [i_e \Delta x] - \left[(\rho q) + \frac{\partial(\rho q)}{\partial x} \frac{\Delta x}{2} \right] = \frac{\partial y}{\partial t} \rho \Delta x \tag{6.73}$$

where the first term in square brackets is the inflow into the control volume, the second term is the rainfall excess entering the control volume, the third term is the outflow, and the righthand side of Equation 6.73 is equal to the rate of change of fluid mass within the control volume. Taking the density as being constant and simplifying yields

$$\frac{\partial y}{\partial t} + \frac{\partial q}{\partial x} = i_e \tag{6.74}$$

This equation contains two unknowns, q and y, and a second relationship between these variables is needed to solve this equation. Normally, the second equation is the momen-

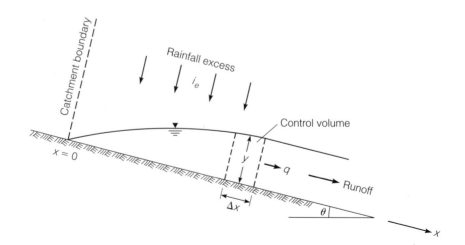

Figure 6.21 ■
One-Dimensional
Approximation of
Surface-Runoff Process

tum equation; however, a unique relationship between the flow rate, q, and the flow depth, y, can be assumed to have the form

$$q = \alpha y^m \tag{6.75}$$

where α is a proportionality constant. The assumption of a relationship such as Equation 6.75 is justified in that equations describing steady-state flow in open channels, such as the Manning and Darcy-Weisbach equations, can be put in the form of Equation 6.75. Combining Equations 6.74 and 6.75 leads to the following differential equation for y

$$\frac{\partial y}{\partial t} + \alpha m y^{m-1} \frac{\partial y}{\partial x} = i_e \tag{6.76}$$

Solution of this equation can be obtained by comparing it with

$$\frac{dy}{dt} = \frac{\partial y}{\partial t} + \frac{dx}{dt}\frac{\partial y}{\partial x} = i_e \tag{6.77}$$

which gives the rate of change of y with respect to t observed by moving at a velocity dx/dt. Consequently, Equation 6.76 is equivalent to the following pair of equations

$$\frac{dx}{dt} = \alpha m y^{m-1} \tag{6.78}$$

$$\frac{dy}{dt} = i_e \tag{6.79}$$

where dx/dt is called the *wave speed*, and Equation 6.77 is called the *kinematic-wave equation*. Solution of Equation 6.79 subject to the boundary condition that $y = 0$ at $t = 0$ yields

$$y = i_e t \tag{6.80}$$

Substituting this result into Equation 6.78 and integrating subject to the boundary condition that $x = 0$ at $t = 0$ yields

$$x = \alpha i_e^{m-1} t^m \tag{6.81}$$

Equations 6.80 and 6.81 are parametric equations describing the water surface illustrated in Figure 6.21, and the discharge at any location along the catchment area can be obtained by combining Equations 6.75 and 6.80 to yield

$$q = \alpha (i_e t)^m \tag{6.82}$$

Defining the time of concentration, t_c, of a catchment as the time required for a kinematic wave to travel the distance L from the catchment boundary to the catchment outlet, then Equation 6.81 gives the time of concentration as

$$t_c = \left(\frac{L}{\alpha i_e^{m-1}} \right)^{\frac{1}{m}} \tag{6.83}$$

If the Manning equation is used to relate the runoff rate to the depth, then Equation 6.83 can be written in the form (ASCE, 1992)

$$t_c = 6.99 \frac{(nL)^{0.6}}{i_e^{0.4} S_o^{0.3}} \tag{6.84}$$

where t_c is in minutes, i_e in mm/h, and L in m, n is the Manning roughness coefficient for overland flow, and S_o is the ground slope. Estimates of the Manning roughness coefficient for overland flow are given in Table 6.15. On the basis of Equation 6.84, the time of concentration for overland flow should be regarded as a function of the rainfall-excess rate (i_e), the catchment-surface roughness (n), the flow length from the catchment boundary to the outlet (L), and the slope of the flow path (S_o).

Equation 6.84 assumes that the surface runoff is described by the Manning equation, which is only valid for turbulent flows; however, at least a portion of the surface runoff will be in the laminar and transition regimes (Wong and Chen, 1997). This limitation associated with using the Manning equation can be addressed by using the Darcy-Weisbach equation, which yields

$$\alpha = \left(\frac{8gS_o}{Cv^k} \right)^{\frac{1}{(2-k)}}, \quad \text{and} \quad m = \frac{3}{2-k} \tag{6.85}$$

where v is the kinematic viscosity of water, C and k are parameters relating the Darcy-Weisbach friction factor, f, to the Reynolds number, Re:

$$f = \frac{C}{\mathrm{Re}^k} \tag{6.86}$$

where $k = 0$ for turbulent flow, $k = 1$ for laminar flow, $0 < k < 1$ for transitional flow, and the Reynolds number is defined by

$$\mathrm{Re} = \frac{q}{v} \tag{6.87}$$

Laminar flow typically occurs where $\mathrm{Re} < 200$, turbulent flow where $\mathrm{Re} > 2{,}000$, and transition flow where $200 < \mathrm{Re} < 2{,}000$. Values of C for overland flow have not been widely published, but Radojkovic and Maksimovic (1987) and Wenzel (1970) indicate

■ Table 6.15
Manning's n for
Overland Flow

Surface Type	Manning n	Range
Concrete/asphalt	0.011	0.01–0.013
Bare sand	0.01	0.01–0.016
Bare clay–loam (eroded)	0.02	0.012–0.033
Gravelled surface	0.02	0.012–0.03
Packed clay	0.03	
Short grass prairie	0.15	0.10–0.20
Light turf	0.20	
Lawns	0.25	0.20–0.30
Dense turf	0.35	
Pasture	0.35	0.30–0.40
Dense shrubbery and forest litter	0.40	
Bluegrass sod	0.45	0.39–0.63

Source: ASCE, 1992. *Design and Construction of Urban Stormwater Mangement Systems.* p. 88. Reprinted by permission of ASCE.

that for concrete surfaces C values of 41.8, 2, and 0.04 are appropriate for laminar, transition, and turbulent flow regimes, respectively. Combining the kinematic-wave expression for t_c, Equation 6.83, with the Darcy-Weisbach equation, Equations 6.85 to 6.87, yields

$$t_c = \left[\frac{0.21(3.6 \times 10^6 \nu)^k C L^{2-k}}{S_o i_e^{1+k}} \right]^{\frac{1}{3}} \tag{6.88}$$

where t_c is in minutes, ν is in m^2/s, L is in meters, and i_e in mm/h. Equation 6.88 can be used to account for various flow regimes in overland flow, and it has been shown that assuming a single flow regime (laminar, transition, or turbulent) will tend to underestimate the time of concentration (Wong and Chen, 1997). Indications are that overland flow is predominantly in the transition regime and that Equation 6.88 may be most applicable using $k \approx 0.5$.

NRCS Method. NRCS (SCS, 1986) proposed that overland flow originates as *sheet flow* that eventually becomes *shallow concentrated flow*. The flow characteristics of sheet flow are sufficiently different from shallow concentrated flow that separate equations should be used. The flow length of the sheet flow regime should be less than 100 m, and the travel time, t_f (in hours), over a flow length, L (in m), is estimated by (SCS, 1986)

$$t_f = 0.0288 \frac{(nL)^{0.8}}{P_2^{0.5} S_o^{0.4}} \tag{6.89}$$

where n is the Manning roughness coefficient for overland flow (Table 6.15), S_o is the land slope, and P_2 is the two-year 24-hour rainfall (in cm). Equation 6.89 is a simplified form of the kinematic-wave solution developed by Overton and Meadows (1976), and the assumptions used in the simplification were: (1) The flow is steady and uniform with a depth of about 3 cm; (2) the rainfall intensity is uniform over the catchment; (3) the rainfall duration is 24 hours; (4) infiltration is neglected; and (5) the maximum flow length is 100 m. In considering the validity of these assumptions, it should be noted that overland flow may be significantly different than 3 cm in many areas, the rainfall duration may differ from 24 hours, and the actual travel time can increase if there is a significant amount of infiltration in the catchment. By limiting the maximum flow length to 100 m, the catchment is necessarily small, and the assumption of a spatially uniform rainfall distribution is reasonable. After a maximum distance of 100 m, sheet flow usually becomes shallow concentrated flow, and the average velocity is taken to be a function of the slope of the flow path and the type of land surface as given in Figure 6.22. The average velocity, V_{sc}, derived from Figure 6.22 is then combined with the flow length, L_{sc}, of shallow concentrated flow to yield the flow time, t_{sc}, as

$$t_{sc} = \frac{L_{sc}}{V_{sc}} \tag{6.90}$$

The average velocity, V_{sc}, plotted in Figure 6.22 is derived directly from the Manning equation

$$V_{sc} = \frac{1}{n} R^{\frac{2}{3}} S_o^{\frac{1}{2}} \tag{6.91}$$

where n is the roughness coefficient, R is the hydraulic radius, and S_o is the slope of the watercourse. For unpaved areas, Figure 6.22 assumes that $n = 0.05$ and $R = 12$ cm;

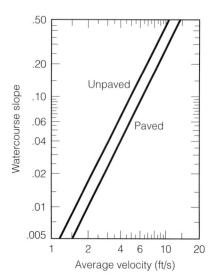

Figure 6.22 ■ Average Velocity for Shallow Concentrated Flow

for paved areas, $n = 0.025$ and $R = 6$ cm. Appropriate modifications can be made for site-specific conditions. The total time of concentration, t_c, of overland flow is taken as the sum of the sheet flow time, t_f, given by Equation 6.89, and the shallow concentrated flow time, t_{sc}, given by Equation 6.90.

Kirpich Equation.   An empirical time of concentration formula that is especially popular is the *Kirpich formula* (Kirpich, 1940) given by

$$t_c = 0.019 \frac{L^{0.77}}{S_o^{0.385}} \qquad (6.92)$$

where t_c is the time of concentration in minutes, L is the flow length in meters, and S_o is the average slope along the flow path. Equation 6.92 was originally developed and calibrated from NRCS data on rural catchments in Tennessee, ranging in size from 1 to 112 ac, with slopes varying from 3% to 10%; it has found widespread use in urban applications to estimate both overland flow and channel flow times. Equation 6.92 is most applicable for natural basins with well-defined channels, bare-earth overland flow, or flow in mowed channels (Debo and Reese, 1995). Rossmiller (1980) has reviewed field applications of the Kirpich equation and suggested that for overland flow on concrete and asphalt surfaces, t_c should be multiplied by 0.4; for concrete channels, multiply t_c by 0.2; and for general overland flow and flow in natural grass channels, multiply t_c by 2. The Kirpich formula is usually considered applicable to small agricultural watersheds with drainage areas less than 80 ha (200 ac).

Izzard Equation.   The Izzard equation (Izzard, 1944) was derived from experiments on pavements and turf where overland flow was dominant. The Izzard equation is given by

$$t_c = \frac{530KL^{1/3}}{i_e^{2/3}}, \qquad \text{where } i_e L < 3.9 \text{ m}^2/\text{h} \qquad (6.93)$$

■ Table 6.16
Values of c_r in the
Izzard Equation

Surface	c_r
Very smooth asphalt	0.007
Tar and sand pavement	0.0075
Crushed-slate roof	0.0082
Concrete	0.012
Tar and gravel pavement	0.017
Closely clipped sod	0.046
Dense bluegrass	0.060

Source: Adapted from Izzard (1944).

where t_c is the time of concentration in minutes, L is the overland flow distance in meters, i_e is the effective rainfall intensity in mm/h, and K is a constant given by

$$K = \frac{2.8 \times 10^{-6} i_e + c_r}{S_o^{1/3}} \qquad (6.94)$$

where c_r is a retardance coefficient that is determined by the catchment surface as given in Table 6.16 and S_o is the catchment slope.

Kerby Equation. The Kerby equation (Kerby, 1959) is given by

$$\boxed{t_c = 1.44(LrS_o^{-0.5})^{0.467}, \qquad \text{where } L < 365 \text{ m}} \qquad (6.95)$$

where t_c is the time of concentration in minutes, L is the length of flow in meters, r is a retardance roughness coefficient given in Table 6.17, and S_o is the slope of the catchment. Catchments with areas less than 10 ac, slopes less than 1%, and retardance coefficients less than 0.8 were used in calibrating the Kerby equation, and application of this equation should also be limited to this range.

■ Example 6.19

An urban catchment with an asphalt surface has an average slope of 0.5%, and the distance from the catchment boundary to the outlet is 90 m. For a 20-min storm with an effective rainfall rate of 75 mm/h, estimate the time of concentration using: (a) the kinematic-wave equation, (b) the NRCS method, (c) the Kirpich equation, (d) the Izzard equation, and (e) the Kerby equation.

■ Table 6.17
Values of r in the
Kerby Equation

Surface	r
Smooth pavements	0.02
Smooth bare packed soil, free of stones	0.10
Poor grass, bare sod	0.30
Average grass	0.40
Deciduous timberland	0.60
Conifer timberland, dense grass	0.80

Source: Adapted from Kerby (1959).

Solution

(a) *Kinematic-wave equation:* If the overland flow is assumed to be fully turbulent, the Manning form of the kinematic-wave equation (Equation 6.84) can be used. From the given data: $L = 90$ m, $i_e = 75$ mm/h, $S_o = 0.005$, and for an asphalt surface Table 6.15 gives $n = 0.011$. According to Equation 6.84

$$t_c = 6.99 \frac{(nL)^{0.6}}{i_e^{0.4} S_o^{0.3}} = 6.99 \frac{(0.011 \times 90)^{0.6}}{(75)^{0.4}(0.005)^{0.3}} = 6 \text{ min}$$

If the overland flow is assumed to be in the transition range (which is more probable), then the Darcy-Weisbach form of the kinematic-wave equation (Equation 6.88) must be used. Substituting $v = 10^{-6}$ m²/s, $k = 0.5$, and $C = 2$ into Equation 6.88 gives

$$t_c = \left[\frac{0.21(3.6 \times 10^6 v)^k CL^{2-k}}{S_o i_e^{1+k}} \right]^{\frac{1}{3}}$$

$$= \left[\frac{0.21(3.6 \times 10^6 \times 10^{-6})^{0.5}(2)(90)^{2-0.5}}{(0.005)(75)^{1+0.5}} \right]^{\frac{1}{3}} = 6 \text{ min}$$

Therefore, both forms of the kinematic-wave equation yield the same result (to the nearest minute).

(b) *NRCS method:* Use $n = 0.011$, $L = 90$ m, $S_o = 0.005$, and take P_2 as $(20/60) \times 75 = 25$ mm $= 2.5$ cm. According to the NRCS equation

$$t_c = 0.0288 \frac{(nL)^{0.8}}{P_2^{0.5} S_o^{0.4}} = 0.0288 \frac{(0.011 \times 90)^{0.8}}{(2.5)^{0.5}(0.005)^{0.4}} = 0.15 \text{ h} = 9 \text{ min}$$

(c) *Kirpich equation:* Use $L = 90$ m and $S_o = 0.005$. According to the Kirpich equation, with a factor of 0.4 to account for the asphalt surface,

$$t_c = 0.4(0.019) \frac{L^{0.77}}{S_o^{0.385}} = 0.4(0.019) \frac{(90)^{0.77}}{(0.005)^{0.385}} = 2 \text{ min}$$

This is an overestimate of the t_c, since P_2 should be the 24-h rainfall rather than the 20-min rainfall.

(d) *Izzard equation:* Use $S_o = 0.005$, $L = 90$ m, $i_e = 75$ mm/h, and a retardance coefficient, c_r, given by Table 6.16 as 0.007. The constant, K, is given by

$$K = \frac{2.8 \times 10^{-6} i_e + c_r}{S_o^{1/3}} = \frac{2.8 \times 10^{-6} \times 75 + 0.007}{(0.005)^{1/3}} = 0.0422$$

The Izzard equation gives the time of concentration as

$$t_c = \frac{530 K L^{1/3}}{i_e^{2/3}} = \frac{530(0.0422)(90)^{1/3}}{(75)^{2/3}} = 6 \text{ min}$$

In this case, $i_e L = (0.075)(90) = 6.75$ m²/h. Therefore, since $i_e L > 3.9$ m²/h, the Izzard equation is not strictly applicable.

(e) *Kerby equation:* Use $L = 90$ m, $S_o = 0.005$, and a retardance coefficient, r, given by Table 6.17 as 0.02. According to the Kerby equation

$$t_c = 1.44[LrS_o^{-0.5}]^{0.467} = 1.44[(90)(0.02)(0.005)^{-0.5}]^{0.467} = 7 \text{ min}$$

The computed times of concentration are summarized in the following table:

Equation	t_c (min)
kinematic wave	6
NRCS	9
Kirpich	2
Izzard	6
Kerby	7

Noting the NRCS method overestimates t_c and that the Kirpich equation would give $t_c = 5$ min if the Rosmiller factor of 0.4 were not applied, the (overland flow) time of concentration of the catchment can be taken to be on the order of 6 min. ■

6.4.1.2 Channel Flow

Channel flow elements in flow paths include street gutters, roadside swales, storm sewers, drainage channels, and small streams. In these cases, it is recommended that velocity-based equations such as Manning and Darcy-Weisbach be used to estimate the flow time in each segment (ASCE, 1992). Here the flow time, t_o, is estimated by the relation

$$t_o = \frac{L}{V_o} \tag{6.96}$$

where L is the length of the flow path in the channel and V_o is the estimated velocity of flow. The time of concentration of the entire catchment is equal to the sum of the time of concentration for overland flow and the channel flow time, t_o.

■ Example 6.20

A catchment consists of an asphalt pavement that drains into a rectangular concrete channel. The asphalt surface has an average slope of 0.6%, and the distance from the catchment boundary to the drain is 50 m. The drainage channel is 30 m long, 25 cm wide, 20 cm deep, and has a slope of 0.8%. For an effective rainfall rate of 60 mm/h, the flowrate in the channel is estimated to be 0.025 m³/s. Estimate the time of concentration of the catchment.

Solution

The flow consists of both overland flow and channel flow. Use the kinematic-wave equation to estimate the time of concentration of overland flow, t_1, where $L = 50$ m, $i_e = 60$ mm/h, $S_o = 0.006$, and for an asphalt surface Table 6.15 gives $n = 0.011$. The kinematic-wave equation (Equation 6.84) gives

$$t_1 = 6.99 \frac{(nL)^{0.6}}{i_e^{0.4} S_o^{0.3}} = 6.99 \frac{(0.011 \times 50)^{0.6}}{(60)^{0.4}(0.006)^{0.3}} = 5 \text{ min}$$

The flow area, A, in the drainage channel can be calculated using the Manning equation

$$Q = \frac{1}{n} A R^{2/3} S_o^{1/2} = \frac{1}{n} A \left(\frac{A}{P}\right)^{2/3} S_o^{1/2} = \frac{1}{n} \frac{A^{5/3}}{P^{2/3}} S_o^{1/2}$$

where $Q = 0.025$ m³/s, $n = 0.013$ (for concrete), and $S_o = 0.008$. The area, A, and wetted perimeter, P, can be written in terms of the depth of flow, d, in the drainage channel as $A = 0.25d$ and $P = 2d + 0.25$. Therefore, the Manning equation can be put in the form

$$0.025 = \frac{1}{0.013} \frac{(0.25d)^{5/3}}{(2d + 0.25)^{2/3}} (0.008)^{1/2}$$

Solving iteratively for d yields

$$d = 0.10 \text{ m} = 10 \text{ cm}$$

The flow area, A, in the drainage channel is given by

$$A = 0.25d = (0.25)(0.10) = 0.025 \text{ m}^2$$

Therefore, the flow velocity, V_o, is given by

$$V_o = \frac{Q}{A} = \frac{0.025}{0.025} = 1.0 \text{ m/s}$$

Since the length, L, of the drainage channel is 30 m, the flow time, t_2, in the channel is given by

$$t_2 = \frac{L}{V_o} = \frac{30}{1.0} = 30 \text{ s} = 0.5 \text{ min}$$

The time of concentration, t_c, of the entire catchment area is equal to the overland flow time, t_1, plus the channel flow time, t_2. Hence

$$t_c = t_1 + t_2 = 5 + 0.5 = 5.5 \text{ min}$$

Because of uncertainties in estimating t_c, the time of concentration of the catchment can reasonably be taken as 6 minutes.   ■

6.4.1.3   Accuracy of Estimates

McCuen and colleagues (1984) compared 11 equations for estimating t_c in 48 urban catchments in the United States. The catchments used in the study all had areas less than 1,600 ha (4,000 ac), average impervious areas of approximately 29%, and times of concentration from 0.21 to 6.14 h, with an average time of concentration of 1.5 h. The equations used in these studies included, among others, the kinematic-wave equation (Equation 6.84), the Kirpich equation (Equation 6.92), and the Kerby equation (Equation 6.95). The results of this study indicated that the error in the estimated value of t_c exceeded 0.5 h for more than 50% of the catchments, with the standard deviation of the errors ranging from 0.37 to 2.27 h. These results indicate that relatively large errors can be expected in estimating t_c from commonly used equations, and Singh (1989) has noted that this can lead to significant errors in design discharges. In spite of this note of caution, it is generally accepted that estimates of travel time in drainage channels are usually much more accurate than estimates of overland flow; therefore, in catchments where flow in drainage channels constitute a significant portion of the travel time, estimates of the time of concentration can be estimated more accurately.

6.4.2 ■ Peak-Runoff Models

Peak-runoff models estimate only the peak runoff, not the entire runoff hydrograph. Peak runoff is typically sufficient for the design of stormwater conveyance structures such as sewer pipes and culverts.

6.4.2.1 The Rational Method

The *rational method*, sometimes called the Lloyd-Davies method in Great Britain (after Lloyd-Davies, 1906), is the most widely used peak runoff method in urban hydrology. This method has been used by engineers since the nineteenth century (Mulvaney, 1850; Kuichling, 1889). The rational method relates the peak runoff rate, Q_p, to the rainfall intensity, i, by the relation

$$Q_p = CiA \qquad (6.97)$$

where C is the runoff coefficient and A is the area of the catchment. Application of the rational method assumes that (1) the entire catchment area is contributing to the runoff, in which case the duration of the storm must equal or exceed the time of concentration of the catchment; (2) the rainfall is distributed uniformly over the catchment area; and (3) all catchment losses are incorporated into the runoff coefficient, C. The runoff mechanism assumed by the rational method is illustrated in Figure 6.23, where the runoff from a storm gradually increases until an equilibrium is reached in which the runoff rate, Q, is equal to the effective rainfall rate over the entire catchment area. This condition occurs at the time of concentration, t_c. Under these circumstances, the duration of the rainfall does not affect the peak runoff as long as the duration equals or exceeds t_c. Since the storm duration must equal or exceed the time of concentration for the rational method to be applicable and the average intensity of a storm is inversely proportional to the duration of the storm, then the rainfall intensity used in the rational method to generate the largest peak runoff must correspond to a storm whose duration is equal to the time of concentration. The rational method implicitly assumes that the rainfall excess, i_e, is related to the rainfall rate, i, by the relation

$$i_e = Ci \qquad (6.98)$$

which means that the rainfall-excess rate is a constant fraction, C, of the rainfall rate. The assumption that C is a constant that is independent of the rainfall rate and antecedent moisture conditions makes the application of the rational method questionable from a physical viewpoint, but this approximation improves as the imperviousness of the catchment increases. In addition to this flaw in the underlying theory of the rational

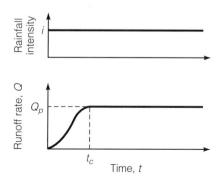

Figure 6.23 ■ Runoff Mechanism Assumed by Rational Method

method, the assumed uniformity of the rainfall distribution over the catchment area and the assumption of equilibrium conditions indicate that the rational method should not be used on areas larger than about 80 ha (200 ac) (ASCE, 1992), which typically have times of concentrations less than 20 min (Wanielista et al., 1997). Debo and Reese (1995) recommend that the rational method be limited to areas less than 40 ha (100 ac), and some regulations prohibit the use of the rational method for areas larger than 8 ha (20 ac) (South Carolina, 1992). Use of the rational method is not recommended in any catchment where ponding of stormwater might affect the peak discharge, or where the design and operation of large drainage facilities are to be undertaken.

Application of the rational method requires a simultaneous solution of the IDF equation and the time of concentration expression to determine the rainfall intensity and time of concentration. Typically, the IDF curve is of the form

$$i = f(t_c) \qquad (6.99)$$

and an expression for the time of concentration is of the form

$$t_c = g(i) \qquad (6.100)$$

where f and g are known functions; Equation 6.99 implicitly assumes that the storm duration is equal to the time of concentration. Typical runoff coefficients for rainfall intensities with 2- to 10-year return periods are shown in Table 6.18. Higher coefficients should be used for less frequent storms having higher return periods, and the runoff coefficients can be reasonably increased by 10%, 20%, and 25% for 20-, 50-, and 100-year storms, respectively (Akan, 1993). Obviously the increased runoff coefficients must not be greater than 1.0.

■ **Example 6.21**

A new 1.2-ha suburban residential development is to be drained by a storm sewer that connects to the municipal drainage system. The development is characterized by an average runoff coefficient of 0.4, a Manning n for overland flow of 0.20, an average overland flow length of 70 m, and an average slope of 0.7%. The time of concentration can be estimated by the kinematic-wave equation. Local drainage regulations require the sewer pipe to be sized for the peak runoff resulting from a 10-year rainfall event. The 10-year IDF curve is given by

$$i = \frac{315.5}{t^{0.81} + 6.19}$$

where i is the rainfall intensity in cm/h and t is the duration in minutes. Local drainage regulations further require a minimum time of concentration of 5 minutes. Determine the peak runoff rate to be handled by the storm sewer.

Solution
The time of concentration, t_c, is estimated by the kinematic-wave equation (Equation 6.84) as

$$t_c = 6.99 \frac{(nL)^{0.6}}{i_e^{0.4} S_o^{0.3}}$$

■ **Table 6.18**
Typical Runoff Coefficients
(2-year to 10-year
Return Periods)

Description of area	Runoff coefficient
Business:	
Downtown areas	0.70–0.95
Neighborhood areas	0.50–0.70
Residential:	
Single family areas	0.30–0.50
Multiunits, detached	0.40–0.60
Multiunits, attached	0.60–0.75
Residential, suburban	0.25–0.40
Apartment dwelling areas	0.50–0.70
Industrial:	
Light areas	0.50–0.80
Heavy areas	0.60–0.90
Parks, cemeteries	0.10–0.25
Railroad yard areas	0.20–0.35
Unimproved areas	0.10–0.30
Pavement:	
Asphalt or concrete	0.70–0.95
Brick	0.70–0.85
Roofs	0.75–0.95
Lawns, sandy soil:	
Flat, 2%	0.05–0.10
Average, 2%–7%	0.10–0.15
Steep, 7% or more	0.15–0.20
Lawns, heavy soil:	
Flat, 2%	0.13–0.17
Average, 2%–7%	0.18–0.22
Steep, 7% or more	0.25–0.35

Source: ASCE, 1992. *Design and Construction of Urban Stormwater Management Systems.* p. 91–92. Reprinted by permission of ASCE.

where $n = 0.20$, $L = 70$ m, and $S_o = 0.007$, and therefore

$$t_c = 6.99 \frac{(0.20 \times 70)^{0.6}}{i_e^{0.4}(0.007)^{0.3}} = \frac{151}{i_e^{0.4}} \text{ min}$$

where i_e is in mm/h. Equating the storm duration to the time of concentration, then the effective rainfall rate, i_e, for a 10-year storm is given by the IDF relation as

$$i_e = Ci = \frac{3155C}{t_c^{0.81} + 6.19} \text{ mm/h}$$

Combining the latter two equations with $C = 0.4$ yields

$$i_e = \frac{3155(0.4)}{\left(\frac{151}{i_e^{0.4}}\right)^{0.81} + 6.19} = \frac{1262 i_e^{0.324}}{58.2 + 6.19 i_e^{0.324}}$$

Solving by trial and error yields $i_e = 58$ mm/h and a corresponding time of concentration of 30 min, which exceeds the minimum allowable time of concentration of 5 minutes. The peak runoff, Q_p, from the residential development is given by the rational formula as

$$Q_p = CiA = i_e A$$

where $i_e = 58$ mm/h $= 1.61 \times 10^{-5}$ m/s, and $A = 1.2$ ha $= 1.2 \times 10^4$ m². Therefore

$$Q_p = (1.61 \times 10^{-5})(1.2 \times 10^4) = 0.19 \text{ m}^3/\text{s}$$

The storm sewer serving the residential area should be sized to accommodate 0.19 m³/s when flowing full. ■

In using the rational method to calculate the peak discharge from a composite catchment area, the rational formula should generally be applied in two ways (ASCE, 1992): (1) using the entire drainage area, and (2) using the most densely developed hydraulically connected area. In using the entire drainage area, the rainfall duration is equated to the time of concentration of the entire catchment, the average intensity corresponding to a duration equal to the time of concentration is derived from the IDF curve, the average runoff coefficient is an area-weighted average, and the resulting peak runoff is calculated by substituting these values into the rational formula, Equation 6.97. Using only the portion of the catchment that is most densely developed and connected to the catchment outlet, a smaller (sub)catchment area results in a shorter time of concentration and higher average rainfall intensity, a higher average runoff coefficient because of the more impervious nature of the subcatchment, and the peak runoff from this area is obtained by substituting these values into Equation 6.97. If the increased average rainfall intensity and runoff coefficient are sufficient to offset the smaller catchment area, then the peak runoff from the densely developed hydraulically connected area will be greater than the peak runoff from the entire catchment and will govern the design of the downstream drainage structures.

■ Example 6.22

Consider the case where the residential development described in Example 6.21 contains 0.4 ha of impervious area that is directly connected to the storm sewer. If the runoff coefficient of the impervious area is 0.9, the Manning n for overland flow on the impervious surface is 0.03, the average flow length is 20 m, and the average slope is 0.1%, estimate the design runoff to be handled by the storm sewer.

Solution

The time of concentration, t_c, of the impervious area is given by

$$t_c = 6.99 \frac{(nL)^{0.6}}{i_e^{0.4} S_o^{0.3}}$$

where $n = 0.03$, $L = 20$ m, and $S_o = 0.001$. Therefore

$$t_c = 6.99 \frac{(0.03 \times 20)^{0.6}}{i_e^{0.4}(0.001)^{0.3}} = \frac{40.9}{i_e^{0.4}} \text{ min}$$

where i_e is in mm/h. Taking the storm duration as t_c, the 10-year effective rainfall rate, i_e, is given by the IDF curve as

$$i_e = Ci = \frac{3155C}{t_c^{0.81} + 6.19} \text{ mm/h}$$

Combining the latter two equations with $C = 0.9$ yields

$$i_e = \frac{3155(0.9)}{\left(\frac{40.9}{i_e^{0.4}}\right)^{0.81} + 6.19} = \frac{2840i_e^{0.324}}{20.2 + 6.19i_e^{0.324}}$$

Solving by trial and error yields $i_e = 303$ mm/h and a corresponding time of concentration of 4.2 min, which is less than the minimum allowable time of concentration of 5 minutes. Taking $t_c = 5$ min, the IDF relation yields

$$i_e = \frac{3155(0.9)}{5^{0.81} + 6.19} = 288 \text{ mm/h}$$

The peak runoff, Q_p, from the impervious area is given by the rational formula as

$$Q_p = CiA = i_e A$$

where $i_e = 288$ mm/h $= 8 \times 10^{-5}$ m/s, and $A = 0.4$ ha $= 4,000$ m^2, therefore

$$Q_p = (8 \times 10^{-5})(4000) = 0.32 \text{ m}^3/\text{s}$$

Example 6.21 indicated that the peak runoff from the entire composite catchment is 0.19 m^3/s, and this example shows that the peak runoff from the directly connected impervious area is 0.32 m^3/s. The storm sewer serving this residential development should therefore be designed to accommodate 0.32 m^3/s. ■

6.4.2.2 NRCS-TR55 Method

The Natural Resources Conservation Service (NRCS) computed the runoff from many small and midsize catchments using the NRCS unit hydrograph method (SCS, 1983) and, on the basis of these results, proposed the *TR-55 method* for estimating peak runoff. The TR-55 method, which is named after the technical report in which it is described (SCS, 1986), expresses the peak runoff, q_p, in m^3/s as

$$q_p = q_u A Q F_p \tag{6.101}$$

where q_u is the unit peak discharge in m^3/s per cm of runoff per km^2 of catchment area, A is the catchment area in km^2, Q is the runoff in centimeters from a 24-h storm with a given return period, and F_p is the pond and swamp adjustment factor (dimensionless). The runoff, Q, is derived directly from NRCS curve number model, Equation 6.59, using the 24-h precipitation. F_p is derived from Table 6.19, assuming that the ponds and/or swampy areas are distributed throughout the catchment, and the unit peak discharge, q_u, is obtained using the empirical relation

$$\log(q_u) = C_o + C_1 \log t_c + C_2(\log t_c)^2 - 2.366 \tag{6.102}$$

where C_o, C_1, and C_2 are obtained from Table 6.20, and t_c is in hours. Values of t_c in Equation 6.102 must be between 0.1 h and 10 h (calculated using the NRCS method described in Section 6.4.1.1). Values of I_a in Table 6.20 are derived using

$$I_a = 0.2S \tag{6.103}$$

where S is obtained from the curve number in accordance with Equation 6.63. If $I_a/P < 0.1$, where P corresponds to the 24-h precipitation for the given return period, values of

■ Table 6.19
Pond and Swamp
Adjustment Factor, F_p

Percentage of pond and swamp areas	F_p
0	1.00
0.2	0.97
1.0	0.87
3.0	0.75
5.0*	0.72

*If the percentage of pond and swamp areas exceeds 5%, then consideration should be given to routing the runoff through these areas.

C_o, C_1, and C_2 corresponding to $I_a/P = 0.1$ should be used, and if $I_a/P > 0.5$, values of C_o, C_1, and C_2 corresponding to $I_a/P = 0.5$ should be used. These approximations result in reduced accuracy of the peak discharge estimates (SCS, 1986). The Federal Highway Administration (USFHWA, 1995) and the NRCS (SCS, 1986) have both recommended that the NRCS-TR55 method be used only with homogenous catchments, where the curve numbers vary within ±5 between zones, CN of the catchment should be greater

■ Table 6.20
Parameters Used to Estimate Unit Peak Discharge, q_u

Rainfall type	I_a/P	C_o	C_1	C_2
I	0.10	2.30550	−0.51429	−0.11750
	0.20	2.23537	−0.50387	−0.08929
	0.25	2.18219	−0.48488	−0.06589
	0.30	2.10624	−0.45695	−0.02835
	0.35	2.00303	−0.40769	0.01983
	0.40	1.87733	−0.32274	0.05754
	0.45	1.76312	−0.15644	0.00453
	0.50	1.67889	−0.06930	0.0
IA	0.10	2.03250	−0.31583	−0.13748
	0.20	1.91978	−0.28215	−0.07020
	0.25	1.83842	−0.25543	−0.02597
	0.30	1.72657	−0.19826	0.02633
	0.50	1.63417	−0.09100	0.0
II	0.10	2.55323	−0.61512	−0.16403
	0.30	2.46532	−0.62257	−0.11657
	0.35	2.41896	−0.61594	−0.08820
	0.40	2.36409	−0.59857	−0.05621
	0.45	2.29238	−0.57005	−0.02281
	0.50	2.20282	−0.51599	−0.01259
III	0.10	2.47317	−0.51848	−0.17083
	0.30	2.39628	−0.51202	−0.13245
	0.35	2.35477	−0.49735	−0.11985
	0.40	2.30726	−0.46541	−0.11094
	0.45	2.24876	−0.41314	−0.11508
	0.50	2.17772	−0.36803	−0.09525

than 40, t_c should be between 0.1 and 10 h, and t_c should be approximately the same for all main channels.

■ **Example 6.23**

A 2.25-km² catchment with 0.2% pond area is estimated to have a curve number of 85, and a time of concentration of 2.4 h, and a 24-h Type III precipitation of 13 cm. Estimate the peak runoff from the catchment.

Solution

For CN = 85, the storage, S, is given by

$$S = \frac{1}{0.0394}\left(\frac{1000}{CN} - 10\right) = \frac{1}{0.0394}\left(\frac{1000}{85} - 10\right) = 45 \text{ mm}$$

For $P = 130$ mm, the runoff, Q, is given by

$$Q = \frac{[P - 0.2S]^2}{P + 0.8S} = \frac{[130 - 0.2(45)]^2}{130 + 0.8(45)} = 88 \text{ mm} = 8.8 \text{ cm}$$

From Table 6.19, $F_p = 0.97$. By definition

$$\frac{I_a}{P} = \frac{0.2S}{P} = \frac{(0.2)(45)}{130} = 0.069$$

Since $I_a/P < 0.1$, Table 6.20 gives $C_o = 2.47317$, $C_1 = -0.51848$, $C_2 = -0.17083$, and since $t_c = 2.4$ h, Equation 6.102 gives

$$\log(q_u) = C_o + C_1 \log t_c + C_2(\log t_c)^2 - 2.366$$
$$= 2.47317 - 0.51848\log(2.4) - 0.17083(\log 2.4)^2 - 2.366$$
$$= -0.116$$

which leads to

$$q_u = 0.765 \text{ (m}^3\text{/s)/cm/km}^2$$

Therefore, according to the TR-55 method, the peak discharge, q_p, is given by Equation 6.101 as

$$q_p = q_u AQF_p = 0.765(2.25)(8.8)(0.97) = 14.7 \text{ m}^3\text{/s}$$ ■

6.4.3 ■ Continuous-Runoff Models

Continuous-runoff models estimate the entire runoff hydrograph from the rainfall excess remaining after initial abstraction, infiltration, and depression storage have been taken into account. A good example of a method for calculating the rainfall excess is the Green-Ampt model discussed earlier. Since surface conditions tend to be spatially variable within a catchment area, the rainfall excess will also be variable in both space and time, and the runoff hydrograph at the catchment outlet will depend on such factors as the topography and the surface roughness of the catchment. Four types of models commonly used in engineering practice to estimate the runoff hydrograph are: (1) unit-hydrograph models, (2) time-area curves, (3) kinematic-wave models, and (4) nonlinear reservoir models.

6.4.3.1   Unit-Hydrograph Models

The idea of using a unit hydrograph to describe the response of a catchment to a given rainfall excess was first introduced by Sherman (1932) and is currently the most widely used method of estimating runoff hydrographs in midsize catchments (Viessman and Lewis, 1996). A *unit hydrograph* is defined as the temporal distribution of runoff resulting from a unit depth (1 cm or 1 in.) of rainfall excess occurring over a given duration, and distributed uniformly in time and space over the catchment area.

To demonstrate the application of unit-hydrograph models, consider the unit hydrograph, $u(t)$, for a given catchment and rainfall duration, Δt, illustrated in Figure 6.24(a). Assuming that the catchment response is linear, then the runoff hydrograph, $Q(t)$, for a rainfall excess P occurring over a duration Δt is given by

$$Q(t) = Pu(t) \tag{6.104}$$

where the ordinates of $Q(t)$ are equal to P times the ordinates of the unit hydrograph, $u(t)$. This is illustrated in Figure 6.24(b). If the rainfall excess, P, occurs over a duration equal to an integral multiple of Δt, say $n\Delta t$, and assuming the catchment response is linear, the response of the catchment is equal to that of n storms occurring sequentially, with the rainfall excess in each storm equal to P/n. The runoff hydrograph, $Q(t)$, is then given by

$$Q(t) = \frac{P}{n} \sum_{i=0}^{n-1} u(t - i\Delta t) \tag{6.105}$$

where the response of the catchment is equal to the summation of the responses to the incremental rainfall excesses of duration Δt. This is illustrated in Figure 6.24(c).

■ Example 6.24

The 10-min unit hydrograph for a 2.25 km^2 urban catchment is given by

Time (min)	0	30	60	90	120	150	180	210	240	270	300	330	360	390
Runoff (m^3/s)	0	1.2	2.8	1.7	1.4	1.2	1.1	0.91	0.74	0.61	0.50	0.28	0.17	0

(a) Verify that the unit hydrograph is consistent with a 1 cm rainfall excess, (b) estimate the runoff hydrograph for a 10-min rainfall excess of 3.5 cm, and (c) estimate the runoff hydrograph for a 20-min rainfall excess of 8.5 cm.

Solution

(a) The area, A_h, under the unit hydrograph can be estimated by numerical integration as

$$A_h = (1800)(1.2 + 2.8 + 1.7 + 1.4 + 1.2 + 1.1 + 0.91 + 0.74 + 0.61 + 0.50$$
$$+ 0.28 + 0.17)$$
$$= 22{,}698 \text{ m}^3$$

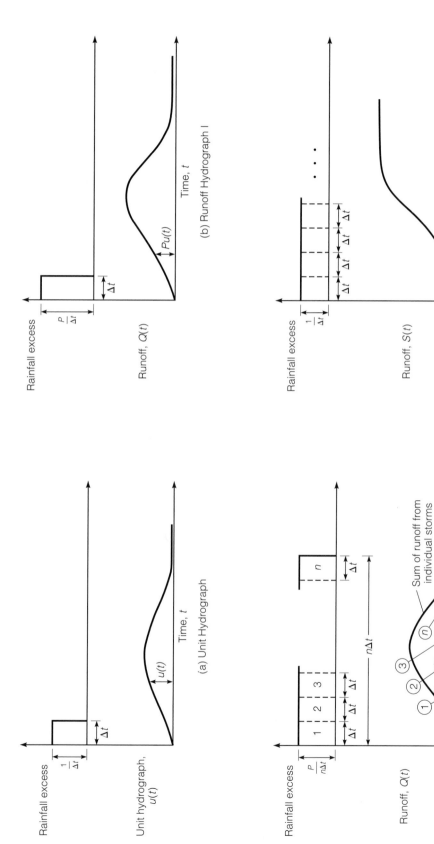

Figure 6.24 ■ Applications of the Unit Hydrograph

367

where the time increment between the hydrograph ordinates is 30 min = 1,800 s. Since the area of the catchment is 2.25 km² = 2.25 × 10⁶ m², then the depth, h, of rainfall excess is given by

$$h = \frac{22698}{2.25 \times 10^6} = 0.01 \text{ m} = 1 \text{ cm}$$

Since the depth of rainfall excess is 1 cm, the given hydrograph qualifies as a unit hydrograph.

(b) For a 10-min rainfall excess of 3.5 cm, the runoff hydrograph is estimated by multiplying the ordinates of the unit hydrograph by 3.5. This yields the following runoff hydrograph:

Time (min)	0	30	60	90	120	150	180	210	240	270	300	330	360	390
Runoff (m³/s)	0	4.2	9.8	6.0	4.9	4.2	3.9	3.2	2.6	2.1	1.8	0.98	0.60	0

(c) The runoff hydrograph for a 20-min rainfall excess of 8.5 cm is calculated by adding the runoff hydrographs from two consecutive 10-min events, with each event corresponding to a rainfall excess of 4.25 cm. The unit hydrograph is first interpolated for 10-minute intervals and multiplied by 4.25 to give the runoff from a 10-min 4.25-cm event. This hydrograph is then added to the same hydrograph shifted by 10 minutes. The computations are summarized in Table 6.21, where Runoff 1 and Runoff 2 are the runoff hydrographs of the two 10-min 4.25-cm events, respectively. ■

For storms in which the duration of the rainfall excess is not an integral multiple of Δt, the runoff is derived from the *S-hydrograph*, $S(t)$, which is defined as the response to a storm of infinite duration as illustrated in Figure 6.24(d). The S-hydrograph is related to the unit hydrograph by

$$S(t) = \sum_{i=0}^{\infty} u(t - i\Delta t) \tag{6.106}$$

The S-hydrograph defined by Equation 6.106 converges quickly, since the ordinates of the individual-storm unit hydrographs, $u(t)$, are only nonzero for a finite interval of time. The S-hydrograph is the response of a catchment to a rainfall excess of intensity $1/\Delta t$ beginning at $t = 0$ and continuing to $t = \infty$. If the rainfall excess were to begin at $t = \tau$, the response of the catchment would be identical to the response to a rainfall excess beginning at $t = 0$, except that the S-hydrograph would begin at time $t = \tau$ instead of $t = 0$. In fact, linearity of the catchment response requires that the difference between the runoff hydrograph resulting from an infinite rainfall excess of intensity $1/\Delta t$ beginning at $t = 0$ and the runoff hydrograph resulting from an infinite rainfall excess of intensity $1/\Delta t$ beginning at $t = \tau$ must be equal to the runoff hydrograph resulting from a rainfall excess of intensity $1/\Delta t$ beginning at $t = 0$ and ending at $t = \tau$. This is illustrated in Figure 6.25.

The unit hydrograph for a rainfall excess of duration τ, denoted by u_τ, is given in terms of the S-hydrograph by

$$u_\tau = \frac{\Delta t}{\tau}[S(t) - S(t - \tau)] \tag{6.107}$$

	Time (min)	Runoff 1 (m³/s)	Runoff 2 (m³/s)	Total runoff (m³/s)
■ Table 6.21 Hydrograph Computation	0	0	0	0
	10	1.70	0	1.70
	20	3.40	1.70	5.10
	30	5.10	3.40	8.50
	40	7.35	5.10	12.45
	50	9.65	7.35	17.00
	60	11.90	9.65	21.55
	70	10.33	11.90	22.23
	80	8.80	10.33	19.13
	90	7.23	8.80	16.03
	100	6.80	7.23	14.03
	110	6.38	6.80	13.18
	120	5.95	6.38	12.33
	130	5.65	5.95	11.60
	140	5.40	5.65	11.05
	150	5.10	5.40	10.50
	160	4.97	5.10	10.07
	170	4.80	4.97	9.77
	180	4.68	4.80	9.48
	190	4.42	4.68	9.10
	200	4.12	4.42	8.54
	210	3.87	4.12	7.99
	220	3.61	3.87	7.48
	230	3.40	3.61	7.01
	240	3.15	3.40	6.55
	250	2.98	3.15	6.13
	260	2.76	2.98	5.74
	270	2.59	2.76	5.35
	280	2.42	2.59	5.01
	290	2.30	2.42	4.72
	300	2.13	2.30	4.43
	310	1.83	2.13	3.96
	320	1.49	1.83	3.32
	330	1.19	1.49	2.68
	340	1.02	1.19	2.21
	350	0.89	1.02	1.91
	360	0.72	0.89	1.61
	370	0.47	0.72	1.19
	380	0.26	0.47	0.73
	390	0	0.26	0.26
	400	0	0	0

and the catchment response, $Q(t)$, to a rainfall excess P with duration τ is

$$Q(t) = Pu_\tau = \frac{P\Delta t}{\tau}[S(t) - S(t - \tau)] \tag{6.108}$$

Since the S-hydrograph is derived from the unit hydrograph of the catchment, Equation 6.108 verifies that the runoff hydrograph for a rainfall excess of any amount and

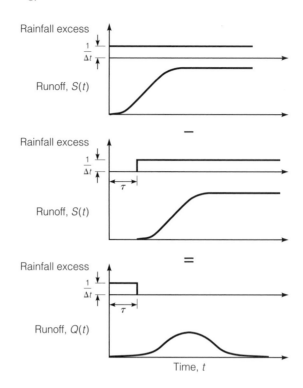

Figure 6.25 ■ Estimation of Unit Hydrograph from S-Hydrograph

duration can be estimated from the unit hydrograph of the catchment, provided that the assumptions of the unit-hydrograph method are approximately valid.

The basic assumptions of the unit-hydrograph model are (Chow et al., 1988): (1) The rainfall excess has a constant intensity within the effective duration; (2) the rainfall excess is uniformly distributed throughout the catchment; (3) the *base time* (= duration) of the runoff hydrograph resulting from a rainfall excess of given duration is constant; (4) the ordinates of all the runoff hydrographs from the catchment are directly proportional to the amount of rainfall excess; and (5) for a given catchment, the hydrograph resulting from a given rainfall excess reflects the unchanging characteristics of the catchment. This assumption is sometimes called the *principle of invariance*. The assumption of a uniformly distributed rainfall excess in space and time limits application of the unit hydrograph methods to relatively small catchment areas, typically in the range of 0.5 ha to 25 km^2 (1 ac to 10 mi^2) although even smaller catchment limitations, on the order of 2.5 km^2 (1 mi^2), are frequently used. If the catchment area is too large for the unit-hydrograph model to be applied over the entire catchment, then the area should be divided into smaller areas that are analyzed separately. The assumption of linearity has been found to be satisfactory in many practical cases (Chow et al., 1988); however, unit-hydrograph models are applicable only when channel conditions remain unchanged and catchments do not have appreciable storage.

■ Example 6.25

The 30-min unit hydrograph for a 2.25 km^2 catchment is given by

Time (min)	0	30	60	90	120	150	180	210	240	270	300	330	360	390
Runoff (m^3/s)	0	1.2	2.8	1.7	1.4	1.2	1.1	0.91	0.74	0.61	0.50	0.28	0.17	0

(a) Determine the S-hydrograph, (b) calculate the 40-min unit hydrograph for the catchment, and (c) verify that the 40-min unit hydrograph corresponds to a 1-cm rainfall excess.

Solution

(a) The S-hydrograph is calculated by summing the lagged unit hydrographs (Equation 6.106), and these computations are summarized in the following table.

Time (min)	0	30	60	90	120	150	180	210	240	270	300	330	360	390	420
Runoff 1 (m^3/s)	0	1.2	2.8	1.7	1.4	1.2	1.1	0.91	0.74	0.61	0.50	0.28	0.17	0	0
Runoff 2 (m^3/s)	0	0	1.2	2.8	1.7	1.4	1.2	1.1	0.91	0.74	0.61	0.50	0.28	0.17	0
Runoff 3 (m^3/s)	0	0	0	1.2	2.8	1.7	1.4	1.2	1.1	0.91	0.74	0.61	0.50	0.28	0.17
Runoff 4 (m^3/s)	0	0	0	0	1.2	2.8	1.7	1.4	1.2	1.1	0.91	0.74	0.61	0.50	0.28
Runoff 5 (m^3/s)	0	0	0	0	0	1.2	2.8	1.7	1.4	1.2	1.1	0.91	0.74	0.61	0.50
Runoff 6 (m^3/s)	0	0	0	0	0	0	1.2	2.8	1.7	1.4	1.2	1.1	0.91	0.74	0.61
Runoff 7 (m^3/s)	0	0	0	0	0	0	0	1.2	2.8	1.7	1.4	1.2	1.1	0.91	0.74
Runoff 8 (m^3/s)	0	0	0	0	0	0	0	0	1.2	2.8	1.7	1.4	1.2	1.1	0.91
Runoff 9 (m^3/s)	0	0	0	0	0	0	0	0	0	1.2	2.8	1.7	1.4	1.2	1.1
Runoff 10 (m^3/s)	0	0	0	0	0	0	0	0	0	0	1.2	2.8	1.7	1.4	1.2
Runoff 11 (m^3/s)	0	0	0	0	0	0	0	0	0	0	0	1.2	2.8	1.7	1.4
Runoff 12 (m^3/s)	0	0	0	0	0	0	0	0	0	0	0	0	1.2	2.8	1.7
Runoff 13 (m^3/s)	0	0	0	0	0	0	0	0	0	0	0	0	0	1.2	2.8
Runoff 14 (m^3/s)	0	0	0	0	0	0	0	0	0	0	0	0	0	0	1.2
Runoff 15 (m^3/s)	0	0	0	0	0	0	0	0	0	0	0	0	0	0	0
S-hydrograph (m^3/s)	0	1.2	4.0	5.7	7.1	8.3	9.4	10.3	11.1	11.7	12.2	12.4	12.6	12.6	12.6

(b) To estimate the 40-min unit hydrograph, the S-hydrograph must be shifted by 40 minutes and subtracted from the original S-hydrograph (Equation 6.107). These computations are summarized in columns 2 to 4 in the following table.

(1) t (min)	(2) $S(t)$ (m^3/s)	(3) $S(t-40)$ (m^3/s)	(4) $S(t) - S(t-40)$ (m^3/s)	(5) 40-min UH (m^3/s)
0	0	0	0	0
30	1.2	0	1.2	0.9
60	4.0	0.8	3.2	2.4
90	5.7	3.1	2.6	2.0
120	7.1	5.1	2.0	1.5
150	8.3	6.6	1.7	1.3
180	9.4	7.9	1.5	1.1
210	10.3	9.0	1.3	1.0
240	11.1	10.0	1.1	0.8
270	11.7	10.8	0.9	0.7
300	12.2	11.5	0.7	0.5
330	12.4	12.0	0.4	0.3
360	12.6	12.3	0.3	0.2
390	12.6	12.5	0.1	0.1
420	12.6	12.6	0	0

Note that values of $S(t-40)$ must be interpolated. From the given data, $\Delta t = 30$ min and $\tau = 40$ min, therefore $\Delta t/\tau = 30/40 = 0.75$, and the 40-min unit hydrograph is obtained by multiplying column 4 by 0.75 (see Equation 6.107). The 40-min unit hydrograph is given in column 5.

(c) The area under the computed 40-min unit hydrograph is given by

$$\text{Area} = (30)(60)(0.9 + 2.4 + 2.0 + 1.5 + 1.3 + 1.1 + 1.0 + 0.8 + 0.7 + 0.5$$
$$+ 0.3 + 0.2 + 0.1)$$
$$= 23040 \text{ m}^3$$

Since the catchment area is $2.25 \text{ km}^2 = 2.25 \times 10^6 \text{ m}^2$, then the depth of rainfall is given by

$$\text{Depth of rainfall} = \frac{23040}{2.25 \times 10^6} = 0.01 \text{ m} = 1 \text{ cm}$$

The depth of rainfall is equal to 1 cm, and therefore the 40-min unit hydrograph calculated in (b) is a valid unit hydrograph.   ■

Unit hydrographs can also be used to determine the runoff in cases where the rainfall excess is nonuniform in time. In these cases, the rainfall excess is discretized into discrete uniform events over time intervals Δt, and the runoffs from each of these discrete events are added together to give the runoff from the entire event. Hence, if the rainfall excess is divided into N discrete events, with rainfall excesses $P_i, i = 1 \ldots N$, then the runoff, $Q(t)$, from the entire event is given by

$$Q(t) = \sum_{i=1}^{N} P_i u(t - i\Delta t + \Delta t) \tag{6.109}$$

where $u(t)$ is the unit hydrograph for a unit rainfall excess of duration Δt.

■ Example 6.26

The 30-min unit hydrograph for a catchment is given by

Time (min)	0	30	60	90	120	150	180	210	240	270	300	330	360	390
Runoff (m^3/s)	0	1.2	2.8	1.7	1.4	1.2	1.1	0.91	0.74	0.61	0.50	0.28	0.17	0

Estimate the runoff resulting from the following 90-min storm:

Time (min)	Rainfall Excess (cm)
0–30	3.1
30–60	2.5
60–90	1.7

Solution

The given 90-min storm can be viewed as three consecutive 30-min storms with rainfall excess amounts of 3.1 cm, 2.5 cm, and 1.7 cm. The runoff from each storm is estimated by multiplying the 30-min unit hydrograph by 3.1, 2.5, and 1.7, respectively. The hydrographs are then lagged by 30 minutes and summed to estimate the total runoff from the storm (see Equation 6.109). These computations are summarized in the following table:

Time (min)	UH × 3.1 (m³/s)	UH × 2.5 (m³/s)	UH × 1.7 (m³/s)	Total runoff (m³/s)
0	0	0	0	0
30	3.7	0	0	3.7
60	8.7	3.0	0	11.7
90	5.3	7.0	2.0	14.3
120	4.3	4.3	4.8	13.4
150	3.7	3.5	2.9	10.1
180	3.4	3.0	2.4	8.8
210	2.8	2.8	2.0	7.6
240	2.3	2.3	1.9	6.4
270	1.9	1.9	1.5	5.3
300	1.6	1.5	1.3	4.4
330	0.9	1.3	1.0	3.2
360	0.5	0.7	0.9	2.1
390	0	0.4	0.5	0.9
420	0	0	0.3	0.3
450	0	0	0	0

■

Unit hydrographs are estimated in practice from either contemporaneous rainfall and runoff measurements or by using empirical relationships. The estimation of unit hydrographs from measured rainfall and runoff hydrographs is seldom an option in urban hydrology, where runoff is to be estimated from planned developments. The most common method of estimating the unit hydrograph in urban applications is to use empirical relationships to construct a *synthetic unit hydrograph*, where the shape of the unit hydrograph is related to the catchment characteristics and the duration of the rainfall excess. The shape parameters are usually the peak runoff, time to peak, and time base of the unit hydrograph. A wide variety of synthetic unit hydrograph methods have been developed over the years, but only some of these methods are appropriate for use in designing urban stormwater systems (ASCE, 1992).

Espey-Altman 10-min Unit Hydrograph.   Espey and Altman (1978) studied the unit hydrographs resulting from 10-min rainfall excesses over 41 small watersheds ranging in size from 0.036 to 39 km² (0.014 to 15 mi²), and impervious fractions ranging from 2% to 100%. Of the 41 watersheds studied, 16 are in Texas, nine in North Carolina, six in Kentucky, four in Indiana, two in Colorado, two in Mississippi, one in Tennessee, and one in Pennsylvania. The characterictics of the catchment areas were measured by the following five parameters: (1) size of the drainage area, A; (2) main channel length, L, defined as the total distance along the main channel from the catchment outlet to the upstream boundary of the catchment; (3) main channel slope, S, defined as $H/0.8L$, where H is the difference in elevation between a point on the channel bottom at a distance of $0.2L$ downstream from the upstream catchment boundary and a point on the channel bottom at the catchment outlet; (4) impervious cover, I (%), which is assumed to be 5% for undeveloped catchments; and (5) dimensionless catchment conveyance factor, ϕ, which is a function of the impervious cover, I, and the Manning n of the main channel, as shown in Figure 6.26.

The observed 10-min unit hydrographs were described by the following five parameters: (1) time of rise to the peak of the unit hydrograph, T_p, measured from the beginning of the runoff; (2) peak flow of the unit hydrograph, Q_p; (3) time base of the unit hydrograph, T_B; (4) width of the hydrograph at 50% of Q_p, W_{50}; and (5) width of

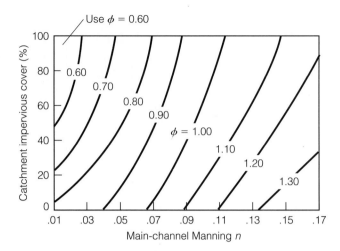

Figure 6.26 ■ Catchment
Conveyance Factor
Source: Espey and Altman
(1978).

the unit hydrograph at 75% of Q_p, W_{75}. The empirical relationships between these five
unit-hydrograph parameters and the catchment characteristics are given by Equations
6.110 to 6.114

$$T_p = 4.1 \frac{L^{0.23} \phi^{1.57}}{S^{0.25} I^{-0.18}} \tag{6.110}$$

$$Q_p = 359 \frac{A^{0.96}}{T_p^{1.07}} \tag{6.111}$$

$$T_B = 1645 \frac{A}{Q_p^{0.95}} \tag{6.112}$$

$$W_{50} = 252 \frac{A^{0.93}}{Q_F^{0.92}} \tag{6.113}$$

$$W_{75} = 95 \frac{A^{0.79}}{Q_p^{0.78}} \tag{6.114}$$

where L is in meters, I is in %, A is in km^2, T_p, T_B, W_{50}, and W_{75} are in minutes, Q_p is
in m^3/s, and S and ϕ are dimensionless. The application of Equations 6.110 to 6.114 is
illustrated in Figure 6.27, where the widths W_{50} and W_{75} are normally positioned such
that one-third lies on the rising side and two-thirds on the recession limb of the unit
hydrograph.

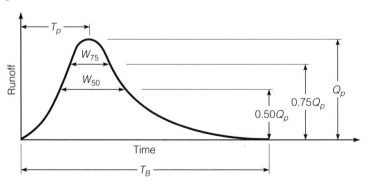

Figure 6.27 ■ Espey-
Altman Unit
Hydrograph

■ **Example 6.27**

A 2.25 km² urban catchment has a main channel with a slope of 0.5% and Manning n of 0.06, the catchment is 40% impervious, and the distance along the main channel from the catchment boundary to the outlet is 1,680 m. Calculate the 10-min unit hydrograph using the Espey-Altman method.

Solution

From the given data: $A = 2.25$ km², $S = 0.005$, $I = 40\%$, and $L = 1,680$ m. For $n = 0.06$, Figure 6.26 gives $\phi = 0.85$. Substituting these data into Equations 6.110 to 6.114 yields

$$T_p = 4.1\frac{L^{0.23}\phi^{1.57}}{S^{0.25}I^{-0.18}} = 4.1\frac{(1680)^{0.23}(0.85)^{1.57}}{(0.005)^{0.25}(40)^{-0.18}} = 128 \text{ min} = 2.13 \text{ h}$$

$$Q_p = 359\frac{A^{0.96}}{T_p^{1.07}} = 359\frac{(2.25)^{0.96}}{(128)^{1.07}} = 4.35 \text{ m}^3/\text{s}$$

$$T_B = 1645\frac{A}{Q_p^{0.95}} = 1645\frac{(2.25)}{(4.35)^{0.95}} = 916 \text{ min} = 15.3 \text{ h}$$

$$W_{50} = 252\frac{A^{0.93}}{Q_p^{0.92}} = 252\frac{(2.25)^{0.93}}{(4.35)^{0.92}} = 139 \text{ min} = 2.32 \text{ h}$$

$$W_{75} = 95\frac{A^{0.79}}{Q_p^{0.78}} = 95\frac{(2.25)^{0.79}}{(4.35)^{0.78}} = 57.3 \text{ min} = 0.96 \text{ h}$$

The hydrograph corresponding to these parameters is shown in Figure 6.28, where all intermediate points have been linearly interpolated. The hydrograph coordinates are:

Time (h)	0	1.36	1.81	2.13	2.77	3.68	15.3
Runoff (m³/s)	0	2.18	3.26	4.35	3.26	2.18	0.0

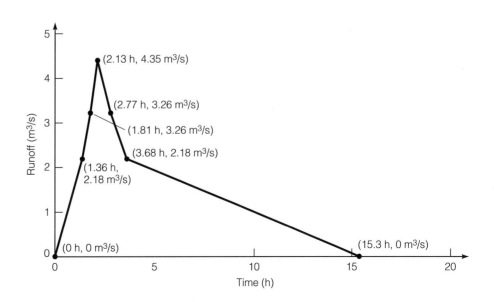

Figure 6.28 ■ Espey-Altman Runoff Hydrograph

For the estimated hydrograph to qualify as a unit hydrograph, it must correspond to a rainfall excess of 1 cm. This can be verified by dividing the area under the estimated hydrograph (= volume of runoff) by the catchment area. The area under the estimated hydrograph is given by (see Figure 6.28):

$$\text{Hydrograph area} = 3600\left[\frac{1}{2}(1.36)(2.18) + \frac{1}{2}(2.18+3.26)(1.81-1.36)\right.$$

$$+ \frac{1}{2}(3.26+4.35)(2.13-1.81) + \frac{1}{2}(4.35+3.26)(2.77-2.13)$$

$$\left.+ \frac{1}{2}(3.26+2.18)(3.68-2.77) + \frac{1}{2}(2.18+0.0)(15.3-3.68)\right]$$

$$= 77{,}436 \text{ m}^3$$

The catchment area is $2.25 \text{ km}^2 = 2.25 \times 10^6 \text{ m}^2$, and therefore the runoff depth is given by

$$\text{Runoff depth} = \frac{77436 \text{ m}^3}{2.25 \times 10^6 \text{ m}^2} = 0.0344 \text{ m} = 3.44 \text{ cm}$$

Since the estimated hydrograph corresponds to a runoff depth of 3.4 cm and a unit hydrograph corresponds to a depth of 1 cm, the estimated hydrograph coordinates must be adjusted by a factor of $1/3.44 = 0.291$. Applying this factor to the estimated hydrograph coordinates yields the following Espey-Altman unit hydrograph:

Time (h)	0	1.36	1.81	2.13	2.77	3.68	15.3
Runoff (m³/s)	0	0.634	0.949	1.27	0.949	0.634	0.0

NRCS Dimensionless Unit Hydrograph.   The Natural Resources Conservation Service (SCS, 1985; SCS, 1986) developed a dimensionless unit hydrograph that represents the average shape of a large number of unit hydrographs from natural catchments of different characteristics. This dimensionless unit hydrograph is frequently applied to urban catchments. The NRCS dimensionless unit hydrograph is illustrated in Figure 6.29(a) and expresses the normalized runoff, Q/Q_p, as a function of the normalized time, t/T_p,

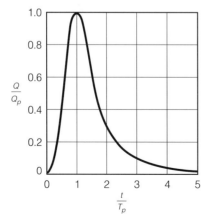
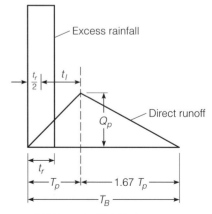

Figure 6.29 ■ NRCS Unit Hydrographs

(a) Dimensionless Unit Hydrograph

(b) Triangular Unit Hydrograph

	t/T_p	Q/Q_p
■ **Table 6.22**	0.0	0.000
NRCS Dimensionless Unit	0.2	0.100
Hydrograph	0.4	0.310
	0.6	0.660
	0.8	0.930
	1.0	1.000
	1.2	0.930
	1.4	0.780
	1.6	0.560
	1.8	0.390
	2.0	0.280
	2.2	0.207
	2.4	0.147
	2.6	0.107
	2.8	0.077
	3.0	0.055
	3.4	0.029
	4.2	0.010
	4.6	0.003
	5.0	0.000

where Q_p is the peak runoff and T_p is the time to the peak of the hydrograph from the beginning of the rainfall excess. The coordinates of the NRCS dimensionless unit hydrograph are given in Table 6.22.

The NRCS dimensionless unit hydrograph can be converted to an actual hydrograph by multiplying the abcissa by T_p and the ordinate by Q_p. The time, T_p, is estimated using

$$T_p = \frac{1}{2}t_r + t_l \tag{6.115}$$

where t_r is the duration of the rainfall excess and t_l is the time lag from the centroid of the rainfall excess to the peak of the runoff hydrograph. NRCS recommends that the specified value of t_r not exceed two-tenths of t_c or three-tenths of T_p for the NRCS dimensionless unit hydrograph to be valid. The time lag, t_l, can be estimated by

$$t_l = 0.6t_c \tag{6.116}$$

for rural catchments and, if justified by available data, can also be estimated using the more detailed relation

$$t_l = \frac{L^{0.8}(S + 1)^{0.7}}{1900Y^{0.5}} \tag{6.117}$$

where t_l is in hours, L is the length to the catchment divide (ft), and S is the potential maximum retention (in.) given by

$$S = \frac{1000}{\text{CN}} - 10 \tag{6.118}$$

where CN is the curve number and Y is the average catchment slope (%). The curve number in Equation 6.118 should be based on antecedent runoff condition II (AMCII) since it is being used as a measure of the surface roughness and not runoff potential (Haan et al., 1994). Equation 6.117 is applicable to curve numbers between 50 and 95, and catchment areas less than 8 km^2 (2,000 ac). Otherwise, Equation 6.116 should be used to estimate the basin lag, t_l (Ponce, 1989). If the watershed is in an urban area, then t_l given by Equation 6.117 is adjusted for imperviousness and/or improved water courses by the factor M, where

$$M = 1 - I(-6.8 \times 10^{-3} + 3.4 \times 10^{-4}CN - 4.3 \times 10^{-7}CN^2 - 2.2 \times 10^{-8}CN^3) \quad (6.119)$$

and I is either the percentage impervious or the percentage of the main watercourse that is hydraulically improved from natural conditions. If part of the area is impervious and part of the channel is improved, then two values of M are determined and both are multiplied by t_l. Once t_l is determined, then the time to peak, T_p, is estimated using Equation 6.115. The time base of the unit hydrograph, T_B, is equal to $5T_p$, and both T_p and T_B should be rounded to the nearest whole number multiple of t_r. The peak runoff, Q_p, from the catchment can be estimated using the empirical relation

$$Q_p = 2.08 \frac{A}{T_p} \quad (6.120)$$

where Q_p is in (m^3/s)/cm, A is the catchment area in km^2, and T_p is in hours. The coefficient of 2.08 in Equation 6.120 is appropriate for the average rural experimental watersheds used in calibrating the formula, but this coefficient should be increased by about 20% for steep mountainous conditions and decreased by about 30% for flat swampy conditions.

The NRCS dimensionless unit hydrograph is sometimes approximated by the triangular unit hydrograph illustrated in Figure 6.29(b). In this case, both Q_p and T_p are still estimated by Equations 6.120 and 6.115, respectively, and the time base of the unit hydrograph, T_B, is equal to $2.67T_p$ rounded to the nearest whole number multiple of t_r. The coefficient of 2.67 should be decreased by 20% for mountainous conditions or increased by 30% for flat swampy conditions. The triangular unit hydrograph incorporates key properties of the dimensionless unit hydrograph in that (1) the total volume under the dimensionless unit hydrograph is the same, (2) the volume under the rising limb is the same, and (3) the peak discharge is the same. According to Debo and Reese (1995), the triangular unit hydrograph approximation produces results that are sufficiently accurate for most stormwater management facility designs, including curbs, gutters, storm drains, channels, ditches, and culverts.

■ Example 6.28

A 2.25 km^2 urban catchment is estimated to have an average curve number of 70, an average slope of 0.5%, and a flow length from the catchment boundary to the outlet of 1,680 m. If the imperviousness of the catchment is estimated at 40%, determine the NRCS unit hydrograph for a 30-minute rainfall excess. Determine the approximate NRCS triangular unit hydrograph, and verify that it corresponds to a rainfall excess of 1 cm.

Solution

The time lag of the catchment can be estimated using Equation 6.117, where $L = 1680$ m $= 5,512$ ft, $Y = 0.5\%$, and S is derived from CN $= 70$ using Equation 6.118:

$$S = \frac{1000}{CN} - 10 = \frac{1000}{70} - 10 = 4.3 \text{ in.}$$

The time lag, t_l, is therefore given by

$$t_l = \frac{L^{0.8}(S+1)^{0.7}}{1900Y^{0.5}} = \frac{(5512)^{0.8}(4.3+1)^{0.7}}{1900(0.5)^{0.5}} = 2.35 \text{ h}$$

For a 40% impervious catchment, $I = 40\%$ and $CN = 70$, Equation 6.119 gives the correction factor, M, as

$$M = 1 - I[-6.8 \times 10^{-3} + 3.4 \times 10^{-4}CN - 4.3 \times 10^{-7}CN^2 - 2.2 \times 10^{-8}CN^3]$$

$$= 1 - (40)[-6.8 \times 10^{-3} + 3.4 \times 10^{-4}(70) - 4.3 \times 10^{-7}(70)^2 - 2.2 \times 10^{-8}(70)^3]$$

$$= 0.706$$

Applying the correction factor, M, to calculate the adjusted time lag yields

$$t_l = 0.706 \times 2.35 = 1.66 \text{ h}$$

For a 30-minute rainfall excess, $t_r = 0.5$ h and the time to peak, T_p, is given by Equation 6.115 as

$$T_p = \frac{1}{2}t_r + t_l = \frac{1}{2}(0.5) + 1.66 = 1.91 \text{ h}$$

The NRCS unit hydrograph is limited to cases where $t_r/T_p \leq 0.3$; in this case, $t_r/T_p = 0.5/1.91 = 0.26$. The NRCS unit hydrograph is therefore applicable. The peak of the unit hydrograph, Q_p, is estimated using Equation 6.120, where $A = 2.25$ km² and

$$Q_p = 2.08\frac{A}{T_p} = 2.08\frac{2.25}{1.91} = 2.45 \text{ (m}^3\text{/s)/cm}$$

With $T_p = 1.91$ h and $Q_p = 2.45$ (m³/s)/cm, the NRCS 30-min unit hydrograph is obtained by multiplying t/T_p in Table 6.22 by 1.91 h and Q/Q_p by 2.45 (m³/s)/cm to yield

t (h)	Q [(m³/s)/cm]	t (h)	Q [(m³/s)/cm]
0.0	0.0	3.82	0.69
0.38	0.25	4.20	0.51
0.76	0.76	4.58	0.36
1.15	1.62	4.97	0.26
1.53	2.28	5.35	0.19
1.91	2.45	5.73	0.14
2.29	2.28	6.49	0.07
2.67	1.91	8.02	0.03
3.06	1.37	8.79	0.01
3.44	0.96	9.55	0.00

For the approximate triangular unit hydrograph, T_p and Q_p are the same, but the base of the hydrograph is $2.67T_p = 2.67 \times 1.91 = 5.10$ h ($= 18,360$ s). The area under the triangular hydrograph ($=$ volume of runoff) is given by

$$\text{Area} = \frac{1}{2}\text{base} \times \text{height} = \frac{1}{2}18360 \times 2.45 = 22,500 \text{ m}^3$$

Since the catchment area is 2.25 km² = 2.25 × 10⁶ m², the depth of rainfall is equal to the volume of runoff divided by the area of the catchment. Hence

$$\text{Depth of rainfall} = \frac{22500 \text{ m}^3}{2.25 \times 10^6 \text{ m}^2} = 0.01 \text{ m} = 1 \text{ cm}$$

Since the runoff hydrograph corresponds to a unit depth (1 cm) of rainfall, it is a valid unit hydrograph.   ■

6.4.3.2   Time-Area Curves

Time-area curves describe the relationship between the travel time to the catchment outlet and the location within the catchment. This relationship is based on the estimated velocity of direct runoff and neglects storage effects. The time-area method can be illustrated by considering the runoff isochrones illustrated in Figure 6.30, and the corresponding histogram of contributing area versus time shown in Figure 6.31. These figures indicate that the incremental area contributing to runoff varies throughout a storm, with full contribution from all areas for $t \geq t_5$. If the rainfall excess is expressed as the average rainfall intensity over the time increments in the time-area histogram, then the runoff at the end of time increment j, Q_j, can be expressed in terms of the contributing areas by the relation

$$Q_j = i_j A_1 + i_{j-1} A_2 + \cdots + i_1 A_j$$

or

$$Q_j = \sum_{k=1}^{j} i_{j-k+1} A_k \qquad (6.121)$$

where i_j and A_j are the average rainfall intensity and contributing area, respectively, during time increment j.

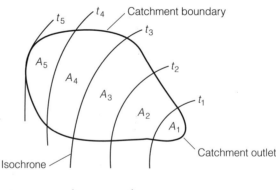

Figure 6.30 ■
Runoff Isochrones

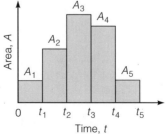

Figure 6.31 ■ Time-Area
Histogram

The difficulty is estimating isochronal lines and the need to account for storage effects make the time-area method difficult to apply in practice. The time-area method uses a simple routing procedure that does not consider attenuation of the runoff hydrograph associated with storage effects, and is therefore limited to small watersheds with areas less than 200 ha (500 ac) (Ward, 1995).

■ **Example 6.29**

A 100-ha catchment is estimated to have the following time-area relationship:

Time (min)	Contributing area (ha)
0	0
5	3
10	9
15	25
20	51
25	91
30	100

If the rainfall-excess distribution is given by

Time (min)	Average intensity (mm/h)
0–5	132
5–10	84
10–15	60
15–20	36

estimate the runoff hydrograph using the time-area method.

Solution

First, tabulate the contributing area for each of the given time intervals:

Time interval (min)	Contributing area (ha)
0–5	3
5–10	6
10–15	16
15–20	26
20–25	40
25–30	9

The runoff hydrograph is calculated using Equation 6.121. The computations are given in the following table (a factor of 0.002778 is used to convert ha·mm/h to m³/s):

Time (min)		Runoff (m³/s)
0		0
5	$[132(3)](0.002778) =$	1.1
10	$[84(3) + 132(6)](0.002778) =$	4.0
15	$[60(3) + 84(6) + 132(16)](0.002778) =$	7.8
20	$[36(3) + 60(6) + 84(16) + 132(26)](0.002778) =$	14.6
25	$[36(6) + 60(16) + 84(26) + 132(40)](0.002778) =$	24.0
30	$[36(16) + 60(26) + 84(40) + 132(9)](0.002778) =$	18.6
35	$[36(26) + 60(40) + 84(9)](0.002778) =$	11.4
40	$[36(40) + 60(9)](0.002778) =$	5.5
45	$[36(9)](0.002778) =$	0.9
50	$0 =$	0

■

6.4.3.3   Kinematic-Wave Model

The kinematic-wave model describes runoff by solving the one-dimensional continuity equation

$$\frac{\partial y}{\partial t} + \frac{\partial q}{\partial x} = i_e \tag{6.122}$$

and a momentum equation of the form

$$q = \alpha y^m \tag{6.123}$$

where y is the flow depth, q is the flow per unit width of the catchment, and i_e is the effective rainfall. Combining Equations 6.122 and 6.123 yields the following kinematic-wave equation for overland flow

$$\boxed{\frac{\partial y}{\partial t} + \alpha m y^{m-1} \frac{\partial y}{\partial x} = i_e} \tag{6.124}$$

The kinematic-wave model solves Equation 6.124 for the flow depth, y, as a function of x and t for given initial and boundary conditions. This solution is then substituted into Equation 6.123 to determine the runoff, q, as a function of space and time. In cases where the rainfall excess is independent of x and t, then the solution was derived previously and is given by Equation 6.82. In the more general case with variable effective rainfall, a numerical solution of Equation 6.124 is required. In cases where overland flow enters a drainage channel, then flow in the channel can be described by

$$\frac{\partial A}{\partial t} + \frac{\partial Q}{\partial x} = q_0 \tag{6.125}$$

and

$$Q = \alpha A^m \tag{6.126}$$

where A is the cross-sectional flow area, Q is the flowrate in the channel, and q_0 is the overland inflow entering the channel per unit length of the channel. Combining Equations 6.125 and 6.126 yields the following kinematic-wave equation for channel flow

$$\frac{\partial A}{\partial t} + \alpha m A^{m-1} \frac{\partial A}{\partial x} = q_0 \tag{6.127}$$

Because this equation does not allow for hydrograph diffusion, the application of the kinematic-wave equation is limited to flow conditions that do not demonstrate appreciable hydrograph attenuation. Accordingly, the kinematic-wave approximation works best when applied short well-defined channel reaches as found in urban drainage applications (ASCE, 1996b). A typical kinematic-wave model divides the catchment into overland flow planes that feed collector channels. The kinematic-wave model is used only in numerical models, and the documentation accompanying commercial software packages that implement the kinematic-wave model generally provide detailed descriptions of the numerical procedures used to solve the kinematic-wave equation.

6.4.3.4   Nonlinear Reservoir Model

The nonlinear resevoir model views the catchment as a very shallow reservoir, where the inflow is equal to the rainfall excess, the outflow is a (nonlinear) function of the depth of flow over the catchment, and the difference between the inflow and outflow is equal to the rate of change of storage within the catchment. The nonlinear reservoir model can be stated as

$$A \frac{dy}{dt} = A i_e - Q \tag{6.128}$$

where A is the surface area of the catchment and Q is the surface runoff at the catchment outlet. The model assumes uniform overland flow at the catchment outlet at a depth equal to the difference between the water depth, y, and the (constant) depression storage, y_d; therefore, Q can be expressed as a function of y. Equation 6.128 is an ordinary differential equation in y that can be solved, given an initial condition within the catchment, and the variation of i_e as a function of time.

A particular application of the nonlinear reservoir model is described by Huber and Dickinson (1988), where the runoff at the catchment outlet is given by the Manning relation

$$Q = \frac{CW}{n}(y - y_d)^{5/3} S_o^{1/2} \tag{6.129}$$

where C is a constant to account for the units ($C = 1$ for SI units, $C = 1.49$ for U.S. Customary units), W is a representative width of the catchment, n is an average value of the Manning roughness coefficient for the catchment, and S_o is the average slope of the catchment. Estimates of the Manning n for overland flow are given in Table 6.15. The combination of Equations 6.128 and 6.129 can be put in the simple finite-difference form

$$\frac{y_2 - y_1}{\Delta t} = \bar{i}_e - \frac{CW S_o^{1/2}}{An}\left(\frac{y_1 + y_2}{2} - y_d\right)^{5/3} \tag{6.130}$$

where Δt is the time step, y_1 and y_2 are the depths at the beginning and end of the time step, and $\bar{i}_e$ is the average effective rainfall over the time step. Equation 6.130 must be solved iteratively for y_2, which is then substituted into Equation 6.129 to determine the runoff, Q.

■ Example 6.30

A 1-ha catchment consists of mostly light turf with an average slope of 0.8%. The width of the catchment is approximately 100 m, and the depression storage is estimated to be 5 mm. Use the nonlinear reservoir model to calculate the runoff from the following 15-min rainfall excess:

Time (min)	Effective rainfall (mm/h)
0–5	120
5–10	70
10–15	50

Solution

The nonlinear reservoir model is given by Equation 6.130, where $C = 1$, $W = 100$ m, $S_o = 0.008$, $A = 1$ ha $= 10^4$ m^2, $n = 0.20$ (from Table 6.15), and $y_d = 5$ mm $= 0.005$ m. Taking $\Delta t = 5$ min $= 300$ s and substituting the given parameters into Equation 6.130 gives

$$\frac{y_2 - y_1}{300} = \bar{i}_e - \frac{(1)(100)(0.008)^{1/2}}{(10^4)(0.20)} \left(\frac{y_1 + y_2}{2} - 0.005 \right)^{5/3}$$

which simplifies to

$$y_2 = y_1 + 300 \left[2.78 \times 10^{-7}\bar{i}_e - 0.00447 \left(\frac{y_1 + y_2}{2} - 0.005 \right)^{5/3} \right]$$

where the factor 2.78×10^{-7} is introduced to convert $\bar{i}_e$ in mm/h to m/s. For given values of y_1, this equation is solved iteratively for $y = y_2$ as a function of time. The corresponding runoff, Q, is given by Equation 6.129 as

$$Q = \frac{CW}{n}(y - y_d)^{5/3} S_o^{1/2} = \frac{(1)(100)}{0.20}(y - 0.005)^{5/3}(0.008)^{1/2} = 44.7(y - 0.005)^{5/3}$$

Starting with $y_1 = 0$ m at $t = 0$, the runoff computations are tabulated for $t = 0$ to 60 min as follows:

t (min)	$\bar{i}_e$ (mm/h)	y (m)	Q (m^3/s)
0		0	0
	120		
5		0.0100	0.00654
	70		
10		0.0154	0.0221
	50		
15		0.0187	0.0351
	0		
20		0.0177	0.0309
	0		
25		0.0168	0.0273
	0		

t (min)	$\bar{i}_e$ (mm/h)	y (m)	Q (m^3/s)
30		0.0160	0.0243
	0		
35		0.0153	0.0218
	0		
40		0.0147	0.0197
	0		
45		0.0141	0.0177
	0		
50		0.0136	0.0161
	0		
55		0.0131	0.0146
	0		
60		0.0127	0.0134

Beyond $t = 60$ min, the runoff, Q, decreases gradually to zero. ■

6.4.3.5 Santa Barbara Urban Hydrograph Model

The Santa Barbara Urban Hydrograph (SBUH) model was developed for the Santa Barbara County Flood Control and Water Conservation District in California (Stubchaer, 1975) and has been adopted by other water-management districts in the United States (for example, the South Florida Water Management District, 1994). In the SBUH model, the impervious portion of the catchment is assumed to be directly connected to the drainage system, abstractions from rain falling on impervious surfaces are neglected, and abstractions from pervious areas are accounted for by using either the NRCS curve-number method or a similar technique. The SBUH combines the runoff from impervious and pervious areas to develop a runoff hydrograph that is routed through an imaginary reservoir that causes a time delay equal to the time of concentration of the catchment. The computations proceed through consecutive time intervals, Δt, during which the instantaneous runoff, I, is calculated using the relation

$$I = [ix + i_e(1.0 - x)]A \tag{6.131}$$

where i is the rainfall rate, x is the fraction of the catchment that is impervious, i_e is the effective rainfall rate, and A is the area of the catchment. The runoff hydrograph is then obtained using the following routing equation

$$Q_j = Q_{j-1} + K_r(I_{j-1} + I_j - 2Q_{j-1}) \tag{6.132}$$

where Q_j and Q_{j-1} are the ordinates of the runoff hydrograph at times $j\Delta t$ and $(j - 1)\Delta t$, I_j and I_{j-1} are the instantaneous runoff rates at $j\Delta t$ and $(j - 1)\Delta t$, and K_r is a routing constant given by

$$K_r = \frac{\Delta t}{2t_c + \Delta t} \tag{6.133}$$

where t_c is the time of concentration of the catchment. An important limitation of the SBUH method is that the calculated peak discharge cannot occur after precipitation ceases. In reality, for short-duration storms over flat and large watersheds, the peak discharge can actually occur after rainfall ends.

■ Example 6.31

A 2.25-km^2 catchment is estimated to have a time of concentration of 45 minutes, and 45% of the catchment is impervious. Estimate the runoff hydrograph for the following rainfall event:

Time (min)	Rainfall (mm/h)	Rainfall excess (mm/h)
0	0	0
10	210	150
20	126	102
30	78	66

Use a time increment of 10 minutes to calculate the runoff hydrograph.

Solution

From the given data, $\Delta t = 10$ min and $t_c = 45$ min, therefore Equation 6.133 gives

$$K_r = \frac{\Delta t}{2t_c + \Delta t} = \frac{10}{2(45) + 10} = 0.10$$

Also, from the given data, $x = 0.45$, $A = 2.25$ km^2, and Equation 6.131 gives the instantaneous runoff, I, as

$$I = [ix + i_e(1.0 - x)]A = [i(0.45) + i_e(1.0 - 0.45)](2.25)(0.278)$$

$$= 0.625[0.45i + 0.55i_e] \tag{6.134}$$

where the factor 0.278 is required to give I in m^3/s. The catchment runoff is given by Equation 6.132 as

$$Q_j = Q_{j-1} + K_r(I_{j-1} + I_j - 2Q_{j-1}) = Q_{j-1} + (0.10)(I_{j-1} + I_j - 2Q_{j-1}) \tag{6.135}$$

Beginning with $I_1 = Q_1 = 0$, Equation 6.134 is applied to calculate I at each time step, and Equation 6.135 is applied to calculate the runoff hydrograph. These calculations are given in the following table:

t (min)	i (mm/h)	i_e (mm/h)	I (m^3/s)	Q (m^3/s)
0	0	0	0	0
10	210	150	111	11.1
20	126	102	70.5	27.0
30	78	66	44.6	33.1
40	0	0	0	30.9
50	0	0	0	24.7
60	0	0	0	19.8
70	0	0	0	15.8
80	0	0	0	12.7
90	0	0	0	10.1
100	0	0	0	8.1

t (min)	i (mm/h)	i_e (mm/h)	I (m³/s)	Q (m³/s)
110	0	0	0	6.5
120	0	0	0	5.2
130	0	0	0	4.1
140	0	0	0	3.3
150	0	0	0	2.7
160	0	0	0	2.1
170	0	0	0	1.7
180	0	0	0	1.4
190	0	0	0	1.1
200	0	0	0	0.9

The runoff, Q, as a function of time reaches a peak at the time the rainfall ends and decreases exponentially thereafter. ■

6.5 Routing Models

Within a hydrologic context, *routing* is the process of determining the spatial and temporal variations in flow rate along a watercourse. Routing is sometimes described as *flow routing* or *flood routing*. Routing models are classified as either *lumped* or *distributed* models. In lumped models, the channel and flow characteristics between the upstream and downstream sections in a watercourse are described by composite parameters, and the flow hydrograph is calculated at the downstream section directly from a given flow hydrograph at the upstream section. In distributed models, the channel and flow characteristics are determined at points in between the upstream and downstream sections, incorporating a more complete description of the flow as it moves from the upstream to the downstream section. Flow routing using lumped parameter models is commonly called *hydrologic routing*, and flow routing using distributed parameter models is called *hydraulic routing*.

6.5.1 ■ Hydrologic Routing

The basic equation used in hydrologic routing is the continuity equation

$$\frac{dS}{dt} = I(t) - O(t) \qquad (6.136)$$

where S is the storage between the upstream and downstream sections, t is time, $I(t)$ is the inflow at the upstream section and $O(t)$ is the outflow at the downstream section. Hydrologic routing is frequently applied to storage reservoirs and stormwater detention basins, where $I(t)$ is the inflow to the storage reservoir, typically from rivers, streams, or drainage channels; $O(t)$ is the outflow from the reservoir, typically over spillways, weirs, or orifice-type outlets; and S is the storage in the reservoir. Although the application of hydrologic routing to rivers and drainage channels is also well established, in many of these cases hydraulic routing is preferable. The procedure for hydrologic routing depends on the particular system being modeled. In the case of storage reservoirs, the storage, S, is typically a function of the outflow, $O(t)$, and the *storage indication method*, also called the *modified Puls method* (Puls, 1928), is preferred. In the case of channel routing, S is typically related to both the upstream inflow, $I(t)$, and downstream outflow, $O(t)$, and other methods such as the *Muskingum method* are preferred.

6.5.1.1 Storage Indication Method

Over the finite interval of time between t and $t + \Delta t$, Equation 6.136 can be written in the finite-difference form

$$\frac{S_2 - S_1}{\Delta t} = \left(\frac{I_1 + I_2}{2}\right) - \left(\frac{O_1 + O_2}{2}\right) \tag{6.137}$$

where the subscripts 1 and 2 refer to the values of the variables at times t and $t + \Delta t$, respectively. For computational convenience, Equation 6.137 can be put in the form

$$\boxed{(I_1 + I_2) + \left(\frac{2S_1}{\Delta t} - O_1\right) = \left(\frac{2S_2}{\Delta t} + O_2\right)} \tag{6.138}$$

In using this form of the continuity equation to route the hydrograph through a storage reservoir, it is important to recognize what is known and what must be determined. Since the inflow hydrograph is generally a given, then I_1 and I_2 are typically known at every time step. For ungated spillways, orifice-type outlets, and weir-type outlets, the discharge, O, is a known function of the water-surface elevation (stage) in the reservoir. Since the storage, S, in the reservoir is also a function of the water stage in the reservoir, then $2S/\Delta t + O$ is typically a known function of the stage in the reservoir, and the following procedure can be used to route the inflow hydrograph through the reservoir:

1. Substitute known values of I_1, I_2, and $2S_1/\Delta t - O_1$ into the lefthand side of Equation 6.138. This gives the value of $2S_2/\Delta t + O_2$.
2. From the reservoir characteristics, determine the discharge, O_2, corresponding to the calculated value of $2S_2/\Delta t + O_2$.
3. Subtract $2O_2$ from $2S_2/\Delta t + O_2$ to yield $2S_2/\Delta t - O_2$ at the end of this time step.
4. Repeat steps 1 to 3 until the entire outflow hydrograph, $O(t)$ is calculated.

In performing these computations, it is recommended to select a time step, Δt, such that there are five or six points on the rising side of the inflow hydrograph, one of which coincides with the inflow peak. In computer-aided computations, many more points can be used.

■ Example 6.32

A stormwater detention basin is estimated to have the following storage characteristics:

Stage (m)	Storage (m^3)
5.0	0
5.5	694
6.0	1,525
6.5	2,507
7.0	3,652
7.5	4,973
8.0	6,484

The discharge weir from the detention basin has a crest elevation of 5.5 m, and the weir discharge, Q, is given by

$$Q = 1.83H^{\frac{3}{2}}$$

where H is the height of the water surface above the crest of the weir. The catchment runoff hydrograph is given by

Time (min)	0	30	60	90	120	150	180	210	240	270	300	330	360	390
Runoff (m³/s)	0	2.4	5.6	3.4	2.8	2.4	2.2	1.8	1.5	1.2	1.0	0.56	0.34	0

If the prestorm stage in the detention basin is 5.0 m, estimate the discharge hydrograph from the detention basin.

Solution

This problem requires that the runoff hydrograph be routed through the detention basin. Using $\Delta t = 30$ min, the storage and outflow characteristics can be put in the convenient tabular form:

Stage (m)	S (m³)	O (m³/s)	$2S/\Delta t + O$ (m³/s)
5.0	0	0	0
5.5	694	0	0.771
6.0	1,525	0.647	2.34
6.5	2,507	1.83	4.62
7.0	3,652	3.36	7.42
7.5	4,973	5.18	10.7
8.0	6,484	7.23	14.4

and the computations in the routing procedure are summarized in the following table:

(1) Time (min)	(2) I (m³/s)	(3) $2S/\Delta t - O$ (m³/s)	(4) $2S/\Delta t + O$ (m³/s)	(5) O (m³/s)
0	0	0	0	0
30	2.4	1.04	2.4	0.68
60	5.6	0.52	9.04	4.3
90	3.4	0.46	9.52	4.5
120	2.8	0.78	6.66	2.9
150	2.4	1.41	5.98	2.6
180	2.2	0.83	6.01	2.6
210	1.8	0.95	4.83	1.9
240	1.5	0.97	4.25	1.6
270	1.2	0.99	3.67	1.3
300	1.0	1.01	3.19	1.1
330	0.56	1.03	2.57	0.77
360	0.34	0.97	1.93	0.48
390	0	0.87	1.31	0.22
420	0	0.79	0.87	0.04
450	0	0.77	0.79	0.01
480	0	0.77	0.79	0

The sequence of computations begins with tabulating the inflow hydrograph in columns 1 and 2, and the first row of the table ($t = 0$ min) is given by the initial conditions. At $t = 30$ min, $I_1 = 0$ m³/s, $I_2 = 2.4$ m³/s, $2S_1/\Delta t - O_1 = 0$ m³/s, and Equation 6.138 gives

$$\frac{2S_2}{\Delta t} + O_2 = (I_1 + I_2) + \left(\frac{2S_1}{\Delta t} - O_1\right) = 0 + 2.4 + 0 = 2.4 \text{ m}^3\text{/s}$$

This value of $2S/\Delta t + O$ is written in column 4 (at $t = 30$ min), the corresponding outflow, O, from the detention basin is interpolated from the reservoir properties as 0.68 m³/s, and this value is written in column 5 (at $t = 30$ min), and the corresponding value of $2S/\Delta t - O$ is given by

$$\frac{2S}{\Delta t} - O = \left(\frac{2S}{\Delta t} + O\right) - 2O = 2.4 - 2(0.68) = 1.04 \text{ m}^3\text{/s}$$

which is written in column 3 (at $t = 30$ min). The variables at $t = 30$ min are now all known, and form a new set of initial conditions to repeat the computation procedure for subsequent time steps.

The outflow hydrograph, given in column 5, indicates a peak discharge of 4.5 m³/s from the detention basin. This is a reduction of 20% from the peak catchment runoff (= basin inflow) of 5.6 m³/s.   ■

6.5.1.2   Muskingum Method

The Muskingum method for flood routing, developed for the Muskingum Conservancy District (Ohio) flood control study in the 1930s (McCarthy, 1938), is used for routing flows in drainage channels, including rivers and streams, and is the most widely used method of hydrologic stream channel routing (Ponce, 1989). The Muskingum method approximates the storage volume in a channel by a combination of *prism storage* and *wedge storage*, as illustrated in Figure 6.32 for the case in which the inflow exceeds the outflow. A *negative wedge* is produced when the outflow exceeds the inflow, and it occurs as the water level recedes in the channel. The prism storage is a volume of constant cross-section corresponding to uniform flow in a prismatic channel; the wedge storage is generated by the passage of the flow hydrograph.

Assuming that the flow area is directly proportional to the channel flow, the volume of prism storage can be expressed as KO, where K is the travel time through the reach and O is the flow through the prism. The wedge storage can therefore be approximated by $KX(I - O)$, where X is a weighting factor in the range $0 \leq X \leq 0.5$. For reservoir-

Figure 6.32 ■ Muskingum Storage Approximation

Source: Chow, Ven Te et al. *Applied Hydrology.* The McGraw-Hill Companies, New York. Copyright © 1988. Reprinted by permission of The McGraw-Hill Companies.

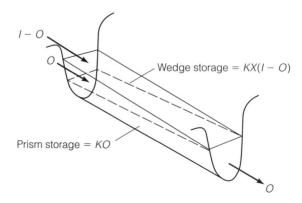

type storage, $X = 0$; for a full wedge, $X = 0.5$. In natural streams, X is usually between 0 and 0.3, with a typical value near 0.2 (Chow et al., 1988). The total storage, S, between the inflow and outflow sections is therefore given by

$$S = KO + KX(I - O) \tag{6.139}$$

or

$$S = K[XI + (1 - X)O] \tag{6.140}$$

Applying Equation 6.140 at time increments of Δt, the storage, S, in the channel between the inflow and outflow sections at times $j\Delta t$ and $(j + 1)\Delta t$ can be written as

$$S_j = K[XI_j + (1 - X)O_j] \tag{6.141}$$

and

$$S_{j+1} = K[XI_{j+1} + (1 - X)O_{j+1}] \tag{6.142}$$

respectively, and the change in storage over Δt is therefore given by

$$S_{j+1} - S_j = K\{[XI_{j+1} + (1 - X)O_{j+1}] - [XI_j + (1 - X)O_j]\} \tag{6.143}$$

The discretized form of the continuity equation, Equation 6.137, can be written as

$$S_{j+1} - S_j = \frac{(I_j + I_{j+1})}{2}\Delta t - \frac{(O_j + O_{j+1})}{2}\Delta t \tag{6.144}$$

Combining Equations 6.143 and 6.144 yields the routing expression

$$\boxed{O_{j+1} = C_1 I_{j+1} + C_2 I_j + C_3 O_j} \tag{6.145}$$

where C_1, C_2, and C_3 are given by

$$C_1 = \frac{\Delta t - 2KX}{2K(1 - X) + \Delta t} \tag{6.146}$$

$$C_2 = \frac{\Delta t + 2KX}{2K(1 - X) + \Delta t} \tag{6.147}$$

$$C_3 = \frac{2K(1 - X) - \Delta t}{2K(1 - X) + \Delta t} \tag{6.148}$$

and it is apparent that $C_1 + C_2 + C_3 = 1$.

The routing equation, Equation 6.145, is applied to a given inflow hydrograph, I_j ($j = 1, J$), and initial outflow, O_1, to calculate the outflow hydrograph, O_j ($j = 2, J$), at a downstream section. The constants in the routing equation, C_1, C_2, and C_3 are expressed in terms of the routing time step, Δt, and the channel parameters K and X. Ideally, subreaches selected for channel routing should be such that the travel time, K, through each subreach is equal to the time step, Δt. If this is not possible, then Δt should be selected such that

$$\Delta t \geq 2KX \tag{6.149}$$

This requirement prevents negative coefficients and instabilities in the routing procedure (ASCE, 1996). Viessman and Lewis (1996) recommend that Δt be assigned any convenient value between $K/3$ and K.

The proportionality factor, K, is a measure of time of travel of a flood wave through the channel reach, and X in natural streams is between 0 and 0.3, with a mean value near 0.2. Values of X in excess of 0.5 produce hydrograph amplification, which is unrealistic. If measured inflow and outflow hydrographs are available for the channel reaches, then K and X can be estimated using the rearranged routing equation (Equation 6.145)

$$K = \frac{0.5\Delta t[(I_{j+1} + I_j) - (O_j + O_{j+1})]}{X(I_{j+1} - I_j) + (1 - X)(O_{j+1} - O_j)} \tag{6.150}$$

where I_j ($j = 1, J$), O_j ($j = 1, J$), and Δt are known from field measurements. For assumed values of X, the numerator is plotted versus the denominator for each time step. The resulting curve typically has the form of a loop, and the value of X that yields a loop that is closest to a straight line is the value of X for the channel reach, and the slope of the line is the corresponding value of K. Furthermore, since K is defined as the time required for an incremental flood wave to move through the channel reach, K can also be estimated by the observed time for the hydrograph peak to move through the channel reach. In the absence of measured data, K is usually taken as the estimated mean travel time between the inflow and ouflow section and X is taken as 0.2.

A more refined method for estimating K and X was proposed by Cunge (1967), where K is estimated by the relation

$$K = \frac{L}{c} \tag{6.151}$$

where L is the distance between the inflow and outflow section and c is the wave celerity defined by

$$c = \frac{5}{3}v \tag{6.152}$$

where v is the average velocity at the bankfull discharge. The coefficient $\frac{5}{3}$ in Equation 6.152 is derived from the Manning equation (Viessman and Lewis, 1996). The value of X suggested by Cunge (1967) is

$$X = \frac{1}{2}\left(1 - \frac{q_o}{S_o c L}\right) \tag{6.153}$$

where q_o is the flow per unit width, calculated at the peak flow rate and S_o is the slope of the channel. In cases where the Muskingum method is applied using values of K and X estimated from Equations 6.151 to 6.153, the approach is referred to as the *Muskingum-Cunge method*. According to Chow et al. (1988), great accuracy in determining X may not be necessary because the results of the Muskingum method are relatively insensitive to this parameter. In cases where there are significant backwater effects and where the channel is either very steep or very flat, then dynamic effects may be significant and hydraulic routing is preferred over hydrologic routing.

■ **Example 6.33**

The flow hydrograph at a channel section is given by:

Time (min)	Flow (m³/s)
0	10.0
30	10.0
60	25.0
90	45.0
120	31.3
150	27.5
180	25.0
210	23.8
240	21.3
270	19.4
300	17.5
330	16.3
360	13.5
390	12.1
420	10.0
450	10.0
480	10.0

Use the Muskingum method to estimate the hydrograph 1,200 m downstream from the channel section. Assume that $X = 0.2$ and $K = 40$ min.

Solution

In accordance with Equation 6.149, select Δt such that

$$\Delta t \geq 2KX = 2(40)(0.2) = 16 \text{ min}$$

According to Viessman and Lewis (1996)

$$\frac{K}{3} \leq \Delta t \leq K$$

or

$$\frac{40}{3} \text{ min} \leq \Delta t \leq 40 \text{ min}$$

Taking $\Delta t = 30$ min, the Muskingum constants are given by Equations 6.146 to 6.148 as

$$C_1 = \frac{\Delta t - 2KX}{2K(1 - X) + \Delta t} = \frac{30 - 2(40)(0.2)}{2(40)(1 - 0.2) + 30} = 0.149$$

$$C_2 = \frac{\Delta t + 2KX}{2K(1 - X) + \Delta t} = \frac{30 + 2(40)(0.2)}{2(40)(1 - 0.2) + 30} = 0.489$$

$$C_3 = \frac{2K(1 - X) - \Delta t}{2K(1 - X) + \Delta t} = \frac{2(40)(1 - 0.2) - 30}{2(40)(1 - 0.2) + 30} = 0.362$$

These results can be verified by taking $C_1 + C_2 + C_3 = 0.149 + 0.489 + 0.362 = 1$. The Muskingum routing equation, Equation 6.145, is therefore given by

$$O_{j+1} = C_1 I_{j+1} + C_2 I_j + C_3 O_j = 0.149 I_{j+1} + 0.489 I_j + 0.362 O_j$$

This routing equation is applied repeatedly to the given inflow hydrograph, and the results are as follows:

(1) Time (min)	(2) I (m^3/s)	(3) O (m^3/s)
0	10.0	10.0
30	10.0	10.0
60	25.0	12.2
90	45.0	23.4
120	31.3	35.1
150	27.5	32.1
180	25.0	28.8
210	23.8	26.2
240	21.3	24.3
270	19.4	22.1
300	17.5	20.1
330	16.3	18.3
360	13.5	16.6
390	12.1	14.4
420	10.0	12.6
450	10.0	11.0
480	10.0	10.3
510	10.0	10.1
540	10.0	10.1
570	10.0	10.0
600	10.0	10.0

The computations begin with the inflow hydrograph in columns 1 and 2 and the initial outflow, O, at $t = 0$ min. At $t = 30$ min, $I_1 = 10$ m^3/s, $I_2 = 10$ m^3/s, $O_1 = 10$ m^3/s, and the Muskingum method gives the outflow, O_2, at $t = 30$ min as

$$O_{j+1} = 0.149I_{j+1} + 0.489I_j + 0.362O_j = 0.149(10) + 0.489(10) + 0.362(10) = 10 \text{ m}^3/\text{s}$$

This procedure is repeated for subsequent times, and the outflow hydrograph is shown in column 3.   ■

6.5.2 ■ Hydraulic Routing

In hydraulic routing, the one-dimensional continuity and momentum equations are solved along the channel continuum. This approach is in contrast to hydrologic routing, which is based on the finite-difference approximation to the continuity equation plus a second equation that relates the channel storage to the flowrate. The one-dimensional continuity and momentum equations in open channels, first presented by Barre de Saint-Venant (1871), are commonly called the Saint-Venant equations, and can be written as

$$\frac{\partial Q}{\partial x} + \frac{\partial A}{\partial t} = 0 \qquad \text{(1-D continuity)} \tag{6.154}$$

$$\frac{1}{A}\frac{\partial Q}{\partial t} + \frac{1}{A}\frac{\partial}{\partial x}\left(\frac{Q^2}{A}\right) + g\frac{\partial y}{\partial x} - g(S_o - S_f) = 0 \qquad \text{(1-D momentum)} \tag{6.155}$$

where Q is the volumetric flowrate, x is the distance along the channel, t is time, A is the cross-sectional area, g is gravity, y is the depth of flow, S_o is the slope of the channel, and S_f is the slope of the energy grade line, which can be estimated using the Manning equation. The assumptions inherent in the Saint-Venant equations are: (1) The flow is one-dimensional (i.e., the depth and velocity vary only in the longitudinal direction of the channel); (2) the flow is gradually varied along the channel (i.e., the vertical pressure distribution is hydrostatic); (3) the longitudinal axis of the channel is straight; (4) the bottom slope of the channel is small and the channel bed is fixed (i.e., the effects of scour and deposition are negligible); and (5) the friction coefficients for steady uniform flow are applicable.

In some cases, inertial and pressure effects are much smaller than friction and gravity effects, and Equation 6.155 is closely approximated by the simple relation

$$S_o = S_f \tag{6.156}$$

A model that describes the channel flow by Equations 6.154 and 6.156 is called a *kinematic-wave model* and is frequently used for overland-flow routing. With the exception of a few simple cases, the Saint-Venant equations, given by Equations 6.154 and 6.155, cannot be solved analytically; they are usually solved with numerical models that use implicit or explicit finite-difference algorithms (Amein and Fang, 1969) or the method of characteristics (Amein, 1966).

■ **Example 6.34**

Write a simple finite-difference model for solving the Saint-Venant equations.

Solution

The dependent variables in the Saint-Venant equations are the flowrate, Q, and the depth of flow, y; and the independent variables are the distance along the channel, x, and the time, t. In a finite-difference approximation, the solution is calculated at discrete intervals, Δx, along the x-axis, and at discrete intervals, Δt, in time. Denoting the distance index by i, and the time index by j, then the following convention for specifying variables at index locations is adopted:

$$Q_{i,j} = Q(i\Delta x, j\Delta t)$$
$$y_{i,j} = y(i\Delta x, j\Delta t)$$
$$A_{i,j} = A(y_{i,j})$$

where $A_{i,j}$ is the flow area corresponding to $y_{i,j}$.

The (explicit) finite-difference approximation to the one-dimensional continuity equation (Equation 6.154) can be written as

$$\frac{Q_{i+1,j} - Q_{i,j}}{\Delta x} + \frac{A_{i,j+1} - A_{i,j}}{\Delta t} = 0$$

and the (explicit) finite-difference approximation to the one-dimensional momentum equation (Equation 6.155) can be written as

$$\frac{1}{A_{i,j}}\frac{Q_{i,j+1} - Q_{i,j}}{\Delta t} + \frac{1}{A_{i,j}}\frac{1}{\Delta x}\left(\frac{Q_{i+1,j}^2}{A_{i+1,j}} - \frac{Q_{i,j}^2}{A_{i,j}}\right) + g\frac{y_{i+1,j} - y_{i,j}}{\Delta x} - g(S_o - S_{i,j}) = 0$$

where $S_{i,j}$ is the friction slope calculated using the Manning equation with $Q_{i,j}$ and $y_{i,j}$. These finite-difference equations, along with given initial and boundary conditions, can be used to solve for Q and y at the discrete points $i\Delta x$ and $j\Delta t$ along the x and t axes, respectively.

It should be noted that this numerical scheme requires some constraints on the choice of the discrete intervals Δx and Δt for the scheme to produce accurate and stable solutions (Ames, 1977).   ■

6.6   Water Quality Models

The quality of surface runoff is generally site-specific and can only be predicted with order-of-magnitude accuracy. Field studies have demonstrated that, in many cases, urban stormwater runoff is a major source of pollution in receiving waters. Typical characteristics of urban stormwater runoff are shown in Table 6.23, which includes seven conventional pollutants and three metals. These pollutants originate from a variety of sources, including motor vehicles, construction activities, chemicals from lawns, solid waste litter, and fecal droppings from animals. The pollutants listed in Table 6.23 were selectively drawn from the following categories: solids, oxygen-consuming constituents, nutrients, and heavy metals. Most pollutants in stormwater runoff have a strong affinity to suspended solids (TSS), and the removal of TSS often removes many other pollutants found in urban stormwater (Urbonas and Stahre, 1993). According to ASCE (1992), for general-purpose planning the concentrations of pollutants in runoff from large residential and commercial areas can be assumed to be in the range of the concentrations shown in Table 6.23, with central business districts usually having the highest pollutant loadings per unit area. In the United States, water-quality criteria for receiving waterbodies are set by the states in accordance with minimum levels specified by the federal government. The deleterious effect of discharging stormwater directly into open waters has prompted the United States to require all municipalities with populations exceeding 100,000 to obtain stormwater discharge permits. Such permits are called National Pollutant Discharge Elimination System (NPDES) permits.

The water quality in drainage channels and conduits, including storm sewers, is determined by both runoff-related sources and nonstormwater sources. Nonstormwater sources of pollution in storm sewers include illicit connections, interactions with sewage systems, improper disposal, spills, malfunctioning septic tanks, and infiltration of contaminated water. The removal of nonstormwater sources of pollutants can result in a dramatic improvement in the quality of water discharged from storm sewers. Runoff-related sources of pollution in storm sewers originate from two main sources:

■ **Table 6.23**
Typical Water Quality of Runoff from Residential and Commercial Areas

Constituent	Typical concentrations
Total suspended solids (TSS)	180–548 mg/L
Biochemical oxygen demand (BOD)	12–19 mg/L
Chemical oxygen demand (COD)	82–178 mg/L
Total phosphorous, as P (TP)	0.42–0.88 mg/L
Soluble phosphorous, as P (SP)	0.15–0.28 mg/L
Total Kjeldahl nitrogen, as N (TKN)	1.90–4.18 mg/L
Nitrite + Nitrate, as N (NO2+NO3)	0.86–2.2 mg/L
Total copper (Cu)	43–118 μg/L
Total lead (Pb)	182–443 μg/L
Total zinc (Zn)	202–633 μg/L

Source: USEPA (1983).

(1) land surface, and (2) catch basins. On land surfaces, pollutants accumulate much more rapidly on impervious areas than they do on pervious areas. In urban areas, the primary areas of accumulation are streets and gutters. Initial planning estimates of pollutant levels in stormwater effluents can be estimated using either the concentrations given in Table 6.23 or regression equations for pollutant loads that relate the amount of pollutants in the runoff to the rainfall amount. The most widely used regression equations for pollutant loads are incorporated in the USGS model and the EPA model.

6.6.1 ■ USGS Model

The U.S. Geological Survey (Driver and Tasker, 1988; 1990) analyzed the pollutant loads resulting from 2,813 storms at 173 urban stations in 13 metropolitan areas and developed empirical equations to estimate the annual pollutant loads in terms of the rainfall and catchment characteristics. Pollutant loads refer to the amount of pollutant per unit time, usually a year and, for water managers, pollutant loads are usually of greater interest than runoff concentrations in individual storms (Black, 1996). The USGS regression equations for annual load have the form

$$Y = 0.454(N)(BCF)10^{[a+b\sqrt{(DA)}+c(IA)+d(MAR)+e(MJT)+f(X2)]} \tag{6.157}$$

where Y is the pollutant load (kg) for the pollutants listed in Table 6.24, N is the average number of storms in a year, BCF is the bias correction factor given in Table 6.25, DA is the total contributing drainage area (ha), IA is the impervious area as a percentage of the total contributing area (%), MAR is the mean annual rainfall (cm), MJT is the mean minimum January temperature (°C), and $X2$ is an indicator variable that is equal to 1.0 if commercial land use plus industrial land use exceeds 75% of the total contributing drainage area and is zero otherwise. A storm is defined as a rainfall event with at least 0.05 in. (1.3 mm) of rain, and storms are separated by at least 6 consecutive hours of zero rainfall. The regression constants in Equation 6.157 depend on the type of pollutant and are given in Table 6.25. As in the case of most empirical relationships, application of Equation 6.157 is not recommended beyond the ranges of variables used in developing the equation. The ranges of variables used in developing Equation 6.157 are given in Table 6.26.

■ Example 6.35

The quality of surface runoff from a 100-ha planned development in Miami, Florida, is to be assessed. The planned development will be 40% impervious, with 50% commer-

■ Table 6.24
Pollutants in USGS Formula

Y	Pollutant
COD	chemical oxygen demand
SS	total suspended solids
DS	dissolved solids
TN	total nitrogen
AN	total ammonia plus organic nitrogen
TP	total phosphorous
DP	dissolved phosphorous
CU	total recoverable copper
PB	total recoverable lead
ZN	total recoverable zinc

■ **Table 6.25**
Regression Constants
for USGS Pollutant
Load Equation

Y	a	b	c	d	e	f	BCF
COD	1.1174	0.1427	0.0051	—	—	—	1.298
SS	0.5926	0.0988	—	0.0104	−0.0535	—	1.521
DS	1.1025	0.1583	—	—	−0.0418	—	1.251
TN	−0.2433	0.1018	0.0061	—	—	−0.4442	1.345
AN	−1.4002	0.1002	0.0064	0.00890	−0.0378	−0.4345	1.277
TP	−2.0700	0.1294	—	0.00921	−0.0383	—	1.314
DP	−1.3661	0.0867	—	—	—	—	1.469
CU	−1.9336	0.1136	—	—	−0.0254	—	1.403
PB	−1.9679	0.1183	0.0070	0.00504	—	—	1.365
ZN	−1.6302	0.1267	0.0072	—	—	—	1.322

Source: Driver and Tasker (1990).

cial and industrial use. According to Winsberg (1990), there are typically 84 storms per year in Miami, the mean annual rainfall is 147 cm, and the mean minimum January temperature is 14.4°C. Estimate the annual load of total phosphorous contained in the runoff.

Solution

From the given data: $N = 84$ storms, $BCF = 1.314$ (Table 6.25 for TP), $DA = 100$ ha, $IA = 40\%$, $MAR = 147$ cm, $MJT = 14.4°C$, and $X2 = 0$ (since commercial plus industrial use is less than 75%). These variables are within the ranges given in Table 6.26, and therefore the USGS regression equation, Equation 6.157 can be used. The regression constants taken from Table 6.25 (for TP) are $a = -2.0700$, $b = 0.1294$, $d = 0.00921$, $e = -0.0383$, and $c = f = 0$. Substituting data into Equation 6.157 gives

$$Y = 0.454(N)(BCF)10^{[a+b\sqrt{(DA)}+c(IA)+d(MAR)+e(MJT)+f(X2)]}$$

$$= 0.454(84)(1.314)10^{[-2.0700+0.1294\sqrt{(100)}+0.00921(147)-0.0383(14.4)]}$$

$$= 53 \text{ kg}$$

The annual load of total phosphorous contained in the runoff from the planned development is expected to be on the order of 53 kg.   ■

■ **Table 6.26**
Ranges of Variables Used in
Developing USGS Equation

Y	DA (ha)	IA (%)	MAR (cm)	MJT °C
COD	4.9–183	4–100	21.3–157.5	−16.0–14.8
SS	4.9–183	4–100	21.3–125.4	−16.0–10.1
DS	5.2–117	19–99	26.0–95.5	−11.4–2.1
TN	4.9–215	4–100	30.0–157.5	−16.0–14.8
AN	4.9–183	4–100	21.3–157.5	−16.0–14.8
TP	4.9–215	4–100	21.3–157.5	−16.0–14.8
DP	5.2–183	5–99	21.3–117.3	−11.8–2.1
CU	3.6–215	6–99	21.3–157.5	−9.3–14.8
PB	4.9–215	4–100	21.3–157.5	−16.0–14.8
ZN	4.9–215	13–100	21.3–157.5	−11.4–14.8

Source: Driver and Tasker (1990).

6.6.2 ■ EPA Model

The U.S. Environmental Protection Agency (Heany et al., 1977) has also developed a set of empirical formulae that can be used to estimate the average annual pollutant loads in urban stormwater runoff. The empirical equation for urban areas having separate storm sewer systems is given by

$$M_s = 0.0442\alpha P f s \qquad (6.158)$$

where M_s is the amount of pollutant (kg) generated per hectare of land per year, α is a pollutant loading factor given in Table 6.27 for various pollutants (BOD_5 = five-day biochemical oxygen demand, SS = suspended solids, VS = volatile solids, PO_4 = phosphate, and N = nitrogen), P is the precipitation (cm/year), f is a population density function, and s is a street sweeping factor. The population density function, f, for residential areas is given by

$$f = 0.142 + 0.134D^{0.54} \qquad (6.159)$$

where D is the population density in persons per hectare. For commercial and industrial areas, the population density function, f, is equal to 1.0; for other types of developed areas, such as parks, cemeteries, and schools, f is taken as equal to 0.142. The street sweeping factor, s, depends on the sweeping interval, N_s (days), where if $N_s > 20$ days then $s = 1.0$, and if $N_s \leq 20$ days, then s is given by

$$s = \frac{N_s}{20} \qquad (6.160)$$

The average annual pollutant concentration can be derived from the annual pollutant load by dividing the annual pollutant load by the annual runoff. Heany and colleagues (1977) suggested that the annual runoff, R (cm), can be estimated using the formula

$$R = \left[0.15 + 0.75 \left(\frac{I}{100} \right) \right] P - 3.004 d^{0.5957} \qquad (6.161)$$

where I is the imperviousness of the catchment (%), P is the annual rainfall (cm), and d is the depression storage (cm), which can be estimated using the relation

$$d = 0.64 - 0.476 \left(\frac{I}{100} \right) \qquad (6.162)$$

■ **Table 6.27**
Pollutant Loading Factor, α

Land Use	BOD_5	SS	VS	PO_4	N
Residential	0.799	16.3	9.4	0.0336	0.131
Commercial	3.200	22.2	14.0	0.0757	0.296
Industrial	1.210	29.1	14.3	0.0705	0.277
Other	0.113	2.7	2.6	0.0099	0.060

Source: Heany et al. (1977).

■ **Example 6.36**

A 100-ha residential development has a population density of 15 persons per hectare, streets are swept every two weeks, and the area is 40% impervious. The average annual rainfall is 147 cm. Estimate the annual load of phosphate (PO_4) expected in the runoff.

Solution

From the given data: $\alpha = 0.0336$ (Table 6.27: PO_4, Residential), $P = 147$ cm, $D = 15$ persons/ha, $N_s = 14$ days, Equation 6.159 gives

$$f = 0.142 + 0.134D^{0.54} = 0.142 + 0.134(15)^{0.54} = 0.72$$

and Equation 6.160 gives

$$s = \frac{N_s}{20} = \frac{14}{20} = 0.7$$

Substituting these data into the EPA model, Equation 6.158, gives

$$M_s = 0.0442\alpha Pfs = 0.0442(0.0336)(147)(0.72)(0.7) = 0.11 \text{ kg/ha}$$

For a 100-ha development, the annual phosphate load is 100 ha × 0.11 kg/ha = 11 kg.

The average concentration in the runoff is obtained by dividing the annual load (= 11 kg) by the annual runoff. Equation 6.162 gives the depression storage, d, as

$$d = 0.64 - 0.476 \left(\frac{I}{100} \right)$$

$$= 0.64 - 0.476 \left(\frac{40}{100} \right)$$

$$= 0.45 \text{ cm}$$

and Equation 6.161 gives the runoff, R, as

$$R = \left[0.15 + 0.75 \left(\frac{I}{100} \right) \right] P - 3.004d^{0.5957}$$

$$= \left[0.15 + 0.75 \left(\frac{40}{100} \right) \right] (147) - 3.004(0.45)^{0.5957}$$

$$= 64 \text{ cm}$$

Since the catchment area is 100 ha = 10^6 m^2, then the volume, V, of annual runoff is 0.64×10^6 m^3 and the average concentration, c, is given by

$$c = \frac{11 \text{ kg}}{0.64 \times 10^6 \text{ m}^3} = 1.7 \times 10^{-5} \text{ kg/m}^3 = 0.017 \text{ mg/L} \qquad ■$$

6.7 Design of Stormwater Management Systems

Stormwater management systems are designed to control the quantity, quality, timing, and distribution of runoff resulting from storm events. Other objectives in the design of stormwater management systems include: erosion control, reuse storage, and ground-water recharge. A typical urban stormwater management system has two distinct components: a minor system and a major system. The minor system consists of storm sewers that route the design runoff to receiving waters, and it is typically designed to handle runoff events with return periods of 2 to 10 years. Typical return periods for various types of areas are given in Table 6.4 on page 314. The major system consists of the above-ground conveyance routes that transport stormwater from larger runoff events with return periods from 25 to 100 years. Major urban conveyance systems that are covered by the National Flood Insurance Program (in the United States) are typically designed for a runoff with a 100-year return period.

6.7.1 ■ Minor System

Most minor systems are designed for urban environments, and the principal hydraulic elements of the minor system are shown in Figure 6.33. The (minor) stormwater-management system collects surface runoff via inlets in road pavements, and the surface runoff is routed to a treatment unit and/or receiving waterbody, usually through underground pipes called *storm sewers*. In some cases, the surface runoff is discharged directly into a receiving body of water such as a drainage canal. In some older U.S. and European cities, storm and sanitary sewers are combined into a single system; these systems are called *combined-sewer systems*.

6.7.1.1 Storm Sewers

Storm sewers are typically located a short distance behind the curb, or in the roadway near the curb. These sewers should be straight between manholes (where possible); where curves are necessary to conform to street layout, the radius of curvature should not be less than 30 m. There should be at least 0.9 m (3 ft) of cover over the crowns of the sewer pipes to prevent excessive loading on the pipe, and crossings with underground utilities should be avoided whenever possible, but, if necessary, should be at an angle greater than 45°. Manholes, also called *clean-out structures* (ASCE, 1992), are placed along the sewer pipeline to provide convenient access for inspection, maintenance, and repair of storm drainage systems; they are normally located at the junctions of sewers and at changes in grade or alignment. Typical manhole spacings depend on the pipe sizes and are given in Table 6.28.

The rational method is commonly used to determine the peak flows to be handled by storm sewers. The flow calculations proceed from the most upstream pipe in the system and, with each new inlet (inflow), the pipe immediately downstream of the inlet is expected to carry the runoff from a storm of duration equal to the time of concentration of the contributing area. Two separate contributing areas must be considered: (1) the entire contributing area, and (2) the impervious area directly connected to the inlets. Directly connected impervious areas must be considered separately since they have considerably shorter times of concentration than the entire catchment, resulting in higher design-rainfall intensities and possibly higher peak runoff rates than the entire catchment. It should be noted that the minimization of a directly connected impervious area is by far the most effective method of controlling the quality of surface runoff (ASCE, 1992). Very short times of concentration will lead to unrealistically high rainfall intensities, and a minimum time of concentration such as 5 min is generally adopted. The procedure for calculating the design flows in storm sewers is illustrated in the following example.

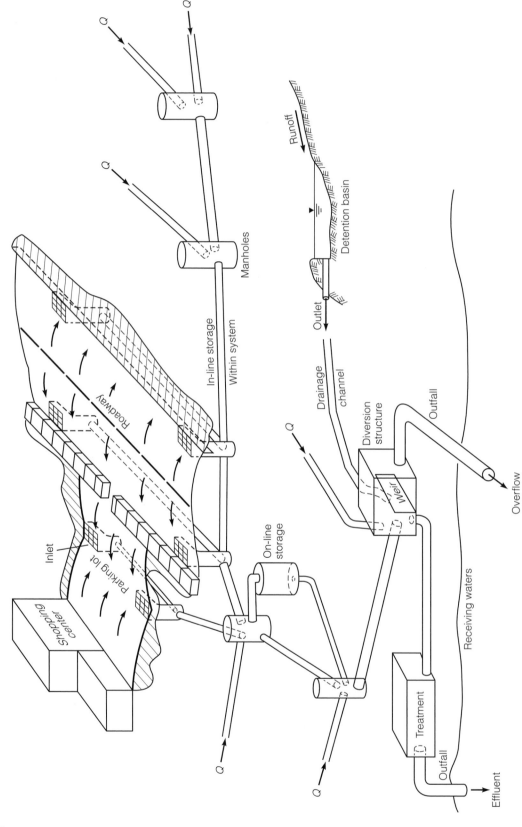

Figure 6.33 ■ Principal Hydraulic Elements in (Minor) Stormwater Management System
Source: ASCE, 1992. Design and Construction of Urban Stormwater Management Systems. p. 114. Reprinted by permission of ASCE.

■ **Table 6.28**
Typical Manhole Spacings

Pipe Size	Maximum Spacing
38 cm or less	122 m
46 cm to 91 cm	152 m
107 cm or greater	183 m

Source: Boulder County (1984).

■ **Example 6.37**

Consider the two inlets and two pipes shown in Figure 6.34. Catchment A has an area of 1 ha and is 50% impervious; catchment B has an area of 2 ha and is 10% impervious. All impervious areas are directly connected to the sewer inlets. The runoff coefficient, C; length of overland flow, L; roughness coefficient, n; and average slope, S_0, of the pervious and impervious surfaces in both catchments are given in Table 6.29. The design storm has a return period of 10 years. The 10-year IDF curve can be approximated by

$$i = \frac{7620}{t + 36}$$

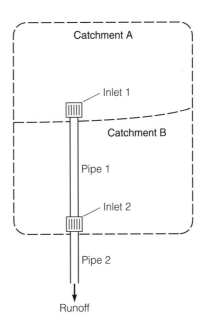

Figure 6.34 ■ Computation of Peak Inlet and Pipe Flows
Adapted from ASCE (1992).

■ **Table 6.29**
Catchment Characteristics

Catchment	Surface	C	L (m)	n	S_0
A	pervious	0.2	80	0.2	0.01
	impervious	0.9	60	0.1	0.01
B	pervious	0.2	140	0.2	0.01
	impervious	0.9	65	0.1	0.01

where i is the average rainfall intensity in mm/h and t is the duration of the storm in minutes. Calculate the peak flows to be handled by the inlets and pipes.

Solution

Using the given IDF curve, the effective rainfall rate, i_e, is given by the rational formula as

$$i_e = Ci = C\frac{7620}{t_c + 36} \tag{6.163}$$

where C is the runoff coefficient. The storm duration, t, is taken to be equal to the time of concentration, t_c, given by Equation 6.84 as

$$t_c = 6.99\frac{(nL)^{0.6}}{i_e^{0.4}S_0^{0.3}} \tag{6.164}$$

Simultaneous solution of Equations 6.163 and 6.164 using the catchment characteristics in Table 6.29 leads to the following times of concentration, t_c:

Catchment	Surface	t_c (min)
A	pervious area	46
	impervious area	11
B	pervious	71
	impervious	12

Consider now the flows at specific locations.

Inlet 1 and Pipe 1: When the entire catchment A is contributing, the time of concentration is 46 min (this is the time for both pervious and impervious areas to be fully contributing), the average rainfall rate, i, from the IDF curve is 92.9 mm/h ($= 2.58 \times 10^{-5}$ m/s), the weighted average runoff coefficient, $\bar{C}$ is given by

$$\bar{C} = 0.5(0.9) + 0.5(0.2) = 0.55$$

Since the area of the catchment is 1 ha ($= 10,000$ m^2), then the peak runoff rate, Q_p, from the catchment is given by the rational formula as

$$Q_p = \bar{C}iA = (0.55)(2.58 \times 10^{-5})(10000) = 0.142 \text{ m}^3/\text{s}$$

Considering only the impervious portion of the catchment, the time of concentration is 11 min, the average rainfall rate, i, from the IDF curve is 162 mm/h ($= 4.50 \times 10^{-5}$ m/s), the runoff coefficient, C, is 0.9, the contributing area is 0.5 ha ($= 5,000$ m^2), and the peak runoff rate, Q_p, is given by

$$Q_p = CiA = (0.9)(4.50 \times 10^{-5})(5000) = 0.203 \text{ m}^3/\text{s}$$

The calculated peak runoff from the directly connected impervious area is greater than the calculated runoff from the entire area, and therefore the design flow to be handled by Inlet 1 and Pipe 1 is controlled by the directly connected impervious area and is equal to 0.203 m^3/s.

Inlet 2: When the entire catchment B is contributing, the time of concentration is 71 min, the average rainfall rate, i, from the IDF curve is 71.2 mm/h ($= 1.98 \times 10^{-5}$ m/s), the weighted average runoff coefficient, $\bar{C}$ is given by

$$\bar{C} = 0.1(0.9) + 0.9(0.2) = 0.27$$

Since the area of the catchment is 2 ha ($= 20{,}000$ m^2), then the peak runoff rate, Q_p, from the catchment is given by the rational formula as

$$Q_p = \bar{C}iA = (0.27)(1.98 \times 10^{-5})(20000) = 0.107 \text{ m}^3/\text{s}$$

Considering only the impervious portion of the catchment, the time of concentration is 12 min, the average rainfall rate, i, from the IDF curve is 159 mm/h ($= 4.41 \times 10^{-5}$ m/s), the runoff coefficient, C, is 0.9, the contributing area is 0.2 ha ($= 2{,}000$ m^2), and the peak runoff rate, Q_p, is given by

$$Q_p = CiA = (0.9)(4.41 \times 10^{-5})(2000) = 0.079 \text{ m}^3/\text{s}$$

The calculated peak runoff from the directly connected impervious area is less than the calculated runoff from the entire area, and therefore the design flow to be handled by Inlet 2 is controlled by the entire catchment and is equal to 0.107 m^3/s.

Pipe 2: First consider the case where the entire tributary area of 3 ha ($= 30{,}000$ m^2) is contributing runoff to pipe 2. The time of concentration of catchment A is equal to 46 min plus the time of flow in pipe 1, which, in lieu of hydraulic calculations, can be taken as 2 min. Therefore, the time of concentration of catchment A is 48 min. The time of concentration of catchment B is 71 min, and therefore the time of concentration of the entire tributary area to pipe 2 (including both catchments A and B) is equal to 71 min. The average rainfall intensity corresponding to this duration (from the IDF curve) is 71.2 mm/h ($= 1.98 \times 10^{-5}$ m/s); the area-weighted runoff coefficient, $\bar{C}$, is given by

$$\bar{C} = \frac{1}{3}[(0.5 + 0.2)(0.9) + (0.5 + 1.8)(0.2)] = 0.36$$

and the rational formula gives the peak runoff rate, Q_p, as

$$Q_p = CiA = (0.36)(1.98 \times 10^{-5})(30000) = 0.214 \text{ m}^3/\text{s}$$

Considering only the impervious portions of catchments A and B, the contributing area is 0.7 ha ($= 7{,}000$ m^2), the time of concentration is 13 min (equal to the time of concentration for Inlet 1 plus travel time of 2 min in pipe), the corresponding average rainfall intensity from the IDF curve is 156 mm/h ($= 4.32 \times 10^{-5}$), the runoff coefficient is 0.9, and the rational formula gives a peak runoff, Q_p, of

$$Q_p = CiA = (0.9)(4.32 \times 10^{-5})(7000) = 0.272 \text{ m}^3/\text{s}$$

Therefore, the peak runoff rate calculated by using the entire catchment is less than the peak runoff rate calculated by considering only the directly connected impervious portion of the tributary area. The design flow to be handled by pipe 2 is therefore controlled by the directly connected impervious area and is equal to 0.272 m^3/s. ■

Most storm sewers are sized to flow full at the design discharge, although where ground elevations are sufficient, a limited surcharge above the pipe crown may be permitted (ASCE, 1992). To prevent deposition of suspended materials, the minimum slope of the sewer should produce a velocity of at least 60 to 90 cm/s (2 to 3 ft/s) when the sewer is flowing full; to prevent scouring, the velocity should be less than 3 to 4.5 m/s (10 to 15 ft/s). The appropriate flow equation for sizing storm sewers is the Darcy-Weisbach equation, but it is also common practice to use the Manning equation. Caution should be exercised when using the Manning equation, which is only valid for rough-pipe (fully turbulent) flow and is only appropriate when the following condition is satisfied (French, 1985)

$$n^6 \sqrt{RS_o} \geq 10^{-13} \tag{6.165}$$

	Material	n
■ **Table 6.30**	Asbestos-cement pipe	0.011–0.015
Manning Coefficient in	Brick	0.013–0.017
Closed Conduits	Cast-iron pipe (cement lined and seal coated)	0.011–0.015
	Concrete (monolithic)	
	Smooth forms	0.012–0.014
	Rough forms	0.015–0.017
	Concrete pipe	0.011–0.015
	Corrugated-metal pipe (1.3 cm × 6.4 cm corrugations)	
	Plain	0.022–0.026
	Paved invert	0.018–0.022
	Spun asphalt lined	0.011–0.015
	Plastic pipe (smooth)	0.011–0.015
	Vitrified clay	
	Pipes	0.011–0.015
	Liner plates	0.013–0.017

Source: ASCE (1982). Reprinted by permission of the American Society of Civil Engineers.

where n is the Manning roughness coefficient, R is the hydraulic radius of the pipe (in meters), and S_o is the slope of the pipe. Roughness coefficients recommended for closed-conduit flow are given in Table 6.30.

In the hydraulic design of sewer pipes, the basic objective is to calculate the size and slope of the pipes that will carry the design flows at velocities that are within a specified range and with flow depths that are less than or equal to the diameter of the pipes. In most situations, it can be assumed that the flow is uniform and any losses other than pipe friction can be accounted for by assuming point losses at each manhole. In calculating the diameter, D, of storm sewers, the Manning equation can be put in the convenient form

$$D = \left(\frac{3.21 Q n}{\sqrt{S_o}} \right)^{\frac{3}{8}}$$
(6.166)

where Q is the design flowrate in m^3/s and D is in meters. If the Darcy-Weisbach equation is used in design, the convenient form is

$$D = \left(\frac{0.811 f Q^2}{g S_o} \right)^{\frac{1}{5}}$$
(6.167)

where f is the friction factor. The actual size of the pipe to be used should be the next larger commercial size than calculated using either Equation 6.166 or 6.167. Concrete, asbestos-cement, and clay pipes are commonly used for diameters between 100 mm (4 in.) and 60 cm (25 in.), with reinforced or prestressed concrete pipes commonly used for diameters larger than 60 cm (Novotny et al., 1989). It is generally recommended to choose a pipe diameter larger than 30 cm (12 in.) to prevent clogging and facilitate maintenance (Gribbin, 1997).

Junction and manhole losses usually have a significant effect on flows in sewers. Junctions are locations where two or more pipes join together and enter another pipe or channel, and these transitions need to be smooth to avoid high head losses. Conditions that promote turbulent flow and high losses include a large angle between the incoming pipes ($> 60°$), a large vertical distance between the pipes (> 15 cm between the two inverts), and the absence of a semicircular channel at the bottom of the manhole (ASCE, 1992). Manholes are generally placed at sewer junctions, as well as at other locations, to permit access to the entire pipeline system. The head loss, h_m, in manholes can be estimated using the equation

$$h_m = K_c \frac{V^2}{2g} \qquad (6.168)$$

where K_c is a head loss coefficient and V is the average velocity in the pipe. Head-loss coefficients for manholes with single inflow and outflow pipes aligned opposite to each other vary between 0.12 and 0.32, while the loss coefficients vary between 1.0 and 1.8 for inflow and outflow pipes at $90°$ to each other (ASCE, 1992).

■ Example 6.38

A concrete sewer pipe is to be laid parallel to the ground surface on a slope of 0.5% and is to be designed to carry 0.43 m^3/s of stormwater runoff. Estimate the required pipe diameter using (a) the Manning equation, and (b) the Darcy-Weisbach equation. If service manholes are placed along the pipeline, estimate the head loss at each manhole.

Solution

From the given data: $S_o = 0.005$, $Q = 0.43$ m^3/s, and $n = 0.013$ (Table 6.30, average for concrete pipe).

(a) The Manning equation (Equation 6.166) gives

$$D = \left[\frac{3.21Qn}{\sqrt{S_o}} \right]^{\frac{3}{8}} = \left[\frac{3.21(0.43)(0.013)}{\sqrt{0.005}} \right]^{\frac{3}{8}} = 0.60 \text{ m}$$

Use the limitation given by Equation 6.165 to check whether the Manning equation is valid:

$$n^6 \sqrt{RS_o} = (0.013)^6 \sqrt{(0.6/4)(0.005)} = 1.3 \times 10^{-13} \geq 10^{-13}$$

Therefore, the Manning equation is valid. Using a 60-cm pipe, the flow velocity, V, is given by

$$V = \frac{Q}{A} = \frac{0.43}{\frac{\pi}{4}(0.6)^2} = 1.52 \text{ m/s}$$

This velocity exceeds the minimum velocity to prevent sedimentation (0.60 to 0.90 m/s) and is less than the maximum velocity to prevent excess scour (3 to 4.5 m/s). According to the Manning equation, the pipe should have a diameter of 60 cm.

(b) The Darcy-Weisbach equation (Equation 6.167) gives

$$D = \left[\frac{0.811 f Q^2}{g S_o} \right]^{\frac{1}{5}}$$

where f depends on D via the Colebrook equation (Equation 3.35). Assuming $f = 0.020$ (a typical median value), then

$$D = \left[\frac{0.811(0.020)(0.43)^2}{(9.81)(0.005)}\right]^{\frac{1}{5}} = 0.57 \text{ m}$$

The assumed value of f must now be verified. A pipe diameter of 0.57 m carrying 0.43 m m³/s has a velocity, V, given by

$$V = \frac{Q}{A} = \frac{0.43}{\frac{\pi}{4}(0.57)^2} = 1.69 \text{ m/s}$$

The equivalent sand roughness, k_s, of concrete is in the range 0.3–3.0 mm (Table 3.1) and can be taken as $k_s = 1.7$ mm. Assuming that the temperature of the water is 20°C, then the kinematic viscosity, ν, is equal to 1.00×10^{-6} m/s², and the Reynolds number, Re, is given by

$$\text{Re} = \frac{VD}{\nu} = \frac{(1.69)(0.57)}{1.00 \times 10^{-6}} = 9.63 \times 10^5$$

According to the Jain approximation to the Colebrook equation (Equation 3.38):

$$\frac{1}{\sqrt{f}} = -2\log\left[\frac{k_s/D}{3.7} + \frac{5.74}{\text{Re}^{0.9}}\right] = -2\log\left[\frac{0.17/57}{3.7} + \frac{5.74}{(9.63 \times 10^5)^{0.9}}\right] = 6.16$$

which leads to

$$f = 0.0263$$

Since this value of f differs from the assumed value of 0.020, the computations must be repeated until the assumed value of f and the value of f derived from the Jain equation are the same. These computations are summarized in the following table:

Assumed f	D (m)	V (m/s)	Re	Computed f
0.020	0.57	1.69	9.63×10^5	0.0263
0.0263	0.60	1.52	9.12×10^5	0.0263

Therefore, the Darcy-Weisbach equation requires that the sewer pipe be at least 60 cm in diameter.

The service manholes placed along the pipe will each cause a head loss, h_m, where

$$h_m = K_c \frac{V^2}{2g}$$

For inflow and outflow pipes aligned opposite to each other, K_c is between 0.12 and 0.32 and can be assigned an average value of $K_c = 0.22$. Since $V = 1.52$ m/s, the head loss, h_m, is therefore given by

$$h_m = 0.22\frac{(1.52)^2}{2(9.81)} = 0.026 \text{ m}$$

This head loss must be accounted for in computing the energy grade line in the sewer system.   ■

6.7.1.2 Street Gutters and Inlets

Surface runoff from urban streets are typically routed to sewer pipes through street gutters and inlets. To facilitate drainage, urban roadways are designed with both cross-slopes and longitudinal slopes. The cross-slope directs the incident rainfall to the sides of the roadway, where the pavement intersects the curb and forms an open channel called a *gutter*. Longitudinal slopes direct the flow in the gutters to stormwater *inlets* that direct the flow into sewer pipes. Typical cross-slopes on urban roadways are in the range of 1.5% to 6%, depending on the type of pavement surface (Easa, 1995), and typical longitudinal slopes are in the range of 0.5% to 5%, depending on the topography. The spacing between stormwater inlets depends on several criteria, but it is usually controlled by the rate of flow and the allowable water spread toward the crown of the street.

The flowrate in a triangular curb gutter, illustrated in Figure 6.35, can be derived from the Manning equation (ASCE, 1992) as

$$Q = 0.38 \left(\frac{1}{nS_x} \right) d^{8/3} S_o^{1/2} \qquad (6.169)$$

where S_x is the street cross-slope, n is the Manning roughness coefficient, S_o is the longitudinal slope of the street, and d is the depth of flow at the curb. The conventional Manning equation has been modified in deriving Equation 6.169 because the hydraulic radius does not adequately describe the gutter cross-section, particularly when the top-width T exceeds 40 times the depth at the curb (ASCE, 1992). The depth and top-width of gutter flow are related by

$$d = TS_x \qquad (6.170)$$

Typical Manning n values for street and pavement gutters are given in Table 6.31 (USFHWA, 1984b). To facilitate proper drainage, it is recommended that the gutter grade exceed 0.4% and the street cross-slope exceed 2% (ASCE, 1992). The gutter, together with a curb, should be at least 15 cm (6 in.) deep and 60 cm (2 ft) wide, with the deepest portion adjacent to the curb. The maximum allowable width of street flooding depends on the type of street and is usually specified separately for minor and major design runoff events. Typical guidelines for allowable pavement encroachment are given in Table 6.32. Allowable pavement encroachment is the basis for computing the street drainage capacity using the modified Manning equation.

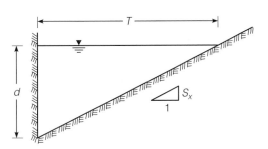

Figure 6.35 ■ Triangular Curb Gutter

■ Table 6.31
Typical Manning *n* Values
for Street and Pavement
Gutters

Type of gutter or pavement	Manning *n*
Concrete gutter, troweled finish	0.012
Asphalt pavement	
Smooth texture	0.013
Rough texture	0.016
Concrete gutter with asphalt pavement	
Smooth	0.013
Rough	0.015
Concrete pavement	
Float finish	0.014
Broom finish	0.016

Source: USFHWA (1984b).

■ Table 6.32
Pavement Encroachment
Guidelines

Street type	Minor storm runoff	Major storm runoff
Local*	no curb overtopping[†]; flow may spread to crown of street	Residential dwellings, public, commercial and industrial buildings shall not be inundated at the ground line, unless buildings are flood-proofed. The depth of water over the gutter flow line shall not exceed an amount specified by local regulation, often 30 cm (12 in.).
Collector[‡]	no curb overtopping[†]; flow spread must leave at least one lane free of water	Same as for local streets.
Arterial[§]	no curb overtopping[†]; flow spread must leave at least one lane free of water in each direction	Residential dwellings, public, commercial, and industrial buildings shall not be inundated at the ground line, unless buildings are flood-proofed. Depth of water at the street crown shall not exceed 15 cm (6 in.) to allow operation of emergency vehicles. The depth of water over the gutter flow line shall not exceed a locally prescribed amount.
Freeway[¶]	no encroachment allowed on any traffic lanes	Same as for arterial streets.

Source: Denver Regional Urban Storm Drainage Criteria Manual (1984).
*A local street is a minor traffic carrier within a neighborhood characterized by one or two moving lanes and parking along curbs. Traffic control may be by stop or yield signs.
[†]Where no curb exists, encroachment onto adjacent property should not be permitted.
[‡]A collector street collects and distributes traffic between arterial and local streets. There may be two or four moving traffic lanes and parking may be allowed adjacent to curbs.
[§]An arterial street permits rapid and relatively unimpeded traffic movement. There may be four to six lanes of traffic, and parking adjacent to curbs may be prohibited. The arterial traffic normally has the right-of-way over collector streets. An arterial street will often include a median strip with traffic channelization and signals at numerous intersections.
[¶]Freeways permit rapid and unimpeded movement of traffic through and around a city. Access is normally controlled by interchanges at major arterial streets. There may be eight or more traffic lanes, frequently separated by a median strip.

■ **Example 6.39**

A four-lane collector roadway is to be constructed with 3.66 m (12 ft) lanes, a cross-slope of 2%, a longitudinal slope of 0.5%, and pavement made of rough asphalt. If the (minor) roadway drainage system is to be designed for a rainfall intensity of 150 mm/h, determine the spacing of the inlets.

Solution

For a collector street, Table 6.32 indicates that at least one lane must be free of water. However, since the roadway has four lanes, the drainage system must necessarily be designed to leave two lanes free of water (one on each side of the crown). Since each lane is 3.66 m wide, the allowable top width, T, is 3.66 m, with $n = 0.016$ (rough asphalt), $S_x = 0.02$, and $S_o = 0.005$. The maximum allowable depth of flow, d, at the curb is given by

$$d = TS_x = (3.66)(0.02) = 0.0732 \text{ m} = 7.32 \text{ cm}$$

and the maximum allowable flowrate, Q, in the gutter is given by the Manning equation (Equation 6.169) as

$$Q = 0.38 \left[\frac{1}{nS_x} \right] d^{8/3} S_o^{1/2} = 0.38 \left[\frac{1}{(0.016)(0.02)} \right] (0.0732)^{8/3} (0.005)^{1/2} = 0.0787 \text{ m}^3/\text{s}$$

Since the design-rainfall intensity is 150 mm/h $= 4.17 \times 10^{-5}$ m/s, then the contributing area, A, required to produce a runoff of 0.0787 m³/s is given by

$$A = \frac{0.0787}{4.17 \times 10^{-5}} = 1887 \text{ m}^2$$

The roadway has two lanes contributing runoff to each gutter. Therefore, the width of the contributing area is $2 \times 3.66 = 7.32$ m, and the length, L, of roadway required for a contributing area of 1,887 m² is given by

$$L = \frac{1887}{7.32} = 258 \text{ m}$$

The required spacing of inlets is therefore 258 m. This is a rather long spacing, and the required placement of inlets at roadway intersections may govern the location of inlets.

■

Stormwater inlets can take many forms but are usually classified as curb inlets, gutter inlets, or slotted drains. The various inlet types are illustrated in Figure 6.36. These inlets are typically located at low points (sumps), directly upstream from intersections, or at intermediate locations. Municipalities sometimes specify the manufacturer and specific inlet types that are acceptable within their jurisdiction.

Curb Inlets. Curb inlets are vertical openings in the curb covered by a top slab. The capacity of a curb inlet depends on the amount of debris blockage, the amount of the total flow that can be intercepted by the inlet, whether the inlet throat is depressed, and whether deflectors are used. Details of the performance of curb inlets can be found in regulatory manuals such as the *Denver Regional Urban Storm Drainage Criteria Manual* (1984). According to ASCE (1992), curb inlets without local depressions are very inefficient, the final design should show roughly one 1.2-m long inlet for every 0.08 m³/s on a street with a longitudinal slope of 2% or less, and opening heights should not exceed 15 cm (6 in.) to reduce risks to children. Curb openings act as weirs up to a depth equal

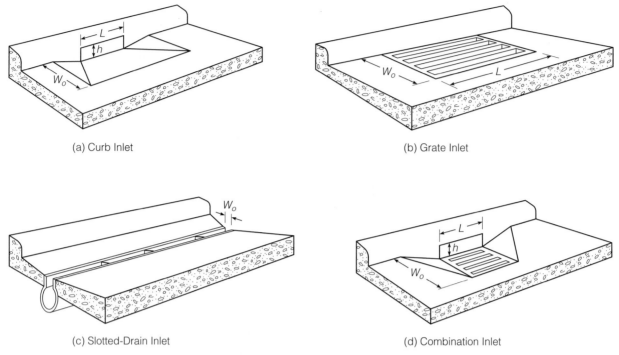

(a) Curb Inlet

(b) Grate Inlet

(c) Slotted-Drain Inlet

(d) Combination Inlet

Figure 6.36 ■ Stormwater Inlets
Source: FHWA (1984a).

to the opening height, and the inlet operates as an orifice when the water depth is greater than 1.4 times the opening height. The weir flow equation gives the flowrate, Q_i (m^3/s), into the curb inlet as (Wanielista et al., 1997)

$$Q_i = 1.27(L + 1.8W_o)d^{1.5} \tag{6.171}$$

where L is the length of the curb opening (m), W_o is the width of the inlet depression (m), d is the depth of flow at the curb, where d is less than the depression depth and h is the height of the curb opening. Without a depressed gutter, the inflow to a curb inlet is given by

$$Q_i = 1.27Ld^{1.5} \qquad d \leq h \tag{6.172}$$

When the flow depth, d, exceeds 1.4 times the opening height, the inflow to the curb inlet is given by the orifice equation

$$Q_i = 0.67A \left[2g \left(d - \frac{h}{2} \right) \right]^{\frac{1}{2}} \tag{6.173}$$

where A is the area of the curb opening ($= hL$).

■ Example 6.40

A roadway has a maximum allowable flow depth at the curb of 8 cm and a corresponding flowrate in the gutter of 0.08 m^3/s. Determine the length of a 15-cm (6-in.) high curb inlet that is required to remove all the water from the gutter. Consider the cases where (a) the width of the inlet depression is 0.4 m, and (b) there is no inlet depression.

Solution

(a) Since the flow depth (8 cm) is less than the height of the inlet (15 cm), the curb inlet acts as a weir. In this case, $Q_i = 0.08$ m³/s, $W_o = 0.4$ m, $d = 0.08$ m, and the weir equation (Equation 6.171) can be put in the form

$$L = \frac{Q_i}{1.27d^{1.5}} - 1.8W_o$$

The required weir length, L, is therefore given by

$$L = \frac{0.08}{1.27(0.08)^{1.5}} - 1.8(0.4) = 2.06 \text{ m}$$

With an inlet depression, the length of the curb opening should be at least 2.06 m.

(b) In the case of no inlet depression, the inlet equation (Equation 6.172) can be put in the form

$$L = \frac{Q_i}{1.27d^{1.5}}$$

and the required weir length is given by

$$L = \frac{0.08}{1.27(0.08)^{1.5}} = 2.78 \text{ m}$$

The presence of an inlet depression reduces the required curb length from 2.78 m to 2.06 m, a reduction of 26%. ■

Grate Inlets.   Grate inlets consist of an opening in the gutter covered by one or more grates. Grate inlets are flush mounted to the pavement and function best when located in a sump. The main disadvantage of using grated inlets is their interference with bicycles and the tendency for debris blockage. If clogging due to debris is not expected, then a grate or grate/curb combination type inlet will provide more capacity than a curb inlet. For depths of water not exceeding 12 cm, the following weir equation can be used to calculate the capacity, Q_i (m³/s), of the grate inlet

$$Q_i = 1.66Pd^{1.5} \tag{6.174}$$

where P is the perimeter of the grate opening (m) and d is the depth of flow above the grate (m). If the grate is adjacent to a curb, then that side of the grate is not counted in the perimeter. If the flow depth over the grate exceeds 43 cm, then the following orifice equation is used to compute the capacity, Q_i, of the grate inlet

$$Q_i = 0.6A\sqrt{2gd} \tag{6.175}$$

where A is the total area of the opening and d is the depth of flow above the grate. For depths of flow between 12 cm and 43 cm, the capacity of the grate is somewhere between that calculated by Equations 6.174 and 6.175. The minimum length, L (m), of clear opening parallel to the direction of flow to allow the flow to fall through the opening and clear the downstream end of the bars can be estimated by (ASCE, 1992)

$$L = 0.91V(t + d)^{0.5} \qquad (6.176)$$

where V is the average velocity of the water approaching the grate inlet (m/s), d is the depth of flow above the grate (m), and t is the thickness of the grate (m). In typical grated inlets, all of the frontal flow, and a portion of the side flow, is intercepted. The ratio, R, of the side flow intercepted to the total side flow is called the side flow interception efficiency, and can be estimated using

$$R = \frac{1}{1 + \frac{0.15V^{1.8}}{S_x L^{2.3}}} \qquad (6.177)$$

where S_x is the cross-slope of the gutter. A more detailed discussion of the capacity calculation for grate inlets is given by the Federal Highway Administration (USFHWA, 1984a).

■ Example 6.41

A roadway has a cross-slope of 2%, a maximum allowable flow depth at the curb of 8 cm, and a corresponding flowrate in the gutter of 0.08 m³/s. The gutter flow is to be removed by a 1.5-cm thick grate inlet that is mounted flush with the curb. Calculate the minimum dimensions of the grate inlet.

Solution

Since the depth of flow is less than 12 cm, the inflow to the inlet is given by the weir equation (Equation 6.174), which can be put in the form

$$P = \frac{Q_i}{1.66d^{1.5}}$$

where P is the grate-inlet perimeter not including the side adjacent to the curb, $Q_i = 0.08$ m³/s, $d = 0.08$ m, and

$$P = \frac{0.08}{1.66(0.08)^{1.5}} = 2.13 \text{ m}$$

The minimum length, L, of the grate inlet is given by Equation 6.176, where $t = 0.015$ m, $d = 0.08$ m, and

$$V = \frac{Q_i}{A}$$

where A is the flow area in the gutter, given by

$$A = \frac{1}{2}d\left(\frac{d}{S_x}\right) = \frac{1}{2}(0.08)\left(\frac{0.08}{0.02}\right) = 0.16 \text{ m}^2$$

The flow velocity, V, in the gutter is therefore given by

$$V = \frac{Q_i}{A} = \frac{0.08}{0.16} = 0.5 \text{ m/s}$$

and the minimum length of the grate inlet is

$$L = 0.91V(t + d)^{0.5} = 0.91(0.5)(0.015 + 0.08)^{0.5} = 0.14 \text{ m}$$

Therefore, the grate inlet must have a minimum length of 14 cm and a minimum perimeter of 213 cm.   ■

Slotted-Drain Inlets.   Slotted drains are used on both curbed and uncurbed roadways and function much like curb inlets, although they are generally much longer than curb inlets (Loganathan et al., 1996). For slotted-drain inlets with slot widths greater than 4.45 cm (1.75 in.), the length, L, of drain required to intercept a flow Q_i is given by (Wanielista et al., 1997)

$$L = 0.6 Q_i^{0.42} S_o^{0.3} \left(\frac{1}{n S_x} \right)^{0.6} \tag{6.178}$$

where S_o is the longitudinal slope of the drain, n is the Manning roughness coefficient of the drain pipe, and S_x is the cross-slope of the gutter.

■ Example 6.42

A roadway has a cross-slope of 2.5%, a longitudinal slope of 1.5%, and a flowrate in the gutter of 0.1 m³/s. The flow is to be removed by a slotted drain with a slot width of 5 cm, and a Manning n quoted by the manufacturer as 0.015. Estimate the minimum length of slotted drain that can be used.

Solution

From the given data, $Q_i = 0.1$ m³/s, $S_o = 0.015$, $n = 0.015$, $S_x = 0.025$, and Equation 6.178 gives the length of the drain as

$$L = 0.6 Q_i^{0.42} S_o^{0.3} \left[\frac{1}{n S_x} \right]^{0.6} = 0.6(0.1)^{0.42}(0.015)^{0.3} \left[\frac{1}{(0.015)(0.025)} \right]^{0.6} = 7.35 \text{ m}$$

Hence the slotted drain must be at least 7.35 m long to remove the gutter flow.   ■

6.7.2 ■ Runoff Controls

Urban stormwater-management systems are designed to control both the quantity and quality of stormwater runoff. Quantity control usually requires that peak postdevelopment runoff rates do not exceed peak predevelopment runoff rates (for a design rainfall), and quality control usually requires a defined level of treatment, such as a specified detention time in a sedimentation basin or the retention of a specified volume of initial runoff. The runoff events used to design flood control and water-quality control systems are generally different. Flood-control systems are designed for large infrequent runoff events with return periods of 10 to 100 years, while quality-control systems are designed for small frequent events with return periods of less than one year. On a long-term basis, most of the pollutant load in stormwater runoff is contained in the smaller, more frequent storms.

Runoff controls can be either on-site or regional controls. *On-site controls* handle runoff from individual developments, while *regional controls* handle runoff from several developments. The main advantage of on-site facilities is that developers can be required to build them, while the major disadvantage is the larger overall land area that is required compared with regional controls. The main advantage of regional facilities is that they provide more storage and can be designed for longer release periods, while their major disadvantages are the complex arrangements that are necessary to collect funds from developers and use those funds efficiently for the intended stormwater management facilities. The minimization of directly connected impervious areas remains one of the most effective source controls that can be implemented to reduce the quantity and improve the quality of runoff at the source, and it can significantly reduce the capacity requirements in other runoff controls. Under ideal conditions, the minimization of directly connected impervious areas can virtually eliminate surface runoff from storms with less than 13 mm (0.5 in.) of precipitation (Urbonas and Stahre, 1993).

6.7.2.1 Stormwater Impoundments

Stormwater impoundments are facilities (basins) that collect surface runoff and release the runoff either at a reduced rate through an outlet or by infiltration into the ground. These impoundments are frequently used for both flood control and water-quality control purposes; the same impoundment used for both purposes is called *dual-purpose*. The two major types of impoundments are detention and retention impoundments (basins). *Detention basins* are water-storage areas where the stored water is released through an uncontrolled outlet and where the peak discharge from the outlet structure is generally less than the peak runoff rate entering the impoundment. *Retention basins* are water-storage areas where there is either no outlet or the impounded water is stored for a prolonged period of time. Infiltration basins and ponds that maintain water permanently, with freeboard provided for flood storage, are the most common types of retention basins (ASCE, 1992). Storage impoundments where a water body forms the base of the storage area are called *wet* basins; impoundments where the ground surface forms the base of the storage area are called *dry* basins. An infiltration basin is an example of a dry retention basin.

Most detention basins are constructed by a combination of cut and fill and must have at least one service outlet and an emergency spillway. In some cases, more than one outlet at different elevations are used to facilitate the discharge from the basin under multiple design storms with return periods between 20 and 50 years (ASCE, 1992). A schematic diagram of a detention basin is given in Figure 6.37. The most common types of outlets from detention basins are orifice-type and weir-type outlets.

Orifice-Type Outlets.   The most common orifice-type structures are culverts, where the culvert entrance is mostly submerged when the detention basin is filled with water during and immediately after a storm. The culvert outlet is usually free, and the hydraulic equations for culverts discussed in Section 4.4.4 can be used to express the discharge in terms of the headwater elevation. The discharge, Q, from orifice-type outlets

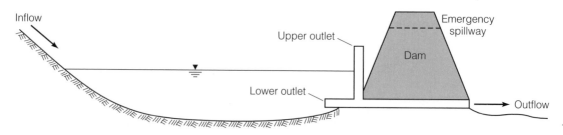

Figure 6.37 ■ Typical Detention Basin
Source: Akam, Osman, 1993. *Urban Stormwater Hydrology—A Guide to Engineering Calculations.* Technomic Publishing Company, Inc. Lancaster, PA.

are typically of the form

$$Q = C_d A \sqrt{2gh}$$

(6.179)

where C_d is a discharge coefficient, A is the cross-sectional area of the culvert (orifice), and h is the elevation of the water surface in the detention basin above the center of the culvert entrance.

Weir-Type Outlets. Weir-type outlets include rectangular broad-crested weirs and overflow spillways. The discharge over the weir can be expressed in terms of the head-water elevation using the hydraulic equations in Section 4.4.1. The discharge, Q, from weir-type outlets is usually of the form

$$Q = \frac{2}{3} C_d \sqrt{2g} b h^{\frac{3}{2}}$$

(6.180)

where C_d is a discharge coefficient, b is the crest length of the weir, and h is the elevation of the water surface in the detention basin above the crest of the weir.

■ Example 6.43

A 30-cm diameter concrete culvert is being used to drain a detention basin, and the culvert entrance is shaped such that $C_d = 0.90$. Estimate the basin discharge when the water in the basin is 95 cm above the center of the culvert entrance, and the culvert is discharging freely into a drainage channel. Assume that the culvert behaves like an orifice.

Solution

Under the given conditions, the culvert discharge, Q, is given by Equation 6.179 as

$$Q = C_d A \sqrt{2gh}$$

From the given data, $C_d = 0.90$, $h = 0.95$ m, $D = 0.30$ m, $A = \pi D^2/4 = \pi(0.30)^2/4 = 0.0707$ m^2, and therefore

$$Q = (0.90)(0.0707)\sqrt{2(9.81)(0.95)} = 0.27 \text{ m}^3/\text{s} \qquad ■$$

■ Example 6.44

A detention basin is drained by a weir with a length of 30 cm, and a discharge coefficient, C_d, of 0.62. Estimate the basin discharge when the water in the basin is 95 cm above the crest of the weir.

Solution

From the given data: $C_d = 0.62$, $b = 0.30$ m, $h = 0.95$ m, and the weir discharge, Q, is given by Equation 6.180 as

$$Q = \frac{2}{3} C_d \sqrt{2g} b h^{\frac{3}{2}} = \frac{2}{3}(0.62)\sqrt{2(9.81)}(0.30)(0.95)^{\frac{3}{2}} = 0.51 \text{ m}^3/\text{s} \qquad ■$$

6.7.2.2 Flood Control

Both detention and retention basins are used for flood control. The design of retention basins for flood control consists first of providing sufficient freeboard in an impoundment area to store the runoff volume resulting from the design runoff event and then of verifying that the infiltration capacity of the impoundment is sufficient to remove the stored water in a reasonable amount of time, usually a typical interstorm period. The design runoff event is usually equal to the runoff resulting from a storm of specified duration and frequency, such as a 24-hour storm with a 25-year return period. The design of detention basins for flood control consists of the following steps: (1) select a drainage basin configuration; (2) select an outlet structure; (3) route the design runoff hydrograph through the detention basin and determine the peak discharge from the detention basin and the maximum water elevation in the basin; and (4) repeat steps 1 to 3 until the peak discharge and maximum water surface elevation are acceptable. An acceptable peak discharge is usually one that is less than or equal to the predevelopment peak discharge, and an acceptable maximum water surface elevation maintains the stored runoff within the confines of the detention basin. Most drainage ordinances require that postdevelopment discharges not exceed predevelopment discharges for multiple events, such as the 2-year, 10-year, 25-year, and 50-year storms (ASCE, 1996a). Detention facilities should be designed to drain within the typical interstorm period, usually on the order of 72 hours (Debo and Reese, 1995). Some guidelines in determining the required size of a detention basin for flood control are given below.

Step 1: Make a preliminary selection of a drainage basin. A preliminary estimate of the required drainage-basin volume can be obtained by subtracting the predevelopment runoff volume from the postdevelopment runoff volume. This volume represents the approximate storage requirement to ensure that the postdevelopment peak discharge rate is less than or equal to the predevelopment peak runoff rate.

Step 2: Make a preliminary selection of an outlet structure. Use the required volume of the detention basin estimated in step 1 with the storage-elevation function for the site to estimate the maximum headwater elevation at the outlet location. Select an outlet structure that will pass the maximum allowable outflow at this headwater elevation.

Step 3: Route the runoff hydrograph through the detention basin. The runoff hydrograph is routed through the detention basin using the storage-indication method described in Section 6.5.1.1. This procedure yields the discharge hydrograph from the detention basin.

Step 4: Assess the performance of the detention basin. Determine whether the peak discharge from the detention basin is less than or equal to the predevelopment peak discharge. If it is, the required detention volume corresponds to the maximum stage in the detention basin. If the peak discharge from the detention basin exceeds the predevelopment peak discharge, then adjust the outlet structure to discharge the predevelopment peak at the maximum stage in the detention basin, and repeat steps 3 and 4.

■ Example 6.45

The estimated runoff hydrographs from a site before and after development are as follows:

Time (min)	0	30	60	90	120	150	180	210	240	270	300	330	360	390
Before (m³/s)	0	1.2	1.7	2.8	1.4	1.2	1.1	0.91	0.74	0.61	0.50	0.28	0.17	0
After (m³/s)	0	2.2	7.7	1.9	1.1	0.80	0.70	0.58	0.38	0.22	0.11	0	0	0

The postdevelopment detention basin is to be a wet detention reservoir drained by an outflow weir. The elevation versus storage in the detention basin is

Elevation (m)	Storage (m³)
0	0
0.5	5,544
1.0	12,200
1.5	20,056

where the weir crest is at elevation 0 m, which is also the initial elevation of the water in the detention basin prior to runoff. The performance of the weir is given by

$$Q = 1.83bh^{\frac{3}{2}}$$

where Q is the overflow rate (m³/s), b is the crest length (m), and h is the head on the weir (m). Determine the required crest length of the weir for the detention basin to perform its desired function. What is the maximum water surface elevation expected in the detention basin?

Solution

The required detention basin volume is first estimated by subtracting the predevelopment runoff volume, V_1, from the postdevelopment runoff volume, V_2. From the given hydrographs:

$$V_1 = (30)(60)[1.2 + 1.7 + 2.8 + 1.4 + 1.2 + 1.1 + 0.91 + 0.74 + 0.61 + 0.50$$

$$+ 0.28 + 0.17]$$

$$= 22{,}698 \text{ m}^3$$

and

$$V_2 = (30)(60)[2.2+7.7+1.9+1.1+0.80+0.70+0.58+0.38+0.22+0.11] = 28{,}242 \text{ m}^3$$

A preliminary estimate of the required volume, V, of the detention basin is

$$V = V_2 - V_1 = 28242 - 22698 = 5{,}544 \text{ m}^3$$

From the storage-elevation function, the head h corresponding to a storage volume of 5,544 m³ is 0.50 m. The maximum predevelopment runoff, Q, is 2.8 m³/s, and the weir equation gives

$$Q = 1.83bh^{\frac{3}{2}}$$

or

$$b = \frac{Q}{1.83h^{3/2}} = \frac{2.8}{1.83(0.50)^{3/2}} = 4.33 \text{ m}$$

Based on this preliminary estimate of the crest length, use a trial length of 4.25 m. The corresponding weir discharge equation is

$$Q = 1.83bh^{\frac{3}{2}} = 1.83(4.25)h^{\frac{3}{2}} = 7.78h^{\frac{3}{2}}$$

The postdevelopment runoff hydrograph can be routed through the detention basin using $\Delta t = 30$ min. The storage and outflow characteristics of the detention basin can be put in the following form

Elevation (m)	Storage, S (m³)	Outflow, O (m³/s)	$2S/\Delta t + O$ (m³/s)
0	0	0	0
0.5	5,544	2.75	8.91
1.0	12,200	7.78	21.34
1.5	20,056	14.29	36.58

The routing computations (using the storage-indication method) are summarized in the following table:

Time (min)	Inflow, I (m³/s)	$2S/\Delta t - O$ (m³/s)	$2S/\Delta t + O$ (m³/s)	O (m³/s)
0	0	0	0	0
30	2.2	0.84	2.2	0.68
60	7.7	3.76	10.74	3.49
90	1.9	4.26	13.36	4.55
120	1.1	2.78	7.26	2.24

From these results, it is already clear that the maximum outflow from the detention basin is 4.55 m³/s, which is higher than the predevelopment peak of 2.8 m³/s and is therefore unacceptable. Decreasing the crest length (= 4.25 m) of the weir by the factor $2.8/4.55 = 0.62$ gives a new crest length, b, of $0.62(4.25) = 2.64$ m. Using a rounded number of 2.50 m, the revised weir equation is

$$Q = 1.83bh^{\frac{3}{2}} = 1.83(2.5)h^{\frac{3}{2}} = 4.58h^{\frac{3}{2}}$$

The revised storage-outflow characteristics of the detention basin are:

Elevation (m)	Storage, S (m³)	Outflow, O (m³/s)	$2S/\Delta t + O$ (m³/s)
0	0	0	0
0.5	5,544	1.62	7.78
1.0	12,200	4.58	18.14
1.5	20,056	8.41	30.70

The routing computations are summarized in the following table:

Time (min)	Inflow, I (m³/s)	$2S/\Delta t - O$ (m³/s)	$2S/\Delta t + O$ (m³/s)	O (m³/s)
0	0	0	0	0
30	2.2	1.28	2.2	0.46
60	7.7	6.00	11.18	2.59
90	1.9	7.90	15.60	3.85
120	1.1	5.88	10.90	2.51

From these results, it is already clear that the maximum outflow from the detention basin is 3.85 m³/s, which is higher than the predevelopment peak of 2.8 m³/s and is therefore unacceptable. The crest length must be further decreased until the maximum weir discharge is less than or equal to 2.8 m³/s. This occurs when the crest length, b, is decreased to 1.30 m. The revised weir discharge equation is then given by

$$Q = 1.83bh^{\frac{3}{2}} = 1.83(1.30)h^{\frac{3}{2}} = 2.38h^{\frac{3}{2}}$$

and the revised storage-outflow characteristics of the reservoir are:

Elevation (m)	Storage, S (m^3/s)	Outflow, O (m^3/s)	$2S/\Delta t + O$ (m^3/s)
0	0	0	0
0.5	5,544	0.84	7.00
1.0	12,200	2.38	15.94
1.5	20,056	4.37	26.66

The routing computations are summarized in the following table:

Time (min)	Inflow, I (m^3/s)	$2S/\Delta t - O$ (m^3/s)	$2S/\Delta t + O$ (m^3/s)	O (m^3/s)
0	0	0	0	0
30	2.2	1.68	2.2	0.26
60	7.7	8.32	11.58	1.63
90	1.9	12.42	17.92	2.75
120	1.1	10.84	15.42	2.29
150	0.80	9.08	12.74	1.83
180	0.70	7.66	10.58	1.46
210	0.58	6.60	8.94	1.17
240	0.38	5.68	7.56	0.94
270	0.22	4.78	6.28	0.75
300	0.11	3.89	5.11	0.61
330	0	3.04	4.00	0.48
360	0	2.32	3.04	0.36
390	0	1.76	2.32	0.28
420	0	1.34	1.76	0.21
450	0	1.02	1.34	0.16
480	0	0.78	1.02	0.12
510	0	0.60	0.78	0.09
540	0	0.46	0.60	0.07
570	0	0.34	0.46	0.06
600	0	0.26	0.34	0.04
630	0	0.20	0.26	0.03
660	0	0.16	0.20	0.02
690	0	0.12	0.16	0.02
720	0	0.10	0.12	0.01

For a crest length of 1.30 m, the maximum postdevelopment discharge is 2.75 m^3/s and is therefore acceptable. The maximum water level in the detention basin corresponds to a weir overflow rate of 2.75 m^3/s; from the weir-discharge equation, this corresponds to $h = 1.10$ m. If this water elevation is excessive, the engineer could consider expanding the proposed detention basin. ■

Flood control systems are primarily designed to ensure that postdevelopment peak discharge rates do not exceed predevelopment peak discharge rates. Using detention basins to accomplish this goal generally results in a postdevelopment discharge hydrograph that is shifted in time and has an overall greater volume compared to the predevelopment discharge hydrograph. Consequently, the postdevelopment runoff hydrograph generally has higher off-peak discharge rates than the predevelopment hydrograph. Effects of increased runoff volumes from developed areas include (1) prolonged rise in the water surface downstream of the site, which might affect the slope and stability of channels; (2) the increase in runoff volume represents the amount of ground-water recharge that is no longer being absorbed onsite; and (3) an increased volume of water is fed into downstream detention ponds (Haestad, 1997a).

6.7.2.3   Water-Quality Control

The quality of urban runoff is determined principally by source controls and treatment controls. *Source controls* (also called "nonstructural" controls) are practices that keep pollutants from entering the runoff, while *treatment controls* (also called *structural controls*) are practices that remove pollutants from the runoff. Examples of source controls include street sweeping and household hazardous waste recycling programs, while examples of treatment controls include stormwater detention reservoirs and infiltration basins. Water-quality control regulations usually require a defined level of treatment, such as a specified detention time in a sedimentation basin and/or the retention of a specified volume of initial runoff. The most effective stormwater-management facilities are designed to satisfy both detention and retention criteria. Detention basins are commonly used for sedimentation purposes, and infiltration basins or underground infiltration trenches are used for retention purposes. The specified retention volume that is used in the design of water-quality control systems is usually less than or equal to the runoff volume in at least 90% of the annual runoff events, and typically corresponds to a runoff depth on the order of 1.3 cm (0.5 in.).

6.7.2.4   Detention Facilities

The design of detention basins is fundamentally different for water-quality control than for flood control because flood-control basins are designed to attenuate the peak runoff rates from large design storms (with long return periods), whereas water-quality basins are designed to provide sufficient detention time for the sedimentation of pollutant loads in smaller, more frequent storms that usually contain much higher concentrations of pollutants than large rainfall events. Processes such as natural die-off of bacteria and plant uptake of soluble nitrogen and phosphorous also occur in detention basins, but sedimentation is the principal process of pollutant removal. A common practice is to use *dual-purpose basins* designed to control both peak discharges and pollution from stormwater runoff.

The main design criteria for water-quality detention basins are (Akan, 1993): (1) detain the design runoff long enough to provide the targeted level of treatment, and (2) evacuate the design runoff soon enough to provide available storage for the next runoff event. The required detention time is determined by the settling velocities of the pollutants in the runoff. According to Whipple and Randall (1983), a mean detention time of about 18 hours is usually sufficient to settle out 60% of total suspended solids, lead, and hydrocarbons and 45% of total BOD, copper, and phosphates from urban storm runoff. A typical distribution of settling velocities in urban runoff is given in Table 6.33 (USEPA, 1986a). The data given in Table 6.33 were derived from approximately 50 different runoff samples at seven urban sites and can be used for design purposes in the absence of local data. The required evacuation time for a water-quality detention basin is based on the average time between design runoff events and is usually specified by local

Table 6.33	Percent of Particle Mass (%)	Average Settling Velocity (m/h)
Distribution of Settling Velocities in Urban Runoff	0–20	0.009
	20–40	0.091
	40–60	0.46
	60–80	2.1
	80–100	20

Source: USEPA (1986a).

regulatory requirements. According to the Environmental Protection Agency (USEPA, 1986a), the average interval between storms in most parts of the United States is between 73 and 108 hours, while the average time between storms in the southwestern part of the United States is much higher, on the order of 277 hours (USEPA, 1986a). Florida requires that detention basins should empty within 72 hours after a storm event, and Delaware requires that 90% of the runoff be evacuated in 36 hours or in 18 hours for residential areas (Akan, 1993).

The design of a detention basin for given detention and evacuation times is a reservoir routing problem. *Wet detention basins* contain a permanent pool of water, and the detention time, t_d, is estimated from the outflow hydrograph by the relation

$$\int_0^{t_d} O(t) \, dt = V \tag{6.181}$$

where $O(t)$ is the outflow from the detention basin as a function of time and V is the average volume of the detention basin. Detention basins without a permanent pool of water are called *dry detention basins*. Haan and colleagues (1994) define the detention time in dry detention basins as the time difference between the centroids of the inflow and outflow hydrographs. The evacuation time is defined as the interval between when inflow first enters the detention basin and when outflow from the basin ceases. Various configurations of the detention basin can be tried until all the design criteria are met. Because in many cases the same detention basin is used for flood-control purposes, the ability of the detention basin to attenuate the peak runoff from a larger storm must also be investigated.

■ Example 6.46

A runoff hydrograph is routed through a wet detention basin, and the results are given in the following table:

Time (min)	Inflow (m³/s)	Storage (m³/s)	Outflow (m³/s)
0	0	10,000	0
30	2.2	11,746	0.26
60	7.7	18,955	1.63
90	1.9	23,653	2.75
120	1.1	21,817	2.29
150	0.8	19,819	1.83
180	0.7	18,208	1.46

Time (min)	Inflow (m^3/s)	Storage (m^3/s)	Outflow (m^3/s)
210	0.58	16,993	1.17
240	0.38	15,958	0.94
270	0.22	14,977	0.75
300	0.11	14,050	0.61
330	0	13,168	0.48
360	0	12,412	0.36
390	0	11,836	0.28
420	0	11,395	0.21
450	0	11,062	0.16
480	0	10,810	0.12
510	0	10,621	0.09
540	0	10,477	0.07
570	0	10,360	0.06
600	0	10,270	0.04
630	0	10,207	0.03
660	0	10,162	0.02
690	0	10,126	0.02
720	0	10,099	0.01

Estimate the detention time and evacuation time of the detention basin.

Solution

The average volume, V, of the detention basin is determined by averaging the storage over the duration of the discharge (0 to 720 min), which yields $V = 13{,}567$ m^3. The detention time, t_d, is defined by Equation 6.181, and t_d is the time when the cumulative outflow is equal to V. The cumulative outflow as a function of time is tabulated as follows:

Time (min)	Outflow (m^3/s)	Cumulative outflow (m^3)	Time (min)	Outflow (m^3/s)	Cumulative outflow (m^3/s)
0	0	0	390	0.28	26,406
30	0.26	234	420	0.21	26,847
60	1.63	1,935	450	0.16	27,180
90	2.75	5,877	480	0.12	27,432
120	2.29	10,413	510	0.09	27,621
150	1.83	14,121	540	0.07	27,765
180	1.46	17,082	570	0.06	27,882
210	1.17	19,449	600	0.04	27,972
240	0.94	21,348	630	0.03	28,035
270	0.75	22,869	660	0.02	28,080
300	0.61	24,093	690	0.02	28,116
330	0.48	25,074	720	0.01	28,143
360	0.36	25,830			

From these results, $t_d = 146$ min $= 2.4$ h. The evacuation time is equal to the time from when inflow first enters the detention basin to when outflow from the basin ceases. Inflow begins at $t = 0$ min and ceases at about $t = 720$ min. Therefore, the evacuation time is equal to 720 min or 12 h.   ■

■ **Example 6.47**

If the inflow and outflow hydrographs given in the previous example were from a dry detention basin, estimate the detention time and evacuation time.

Solution

The detention time can be approximated by the difference between the centroids of the inflow and outflow hydrographs. Denoting the points on the inflow hydrograph by $I_i = I(t_i)$, and the outflow hydrograph by $O_i = O(t_i)$, then the centroids of the inflow and outflow hydrographs, t_I and t_O, are defined as

$$t_I = \frac{\sum_{i=1}^{N_I} t_i I_i}{\sum_{i=1}^{N_I} I_i}$$

and

$$t_O = \frac{\sum_{i=1}^{N_O} t_i O_i}{\sum_{i=1}^{N_O} O_i}$$

where N_I and N_O are the number of points on the inflow and outflow hydrographs, respectively. The computations of t_I and t_O are summarized in the following table:

t_i (min)	I_i (m^3/s)	$t_i I_i$ (m^3)	O_i (m^3/s)	$t_i O_i$ (m^3)
0	0	0	0	0
30	2.2	3,960	0.26	468
60	7.7	27,720	1.63	5,868
90	1.9	10,260	2.75	14,850
120	1.1	7,920	2.29	16,488
150	0.80	7,200	1.83	16,470
180	0.70	7,560	1.46	15,768
210	0.58	7,308	1.17	14,742
240	0.38	5,472	0.94	13,536
270	0.22	3,564	0.75	12,150
300	0.11	1,980	0.61	10,980
330	0	0	0.48	9,504
360	0	0	0.36	7,776
390	0	0	0.28	6,552
420	0	0	0.21	5,292
450	0	0	0.16	4,320
480	0	0	0.12	3,456
510	0	0	0.09	2,754
540	0	0	0.07	2,268
570	0	0	0.06	2,052
600	0	0	0.04	1,440
630	0	0	0.03	1,134
660	0	0	0.02	792
690	0	0	0.02	828
720	0	0	0.01	432
Total	15.69	82,944	15.64	169,920

Based on these results:

$$t_I = \frac{82944}{15.69} = 88 \text{ min}$$

$$t_O = \frac{169920}{15.64} = 181 \text{ min}$$

The detention time, t_d, is therefore given by

$$t_d = t_O - t_I = 181 - 88 = 93 \text{ min} = 1.6 \text{ h}$$

The evacuation time is equal to 720 min = 12 h, as in the previous example.   ■

Wet Detention Basins.   Wet detention basins are the most common type of detention basin used in stormwater management. *Wet detention basins*, also called *detention ponds*, remove pollutants by physical, chemical, and biological processes. In addition to sedimentation, chemical flocculation occurs when heavier sediment particles overtake and coalesce with smaller (lighter) particles; biological removal is accomplished by the uptake of pollutants by aquatic plants and metabolism by phytoplankton and microorganisms. The removal of dissolved pollutants primarily occurs between storms.

The three most important factors in determining the removal efficiency of detention ponds are: (1) the volume of the permanent pool, (2) the depth of the permanent pool, and (3) the presence of a shallow littoral zone. The volume of the permanent pool should be sufficient to provide two to four weeks of detention time so that algae can grow, and the ratio of the volume of the detention pond to the detained volume should be at least 4 to achieve total suspended sediment removal rates of 80% to 90% (ASCE, 1998). The depth of the permanent pool should be greater than 1 to 2 meters, to minimize sunlight penetration to the bottom of the pond and to reduce bottom-weed growth; the depth of the pond should be less than 3 to 5 meters so that the water remains well mixed and the bottom sediment remains aerobic. An anaerobic condition in the bottom of the pond will mobilize nutrients and metals into the water column and significantly reduce the effectiveness of the detention pond. In Florida, detention ponds up to 9 meters deep have been used successfully when excavated in high ground water areas, probably because of the improved circulation at the bottom of the pond as a result of ground water moving through the pond (ASCE, 1998). If the pond has more than 0.8 ha of water surface, a mean depth of at least 2 m will protect the detention pond against wind-generated resuspension of sediments (ASCE, 1998). The presence of a littoral zone is essential to the proper performance of a wet detention pond, since it is the aquatic plants in the littoral zone that provide much of the biological assimilation of the dissolved stormwater pollutants. The littoral zone should cover 25% to 50% the surface area of the detention pond (ASCE, 1998) and have a slope of 6:1 (H:V) or less to a depth of 60 cm (2 ft) below the permanent pond elevation. Flat side slopes provide a measure of public safety, especially for children. Urbonas and Stahre (1993) recommend that the flow length be extended as much as possible between the inlet and outlet structures, and that the outlet structure from the wet detention pond be designed such that an average annual runoff event, captured as a surcharge above the permanent pool, be drained in approximately 72 hours, or whatever is the local interstorm duration. Yousef and Wanielista (1989) recommend that the length-to-width ratio of detention ponds should be greater than 4:1, but ASCE (1998) suggests that a length-to-width ratio of 3:1 is preferable. This requirement is intended to minimize short circuiting, enhance sedimentation, and prevent vertical stratification within the permanent pool (Hartigan, 1989). Overflow from wet

Pollutant	Removal Rate
TSS	50%–70%
TP	10%–20%
Nitrogen	10%–20%
Organic matter	20%–40%
Pb	75%–90%
Zn	30%–60%
Hydrocarbons	50%–70%
Bacteria	50%–90%

■ **Table 6.34**
Typical Removal Rates in
Dry Detention Basins

Source: Adapted from Urbonas and Stahre (1993).

detention ponds are allowed for larger rainfall events, with appropriate restrictions for flood control.

Dry Detention Basins. Dry detention basins remove pollutants primarily by sedimentation. The removal efficiency of these basins is regarded as poor for detention times shorter than 12 hours and good for detention times longer than 24 hours. The design of dry detention basins for pollutant removal is much less scientific than for wet detention ponds. Although sedimentation is still the primary pollutant-removal process, the estimation of basin performance is based mostly on empirical results. Grizzard and colleagues (1986) recommend that a dry detention basin should have a volume at least equal to the average runoff event during the year; in addition, Urbonas and Stahre (1993) recommend that this volume be drained in no less than 40 hours. However, basins with long drain times tend to have "boggy" bottoms with marshy vegetation and can be difficult to maintain and clean. Typical removal rates for properly designed dry detention basins are given in Table 6.34.

In designing dry detention basins, the volume of the basin can be taken as equal to the runoff volume to be treated, provided the catchment area is less than 100 ha. If the catchment area is larger than 100 ha, reservoir routing is necessary to determine the volume of the basin. The calculated volume should be increased by 20% to account for sediment accumulation (ASCE, 1998). The outlet structure, such as a V-notch weir or perforated riser, should be designed to drain the detention basin in the specified design period. To ensure that small runoff events will be adequately detained, ASCE (1998) recommends that the outlet empty less than 50% of the design volume in the first one-third of the design emptying period. The shape of dry detention basins should be such that they gradually expand from the inlet and contract toward the outlet to reduce short circuiting. A length-to-width ratio of 2 or greater, preferably up to a ratio of 4, is recommended (ASCE, 1998). Basin side slopes of 4:1 or greater provide for facility maintenance and safety concerns, and a forebay with a volume equal to approximately 10% of the total design volume can help with the maintenance of the basin by facilitating sediment deposition near the inflow, thereby extending the service life of the remainder of the basin. Embankments for small on-site basins should be protected from at least the 100-year flood; whenever possible, dry detention basins should be incorporated within larger flood-control facilities.

■ **Example 6.48**

A dry detention basin is to be designed for a 60-ha (150-ac) residential development. The runoff coefficient of the area is estimated as 0.3, and the average rainfall depth during the year is 3.3 cm (1.3 in.). Design the detention basin and estimate the suspended solids removal efficiency.

Solution

The average runoff volume, V, is given by

$$V = CAD$$

where C is the runoff coefficient ($= 0.3$), A is the catchment area ($= 60$ ha), and D is the average rainfall depth ($= 3.3$ cm). Hence,

$$V = (0.30)(60 \times 10^4)(3.3 \times 10^{-2}) = 5{,}940 \text{ m}^3$$

The required volume of the dry detention basin should be increased by 20% to account for sediment accumulation and is therefore equal to $1.2(5940) = 7{,}130$ m^3. The outlet between the base of the detention basin and the elevation corresponding to a basin storage of 7130 m^3 should be designed to discharge 7,130 m^3 in approximately 40 hours. An appropriate outlet structure could be a perforated riser or a V-notch weir, and overflow from the detention basin should be accommodated when the storage volume exceeds 7,130 m^3. In accordance with Table 6.34, the suspended solids removal in the basin is expected to be in the range of 50% to 70%.   ■

In deciding whether to use a wet or dry detention basin as a water-quality control, an important consideration is whether nutrient removal (nitrogen and phosphorous) is an important requirement. This is particularly the case when the quality of the receiving water is sensitive to nutrient loadings. Properly designed wet detention basins generally provide much better nutrient removal than dry detention basins, since many of the nutrients in surface runoff are in dissolved form and are not significantly affected by the sedimentation process in dry detention basins (Hartigan, 1989). This functional advantage of wet detention basins must usually be balanced against the greater land requirements for wet detention basins versus dry detention basins. For example, the permanent pool of a wet detention basin can require anywhere from two to seven times more storage than the alternative dry detention basin.

Retention Facilities.   Stormwater retention is the most effective quality-control method, but it can only be used in situations where the captured volume of water can infiltrate into the ground before the next storm. The most common retention facilities are infiltration basins, swales, and below-ground exfiltration trenches.

Infiltration Basins.   *Infiltration basins* retain stormwater runoff, which then infiltrates into the ground. These basins typically serve areas ranging from front yards to 20 ha (ASCE, 1992) and are typically designed to capture either a given depth of runoff, or the runoff from a given depth of rainfall. For example, an infiltration basin may be designed to retain the first (0.5 in.) 1.3 cm of runoff, or the runoff from the first (1 in.) 2.5 cm of rainfall. The initial runoff is commonly called the *first flush*, and infiltration basins are classified as either on-line or off-line. *On-line detention basins* retain a specified first flush of runoff; when a larger runoff occurs, it overflows the basin, which then acts as a detention pond for the larger event. Some drainage systems divert the first flush of stormwater out of the normal drainage path and into *off-line detention basins* that hold it for later water-quality treatment. Experience in Florida indicates that on-line basins are not as efficient as off-line infiltration basins.

Infiltration basins must be located in soils that allow the runoff to infiltrate within 72 hours, or within 24 to 36 hours for infiltration areas that are planted with grass. The seasonal high-water table should be at least 1.2 m (4 ft) below the ground surface in the infiltration basin to assure that the pollutants in the runoff are removed by the vegetation, soil, and microbes before reaching the water table (Urbonas and Stahre, 1993). Infiltration rates can be calculated using the Green-Ampt or similar models (see Section 6.3.3.3), but soils with saturated infiltration rates less than 8 mm/h are not suitable for infiltration basins (ASCE, 1998). Design guidelines suggested by ASCE (1998) are that water ponding in infiltration basins be less than 0.3 m during the design storm and that the design infiltration rate be limited to a maximum of 50 mm/h to account for clogging. In urban settings, infiltration basins are commonly integrated into park lands and open spaces, while in highway drainage they may be located in rights-of-way or in open space within freeway interchange loops.

▪ Example 6.49

An infiltration basin is to be designed to retain the first 1.3 cm of runoff from a 10 ha catchment. The area to be used for the infitration basin is turfed, and field measurements indicate that the native soil has a minimum infiltration rate of 150 mm/h. If the retained runoff is to infiltrate within 24 hours, determine the surface area that must be set aside for the basin.

Solution

The volume, V, of the runoff corresponding to a depth of 1.3 cm = 0.013 m on an area of 10 ha = 10^5 m^2 is given by

$$V = (0.013)(10^5) = 1,300 \text{ m}^3$$

The native soil has a minimum infiltration rate of 150 mm/h. To account for clogging, however, the design infiltration rate will be taken as 50 mm/h = 0.05 m/h. The area, A, of infiltration basin required to infiltrate 1,300 m^3 in 24 h is given by

$$A = \frac{1300}{(0.05)(24)} = 1080 \text{ m}^2 = 0.11 \text{ ha} \qquad ▪$$

Swales.   Swales are vegetated open channels that transport and infiltrate runoff from adjacent land areas. Sometimes referred to as *borrow ditches*, swales are commonly for roadside drainage on rural roads. Typically, swales are designed as free-flowing open channels with inclined slopes, but they can also be designed to be nonflowing, with all the surface runoff retained and infiltrated into the ground. If the infiltration rate is equal to the runoff rate, Q, then the length, L, of swale required for the infiltration rate to be equal to the runoff rate is given by

$$L = \frac{Q}{fP} \qquad (6.182)$$

where f is equal to the infiltration capacity of the soil and P is the wetted perimeter of the swale. For any cross-sectional shape, the wetted perimeter can be expressed in terms of the runoff rate via the Manning equation. For triangular-shaped swales, Wanielista and colleagues (1997) have shown that Equation 6.182 can be combined with the Manning equation to yield

$$L = \frac{151400Q^{5/8}m^{5/8}S^{3/16}}{n^{3/8}(1+m^2)^{5/8}f} \qquad (6.183)$$

where L is the swale length in meters, Q is the runoff rate in m³/s, m is the side slope, S is the longitudinal slope, n is the Manning roughness coefficient, and f is the infiltration rate in cm/h. Typical values of the Manning roughness coefficient are $n = 0.20$ for routinely mowed swales and $n = 0.24$ for infrequently mowed swales (ASCE, 1998). In the case of trapezoidal sections

$$L = \frac{360000Q}{\left\{b + 2.38\left[\frac{Qn}{(2\sqrt{1+m^2}-m)S^{1/2}}\right]^{3/8}\sqrt{1+m^2}\right\}f} \qquad (6.184)$$

where b is the bottom width of the swale in meters. Equation 6.184 applies to the *best trapezoidal section*, where the perimeter, P, is related to the flow depth, y, and the side slope m by

$$P = 4y\sqrt{1+m^2} - 2my \qquad (6.185)$$

Urbonas and Stahre (1993) recommend that the longitudinal grade should be set as flat as possible to promote infiltration, never steeper than 3%, and that side slopes should be flatter than 4:1 (H:V) to maximize the contact area. Urbonas and Stahre (1993) also recommend that swales be designed for 2-year storms and that the runoff volume in nonflowing or slow-moving swales, be infiltrated within 36 hours. In cases where the length of the swale required to infiltrate the runoff is excessive, then a swale block (or berm) can be used to store the runoff within the swale, in which case the depth in the swale should not exceed 0.5 m.

■ **Example 6.50**

A triangular-shaped swale is to retain the runoff from a catchment with design runoff rate of 0.02 m³/s. The longitudinal slope of the swale is to be 3% with side slopes of 4:1 (H:V). If the grassed swale has a Manning n of 0.24 (infrequently mowed grass) and a minimum infiltration rate of 150 mm/h, determine the length of swale required.

Solution
From the given data: $Q = 0.02$ m³/s, $m = 4$, $S = 0.03$, $n = 0.24$, and $f = 150$ mm/h = 15 cm/h. Equation 6.183 gives the required swale length, L, as

$$L = 151400\frac{Q^{5/8}m^{5/8}S^{3/16}}{n^{3/8}(1+m^2)^{5/8}f}$$

$$= 151400\frac{(0.02)^{5/8}(4)^{5/8}(0.03)^{3/16}}{(0.24)^{3/8}(1+4^2)^{5/8}15}$$

$$= 314 \text{ m}$$

This length of swale is quite long. A ponded infiltration basin would require less area and would probably be more cost effective.   ■

In Example 6.50, the swale was designed to retain and infiltrate the entire surface runoff. In cases where the swale is to provide a reduced level of treatment of the surface runoff, primarily through trapping a portion of the sediment and organic biosolids in the vegetative cover, the swale is designed as a *biofilter*. For the swale to perform adequately as a biofilter, the following design criteria are recommended (ASCE, 1998):

- Minimum hydraulic residence time of 5 minutes
- Maximum flow velocity of 0.3 m/s
- Maximum bottom width of 2.4 m
- Minimum bottom width of 0.6 m
- Maximum depth of flow no greater than one-third of the gross or emergent wetland vegetation height for infrequently mowed swales, or no greater than one-half of the vegetation height for regularly mowed swales, up to a maximum of approximately 75 mm for grass and approximately 50 mm below the normal height of the shortest wetland plant species
- Minimum length of 30 m

For swales designed as biofilters, ASCE (1998) recommends longitudinal slopes of 1% to 2%, with a minimum of 0.5% and a maximum of 6%. When the longitudinal slope is less than 1% to 2%, perforated underdrains should be installed or, if there is adequate moisture, wetland species should be established. If the slope is greater than 2%, check dams should be used to reduce the effective slope to approximately 2%. Using these guidelines, the following design procedure is proposed (ASCE, 1998):

1. Estimate the runoff rate for the design event and limit the discharge to approximately 0.03 m^3/s by dividing the flow among several swales, installing upstream detention to control release rates, or reducing the developed surface area to reduce the runoff coefficient and gain space for biofiltration.
2. Establish the slope of the swale.
3. Select a vegetation cover suitable for the site.
4. Estimate the height of vegetation that is expected to occur during the storm runoff season. The design flow depth should be at least 50 mm less than this vegetation height and a maximum of approximately 75 mm in swales and 25 mm in filter strips (discussed in following section).
5. Estimate the Manning n.
6. Typically, swales are designed as trapezoidal channels (skip this step for filter strip design). When using a rectangular section, provide reinforced vertical walls.
7. Use the Manning equation to approximate the initial dimensions of the swale. For a trapezoidal cross-section, select a side slope that is no steeper than 3:1, with 4:1 or flatter preferred. Set the bottom width to be between 0.6 m and 2.5 m.
8. Compute the cross-sectional area and flow velocity. Limit the design velocity to less than 0.3 m/s. Iterate until the estimated Manning n is consistent with the relation between the Manning n and the product of the velocity and hydraulic radius given in Figure 4.33 in Chapter 4.
9. Compute the swale length using the design velocity from step 8 and an assumed hydraulic detention time, preferably no less than 5 minutes. If the computed swale length is less than 30 m, increase the swale length to 30 m and adjust the bottom width.

■ Example 6.51

A swale is to be excavated on a 1% slope to handle a design runoff of 0.03 m^3/s. During the wet season, the swale is expected to be covered with grass having an average height of 130 mm with type E retardance. Design the swale as a biofilter.

Solution

From the given data: $Q = 0.03$ m^3/s, $S_o = 0.01$, and the average height of the vegetation is 130 mm. This given data covers the specifications in steps 1 to 3 of the design procedure.

Step 4: The design depth in the swale should be at least 50 mm below the height of the vegetation (130 mm $-$ 50 mm $=$ 80 mm), with a maximum height of 75 mm. Therefore, in this case, the design flow depth is taken as 75 mm.

Step 5: A preliminary estimate of the Manning n is 0.24, which corresponds to infrequently mowed grass in deep swales.

Step 6: Use a trapezoidal section for the swale.

Step 7: Use side slopes of 4:1 ($m = 4$), a bottom width b, and a depth $y = 75$ mm ($= 0.075$ m). The flow area, A, wetted perimeter, P, and hydraulic radius, R, are given by

$$A = by + my^2 = b(0.075) + (4)(0.075)^2 = 0.075b + 0.0225$$

$$P = b + 2\sqrt{1 + m^2}\,y = b + 2\sqrt{1 + 4^2}(0.075) = b + 0.618$$

$$R = \frac{A}{P} = \frac{0.075b + 0.0225}{b + 0.618}$$

where 0.6 m $< b <$ 2.5 m.

Step 8: The Manning equation requires that

$$Q = \frac{1}{n}AR^{2/3}S_o^{1/2} = \frac{1}{n}\frac{A^{5/3}}{P^{2/3}}S_o^{1/2}$$

In this case,

$$0.03 = \frac{1}{n}\frac{(0.075b + 0.0225)^{5/3}}{(b + 0.618)^{2/3}}(0.01)^{1/2}$$

or

$$\frac{1}{n^3}\frac{(0.075b + 0.0225)^5}{(b + 0.618)^2} = 0.027$$

This equation can be used to find values of b for assumed values of n. The values of n must be consistent with the product of the velocity, V ($= Q/A$), and the hydraulic radius R, as given in Figure 4.33 for type E retardance, and the computations are summarized in the following table:

Assumed n	b (m)	A (m^2)	P (m)	R (m)	V (m/s)	VR (m^2/s)	VR (ft^2/s)	n (Fig. 4.33)
0.024	0.394	0.0521	1.01	0.0514	0.576	0.0296	0.32	0.052
0.052	1.05	0.101	1.67	0.0607	0.296	0.0180	0.19	0.063
0.063	1.30	0.120	1.92	0.0626	0.250	0.0156	0.17	0.063

Hence a swale with a bottom width of 1.30 m and side slopes of 4:1 will accommodate the design flow at a velocity of 0.25 m/s, which is less than the limiting velocity of 0.30 m/s.

Step 9: Using a detention time of 5 minutes with the design velocity of 0.25 m/s gives the length, *L*, of the swale as

$$L = Vt = (0.25)(5 \times 60) = 75 \text{ m}$$

which is longer than the minimm length of 30 m.

In summary, the swale is designed to have a trapezoidal cross-section with a bottom width of 1.30 m, side slopes of 4:1, and be 75 m long. ■

The total depth of the swale is determined by ensuring that the swale has sufficient capacity to accomodate the highest expected flow in the swale, typically resulting from a 10- to 100-year design storm. Under this high-flow condition, the depth of flow in the swale is calculated using the procedures described in Chapter 4, Section 4.5.4. The depth of the swale should be equal to the maximum expected depth of flow plus 0.3 m freeboard.

Filter Strips. Filter strips perform in a similar manner to swales but are not channels. *Filter strips* (also called *buffer strips*) are mildly sloping vegetated surfaces that are located adjacent to impervious surfaces. These areas are designed to slow the runoff velocity from the impervious area, increasing the opportunities for infiltration and trapping the pollutants. The design procedure for filter strips is the same as that for swales, with the additional constraints that the average depth of flow be no more than 25 mm and the hydraulic radius be taken equal to the flow depth (ASCE, 1998). The width of the filter strip should be sufficiently limited to achieve a uniform-flow distribution.

Exfiltration Trenches. Exfiltration trenches, also called *percolation trenches* and *french drains*, are common in urban areas with large impervious areas and high land costs. An exfiltration trench typically consists of a long narrow excavation, ranging from 1 to 4 m in depth, backfilled with gravel aggregate (2.5 to 7.6 cm) and surrounded by a filter fabric to prevent the migration of fine soil particles into the trench (Harrington, 1989). The maximum trench depth is limited by trench-wall stability, seasonal high ground-water levels, and the depth to any impervious soil layer. Exfiltration trenches 1 m wide and 1 to 2 m deep seem to be most efficient (ASCE, 1998). In some cases, the surface runoff enters the trench through a perforated pipe that is centrally located within the aggregate that fills the trench; in other cases, the trench is constructed in swales to accept infiltrated flow directly from the ground surface.

Exfiltration trenches are considered off-line systems, and the purpose of these trenches are to store and exfiltrate the runoff from frequent storms into the ground. Exfiltration trenches are typically designed to serve single-family residential areas up to 4 ha (10 ac) and commercial areas up to 2 ha (5 ac) in size. These trenches must be carefully designed, installed, and maintained since they are very susceptible to clogging (especially during construction); once they are clogged, rehabilitation is difficult. It is usually desirable to have a minimum 6-m grass buffer to remove larger sediment particles prior to runoff entering the trench (Harrington, 1989).

Exfiltration trenches can only be used at sites with porous soils, favorable site geology, and proper ground-water conditions. Site conditions that are favorable to exfiltration trenches are (Stahre and Urbonas, 1989; Harrington, 1989; ASCE, 1998):

1. The hydraulic conductivity surrounding the trench exceeds 2 m/d.
2. The distance between the bottom of the trench and the seasonal high-water table or bedrock exceeds 1.2 m.
3. Water-supply wells are more than 30 m from the trench (to prevent possible contamination).

4. The trench is located at least 6 m from building foundations to avoid possible hydrostatic pressures on foundations or basements.

5. The ground slope downstream of the trench does not exceed 20%, which would increase the chance of downstream seepage and slope failure.

Exfiltration trenches are generally designed to retain a volume equal to the difference between the runoff volume and the volume of water exfiltrated during a storm (ASCE, 1998). Assuming that the water in the trench percolates through one-half of the trench height, and that there is negligible outflow from the bottom of the trench (due to clogging), the total outflow rate, Q_{out}, from the two (long) sides of the trench is given by Darcy's law (see Chapter 7, Section 7.2.1) as

$$Q_{out} = 2\left(K_t \frac{H}{2} L\right) = K_t H L \tag{6.186}$$

where K_t is the trench hydraulic conductivity, H is the height of the trench, L is the length of the trench, and the hydraulic gradient is assumed equal to unity. Taking t as the duration of the storm, the volume exfiltrated in time t, V_{out}, is given by

$$V_{out} = Q_{out}t = K_t H L t \tag{6.187}$$

The runoff volume into the trench during time t, V_{in}, is given by the rational formula (see Section 6.4.2.1) as

$$V_{in} = CiAt \tag{6.188}$$

where C is the runoff coefficient, i is the average intensity of a storm with duration t, and A is the area of the catchment contributing flow to the exfiltration trench. The storage capacity of the trench, V_{stor}, is given by

$$V_{stor} = nWHL \tag{6.189}$$

where n is the porosity in the trench, typically taken as 40% (ASCE, 1998). The trench dimensions must be such that

$$V_{in} = V_{stor} + V_{out} \tag{6.190}$$

Combining Equations 6.187 to 6.190 yields

$$CiAt = nWHL + K_t H L t \tag{6.191}$$

Solving for the trench length, L, gives

$$\boxed{L = \frac{CiAt}{(nW + K_t t)H}} \tag{6.192}$$

The rainfall intensity, i, is related to the storm duration, t, by an intensity-duration-frequency (IDF) curve that typically has the form

$$i = \frac{a}{(t + b_1)^{c_1}} \tag{6.193}$$

where a, b_1, and c_1 are constants (see Section 6.2.1). The combination of Equations 6.192 and 6.193 indicates that the required trench length, L, varies as a function of the storm duration, t, and the design length should be chosen as the maximum length required for any storm duration (with a given return period). In some cases, the runoff from a specified rainfall depth is required to be handled by the trench, and in these cases ASCE (1998) recommends that the trench be designed for an IDF curve with a return period in which the specified rainfall depth falls in 1 hour.

▪ Example 6.52

An exfiltration trench is to be designed to handle the runoff from a 1-ha commercial area with an average runoff coefficient of 0.7. The IDF curve for the design rainfall is

$$i = \frac{548}{(t + 7.24)^{0.73}}$$

where i is the average rainfall intensity in mm/h and t is the storm duration in minutes. The trench hydraulic conductivity estimated from field tests is 15 m/d, the seasonal high water table is 5 m below the ground surface, and local regulations require that a safety factor of 2 be applied to the trench hydraulic conductivity to account for clogging. Design the exfiltration trench.

Solution

From the given data: $C = 0.7$, $A = 1$ ha $= 10,000$ m^2, and $K_t = 15/2 = 7.5$ m/d $= 0.31$ m/h (using a safety factor of 2). According to ASCE (1998) guidelines, the porosity, n, of the gravel pack can be taken as 40% ($n = 0.4$), and a trench width, W, of 1 m and height, H, of 2 m can be expected to perform efficiently. Substituting these values into Equation 6.192 gives

$$L = \frac{CiAt}{(nW + K_t t)H} = \frac{(0.7)i(10000)t}{(0.4 \times 1 + 0.31t)(2)} = \frac{7000it}{0.8 + 0.62t} \qquad (6.194)$$

where i in m/h, and t in hours are related by

$$i = \frac{0.548}{(60t + 7.24)^{0.73}} \text{ m/h} \qquad (6.195)$$

Combining Equations 6.194 and 6.195 gives the required trench length, L, as a function of the storm duration, t, as

$$L = \frac{3836t}{(0.8 + 0.62t)(60t + 7.24)^{0.73}}$$

and values of L versus t are given in the following table:

t (h)	0.0	0.1	0.2	0.3	0.4	0.5	0.6	0.7	0.8	0.9	1.0	1.1	1.2	1.3	1.4	1.5
L (m)	0	68	96	111	119	123	126	127	127	126	126	124	123	121	120	118

On the basis of these results, a trench length of 127 m gives the trench volume required to handle the design storm without causing surface ponding. Since the seasonal high-water table is 5 m below the ground surface and the minimum allowable spacing between the bottom of the trench and the water table is 1.2 m, a (maximum) trench height of 5 m − 1.2 m = 3.8 m would still be adequate and would yield the shortest possible trench length (for the specified trench width of 1 m). Since the trench length is inversely proportional to the trench height (see Equation 6.192), using a trench height, H, of 3.8 m would give a required trench length, L, of

$$L = 127 \text{ m} \times \frac{2}{3.8} = 67 \text{ m}$$

Hence, a trench capable of handling the design storm is 1 m wide, 3.8 m high, and 67 m long. If the (vertical) side slopes are not stable for a trench of this depth, then the trench height and length can be adjusted according to their inverse proportionality. ■

6.7.3 ■ Major System

The major drainage system includes features such as natural and constructed open channels, streets, and drainage easements such as floodplains. The major drainage system handles runoff events that exceed the capacity of the minor drainage system and must be planned concurrently with the design of the minor system. The design of open channels has been discussed extensively in Chapter 4, Section 4.5. In major urban drainage systems, concrete- and grass-lined channels are the most common, with grass lining usually preferred for aesthetic reasons. Concrete-lined channels have smaller roughness coefficients, require smaller flow areas, and are used when hydraulic, topographic, and right-of-way needs are important considerations.

6.8 Evapotranspiration

The abstraction processes of evaporation and transpiration do not contribute significantly to the water budget over time scales of individual storms, but over longer time periods (weeks and months) these processes are major components in the terrestrial water budget. On an annual basis, approximately 70% of the rainfall in the United States is returned to the atmosphere via evaporation and transpiration. Evaporation and transpiration processes are of primary interest in the design of irrigation and surface-water storage systems. *Evaporation* is the process by which water is transformed from the liquid phase to the vapor phase, and *transpiration* is the process by which water moves through plants and evaporates through leaf stomatae. In cases where the ground surface is covered by vegetation, it is usually not feasible to differentiate between evaporation from the ground surface and transpiration through plants; these combined processes are collectively called *evapotranspiration*. The estimation of evapotranspiration is the basis for designing most irrigation projects and water-resource developments.

Three standard evapotranspiration rates are commonly used in practice: (1) potential evapotranspiration, (2) reference-crop evapotranspiration, and (3) actual evapotranspiration. *Potential evapotranspiration* was defined by Penman (1956) as the amount of water transpired in unit time by a short green crop, completely shading the ground, of uniform height and never short of water. Potential evapotranspiration is used synonymously with the term *potential evaporation*, which is commonly defined as the quantity of water evaporated per unit area, per unit time from an idealized, extensive free water surface under existing atmospheric conditions. *Reference-crop evapotranspiration* is defined as the rate of evaporation from an area planted with a specific (reference) crop, where water availability is not a limiting factor. Shuttleworth (1993) defines a standard reference crop as an idealized grass crop with a fixed height of 12 cm, an albedo of 0.23, and a surface resistance of 69 s m^{-1}. *Surface resistance* is a parameter of the Penman-Monteith equation and is defined in Section 6.8.1. A more widely cited definition of a reference crop was given by Doorenbos and Pruitt (1977) as an extensive surface of 8-cm to 15-cm tall green grass cover of uniform height, actively growing, completely shading the ground and not short of water. The *actual evapotranspiration* is defined as the evapotranspiration that occurs under actual soil, ground cover, and water-availability conditions. Potential evaporation is typically used as a measure of the meteorological control on evaporation from an open water surface, such as a lake or reservoir, while reference-crop evapotranspiration is used as a measure of evapotranspiration from a standard vegetated surface.

Over the past several decades, hundreds of empirical and semi-empirical methods of estimating evapotranspiration have evolved. These methods can be broadly classified as *combination*, *radiation*, *temperature*, or *evaporation-pan* methods. Combination methods are the most comprehensive and account for both energy utilization as well as the processes required to remove the vapor, once it has evaporated into the atmosphere. Radiation and temperature methods relate evapotranspiration solely to net radiation (solar plus long-wave) and air temperature, respectively. In contrast to the combination methods, the radiation and temperature methods neglect the wind velocity and specific-humidity gradient above the evaporating surface, both of which relate to the ability of the air to transport water vapor away from the evaporative surface. Evaporation-pan methods are empirical formulations that relate the evapotranspiration to measured water evaporation from standardized open pans. Several evapotranspiration methods of various types are listed in Table 6.35. However, there seems to be a consensus among present-day hydrologists that the physically based Penman-Monteith equation provides the best description of the evaporation and evapotranspiration processes. This assertion is supported by several quantitative comparisons between measured and predicted evapotranspiration at several test sites around the world (ASCE, 1990).

■ **Table 6.35**
Methods for Estimating Evapotranspiration

Classification	Method	References
Combination	Penman-Monteith	Monteith (1965); Allen (1986); Allen et al. (1989)
	Penman	Penman (1963)
	1972 Kimberly-Penman	Wright and Jensen (1972)
	1982 Kimberly-Penman	Wright (1982)
	FAO-24 Penman	Doorenbos and Pruitt (1975; 1977)
	FAO-PPP-17 Penman	Frére and Popov (1979)
	Businger-van Bavel	Businger (1956); van Bavel (1966)
Radiation	Jensen-Haise	Jensen and Haise (1963); Jensen et al. (1971)
	FAO-24 Radiation	Doorenbos and Pruitt (1975; 1977)
	Priestly-Taylor	Priestly and Taylor (1972)
	Turc	Turc (1961); Jensen (1966)
Temperature	NRCS Blaney-Criddle	USDA (1970)
	FAO-24 Blaney-Criddle	Doorenbos and Pruitt (1977); Allen and Pruitt (1986)
	Hargreaves	Hargreaves et al. (1985); Hargreaves and Samani (1985)
	Thornthwaite	Thornthwaite (1948); Thornthwaite and Mather (1955)
Evaporation-pan	Christiansen	Christiansen (1968); Christiansen and Hargreaves (1969)
	FAO-24 Pan	Doorenbos and Pruitt (1977)

Source: ASCE, 1990. *Evapotranspiration and Irrigation Water Requirements.* p. 165. Reprinted by permission of ASCE.

6.8.1 ■ The Penman-Monteith Equation

The *Penman-Monteith method* (Monteith, 1981) is the most widely recommended approach to estimating the evapotranspiration, ET, from meteorological data. The basic hypothesis of the Penman-Monteith approach is that the transpiration of water through leaves is composed of three serial processes: the transport of water through the surface of the leaves against a canopy resistance, r_c; molecular diffusion against a molecular boundary layer resistance, r_b; and turbulent transport against an aerodynamic resistance, r_a, between the layer in the immediate vicinity of the canopy surface and the planetary boundary layer. The boundary layer resistance, r_b, for water vapor is usually much smaller than the aerodynamic resistance, r_a, or the canopy resistance, r_c, and can be ignored in comparison with both other resistances. The Penman-Monteith equation for estimating the evapotranspiration, ET, from vegetated surfaces is given by

$$ET = \frac{1}{\rho_w \lambda} \left[\frac{\Delta A + \rho_a c_p D / r_a}{\Delta + \gamma(1 + r_c / r_a)} \right] \tag{6.196}$$

where ρ_w is the density of water; λ is the latent heat of vaporization of water; Δ is the gradient of the saturated vapor pressure versus temperature curve; A is the available energy (solar plus long wave); ρ_a is the density of moist air; c_p is the specific heat of moist air ($= 1.013$ kJ/(kg°C)); D is the vapor-pressure deficit, equal to the saturation vapor pressure minus the actual vapor pressure; r_a is the aerodynamic resistance to vapor and heat diffusion; γ is the psychrometric constant; and r_c is the bulk stomatal (canopy) resistance. Equation 6.196 is dimensionally homogeneous, and any consistent set of units can be used. Methods that are commonly used to calculate the parameters in the Penman-Monteith equation are described here.

Aerodynamic Resistance, r_a. In the absence of significant thermal stratification, the aerodynamic resistance between the canopy surface and a height z above the surface is given by (Calder, 1993)

$$r_a = \frac{\left[\ln\left(\frac{z-d}{z_o}\right) \right]^2}{k^2 u(z)} \tag{6.197}$$

where $u(z)$ is the wind speed at height z, k is von Kármán's constant ($= 0.41$), d is the zero-plane displacement height, and z_o is the roughness height. The height, z, above the canopy at which the wind speed, $u(z)$, is measured should be at least 2 m. According to Calder (1993), d and z_o can be estimated from the relations

$$d = 0.75h \tag{6.198}$$

and

$$z_o = 0.1h \tag{6.199}$$

where h is the average height of the surface vegetation.

Canopy Resistance, r_c. The canopy resistance, r_c, varies as leaf stomatae open and close in response to various micrometeorological conditions and is dependent on the particular plant species. It is generally assumed that the canopy resistances of trees are greater than those of shorter vegetation, since trees tend to have stomatal control, while shorter vegetation does not (Olmsted, 1978). Allen and colleagues (1989) suggested that values of r_c can be estimated from the *leaf-area index*, LAI, using the relation

$$r_c = \frac{100}{0.5\text{LAI}}$$

(6.200)

where LAI is defined as the ratio of the surface area of the leaves to the projection of the vegetation on the ground surface. The surface area of only one side of each leaf is counted in calculating the leaf area index. Typical short-grass vegetation has a canopy resistance on the order of 70 s m^{-1}.

Latent Heat of Vaporization, λ.   The latent heat of vaporization, λ, can be expressed as a function of the water-surface temperature, T_s, using the empirical equation (Harrison, 1963)

$$\lambda = 2.501 - 0.002361T_s$$

(6.201)

where λ is in MJ/kg and T_s is in °C.

Vapor Pressure Gradient, Δ.   The gradient of the saturated vapor pressure versus temperature curve, Δ, can be estimated using the equation (Shuttleworth, 1993)

$$\Delta = \frac{4098e_s}{(237.3 + T)^2}$$

(6.202)

where Δ is in kPa/°C, e_s is the saturation vapor pressure (kPa), and T is the air temperature (°C). Since e_s is determined by the air temperature, T, (Equation 6.216), then Δ can be calculated directly from the air temperature.

Available Energy, A.   Neglecting energy storage, the available energy, A, is approximately equal to the net solar (short-wave) radiation, S_n, plus the net long-wave radiation, L_n, hence

$$A = S_n + L_n$$

(6.203)

According to Shuttleworth (1993), the net short-wave radiation can be estimated by

$$S_n = (1 - \alpha)\left(a_s + b_s\frac{n}{N}\right)S_0$$

(6.204)

where α is the *albedo*, defined as the fraction of short-wave radiation reflected at the surface; a_s and b_s are empirical constants; n is the number of bright-sunshine hours per day (h), N is the total number of daylight hours in the day (h); and S_0 is the extraterrestrial radiation (MJ m^{-2}d^{-1}). Typical albedos for various surfaces are given in Table 6.36; a more detailed list of albedos can be found in Ponce and colleagues (1997).

   The empirical constants a_s and b_s are usually estimated from field measurements of the incident solar radiation. However, in the absence of experimental data, the following values are recommended (Shuttleworth, 1993)

$$a_s = 0.25, \qquad b_s = 0.50$$

(6.205)

■ **Table 6.36**
Typical Albedos

Land Cover	Albedo, α
Open water	0.08
Tall forest	0.11–0.16
Tall farm crops	0.15–0.20
Cereal crops	0.20–0.26
Short farm crops	0.20–0.26
Grass and pasture	0.20–0.26
Bare soil	0.10 (wet)–0.35 (dry)

Source: Shuttleworth, James M. and Maidement, David R. 1993. *Handbook of Hydrology*. The McGraw-Hill Companies. Reprinted by permission of The McGraw-Hill Companies.

The number of bright-sunshine hours per day can be estimated by a combination of theoretical and measured results. Theoretically, the maximum possible number of daylight hours, n_{max} is given by (Duffie and Beckman, 1980)

$$n_{max} = \frac{24}{\pi}\omega_s \qquad (6.206)$$

where ω_s is the sunset hour angle (radians) given by

$$\omega_s = \cos^{-1}(-\tan\phi\tan\delta) \qquad (6.207)$$

where ϕ is the average latitude of the site, and δ is the solar declination given by

$$\delta = 0.4093\sin\left(\frac{2\pi}{365}J - 1.405\right) \qquad (6.208)$$

where δ is in radians, and J is the Julian day number. Cloudy skies generally result in a reduction in the number of hours of bright sunshine. The amount of sunshine as a percentage of the maximum possible amount of sunshine, p_s, is combined with the maximum number of daylight hours, n_{max}, to yield the number of bright-sunshine hours per day, n, using the relation

$$n = p_s n_{max} \qquad (6.209)$$

The extraterrestrial solar radiation, S_0, can be estimated using the following equation (Duffie and Beckman, 1980)

$$S_0 = 37.7d_r(\omega_s\sin\phi\sin\delta + \cos\phi\cos\delta\sin\omega_s) \qquad (6.210)$$

where S_0 is in MJ/(m^{2}d), and d_r is the relative distance between the earth and the sun given by

$$d_r = 1 + 0.033\cos\left(\frac{2\pi}{365}J\right) \qquad (6.211)$$

In addition to the short-wave (0.3 μm to 3 μm) solar energy that is added to the surface vegetation, there is also long-wave radiation (3 μm to 100 μm) that is emitted by both the atmosphere and the ground. According to Shuttleworth (1993), the net incoming long-wave radiation, L_n, can be estimated using the equation

$$L_n = -f\epsilon'\sigma(T + 273.2)^4 \tag{6.212}$$

where L_n is in MJ m^{-2}d^{-1}, f is a cloudiness factor, ϵ' is the net emissivity between the atmosphere and the ground, σ is the Stefan-Boltzmann constant (4.903×10^{-9} MJ m^{-2}K^{-4}d^{-1}), and T is the mean air temperature (°C). The cloudiness factor, f, can be estimated by the equation

$$f = \left(a_c\frac{b_s}{a_s + b_s}\right)\frac{n}{N}\left(b_c + \frac{a_s}{a_s + b_s}a_c\right) \tag{6.213}$$

where a_c, b_c, a_s, and b_s are empirical constants; n is the number of hours of bright sunshine per day; and N is total number of daylight hours in one day. Suggested values for the empirical constants are

$$a_c = 1.00, \qquad b_c = 0.00, \qquad a_s = 0.25, \qquad b_s = 0.50 \tag{6.214}$$

The net emissivity, ϵ', can be estimated using the equation (Brunt, 1932; Allen et al., 1989)

$$\epsilon' = a_e + b_e\sqrt{e_d} \tag{6.215}$$

where a_e and b_e are empirical constants and e_d is the vapor pressure (kPa) of water in the atmosphere. The average vapor pressure for each month of the year can be estimated using average (monthly) temperatures and relative humidities. Using these average temperatures, the saturation vapor pressure, e_s, can be estimated from the following equation (Bosen, 1960)

$$e_s \approx 3.38639[(0.00738T + 0.8072)^8 - 0.000019|1.8T + 48| + 0.001316] \tag{6.216}$$

where e_s is in kPa and the ambient temperature, T, is in °C. The ambient vapor pressure, e_d, can then be estimated from the saturation vapor pressure, e_s, and relative humidity, RH, by the relation

$$e_d = \frac{\text{RH}}{100}e_s \tag{6.217}$$

The empirical constants a_e and b_e can be estimated by (Doorenbos and Pruitt, 1975; 1977)

$$a_e = 0.34, \qquad b_e = -0.14 \tag{6.218}$$

Combining Equations 6.212 to 6.218 yields estimates of the net long-wave radiation, L_n, in MJ m^{-2} d^{-1}, which are converted to mm/d by dividing by $\rho_w\lambda$. Negative values of L_n are possible and indicate that there is a net loss of long-wave energy, thereby reducing the energy flux that is available for ET. The short-wave and long-wave radiation, S_n and L_n, are combined to yield the available energy, A, according to Equation 6.203.

Air Density, ρ_a.   The density of moist air, ρ_a, is a function of the temperature and pressure of the air, and can be estimated using the relation

$$\rho_a = 3.486\frac{p}{275 + T} \text{ kg/m}^3 \tag{6.219}$$

where the atmospheric pressure, p, can be assumed to equal 101.32 kPa.

Vapor-Pressure Deficit, D. The vapor-pressure deficit, D, is defined by the relation

$$D = e_s - e_d \qquad (6.220)$$

and can also be written in the form

$$D = e_s\left(1 - \frac{\text{RH}}{100}\right) \qquad (6.221)$$

where e_s is the saturation vapor pressure, e_d is the actual vapor pressure, and RH is the relative humidity. The saturation vapor pressure, e_s, can be derived from the air temperature using Equation 6.216, and the relative humidity, RH, is derived from averaged measurements at the site under investigation. Substituting the values for e_s and RH into Equation 6.221 yields the vapor-pressure deficit.

Psychrometric Constant, γ. The psychrometric constant, γ, depends on the atmospheric pressure, p, and the latent heat of vaporization, λ, in accordance with

$$\gamma = \frac{c_p p}{\epsilon \lambda} \times 10^{-3} = 0.0016286\frac{p}{\lambda} \text{ kPa/°C} \qquad (6.222)$$

where the specific heat of moist air, c_p, is taken as 1.013 kJ/(kg°C), and ϵ is the ratio of the molecular weight of water vapor to the molecular weight of dry air ($= 0.622$). Assuming an atmospheric pressure of 101.32 kPa, and a latent heat of vaporization of 2.444 MJ/kg, then the psychrometric constant, γ, is equal to 0.06752 kPa/°C.

■ Example 6.53

The monthly-averaged measurements of rainfall, humidity, wind speed, air temperature, and fraction of daylight that is bright sunshine for Miami, Florida, are given in the following table.

Month	Rainfall (mm)	Humidity (%)	Wind speed (m/s)	Air temperature (°C)	Daylight fraction
Jan	51	83	4.0	19	0.69
Feb	58	82	3.6	15	0.68
Mar	56	79	4.0	18	0.77
Apr	94	77	4.0	21	0.78
May	150	76	3.6	23	0.71
Jun	230	83	3.1	25	0.74
Jul	190	80	3.1	26	0.76
Aug	210	80	2.7	26	0.75
Sep	220	82	3.1	26	0.72
Oct	150	83	3.1	26	0.72
Nov	64	83	4.0	21	0.67
Dec	43	83	3.6	18	0.65
Total	1,516				

Miami is located at approximately 26°N latitude. Compute the monthly reference-crop ET using the Penman method, compare the annual ET with the annual rainfall, and assess the implications of this differential on available water resources.

Solution

The Penman equation is given by Equation 6.196, and the parameters of this equation are either given or can be computed from the given data. The computational procedure will be illustrated for the month of January. The Penman equation estimates the evapotranspiration, ET, by

$$ET = \frac{1}{\rho_w \lambda} \left[\frac{\Delta A + \rho_a c_p D / r_a}{\Delta + \gamma (1 + r_c / r_a)} \right]$$

and the computation of the parameters in the Penman equation for January are as follows:

ρ_w: The density of water, ρ_w, can be taken to be equal to 998 kg/m³.

λ: The latent heat of vaporization, λ, is given by Equation 6.201 as

$$\lambda = 2.501 - 0.002361 T_s = 2.501 - 0.002361(19) = 2.46 \text{ MJ/kg}$$

Δ: The slope of the vapor pressure versus temperature curve, Δ, is given by Equation 6.202 as

$$\Delta = \frac{4098 e_s}{(237.3 + T)^2}$$

where e_s is the saturation vapor pressure given by Equation 6.216 as

$$e_s \approx 3.38639[(0.00738 T + 0.8072)^8 - 0.000019|1.8T + 48| + 0.001316]$$

$$\approx 3.38639[(0.00738(19) + 0.8072)^8 - 0.000019|1.8(19) + 48| + 0.001316]$$

$$\approx 2.20 \text{ kPa}$$

and therefore Δ is given by

$$\Delta = \frac{4098(2.20)}{(237.3 + 19)^2} = 0.137 \text{ kPa/°C}$$

A: The available energy, A, is equal to the sum of the net short-wave and long-wave radiation according to Equation 6.203, where

$$A = S_n + L_n$$

The net short-wave radiation can be estimated using Equation 6.204 as

$$S_n = (1 - \alpha) \left(a_s + b_s \frac{n}{N} \right) S_0$$

where the albedo, α, is equal to 0.23 for the reference crop; a_s and b_s can be taken as 0.25 and 0.5, respectively; and the fraction of bright sunshine in the daylight hours is given for January as 0.69. The net short-wave radiation can therefore be written as

$$S_n = (1 - 0.23)[0.25 + 0.5(0.69)]S_0 = 0.46 S_0 \tag{6.223}$$

S_0 is the extraterrestrial radiation given by Equation 6.210 as

$$S_0 = 37.7 d_r (\omega_s \sin \phi \sin \delta + \cos \phi \cos \delta \sin \omega_s)$$

where d_r is the relative distance between the earth and the sun given by Equation 6.211 as

$$d_r = 1 + 0.033 \cos\left(\frac{2\pi}{365}J\right) = 1 + 0.033 \cos\left(\frac{2\pi}{365}(15)\right) = 1.032$$

where the mean Julian day, J, for January is taken as 15. The solar declination, δ, is given by Equation 6.208 as

$$\delta = 0.4093 \sin\left[\frac{2\pi}{365}J - 1.405\right] = 0.4093 \sin\left[\frac{2\pi}{365}(15) - 1.405\right] = -0.373 \text{ radians}$$

The sunset hour angle, ω_s, is given by Equation 6.207 as

$$\omega_s = \cos^{-1}[-\tan\phi\tan\delta] = \cos^{-1}[-\tan(26°)\tan(-0.373)] = 1.38 \text{ radians}$$

where the latitude, ϕ, is given as $26°$ (= 0.454 rad). Substituting the calculated values of d_r, ω_s, and δ into the expression for S_0 yields

$$S_0 = 37.7(1.032)[(1.38)\sin(0.454)\sin(-0.373) + \cos(0.454)\cos(-0.373)\sin(1.38)]$$
$$= 26 \text{ MJ/(m}^2\text{d)}$$

The net short-wave radiation, S_n, for January is therefore given by Equation 6.223 as

$$S_n = 0.46S_0 = 0.46(26) = 12 \text{ MJ/(m}^2\text{d)}$$

The net long-wave radiation, L_n, is given by Equation 6.212 as

$$L_n = -f\epsilon'\sigma(T + 273.2)^4$$

The cloudiness factor, f, can be estimated using Equation 6.213 as

$$f = \left(a_c\frac{b_s}{a_s + b_s}\right)\frac{n}{N}\left(b_c + \frac{a_s}{a_s + b_s}a_c\right)$$

where it can be assumed that $a_c = 1.00$, $b_c = 0.00$, $a_s = 0.25$, $b_s = 0.50$, and the daylight fraction, n/N, for January is given as 0.69. Therefore,

$$f = \left(1.00\frac{0.50}{0.25 + 0.50}\right)(0.69)\left(0.00 + \frac{0.25}{0.25 + 0.50}1.00\right) = 0.153$$

The net emissivity, ϵ', can be estimated using Equation 6.215 as

$$\epsilon' = a_e + b_e\sqrt{e_d}$$

where a_e and b_e can be taken as 0.34 and -0.14, respectively, and the vapor pressure in the atmosphere, e_d can be estimated as

$$e_d = \frac{\text{RH}}{100}e_s$$

where RH is the given relative humidity for January (= 83%) and e_s is the saturation vapor pressure (= 2.20 kPa). The net emissivity, ϵ', is therefore given by

$$\epsilon' = 0.34 + (-0.14)\sqrt{(0.89)(2.20)} = 0.151$$

The Stefan-Boltzmann constant, σ, is equal to 4.903×10^{-9} MJ m^{-2}K^{-4}d^{-1} and the net long-wave radiation for January can be calculated as

$$L_n = -(0.153)(0.151)(4.903 \times 10^{-9})(19 + 273.2)^4 = -0.827 \text{ MJ/(m}^2\text{d)}$$

where the negative value indicates that the net long-wave radiation in January is away from the earth. The total available energy in January is then equal to the sum of S_n and L_n and is given by

$$A = S_n + L_n = 12 - 0.827 = 11.2 \text{ MJ/(m}^2\text{d)}$$

ρ_a: The density of air, ρ_a, can be estimated using Equation 6.219 as

$$\rho_a = 3.486\frac{p}{275 + T} \text{ kg/m}^3$$

Taking the atmospheric pressure, p, as 101 kPa yields

$$\rho_a = 3.486\frac{101}{275 + 19} = 1.20 \text{ kg/m}^3$$

c_p: The specific heat of moist air, c_p, can be taken as a constant equal to 1.013 kJ/(kg°C).
D: The vapor deficit, D, is defined by Equation 6.220 as

$$D = e_s - e_d$$

and for January

$$D = 2.20 - (0.83)(2.20) = 0.374 \text{ kPa}$$

where the relative humidity in January is given as 83%.
r_a: The aerodynamic resistance, r_a, is given by Equation 6.197 as

$$r_a = \frac{\left[\ln\left(\frac{z-d}{z_o}\right)\right]^2}{k^2 u(z)}$$

It can be assumed that the wind speed is characteristic of a height, $z = 2$ m above the ground surface; the displacement height, d, can be estimated as $0.75\,h$, where h is the height of the reference crop ($= 0.12$ m), and therefore $d = (0.75)(0.12) = 0.094$ m, the roughness height, z_o, can be estimated as $0.1h = (0.1)(0.12) = 0.012$ m, the von Kármán constant, k, can be taken as 0.41, and the average wind speed for January is given as 4 m/s. The aerodynamic resistance, r_a, for January is therefore given by

$$r_a = \frac{\left[\ln\left(\frac{2-0.094}{0.012}\right)\right]^2}{(0.41)^2(4)} = 38 \text{ s/m}$$

γ: The psychrometric constant, γ, can be estimated using Equation 6.222 as

$$\gamma = 0.0016286\frac{p}{\lambda} \text{ kPa/°C}$$

where the atmospheric pressure, p, can be taken as 101 kPa, and the latent heat of vaporization, λ, for January has already been calculated as 2.46 MJ/kg. Therefore, γ is given by

$$\gamma = 0.0016286\frac{101}{2.46} = 0.067 \text{ kPa/°C}$$

r_c: The canopy resistance, r_c, for the reference crop is 69 s/m.

Based on the calculated parameters for the month of January, the reference-crop evapotranspiration estimated by the Penman equation is given by

$$ET = \frac{1}{\rho_w \lambda}\left[\frac{\Delta A + \rho_a c_p D/r_a}{\Delta + \gamma(1 + r_c/r_a)}\right]$$

$$= \frac{1}{(998)(2.46)}\left[\frac{(0.137)(11.2) + (86.4)(1.20)(1.013)(0.374)/(38)}{(0.137) + (0.067)(1 + 69/38)}\right]$$

$$= 0.0032 \text{ m/d}$$

where the conversion factor 86.4 has been used in the numerator to adjust the units. The *ET* in January is more conveniently written as 3.2 mm/d. The computations illustrated here for January are repeated for other months of the year, and the results are tabulated as follows:

Month	λ (MJ/kg)	Δ (kPa/°C)	A (MJ/m^{2}d)	ρ_a (kg/m^3)	D (kPa)	r_a (s/m)	γ (kPa/°C)	ET (mm/d)	ET (mm)
Jan	2.46	0.137	11.0	1.20	0.374	38	0.067	3.2	98
Feb	2.47	0.110	12.5	1.22	0.307	42	0.067	3.0	85
Mar	2.46	0.130	15.3	1.21	0.433	38	0.067	4.1	126
Apr	2.45	0.153	16.7	1.19	0.572	38	0.067	4.9	147
May	2.45	0.170	16.0	1.19	0.674	42	0.067	5.1	160
Jun	2.44	0.189	16.4	1.18	0.538	49	0.068	4.9	148
Jul	2.44	0.199	16.6	1.17	0.672	49	0.068	5.3	166
Aug	2.44	0.199	16.5	1.17	0.672	57	0.068	5.3	164
Sep	2.44	0.199	15.4	1.17	0.605	49	0.068	4.9	147
Oct	2.44	0.199	13.8	1.17	0.571	49	0.068	4.5	138
Nov	2.45	0.153	11.3	1.19	0.423	38	0.067	3.4	103
Dec	2.46	0.130	10.1	1.21	0.351	42	0.067	2.9	90
Total									1,572

The annual *ET* is calculated to be 1,572 mm, and the annual rainfall is given as 1,516 mm. This indicates that for areas composed of the reference crop with adequate supplies of water there will be a water deficit, which could only be maintained by an inflow of water from surrounding areas. ■

It is important to note that the Penman-Monteith equation is most accurate with hourly data, however, the method still gives reasonably reliable estimates of *ET* using meteorological data that are averaged over longer time intervals (ASCE, 1990). In using climatic data averaged over longer time intervals than an hour, the effect of the phase relationship between various climatic variables on the evapotranspiration process is lost.

The Penman-Monteith equation can also be used to estimate the evaporation from large open bodies of water. The albedo of open water is commonly taken as 0.08, and the aerodynamic resistance, r_a, of open water is estimated using the equation (Thom and Oliver, 1977)

$$r_a = \frac{4.72[\ln(z_m/z_0)]^2}{1 + 0.536U_2} \text{ s/m} \tag{6.224}$$

where z_m is the measurement height of meteorological variables above the water surface, z_0 is the aerodynamic roughness of the surface, and U_2 is the wind speed 2m above the water surface. Assuming that meteorological variables are characteristic of conditions

2 m above the water surface and using the recommended value of 1.37 mm for z_0 (Thom and Oliver, 1977), the aerodynamic resistance is determined by the wind speed. These aerodynamic resistances are generally higher than found for either short vegetation or forest and reflect a greater resistance to vapor transport due to lower levels of turbulence. The canopy resistance, r_c, for lakes is zero by definition, indicating that there are no restrictions to water moving between the liquid and vapor phases at the surface of the lake.

6.8.2 ■ Evaporation Pans

Evaporation pans provide a direct measure of evaporation in the field and are used to estimate both the potential evapotranspiration, E_{po}, and the reference-crop evapotranspiration, E_{rc}, by multiplying pan-evaporation measurements by *pan coefficients*. According to Shuttleworth (1993), mean monthly values of E_{po} and E_{rc} can usually be estimated from pan measurements to within 10 percent in most climates. The most commonly used pan design is the U.S. Weather Bureau Class A pan, which is 122 cm (4 ft) in diameter, 25.4 cm (10 in.) deep, and made of either galvanized iron or Monel™ metal. The pan is mounted on a wooden frame 15 cm above ground level and is filled with water to within 5 cm of the rim. The water level should not fall more than 7.5 cm below the rim, and the pan should be surrounded by short grass turf (4 to 10 cm high) that is well irrigated during the dry season. A Class A pan station generally includes an anemometer mounted 15 cm (6 in.) above the pan rim and, in some cases, the pan is covered with a mesh to prevent animals from entering the pan or drinking the water. A 12.5-mm (0.5-in.) mesh screen is commonly used. The mesh lowers the measured evaporation by 5%–20% (Stanhill, 1962; Dagg, 1968). Potential evapotranspiration, E_{po}, is estimated from pan measurements by multiplicative factors called *pan coefficients*. Therefore

$$\boxed{E_{po} = k_p E_p} \tag{6.225}$$

where k_p is the pan coefficient and E_p is the measured pan evaporation.

Pan coefficients are typically in the range of 0.35 to 0.85 (Doorenbos and Pruitt, 1975; 1977); representative pan coefficients for various site conditions are given in Table 6.37. Key factors affecting the pan coefficient are the average wind speed, upwind fetch characteristics, and the ambient humidity. The estimation of evapotranspiration using the pan coefficients in Table 6.37 is commonly called the *FAO-24 pan evaporation method*. This method was published by the Food and Agriculture Organization (FAO) in paper number 24, hence the name. The pan coefficients given in Table 6.37 are applicable to short-irrigated grass turf. For taller and aerodynamically rougher crops, the values of k_p would be higher and vary less with differences in weather conditions (ASCE, 1990). Where a standard pan environment is maintained and where strong dry-wind conditions occur only occasionally, mean monthly evapotranspiration for well-watered short grass should be predictable to within ± 10% or better for most climates.

■ Example 6.54

An evaporation pan indicates a monthly evaporation of 130 mm. The pan is surrounded by a short green crop for approximately 100 m in the upwind direction, the average wind speed is 3 m/s, and the mean relative humidity is 90%. Estimate the potential evapotranspiration for short irrigated grass turf.

■ Table 6.37		Case A: Pan surrounded by short green crop				Case B: Pan Surrounded by dry, bare area			
Pan Coefficients for Various Site Conditions			Mean relative humidity, %				Mean relative humidity, %		
Wind	Upwind fetch of green crop, m	Low < 40	Med 40–70	High > 70	Upwind fetch of dry fallow, m	Low < 40	Med 40–70	High > 70
Light	0	0.55	0.65	0.75	0	0.70	0.80	0.85
(< 1 m/s)	10	0.65	0.75	0.85	10	0.60	0.70	0.80
	100	0.70	0.80	0.85	100	0.55	0.65	0.75
	1000	0.75	0.85	0.85	1000	0.50	0.60	0.70
Moderate	0	0.50	0.60	0.65	0	0.65	0.75	0.80
(2–5 m/s)	10	0.60	0.70	0.75	10	0.55	0.65	0.70
	100	0.65	0.75	0.80	100	0.50	0.60	0.65
	1000	0.70	0.80	0.80	1000	0.45	0.55	0.60
Strong	0	0.45	0.50	0.60	0	0.60	0.65	0.70
(5–8 m/s)	10	0.55	0.60	0.65	10	0.50	0.55	0.65
	100	0.60	0.65	0.70	100	0.45	0.50	0.60
	1000	0.65	0.70	0.75	1000	0.40	0.45	0.55
Very strong	0	0.40	0.45	0.50	0	0.50	0.60	0.65
(> 8 m/s)	10	0.45	0.55	0.60	10	0.45	0.50	0.55
	100	0.50	0.60	0.65	100	0.40	0.45	0.50
	1000	0.55	0.60	0.65	1000	0.35	0.40	0.45

Source: Doorenbos and Pruitt (1977). Reprinted by permission of Food and Agriculture Organization of the United Nations.

Solution

From the given data: $U = 3$ m/s, $F = 100$ m, RH = 0.90, and $E_p = 130$ mm; Table 6.37 gives $k_p = 0.80$. The potential evapotranspiration, E_{po}, is therefore given by

$$E_{po} = k_p E_p = (0.80)(130) = 104 \text{ mm}$$ ■

6.9 Computer Models

Several good computer models are available for simulating surface runoff processes in both urban and rural catchments. In their most general form, these models simulate the rainfall-runoff process and route the runoff through drainage structures such as storm sewers, detention basins, and canals. In engineering practice, the use of computer models to apply the fundamental principles covered in this chapter is usually essential. Typically, there are a variety of models to choose from for a particular application. However, in doing work that is to be reviewed by regulatory bodies, models developed and maintained by agencies of the U.S. government have the greatest credibility and, perhaps more important, are almost universally acceptable in supporting permit applications and defending design protocols on liability issues. A secondary guideline in choosing a model is that the simplest model that will accomplish the design objectives should be given the most serious consideration. Several of the more widely used models that have been developed and endorsed by U.S. government agencies are described briefly here.

■ **SWMM.** This model was developed and is maintained by the U.S. Environmental Protection Agency (USEPA). SWMM consists principally of a Runoff Module, a Transport Module, and a Storage/Treatment Module. The catchment is separated into subareas that are connected to small gutter/pipe elements leading to an outfall location. The Runoff Module simulates runoff quantity and quality by means of kinematic overflow routing within subareas, the Transport Module receives subarea hydrographs and uses the kinematic approximation to route sewer flows and to compute hydraulic head at key junctions, and the Storage/Treatment Module simulates the effects of detention basins and combined sewer overflows.

■ **TR 20.** TR 20 was developed and is maintained by the Natural Resources Conservation Service, U.S. Department of Agriculture. This model represents the catchment as several subbasins with homogeneous properties. The NRCS curve-number method is used to calculate the runoff hydrographs resulting from a single storm; the runoff hydrographs from the subareas are routed through drainage channels and reservoirs using the kinematic method (channel routing) and the storage indication method (reservoir routing). The catchment-outflow hydrograph is the sum of the routed hydrographs at the catchment outlet.

■ **HEC-1/HEC-HMS.** HEC-1 was developed and is maintained by the U.S. Army Corps of Engineers Hydrologic Engineering Center. This model sumulates the catchment as a series of hydraulic and hydrologic components and calculates the runoff from single storms. The user can choose from a variety of submodels to simulate precipitation, infiltration, and runoff, as well as a variety of techniques to route the flows through rivers and reservoirs. Special features of HEC-1 include dam safety and failure analysis, flood damage analysis, and parameter optimization. The program HEC-HMS (Hydrologic Modeling System), which is in the process of superseding HEC-1, has many of the same capabilities as HEC-1 but with improved user, graphics, and reporting facilities.

Only a few of the more widely used computer models have been cited here, and there are certainly many other good models that are capable of performing the same tasks.

Summary

This chapter focuses on the movement and control of surface runoff resulting from rainfall. The material covered here is particularly applicable to the design of urban stormwater-management systems and provides the fundamental background necessary to design stormwater-management systems in accordance with the ASCE Manual of Practice (ASCE, 1992). The chapter begins with the quantitative specification of design rainfall events, which include: return period, duration, depth, temporal distribution, and spatial distribution. After specifying the design rainfall, abstraction models are used to account for interception (Equation 6.14), depression storage (Table 6.8), and infiltration (Horton, Green-Ampt, or NRCS models). Removal of the abstractions from the design rainfall yields the temporal distribution of runoff depth, which is used to calculate either the peak-runoff or continuous-runoff hydrograph. In small catchments, peak runoff rates are adequately calculated using the rational or TR-55 methods, and the computed peak runoff rates are used in sizing gutters, inlets, and storm sewers.

Complete runoff hydrographs are used to size storage basins and are typically computed using a unit-hydrograph model, although the time-area, kinematic-wave, nonlinear-reservoir, and Santa Barbara models are also used in practice.

The quality of stormwater entering storage basins can be estimated using either the USGS or USEPA water-quality models. Stormwater management systems consist of both a minor and a major system. Components of minor stormwater systems include street gutters, inlets, sewers, and storage basins, while the major stormwater system includes drainage pathways when the capacity of the minor system is exceeded. Procedures for designing storm sewers to accommodate peak runoff rates are presented. Stormwater impoundments are designed so that peak postdevelopment runoff is less than or equal to peak predevelopment runoff and so that sufficient detention time is provided for water-quality control. The storage-indication method is used to route runoff hydrographs through storage basins, and detention time and retention volume criteria are used to ensure an adequate level of treatment.

Water consumption by evapotranspiration is of primary importance in designing irrigation systems in agricultural areas and in computing long-term water balances. The Penman-Monteith equation is widely regarded as one of the most accurate and technically sound methods for estimating evapotranspiration, and the application of this method is presented in detail. In engineering practice, many of the techniques presented in this chapter are implemented using readily available computer programs. A few of the more widely used programs that have been developed by U.S. government agencies are SWMM, TR 20, and HEC-1/HEC-HMS.

Problems

6.1. Show that the maximum return period that can be investigated when using the Weibull formula to derive the IDF curve from a n-year annual rainfall series is equal to $n + 1$.

6.2. A rainfall record contains 50 years of measurements at 5-minute intervals. The annual maximum rainfall amounts for intervals of 5 min, 10 min, 15 min, 20 min, 25 min, and 30 min are ranked as follows:

Rank	\multicolumn{6}{c}{Δt in minutes}
	5	**10**	**15**	**20**	**25**	**30**
1	26.2	45.8	60.5	72.4	81.8	89.7
2	25.3	44.0	58.1	69.6	78.6	86.3
3	24.2	42.2	55.8	66.8	75.5	82.8

where the rainfall amounts are in millimeters. Calculate the IDF curve for a return period of 40 years.

6.3. An eight-year rainfall record measures the rainfall increments at 10-minute intervals, and the top five rainfall increments are as follows:

Rank	1	2	3	4	5
10-min rainfall (mm)	39.6	39.4	38.5	37.3	36.5

Estimate the frequency distribution of the annual maxima with return periods greater than two years.

6.4. Assuming that the ranked rainfall increments in Problem 6.2 were derived from a partial-duration series, calculate the annual-series IDF curve for a 40-year return period.

6.5. Use the Chen method to estimate the 10-year IDF curve for Atlanta, Georgia.

6.6. Use the Chen method to estimate the 10-year IDF curves for New York City and Los Angeles. Compare your results with those obtained by Wenzel (1982) and listed in Table 6.2.

6.7. The spatially averaged rainfall is to be calculated for the catchment area shown in Figure 6.7, where the coordinates of the nearby rain gages are as follows:

Gage	x (km)	y (km)
A	1.3	7.0
B	1.0	3.7
C	4.2	4.9
D	3.5	1.4
E	2.1	−1.0

and the coordinates of the grid origin are (0, 0). The measured rainfall at each of the gages during a 1-hour interval are 60 mm at A, 90 mm at B, 65 mm at C, 35 mm at D, and 20 mm at E. Use the reciprocal-distance method with the 1 km × 1 km grid shown in Figure 6.7 to estimate the average rainfall over the catchment during the 1-hour interval. Assume that the rainfall weights are proportional to the inverse distance squared.

6.8. Repeat Problem 6.7 by assigning equal weight to each rain gage. Compare your result with that obtained using the reciprocal-distance squared method. Which result do you think is more accurate?

6.9. The IDF curve for 10-year storms in Atlanta, Georgia, is given by

$$i = \frac{1628}{(t_d + 8.16)^{0.76}}$$

where i is the average intensity in mm/h and t_d is the duration in minutes. Assuming that the mean of the rainfall distribution is equal to 38% of the rainfall duration, estimate the triangular hydrograph for a 40-minute storm.

6.10. The IDF curve for 10-year storms in Santa Fe, New Mexico, is given by

$$i = \frac{818}{(t_d + 8.54)^{0.76}}$$

where i is the average intensity in mm/h and t_d is the duration in minutes. Assuming that the mean of the rainfall distribution is equal to 44% of the rainfall duration, estimate the triangular hydrograph for a 50-minute storm.

6.11. Use the alternating-block method and the IDF curve given in Problem 6.10 to calculate the hyetograph for a 10-year 1-hour storm using 11 time intervals.

6.12. Derive the IDF curve for Boston, Massachusetts, using the Chen method. Use the derived IDF curve with the alternating-block method to determine the hyetograph for a storm with a return period of 10 years and a duration of 40 minutes. Use nine time intervals to construct the hyetograph.

6.13. Use the 10-year IDF curve for Atlanta in Table 6.2 given by Wenzel (1982) and the alternating-block method to estimate the rainfall hyetograph for a 10-year 50-minute storm in Atlanta. Use seven time intervals.

6.14. The alternating-block method assumes that the maximum rainfall for any duration less than or equal to the total storm duration has the same return period. Discuss why this is a conservative assumption.

6.15. The precipitation resulting from a 25-year 24-hour storm in Miami, Florida, is estimated to be 260 mm. Calculate the NRCS 24-hour hyetograph.

6.16. The precipitation resulting from a 25-year 24-hour storm in Atlanta, Georgia, is estimated to be 175 mm. Calculate the NRCS 24-hour hyetograph, and compare this hyetograph with the 25-year 24-hour hyetograph for Miami that is calculated in Problem 6.15.

6.17. The rainfall measured at a rain gage during a 10-hour storm is 193 mm. Estimate the average rainfall over a catchment containing the rain gage, where the area of the catchment is 200 km^2.

6.18. Use Equation 6.11 to show that for long-duration storms the areal-reduction factor becomes independent of the storm duration. Under these asymptotic conditions, determine the relationship between the areal-reduction factor and the catchment area.

6.19. Estimate the maximum amount of rainfall that can be expected in one hour.

6.20. Maximum precipitation amounts can be estimated using Equation 6.12, and observed precipitation maxima are listed in Table 6.6. Compare the predictions given by Equation 6.12 with the observations in Table 6.6.

6.21. A pine forest is to be cleared for a development in which all the trees on the site will be removed. The IDF curve for the area is given by

$$i = \frac{2819}{t + 16}$$

where i is the average rainfall intensity in mm/h and t is the duration in minutes. The storage capacity of the trees in the forest is estimated as 5 mm, the leaf area index is 6, and the evaporation rate during the storm is estimated as 0.3 mm/h. Determine the increase in precipitation reaching the ground during a 30-min storm that will result from clearing the site.

6.22. The storage capacity of a forest canopy covering a catchment is 9 mm, the leaf-area index is 8, and the evaporation rate during a storm is 0.5 mm/h. If a storm has the hyetograph calculated in Problem 6.13, estimate the amount of precipitation intercepted during each of the seven time intervals of the hyetograph. Determine the hyetograph of the rainfall reaching the ground.

6.23. For the pine forest described in Problem 6.21, estimate the interception using a Horton-type empirical equation of the form $I = a + bP^n$, where a and b are constants and P is the precipitation amount.

6.24. A 25-min storm produces 30 mm of rainfall on the following surfaces: steep pavement, flat pavement, impervious surface, lawn, pasture, and forest litter. Assuming that depression storage is the dominant abstraction process, for each surface estimate the fraction of rainfall that becomes surface runoff. On which surface is depression storage most significant, and on which surface is depression storage least important?

6.25. A soil has the Horton infiltration parameters: $f_o = 200$ mm/h, $f_c = 60$ mm/h, and $k = 4$ min^{-1}. If rainfall in excess of 200 mm/h is maintained for 50 min, estimate the infiltration as a function of time. What is the infiltration rate at the end of 50 min? How would this rate be affected if the rainfall rate were less than 200 mm/h?

6.26. Show that the Horton infiltration model given by Equation 6.32 satisfies the differential equation

$$\frac{df_p}{dt} = -k(f_p - f_c)$$

6.27. A catchment has the Horton infiltration parameters: $f_o = 1{,}504$ mm/h, $f_c = 50$ mm/h, and $k = 3$ min^{-1}. The design storm is:

Interval (min)	Average rainfall (mm/h)
0–10	20
10–20	40
20–30	80
30–40	170
40–50	90
50–60	20

Estimate the time when ponding begins.

6.28. An alternating-block analysis indicated the following rainfall distribution for a 70-min storm:

Interval	Rainfall (mm/h)
1	7
2	25
3	45
4	189
5	97
6	44
7	18

where each interval corresponds to 10 minutes. If infiltration can be described by the Horton parameters: $f_o = 600$ mm/h, $f_c = 30$ mm/h, $k = 0.5$ min^{-1}, and the depression storage is 4 mm, use the Horton method to determine the distribution of runoff, and hence the total runoff.

6.29. An area consists almost entirely of sandy loam, which typically has a saturated hydraulic conductivity of 11 mm/h, average suction head of 110 mm, porosity of 0.45, field capacity of 0.190, wilting point of 0.085, and depression storage of 4 mm. The design rainfall is given as:

Interval (min)	Average rainfall (mm/h)
0–10	20
10–20	40
20–30	60
30–40	110
40–50	60
50–60	20

Use the Green-Ampt method to determine the runoff versus time for average initial moisture conditions, and contrast the depth of rainfall with the depth of runoff. Assume that the initial moisture conditions are midway between the field capacity and wilting point.

6.30. Repeat Problem 6.29 for a sandy clay soil with a depression storage of 9 mm. (*Hint:* Use Table 6.10 to estimate the soil properties.)

6.31. Derive the NRCS curve-number model for the infiltration rate given by Equation 6.61. Explain why this infiltration model is unrealistic.

6.32. Drainage facilities are to be designed for a rainfall of return period 10 years and duration 1 hour. The IDF curve is given by

$$i = \frac{203}{(t + 7.24)^{0.73}}$$

where i is the average intensity in cm/h and t is the storm duration in minutes. The minimum infiltration rate is 10 mm/h, and the area to be drained is primarily residential with lot sizes on the order of 0.2 ha (0.5 ac). Use the NRCS method to estimate the total amount of runoff (in cm), assuming the soil is in average condition at the beginning of the design storm.

6.33. Repeat Problem 6.32 for the case in which heavy rainfall occurs within the previous five days and the soil is saturated.

6.34. Use Equation 6.61 to calculate the average infiltration rate during the storm described in Problem 6.32. Compare this calculated infiltration rate with the given minimum infiltration rate of 10 mm/h.

6.35. A proposed 20-ha development includes 5 ha of parking lots, 10 ha of buildings, and 5 ha of grassed area. The runoff from the parking lots and buildings are both routed directly to grassed areas. If the grassed areas contain Type A soil in good condition, estimate the runoff from the site for a 180 mm rainfall event.

6.36. Repeat Problem 6.35 for the case where the runoff from the buildings, specifically the roofs of the buildings, is discharged directly onto the parking lots. Based on this result, what can you infer about the importance of directing roof drains to pervious areas?

6.37. Repeat Problem 6.35 using an area-weighted curve number. How would your result change if the roof drains are directly connected to the parking lot?

6.38. Discuss why it is preferable to route rainfall excesses on composite areas rather than using weighted-average curve numbers.

6.39. An catchment with a grass surface has an average slope of 0.8%, and the distance from the catchment boundary to the outlet is 80 m. For a 30-min storm with an effective rainfall rate of 70 mm/h, estimate the time of concentration using: (a) the kinematic-wave equation, (b) the NRCS method, (c) the Kirpich equation, (d) the Izzard equation, and (e) the Kerby equation.

6.40. What is the maximum flow distance that should be described by overland flow?

6.41. Find α and m in the kinematic wave model (Equation 6.75) corresponding to: (a) the Manning equation, and (b) the Darcy-Weisbach equation.

6.42. An asphalt pavement drains into a rectangular concrete channel. The catchment surface has an average slope of 1.0%, and the distance from the catchment boundary to the drain is 30 m. The drainage channel is 60 m long, 20 cm wide, 25 cm deep, and has a slope of 0.6%. For an effective rainfall rate of 50 mm/h, the flowrate in the channel is estimated to be 0.02 m^3/s. Estimate the time of concentration of the catchment.

6.43. The surface of a 2-ha catchment is characterized by a runoff coefficient of 0.5, a Manning n for overland flow of 0.25, an average overland flow length of 60 m, and an average slope of 0.5%. Calculate the time of concentration using the kinematic-wave equation. The drainage channel is to be sized for the peak runoff resulting from a 10-year rainfall event, and the 10-year IDF curve is given by

$$i = \frac{150}{(t + 8.96)^{0.78}}$$

where i is the average rainfall intensity in cm/h and t is the duration in minutes. The minimum time of concentration is 5 minutes. Determine the peak runoff rate.

6.44. Explain why higher runoff coefficients should be used for storms with longer return periods.

6.45. Suppose that the catchment described in Problem 6.43 contains 0.5 ha of impervious area that is directly connected to the storm sewer. If the runoff coefficient of the impervious area is 0.9, the Manning n for overland flow on the impervious surface is 0.035, the average flow length is 30 m, and the average slope is 0.5%, then estimate the peak runoff.

6.46. A 4.2-km^2 catchment with 0.5% pond area has a curve number of 79, a time of concentration of 3 h, and a 24-h Type II precipitation of 10 cm. Estimate the peak runoff.

6.47. A 1-km^2 catchment with 3% pond area has a curve number of 70, a time of concentration of 1.7 h, and a 24-h Type I precipitation of 13 cm. Estimate the peak runoff.

6.48. The 15-min unit hydrograph for a 2.1-km^2 urban catchment is given by

Time (min)	Runoff (m^3/s)	Time (min)	Runoff (m^3/s)
0	0	210	0.66
30	1.4	240	0.49
60	3.2	270	0.36
90	1.5	300	0.28
120	1.2	330	0.25
150	1.1	360	0.17
180	1.0	390	0

(a) Verify that the unit hydrograph is consistent with a 1-cm rainfall excess; (b) estimate the runoff hydrograph for a

15-min rainfall excess of 2.8 cm; and (c) estimate the runoff hydrograph for a 30-min rainfall excess of 10.3 cm.

6.49. The 15-min unit hydrograph for a 2.1-km^2 catchment is given in Problem 6.48. Determine the S-hydrograph, calculate the 50-min unit hydrograph for the catchment, and verify that your calculated unit hydrograph corresponds to a 1-cm rainfall excess.

6.50. The 15-min unit hydrograph for a 2.1-km^2 catchment is given in Problem 6.48. Estimate the runoff resulting from the following 120-min storm:

Time (min)	Rainfall (cm)
0–30	2.4
30–60	4.5
60–90	2.1
90–120	0.8

6.51. Calculate the Espey-Altman 10-minute unit hydrograph for a 3-km^2 catchment. The main channel has a slope of 0.9% and a Manning n of 0.10, the catchment is 50% impervious, and the distance along the main channel from the catchment boundary to the outlet is 1,100 m.

6.52. If the rainfall excess versus time from a 2.25-km^2 catchment is given by

Time (min)	Rainfall excess (mm)
0–10	0
10–20	10
20–30	15
30–40	0
40–50	0

and the 10-min unit hydrograph derived using the Espey-Altman method is given by

Time (h)	Runoff (m^3/s)
0	0
1.36	0.634
1.81	0.949
2.13	1.27
2.77	0.949
3.68	0.634
15.3	0

Estimate the runoff hydrograph of the catchment.

6.53. A 3-km^2 catchment has an average curve number of 60, an average slope of 0.8%, and a flow length from the catchment boundary to the outlet of 1,000 m. If the catchment is 40% impervious, determine the NRCS unit hydrograph for a 20-minute storm.

6.54. Determine the approximate NRCS triangular unit hydrograph for the catchment described in Problem 6.53, and verify that it corresponds to a rainfall excess of 1 cm.

6.55. Show that any triangular unit hydrograph that has a peak runoff, Q_p, in (m³/s)/cm, given by

$$Q_p = 2.08 \frac{A}{T_p}$$

must necessarily have a runoff duration equal to $2.67 T_p$, where T_p is the time to peak in hours, and A is the catchment area in km².

6.56. A 75-ha catchment has the following time-area relation:

Time (min)	Contributing area (ha)
0	0
5	2
10	7
15	19
20	38
25	68
30	75

Estimate the runoff hydrograph using the time-area method for a 30-min rainfall-excess distribution given by

Time (min)	Average intensity (mm/h)
0–5	120
5–10	70
10–15	50
15–20	30
20–25	20
25–30	10

6.57. A 2-ha catchment consists of pastures with an average slope of 0.5%. The width of the catchment is approximately 250 m, and the depression storage is estimated to be 10 mm. Use the nonlinear reservoir model to calculate the runoff from the following 20-min rainfall excess:

Time (min)	Effective rainfall (mm/h)
0–5	110
5–10	50
10–15	40
15–20	10

6.58. A 1-km² catchment has a time of concentration of 25 minutes and is 45% impervious. Use the Santa Barbara urban hydrograph method to estimate the runoff hydrograph for the following rainfall event:

Time (min)	Rainfall (mm/h)	Rainfall excess (mm/h)
0	0	0
10	50	0
20	200	130
30	103	91
40	52	11

Use a time increment of 10 minutes to calculate the runoff hydrograph.

6.59. A stormwater detention basin has the following storage characteristics:

Stage (m)	Storage (m³)
8.0	0
8.5	1,041
9.0	2,288
9.5	3,761
10.0	5,478
10.5	7,460
11.0	9,726

The discharge weir from the detention basin has a crest elevation of 8.0 m, and the weir discharge, Q (m³/s), is given by

$$Q = 3.29 h^{\frac{3}{2}}$$

where h is the height of the water surface above the crest of the weir in meters. The catchment runoff hydrograph is given by:

Time (min)	Runoff (m³/s)
0	0
30	3.6
60	8.4
90	5.1
120	4.2
150	3.6
180	3.3
210	2.7
240	2.3
270	1.8
300	1.5
330	0.84
360	0.51
390	0

If the prestorm stage in the detention basin is 8.0 m, estimate the discharge hydrograph from the detention basin.

6.60. The flow hydrograph at a channel section is given by:

Time (min)	Flow (m³/s)
0	0.0
30	12.5
60	22.1
90	15.4
120	13.6
150	12.4
180	11.7
210	10.8
240	9.9
270	8.4
300	8.1
330	7.5
360	4.2
390	0.0

Use the Muskingum method to estimate the hydrograph 1,000 m downstream from the channel section. Assume that $X = 0.3$ and $K = 35$ min.

6.61. Write a simple finite-difference model to solve the kinematic-wave equation for channel routing. [*Hint*: The kinematic-wave equation is obtained by combining Equations 6.154 and 6.156.]

6.62. A 70-ha catchment is 50% impervious, with 60% commercial and industrial use. The site is located in a city where there are typically 84 storms per year, the mean annual rainfall is 95 cm and the mean minimum January temperature is 8.1°C. Estimate the annual load of suspended solids contained in the runoff.

6.63. Calculate the annual load of lead in the runoff for the site described in Problem 6.62.

6.64. A 80-ha residential development has a population density of 15 persons per hectare, streets are swept every two weeks, and the area is 25% impervious. If the average annual rainfall is 98 cm, estimate the annual load of phosphate (PO_4) expected in the runoff.

6.65. A stormwater-management system is to be designed for a new 10-ha residential development. The site is located in a region where there are 62 storms per year with at least 1.3 mm of rain, and the IDF curve is given by

$$i = \frac{6000}{t + 20} \text{ mm/h} \qquad (6.226)$$

where i is the average rainfall intensity and t is the duration of the storm in minutes. The mean annual rainfall in the region is 98.5 cm, and the mean January temperature is 9.6°C. The developed site is to have an impervious area of 65%, an

estimated depression storage of 1.5 mm, an estimated population density of 25 persons/hectare, and no street sweeping. Estimate the average concentration in mg/L of suspended sediments in the runoff using: (a) the USGS model, and (b) the EPA model.

6.66. Consider the two inlets and two pipes shown in Figure 6.34. Catchment A has an area of 0.5 ha and is 60% impervious, and catchment B has an area of 1 ha and is 15% impervious. The runoff coefficient, C; length of overland flow, L; roughness coefficient, n; and average slope, S_o, of the pervious and impervious surfaces in both catchments are given in the following table:

Catchment	Surface	C	L (m)	n	S_0
A	pervious	0.2	80	0.2	0.01
	impervious	0.9	60	0.1	0.01
B	pervious	0.2	140	0.2	0.01
	impervious	0.9	65	0.1	0.01

The IDF curve of the design storm is given by

$$i = \frac{8000}{t + 40}$$

where i is the average rainfall intensity in mm/h and t is the duration of the storm in minutes. Calculate the peak flows in the inlets and pipes.

6.67. Within the storm-sewer system for the site described in Problem 6.65, pipe I and pipe II intersect (at a manhole) and flow into Pipe III. All pipes have a slope of 2%. Pipe I and pipe II both drain 1-ha areas that are 65% impervious, with the impervious area directly connected to the inlet to pipe I, but the inlet to pipe II is not directly connected to any impervious area. The runoff coefficients can be taken as 0.3 and 0.9 for the pervious and impervious areas, respectively. The times of concentration of the catchments contributing to pipes I and II are:

Pipe	Area	t_c (min)
I	all	25
	DCIA*	12
II	all	30
	DCIA*	—

*DCIA means Directly Connected Impervious Area.

If the flow time in pipe I and pipe II are 3 min (in each pipe), estimate the design flows in pipes I, II, and III. What diameter of concrete pipe would you use for pipe III?

6.68. A concrete sewer pipe is to be laid on a slope of 0.90% and is to be designed to carry 0.50 m³/s of stormwater runoff.

Estimate the required pipe diameter using the Manning equation.

6.69. Repeat Problem 6.68 using the Darcy-Weisbach equation.

6.70. If service manholes are placed along the pipeline in Problem 6.68, estimate the head loss at each manhole.

6.71. A four-lane collector roadway is to be constructed with 3.66-m (12-ft) lanes, a cross-slope of 1.5%, a longitudinal slope of 0.8%, and pavement of smooth asphalt. If the roadway drainage system is to be designed for a rainfall intensity of 120 mm/h, determine the spacing of the inlets.

6.72. A roadway has a maximum allowable flow depth at the curb of 9 cm and a corresponding flowrate in the gutter of 0.1 m^3/s. Determine the length of a 15-cm (6-in.) high curb inlet that is required to remove all the water from the gutter. Consider the cases where (a) the width of the inlet depression is 0.3 m, and (b) there is no inlet depression.

6.73. A roadway has a cross slope of 1.5%, a maximum allowable flow depth at the curb of 9 cm, and a corresponding flowrate in the gutter of 0.1 m^3/s. The gutter flow is to be removed by a 1.8-cm thick grate inlet that is mounted flush with the curb. Calculate the minimum dimensions of the grate inlet.

6.74. A roadway has a cross-slope of 1.5%, a logitudinal slope of 0.5%, and a flowrate in the gutter of 0.2 m^3/s. The flow is to be removed by a slotted drain with a slot width of 5 cm, and a Manning n quoted by the manufacturer as 0.017. Estimate the minimum length of slotted drain that can be used.

6.75. A 45-cm diameter concrete culvert drains a storage basin, and the culvert entrance is shaped such that $C_d = 0.80$. Estimate the basin discharge when the water in the basin is 1.1 m above the center of the culvert entrance and the culvert is discharging freely into a drainage channel.

6.76. A storage basin is drained by a weir with a crest length of 45 cm and a discharge coefficient, C_d, of 0.60. Estimate the basin discharge when the water in the basin is 1.1 m above the crest of the weir.

6.77. Show that if the bottom of a storage reservoir is a rectangle of dimensions $L \times W$, the longitudinal cross-section is a trapezoid with base W and side slope angle α, and the transverse cross-section is rectangular with base L, then the stage-storage relation is given by

$$S = \frac{L}{\tan \alpha} h^2 + (LW)h$$

where S is the storage, and h is the depth.

6.78. The runoff hydrographs from a site before and after development are as follows:

Time (min)	Before (m³/s)	After (m³/s)
0	0	0
30	2.0	3.5
60	7.5	10.6
90	1.7	7.5
120	0.90	5.1
150	0.75	3.0
180	0.62	1.5
210	0.49	0.98
240	0.30	0.75
270	0.18	0.62
300	0.50	0.51
330	0	0.25
360	0	0.12
390	0	0

The postdevelopment detention basin is to be a wet detention reservoir drained by an outflow weir. The elevation versus storage in the detention basin is

Elevation (m)	Storage (m³)
0	0
0.5	11,022
1.0	24,683
1.5	41,522

where the weir crest is at elevation 0 m, which is also the initial elevation of the detention basin prior to runoff. The performance of the weir is given by

$$Q = 1.83bh^{\frac{3}{2}}$$

where Q is the overflow rate in m^3/s, b is the crest length in m, and h is the head of the weir in m. Determine the crest length of the weir for the detention basin to perform its desired function. What is the maximum water surface elevation expected in the detention basin?

6.79. The runoff from the site described in Problem 6.65 is to be routed to a dual-purpose dry detention reservoir for both flood control and water-quality control. The predevelopment and postdevelopment runoff volumes can be estimated using the USEPA empirical formula (Equation 6.161), where the predevelopment imperviousness and depression storage are 10% and 2.5 mm, respectively. The peak predevelopment runoff is 3 m^3/s, and the stage-storage relation of the detention reservoir is:

Elevation (m)	Storage (m³)
0.0	0
0.5	300
1.0	600
1.5	900

The basin is to be drained by a circular orifice, where the discharge, Q, is related to the area of the opening, A, and the stage, h, by

$$Q = 0.65A\sqrt{2gh} \qquad (6.227)$$

Make a preliminary estimate the required diameter of the orifice opening.

6.80. A runoff hydrograph is routed through a wet detention basin, and the results are given in the following table:

Time (min)	Inflow (m^3/s)	Storage (m^3)	Outflow (m^3/s)
0	0	5,000	0
30	1.1	5,873	0.13
60	3.8	9,478	0.81
90	0.95	11,827	1.37
120	0.55	10,909	1.15
150	0.40	9,910	0.91
180	0.35	9,104	0.73
210	0.29	8,497	0.59
240	0.19	7,979	0.47
270	0.11	7,489	0.37
300	0	7,025	0.30
330	0	6,584	0.24
360	0	6,206	0.17
390	0	5,918	0.14
420	0	5,698	0.11
450	0	5,531	0.08
480	0	5,405	0.07
510	0	5,311	0.06
540	0	5,239	0.05
570	0	5,180	0.04
600	0	5,135	0.03
630	0	5,104	0.02
660	0	5,081	0.01
690	0	5,063	0.01
720	0	5,050	0.01

Estimate the detention time and evacuation time of the detention basin.

6.81. If the inflow and outflow hydrographs given in Problem 6.80 are for a dry detention basin, estimate the detention time and evacuation time in the basin.

6.82. Determine the required volume of a dry detention basin for a 25-ha residential development, where the runoff coefficient is estimated as 0.4 and the average rainfall depth during the year is 4.2 cm. Estimate the suspended solids removal in the basin.

6.83. An infiltration basin to retain the first 2.5 cm of runoff from a 20 ha catchment. The area to be used for the infiltration basin is turfed, and the soil has a minimum infiltration rate of 100 mm/h. If the retained runoff is to infiltrate within 36 hours, determine the surface area to be set aside for the basin.

6.84. Derive Equation 6.183 for a swale with a triangular cross section.

6.85. Derive Equation 6.184 for a swale with a trapezoidal cross section.

6.86. A triangular-shaped swale is to retain the runoff from a catchment with a design runoff rate of 0.01 m^3/s. The longitudinal slope of the swale is to be 1.5% with side slopes of 5:1 (H:V). If the grassed swale has a Manning n of 0.030 and a minimum infiltration rate of 200 mm/h, determine the length of swale required.

6.87. Repeat Problem 6.86 for a trapezoidal swale with a bottom width of 1 m.

6.88. A swale is to be laid on a 2% slope to handle a design runoff of 0.02 m^3/s. During the wet season, the swale is expected to be covered with grass having an average height of 100 mm with type E retardance. Design the swale as a biofilter.

6.89. An exfiltration trench is to be designed to handle the runoff from a 3-ha residential area with a runoff coefficient of 0.5. The IDF curve for the design rainfall is given by

$$i = \frac{403}{(t + 8.16)^{0.69}} \ \text{mm/h}$$

where i is the average rainfall intensity, and t is the storm duration in minutes. The trench hydraulic conductivity is 35 m/d, and the seasonal high-water table is 4.6 m below the ground surface. Assuming a safety factor of 3 for the trench hydraulic conductivity, design the exfiltration trench.

6.90. Monthly-averaged measurements of rainfall, humidity, wind speed, air temperature, and fraction of daylight that is bright sunshine for a location at 30°N are given in the following table:

Month	Rainfall (mm)	Humidity (%)	Wind speed (m/s)	Air temp. (°C)	Daylight fraction
Jan	55	73	3.5	17	0.70
Feb	63	74	3.2	13	0.65
Mar	71	76	3.8	17	0.75
Apr	95	79	3.6	18	0.79
May	120	83	3.1	22	0.73
Jun	180	84	2.8	23	0.76
Jul	190	88	3.1	24	0.79
Aug	210	87	3.2	24	0.82
Sep	210	81	2.9	25	0.73
Oct	120	79	2.8	23	0.75
Nov	70	77	3.7	19	0.79
Dec	40	78	3.5	17	0.61
Total	1,424				

Compute the monthly reference-crop *ET* using the Penman method.

6.91. An evaporation pan measures a monthly evaporation of 152 mm. The pan is surrounded by a short green crop for approximately 300 m in the upwind direction, the average wind speed is 4 m/s, and the mean relative humidity is 80%. Estimate the potential evapotranspiration for short irrigated grass turf.

7 Ground-Water Hydrology

Introduction

Ground-water hydrology is the science dealing with the movement, quantity, quality, and distribution of water below the surface of the earth. A major application of the principles of ground-water hydrology is in the development of water supplies by means of wells and infiltration galleries. Other important applications of ground-water hydrology include the evaluation, mitigation, and remediation of contaminated ground water; the storage of surface waters in underground reservoirs; and the lowering of ground water levels to permit crop growth. Ground water accounts for approximately 30% of all the fresh water on earth, which is second only to the polar ice (69%), and two orders of magnitude greater than the amount of fresh water in lakes and rivers (0.3%).

The subsurface environment consists of a porous medium in which the void spaces have varying degrees of water saturation. Regions where the void spaces are completely filled with water are called *zones of saturation*, and regions where the void spaces are not completely filled with water are called *zones of aeration*. Water in a zone of aeration is sometimes called *vadose* water*, and the zone of aeration is sometimes called the *vadose zone*. Typically, the zone of aeration (vadose zone) lies above the zone of saturation, and the upper boundary of the zone of saturation is called the *phreatic surface*† or *water table*. At the water table, the pressure is equal to atmospheric pressure, and this condition is illustrated in Figure 7.1. Within the zone of aeration are three subzones: the soil-water, intermediate, and capillary zones. The *soil-water zone* is the region containing the roots of surface vegetation, and the maximum moisture content of the soil in the soil-water zone corresponds to the maximum moisture that can be held by the soil against the force of gravity, regardless of the depth of the water table below ground surface. The maximum moisture content in the soil-water zone is called the *field capacity*, and the thickness of the soil-water zone is typically on the order of 3 m (Raudkivi and Callander, 1976). Beneath the soil-water zone is the *intermediate zone*, which extends from the bottom of the soil-water zone to the upper limit of the capillary zone. The *capillary zone* extends from the water table up to the limit of the capillary rise of the water from the zone of saturation (through the pores of the porous formation). The thickness of the capillary zone depends on the pore size, and can vary from 1 cm to several meters.

---

*Vadose is a derivative of the Latin word *vadosus*, which means "shallow."
† *Phreatic* is a derivative of the Greek word *phreatos*, which means "well." In this context, a saturated zone is encountered when a well is dug.

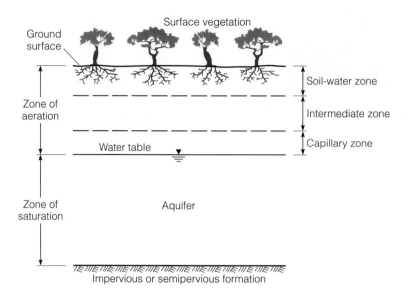

Figure 7.1 ■
Unconfined Aquifer

An *aquifer** is a geologic formation containing water that can be withdrawn in significant amounts. *Aquicludes* contain water but are incapable of transmitting it in significant quantities, and *aquifuges* neither contain nor transmit water. A clay layer is an example of an aquiclude; solid rock is an example of an aquifuge; and, for most practical purposes, aquicludes can be taken as impervious formations. Aquifers are classified as either unconfined or confined. *Unconfined aquifers* are open to the atmosphere, as illustrated in Figure 7.1, and are also called *phreatic aquifers* or *water-table aquifers*. In *confined aquifers*, water in the saturated zone is bounded above by either impervious or semipervious formations. A typical configuration of a confined aquifer is shown in Figure 7.2, where the water in the confined aquifer is recharged by inflows at A (usually from rainfall).

Land surfaces that supply water to aquifers are called *recharge areas*, and maintaining an adequate recharge area (and recharge-water supply) is particularly important in urban areas where ground water is a major source of drinking water. The primary ground-water recharge mechanism is the infiltration of rainfall. *Piezometers* are observation wells that are used to measure the piezometric head, ϕ, which for an incompressible fluid is

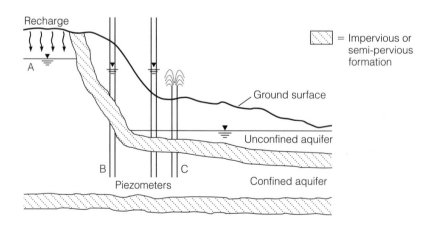

Figure 7.2 ■ Confined,
Unconfined, and Artesian
Aquifers

---

*Aquifer is a derivative of the Latin words *aqua* ("water") and *ferre* ("to bear").

given by

$$\phi = \frac{p}{\gamma} + z \qquad (7.1)$$

where p and z are the pressure and elevation at the base (opening) of the observation well (piezometer) and γ is the specific weight of the ground water. If the confined aquifer in Figure 7.2 is penetrated by piezometers at B and C, then the water levels in these piezometers rise to levels equal to the piezometric heads at B and C, respectively. At B, the piezometric head rises above the top confining layer of the (confined) aquifer, indicating that the water pressure at the top of the confined aquifer is greater than atmospheric pressure. This condition can be contrasted with the top of an unconfined aquifer, where the pressure is equal to atmospheric pressure. The piezometer at C behaves similarly to the piezometer at B, with the difference being that the water level in the piezometer at C rises above the ground surface. In practical terms, this means that if a well were extended from ground surface down into the confined aquifer at C, then the water would flow continuously from the well until the piezometric head was reduced to the elevation at the top of the well. An aquifer that produces flowing water when penetrated by a well from the ground surface is called an *artesian aquifer*.* As indicated in Figure 7.2, a confined aquifer can be an artesian aquifer at some locations, such as at C, and an unconfined aquifer at other locations, such as at A.

The presence of artesian aquifers and the distribution of hydrostatic pressures in aquifers is frequently identified by plotting the areal distribution of piezometric head, and such plots commonly referred to as *piezometric surfaces* or *potentiometric surfaces*. In cases where artesian aquifers intersect the ground surface, concentrated flows of ground water called *springs* are formed. Ground-water inflows into surface-water channels are a common source of perennial discharge in streams, which is commonly referred to as the *base flow* of the stream. Stream flows mostly consist of the base flow plus the flow resulting from stormwater runoff. Both unconfined and confined aquifers can be bounded by semipervious formations called *aquitards*, which are significantly less permeable than the aquifer but are not impervious. Of course, unconfined aquifers can only be bounded by impervious or semipervious layers on the bottom, while confined aquifers are bounded on both the top and bottom by either impervious or semipervious layers. Aquifers bounded by semipervious formations are called *leaky*, and terms such as *leaky-unconfined aquifer* and *leaky-confined aquifer* are used.

A microscopic view of the flow through porous media is illustrated in Figure 7.3, where water flows through the *void space* and around the *solid matrix* within the porous medium. It is difficult to describe the details of the flow field within the void spaces, since this would necessarily require a detailed knowledge of the geometry of the void space within the porous medium. To deal with this problem, it is convenient to work with spatially averaged variables rather than variables at a point (which is a spatial average over a very small volume). Referring to Figure 7.3, a property of the porous medium at P can be taken as the average value of that property within a volume centered at P. The scale of the averaging volume is called the *support scale*. Almost all properties of a porous medium that are relevant to ground-water engineering are associated with a support scale, and in most cases the value of the averaged quantity is independent of the size of the support scale. A case in point is the *porosity*, n, of a porous medium, which is

---

*The name "artesian" is derived from the name of the northern French city of Artois, where wells penetrating artesian aquifers are common.

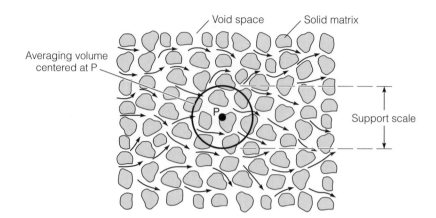

Figure 7.3 ■ Microscopic View of Flow Through Porous Media

defined by the relation

$$n = \frac{\text{volume of voids}}{\text{sample volume}} \qquad (7.2)$$

where the sample volume corresponds to the spherical volume with radius equal to the support scale. A typical relationship between the porosity and support scale is shown in Figure 7.4. When the support scale is very small, the porosity is sensitive to the location and size of the sample volume. Clearly, for sample volumes of the order of the size of the void space, it makes a significant difference whether the sample volume is located within a void space or within the solid matrix. As the support scale gets larger, the porosity becomes less sensitive to the location and size of the sample volume and approaches a constant value that is independent of the support scale. The porosity remains independent of the support scale until the averaging volume becomes so large that it encompasses portions of the porous medium that have significantly different characteristics; under these circumstances, the porosity again becomes dependent on the size of the support scale. The range within which the porosity is independent of the support scale is given by the interval between the scales L_0 and L_1 shown in Figure 7.4. Therefore, as long as the support scale is between L_0 and L_1, the porosity need not be associated with any particular support scale. The sample volume associated with the support scale L_0 is commonly referred to as the *representative elementary volume* (REV). The relationship demonstrated here between the porosity and support scale is typical of the relationship between many other hydrogeologic parameters and support scales, although the REV of other parameters may be different. In general, there is no guarantee that a REV exists for

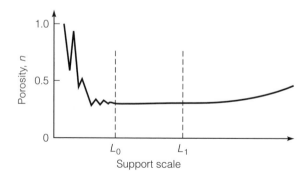

Figure 7.4 ■ Porosity Versus Support Scale

any hydrogeologic parameter, and in the absence of a REV the value of the parameter must be associated wth a support scale.

All earth materials are collectively known as *rocks*. The three main categories of rocks are igneous, sedimentary, and metamorphic rocks. *Igneous rocks*, such as basalt and granite, are formed from molten or partially molten rock (magma) formed deep within the earth; *sedimentary rocks*, such as sand, gravel, sandstone, and limestone, are formed by the erosion of previously existing rocks and/or the deposition of marine sediment; and *metamorphic rocks*, such as schist and shale, are formed through the alteration of igneous or sedimentary rock by extreme heat or pressure or both. Typically, igneous and metamorphic rocks have less pore space and fewer passageways for water than sedimentary deposits. Rocks are further classified as either *consolidated* or *unconsolidated*. Solid masses of rock are referred to as consolidated, while rocks consisting of loose granular material are termed unconsolidated. In consolidated formations, *original porosity* is associated with pore spaces created during the formation of the rock, while *secondary porosity* is associated with pore spaces created after rock formation. Examples of secondary porosity in consolidated formations include fractures, and solution cavities in limestone. Representative values of porosity in consolidated formations are given in Table 7.1. Porosities are considered small when $n < 0.05$, medium when $0.05 \leq n \leq 0.20$, and large when $n > 0.20$ (Kashef, 1986). Between 60% and 90% of all developed aquifers consist of unconsolidated rocks (Lehr et al., 1988; Todd, 1980), where the porosities are associated with the intergranular spaces and are determined by the particle-size distributions of the granular material.

The classification of granular material in unconsolidated formations by particle size is illustrated in Table 7.2, with corresponding values of porosity. The porosities of granular materials tend to decrease with increasing particle size, however, this does not mean that water flows with more resistance through aquifers composed of larger particle sizes. In fact, the opposite is true. The *lithology* of a formation is described by its mineral composition, grain-size distribution, and grain packing; and the *stratigraphy* describes the geometric grouping of the various lithologic units (Gorelick et al., 1993). For example, in sedimentary deposits, the stratigraphy is often horizontally layered.

The most common aquifer materials are unconsolidated sands and gravels, which occur in alluvial valleys, coastal plains, dunes, and glacial deposits (Bouwer, 1978). Consolidated formations that make good aquifers are sandstones, limestones with solution channels, and heavily fractured volcanic and crystalline rocks. Clays, shales, and dense crystalline rocks are the most common materials found in aquitards. Aquifers range in thickness from less than 1 m to several hundred meters, and may be long and narrow

■ **Table 7.1**

Representative Hydrologic Properties in Consolidated Formations

Material	Porosity	Specific yield	Hydraulic conductivity (m/d)
Sandstone	0.05–0.50	0.01–0.40	10^{-5}–3
Limestone	0.05–0.55	0–0.35	10^{-4}–800
Schist	0.01–0.50	0.20–0.35	10^{-4}–0.2
Siltstone	0.20–0.40	0.01–0.35	10^{-6}–0.001
Shale	0.01–0.10	—	10^{-8}–0.004
Till	0.25–0.45	0.05–0.20	10^{-5}–30
Basalt	0.01–0.50	—	10^{-6}–800
Pumice	0.80–0.90	—	—
Tuff	0.10–0.40	0.01–0.45	—

■ **Table 7.2**
Classification and
Representative Hydrologic
Properties in Unconsol-
idated Formations

Material	Particle size* (mm)	Porosity	Specific yield	Hydraulic conductivity (m/d)
Very coarse gravel	32.0–64.0	—	—	–
Coarse gravel	16.0–32.0	0.25–0.40	0.10–0.25	860–8,600
Medium gravel	8.0–16.0	—	0.15–0.45	20–1,000
Fine gravel	4.0–8.0	0.25–0.40	0.15–0.40	—
Very fine gravel	2.0–4.0	—	—	—
Very coarse sand	1.0–2.0	—	—	—
Coarse sand	0.5–1.0	0.20–0.50	0.15–0.45	0.1–860
Medium sand	0.25–0.5	0.30–0.40	0.15–0.45	0.1–50
Fine sand	0.125–0.25	0.25–0.55	0.01–0.45	0.01–40
Very fine sand	0.062–0.125	—	—	—
Silt	0.004–0.062	0.35–0.70	0.01–0.40	10^{-5}–2
Clay	< 0.004	0.35–0.70	0.01–0.20	$< 10^{-2}$

*Morris and Johnson (1967).

as in small alluvial valleys, or they may extend over millions of square kilometers and underlie major portions of states (Bouwer, 1984). The depth from the ground surface to the top of an aquifer may range from 1 m to more than several hundred meters.

7.2   Basic Equations of Ground-Water Flow

7.2.1 ■ Darcy's Law

The study of flow through porous media was pioneered by Darcy (1856) using the experimental setup shown in Figure 7.5.* In this experiment, Darcy investigated the flowrate

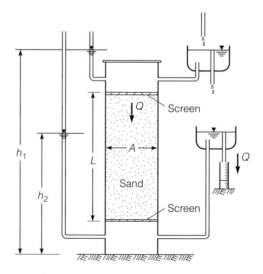

Figure 7.5 ■ Darcy's
Experimental Setup

Source: Bear, Jacob. *Hydraulics
of Groundwater.* Copyright
© 1979. The McGraw-Hill
Companies, New York.
Reprinted by permission of The
McGraw-Hill Companies.

---

*The motivation for Darcy's experiments was to study the performance of the sand filters in the water-supply system for the city of Dijon, France.

of water through a column of sand with cross-sectional area A, and length, L. Darcy (1856) found that the flowrate, Q, of water through the sand column could be described by the relation

$$Q = KA\frac{h_1 - h_2}{L} \tag{7.3}$$

where K is a proportionality constant, and h_1 and h_2 are the piezometric heads at the entrance and exit of the sand column, respectively. Recall that the piezometric head, h, of an incompressible fluid is given by $h = p/\gamma + z$. The piezometric head is sometimes called the *hydraulic head*. Defining the gradient in the piezometric head or *hydraulic gradient*, J, across the sand column by

$$J = \frac{h_2 - h_1}{L} \tag{7.4}$$

and defining the *specific discharge* or *filtration velocity*, q, through the sand column by

$$q = \frac{Q}{A} \tag{7.5}$$

then Equation 7.3 can be written in the form

$$\boxed{q = -KJ} \tag{7.6}$$

which is commonly known as *Darcy's law*. This experimentally validated (phenomeno-logical) relationship states that the specific discharge through a porous medium is lin-early proportional to the gradient in the piezometric head in the direction of flow. The proportionality constant, K, is called the *hydraulic conductivity* of the porous medium. In the field of geotechnical engineering, the term *coefficient of permeability* is commonly used instead of hydraulic conductivity, but both terms refer to the same quantity.

Before proceeding to study the applications of Darcy's law, several important points must be made regarding the validity and application of Darcy's law. First of all, the spe-cific discharge, q, is defined as the volumetric flowrate per unit cross-section of the porous medium and, since the flow only occurs within the pores, the actual velocity of flow through the pores is necessarily higher than the specific discharge. The flowrate through the pores is called the *seepage velocity*, v, and is defined by

$$v = \frac{Q}{A_p} \tag{7.7}$$

where A_p is the area of the pores (normal to the flow direction). Comparing Equations 7.5 and 7.7, the seepage velocity, v, is related to the specific discharge, q, by

$$v = q\frac{A}{A_p} \tag{7.8}$$

The ratio of the pore area, A_p, to the bulk cross-sectional area, A, is defined as the *areal porosity*, and is equal to the volumetric porosity, n, defined by Equation 7.2 (Bear, 1972). In reality, not all of the pore space is connected and available for fluid flow, and there-fore the effective areal porosity is less than the volumetric porosity, n. The ratio of the effective flow area to the bulk cross-sectional area is defined as the *effective porosity*, n_e, where

$$n_e = \frac{A_e}{A} \tag{7.9}$$

where A_e is the effective flow area through the pores. For unconsolidated porous media and for many consolidated rocks, the effective porosity is approximately equal to the volumetric porosity (Todd, 1980). Combining Equations 7.8 and 7.9, and taking A_p equal to A_e, yields the following relationship between the seepage velocity, v, and the specific discharge, q, given by Darcy's law

$$v = \frac{q}{n_e} = -\frac{K}{n_e}J \qquad (7.10)$$

In practice, the effective porosity is sometimes denoted by n rather than n_e, and it is differentiated from the volumetric porosity by the stated definition of n.

▪ Example 7.1

Water flows through a sand aquifer with a piezometric head gradient of 0.01. (a) If the hydraulic conductivity and effective porosity of the aquifer are 2 m/d and 0.3, respectively, estimate the specific discharge and seepage velocity in the aquifer; (b) estimate the volumetric flowrate of the ground water if the aquifer is 15 m deep and 1 km wide; and (c) how long does it take the ground water to move 100 m?

Solution

(a) From the given data: $J = -0.01$; $K = 2$ m/d; $n_e = 0.3$; and, according to Darcy's law, the specific discharge, q, is given by

$$q = -KJ = -(2)(-0.01) = 0.02 \text{ m/d}$$

The corresponding seepage velocity, v, is

$$v = \frac{q}{n_e} = \frac{0.02}{0.3} = 0.067 \text{ m/d}$$

(b) The volumetric flowrate, Q, of ground water across an area $A = 15$ m $\times$ 1 km $=$ 15,000 m² is

$$Q = qA = (0.02)(15000) = 300 \text{ m}^3/\text{d}$$

(c) Ground water flows with the seepage velocity, v, and therefore the time, t, to travel 100 m is given by

$$t = \frac{100 \text{ m}}{v} = \frac{100}{0.067} = 1490 \text{ days} = 4.09 \text{ years}$$

This result is indicative of the slow movement of most ground waters.   ▪

The hydraulic conductivity, K, that appears in Darcy's law is a function of both the fluid properties and the geometry of the solid matrix. This point is intuitively obvious if one considers the water in Darcy's experiment being replaced by oil. Clearly there would be less flow than for water under the same piezometric gradient, thereby indicating a smaller hydraulic conductivity for oil in sand than for water in sand. Also, if clay were used instead of sand in Darcy's experiment, then there would be less flow than for sand under the same hydraulic gradient, indicating a smaller hydraulic conductivity for water in clay than for water in sand. The functional relationship between the hydraulic conductivity and the fluid and solid matrix properties can be extracted using dimensional analysis. If the fluid properties are characterized by the specific weight, γ, and dynamic viscosity, μ; and the solid matrix is characterized by the length scale of the pores, d, then the hydraulic conductvity, K, is related to the fluid and solid matrix properties by the following functional relationship

$$K = f_1(\gamma, \mu, d) \tag{7.11}$$

where f_1 is an undetermined function. Using the Buckingham pi theorem, this functional relationship between four variables containing three dimensions (mass, length, and time) can be expressed in the following form

$$f_2\left(\frac{K\mu}{\gamma d^2}\right) = 0 \tag{7.12}$$

where f_2 is an undetermined function. The relationship given by Equation 7.12 can theoretically be solved for the dimensionless quantity, $K\mu/\gamma d^2$, to yield

$$\frac{K\mu}{\gamma d^2} = \alpha \tag{7.13}$$

where α is a dimensionless constant that incorporates the secondary structural characteristics of the porous medium that are not simply characterized by the pore scale, such as the distribution and shape of the pore sizes. Solving for K in Equation 7.13 yields

$$K = \alpha \frac{\gamma}{\mu} d^2 \tag{7.14}$$

This equation clearly separates the hydraulic conductivity into a term related to the fluid properties, γ/μ, and a term related to the geometry of the solid matrix, αd^2. The *intrinsic permeability*, k, is defined by the relation

$$\boxed{k = \alpha d^2} \tag{7.15}$$

The intrinsic permeability should not be confused with the "coefficient of permeability," a term used by geotechnical engineers to mean the hydraulic conductivity. There have been a number of investigations to estimate the magnitude of α in Equation 7.15, and several of these results are summarized in Table 7.3. The characteristic pore size, d, is most often taken as the 10-percentile grain size, d_{10}, which is also called the *effective grain diameter*, and proportionality constant, α, is commonly considered to decrease as the porosity, n, decreases (Venkataraman and Rao, 1998).

A recent comparative study by Sperry and Pierce (1995) indicates that the Hazen equation for intrinsic permeability will usually provide a good estimate of the intrinsic permeability, except for irregularly shaped particles. Using the definition of the intrinsic permeability given by Equation 7.15, the hydraulic conductivity, K, given by Equation 7.14, can be written in the form

$$\boxed{K = k\frac{\gamma}{\mu}} \tag{7.16}$$

On the basis of Equation 7.16, it is appropriate to associate the intrinsic permeability with the solid matrix, and associate the hydraulic conductivity with the fluid/matrix combination. The fluid properties, γ and μ, and hence the hydraulic conductivity, are significantly influenced by the temperature of the ground water. The temperature of shallow ground water is typically 1 to 1.5°C greater than the mean annual air temper-

	Equation	Reference	Comments
■ **Table 7.3** Equations for Estimating Intrinsic Permeability	$k = 0.617 \times 10^{-3} d^2$	Krumbein and Monk (1942)	d is a measure of the pore size
	$k = 1.02 \times 10^{-3} d_{10}^2$	Hazen (1911)	restricted to uniformity coefficient $(d_{60}/d_{10}) < 5$, and $0.1 \text{ mm} < d_{10} < 3 \text{ mm}$
	$k = 0.654 \times 10^{-3} d_{10}^2$	Harleman et al. (1963)	—
	$k = 0.750 \times 10^{-3} d^2 e^{-1.31\sigma}$	Krumbein and Monk (1943)	d is the geometric mean diameter and σ is the log standard deviation of the size distribution
	$k = C_s \dfrac{n^3}{1-n} D_s^2$	Taylor (1948)	C_s is a shape factor, n is the porosity, and D_s is the equivalent (uniform) spherical diameter of the porous medium $(D_s \approx d_{10})$
	$k = \dfrac{n^3}{180(1-n)^2} d^2$	Kozeny-Carman equation (Kozeny, 1927; Carman 1937, 1956)	d is any representative grain size
	$k = \dfrac{1}{m}\left[\dfrac{(1-n)^2}{n^3}\left(\dfrac{a}{100}\sum\dfrac{P}{d_g}\right)^2\right]^{-1}$	Fair and Hatch (1933)	m is a packing factor (≈ 5 by experiment), a is a sand shape factor which ranges from 6.0 for spherical grains to 7.7 for angular grains, P is the percentage of grains passing one sieve and held on the next, and d_g is the geometric mean of these sieve sizes

Source: Adapted from Walton (1991).

ature. The depth of nearly uniform temperature occurs at about 10 m in the tropics and at about 20 m in polar regions (Todd, 1980) and, above these depths, significant seasonal fluctuations in temperature can occur. The ground-water temperature typically increases with depth by about 2 to 3.5°C/100 m (Warner and Lehr, 1981; Todd, 1980), and the rate of increase in water temperature with depth is commonly referred to as the *geothermal gradient*. In the contiguous United States, ground-water temperatures vary from about 4°C in the north (10°C in the Northwest) to around 20°C in the south (Miller et al., 1962). The intrinsic permeability, k, in Equation 7.16 is generally a very small number when expressed in m^2 and is commonly expressed in darcys, where 1 darcy is equal to 0.987×10^{-12} m^2.

■ Example 7.2

A sand aquifer has a 10-percentile particle size of 0.5 mm. Estimate the intrinsic permeability and hydraulic conductivity at 20°C using the Harleman et al. (1963) equation.

Compare your result with the representative hydraulic conductivity of medium sand given in Table 7.2.

Solution

From the given data: $d_{10} = 0.5$ mm $= 5 \times 10^{-4}$ m. According to the Harleman et al. (1963) relation in Table 7.3, the intrinsic permeability, k, can be estimated by

$$k = 0.654 \times 10^{-3} d_{10}^2 = 0.654 \times 10^{-3} (5 \times 10^{-4})^2 = 1.64 \times 10^{-10} \text{ m}^2 = 1{,}660 \text{ darcys}$$

At 20°C, $\mu = 1.00 \times 10^{-3}$ N·s/m^2, and $\gamma = \rho g = (998)(9.81) = 9790$ N/m^3 (Table 2.1). The hydraulic conductivity, K, is therefore given by Equation 7.16 as

$$K = k\frac{\gamma}{\mu} = (1.64 \times 10^{-10})\frac{9790}{1.00 \times 10^{-3}} = 0.00161 \text{ m/s} = 139 \text{ m/d}$$

This value of hydraulic conductivity is higher than the typical range of 0.1–50 m/d for medium sand given in Table 7.2.   ■

Hydraulic conductivities of natural porous media cover a wide range of values and can be classified as ranging from "very high" to "very low," as shown in Table 7.4.

Darcy's law indicates a linear relationship between the flowrate and the gradient in the piezometric head measured in the direction of flow. This linear relationship is the same as for flow in pipes under low Reynolds number (Hagen-Poiseuille flow) and is symptomatic of the dominance of viscous forces as water flows through the pore spaces (Munson et al., 1994). As the Reynolds number increases, viscous forces become less dominant, and the flowrate deviates from being linearly proportional to the piezometric head gradient. The characteristic Reynolds number, Re, can be defined in terms of the specific discharge, q; pore scale, d; and kinematic viscosity of water, ν, by the relation

$$\text{Re} = \frac{qd}{\nu} \tag{7.17}$$

Experiments indicate that deviations from Darcy's law begin to occur for values of Re > 1, but serious deviations do not occur up to Re $= 10$ (Ahmed and Sunada, 1969). Such conditions are routinely found in the immediate vicinity of large water-supply wells, and in fractured rock formations. Darcy's law has also been found to be invalid at

■ **Table 7.4** Classification of Hydraulic Conductivities	**Hydraulic conductivity, K (m/d)**	**Class**	**Unconsolidated deposits**	**Consolidated rocks**
	$> 1{,}000$	very high	clean gravel	vesicular and scoriaceous basalt and cavernous lime-stone and dolomite
	10–$1{,}000$	high	clean sand, and sand and gravel	clean sandstone and fractured igneous and metamorphic rocks
	0.01–10	moderate	fine sand	laminated sandstone, shale, mudstone
	0.0001–0.01	low	silt, clay, and mixtures of sand, silt, and clay	massive igneous and metamorphic rocks
	< 0.0001	very low	massive clay	

Source: USBR (1977).

very small seepage velocities in compact clays (such as in clay liners beneath landfills) and is not theoretically valid under transient conditions (de Marsily, 1986). However, the effect of transient flows on the validity of Darcy's law can be taken as negligible in most cases of practical interest.

■ Example 7.3

A sand aquifer has a 10-percentile particle size of 0.4 mm and an effective porosity of 0.3. If the temperature of the water in the aquifer is 20°C, estimate the range of seepage velocities for which Darcy's law describes the flow.

Solution

Darcy's law can be taken to be valid when Re < 10,

$$\frac{qd}{\nu} < 10$$

which can be put in the form

$$q < \frac{10\nu}{d}$$

or

$$\nu < \frac{10\nu}{n_e d}$$

where the seepage velocity, q, is related to the seepage velocity, ν, and the effective porosity, n_e, by $\nu = q/n_e$. From the given data, $n_e = 0.3$, $d \approx d_{10} = 0.4$ mm $= 4 \times 10^{-4}$ m, and at 20°C, $\nu = 1.00 \times 10^{-6}$ m^2/s. Hence

$$\nu < \frac{10(1.00 \times 10^{-6})}{(0.3)(4 \times 10^{-4})} = 0.0833 \text{ m/s} = 7{,}200 \text{ m/d}$$

Darcy's law can be applied in the aquifer whenever the seepage velocity is less than 7,200 m/d.   ■

Darcy's law relates the specific discharge to the piezometric gradient in the direction of flow by Equation 7.6. In the case of three-dimensional flow, the components of the specific-discharge vector are related to the corresponding components of the head gradient vector by the relationship

$$q_i = -KJ_i \tag{7.18}$$

where q_i and J_i are the i-components of the specific discharge and head gradient respectively. This formulation assumes that the hydraulic conductivity is *isotropic*, in which case the hydraulic conductivity does not depend on the flow direction. In cases where the hydraulic conductivity depends on the flow direction, the porous medium is called *anisotropic*, the hydraulic conductivity is a tensor, and the specific discharge is related to the head gradient by

$$\boxed{q_i = -K_{ij}J_j} \tag{7.19}$$

where the Einstein summation convention is used. The hydraulic conductivity is a symmetric second-rank tensor with nine components (K_{ij}, $i = 1, 3$; $j = 1, 3$), of which six are independent.

Anisotropy can be caused by a variety of factors, such as solution cavities in carbonate rocks preferentially forming in the horizontal flow direction, or the deposition of flat granular material (tilted slightly upward in the direction of flow) in the formation of alluvial aquifers. In cases where the coordinate axes are chosen to coincide with the *principal axes* of the hydraulic conductivity tensor, Darcy's law can be written in the simple form

$$q_i' = -K_{ii}'J_i' \tag{7.20}$$

where the primed quantities indicate components in the principal directions of the hydraulic conductivity tensor, and the Einstein summation is not used. In other words, if the coordinate axes coincide with the principal axes of the hydraulic conductivity tensor, then the components of the seepage velocity in the principal directions are given by

$$
\begin{aligned}
q_1' &= -K_{11}'J_1' \\
q_2' &= -K_{22}'J_2' \\
q_3' &= -K_{33}'J_3'
\end{aligned}
\tag{7.21}
$$

If coordinate axes other than the principal axes are used, then the components of the hydraulic conductivity tensor can be computed using the relationship (Bear, 1972)

$$\boxed{K_{ij} = K_{11}'\alpha_{i1}\alpha_{j1} + K_{22}'\alpha_{i2}\alpha_{j2} + K_{33}'\alpha_{i3}\alpha_{j3}} \tag{7.22}$$

where α_{ij} is the cosine of the angle between the x_i and x_j' axes. Primed coordinates refer to the principal axes. In two-dimensional aquifers, Equation 7.22 can be written in the form

$$K_{11} = \frac{K_{11}' + K_{22}'}{2} + \frac{K_{11}' - K_{22}'}{2}\cos 2\theta \tag{7.23}$$

$$K_{22} = \frac{K_{11}' + K_{22}'}{2} - \frac{K_{11}' - K_{22}'}{2}\cos 2\theta \tag{7.24}$$

$$K_{12} = K_{21} = -\frac{K_{11}' - K_{22}'}{2}\sin 2\theta \tag{7.25}$$

where θ is the angle between the x_1- and x_1'-axes, measured counterclockwise from the x_1'-axis. These equations can also be used to estimate the principal components of the hydraulic conductivity and the orientation of the principal axes from the hydraulic conductivity tensor, K_{ij}.

In most cases, the principal axes are such that the x_1- and x_2-axes are in the horizontal plane and the x_3-axis is vertically upward. Another common approximation is that the hydraulic conductivity is isotropic in the horizontal plane (i.e., $K_{11} = K_{22}$), and the ratio of the vertical hydraulic conductivity, K_{33}, to the horizontal hydraulic conductivity, K_{11} or K_{22}, is defined as the *anisotropy ratio*. Typical values of the anisotropy ratio are in the range of 0.1 to 0.5 for alluvial aquifers, with values as low as 0.01 where clay layers exist (Bedient et al., 1994). The primary cause of anisotropy on a small scale is the orientation of clay minerals in sedimentary rocks and unconsolidated sediments. In nongranular rocks, the size, shape, orientation, and spacing of fractures and other voids (such as solution cavities) are the primary causes of anisotropy. Porous formations are called *homogeneous* if the hydraulic conductivity is independent of location and *nonhomogeneous* if the hydraulic conductivity varies spatially.

■ **Example 7.4**

The piezometric heads are measured at three locations in an aquifer. Point A is located at (0 km, 0 km), Point B is located at (1 km, −0.5 km), and Point C is located at (0.5 km, −1.2 km), and the piezometric heads at A, B, and C are 2.157 m, 1.752 m, and 1.629 m. (a) Determine the head gradient in the aquifer; (b) if the coordinate locations are measured relative to the principal axes (x', y'), and the principal components of the hydraulic conductivity tensor are $K'_{xx} = 15$ m/d and $K'_{yy} = 5$ m/d, calculate the magnitude and direction of the specific discharge; (c) if the coordinate axes are rotated 30° clockwise from the principal axes, calculate the components of the hydraulic conductivity tensor relative to the new coordinate axes; and (d) verify that the magnitude and direction of the specific discharge calculated in part (b) is not affected by axis rotation.

Solution

(a) The piezometric head distribution, $h(x', y')$, in the triangular region ABC can be assumed to be planar and given by

$$h(x', y') = ax' + by' + c$$

where a, b, and c are constants, and (x', y') are the (principal) coordinate locations. Applying this equation to points A, B, and C (with all linear dimensions in meters) yields

$$2.157 = a(0) + b(0) + c$$

$$1.752 = a(1000) + b(-500) + c$$

$$1.629 = a(500) + b(-1200) + c$$

The solution of these equations is $a = -0.0002337$, $b = 0.0003426$, and $c = 2.157$. From the planar head distribution, it is clear that the components of the head gradient are given by

$$\frac{\partial h}{\partial x'} = a \quad \text{and} \quad \frac{\partial h}{\partial y'} = b$$

Therefore, in this case the components of the head gradient are

$$\frac{\partial h}{\partial x'} = -0.0002337 \quad \text{and} \quad \frac{\partial h}{\partial y'} = 0.0003426$$

which can be written in vector notation as

$$\nabla' h = -0.0002337 i' + 0.0003426 j'$$

where i' and j' are unit vectors in the principal directions.

(b) The components of the specific discharge vector relative to the principal axes are given by

$$q'_x = -K'_{xx} J'_x$$

$$q'_y = -K'_{yy} J'_y$$

where $J'_x = \partial h/\partial x' = -0.0002337$, $J'_y = \partial h/\partial y' = 0.0003426$, $K'_{xx} = 15$ m/d, and $K_{yy} = 5$ m/d. Substituting these values yields

$$q'_x = -15(-0.0002337) = 0.00351 \text{ m/d}$$

$$q'_y = -5(0.0003426) = -0.00171 \text{ m/d}$$

The magnitude of the specific discharge, q', is given by

$$q' = \sqrt{(q'_x)^2 + (q'_y)^2} = \sqrt{0.00351^2 + (-0.00171)^2} = 0.00390 \text{ m/d}$$

and the direction of flow is at an angle η measured clockwise from the principal axes, where

$$\eta = \tan^{-1}\left(\frac{0.00171}{0.00351}\right) = 26.0°$$

(c) If the reference axes are rotated 30° clockwise from the principal axes then, according to Equations 7.23 to 7.25, the components of the hydraulic conductivity tensor become

$$K_{xx} = \frac{K'_{xx} + K'_{yy}}{2} + \frac{K'_{xx} - K'_{yy}}{2}\cos 2\theta = \frac{15+5}{2} + \frac{15-5}{2}\cos 2(-30°) = 12.5 \text{ m/d}$$

$$K_{yy} = \frac{K'_{xx} + K'_{yy}}{2} - \frac{K'_{xx} - K'_{yy}}{2}\cos 2\theta = \frac{15+5}{2} - \frac{15-5}{2}\cos 2(-30°) = 7.5 \text{ m/d}$$

$$K_{xy} = K_{yx} = -\frac{K'_{xx} - K'_{yy}}{2}\sin 2\theta = -\frac{15-5}{2}\sin 2(-30°) = 4.33 \text{ m/d}$$

The components of the piezometric head gradient also change with axis rotation, and the new components, J_x and J_y, are given by

$$J_x = J'_x \cos 30° - J'_y \sin 30°$$
$$= (-0.0002337)\cos 30° - (0.0003426)\sin 30° = -0.0003738$$
$$J_y = J'_x \sin 30° + J'_y \cos 30°$$
$$= (-0.0002337)\sin 30° + (0.0003426)\cos 30° = 0.0001799$$

The components of the specific discharge relative to the new coordinate axes, q_x and q_y, are therefore

$$q_x = -K_{xx}J_x - K_{xy}J_y = -(12.5)(-0.0003738) - (4.33)(0.0001799) = 0.00389 \text{ m/d}$$
$$q_y = -K_{yx}J_x - K_{yy}J_y = -(4.33)(-0.0003738) - (7.5)(0.0001799) = 0.000269 \text{ m/d}$$

The magnitude of the specific discharge, q, is

$$q = \sqrt{q_x^2 + q_y^2} = \sqrt{0.00389^2 + (0.000269)^2} = 0.00390 \text{ m/d}$$

and the specific discharge vector is oriented at an angle ξ measured anti-clockwise from the rotated x-axis, where

$$\xi = \tan^{-1}\left(\frac{0.000269}{0.00389}\right) = 4.0°$$

(d) The magnitude and direction of the specific discharge vector calculated for the rotated axes in (c) is exactly the same as the specific discharge vector calculated in (b). Hence, as expected, the specific discharge vector does not depend on orientation of the coordinate axes.   ■

As with most hydrogeologic parameters, the hydraulic conductivity is associated with a support scale, which corresponds the volume over which the hydraulic conductivity describes the proportionality between the average specific discharge and the average hydraulic gradient. Hydraulic conductivities averaged over a support scale tend to vary in space and can be described as a *random space function* (RSF). The probability distribution of the hydraulic conductivity at any point in space has been found to be log-normally distributed in most cases. Defining the natural logarithm of the hydraulic conductivity, Y, by

$$Y = \ln K \tag{7.26}$$

where K is the hydraulic conductivity, then the log-normal probability distribution of K can be described by the mean and variance of Y, $\langle Y \rangle$ and σ_Y^2, respectively. Alternatively, the mean of $\ln K$ can be expressed in terms of the *geometric mean* of the hydraulic conductivity, K_G where

$$K_G = e^{\langle Y \rangle} \tag{7.27}$$

Besides the mean and variance of $\ln K$ at any point in an aquifer, the spatial variability in $\ln K$ is described by its spatial correlation, which is defined as the correlation between $\ln Ky$ = measured at any two points in space. The spatial correlation of $\ln K$, $\rho_Y(\mathbf{r})$, in a statistically homogeneous porous medium is defined by the relation

$$\rho_Y(\mathbf{r}) = \frac{\langle (Y(\mathbf{x}) - \langle Y \rangle)(Y(\mathbf{x} + \mathbf{r}) - \langle Y \rangle) \rangle}{\sigma_Y^2} \tag{7.28}$$

where $\mathbf{r}$ is the separation vector between the measurements of Y, and $\mathbf{x}$ is the position vector indicating the location of the reference measurement. Statistical homogeneity of the porous medium guarantees that the spatial correlation calculated by Equation 7.28 is independent of the reference location, $\mathbf{x}$. In practice, there are several empirical correlation functions that can be used to describe $\rho_Y(\mathbf{r})$ (Dagan, 1989), the most common of which is the *exponential correlation function* given by

$$\rho_Y(\mathbf{r}) = \exp\left[-(r_1^2/\lambda_1^2 + r_2^2/\lambda_2^2 + r_3^2/\lambda_3^2)^{\frac{1}{2}} \right] \tag{7.29}$$

where (r_1, r_2, r_3) are the components of the separation vector, $\mathbf{r}$, and $(\lambda_1, \lambda_2, \lambda_3)$ are the *correlation length scales* in each of the coordinate directions. The correlation length scales are empirical parameters that measure the separation distances at which the correlation between hydraulic conductivities are equal to e^{-1} ($= 0.368$). The correlation between hydraulic conductivities at zero separation is equal to one, indicating perfect correlation.

The statistical characterization of the hydraulic conductivity provides the most realistic description of the distribution of hydraulic conductivities in porous media. However, the relationship between the statistical properties of the hydraulic conductivity and the support scale must not be overlooked. For example, whenever hydraulic conductivities are measured at small support scales, such as in permeameters in the laboratory where the support scale is on the order of a few centimeters, it is expected that individual measurements of the log-hydraulic conductivity may deviate significantly from the mean,

$\langle Y \rangle$, the variance of the measurements, σ_Y^2, will be high, and the correlation length scales $(\lambda_1, \lambda_2, \lambda_3)$ will be small, and on the order of the support scale. If, in the same porous medium, the hydraulic conductivities are measured over a much larger support scale using pump tests, where the support scale may be on the order of 100 m, then it is expected that individual measurements at different locations in the aquifer would be much closer to the ensemble mean, $\langle Y \rangle$, the variance of the measurements, σ_Y^2, will be small, and the correlation length scale will be large, and on the order of the support scale. The sensitivity of the statistical properties of the hydraulic conductivity to the support scale of the measurements is called the *scale effect*. At any given scale, the mean hydraulic conductivity, K_G, can be used in Darcy's law to yield the average specific discharge over the support scale, and on scales smaller than the support scale, the specific discharge may be expected to deviate from the averaged value. The average hydraulic conductivities over scales much larger than the support scale of individual measurements can be estimated by assuming a log-normal distribution of the hydraulic conductivity, with an exponential correlation function, and for small variances in hydraulic conductivity ($\sigma_Y < 1$), the mean values of the hydraulic conductivity, $\langle K \rangle$, can be estimated by (Dagan, 1989)

$$1\text{-D Flow:} \quad \langle K \rangle = K_G \left(1 - \frac{\sigma_Y^2}{2} \right) \tag{7.30}$$

$$2\text{-D Flow:} \quad \langle K \rangle = K_G \tag{7.31}$$

$$3\text{-D Flow:} \quad \langle K \rangle = K_G \left(1 + \frac{\sigma_Y^2}{6} \right) \tag{7.32}$$

Large-scale averaged hydraulic conductivities, $\langle K \rangle$, are commonly referred to as *macroscopic hydraulic conductivities*.

■ Example 7.5

Ten pump tests have been performed at different locations in an aquifer, and the calculated hydraulic conductivities were found to be: 15, 20, 30, 10, 8, 55, 17, 86, 12, and 35 m/d. The support scale for each of these measurements is around 100 m, and the geologic logs indicate that the aquifer can be considered statistically homogeneous. Estimate the macroscopic hydraulic conductivity (applied over scales of several hundreds of meters) that is appropriate for simulating three-dimensional flows in the aquifer.

Solution
The natural logs of the hydraulic conductivities are given in the following table:

K	$Y (= \ln K)$	Y^2
15	2.71	7.33
20	3.00	8.97
30	3.40	11.57
10	2.30	5.30
8	2.08	4.32
55	4.01	16.06
17	2.83	8.03
86	4.45	19.84
12	2.48	6.17
35	3.56	12.64
Sum	30.82	100.23

The mean and variance of the log-hydraulic conductivity, $\langle Y \rangle$ and σ_Y^2, can be estimated by $\bar{y}$ and S_y^2, respectively, where

$$\bar{Y} = \frac{1}{N} \sum_{i=1}^{N} Y_i$$

and

$$S_Y^2 = \frac{1}{N-1} \sum_{i=1}^{N} (Y_i - \bar{Y})^2$$

$$= \frac{1}{N-1} \sum_{i=1}^{N} Y_i^2 - \frac{N}{N-1} \bar{Y}^2$$

where N is the number of measurements, and Y_i are the log-hydraulic conductivity measurements. From the given data,

$$\bar{Y} = \frac{1}{10}(30.82) = 3.082$$

and

$$S_Y^2 = \frac{1}{10-1} 100.23 - \frac{10}{9} 3.082^2 = 0.583$$

For three-dimensional flows, the mean (large-scale) hydraulic conductivity is given by Equation 7.32 as

$$\langle K \rangle = K_G \left(1 + \frac{\sigma_Y^2}{6} \right)$$

where K_G is the geometric hydraulic conductivity given by

$$K_G = e^{\langle Y \rangle} \approx e^{\bar{Y}} = e^{3.082} = 22 \text{ m/d}$$

Substituting $K_G = 22$ m/d, and $\sigma_Y^2 = 0.583$ into Equation 7.32 yields the macroscopic hydraulic conductivity

$$\langle K \rangle = 22 \left(1 + \frac{0.583}{6} \right) = 24 \text{ m/d} \qquad ■$$

In concluding the discussion of Darcy's law, it should be noted that there are some porous media where Darcy's law is not applicable, notably in fractured media. In these formations, the flow is either modeled in individual fractures or, if the fractures are sufficiently dense, modeled as an *equivalent* porous medium. Techniques for describing flow in fractured media can be found in de Marsily (1986).

7.2.2 ■ General Flow Equation

Consider the control volume shown in Figure 7.6. The net influx of fluid mass into the control volume is given by

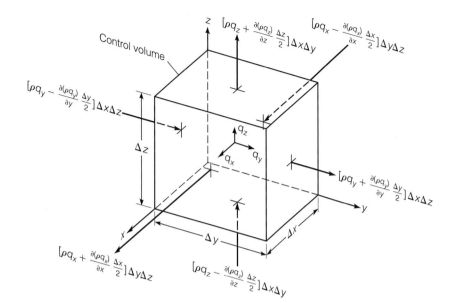

Figure 7.6 ■ Control Volume in a Porous Medium

$$\text{Net mass inflow} = \left[(\rho q_x) - \frac{\partial(\rho q_x)}{\partial x} \frac{\Delta x}{2} \right] \Delta y \Delta z - \left[(\rho q_x) + \frac{\partial(\rho q_x)}{\partial x} \frac{\Delta x}{2} \right] \Delta y \Delta z$$

$$+ \left[(\rho q_y) - \frac{\partial(\rho q_y)}{\partial y} \frac{\Delta y}{2} \right] \Delta x \Delta z - \left[(\rho q_y) + \frac{\partial(\rho q_y)}{\partial y} \frac{\Delta y}{2} \right] \Delta x \Delta z$$

$$+ \left[(\rho q_z) - \frac{\partial(\rho q_z)}{\partial z} \frac{\Delta z}{2} \right] \Delta x \Delta y - \left[(\rho q_z) + \frac{\partial(\rho q_z)}{\partial z} \frac{\Delta z}{2} \right] \Delta x \Delta y$$

$$(7.33)$$

where ρ is the density of the fluid. Combining terms in Equation 7.33 and simplifying leads to

$$\text{Net mass inflow} = - \left[\frac{\partial(\rho q_x)}{\partial x} + \frac{\partial(\rho q_y)}{\partial y} + \frac{\partial(\rho q_z)}{\partial z} \right] \Delta x \Delta y \Delta z \qquad (7.34)$$

In accordance with the law of conservation of mass, the net mass inflow into the control volume is equal to rate of change of mass within the control volume, which is given by

$$\text{Rate of change of mass} = \frac{\partial(n\rho)}{\partial t} \Delta x \Delta y \Delta z \qquad (7.35)$$

where n is the porosity of the porous medium, and t is the time. Combining Equations 7.34 and 7.35 yields the following relation

$$- \left[\frac{\partial(\rho q_x)}{\partial x} + \frac{\partial(\rho q_y)}{\partial y} + \frac{\partial(\rho q_z)}{\partial z} \right] = \frac{\partial(n\rho)}{\partial t} \qquad (7.36)$$

This equation can be further reduced for the cases where the density, ρ, is independent of space and time. Under this condition, Equation 7.36 reduces to

$$- \left(\frac{\partial q_x}{\partial x} + \frac{\partial q_y}{\partial y} + \frac{\partial q_z}{\partial z} \right) = \frac{\partial n}{\partial t} \qquad (7.37)$$

The *specific storage*, S_s, of a porous medium is defined as the volume of water released from storage per unit volume of the porous medium per unit decline in piezometric head. Considering the control volume shown in Figure 7.6, the time rate of change of fluid volume within the control volume can be expressed in terms of the specific storage, S_s, by the following relation

$$\frac{\partial n}{\partial t}\Delta x \Delta y \Delta z = S_s \frac{\partial \phi}{\partial t}\Delta x \Delta y \Delta z \tag{7.38}$$

where ϕ is the piezometric head, defined as $p/\gamma + z$. Equation 7.38 simplifies into

$$\frac{\partial n}{\partial t} = S_s \frac{\partial \phi}{\partial t} \tag{7.39}$$

and combining Equations 7.37 and 7.39 yields

$$-\left(\frac{\partial q_x}{\partial x} + \frac{\partial q_y}{\partial y} + \frac{\partial q_z}{\partial z}\right) = S_s \frac{\partial \phi}{\partial t} \tag{7.40}$$

According to Darcy's law, the components of the specific discharge vector, q_x, q_y, and q_z, can be written in terms of the piezometric head gradients, where

$$q_x = -K_{xx}\frac{\partial \phi}{\partial x}$$
$$q_y = -K_{yy}\frac{\partial \phi}{\partial y} \tag{7.41}$$
$$q_z = -K_{zz}\frac{\partial \phi}{\partial z}$$

and it is assumed that the coordinate axes are in the directions of the principal axes of the hydraulic conductivity. Combining Equations 7.40 and 7.41 leads to

$$\boxed{\frac{\partial}{\partial x}\left(K_{xx}\frac{\partial \phi}{\partial x}\right) + \frac{\partial}{\partial y}\left(K_{yy}\frac{\partial \phi}{\partial y}\right) + \frac{\partial}{\partial z}\left(K_{zz}\frac{\partial \phi}{\partial z}\right) = S_s \frac{\partial \phi}{\partial t}} \tag{7.42}$$

This general equation is applicable to both isotropic and anisotropic formations. In cases where the aquifer is homogeneous and anisotropic, Equation 7.42 becomes

$$\boxed{K_{xx}\frac{\partial^2 \phi}{\partial x^2} + K_{yy}\frac{\partial^2 \phi}{\partial y^2} + K_{zz}\frac{\partial^2 \phi}{\partial z^2} = S_s \frac{\partial \phi}{\partial t} \qquad \text{homogeneous, anisotropic}} \tag{7.43}$$

In cases where the hydraulic conductivity is isotropic, the hydraulic conductivity, K, is independent of the coordinate direction, which means that

$$K_{xx} = K_{yy} = K_{zz} = K \tag{7.44}$$

and Equation 7.43 can be written as

$$\boxed{\frac{\partial^2 \phi}{\partial x^2} + \frac{\partial^2 \phi}{\partial y^2} + \frac{\partial^2 \phi}{\partial z^2} = \frac{S_s}{K}\frac{\partial \phi}{\partial t} \qquad \text{homogeneous, isotropic}} \tag{7.45}$$

or in the more convenient vector form as

$$\nabla^2 \phi = \frac{S_s}{K} \frac{\partial \phi}{\partial t} \qquad \text{homogeneous, isotropic} \qquad (7.46)$$

where $\nabla^2()$ is the Laplacian operator defined by

$$\nabla^2 f = \frac{\partial^2 f}{\partial x^2} + \frac{\partial^2 f}{\partial y^2} + \frac{\partial^2 f}{\partial z^2} \qquad (7.47)$$

where f is a scalar function. In cases where there is significant radial symmetry, it is convenient to use cylindrical coordinates, and the Laplacian in Equation 7.46 is given by

$$\nabla^2 f = \frac{\partial^2 f}{\partial r^2} + \frac{1}{r} \frac{\partial f}{\partial r} + \frac{1}{r^2} \frac{\partial^2 f}{\partial \theta^2} + \frac{\partial^2 f}{\partial z^2} \qquad (7.48)$$

In summary, the governing equation for the flow of a homogeneous fluid in anisotropic nonhomogeneous porous media is given Equation 7.42, which simplifies to Equation 7.43 for homogeneous media and to Equation 7.45 in media that are both homogeneous and isotropic.

■ Example 7.6

Give the equation describing the piezometric head distribution in a homogeneous anisotropic aquifer in which vertical variations in the piezometric head are negligible. Explain why (in this case) the vertical hydraulic conductivity does not influence the head distribution. Is the assumption of negligible vertical variation in piezometric head reasonable in some cases?

Solution

The piezometric head distribution, $\phi(x, y, z, t)$, in a homogeneous anisotropic aquifer is given by Equation 7.43 as

$$K_{xx} \frac{\partial^2 \phi}{\partial x^2} + K_{yy} \frac{\partial^2 \phi}{\partial y^2} + K_{zz} \frac{\partial^2 \phi}{\partial z^2} = S_s \frac{\partial \phi}{\partial t}$$

If vertical variations in piezometric head are negligible, then ϕ does not depend on z, which means that $\partial^2 \phi / \partial z^2 = 0$, and therefore Equation 7.43 becomes

$$K_{xx} \frac{\partial^2 \phi}{\partial x^2} + K_{yy} \frac{\partial^2 \phi}{\partial y^2} = S_s \frac{\partial \phi}{\partial t}$$

Since the vertical hydraulic conductivity, K_{zz}, does not appear in this equation, then K_{zz} has no effect on the head distribution, $\phi(x, y, t)$.

The assumption of negligible vertical variation in the piezometric head is exact when the vertical pressure distribution is hydrostatic. This is certainly reasonable for horizontal ground-water flow. ■

Equivalent Anisotropic/Isotropic Media.   Consider a porous medium that is homogeneous and anisotropic, with hydraulic conductivities (K_{xx}, K_{yy}, K_{zz}); specific storage, S_s; and spatial coordinates (x, y, z). The spatial features of the anisotropic domain, described by the (x, y, z) coordinates can be transformed into a new domain, described by

(x', y', z') coordinates, where

$$x' = \sqrt{\frac{K}{K_{xx}}}x, \qquad y' = \sqrt{\frac{K}{K_{yy}}}y, \qquad z' = \sqrt{\frac{K}{K_{zz}}}z \qquad (7.49)$$

and K is defined as an arbitrary coefficient with dimensions of a hydraulic conductivity. Applying these relationships between (x, y, z) and (x', y', z') yields

$$\frac{\partial \phi}{\partial x'} = \frac{\partial \phi}{\partial x}\frac{dx}{dx'} = \sqrt{\frac{K_{xx}}{K}}\frac{\partial \phi}{\partial x} \qquad (7.50)$$

$$\frac{\partial^2 \phi}{\partial x'^2} = \frac{\partial}{\partial x}\left(\frac{\partial \phi}{\partial x'}\right)\frac{dx}{dx'} = \frac{K_{xx}}{K}\frac{\partial^2 \phi}{\partial x^2} \qquad (7.51)$$

$$\frac{\partial \phi}{\partial y'} = \frac{\partial \phi}{\partial y}\frac{dy}{dy'} = \sqrt{\frac{K_{yy}}{K}}\frac{\partial \phi}{\partial y} \qquad (7.52)$$

$$\frac{\partial^2 \phi}{\partial y'^2} = \frac{\partial}{\partial y}\left(\frac{\partial \phi}{\partial y'}\right)\frac{dy}{dy'} = \frac{K_{yy}}{K}\frac{\partial^2 \phi}{\partial y^2} \qquad (7.53)$$

$$\frac{\partial \phi}{\partial z'} = \frac{\partial \phi}{\partial z}\frac{dz}{dz'} = \sqrt{\frac{K_{zz}}{K}}\frac{\partial \phi}{\partial z} \qquad (7.54)$$

$$\frac{\partial^2 \phi}{\partial z'^2} = \frac{\partial}{\partial z}\left(\frac{\partial \phi}{\partial z'}\right)\frac{dz}{dz'} = \frac{K_{zz}}{K}\frac{\partial^2 \phi}{\partial z^2} \qquad (7.55)$$

Substituting Equations 7.50 to 7.55 into Equation 7.43 gives the following equation for the head distribution in the (x', y', z') domain,

$$\frac{\partial^2 \phi}{\partial x'^2} + \frac{\partial^2 \phi}{\partial y'^2} + \frac{\partial^2 \phi}{\partial z'^2} = \frac{S_s}{K}\frac{\partial \phi}{\partial t} \qquad (7.56)$$

which indicates that the head distribution can be calculated by taking the hydraulic conductivity as homogeneous and isotropic with a value of K in the transformed domain. By further requiring that the calculated flowrates across any boundary in the transformed domain be equal to the flowrate across the corresponding boundary in the real (x, y, z) domain, de Marsily (1986) has shown that K should be taken as

$$K = \sqrt[3]{K_{xx}K_{yy}K_{zz}} \qquad (7.57)$$

This result is extremely useful in practical applications, since it means that solutions to the flow equation in isotropic homogeneous formations can be applied to any anisotropic homogeneous formation by simply specifying the hydraulic conductivity components (K_{xx}, K_{yy}, K_{zz}), and transforming from (x', y', z') coordinates to (x, y, z) coordinates using Equation 7.49.

■ Example 7.7

The steady-state head distribution in a large homogeneous isotropic aquifer caused by pumping at a rate Q from the location (x'_o, y'_o, z'_o) is given by

$$\phi(x', y', z') = -\frac{Q}{4\pi}\frac{1}{\sqrt{(x'-x'_o)^2 + (y'-y'_o)^2 + (z'-z'_o)^2}}$$

Determine the head distribution caused by pumping at a rate Q in a homogeneous anisotropic aquifer where $K_{xx} = 10$ m/d, $K_{yy} = 5$ m/d, and $K_{zz} = 1$ m/d.

Solution

The hydraulic conductivity, K, of the equivalent isotropic aquifer is given by Equation 7.57 as

$$K = \sqrt[3]{K_{xx}K_{yy}K_{zz}} = \sqrt[3]{(10)(5)(1)} = 3.68 \text{ m/d}$$

and the relationships between the coordinates in the anisotropic aquifer and equivalent isotropic aquifer are given by Equation 7.49 as

$$x' = \sqrt{\frac{K}{K_{xx}}}x = \sqrt{\frac{3.68}{10}}x = 0.607x$$

$$y' = \sqrt{\frac{K}{K_{yy}}}y = \sqrt{\frac{3.68}{5}}y = 0.858y$$

$$z' = \sqrt{\frac{K}{K_{zz}}}z = \sqrt{\frac{3.68}{1}}z = 1.92z$$

where the primed coordinates apply to the equivalent isotropic aquifer. Applying the coordinate transformation to the head distribution in the equivalent isotropic aquifer gives the head distribution resulting from pumping in an anisotropic aquifer as

$$\phi(x, y, z) = -\frac{Q}{4\pi} \frac{1}{\sqrt{(0.607)^2(x - x_o)^2 + (0.858)^2(y - y_o)^2 + (1.92)^2(z - z_o)^2}}$$

$$= -\frac{Q}{4\pi} \frac{1}{\sqrt{0.368(x - x_o)^2 + 0.736(y - y_o)^2 + 3.69(z - z_o)^2}}$$

where (x_o, y_o, z_o) is the pumping location in the anisotropic aquifer.   ■

The governing equations for flow in porous media, such as Equation 7.43 or 7.45, contain only one unknown quantity, the piezometric head, ϕ, and the statement of the governing partial differential equation, along with the initial and boundary conditions for the piezometric head represents a complete statement of the problem of flow in aquifers (Bear, 1979). Solving this partial differential equation with associated boundary and initial conditions yields the piezometric head distribution, which is also called the piezometric surface or potentiometric surface. The term *potentiometric surface* is used more often in practice than the term *piezometric surface*. The distribution of seepage velocity and specific discharge in the porous medium can be derived directly from the potentiometric surface using Darcy's law.

7.2.3 ■ Two-Dimensional Approximations

7.2.3.1 Unconfined Aquifers

In the case of unconfined aquifers, the general flow equation can be simplified using an approximation first suggested by Dupuit (1863). Consider the typical flow conditions in an unconfined aquifer shown in Figure 7.7. The surface streamline is coincident with the phreatic surface (water table), and the slope of the streamlines become more horizontal

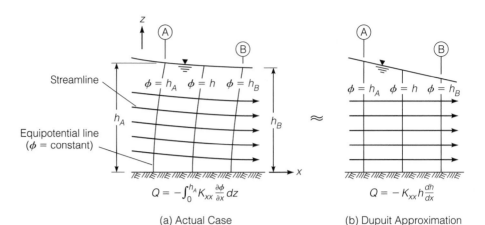

Figure 7.7 ■ Dupuit Approximation in an Unconfined Aquifer

with depth below the phreatic surface. The streamlines eventually become horizontal at the base of the aquifer (which is assumed to be horizontal). Since the equipotential lines are necessarily perpendicular to the streamlines, the equipotential lines are curved, as illustrated in Figure 7.7. The Dupuit (1863) approximation is that the equipotential lines can be assumed to be vertical, making the flowlines horizontal, and therefore the specific discharge components are given by

$$q_x \approx -K_{xx}\frac{\partial h}{\partial x}$$
$$q_y \approx -K_{yy}\frac{\partial h}{\partial y}$$

(7.58)

where h is the depth of the saturated zone. The Dupuit approximation was independently proposed by Forcheimer (1930) and is sometimes called the *Dupuit-Forchheimer* assumption. Clearly, for the Dupuit approximation to be valid, the actual equipotential lines must be nearly vertical, and therefore the slope of the phreatic surface must be nearly horizontal. These conditions are met by most unconfined aquifers, which typically have (phreatic) surface slopes in the range of 0.1% to 1% (Bear, 1979). Analysis of ground-water flow based on the Dupuit approximation is commonly called the *hydraulic approach.*

A general equation for flow in unconfined aquifers can now be derived using the Dupuit approximation. Consider the control volume shown in Figure 7.8, which is bounded on the bottom by the base of the aquifer and on the top by the phreatic surface. In addition to the inflows and outflows from the lateral boundaries of the control volume, there is also a recharge, $N(x, y)$, which accounts for the net inflow from above the aquifer, and is typically associated with infiltrated of water from the ground surface that penetrates the vadose zone. The recharge, $N(x, y)$, has units of volume inflow per unit area per unit time [L/T]. The net inflow of fluid mass into the control volume is given by

$$\text{Net mass inflow} \approx \left[(\rho q_x h) - \frac{\partial(\rho q_x h)}{\partial x}\frac{\Delta x}{2}\right]\Delta y - \left[(\rho q_x h) + \frac{\partial(\rho q_x h)}{\partial x}\frac{\Delta x}{2}\right]\Delta y$$
$$+ \left[(\rho q_y h) - \frac{\partial(\rho q_y h)}{\partial y}\frac{\Delta y}{2}\right]\Delta x - \left[(\rho q_y h) + \frac{\partial(\rho q_y h)}{\partial y}\frac{\Delta y}{2}\right]\Delta x$$
$$+ \rho N \Delta x \Delta y$$

(7.59)

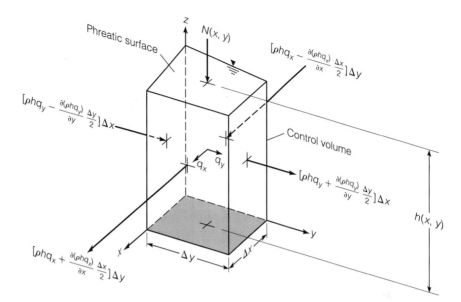

Figure 7.8 ▪ Control Volume in an Unconfined Aquifer

where the net mass inflow is regarded as approximate since vertical flow in the aquifer is neglected. Combining terms in Equation 7.59 and simplifying yields

$$\text{Net mass inflow} \approx -\left[\frac{\partial(\rho q_x h)}{\partial x} + \frac{\partial(\rho q_y h)}{\partial y}\right]\Delta x \Delta y + \rho N \Delta x \Delta y \qquad (7.60)$$

The net mass inflow into the control volume is equal to rate of change of mass within the control volume, which is given by

$$\text{Rate of change of mass} = \frac{\partial(nh\Delta x\Delta y\rho)}{\partial t} \qquad (7.61)$$

Combining Equations 7.60 and 7.61 leads to the Dupuit approximation to the flow equation as

$$-\left[\frac{\partial(\rho q_x h)}{\partial x} + \frac{\partial(\rho q_y h)}{\partial y}\right] + \rho N = \frac{\partial(nh\rho)}{\partial t} \qquad (7.62)$$

For the common case in which the density, ρ, is independent of space and time, Equation 7.62 reduces to

$$-\left[\frac{\partial(q_x h)}{\partial x} + \frac{\partial(q_y h)}{\partial y}\right] + N = \frac{\partial(nh)}{\partial t} \qquad (7.63)$$

The *specific yield*, S_y, of an unconfined aquifer is defined as the volume of water released from storage per unit (plan) area per unit decline in the phreatic surface (water table). Representative values of the specific yield for several aquifer materials are given in Tables 7.1 and 7.2. Considering the control volume shown in Figure 7.8, the time rate of change of water volume within the control volume can be expressed in terms of the specific yield by

$$\frac{\partial(nh)}{\partial t}\Delta x \Delta y = S_y \frac{\partial h}{\partial t}\Delta x \Delta y \qquad (7.64)$$

which simplifies into

$$\frac{\partial(nh)}{\partial t} = S_y \frac{\partial h}{\partial t} \qquad (7.65)$$

Combining Equations 7.63 and 7.65 yields

$$-\left[\frac{\partial(q_x h)}{\partial x} + \frac{\partial(q_y h)}{\partial y}\right] + N = S_y \frac{\partial h}{\partial t} \tag{7.66}$$

Substituting the approximate Darcy relations between the specific discharge and the aquifer depth, Equation 7.58, leads to

$$\boxed{\frac{\partial}{\partial x}\left(K_{xx} h \frac{\partial h}{\partial x}\right) + \frac{\partial}{\partial y}\left(K_{yy} h \frac{\partial h}{\partial y}\right) + N = S_y \frac{\partial h}{\partial t}} \tag{7.67}$$

This equation is applicable to anisotropic nonhomogeneous formations.

Consider the case of a stratified aquifer shown in Figure 7.9, where the aquifer has n layers with hydraulic conductivities (K_{xx}^i, K_{yy}^i), $i = 1, \ldots, n$, and K_{xx}^i and K_{yy}^i are the (principal) components of the hydraulic conductivity in the x- and y-directions in layer i. The flow in the x-direction, Q_x, is equal to the sum of the flows (in the x-direction) in each of the layers. Therefore

$$Q_x = -\sum_{i=1}^{n} K_{xx}^i \frac{\partial h}{\partial x} \Delta z_i \Delta y \tag{7.68}$$

and the effective x-component of the hydraulic conductivity, $\bar{K}_{xx}$, is defined by the relation

$$Q_x = -\bar{K}_{xx} \frac{\partial h}{\partial x} h \Delta y \tag{7.69}$$

Combining Equations 7.68 and 7.69 leads to the following expression for $\bar{K}_{xx}$ in terms of the hydraulic conductivities in the individual layers

$$\bar{K}_{xx} = \frac{1}{h}\sum_{i=1}^{n} K_{xx}^i \Delta z_i \tag{7.70}$$

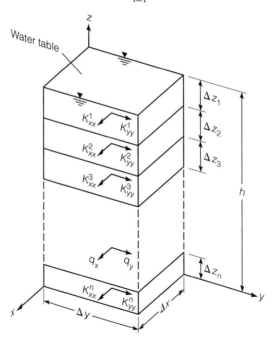

Figure 7.9 ■ Control Volume in a Stratified Unconfined Aquifer

Similarly, the effective hydraulic conductivity in the y-direction is given by

$$\bar{K}_{yy} = \frac{1}{h} \sum_{i=1}^{n} K_{yy}^i \Delta z_i \qquad (7.71)$$

Equations 7.70 and 7.71 yield effective hydraulic conductivities that can be substituted into the Dupuit approximation of the flow equation, Equation 7.67, to determine the distribution of piezometric head in stratified phreatic aquifers. It should be clear that the effective hydraulic conductivities given by Equations 7.70 and 7.71 are both non-homogeneous in space, since the effective hydraulic conductivities are a function of the aquifer depth, h, which is necessarily nonhomogeneous.

■ Example 7.8

The hydraulic conductivity distribution in a 30-m thick stratified surficial aquifer is given in the following table:

Depth (m)	K_{xx} (m/d)	K_{yy} (m/d)
0–5	25	30
5–10	30	33
10–15	40	37
15–20	32	28
20–25	22	19
25–30	13	11

Estimate the effective hydraulic conductivity when the water table is 4 m below the ground surface. Would the effective hydraulic conductivity be the same at a location where the water table is 5 m below the ground surface?

Solution

The effective hydraulic conductivity components, $\bar{K}_{xx}$ and $\bar{K}_{yy}$, are given by Equations 7.70 and 7.71 as

$$\bar{K}_{xx} = \frac{1}{h} \sum_{i=1}^{n} K_{xx}^i \Delta z_i \qquad \text{and} \qquad \bar{K}_{yy} = \frac{1}{h} \sum_{i=1}^{n} K_{yy}^i \Delta z_i$$

When the water table is 4 m below the ground surface, $h = 30 - 4 = 26$ m, and therefore

$$\bar{K}_{xx} = \frac{1}{26}(25 \times 1 + 30 \times 5 + 40 \times 5 + 32 \times 5 + 22 \times 5 + 13 \times 5) = 27.3 \text{ m/d}$$

$$\bar{K}_{yy} = \frac{1}{26}(30 \times 1 + 33 \times 5 + 37 \times 5 + 28 \times 5 + 19 \times 5 + 11 \times 5) = 25.8 \text{ m/d}$$

These calculations account for the fact that the ground water only flows through the top 1 m of the upper layer.

   When the water table is 5 m below the ground surface, $h = 30 - 5 = 25$ m, and therefore

$$\bar{K}_{xx} = \frac{1}{25}(30 \times 5 + 40 \times 5 + 32 \times 5 + 22 \times 5 + 13 \times 5) = 27.4 \text{ m/d}$$

$$\bar{K}_{yy} = \frac{1}{25}(33 \times 5 + 37 \times 5 + 28 \times 5 + 19 \times 5 + 11 \times 5) = 25.6 \text{ m/d}$$

Hence, $\bar{K}_{xx}$ increases slightly and $\bar{K}_{yy}$ decreases slightly when the water table falls from 4 m to 5 m below the ground surface. ■

In cases where the effective hydraulic conductivity of the aquifer is homogeneous, such as when the aquifer is composed of a single homogeneous layer, Equation 7.67 describes the head distribution by

$$K_{xx}\frac{\partial}{\partial x}\left(h\frac{\partial h}{\partial x}\right) + K_{yy}\frac{\partial}{\partial y}\left(h\frac{\partial h}{\partial y}\right) + N = S_y\frac{\partial h}{\partial t} \qquad \text{homogeneous, anisotropic}$$

(7.72)

Noting the identities

$$h\frac{\partial h}{\partial x} = \frac{1}{2}\frac{\partial h^2}{\partial x}, \qquad h\frac{\partial h}{\partial y} = \frac{1}{2}\frac{\partial h^2}{\partial y}$$

(7.73)

then Equation 7.72 can be written in the more compact form

$$K_{xx}\frac{\partial^2 h^2}{\partial x^2} + K_{yy}\frac{\partial^2 h^2}{\partial y^2} + 2N = 2S_y\frac{\partial h}{\partial t}$$

(7.74)

A common assumption is that the hydraulic conductivity is isotropic in the horizontal plane, in which case

$$K_{xx} = K_{yy} = K$$

(7.75)

and Equation 7.74 can be written as

$$K\frac{\partial^2 h^2}{\partial x^2} + K\frac{\partial^2 h^2}{\partial y^2} + 2N = 2S_y\frac{\partial h}{\partial t} \qquad \text{homogeneous, isotropic}$$

(7.76)

or in the more compact vector form as

$$K\nabla^2(h^2) + 2N = 2S_y\frac{\partial h}{\partial t} \qquad \text{homogeneous, isotropic}$$

(7.77)

where $\nabla^2()$ is the Laplacian operator. Equation 7.77, commonly called the *Boussinesq equation*, is an approximate governing equation for flow in unconfined aquifers, derived by invoking the Dupuit approximation. The Boussinesq equation (Equation 7.77) is a nonlinear partial differential equation in one unknown: the aquifer thickness, h. The statement of this equation, along with the initial and boundary conditions for the aquifer thickness, represents a complete statement of the problem of flow in unconfined aquifers. Solving this problem for the distribution of aquifer thickness, h, also yields the distribution of specific discharge in the aquifer, since the specific discharge can be derived directly from the aquifer thickness according to Equation 7.58. There are many problems in ground-water engineering that relate to conditions induced by pumping (or recharging) wells. In these cases, the induced stress on the aquifer is symmetric around the well and it is therefore preferable to work with radial coordinates rather than Cartesian coordinates. The Laplacian operator in radial coordinates is given by

$$\nabla^2() = \frac{\partial^2()}{\partial r^2} + \frac{1}{r}\frac{\partial()}{\partial r} + \frac{1}{r^2}\frac{\partial^2()}{\partial\theta^2}$$

(7.78)

Combining Equations 7.77 and 7.78, the ground-water flow equation in radial coordinates for an isotropic hydraulic conductivity is given by

$$\frac{\partial^2 (h^2)}{\partial r^2} + \frac{1}{r}\frac{\partial (h^2)}{\partial r} + \frac{1}{r^2}\frac{\partial^2 (h^2)}{\partial \theta^2} + \frac{2N}{K} = 2\frac{S_y}{K}\frac{\partial h}{\partial t} \qquad \text{homogeneous, isotropic}$$

(7.79)

■ Example 7.9

Write an equation for the saturated aquifer thickness surrounding a single pumping well in an unconfined isotropic homogeneous aquifer. Assume that the influence of the well is radially symmetric. How would this equation be simplified for steady-state conditions without recharge?

Solution

In this case, it is appropriate to use radial coordinates, and the aquifer thickness, h, can be described by Equation 7.79. Since the influence of the well is radially symmetric around the well, the origin of the radial coordinate system should be taken as the well location and, since h is independent of θ, $\partial^2 (h^2)/\partial \theta^2 = 0$. The governing equation, Equation 7.79, then simplifies to

$$\frac{\partial^2 (h^2)}{\partial r^2} + \frac{1}{r}\frac{\partial (h^2)}{\partial r} + \frac{2N}{K} = 2\frac{S_y}{K}\frac{\partial h}{\partial t}$$

Under steady-state conditions, $\partial h/\partial t = 0$, and in the absence of recharge $N = 0$. Hence, for the steady-state no-recharge case, the aquifer thickness, h, is described by

$$\frac{\partial^2 (h^2)}{\partial r^2} + \frac{1}{r}\frac{\partial (h^2)}{\partial r} = 0 \qquad (7.80)$$

This equation is considerably simpler than the general equation for the aquifer thickness given by Equation 7.79.   ■

7.2.3.2 Confined Aquifers

Consider the case of a confined aquifer of thickness b illustrated in Figure 7.10. In a similar manner to the Dupuit approximation, assuming that the streamlines are horizontal and the seepage velocities are vertically uniform, then the mass inflow into the control volume is given by

$$\text{Net mass inflow} = \left[\rho q_x - \frac{\partial(\rho q_x)}{\partial x}\frac{\Delta x}{2}\right]b\Delta y - \left[\rho q_x + \frac{\partial(\rho q_x)}{\partial x}\frac{\Delta x}{2}\right]b\Delta y$$

$$+ \left[\rho q_y - \frac{\partial(\rho q_y)}{\partial y}\frac{\Delta y}{2}\right]b\Delta x - \left[\rho q_y + \frac{\partial(\rho q_y)}{\partial y}\frac{\Delta y}{2}\right]b\Delta x$$

$$+ \rho N \Delta x \Delta y \qquad (7.81)$$

Combining terms in Equation 7.81 and simplifying yields

$$\text{Net mass inflow} = -\left[\frac{\partial(\rho q_x)}{\partial x} + \frac{\partial(\rho q_y)}{\partial y}\right]b\Delta x \Delta y + \rho N \Delta x \Delta y \qquad (7.82)$$

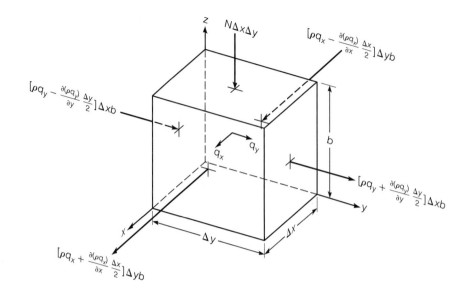

Figure 7.10 ■ Control Volume in a Confined Aquifer

The net mass inflow into the control volume is equal to rate of change of mass within the control volume, which is given by

$$\text{Rate of change of mass} = S\Delta x\Delta y\rho\frac{\partial\phi}{\partial t} \tag{7.83}$$

where S is the *storage coefficient* of the confined aquifer, which is defined as the volume of water released from storage per unit surface area of the aquifer per unit change in piezometric head. The storage coefficient is also called the *storativity*, and typically has magnitudes in the range $10^{-6} < S < 10^{-2}$ (Şen, 1995), which are much smaller than the specific yields in phreatic aquifers. In fact, storage coefficients are typically 1,000 to 10,000 times smaller than corresponding specific yields (de Marsily, 1986) in unconfined aquifers. This indicates that, for the same volume of withdrawal or recharge, changes in piezometric surface elevations are much larger in a confined aquifer than in an unconfined aquifer of the same material. Water released from storage due to declining piezometric heads in confined aquifers is associated with the net effect of the expansion of the water and the reduction in void space associated with the consolidation of the solid matrix. The storage coefficient can be expressed in terms of the elastic properties of the water and porous medium using the following equation (Bear, 1979)

$$S = \rho ngb\left(E_w + \frac{\alpha}{n}\right) \tag{7.84}$$

where ρ is the density of water, b is the thickness of the aquifer, E_w is the compressibility coefficient of water, and α is the compressibility coefficient of the porous matrix. Typically, 40% of the magnitude of S is contributed by the compressibility of water, and 60% by the compressibility of the porous matrix. Lohman (1972) has proposed the following relationship for estimating the storage coefficient

$$S \approx 3 \times 10^{-6}b \tag{7.85}$$

where b is the saturated aquifer thickness in meters. Combining Equations 7.82 and 7.83 leads to

$$-\left[\frac{\partial(\rho q_x)}{\partial x} + \frac{\partial(\rho q_y)}{\partial y}\right] + \frac{\rho N}{b} = \frac{S}{b}\rho\frac{\partial\phi}{\partial t} \qquad (7.86)$$

For the common case in which the density, ρ, is independent of space and time, Equation 7.86 reduces to

$$-\left(\frac{\partial q_x}{\partial x} + \frac{\partial q_y}{\partial y}\right) + \frac{\rho N}{b} = \frac{S}{b}\frac{\partial\phi}{\partial t} \qquad (7.87)$$

According to Darcy's law

$$q_x = -K_{xx}\frac{\partial\phi}{\partial x} \qquad (7.88)$$

$$q_y = -K_{yy}\frac{\partial\phi}{\partial y} \qquad (7.89)$$

and substituting these relations into Equation 7.87 leads to

$$\boxed{\frac{\partial}{\partial x}\left(K_{xx}\frac{\partial\phi}{\partial x}\right) + \frac{\partial}{\partial y}\left(K_{yy}\frac{\partial\phi}{\partial y}\right) + \frac{N}{b} = \frac{S}{b}\frac{\partial\phi}{\partial t}} \qquad (7.90)$$

This equation is applicable to anisotropic nonhomogeneous formations. In cases where the effective hydraulic conductivity of the aquifer is homogeneous, such as when the aquifer is composed of a single homogeneous layer, Equation 7.90 becomes

$$\boxed{K_{xx}\frac{\partial^2\phi}{\partial x^2} + K_{yy}\frac{\partial^2\phi}{\partial y^2} + \frac{N}{b} = \frac{S}{b}\frac{\partial\phi}{\partial t}} \qquad \text{homogeneous, anisotropic} \qquad (7.91)$$

A common assumption is that the hydraulic conductivity is isotropic in the horizontal plane, in which case

$$K_{xx} = K_{yy} = K \qquad (7.92)$$

and Equation 7.91 can be written as

$$\boxed{K\frac{\partial^2\phi}{\partial x^2} + K\frac{\partial^2\phi}{\partial y^2} + \frac{N}{b} = \frac{S}{b}\frac{\partial\phi}{\partial t}} \qquad \text{homogeneous, isotropic} \qquad (7.93)$$

or in the more compact form as

$$\boxed{K\nabla^2\phi + \frac{N}{b} = \frac{S}{b}\frac{\partial\phi}{\partial t}} \qquad \text{homogeneous, isotropic} \qquad (7.94)$$

where $\nabla^2()$ is the two-dimensional Laplacian operator.

■ Example 7.10

If the storage coefficient in a 20-m thick homogeneous isotropic confined aquifer can be estimated using Equation 7.85, determine the volume of water released per m^2 of aquifer when the piezometric head drops by 2 m. Where is this water "released" from?

If the piezometric head distribution reaches a steady state without recharge, state the equation describing the head distribution.

Solution

From the given data: $b = 20$ m, and Equation 7.85 gives the storage coefficient as

$$S \approx 3 \times 10^{-6}b = 3 \times 10^{-6}(20) = 6 \times 10^{-5}$$

This gives the volume of water (in m^3) released per m^2 of aquifer per meter drop in the piezometric head. Hence, for a 2-m drop in piezometric head, the volume of water released per m^2 is $2 \times (6 \times 10^{-5}) = 1.2 \times 10^{-4}$ m^3 per m^2. This water is "released" by the volumetric expansion of water and the compression of the porous matrix (= reduction in pore volume) caused by a decrease in pore pressure.

For a homogeneous isotropic confined aquifer, the piezometric head distribution is given by Equation 7.93 as

$$K\frac{\partial^2 \phi}{\partial x^2} + K\frac{\partial^2 \phi}{\partial y^2} + \frac{N}{b} = \frac{S}{b}\frac{\partial \phi}{\partial t}$$

Under steady-state conditions, $\partial\phi/\partial t = 0$, and without recharge, $N = 0$. In this case, Equation 7.93 becomes

$$K\frac{\partial^2 \phi}{\partial x^2} + K\frac{\partial^2 \phi}{\partial y^2} = 0$$

or

$$\frac{\partial^2 \phi}{\partial x^2} + \frac{\partial^2 \phi}{\partial y^2} = 0$$

which is considerably simpler than the general equation for confined homogeneous aquifers given by Equation 7.93. ■

Basic Equations in Terms of Transmissivity. The transmissivity (T_{xx}, T_{yy}) of a (confined) aquifer is defined as the product of the hydraulic conductivity (K_{xx}, K_{yy}) and the aquifer thickness, b, where

$$T_{xx} = K_{xx}b, \qquad T_{yy} = K_{yy}b \tag{7.95}$$

Hence, the general flow equation applicable to anisotropic nonhomogeneous formations (Equation 7.90) can be written as

$$\boxed{\frac{\partial}{\partial x}\left(T_{xx}\frac{\partial \phi}{\partial x}\right) + \frac{\partial}{\partial y}\left(T_{yy}\frac{\partial \phi}{\partial y}\right) + N = S\frac{\partial \phi}{\partial t}} \tag{7.96}$$

For homogeneous anisotropic formations

$$\boxed{T_{xx}\frac{\partial^2 \phi}{\partial x^2} + T_{yy}\frac{\partial^2 \phi}{\partial y^2} + N = S\frac{\partial \phi}{\partial t}} \quad \text{homogeneous, anisotropic} \tag{7.97}$$

and for homogeneous isotropic formations

$$\boxed{T\frac{\partial^2 \phi}{\partial x^2} + T\frac{\partial^2 \phi}{\partial y^2} + N = S\frac{\partial \phi}{\partial t}} \quad \text{homogeneous, isotropic} \tag{7.98}$$

Many analyses of ground-water flow are conducted using transmissivities rather than hydraulic conductivities and, in these cases, Equations 7.96 to 7.98 are the appropriate governing equations.

■ **Example 7.11**
_____

A homogeneous anisotropic confined aquifer is 25-m thick and has principal hydraulic conductivities of $K_{xx} = 34$ m/d and $K_{yy} = 15$ m/d. Determine the principal transmissivities and state the differential equation describing the piezometric head distribution in the absence of recharge. Would the governing equation be any different if the aquifer were 50-m thick, $K_{xx} = 17$ m/d, and $K_{yy} = 7.5$ m/d?

Solution
From the given data: $b = 25$ m, and the principal transmissivities are given by Equation 7.95 as

$$T_{xx} = K_{xx}b = (34)(25) = 850 \text{ m}^2/\text{d}$$

$$T_{yy} = K_{yy}b = (15)(25) = 375 \text{ m}^2/\text{d}$$

In the absence of recharge, $N = 0$, and the differential equation describing the piezometric head distribution is derived from Equation 7.97 as

$$T_{xx}\frac{\partial^2 \phi}{\partial x^2} + T_{yy}\frac{\partial^2 \phi}{\partial y^2} = S\frac{\partial \phi}{\partial t}$$

or

$$850\frac{\partial^2 \phi}{\partial x^2} + 375\frac{\partial^2 \phi}{\partial y^2} = S\frac{\partial \phi}{\partial t}$$

If the aquifer were 50-m thick, $K_{xx} = 17$ m/d, and $K_{yy} = 7.5$ m/d, then $T_{xx} = (17)(50) = 850 \text{ m}^2/\text{d}$ and $T_{yy} = (7.5)(50) = 375 \text{ m}^2/\text{d}$. Therefore, the governing differential equation would be no different.   ■

Equivalent Anisotropic/Isotropic Media.   Consider a porous medium that is homogeneous and anisotropic, with transmissivities (T_{xx}, T_{yy}), storage coefficient, S, and spatial coordinates (x, y). The spatial features of the anisotropic domain, described by the (x, y) coordinates can be transformed into a new domain, described by (x', y') coordinates, where

$$x' = \sqrt{\frac{T}{T_{xx}}}x, \qquad y' = \sqrt{\frac{T}{T_{yy}}}y \tag{7.99}$$

and T is defined as an arbitrary coefficient with dimensions of a transmissivity $[L^2/T]$. Applying these relationships between (x, y) and (x', y') yields

$$\frac{\partial \phi}{\partial x'} = \frac{\partial \phi}{\partial x}\frac{dx}{dx'} = \sqrt{\frac{T_{xx}}{T}}\frac{\partial \phi}{\partial x} \tag{7.100}$$

$$\frac{\partial^2 \phi}{\partial x'^2} = \frac{\partial}{\partial x}\left(\frac{\partial \phi}{\partial x'}\right)\frac{dx}{dx'} = \frac{T_{xx}}{T}\frac{\partial^2 \phi}{\partial x^2} \tag{7.101}$$

$$\frac{\partial \phi}{\partial y'} = \frac{\partial \phi}{\partial y}\frac{dy}{dy'} = \sqrt{\frac{T_{yy}}{T}}\frac{\partial \phi}{\partial y} \qquad (7.102)$$

$$\frac{\partial^2 \phi}{\partial y'^2} = \frac{\partial}{\partial y}\left(\frac{\partial \phi}{\partial y'}\right)\frac{dy}{dy'} = \frac{T_{yy}}{T}\frac{\partial^2 \phi}{\partial y^2} \qquad (7.103)$$

Substituting Equations 7.100 to 7.103 into Equation 7.97 gives the following equation for the piezometric head distribution in the (x', y') domain,

$$T\frac{\partial^2 \phi}{\partial x'^2} + T\frac{\partial^2 \phi}{\partial y'^2} = S\frac{\partial \phi}{\partial t} \qquad (7.104)$$

which indicates that the head distribution can be calculated by taking the transmissivity as homogeneous and isotropic with a value of T in the transformed domain. By further requiring that the calculated flowrates across any boundary in the transformed domain be equal to the flowrate across the corresponding boundary in the real (x, y) domain, de Marsily (1986) has shown that T should be taken as

$$T = \sqrt{T_{xx}T_{yy}} \qquad (7.105)$$

This result is extremely useful in practical applications, since it means that solutions to the flow equation in isotropic homogeneous formations can be applied to any anisotropic homogeneous formation by simply specifying the transmissivity components (T_{xx}, T_{yy}) and transforming from (x', y') coordinates to (x, y) coordinates using Equation 7.99.

■ Example 7.12

The analytic solution of the ground-water flow equation in an extensive isotropic homogeneous two-dimensional confined aquifer for the case of a single pumping well can be approximated by

$$\phi(x', y') = \phi_o + \frac{Q_w}{2\pi T}\left[0.5772 + \ln\frac{(x'^2 + y'^2)S}{4Tt}\right] \qquad (7.106)$$

where $\phi(x', y')$ is the head distribution, x' and y' are the spatial coordinates, ϕ_o is the initial head distribution in the aquifer (prior to pumping), Q_w is the pumping rate from the well, T is the transmissivity, S is the storage coefficient, and t is the time since pumping began. What would be the piezometric head distribution if the aquifer is anisotropic with transmissivities T_{xx} and T_{yy}.

Solution

Application of the head distribution in an isotropic medium to an anisotropic medium requires a scaling of the spatial coordinates and the transmissivity in accordance with Equations 7.99 and 7.105. These equations yield

$$T = \sqrt{T_{xx}T_{yy}} \Rightarrow \frac{T}{T_{xx}} = \sqrt{\frac{T_{yy}}{T_{xx}}}, \qquad \frac{T}{T_{yy}} = \sqrt{\frac{T_{xx}}{T_{yy}}}$$

$$x'^2 = \frac{T}{T_{xx}}x^2 = \sqrt{\frac{T_{yy}}{T_{xx}}}x^2$$

$$y'^2 = \frac{T}{T_{yy}}y^2 = \sqrt{\frac{T_{xx}}{T_{yy}}}y^2$$

Substituting these relationships into Equation 7.106 gives the piezometric head distribution in an anisotropic homogeneous aquifer as

$$\phi(x, y) = \phi_o + \frac{Q_w}{2\pi\sqrt{T_{xx}T_{yy}}}\left[0.5772 + \ln\frac{(\sqrt{T_{yy}/T_{xx}}x^2 + \sqrt{T_{xx}/T_{yy}}y^2)S}{4\sqrt{T_{xx}T_{yy}}t}\right] \qquad ■$$

This example illustrates the application of an analytic solution to the ground-water flow equation in an isotropic medium to an anisotropic medium. Implicit within this approach is that the initial and boundary conditions in both the isotropic and anisotropic domains are also related by the coordinate transformations. In the following section, several solutions to the ground-water flow equation in isotropic formations are presented for a variety of initial and boundary conditions. These results can be extended to anisotropic formations by using the approach described here.

7.3 Solutions of the Ground-Water Flow Equation

Most engineering applications in ground-water hydrology involve a particular solution of the ground-water flow equation, and these solutions can be either numerical or analytic. Numerical models typically solve the three-dimensional flow equation given by Equation 7.42 and can easily accommodate complex initial and boundary conditions along with complex hydraulic conductivity and specific storage distributions. A complete description of a widely used numerical model is given by McDonald and Harbaugh (1988) and Harbaugh and McDonald (1996). Analytic models typically solve the two-dimensional (Dupuit) approximations to the flow equations and are appropriate whenever the characteristics of the porous formation, as well as the boundary and initial conditions, are particularly simple. Common applications of analytic models are for matching observed aquifer responses with analytic solutions in order to determine aquifer properties. These applications are emphasized in the following sections.

7.3.1 ■ Steady Uniform Flow in a Confined Aquifer

Two-dimensional flow in a confined aquifer is described by Equation 7.94, which in the absence of recharge ($N = 0$) and under steady-state conditions ($\partial\phi/\partial t = 0$) leads to

$$\frac{\partial^2\phi}{\partial x^2} + \frac{\partial^2\phi}{\partial y^2} = 0 \qquad (7.107)$$

For steady flow with seepage velocity v_o in the x-direction, the piezometric head, ϕ, is independent of y, and Equation 7.107 can be written as

$$\frac{d^2\phi}{dx^2} = 0 \qquad (7.108)$$

where the solution of this equation must satisfy the conditions

$$v_o = -\frac{K}{n}\frac{d\phi}{dx} \qquad (7.109)$$

and

$$\phi(0) = \phi_o \tag{7.110}$$

where n is the effective porosity. Integrating Equation 7.108 twice leads to

$$\phi = Ax + B \tag{7.111}$$

where A and B are constants that can be evaluated using the conditions given by Equations 7.109 and 7.110 to yield

$$\phi = -\frac{n v_o}{K} x + \phi_o \tag{7.112}$$

Defining the drawdown s by

$$s = \phi_o - \phi \tag{7.113}$$

then Equation 7.112 can also be written in the form

$$\boxed{s = \frac{n v_o}{K} x} \tag{7.114}$$

This result indicates that in the case of uniform flow the piezometric surface is planar, sloping downward in the direction of flow.

■ Example 7.13

Water flows in a confined aquifer from a fully penetrating river with a piezometric head of 12.00 m toward a river with a piezometric head of 10.00 m located 500 m away. If the aquifer hydraulic conductivity and effective porosity are 3 m/d and 0.15, respectively, estimate the seepage velocity in the aquifer. Determine the drawdown and piezometric head 100 m from the upstream river.

Solution

According to Equation 7.112, the piezometric head, ϕ, varies linearly between the two fully penetrating rivers. The Darcy equation gives the (constant) seepage velocity as

$$v_o = -\frac{K}{n} \frac{d\phi}{dx} = -\frac{K}{n} \frac{\phi_2 - \phi_1}{\Delta x} = -\frac{3}{0.15} \frac{10.00 - 12.00}{500} = 0.08 \text{ m/d}$$

The drawdown, s, 100 m from the upstream river is given by Equation 7.114 as

$$s = \frac{n v_o}{K} x = \frac{(0.15)(0.08)}{3} (100) = 0.40 \text{ m}$$

and the corresponding piezometric head, ϕ, is given by

$$\phi = \phi_1 - s = 12.00 - 0.40 = 11.60 \text{ m} \qquad ■$$

7.3.2 ■ Steady Uniform Flow in an Unconfined Aquifer

Two-dimensional flow in an isotropic unconfined aquifer is described by Equation 7.77, which in the absence of recharge ($N = 0$) and under steady-state conditions

$(\partial h / \partial t = 0)$ leads to

$$\frac{\partial h^2}{\partial x^2} + \frac{\partial h^2}{\partial y^2} = 0 \tag{7.115}$$

Steady flow with constant seepage velocity, v_o, is not possible in unconfined aquifers since the aquifer thickness, h, is also the flow depth that must necessarily decrease in the direction of flow in accordance with Darcy's law. A decreasing flow depth must be accompanied by an increasing velocity by virtue of the law of conservation of mass.

7.3.3 ■ Steady Unconfined Flow Between Two Reservoirs

Consider the case of unconfined flow between two reservoirs shown in Figure 7.11. The flow is from a reservoir with water depth h_L to a reservoir with water depth h_R, through a distance L of stratified phreatic aquifer with hydraulic conductivities K_1 and K_2 over thicknesses b_1 and b_2, respectively. Invoking to the Dupuit approximation, the governing flow equation is given by Equation 7.67, which is rewritten here for convenient reference

$$\frac{\partial}{\partial x}\left(\bar{K}_{xx} h \frac{\partial h}{\partial x}\right) + \frac{\partial}{\partial y}\left(\bar{K}_{yy} h \frac{\partial h}{\partial y}\right) + N = S_y \frac{\partial h}{\partial t} \tag{7.116}$$

Since conditions are uniform in the y-direction (perpendicular to the page), conditions are at steady state, and there is no recharge, then the partial derivatives with respect to y and t are equal to zero, N is equal to zero, and Equation 7.116 reduces to

$$\frac{d}{dx}\left(\bar{K}_{xx} h \frac{dh}{dx}\right) = 0 \tag{7.117}$$

where the partial derivatives have been replaced by total derivatives, since the aquifer thickness, h, depends only on x. The effective hydraulic conductivity, $\bar{K}_{xx}$, derived from Equation 7.70, is given by

$$\bar{K}_{xx} = \frac{1}{h}[K_1(h - b_2) + K_2 b_2] \tag{7.118}$$

and substituting Equation 7.118 into Equation 7.117 yields

$$\frac{d}{dx}\left\{[K_1(h - b_2) + K_2 b_2]\frac{dh}{dx}\right\} = 0 \tag{7.119}$$

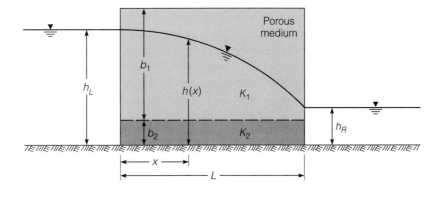

Figure 7.11 ■ Unconfined Flow Between Two Reservoirs

Integrating Equation 7.119 once leads to

$$[K_1(h - b_2) + K_2 b_2]\frac{dh}{dx} = C_1 \qquad (7.120)$$

where C_1 is a constant. Separating terms in Equation 7.120 leads to

$$K_1 h \frac{dh}{dx} + (K_2 - K_1)b_2\frac{dh}{dx} = C_1 \qquad (7.121)$$

which can also be written in the form

$$\frac{K_1}{2}\frac{d(h^2)}{dx} + (K_2 - K_1)b_2\frac{dh}{dx} = C_1 \qquad (7.122)$$

which integrates directly to yield

$$\boxed{\frac{K_1}{2}h^2 + (K_2 - K_1)b_2 h = C_1 x + C_2} \qquad (7.123)$$

The constants C_1 and C_2 can be determined by requiring $h(x)$ in Equation 7.123 to satisfy the boundary conditions: $h(0) = h_L$, and $h(L) = h_R$. Invoking these boundary conditions leads to

$$C_1 = \frac{K_1}{2L}(h_R^2 - h_L^2) + (K_2 - K_1)\frac{b_2}{L}(h_R - h_L) \qquad (7.124)$$

$$C_2 = \frac{K_1}{2}h_L^2 + (K_2 - K_1)b_2 h_L \qquad (7.125)$$

Hence the complete solution to the problem of unconfined flow between two reservoirs is given by Equation 7.123, where the constants C_1 and C_2 are given by Equations 7.124 and 7.125. Once the distribution of aquifer thickness, h, is calculated, then the flow between the two reservoirs, Q, can be calculated directly from Darcy's law and the Dupuit approximation using the relation

$$Q = -K_1(h - b_2)\frac{dh}{dx} - K_2 b_2 \frac{dh}{dx} \qquad (7.126)$$

where dh/dx is determined from the aquifer thickness, h, described by Equation 7.123. The approach used here can be extended to cases where there are more than two layers, and also to cases where a recharge, N, is present.

■ Example 7.14

The Biscayne aquifer is one of the most permeable aquifers in the world, and consists principally of two layers: the Miami Limestone formation and the Fort Thomson formation. In one particular area, the Miami Limestone extends from ground surface at 2.44 m NGVD to −3.00 m NGVD, and the Fort Thomson formation extends from −3.00 m NGVD to −15.24 m NGVD. The hydraulic conductivity of the Miami Limestone formation can be taken as 1,500 m/d, and the hydraulic conductivity of the Fort Thompson formation can be taken as 12,000 m/d. Calculate the shape of the phreatic surface and the flowrate between two fully penetrating canals 1 km apart, when the water elevations in the two canals are 1.07 m NGVD and 1.00 m NGVD.

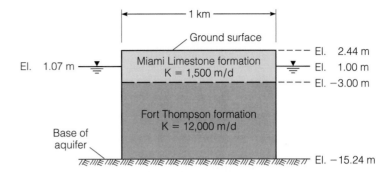

Figure 7.12 ■ Unconfined Flow in the Biscayne Aquifer

Solution

A schematic diagram of the Biscayne aquifer between two fully penetrating canals is shown in Figure 7.12. The phreatic surface is described by Equations 7.123, 7.124, and 7.125, where $h_L = 1.07 \text{ m} - (-15.24 \text{ m}) = 16.31 \text{ m}$, $h_R = 1.00 \text{ m} - (-15.24 \text{ m}) = 16.24 \text{ m}$, $K_1 = 1,500 \text{ m/d}$, $K_2 = 12,000 \text{ m/d}$, $b_2 = -3.00 \text{ m} - (-15.24 \text{ m}) = 12.24 \text{ m}$, and $L = 1,000 \text{ m}$. Substituting these values into Equations 7.123, 7.124, and 7.125 yields

$$750h^2 + 128520h = C_1 x + C_2 \tag{7.127}$$

where

$$C_1 = -10.7 \text{ m}^2/\text{d}, \qquad C_2 = 2,296,000 \text{ m}^3/\text{d} \tag{7.128}$$

Combining these results gives the following (implicit) equation for the phreatic surface

$$h^2 + 171.4h = -0.01428x + 3061 \tag{7.129}$$

The flowrate between the two reservoirs is given by Equation 7.126. Substituting known quantities into this equation yields the following expression for the flowrate, Q,

$$Q = -1500(h - 12.24)\frac{dh}{dx} - 146900\frac{dh}{dx}$$

which can be written in the form

$$Q = -1500h\frac{dh}{dx} - 128500\frac{dh}{dx} \tag{7.130}$$

The slope of the water table, dh/dx, can be obtained by differentiating the equation for the phreatic surface, Equation 7.129, with respect to x, which yields

$$2h\frac{dh}{dx} + 171.4\frac{dh}{dx} = -0.01428$$

or

$$\frac{dh}{dx} = -\frac{0.01428}{2h + 171.4} \tag{7.131}$$

Combining Equations 7.130 and 7.131 yields the following expression for the flowrate as a function of h,

$$Q = \frac{21.42h + 1835}{2h + 171.4} \tag{7.132}$$

At this point we can pick any value of h that exists within the aquifer, including h_L and h_R, and be assured that the value of Q obtained must be the same, since Equation 7.132 must necessarily satisfy the law of conservation of mass. Taking $h = h_L$ yields the desired result

$$Q = 10.7 \text{ m}^2/\text{d}$$ ▪

The problem of flow between two fully penetrating water bodies can also be applied to the cases where the water bodies do not fully penetrate the aquifer. Bear (1979) indicates that beyond a distance on the order of two aquifer depths from a partially penetrating water body, the streamlines are usually very near to being horizontal and the effect of partial penetration is negligible. A practical application of this approximation is in estimating the leakage from partially penetrating open channels. A detailed investigation of this problem by Chin (1991) indicated that a distance of 10 aquifer depths from a partially penetrating open channel may be necessary in order to neglect the effect of partial penetration on leakage.

7.3.4 ▪ Steady Flow to a Well in a Confined Aquifer

In isotropic and homogeneous confined aquifers, the distribution of piezometric head for two-dimensional flows is given by Equation 7.94, which can be written in the form

$$\frac{\partial^2 \phi}{\partial r^2} + \frac{1}{r}\frac{\partial \phi}{\partial r} + \frac{1}{r^2}\frac{\partial^2 \phi}{\partial \theta^2} + \frac{N}{Kb} = \frac{S}{Kb}\frac{\partial \phi}{\partial t} \qquad (7.133)$$

If a well fully penetrates the confined aquifer, and withdraws water at a constant rate uniformly over the aquifer thickness, b, as illustrated in Figure 7.13, then the piezometric head, ϕ, is radially symmetric (independent of θ) and independent of time. Consequently, all derivatives of $\phi(r)$ with respect to θ, and t are zero. If there is no recharge of the confined aquifer, then $N = 0$ and Equation 7.133 reduces to

$$\frac{d^2 \phi}{dr^2} + \frac{1}{r}\frac{d\phi}{dr} = 0 \qquad (7.134)$$

where partial derivatives of ϕ are replaced by total derivatives, since ϕ is only a function of the radial distance from the well, r. Multiplying Equation 7.134 by r yields

$$r\frac{d^2 \phi}{dr^2} + \frac{d\phi}{dr} = 0$$

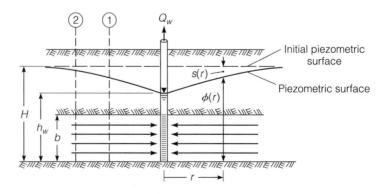

Figure 7.13 ▪ Fully Penetrating Well in a Confined Aquifer

which can be written as

$$\frac{d}{dr}\left(r\frac{d\phi}{dr}\right) = 0 \tag{7.135}$$

This equation is to be solved with the boundary conditions

$$2\pi r_w bK\frac{d\phi}{dr}\bigg|_{r=r_w} = Q_w \tag{7.136}$$

$$\phi(r_w) = h_w \tag{7.137}$$

where r_w is the radius of the well, K is the hydraulic conductivity of the aquifer, Q_w is the flowrate out of the well, and h_w is the depth of the water in the well, which is also equal to the piezometric head at the well ($r = r_w$). Integrating Equation 7.135 with respect to r yields

$$r\frac{d\phi}{dr} = A \tag{7.138}$$

and integrating again with respect to r yields

$$\phi = A \ln r + B \tag{7.139}$$

where A and B are constants to be determined from the boundary conditions, Equations 7.136 and 7.137. Applying the boundary conditions yields

$$A = \frac{Q_w}{2\pi bK}, \qquad B = h_w - A \ln r_w \tag{7.140}$$

Substituting Equation 7.140 into Equation 7.139 yields the following expression for the piezometric head distribution induced by a fully penetrating pumping well in a confined aquifer

$$\phi(r) = h_w + \frac{Q_w}{2\pi Kb} \ln\left(\frac{r}{r_w}\right) \tag{7.141}$$

Although this solution is exact for the flow equation and defined boundary conditions, there is an apparent paradox in that Equation 7.141 indicates that the piezometric head, ϕ, increases without bound as the distance from the well, r, increases. Clearly, this is not realistic, since the piezometric surface should be asymptotic to the initial piezometric surface before pumping, that is, the piezometric surface corresponding to $Q_w = 0$. This is illustrated in Figure 7.13. The reason that an exact solution to the steady-state flow equation does not give a realistic result is that a steady state for this flow condition is in fact impossible, since the water being pumped is constantly being drawn from storage and should result in a continuous decline in the piezometric surface. In spite of the limitation of the steady-state approximation, Equation 7.141 provides a reasonably accurate description of the piezometric surface, as long as $\phi < H$, where H is the elevation of the piezometric surface in the absence of pumping. This limitation in applying Equation 7.141 is frequently made explicit by defining the boundary condition

$$\phi(R) = H \tag{7.142}$$

in lieu of Equation 7.137, where R is commonly referred to as the *radius of influence* of the well. Using Equation 7.142 as a boundary condition, the head distribution is given by

$$\phi(r) = H - \frac{Q_w}{2\pi K b} \ln\left(\frac{R}{r}\right) \qquad (7.143)$$

This equation, originally derived by Thiem (1906), is commonly known as the *Thiem equation*. From a theoretical viewpoint, R must necessarily be transient and cannot be taken as a constant, however, since $\phi(r)$ depends on the logarithm of R, then $\phi(r)$ is not very sensitive to errors in R. The most commonly used (empirical) equation for estimating R (in meters) was originally proposed by Sichard (1927) as

$$R = 3000 s_w \sqrt{K} \qquad (7.144)$$

where $s_w = H - \phi(r_w)$ is the drawdown at the well in meters, and K is the hydraulic conductivity in meters per second.

In many cases, it is convenient to work with the *drawdown*, $s(r)$, instead of the piezometric head, $\phi(r)$, where

$$s(r) = H - \phi(r) \qquad (7.145)$$

Combining Equations 7.145 and 7.141 gives

$$s(r) = s_w - \frac{Q_w}{2\pi T} \ln\left(\frac{r}{r_w}\right), \qquad s(r) \geq 0 \qquad (7.146)$$

where s_w is the drawdown at the well and T is the *transmissivity* of the confined aquifer defined by

$$T = Kb \qquad (7.147)$$

Equation 7.146 describes the drawdown surface surrounding the well, which is commonly referred to as the *cone of depression*. The theoretical drawdown distribution given by Equation 7.146 is used in *pump tests* to determine the transmissivity of the aquifer by measuring the drawdowns resulting from pumpage, and then solving for the transmissivity. Applying Equation 7.146 to estimate the drawdowns at $r = r_1$ and $r = r_2$ and subtracting the result yields

$$s_1 - s_2 = \frac{Q_w}{2\pi T} \ln\left(\frac{r_2}{r_1}\right) \qquad (7.148)$$

where s_1 and s_2 are the drawdowns at r_1 and r_2, respectively. Equation 7.148 can be put in the more useful form

$$T = \frac{Q_w}{2\pi(s_1 - s_2)} \ln\left(\frac{r_2}{r_1}\right) \qquad (7.149)$$

Hence, by measuring the pumping rate, Q_w, and drawdowns, s_1 and s_2, at two monitoring wells located at $r = r_1$ and $r = r_2$, the transmissivity of the aquifer, T, can be estimated directly using Equation 7.149. It is interesting to note that, in reality, even though the steady-state approximation inherent in Equation 7.149 is not strictly correct, the hydraulic gradient, $s_1 - s_2$, in the area surrounding the well approaches a pseudosteady state, and Equation 7.149 provides reasonably accurate estimates of the transmissivity.

■ **Example 7.15**

A pump test was conducted in a confined aquifer where the initial piezometric surface was at elevation 14.385 m, and well logs indicate that the thickness of the aquifer is 25 m. The well was pumped at 31.54 L/s, and after one day the piezometric levels at 50 m and 100 m from the pumping well were measured as 13.585 m and 14.015 m, respectively. Assuming steady-state conditions, estimate the transmissivity and hydraulic conductivity of the aquifer.

Solution

The equation to be used to estimate the transmissivity is given by Equation 7.149 as

$$T = \frac{Q_w}{2\pi(s_1 - s_2)} \ln\left(\frac{r_2}{r_1}\right)$$

where $Q_w = 31.54 \text{L/s} = 2{,}725 \ \text{m}^3/\text{d}$, $r_1 = 50$ m, $r_2 = 100$ m, $s_1 = 14.385$ m − 13.585 m = 0.800 m, and $s_2 = 14.385$ m − 14.015 m = 0.370 m. Substituting these values yields

$$T = \frac{2725}{2\pi(0.800 - 0.370)} \ln\left(\frac{100}{50}\right) = 699 \ \text{m}^2/\text{d}$$

The transmissivity, T, is related to the hydraulic conductivity, K, by

$$T = Kb$$

where b is the aquifer thickness. Since $b = 25$ m,

$$K = \frac{T}{b} = \frac{699}{25} = 28.0 \ \text{m/d}$$

Therefore, the results of the pump test indicate that the transmissivity of the aquifer is 699 m²/d, and the hydraulic conductivity is 28.0 m/d.

As a reference point, aquifers with transmissivity values less than 10 m²/d can supply only enough water for domestic wells and other low-yield uses (Lehr et al., 1988). ■

7.3.5 ■ Steady Flow to a Well in an Unconfined Aquifer

In isotropic and homogeneous unconfined aquifers, the distribution of aquifer thickness, h, can be estimated by assuming two-dimensional flow (Dupuit approximation) and describing the distribution of aquifer thickness by Equation 7.79. The case of a fully penetrating pumping well in an unconfined aquifer is illustrated in Figure 7.14. Since the aquifer response to pumping is radially symmetric around the well, then derivatives with respect to θ are equal to zero and Equation 7.79 can be written as

$$\frac{d^2(h^2)}{dr^2} + \frac{1}{r}\frac{d(h^2)}{dr} = 0 \tag{7.150}$$

where partial derivatives have been replaced by total derivatives (since h is a function of r only), the recharge, N, has been set equal to zero, and the derivative of h with respect to time has been set equal to zero (since steady state is assumed). Equation 7.150 can be

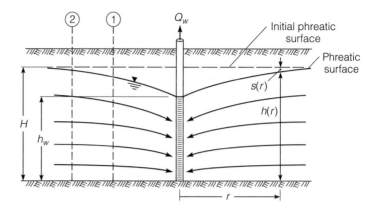

Figure 7.14 ■ Fully Penetrating Well in an Unconfined Aquifer

written as

$$\frac{d}{dr}\left(r\frac{dh^2}{dr}\right) = 0 \qquad (7.151)$$

This equation is to be solved with the boundary conditions

$$2\pi r_w h K \frac{dh}{dr}\bigg|_{r=r_w} = Q_w \qquad (7.152)$$

$$h(r_w) = h_w \qquad (7.153)$$

where r_w is the radius of the well, K is the hydraulic conductivity of the aquifer, Q_w is the pumping rate, and h_w is the depth of the water in the well. Integrating Equation 7.151 twice with respect to r yields

$$h^2 = A \ln r + B \qquad (7.154)$$

where A and B are constants to be determined using the boundary conditions. Applying the boundary conditions yields

$$A = \frac{Q_w}{\pi K}, \qquad B = h_w^2 - A \ln r_w \qquad (7.155)$$

Substituting Equation 7.155 into Equation 7.154 yields the following expression for the piezometric head distribution in an unconfined aquifer surrounding a pumping well

$$h^2 = h_w^2 + \frac{Q_w}{\pi K} \ln\left(\frac{r}{r_w}\right) \qquad (7.156)$$

As was the case for confined aquifers, this solution is apparently exact for the assumed flow equation and defined boundary conditions. However, there is an apparent paradox in that Equation 7.156 indicates that the aquifer thickness, h, increases without bound as the distance from the well, r, increases. Clearly, this is not realistic, since the water table should be asymptotic to the water table prior to pumping, that is, when the pumping rate, Q_w, is equal to zero. The reason that an exact solution to the steady-state flow equation does not give a realistic result is that steady state for this flow condition is impossible. As in the case of a confined aquifer, the pumpage must continuously be drawn from aquifer storage, and therefore there can be no steady state. In spite of this paradox, Equation 7.156 provides a reasonably accurate description of the phreatic surface, as long as $h < H$, where H is the aquifer thickness in the absence of pumping. The *drawdown*, s, is defined by

$$s(r) = H - h(r) \tag{7.157}$$

in which case Equation 7.156 can be written as

$$(H - s)^2 = (H - s_w)^2 + \frac{Q_w}{\pi K} \ln\left(\frac{r}{r_w}\right) \tag{7.158}$$

where s_w is the drawdown at the well. The drawdown distribution given by Equation 7.158 is most commonly used in pump tests to determine the hydraulic conductivity of the aquifer by measuring the drawdowns resulting from pumping water out of the well, and then solving for the hydraulic conductivity. Applying Equation 7.158 to estimate the drawdowns at $r = r_1$ and $r = r_2$ and subtracting the result yields

$$(H - s_2)^2 - (H - s_1)^2 = \frac{Q_w}{\pi K} \ln\left(\frac{r_2}{r_1}\right)$$

which simplifies to

$$\left(s_1 - \frac{s_1^2}{2H}\right) - \left(s_2 - \frac{s_2^2}{2H}\right) = \frac{Q_w}{2\pi KH} \ln\left(\frac{r_2}{r_1}\right) \tag{7.159}$$

Defining the *modified drawdowns*, s_1' and s_2' by the relations

$$s_1' = s_1 - \frac{s_1^2}{2H}, \qquad s_2' = s_2 - \frac{s_2^2}{2H} \tag{7.160}$$

then Equation 7.159 can be written as

$$s_1' - s_2' = \frac{Q_w}{2\pi T} \ln\left(\frac{r_2}{r_1}\right) \tag{7.161}$$

where T is the transmissivity of the unconfined aquifer defined by

$$T = KH \tag{7.162}$$

Making T the subject of the formula in Equation 7.161 yields the useful equation

$$\boxed{T = \frac{Q_w}{2\pi(s_1' - s_2')} \ln\left(\frac{r_2}{r_1}\right)} \tag{7.163}$$

which is exactly the same result that was derived for a confined aquifer (see Equation 7.149), except that modified drawdowns are used. It should be noted, however, that in the common case where the drawdowns are small compared with the aquifer thickness, or specifically

$$s_1^2 \ll 2H \qquad \text{and} \qquad s_2^2 \ll 2H$$

then the modified drawdowns, s_1' and s_2', can be replaced by the actual drawdowns, s_1 and s_2. On the basis of Equation 7.163, the transmissivity of the aquifer can be estimated from measurements of the pumping rate, Q_w, and drawdowns, s_1 and s_2, at two monitoring wells located at $r = r_1$ and $r = r_2$. The hydraulic conductivity can then be estimated using Equation 7.162.

■ **Example 7.16**

A pump test has been conducted in an unconfined aquifer of saturated thickness 15 m. The well was pumped at a rate of 100 L/s, and the drawdowns at 50 m and 100 m from the pumping well after one day of pumping were 0.412 m and 0.251 m, respectively. Estimate the transmissivity and hydraulic conductivity of the aquifer. Assume steady state.

Solution

The steady-state estimate of the transmissivity, T, is given by Equation 7.163 as

$$T = \frac{Q_w}{2\pi(s_1' - s_2')} \ln\left(\frac{r_2}{r_1}\right)$$

where $Q_w = 100$ L/s $= 8,640$ m^3/d, $r_1 = 50$ m, $r_2 = 100$ m, $s_1 = 0.412$ m, $s_2 = 0.251$ m, $H = 15$ m, and s_1' and s_2' are given by Equation 7.160 as

$$s_1' = s_1 - \frac{s_1^2}{2H} = 0.412 - \frac{0.412^2}{2(15)} = 0.406 \text{ m}$$

and

$$s_2' = s_2 - \frac{s_2^2}{2H} = 0.251 - \frac{0.251^2}{2(15)} = 0.249 \text{ m}$$

The transmissivity is therefore estimated as

$$T = \frac{8640}{2\pi(0.406 - 0.249)} \ln\left(\frac{100}{50}\right) = 6,070 \text{ m}^2/\text{d}$$

The hydraulic conductivity, K, is given by

$$K = \frac{T}{H} = \frac{6070}{15} = 405 \text{ m/d}$$

Therefore, the transmissivity and hydraulic conductivity of the aquifer are estimated from the pump test data to be 6,070 m^2/d and 405 m/d, respectively. ■

7.3.6 ■ Steady Flow to a Well in a Leaky Confined Aquifer

In isotropic and homogeneous semiconfined (leaky) aquifers, the distribution of piezometric head for two-dimensional flows is given by Equation 7.94, which can be written in the form

$$\frac{\partial^2\phi}{\partial r^2} + \frac{1}{r}\frac{\partial\phi}{\partial r} + \frac{1}{r^2}\frac{\partial^2\phi}{\partial\theta^2} + \frac{N}{Kb} = \frac{S}{Kb}\frac{\partial\phi}{\partial t} \tag{7.164}$$

If a well fully penetrates the semiconfined aquifer and withdraws water at a constant rate uniformly over the aquifer thickness, b, as illustrated in Figure 7.15, then the piezometric head, ϕ, is radially symmetric (independent of θ) and, under steady-state conditions, is independent of time. Therefore, all derivatives with respect to θ and t are zero, and Equation 7.164 can be written in the form

$$\frac{d^2\phi}{dr^2} + \frac{1}{r}\frac{d\phi}{dr} + \frac{N}{Kb} = 0 \tag{7.165}$$

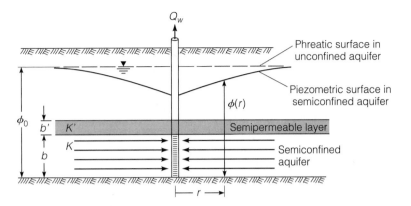

Figure 7.15 ■ Fully Penetrating Well in a Semiconfined Aquifer

where partial derivatives have replaced total derivatives, since ϕ depends only on r. Leakage into the semiconfined aquifer can be calculated by applying Darcy's law across the semipermeable layer, in which case

$$N = -K' \frac{\phi - \phi_o}{b'} \tag{7.166}$$

where K' is the hydraulic conductivity of the semipermeable layer, b' is the thickness of the semipermeable layer, and ϕ_o is the piezometric head above the semipermeable layer, assumed to be constant. Combining Equations 7.165 and 7.166 leads to

$$\frac{d^2\phi}{dr^2} + \frac{1}{r}\frac{d\phi}{dr} + \frac{K'(\phi_o - \phi)}{Kbb'} = 0 \tag{7.167}$$

Defining the parameter λ by the expression

$$\boxed{\lambda^2 = \frac{Kbb'}{K'}} \tag{7.168}$$

then Equation 7.167 can be written as

$$\frac{d^2\phi}{dr^2} + \frac{1}{r}\frac{d\phi}{dr} + \frac{\phi_o - \phi}{\lambda^2} = 0 \tag{7.169}$$

The quantity λ defined by Equation 7.168 is commonly used to parameterize the leakage in semiconfined aquifers, and is called the *leakage factor*. Defining new variables ϕ' and r' by

$$\phi' = \phi_o - \phi \tag{7.170}$$

$$r' = \frac{r}{\lambda} \tag{7.171}$$

then Equation 7.169 can be written in the form

$$r'^2 \frac{d^2\phi'}{dr'^2} + r' \frac{d\phi'}{dr'} - r'^2\phi' = 0 \tag{7.172}$$

This equation is a modified Bessel's equation of order zero, discussed in Appendix F, and has the solution

$$\phi' = AI_o(r') + BK_o(r') \tag{7.173}$$

or

$$\phi_o - \phi(r) = AI_o(r/\lambda) + BK_o(r/\lambda) \tag{7.174}$$

where $I_o(x)$ and $K_o(x)$ are modified Bessel functions of the first and second kind (respectively) of order zero, and A and B are constants to be determined from the boundary conditions. In this case, the boundary conditions are

$$\phi(\infty) = \phi_o \tag{7.175}$$

and

$$2\pi r_w bK \left.\frac{d\phi}{dr}\right|_{r=r_w} = Q_w \tag{7.176}$$

Noting the following Bessel function properties

$$I_o(\infty) = \infty \tag{7.177}$$

and

$$K_o(\infty) = 0 \tag{7.178}$$

then the boundary condition given by Equation 7.175 requires that $A = 0$ in Equation 7.174, which now becomes

$$\phi_o - \phi(r) = BK_o(r/\lambda) \tag{7.179}$$

Application of the second boundary condition, Equation 7.176, requires using the Bessel function identity

$$\frac{d}{dr'}K_o(r') = -K_1(r') \tag{7.180}$$

where $K_1(r')$ is the modified Bessel function of the second kind of order one. Differentiating Equation 7.179 and applying the boundary condition given by Equation 7.176 leads to the following expression for B

$$B = \frac{Q_w}{2\pi T(r_w/\lambda)K_1(r_w/\lambda)} \tag{7.181}$$

where $T(= Kb)$ is the transmissivity of the semiconfined aquifer. Substituting Equation 7.181 into Equation 7.179 yields the following expression for the piezometric head distribution

$$\phi_o - \phi(r) = \frac{Q_w}{2\pi T}\frac{K_o(r/\lambda)}{(r_w/\lambda)K_1(r_w/\lambda)} \tag{7.182}$$

Defining the drawdown, $s(r)$, by

$$s(r) = \phi_o - \phi(r) \tag{7.183}$$

then Equations 7.182 and 7.183 yield the following expression for the drawdown

$$s(r) = \frac{Q_w}{2\pi T}\frac{K_o(r/\lambda)}{(r_w/\lambda)K_1(r_w/\lambda)} \tag{7.184}$$

In this case, there are two aquifer properties: transmissivity, T, and leakage factor, λ. These aquifer properties can be estimated using Equation 7.184 with drawdown measurements at two locations. Denoting the drawdown measurements by s_1 and s_2, at r_1 and r_2, respectively, then Equation 7.184 yields

$$\boxed{\frac{K_o(r_1/\lambda)}{K_o(r_2/\lambda)} = \frac{s_1}{s_2}} \tag{7.185}$$

and

$$\boxed{T = \frac{Q_w}{2\pi s_1} \frac{K_o(r_1/\lambda)}{(r_w/\lambda)K_1(r_w/\lambda)}} \tag{7.186}$$

Equation 7.185 can be used to estimate the leakage factor from the drawdown measurements. This leakage factor can be used in Equation 7.186 to calculate the transmissivity of the semiconfined aquifer.

■ Example 7.17

A pump test is conducted in a leaky confined aquifer in which the semiconfining layer is estimated to have a thickness of 2 m, and the aquifer has a thickness of 20 m. The pumping rate is 50 L/s from a well of radius 0.5 m, and the steady-state drawdowns at 50 m and 100 m from the well are 0.3 m and 0.1 m, respectively. Estimate the transmissivity, hydraulic conductivity, and leakage factor of the aquifer, and the hydraulic conductivity of the semiconfining layer.

Solution

The leakage factor, λ, can be estimated using Equation 7.185, where

$$\frac{K_o(r_1/\lambda)}{K_o(r_2/\lambda)} = \frac{s_1}{s_2}$$

Substituting $r_1 = 50$ m, $r_2 = 100$ m, $s_1 = 0.3$ m, and $s_2 = 0.1$ m leads to

$$\frac{K_o(50/\lambda)}{K_o(100/\lambda)} = \frac{0.3}{0.1} = 3$$

This equation is an implicit equation for λ which can be easily solved by trial and error using the (modified) Bessel functions tabulated in Appendix F. The result is

$$\lambda = 62.7 \text{ m}$$

The transmissivity of the semiconfined aquifer can be estimated using Equation 7.186, where

$$T = \frac{Q_w}{2\pi s_1} \frac{K_o(r_1/\lambda)}{(r_w/\lambda)K_1(r_w/\lambda)}$$

Substituting $Q_w = 50$ L/s $= 4320$ m^3/d, and $r_w = 0.5$ m leads to

$$T = \frac{4320}{2\pi(0.3)} \frac{K_o(50/62.7)}{(0.5/62.7)K_1(0.5/62.7)} = 1{,}300 \text{ m}^2/\text{d}$$

Since the thickness of the aquifer is 20 m, the hydraulic conductivity, K, is

$$K = \frac{1300}{20} = 65 \text{ m/d}$$

Recall the definition of the leakage factor, where

$$\lambda^2 = \frac{Kbb'}{K'}$$

where b is the thickness of the aquifer, b' is the thickness of the semiconfining layer, and K' is the hydraulic conductivity of the semiconfining layer. In the present case, $\lambda = 62.7$ m, $K = 65$ m/d, and $b' = 2$ m, therefore

$$K' = \frac{Kbb'}{\lambda^2} = \frac{(65)(20)(2)}{(62.7)^2} = 0.66 \text{ m/d} \qquad \blacksquare$$

An interesting approximation to the drawdown equation (Equation 7.184) is obtained when $r_w/\lambda \ll 1$, in which case $(r_w/\lambda)K_1(r_w/\lambda) \approx 1$ and Equation 7.184 can be approximated by

$$s(r) = \frac{Q_w}{2\pi T} K_o(r/\lambda) \qquad (7.187)$$

which indicates that the drawdown distribution is independent of the radius of the well, r_w. According to Bear (1979), Equation 7.187 is accurate to within 1% of Equation 7.184 whenever $r_w/\lambda < 0.02$.

■ **Example 7.18**

Repeat the previous problem using the approximation given by Equation 7.187.

Solution

Equation 7.187 gives the following relation for estimating the leakage factor, λ:

$$\frac{K_o(r_1/\lambda)}{K_o(r_2/\lambda)} = \frac{s_1}{s_2}$$

where $r_1 = 50$ m, $r_2 = 100$ m, $s_1 = 0.3$ m, and $s_2 = 0.1$ m. Hence

$$\frac{K_o(50/\lambda)}{K_o(100/\lambda)} = \frac{0.3}{0.1} = 3$$

This is the same expression for λ used in the previous example, and the result is given by

$$\lambda = 62.7 \text{ m}$$

This value of λ indicates that

$$\frac{r_w}{\lambda} = \frac{0.5}{62.7} = 0.008 \ll 0.02$$

and therefore using the approximation given by Equation 7.187 is justified. Equation 7.187 gives the transmissivity, T, as

$$T = \frac{Q_w}{2\pi s_1} K_o(r_1/\lambda)$$

■ Table 7.5	Leakage factor, λ (m)	Condition
Classification of Leakage		
	< 1000	high leakage
	1,000–5,000	moderate leakage
	5,000–10,000	low leakage
	> 10,000	negligible leakage

Source: Şen (1995).

where $Q_w = 50$ L/s $= 4,320$ m^3/d, and therefore

$$T = \frac{4320}{2\pi(0.3)} K_o(50/62.7) = 1,300 \text{ m}^2/\text{d}$$

This transmissivity is the same as was obtained in the previous example. Calculations of both the hydraulic conductivity of the aquifer and semiconfining layer are the same as in the previous example, yielding

$$K = 65 \text{ m/d} \quad \text{and} \quad K' = 0.66 \text{ m/d}$$

The results of this example support the assertion that Equation 7.187 is an adequate approximation to Equation 7.184 when $r_w/\lambda < 0.02$. ■

The derived leakage factor can be used to classify the "leakiness" of the semiconfining layer using the classification system in Table 7.5. Aquifers with leakage factors exceeding 10,000 m can be treated as confined aquifers.

The equations derived in this section assume that the flow is horizontal in the aquifer and vertical in the semiconfining (leaky) layer. Errors introduced by these assumptions are generally less than 5%, provided that the hydraulic conductivity of the aquifer is more than two orders of magnitude greater than that of the semiconfining layer (Neuman and Witherspoon, 1969a).

7.3.7 ■ Steady Flow to a Well in an Unconfined Aquifer with Recharge

In isotropic and homogeneous unconfined aquifers, the distribution of aquifer thickness can be estimated by assuming two-dimensional flow (Dupuit approximation). Under these circumstances, the flow is described by Equation 7.79 which is given by

$$\frac{\partial^2(h^2)}{\partial r^2} + \frac{1}{r}\frac{\partial(h^2)}{\partial r} + \frac{1}{r^2}\frac{\partial^2(h^2)}{\partial \theta^2} + \frac{2N}{K} = 2\frac{S_y}{K}\frac{\partial h}{\partial t} \qquad (7.188)$$

Since the solution is radially symmetric and steady state, h is only a function of r and Equation 7.188 becomes

$$\frac{d^2(h^2)}{dr^2} + \frac{1}{r}\frac{d(h^2)}{dr} + \frac{2N}{K} = 0 \qquad (7.189)$$

where the partial derivatives have been replaced by total derivatives. Equation 7.189 can also be written in the more compact form

$$\frac{d}{dr}\left(r\frac{dh^2}{dr}\right) + r\frac{2N}{K} = 0 \qquad (7.190)$$

This equation is to be solved using the boundary conditions

$$2\pi r_w hK \left.\frac{dh}{dr}\right|_{r=r_w} = Q_w \tag{7.191}$$

$$h(r_w) = h_w \tag{7.192}$$

where r_w is the radius of the well and Q_w is the flowrate out of the well. Since the governing differential equation is in terms of h^2, it is convenient to write the boundary conditions in terms of h^2 as

$$\pi r_w K \left.\frac{dh^2}{dr}\right|_{r=r_w} = Q_w \tag{7.193}$$

and

$$h^2(r_w) = h_w^2 \tag{7.194}$$

Integrating Equation 7.190 with respect to r yields

$$\frac{dh^2}{dr} = -\frac{Nr}{K} + \frac{A}{r} \tag{7.195}$$

and integrating again with respect to r yields

$$h^2 = -\frac{Nr^2}{2K} + A \ln r + B \tag{7.196}$$

where A and B are integration constants. Applying the boundary conditions yields the following expression for the distribution of aquifer thickness

$$\boxed{h^2 = h_w^2 + \frac{N}{2K}(r_w^2 - r^2) + A \ln\left(\frac{r}{r_w}\right)} \tag{7.197}$$

where

$$A = \frac{Q_w}{\pi K} + \frac{Nr_w^2}{K} \tag{7.198}$$

■ Example 7.19

A well pumps at the rate of 20 L/s from an unconfined aquifer with an average surficial recharge of 0.5 m/d. If the radius of the well is 0.5 m, the hydraulic conductivity of the aquifer is 20 m/d, and the aquifer thickness 100 m from the well is 25 m, estimate the aquifer thickness at the well. Determine the additional drawdown at the well in the absence of surficial recharge.

Solution

From the given data: $Q_w = 20$ L/s $= 1{,}728$ m³/d; $N = 0.5$ m/d; $r_w = 0.5$ m; $K = 20$ m/d; and at $r = 100$ m, $h = 25$ m. Equation 7.197 gives the aquifer thickness at the well, h_w, as

$$h_w = \sqrt{h^2 - \frac{N}{2K}(r_w^2 - r^2) - A \ln\left(\frac{r}{r_w}\right)}$$

$$= \sqrt{25^2 - \frac{0.5}{2(20)}(0.5^2 - 100^2) - A \ln\left(\frac{100}{0.5}\right)} = \sqrt{750 - 5.30A}$$

The constant A is given by Equation 7.198 as

$$A = \frac{Q_w}{\pi K} + \frac{N r_w^2}{K} = \frac{1728}{\pi(20)} + \frac{(0.5)(0.5)^2}{20} = 27.5 \text{ m}^2$$

Hence, the aquifer thickness, h_w, at the well is given by

$$h_w = \sqrt{750 - 5.30(27.5)} = 24.6 \text{ m}$$

In the absence of surficial recharge, $N = 0$ m/d, and the aquifer thickness at the well, h_w, is given by Equation 7.197 as

$$h'_w = \sqrt{h^2 - A \ln\left(\frac{r}{r_w}\right)} = \sqrt{25^2 - A \ln\left(\frac{100}{0.5}\right)} = \sqrt{625 - 5.30A}$$

where

$$A = \frac{Q_w}{\pi K} = \frac{1728}{\pi(20)} = 27.5 \text{ m}^2$$

Therefore,

$$h'_w = \sqrt{625 - 5.30(27.5)} = 21.9 \text{ m}$$

and the additional drawdown, Δs, in the absence of recharge is given by

$$\Delta s = h_w - h'_w = 24.6 - 21.9 = 2.7 \text{ m} \qquad \blacksquare$$

7.3.8 ■ Unsteady Flow to a Well in a Confined Aquifer

In isotropic and homogeneous confined aquifers, the distribution of piezometric head for two-dimensional flows is given by Equation 7.94, where

$$\frac{\partial^2 \phi}{\partial r^2} + \frac{1}{r}\frac{\partial \phi}{\partial r} + \frac{1}{r^2}\frac{\partial^2 \phi}{\partial \theta^2} + \frac{N}{Kb} = \frac{S}{Kb}\frac{\partial \phi}{\partial t} \qquad (7.199)$$

If a well fully penetrates the confined aquifer, and withdraws water uniformly over the aquifer thickness, b, then the piezometric head, ϕ, is radially symmetric and therefore independent of θ. If the leakage, N, through the confining layers is equal to zero, then Equation 7.199 becomes

$$\frac{\partial^2 \phi}{\partial r^2} + \frac{1}{r}\frac{\partial \phi}{\partial r} = \frac{S}{T}\frac{\partial \phi}{\partial t} \qquad (7.200)$$

where $T(= Kb)$ is the transmissivity of the aquifer. Defining the drawdown in the aquifer, $s(r, t)$, by the relation

$$s(r, t) = \phi_o - \phi(r, t) \qquad (7.201)$$

where ϕ_o is the piezometric head in the confined aquifer prior to pumping, then Equation 7.200 can be written in terms of drawdown as

$$\frac{\partial^2 s}{\partial r^2} + \frac{1}{r}\frac{\partial s}{\partial r} = \frac{S}{T}\frac{\partial s}{\partial t} \qquad (7.202)$$

This equation is to be solved subject to the following initial and boundary conditions:

$$s(r, 0) = 0 \qquad (7.203)$$

$$s(\infty, t) = 0 \qquad (7.204)$$

$$\lim_{r \to 0} r \frac{\partial s}{\partial r} = -\frac{Q_w}{2\pi T} \qquad (7.205)$$

where Q_w is the pumping rate out of the well. Solution of this problem is facilitated by changing variables from (r, t) to (r, u), where

$$u = \frac{r^2 S}{4Tt} \qquad (7.206)$$

and u is sometimes called the *Boltzman variable*. Using the chain rule on Equation 7.206, the following relationships can be derived

$$\frac{\partial s}{\partial r} = \frac{2u}{r} \frac{\partial s}{\partial u} \qquad (7.207)$$

$$\frac{\partial^2 s}{\partial r^2} = \frac{2u}{r} \left[\frac{2u}{r} \frac{\partial^2 s}{\partial u^2} + \frac{1}{r} \frac{\partial s}{\partial u} \right] \qquad (7.208)$$

$$\frac{\partial s}{\partial t} = -\frac{u}{t} \frac{\partial s}{\partial u} \qquad (7.209)$$

Substituting Equations 7.207 to 7.209 into Equation 7.202 yields

$$\frac{\partial^2 s}{\partial u^2} + \left(1 + \frac{1}{u}\right) \frac{\partial s}{\partial u} = 0 \qquad (7.210)$$

The striking result here is that the variable t has dropped out, and the solution, s, of Equation 7.210 depends only on the Boltzman variable, u. Consequently, the partial derivatives can be replaced by total derivatives, and Equation 7.210 can be written as

$$\frac{d^2 s}{du^2} + \left(1 + \frac{1}{u}\right) \frac{ds}{du} = 0 \qquad (7.211)$$

The boundary conditions, with respect to u, can be derived from Equations 7.203 to 7.205, which yield

$$s(\infty) = 0 \qquad (7.212)$$

$$\lim_{u \to 0} u \frac{ds}{du} = -\frac{Q_w}{4\pi T} \qquad (7.213)$$

Hence, the problem of unsteady flow to a well in a confined aquifer is defined by Equation 7.211, subject to the boundary conditions given by Equations 7.212 and 7.213. To solve Equation 7.211, let

$$p = \frac{ds}{du} \qquad (7.214)$$

then Equation 7.211 can be written as

$$\frac{dp}{du} + \left(1 + \frac{1}{u}\right) p = 0 \qquad (7.215)$$

which can be written as

$$\frac{dp}{p} = -\left(1 + \frac{1}{u}\right) du \tag{7.216}$$

and integrating this equation gives

$$\ln p = -u - \ln u + A' \tag{7.217}$$

where A' is a constant to be determined from the boundary conditions. Rearranging Equation 7.217 leads to

$$up = e^{A'-u}$$
$$= Ae^{-u} \tag{7.218}$$

where A is a constant $(= e^{A'})$. Substituting the definition of p from Equation 7.214 leads to

$$u\frac{ds}{du} = Ae^{-u} \tag{7.219}$$

Applying the boundary condition given by Equation 7.213 to find A leads to

$$u\frac{ds}{du} = -\frac{Q_w}{4\pi T}e^{-u} \tag{7.220}$$

which can be put in the form

$$ds = -\frac{Q_w}{4\pi T}\frac{e^{-u}}{u}du \tag{7.221}$$

or

$$s(u) = -\frac{Q_w}{4\pi T}\int_a^u \frac{e^{-x}}{x}dx + B \tag{7.222}$$

where a and B are constants. Applying the boundary condition given in Equation 7.212, leads to

$$B = \frac{Q_w}{4\pi T}\int_a^\infty \frac{e^{-x}}{x}dx \tag{7.223}$$

Substituting Equation 7.223 into Equation 7.222 yields

$$s(u) = \frac{Q_w}{4\pi T}\left[\int_a^\infty \frac{e^{-x}}{x}dx - \int_a^u \frac{e^{-x}}{x}dx\right]$$
$$= \frac{Q_w}{4\pi T}\int_u^\infty \frac{e^{-x}}{x}dx \tag{7.224}$$

This equation is sufficient to describe the transient drawdown in response to a fully penetrating well in a confined aquifer. The integral in Equation 7.224 is commonly referred to as the *well function*, $W(u)$, where

$$\boxed{W(u) = \int_u^\infty \frac{e^{-x}}{x}dx} \tag{7.225}$$

u	1.0	2.0	3.0	4.0	5.0	6.0	7.0	8.0	9.0
$\times 1$	.2194	.0489	.0130	.0038	.0011	.0004	.0001	.0000	.0000
$\times 10^{-1}$	1.8229	1.2227	.9057	.7024	.5598	.4544	.3738	.3106	.2602
$\times 10^{-2}$	4.0379	3.3547	2.9591	2.6813	2.4679	2.2953	2.1508	2.0269	1.9187
$\times 10^{-3}$	6.3315	5.6394	5.2349	4.9482	4.7261	4.5448	4.3916	4.2591	4.1423
$\times 10^{-4}$	8.6332	7.9401	7.5348	7.2472	7.0242	6.8420	6.6879	6.5545	6.4368
$\times 10^{-5}$	10.9357	10.2428	9.8372	9.5494	9.3264	9.1440	8.9899	8.8564	8.7386

■ Table 7.6
Well Function, $W(u)$

and Equation 7.224 can be written as

$$s(u) = \frac{Q_w}{4\pi T} W(u) \qquad (7.226)$$

This equation is widely used in ground-water engineering and is called the *Theis equation*, after Theis (1935) who originally derived this equation. The well function is shown for several values of u in Table 7.6. The well function can also be expressed in terms of an infinite series as

$$W(u) = -0.5772 - \ln u + u - \frac{u^2}{2 \cdot 2!} + \frac{u^3}{3 \cdot 3!} - \frac{u^4}{4 \cdot 4!} + \cdots \qquad (7.227)$$

The utility of this series expression is that for small values of u, the higher-order terms can be neglected, yielding an analytic expression for $W(u)$. In fact, for values of u less than or equal to 0.004, the first two terms in Equation 7.227 gives values of $W(u)$ that are accurate to within 0.1%, and therefore the following approximation is appropriate

$$W(u) = -0.5772 - \ln u, \qquad u \leq 0.004 \qquad (7.228)$$

This approximation was first applied to the analysis of ground-water flow by Cooper and Jacob (1946) and is commonly referred to as the *Cooper-Jacob approximation*.

The utility of the Theis equation, Equation 7.226, and the approximation given by Equation 7.228, lies primarily in the determination of the aquifer properties, T and S, from pump-drawdown tests. To use the Theis equation to estimate T and S, the relationship between the drawdown, $s(r, t)$, and r^2/t must be measured in the aquifer for a given pumping rate, Q_w. The relationship between s and r^2/t can be obtained from measurements of drawdown, s, versus time, t, at a single monitoring well at a known distance r from the pumping well, or the relationship between s and r^2/t can be obtained from drawdown measurements at several monitoring wells, at varying distances from the pumping well, at a single instance of time, t. To demonstrate how the aquifer properties can be derived from measured values of drawdown, s, versus r^2/t, consider the definition of the Boltzman variable, Equation 7.206, and the logarithm of the Theis equation (Equation 7.226), which yield

$$\ln\left(\frac{r^2}{t}\right) = \ln u + \alpha \qquad (7.229)$$

$$\ln s = \ln W(u) + \beta \qquad (7.230)$$

where α and β are constants given by

$$\alpha = \ln\left(\frac{4T}{S}\right) \qquad (7.231)$$

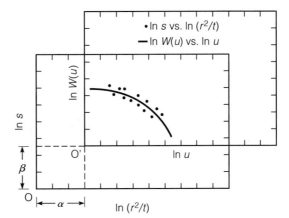

Figure 7.16 ■ Estimation of Aquifer Properties Using Theis Equation and Well Function

$$\beta = \ln\left(\frac{Q_w}{4\pi T}\right) \tag{7.232}$$

Based on Equations 7.229 and 7.230, it is apparent that the theoretical relationship between $\ln s$ and $\ln(r^2/t)$ can be derived directly from the relationship between $\ln W(u)$ and $\ln u$, simply by adding α to $\ln u$ and β to $\ln W(u)$. This relationship is illustrated in Figure 7.16, where the plot of $\ln W(u)$ versus $\ln u$ is commonly referred to as a *type curve*.

To determine α and β from field measurements, $\ln s$ versus $\ln(r^2/t)$ is plotted on axes with the same origin and scale as a reference plot of $\ln W(u)$ versus $\ln u$. The plots are then superimposed such that the axes of the plots are parallel, and the curve of $\ln W(u)$ versus $\ln u$ closely matches the curve of $\ln s$ versus $\ln(r^2/t)$. These matched curves should resemble those shown in Figure 7.16. After matching the curves, then the constants α and β are derived from the coordinates of the origin of the $\ln W(u)$ versus $\ln u$ curve (O' in Figure 7.16) on the $\ln s$ versus $\ln(r^2/t)$ curve. Using the definition of β given by Equation 7.232, the transmissivity, T, of the aquifer is given by

$$T = \frac{Q_w}{4\pi}e^{-\beta} \tag{7.233}$$

and once T is determined, S can be derived from the definition of α given by Equation 7.231 as

$$S = 4Te^{-\alpha} \tag{7.234}$$

The value of the storage coefficient, S, can be used to confirm the assumption that the aquifer is confined. For example, if the calculated storage coefficient exceeds 0.1, the aquifer is most likely unconfined (Gorelick et al., 1993).

■ Example 7.20

A confined aquifer of thickness 30 m is pumped at a rate of 75 L/s from a well of radius 0.3 m. The recorded drawdowns in a monitoring well located 100 m from the pumping well are given in Table 7.7. Use the Theis equation to estimate the hydraulic conductivity, transmissivity, and storage coefficient of the aquifer.

■ Table 7.7 Drawdowns in Monitoring Well	Time (s)	Drawdown (cm)
	1	0.00
	10	0.04
	100	1.98
	1,000	62.80
	10,000	183.20
	100,000	313.80

Solution

The Theis equation is used to extract the transmissivity and storage coefficient by matching the standard curve of $\ln W(u)$ versus $\ln u$, to the measured data in the form of $\ln s$ versus $\ln(r^2/t)$, where s is the drawdown, r is the distance of the monitoring well from the pumping well and t is the time since the beginning of pumping. In this case, $r = 100$ m, and the relationship between s and r^2/t is given in Table 7.8, where t is the time in days, and s is the drawdown in meters. The superposition of the $\ln W(u)$ versus $\ln u$ curve onto the $\ln s$ versus $\ln r^2/t$ curve is illustrated in Figure 7.17 and the axis displacement indicates that $\alpha = 15.1$, and $\beta = -0.557$. Since $Q_w = 75$ L/s $= 6{,}480$ m^3/d, then Equation 7.233 yields

$$T = \frac{Q_w}{4\pi}e^{-\beta} = \frac{6480}{4\pi}e^{-(-0.557)} = 900 \text{ m}^2/\text{d}$$

■ Table 7.8 s versus r^2/t	r^2/t (m^2/d)	s (m)
	8.64×10^8	0.000
	8.64×10^7	0.0004
	8.64×10^6	0.0198
	8.64×10^5	0.628
	8.64×10^4	1.832
	8.64×10^3	3.138

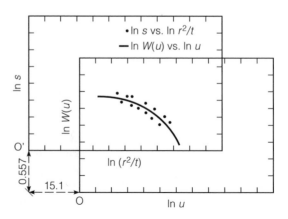

Figure 7.17 ■ Application of Theis Equation and Well Function

and Equation 7.234 yields

$$S = 4Te^{-\alpha} = 4(900)e^{-15.1} = 0.0010$$

The hydraulic conductivity, K, of the aquifer is given by

$$K = \frac{T}{b} = \frac{900}{30} = 30 \text{ m/d} \qquad \blacksquare$$

The Theis equation (Equation 7.226) can also be applied in anisotropic formations using the coordinate transformation technique described in Section 7.2.3.2, which also contains a specific example on the application of the Cooper-Jacob approximation to the Theis equation in anisotropic formations. An important limitation of the Theis equation is that it neglects storage in the well, by assuming that all pumpage is extracted from the aquifer. This assumption has been investigated by Papadopulos and Cooper (1967), who showed that well-storage effects are negligible when

$$t > 2500 \frac{r_c^2}{T} \tag{7.235}$$

where t is the duration of pumping, r_c is the diameter of the well casing and T is the transmissivity of the aquifer. In most practical cases, Equation 7.235 is satisfied and the Theis equation is applicable. In cases where well-storage effects are significant, the analytical formulation developed by Papadopulos and Cooper (1967) should be used in lieu of the Theis equation. A second limitation of the Theis equation relates to the assumption of an infinite aquifer. This assumption is justified as long as the cone of depression does not intersect any boundaries.

7.3.9 ■ Unsteady Flow to a Well in an Unconfined Aquifer

In isotropic and homogeneous unconfined aquifers, the distribution of aquifer thickness, h, can be estimated by invoking the Dupuit approximation of two-dimensional flow, in which case the distribution of aquifer thickness is given by Equation 7.79. Since the drawdown induced by a single pumping well will be radially symmetric, and taking the recharge, N, equal to zero, Equation 7.79 can be written as

$$\frac{\partial^2(h^2)}{\partial r^2} + \frac{1}{r}\frac{\partial(h^2)}{\partial r} = 2\frac{S_y}{K}\frac{\partial h}{\partial t} \tag{7.236}$$

Defining the drawdown, $s(r, t)$, by the relation

$$s(r, t) = H - h(r, t) \tag{7.237}$$

where H is the aquifer thickness prior to pumping, and substituting Equation 7.237 into 7.236 leads to

$$\frac{\partial^2}{\partial r^2}\left(s - \frac{s^2}{2H}\right) + \frac{1}{r}\frac{\partial}{\partial r}\left(s - \frac{s^2}{2H}\right) = \frac{S_y}{KH}\frac{\partial s}{\partial t} \tag{7.238}$$

Noting that

$$s - \frac{s^2}{2H} = s\left(1 - \frac{s}{2H}\right) \tag{7.239}$$

then whenever $s \ll 2H$,

$$s - \frac{s^2}{2H} \approx s \tag{7.240}$$

and Equation 7.238 can be written as

$$\frac{\partial^2 s}{\partial r^2} + \frac{1}{r}\frac{\partial s}{\partial r} = \frac{S_y}{T}\frac{\partial s}{\partial t}, \qquad s \ll 2H \tag{7.241}$$

where T is the transmissivity defined by

$$T = KH \tag{7.242}$$

The initial and boundary conditions for unsteady flow to a well in an unconfined infinite aquifer are

$$s(r, 0) = 0 \tag{7.243}$$

$$s(\infty, t) = 0 \tag{7.244}$$

$$\lim_{r \to 0} r\frac{\partial s}{\partial r} = -\frac{Q_w}{2\pi T} \tag{7.245}$$

Therefore, under the restriction that $s \ll 2H$, the governing equation and associated initial and boundary conditions are exactly the same as the governing equation and associated initial and boundary conditions that describe unsteady drawdown in a confined aquifer, with the exception that the storage coefficient, S, in the confined aquifer case is replaced by the specific yield, S_y, in unconfined aquifer case. The small-drawdown approximation is justified when the drawdown, s, is less than 5% of the original aquifer thickness, H (Boonstra, 1998a). The similarity between the confined and unconfined cases can be observed by comparing Equations 7.241 and 7.243 to 7.245 with Equations 7.202 to 7.205. Consequently, the Theis equation describing the drawdown as a function of time in confined aquifers, Equation 7.226, can also be used in unconfined aquifers, as can the curve-matching approach for determining the hydraulic properties of the aquifer. However, the curve-matching approach in unconfined aquifers produces the specific yield, S_y, instead of the storage coefficient, S, for confined aquifers.

■ Example 7.21

A pump test is conducted at 80 L/s in an unconfined aquifer, and comparison of the drawdown curve with the well function indicates that the best match occurs when $\alpha = 9.9$ and $\beta = -0.60$. If the maximum drawdown during the pump test is 1.5 m and the aquifer thickness prior to pumping is 20 m, estimate the transmissivity and specific yield of the aquifer.

Solution
Since the maximum drawdown is 1.5 m and the aquifer thickness prior to pumping is 20 m, then

$$\frac{s}{2H} \le \frac{1.5}{2(20)} = 0.0375$$

Since $s/2H \ll 1$, the confined aquifer approximation can be used to analyze the data from the pump test. From the given data: $Q_w = 80$ L/s $= 6{,}912$ m³/d, $\alpha = 9.9$, $\beta =$

−0.60, and Equation 7.233 gives

$$T = \frac{Q_w}{4\pi}e^{-\beta} = \frac{6912}{4\pi}e^{-(-0.60)} = 1000 \text{ m}^2/\text{d}$$

and Equation 7.234 gives

$$S_y = 4Te^{-\alpha} = 4(1000)e^{-9.9} = 0.20 \qquad \blacksquare$$

Application of the Theis equation to unconfined aquifers assumes that the ground water drains instantaneously from above the water table. However, finer-grained aquifer materials release drainable waters slowly, and it may take weeks for such formations to drain fully (Gorelick et al., 1993). In such cases, there is said to be a *delayed yield*, and the methodology proposed by Neuman (1972) for such cases should be used to analyze the transient drawdown.

7.3.10 ■ Unsteady Flow to a Well in a Leaky Confined Aquifer

In isotropic and homogeneous semiconfined (leaky) aquifers, the distribution of piezometric head for two-dimensional flows is given by Equation 7.94. If a well fully penetrates the semiconfined aquifer and withdraws water uniformly over the aquifer thickness, b, then the piezometric head, ϕ, is radially symmetric (independent of θ), all derivatives with respect to θ are zero, and Equation 7.94 can be written in the form

$$\frac{\partial^2 \phi}{\partial r^2} + \frac{1}{r}\frac{\partial \phi}{\partial r} + \frac{N}{Kb} = \frac{S}{T}\frac{\partial \phi}{\partial t} \tag{7.246}$$

The leakage into the semiconfined aquifer can be calculated by applying Darcy's law across the semiconfining layer, in which case

$$N = -K'\frac{\phi - \phi_o}{b'} \tag{7.247}$$

where K' is the hydraulic conductivity of the confining layer, b' is the thickness of the confining layer, and ϕ_o is the piezometric head above the confining layer, assumed to be constant. Combining Equations 7.246 and 7.247 leads to

$$\frac{\partial^2 \phi}{\partial r^2} + \frac{1}{r}\frac{\partial \phi}{\partial r} + \frac{\phi_o - \phi}{\lambda^2} = \frac{S}{T}\frac{\partial \phi}{\partial t} \tag{7.248}$$

where λ is the leakage factor defined by Equation 7.168. If the drawdown, $s(r, t)$, is defined by the relation

$$s(r, t) = \phi_o - \phi(r, t) \tag{7.249}$$

then Equation 7.248 can be written in terms of drawdown, s, as

$$\frac{\partial^2 s}{\partial r^2} + \frac{1}{r}\frac{\partial s}{\partial r} - \frac{s}{\lambda^2} = \frac{S}{T}\frac{\partial s}{\partial t} \tag{7.250}$$

The initial and boundary conditions in the case of unsteady flow to a well in an infinite semi-confined aquifer are as follows

$$s(r, 0) = 0 \qquad (7.251)$$

$$s(\infty, t) = 0 \qquad (7.252)$$

$$\lim_{r \to 0} r \frac{\partial s}{\partial r} = -\frac{Q_w}{2\pi T} \qquad (7.253)$$

The solution to this problem has been derived by Hantush and Jacob (1955) as

$$s(r, t) = \frac{Q_w}{4\pi T} W\left(u, \frac{r}{\lambda}\right) \qquad (7.254)$$

where $W(u, r/\lambda)$ is called the *Hantush and Jacob well function for leaky aquifers* (Hantush and Jacob, 1955), or simply the *leaky well function* (Freeze and Cherry, 1979), and is defined as

$$W\left(u, \frac{r}{\lambda}\right) = \int_u^\infty \frac{1}{y} \exp\left(-y - \frac{r^2}{4\lambda^2 y}\right) dy \qquad (7.255)$$

Values of $W(u, r/\lambda)$ for several values of u and r/λ are tabulated in Table 7.9. For $r/\lambda = 0$, $W(u, r/\lambda)$ is equal to $W(u)$. The well function, $W(u, r/\lambda)$, in leaky confined aquifers

■ **Table 7.9**

Well Function for Leaky Aquifer, $W(u, r/\lambda)$

$u \backslash r/\lambda$	0.00	0.002	0.004	0.007	0.01	0.02	0.04	0.06	0.08	0.10
0.00		12.6611	11.2748	10.1557	9.4425	8.0569	6.6731	5.8456	5.2950	4.8541
1×10^{-6}	13.2383	12.4417	11.2711	10.1557						
2×10^{-6}	12.5451	12.1013	11.2259	10.1554						
5×10^{-6}	11.6289	11.4384	10.9642	10.1290	9.4425					
8×10^{-6}	11.1589	11.0377	10.7151	10.0602	9.4313					
1×10^{-5}	10.9357	10.8382	10.5725	10.0034	9.4176	8.0569				
2×10^{-5}	10.2426	10.1932	10.0522	9.7126	9.2961	8.0558				
5×10^{-5}	9.3263	9.3064	9.2480	9.0957	8.8827	8.0080	6.6730			
7×10^{-5}	8.9899	8.9756	8.9336	8.8224	8.6625	7.9456	6.6726			
1×10^{-4}	8.6332	8.6233	8.5937	8.5145	8.3983	7.8375	6.6693	5.8658	5.2950	4.8541
2×10^{-4}	7.9402	7.9352	7.9203	7.8800	7.8192	7.4472	6.6242	5.8637	5.2949	4.8541
5×10^{-4}	7.0242	7.0222	7.0163	6.9999	6.9750	6.8346	6.3626	5.8011	5.2848	4.8530
7×10^{-4}	6.6879	6.6865	6.6823	6.6706	6.6527	6.5508	6.1917	5.7274	5.2618	4.8478
1×10^{-3}	6.3315	6.3305	6.3276	6.3194	6.3069	6.2347	5.9711	5.6058	5.2087	4.8292
2×10^{-3}	5.6394	5.6389	5.6374	5.6334	5.6271	5.5907	5.4516	5.2411	4.9848	4.7079
5×10^{-3}	4.7261	4.7259	4.7253	4.7237	4.7212	4.7068	4.6499	4.5590	4.4389	4.2990
7×10^{-3}	4.3916	4.3915	4.3910	4.3899	4.3882	4.3779	4.3374	4.2719	4.1839	4.0771
1×10^{-2}	4.0379	4.0378	4.0375	4.0368	4.0351	4.0285	4.0003	3.9544	3.8920	3.8190
2×10^{-2}	3.3547	3.3547	3.3545	3.3542	3.3536	3.3502	3.3365	3.3141	3.2832	3.2442
5×10^{-2}	2.4679	2.4679	2.4678	2.4677	2.4675	2.4662	2.4613	2.4531	2.4416	2.4271
7×10^{-2}	2.1508	2.1508	2.1508	2.1507	2.1506	2.1497	2.1464	2.1408	2.1331	2.1232
1×10^{-1}	1.8229	1.8229	1.8229	1.8228	1.8227	1.8222	1.8220	1.8164	1.8114	1.8050
2×10^{-1}	1.2227	1.2226	1.2226	1.2226	1.2226	1.2224	1.2215	1.2201	1.2181	1.2155
5×10^{-1}	0.5598	0.5598	0.5598	0.5598	0.5598	0.5597	0.5595	0.5592	0.5587	0.5581
7×10^{-1}	0.3738	0.3738	0.3738	0.3738	0.3738	0.3737	0.3736	0.3734	0.3732	0.3729
1.00	0.2194	0.2194	0.2194	0.2194	0.2194	0.2194	0.2193	0.2192	0.2191	0.2190
2.00	0.0489	0.0489	0.0489	0.0489	0.0489	0.0489	0.0489	0.0489	0.0489	0.0488
5.00	0.0011	0.0011	0.0011	0.0011	0.0011	0.0011	0.0011	0.0011	0.0011	0.0011
7.00	0.0001	0.0001	0.0001	0.0001	0.0001	0.0001	0.0001	0.0001	0.0001	0.0001
8.00	0.0000	0.0000	0.0000	0.0000	0.0000	0.0000	0.0000	0.0000	0.0000	0.0000

is used in the same way as the well function, $W(u)$, in confined aquifers to determine the transmissivity, T, and storage coefficient, S, from pump tests. In the case of leaky confined aquifers, a curve of $\ln s$ versus $\ln(r^2/t)$ is overlain on a curve of $\ln W(u, r/\lambda)$ versus $\ln u$ (plotted for several different values of r/λ). The displacement of the origin of the $\ln W(u, r/\lambda)$ versus $\ln u$ curve relative to the $\ln s$ versus $\ln(r^2/t)$ curve yields the displacement coordinates (α', β'), and the shape of the $\ln s$ versus $\ln(r^2/t)$ curve yields the best-fit value of r/λ, and hence the leakage factor λ can be derived from the known distance, r, of the monitoring well from the pumping well. The transmissivity, T, and storage coefficient, S, can then be derived from the displacement coordinates using the relations

$$T = \frac{Q_w}{4\pi} e^{-\beta'} \tag{7.256}$$

and

$$S = 4Te^{-\alpha'} \tag{7.257}$$

■ **Example 7.22**

A leaky confined aquifer of thickness 30 m is pumped at a rate of 75 L/s from a fully penetrating well of radius 0.3 m. The drawdown curve is measured at an observation well 10 m from the pumping well, and matched with the well function for a leaky confined aquifer. The best match is obtained for $\alpha' = 15$, $\beta' = -0.60$, and $r/\lambda = 0.04$. If the thickness of the semi-confining layer overlying the aquifer is 2 m, estimate the transmissivity and storativity of the aquifer and the hydraulic conductivity of the semi-confining layer.

Solution

From the given data: $Q_w = 75$ L/s $= 6,480$ m^3/d, $\alpha' = 15$, $\beta' = -0.60$, and Equation 7.256 gives the transmissivity, T, of the aquifer as

$$T = \frac{Q_w}{4\pi} e^{-\beta'} = \frac{6480}{4\pi} e^{-(-0.60)} = 940 \text{ m}^2/\text{d}$$

Equation 7.257 gives the storage coefficient, S, of the aquifer as

$$S = 4Te^{-\alpha'} = 4(940)e^{-15} = 0.0012$$

Since the leaky well function that matches the drawdown data has the parameter $r/\lambda = 0.04$, and the distance, r, of the observation well from the pumping well is 10 m, then

$$\lambda = \frac{r}{0.04} = \frac{10}{0.04} = 250 \text{ m}$$

The leakage factor, λ, is defined by Equation 7.168 as

$$\lambda = \sqrt{\frac{Kbb'}{K'}} = \sqrt{\frac{Tb'}{K'}}$$

which can be put in the form

$$K' = \frac{Tb'}{\lambda^2}$$

where K' and b' are the hydraulic conductivity and thickness of the semi-confining layer respectively. In this case, $T = 940$ m²/d, $b' = 2$ m, $\lambda = 250$ m, and therefore

$$K' = \frac{(940)(2)}{(250)^2} = 0.03 \text{ m/d} \qquad ■$$

This analysis of unsteady flow to a well in a leaky confined aquifer neglects the effect of storage in the semi-confining (leaky) layer and also neglects the drawdown in the overlying aquifer. In some cases, these approximations can lead to significant errors. Neuman and Witherspoon (1969b) have shown that the following condition justifies neglecting the storage in the semiconfining layer

$$\frac{r}{4b}\left(\frac{K'S'}{KS}\right) < 0.01 \qquad (7.258)$$

where S' is the storage coefficient of the semi-confining layer. Neuman and Witherspoon (1969a) have also shown that the drawdown in the overlying aquifer can be neglected when

$$T_s > 100T \qquad (7.259)$$

where T_s is the transmissivity of the overlying aquifer. Whenever storage effects in the semi-confining layer are significant and/or there is appreciable drawdown in the overlying aquifer, the analytical formulation developed by Neuman and Witherspoon (1968; 1969a; 1969b; 1972) should be applied to describe the aquifer response.

7.3.11 ■ Partially Penetrating Wells

The analytic solutions presented in the previous sections have all assumed that the pumping wells fully penetrate the aquifer and that the flow induced by the pumping well is approximately two dimensional. In cases where the pumping well does not fully penetrate the aquifer, in the immediate vicinity of the well the flow pattern is significantly three dimensional and the two-dimensional approximation is not valid. This effect is illustrated in Figure 7.18 for the case of a confined aquifer. The average flowpath induced by a partially penetrating well is longer than for a fully penetrating well; consequently, for the same pumping rate, the drawdown induced by a partially penetrating well is greater than for a fully penetrating well. This condition is illustrated in Figure 7.18(a), where the drawdown, s_p, at the partially penetrating well can be expressed in the form

$$\boxed{s_p = s + \Delta s} \qquad (7.260)$$

where s is the drawdown induced if the well were fully penetrating and Δs is the additional drawdown caused by partial penetration. Figure 7.18(b) illustrates two cases of partial penetration: a case in which the well penetrates from the top of the aquifer and a case in which the well is centered within the aquifer. In the case where the well penetrates from the top of the aquifer, the additional drawdown associated with partial penetration, Δs, can be estimated by (Todd, 1980)

$$\Delta s = \frac{Q_w}{2\pi T}\frac{1-p}{p}\ln\frac{(1-p)h_s}{r_w} \qquad (7.261)$$

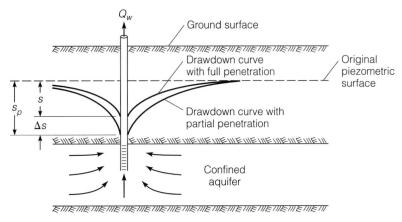

(a) Flow to a Partially Penetrating Well

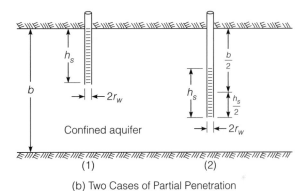

Figure 7.18 ■ Partially Penetrating Well in a Confined Aquifer

Source: Todd, David Keith. *Groundwater Hydrology.* Copyright © 1980 John Wiley & Sons, Inc. Reprinted by permission of John Wiley & Sons, Inc.

(b) Two Cases of Partial Penetration

where Q_w is the pumping rate, T is the transmissivity of the aquifer, p is the *penetration fraction* defined as

$$p = \frac{h_s}{b} \tag{7.262}$$

where h_s is the length of the screen, b is the thickness of the aquifer, and r_w is the radius of the well. Equation 7.261 applies whenever $p > 0.2$. In the cases where the well is screened in the center of the confined aquifer, then the drawdown correction, Δs, is given by (Todd, 1980)

$$\Delta s = \frac{Q_w}{2\pi T} \frac{1-p}{p} \ln \frac{(1-p)h_s}{2r_w} \tag{7.263}$$

The drawdown, s_p, at a partially penetrating well in an unconfined aquifer can be expressed in the form

$$\boxed{s_p^2 = s^2 + 2h_w \Delta s} \tag{7.264}$$

where s is the drawdown induced if the well were fully penetrating, h_w is the saturated thickness of the aquifer at the well, and Δs is a drawdown increment caused by partial penetration. In the case where the well penetrates the top of the unconfined aquifer

$$\Delta s = \frac{Q_w}{2\pi K h_w} \frac{1-p}{p} \ln \frac{(1-p)h_s}{r_w} \tag{7.265}$$

and in the case where the well is centered in the unconfined aquifer

$$\Delta s = \frac{Q_w}{2\pi K h_w} \frac{1-p}{p} \ln \frac{(1-p)h_s}{2r_w} \tag{7.266}$$

As the distance from a partially penetrating well increases, the flow becomes more two dimensional and the effect of partial penetration on the drawdown decreases. In homogeneous isotropic aquifers, this distance is estimated to be 1.5 to 2 times the saturated thickness of the aquifer. Wells that penetrate more than 85% of the saturated thickness of an aquifer can usually be assumed to be fully penetrating (USBR, 1995).

■ **Example 7.23**

A partially penetrating well with a screen length of 8 m is to be installed in a 20-m thick confined aquifer. The hydraulic conductivity of the aquifer is 100 m/d, the radius of the well is 0.3 m, and the pumping rate is 40 L/s. Compare the drawdown at the well with the drawdown at a fully penetrating well where (a) the well penetrates from the top of the aquifer, and (b) the well is screened in the center of the aquifer.

Solution

(a) From the given data: $Q_w = 40$ L/s $= 3{,}456$ m^3/d; $K = 100$ m/d; $b = 20$ m; $h_s = 8$ m; $p = h_s/b = 8/20 = 0.4$; and $r_w = 0.3$ m. Equation 7.261 gives the additional drawdown, Δs, when the well penetrates from the top of the aquifer, as

$$\Delta s = \frac{Q_w}{2\pi T} \frac{1-p}{p} \ln \frac{(1-p)h_s}{r_w} = \frac{3456}{2\pi(100 \times 20)} \frac{1-0.4}{0.4} \ln \frac{(1-0.4)8}{0.3} = 1.14 \text{ m}$$

(b) In the case where the well is screened in the center of the aquifer, Equation 7.263 gives the additional drawdown as

$$\Delta s = \frac{Q_w}{2\pi T} \frac{1-p}{p} \ln \frac{(1-p)h_s}{2r_w} = \frac{3456}{2\pi(100 \times 20)} \frac{1-0.4}{0.4} \ln \frac{(1-0.4)8}{2(0.3)} = 0.86 \text{ m}$$

Therefore, centering the screen in the aquifer will reduce the drawdown at the well by $1.14 - 0.86 = 0.28$ m.   ■

■ **Example 7.24**

Repeat the previous problem for the case of an unconfined aquifer, where the drawdown at the fully penetrating well is 1 m.

Solution

(a) In the case where the well penetrates the top of the aquifer, Equation 7.265 gives

$$\Delta s = \frac{Q_w}{2\pi K h_w} \frac{1-p}{p} \ln \frac{(1-p)h_s}{r_w} = \frac{3456}{2\pi(100)(19)} \frac{1-0.4}{0.4} \ln \frac{(1-0.4)8}{0.3} = 1.20 \text{ m}$$

where $h_w = 20$ m $- 1$ m $= 19$ m is the saturated thickness at the well under full penetration. According to Equation 7.264, the drawdown, s_p, under partially penetrating conditions is given by

$$s_p = \sqrt{s^2 + 2h_w \Delta s}$$

where $s = 1$ m, $h_w = 19$ m, and $\Delta s = 1.20$ m, hence

$$s_p = \sqrt{1^2 + 2(19)(1.20)} = 6.83 \text{ m}$$

Therefore, the additional drawdown due to partial penetration is 6.83 m − 1 m = 5.83 m.

(b) In the case where the well is screened in the center of the aquifer, Equation 7.266 gives

$$\Delta s = \frac{Q_w}{2\pi K h_w} \frac{1-p}{p} \ln \frac{(1-p)h_s}{2r_w} = \frac{3456}{2\pi(100)(19)} \frac{1-0.4}{0.4} \ln \frac{(1-0.4)8}{2(0.3)} = 0.90 \text{ m}$$

and the corresponding total drawdown, s_p, is

$$s_p = \sqrt{s^2 + 2h_w \Delta s} = \sqrt{1^2 + 2(19)(0.90)} = 5.93 \text{ m}$$

Hence the additional drawdown due to partial penetration is 5.93 m − 1 m = 4.93 m. Placing the well in the center of the aquifer rather than at the top of the aquifer, reduces the pumping head by 5.83 m − 4.93 m = 0.90 m. ■

7.4 Principle of Superposition

The principle of superposition can be stated as follows (Bear, 1979): If $s_1, s_2, \ldots, s_n$ are n general solutions of a homogeneous linear partial differential equation

$$L(s) = 0 \tag{7.267}$$

where L represents a linear operator, then

$$s = \sum_{i=1}^{n} C_i s_i \tag{7.268}$$

is also a solution of Equation 7.267, and C_i, $i = [1, n]$ are constants determined by the boundary conditions.

7.4.1 ■ Multiple Wells

The principle of superposition can be elucidated by considering the case of two wells fully penetrating an isotropic and homogeneous confined aquifer, with one well located at (x_1, y_1) and the other well located at (x_2, y_2). Assuming two-dimensional flow, the governing equation for the drawdown, $s(r, t)$ induced by these wells is given by Equation 7.202 as

$$\frac{\partial^2 s}{\partial r^2} + \frac{1}{r}\frac{\partial s}{\partial r} = \frac{S}{T}\frac{\partial s}{\partial t} \tag{7.269}$$

This equation can be written in the form

$$L(s) = 0 \tag{7.270}$$

where $L()$ is the linear operator

$$L() = \frac{\partial^2 ()}{\partial r^2} + \frac{1}{r}\frac{\partial ()}{\partial r} - \frac{S}{T}\frac{\partial ()}{\partial t} \tag{7.271}$$

In the case of unsteady flow to wells in an infinite confined aquifer, Equation 7.269 is to be solved subject to the following initial and boundary conditions

$$s(r, 0) = 0 \tag{7.272}$$

$$s(\infty, t) = 0 \tag{7.273}$$

$$\lim_{r_1 \to 0} r_1 \frac{\partial s}{\partial r_1} = -\frac{Q_1}{2\pi T} \tag{7.274}$$

$$\lim_{r_2 \to 0} r_2 \frac{\partial s}{\partial r_2} = -\frac{Q_2}{2\pi T} \tag{7.275}$$

where r_1 and r_2 are radial coordinates measured from (x_1, y_1) and (x_2, y_2), respectively, and Q_1 and Q_2 are the respective pumping rates from the two wells. To justify solving this problem by superposition, it is noted that if $s_1(r, t)$ and $s_2(r, t)$ are the drawdowns induced by each well operating by itself, then the following equations must be satisfied

$$L(s_1) = 0 \tag{7.276}$$

and

$$L(s_2) = 0 \tag{7.277}$$

with boundary conditions

$$s_1(r_1, 0) = 0 \tag{7.278}$$

$$s_1(\infty, t) = 0 \tag{7.279}$$

$$\lim_{r_1 \to 0} r_1 \frac{\partial s_1}{\partial r_1} = -\frac{Q_1}{2\pi T} \tag{7.280}$$

and

$$s_2(r_2, 0) = 0 \tag{7.281}$$

$$s_2(\infty, t) = 0 \tag{7.282}$$

$$\lim_{r_2 \to 0} r_2 \frac{\partial s_2}{\partial r_2} = -\frac{Q_2}{2\pi T} \tag{7.283}$$

Since both s_1 and s_2 satisfy the governing (homogeneous) differential equation, Equation 7.270, then by the principle of superposition

$$s = C_1 s_1 + C_2 s_2 \tag{7.284}$$

also satisfies the governing differential equation. The initial condition, Equation 7.272, and the boundary condition given by Equation 7.273 are both satisfied by Equation 7.284 by virtue of the fact that s_1 and s_2 satisfy Equations 7.278, 7.279, 7.281, and 7.282. The boundary condition given by Equation 7.274 requires that $C_1 = 1$, by virtue of

Equation 7.280 and the fact that the continuity equation guarantees that

$$\lim_{r_1 \to 0} r_1 \frac{\partial s_2}{\partial r_1} = 0 \tag{7.285}$$

The boundary condition given by Equation 7.275 requires that $C_2 = 1$, by virtue of Equation 7.283 and the fact that the continuity equation guarantees that

$$\lim_{r_2 \to 0} r_2 \frac{\partial s_1}{\partial r_2} = 0 \tag{7.286}$$

Therefore, in the case of two pumping wells in an infinite confined aquifer, it has been demonstrated that the drawdowns induced by the two wells, $s(r, t)$, is equal to the sum of the drawdowns induced by each of the wells operating individually, in which case

$$s(r, t) = s_1(r, t) + s_2(r, t) \tag{7.287}$$

This analysis can be extended to the case of multiple wells, in which case it can be demonstrated that the induced drawdown is simply the sum of the drawdowns induced by each of the wells. Although the example provided here extends only to multiple wells in confined aquifers of infinite extent, the principle of superposition applies also to steady flows in infinite unconfined aquifers, where the governing equation is linear in h^2, and also to unsteady flows in infinite unconfined aquifers, provided the governing equation is linearized by assuming that the drawdowns are small relative to the aquifer thickness.

■ Example 7.25

A municipal wellfield consists of five water-supply wells, each rated at 16 L/s. The wells are located 100 m apart along a north-south line and fully penetrate a 20-m thick confined aquifer that has a transmissivity of 1,000 m^2/d and a storage coefficient of 6×10^{-5}. Estimate the drawdown 500 m west of the center well after one week, one month, one year, and one decade of wellfield operation.

Solution

In accordance with the principle of superposition, the drawdown s_P at a location, P, 500 m west of the center well is given by

$$s_P = \sum_{i=1}^{5} s_i$$

where s_i is the drawdown at P induced by well i. The drawdown in the aquifer induced by each well is given by the Theis equation (Equation 7.226), where

$$s_i = \frac{Q_w}{4\pi T} W(u_i)$$

and hence

$$s_P = \frac{Q_w}{4\pi T} \sum_{i=1}^{5} W(u_i)$$

where

$$u_i = \frac{r_i^2 S}{4Tt}$$

From the given data: $Q_w = 16$ L/s $= 1{,}382$ m³/d, $T = 1{,}000$ m²/d, $S = 0.00006$, $r_1 = 500$ m, $r_2 = r_3 = \sqrt{500^2 + 100^2} = 509.9$ m, $r_4 = r_5 = \sqrt{500^2 + 200^2} = 538.5$ m, and therefore

$$s_P = \frac{1382}{4\pi(1000)} \left[W\left(\frac{500^2 \times 0.00006}{4t(1000)} \right) + 2W\left(\frac{509.9^2 \times 0.00006}{4t(1000)} \right) \right.$$

$$\left. + 2W\left(\frac{538.5^2 \times 0.00006}{4t(1000)} \right) \right]$$

$$= 0.110 \left[W\left(\frac{0.00375}{t} \right) + 2W\left(\frac{0.00390}{t} \right) + 2W\left(\frac{0.00435}{t} \right) \right]$$

Applying this result for $t = 7$ days ($= 1$ week), 30 days ($= 1$ month), 365 days ($= 1$ year), and 3,650 days ($= 10$ years) gives

t	s_P (m)
1 week	3.78
1 month	4.58
1 year	5.96
1 decade	7.22

■

7.4.2 ■ Well in Uniform Flow

In the previous application, the drawdown in the aquifer is zero whenever the pumping rate is equal to zero. In many cases, there is a regional mean flow, which can be characterized by a uniform seepage velocity, v_o. In this case, the drawdown is related to v_o by Darcy's law, and for confined aquifers it is given by Equation 7.114 as

$$s = \frac{nv_o}{K} x \tag{7.288}$$

where n is the effective porosity, K is the hydraulic conductivity of the aquifer, and x is the distance downstream from where the drawdown is zero. This fundamental solution to the ground-water flow equation can be used with the principle of superposition in cases where a regional flow exists. The drawdown, $s'(x, y)$, induced by a single well in an infinite aquifer is given by Equation 7.146 as

$$s'(x, y) = s_w - \frac{Q_w}{2\pi T} \ln\left(\frac{\sqrt{(x - x_o)^2 + (y - y_o)^2}}{r_w} \right), \qquad s'(r) \geq 0 \tag{7.289}$$

where s_w is the drawdown at the pumping well located at (x_o, y_o), Q_w is the pumping rate, T is the transmissivity, and r_w is the radius of the well. The drawdown, $s''(x, y)$, associated with a steady regional flow at seepage velocity v_o is given by Equation 7.114 as

$$s''(x, y) = \frac{nv_o}{K} x \tag{7.290}$$

where the origin of the x-axis is located where the drawdown is zero. Since both Equations 7.289 and 7.290 satisfy the governing flow equation, then by the principle of superposition, the sum of these solutions also satisfy the governing flow equation. Second, the required boundary conditions are that

$$\lim_{r \to r_w} 2\pi r b K \frac{\partial s}{\partial r} = Q_w \tag{7.291}$$

and that as $x \to \pm\infty$ the drawdown approaches the uniform flow solution. If we consider the drawdown, $s(x, y)$, defined by

$$s(x, y) = s'(x, y) + s''(x, y) \tag{7.292}$$

then it is clear that $s(x, y)$ satisfies both the boundary conditions, and the drawdown induced by a pumping well in an aquifer with a regional flow is given by

$$\boxed{s(x, y) = s_w + \frac{nv_o}{K} x - \frac{Q_w}{2\pi T} \ln\left(\frac{\sqrt{(x - x_o)^2 + (y - y_o)^2}}{r_w}\right)} \tag{7.293}$$

■ Example 7.26

A 0.4-m diameter fully penetrating well pumps 15 L/s from a 25-m thick confined aquifer with a hydraulic conductivity of 30 m/d and a porosity of 0.2. The regional mean flow has a seepage velocity of 1 m/d. If the drawdown at the well is 2 m, estimate the drawdown 50 m upstream of the well with and without the regional mean flow. At what distance downstream of the well is the seepage velocity equal to zero?

Solution

From the given data: $s_w = 2$ m, $n = 0.2$, $v_o = 1$ m/d, $K = 30$ m/d, $Q_w = 15$ L/s = 1,300 m³/d, $b = 25$ m, $T = Kb = (30)(25) = 750$ m²/d, and $r_w = 0.2$ m. Taking $(x_o, y_o) = (0$ m, 0 m$)$, then at 50 m upstream of the well $(x, y) = (-50$ m, 0 m$)$ and the drawdown, $s(-50, 0)$ is given by Equation 7.293 as

$$s(x, y) = s_w + \frac{nv_o}{K} x - \frac{Q_w}{2\pi T} \ln\left(\frac{\sqrt{(x - x_o)^2 + (y - y_o)^2}}{r_w}\right)$$

$$= 2 + \left[\frac{(0.2)(1)}{30}(-50)\right] - \frac{1300}{2\pi(750)} \ln\left(\frac{\sqrt{50^2 + 0^2}}{0.2}\right)$$

$$= 0.48 + [-0.33]$$

where the drawdown in square brackets represents the contribution of the regional mean flow. Therefore, in the absence of a regional mean flow, the drawdown 50 m upstream of the well is 0.48 m, and with the regional mean flow the drawdown is 0.48 m − 0.33 m = 0.15 m.

The seepage velocity is equal to zero where $ds/dx = 0$. Differentiating Equation 7.293 with respect to x and taking $y = y_o = 0$ and $x_o = 0$ gives

$$\frac{ds}{dx} = \frac{nv_o}{K} - \frac{Q_w}{2\pi Tx}$$

Taking $ds/dx = 0$ and rearranging yields

$$x = \frac{Q_w K}{2\pi\, T n v_o}$$

Substituting parameter values gives

$$x = \frac{(1300)(30)}{2\pi (750)(0.2)(1)} \approx 41.4 \text{ m}$$

Therefore, at a distance 41.4 m downstream of the well the seepage velocity is equal to zero. At this location, the direction of the seepage velocity changes from toward the well to away from the well. A practical result is that contaminant sources farther than 41.4 m downstream of the well cannot impact the well. ■

This analysis of a single well in a uniform flow can be extended to multiple wells in a uniform flow by superimposing the drawdowns induced by additional wells.

7.5   Method of Images

The method of images is a practical application of the principle of superposition in which several fundamental solutions to the governing (linear) flow equation are combined to yield a solution to the flow equation that satisfies some prescribed boundary conditions. This approach is illustrated in the following sections.

7.5.1   ■   Constant-Head Boundary

Consider the well shown in Figure 7.19, located at a distance L away from a constant-head boundary such as a fully penetrating reservoir. Under steady-state conditions, a solution is sought that satisfies the linearized flow equation given by

$$\nabla^2 s = 0 \tag{7.294}$$

where $s(x, y)$ is the drawdown defined by

$$s(x, y) = H - h(x, y) \tag{7.295}$$

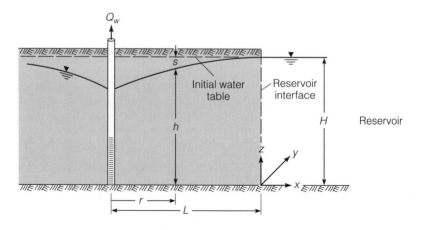

Figure 7.19 ■ Well Near a Constant-Head Boundary

and $\nabla^2()$ is the Laplacian operator. The boundary conditions to be satisfied in this case are

$$s(0, y) = 0 \tag{7.296}$$

$$\lim_{x \to -\infty} s(x, y) = 0 \tag{7.297}$$

$$\lim_{r \to r_w} r \frac{ds}{dr} = -\frac{Q_w}{2\pi T} \tag{7.298}$$

where r is the radial coordinate originating from the location of the pumping well. If the location of the pumping well is (x_o, y_o), then r is related to the Cartesian coordinates by

$$r^2 = (x - x_o)^2 + (y - y_o)^2 \tag{7.299}$$

Although the solution of the governing equation (Equation 7.294) and associated boundary conditions (Equations 7.296 to 7.298) seems challenging, the solution is actually quite simple using the method of images. Consider the situation illustrated in Figure 7.20, where the fully penetrating reservoir is replaced by a well that is identical to the pumping well and placed symmetrically about the reservoir interface. This is called an *image well* and differs from the pumping well only in that the image well pumps water into the aquifer at a rate Q_w, and therefore the magnitude of the buildup of the water table caused by the image well is exactly the same as the drawdown caused by the pumping well. The proof of this assertion is fairly straightforward and is given as a practice problem at the end of the chapter.

Consider now the superposition of the drawdowns caused by the pumping well and image well. According to the superposition principle, the resulting drawdown distribution, $s(x, y)$, is given by

$$s(x, y) = s_p(x, y) + s_i(x, y) \tag{7.300}$$

where s_p and s_i are the drawdowns caused by the pumping and image wells, respectively. The drawdown $s(x, y)$ defined by Equation 7.300 satisfies the boundary conditions required for a well adjacent to a fully penetrating reservoir, given by Equations 7.296 to 7.298. Equation 7.296 is satisfied by virtue of the fact that the drawdown induced by the pumping well and the buildup induced by the recharge well along the reservoir bound-

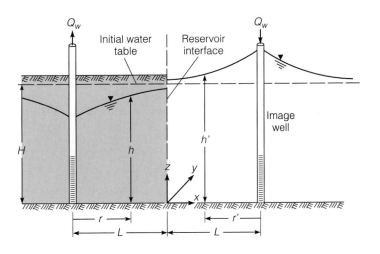

Figure 7.20 ■ Pumping Well and Image Well Relative to a Constant-Head Boundary

ary at $x = 0$ are equal, and therefore the total drawdown, s, given by Equation 7.300, is zero along $x = 0$, thereby meeting the required "zero-drawdown" boundary condition. The second boundary condition, Equation 7.297, is met by virtue of the fact that both s_p and s_i approach zero as $x \to -\infty$, and the third boundary condition, Equation 7.298, is met by virtue of the fact that

$$\lim_{r \to r_w} r \frac{ds_p}{dr} = -\frac{Q_w}{2\pi T} \tag{7.301}$$

and

$$\lim_{r \to r_w} r \frac{ds_i}{dr} = 0 \quad \text{(Continuity)} \tag{7.302}$$

and therefore

$$\lim_{r \to r_w} r \frac{d}{dr}(s_p + s_i) = -\frac{Q_w}{2\pi T} \tag{7.303}$$

which is exactly the requirement of Equation 7.298. The expression for the drawdown induced by a single pumping well in an infinite unconfined aquifer is

$$s_p = \frac{Q_w}{2\pi T} \ln\left(\frac{R}{r}\right) \tag{7.304}$$

and for the image (recharge) well the drawdown, s_i, is equal to

$$s_i = -\frac{Q_w}{2\pi T} \ln\left(\frac{R}{r'}\right) \tag{7.305}$$

where r' is measured relative to the recharge well. According to the analysis presented here, the drawdown distribution, s, induced by a pumping well located at a distance L from a fully penetrating reservoir is given by

$$s = s_p + s_i = \frac{Q_w}{2\pi T}\left[\ln\left(\frac{R}{r}\right) - \ln\left(\frac{R}{r'}\right)\right] \tag{7.306}$$

or

$$\boxed{s = \frac{Q_w}{2\pi T} \ln\left(\frac{r'}{r}\right)} \tag{7.307}$$

which can be written in Cartesian coordinates as

$$s = \frac{Q_w}{4\pi T} \ln\left[\frac{(x - x_o - 2L)^2 + (y - y_o)^2}{(x - x_o)^2 + (y - y_o)^2}\right] \tag{7.308}$$

The example presented here illustrates that the method of images simply consists of the superposition of fundamental solutions to the flow equation, where the fundamental solutions are combined in such a way that certain prescribed boundary conditions are met.

■ **Example 7.27**

A well is located 100 m west of a fully penetrating river that runs in a north-south direction. If the well pumping rate is 20 L/s, the aquifer is 20 m thick, and the hydraulic conductivity is 28 m/d, estimate the drawdowns 30 m north, south, east, and west of the well. What is the leakage out of the river per unit length of river at the section closest to the well?

Solution

From the given data: $Q_w = 20$ L/s $= 1{,}728$ m³/d, $K = 28$ m/d, $b = 20$ m, $T = Kb = (28)(20) = 560$ m²/d, $(x_o, y_o) = (0$ m, 0 m$)$, and $L = 100$ m. The drawdown at (x, y), $s(x, y)$, is given by Equation 7.308 as

$$s = \frac{Q_w}{4\pi T} \ln \left[\frac{(x - x_o - 2L)^2 + (y - y_o)^2}{(x - x_o)^2 + (y - y_o)^2} \right]$$

$$= \frac{1728}{4\pi(560)} \ln \left[\frac{(x - 0 - 2 \times 100)^2 + (y - 0)^2}{(x - 0)^2 + (y - 0)^2} \right] = 0.246 \ln \left[\frac{(x - 200)^2 + y^2}{x^2 + y^2} \right]$$

and the drawdowns 30 m from the well are given in the following table:

Location	x (m)	y (m)	s (m)
North	0	30	0.94
South	0	−30	0.94
East	30	0	0.85
West	−30	0	1.00

The leakage out of the river per unit length (of river), q, is given by

$$q = -Kb \frac{ds}{dx} \bigg|_{x=100, y=0}$$

$$= -T \frac{d}{dx} \left[\frac{Q_w}{4\pi T} \ln \frac{(x - 200)^2}{x^2} \right]_{x=100} = -\frac{Q_w}{4\pi} \left[\frac{2x(x - 200) - 2(x - 200)^2}{x(x - 200)^2} \right]_{x=100}$$

$$= -\frac{1728}{4\pi} \left[\frac{2(100)(100 - 200) - 2(100 - 200)^2}{100(100 - 200)^2} \right]_{x=100} = 5.5 \text{ (m}^3\text{/d)/m}$$

Hence the leakage rate out of the fully penetrating river is 5.5 (m³/d)/m at the section closest to the well. The leakage rates at other river sections are less than this value.   ■

7.5.2 ■ Impermeable Boundary

Consider the pumping well shown in Figure 7.21, located a distance L away from a fully penetrating impermeable boundary. The impermeable boundary could be due to a fault or could simply result from a lateral change in aquifer material. Under steady-state conditions, the equation to be solved is

$$\nabla^2 s = 0 \tag{7.309}$$

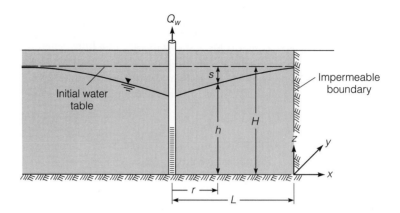

Figure 7.21 ■ Well Near an Impermeable Boundary

where $s(x, y)$ is the drawdown defined by

$$s(x, y) = H - h(x, y) \tag{7.310}$$

and the boundary conditions to be satisfied are

$$\left.\frac{\partial s}{\partial x}\right|_{x=0} = 0 \tag{7.311}$$

$$\lim_{x \to -\infty} s(x, y) = 0 \tag{7.312}$$

$$\lim_{r \to r_w} r\frac{ds}{dr} = -\frac{Q_w}{2\pi T} \tag{7.313}$$

Consider the superposition of the drawdowns induced by a pumping well and an identical image well located symmetrically about the impermeable boundary as illustrated in Figure 7.22. If $s_p(x, y)$ and $s_i(x, y)$ are the drawdowns caused by the real and image wells, respectively, in an infinite aquifer, then it can be shown that the sum of the drawdowns, $s(x, y)$, given by

$$s(x, y) = s_p(x, y) + s_i(x, y) \tag{7.314}$$

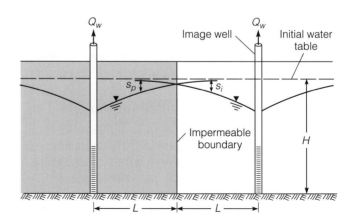

Figure 7.22 ■ Pumping Well and Image Well Relative to an Impermeable Boundary

is the solution of Equation 7.309 and satisfies the boundary conditions given by Equations 7.311 to 7.313. This can be demonstrated by noting that both s_p and s_i satisfy Equation 7.309; therefore, the sum of s_p and s_i also satisfies Equation 7.309. Since the drawdowns induced by the real and image wells are symmetric relative to the location of the impermeable boundary, then the sum of the drawdowns is also symmetric, and the slope of the drawdown curve is equal to zero at the impermeable boundary. Therefore, the boundary condition given by Equation 7.311 is satisfied by the sum of the drawdowns. Since the drawdowns induced by both the real and image well approach zero as the distance from the well increases, then the boundary condition given by Equation 7.312 is satisfied by the sum of the drawdowns. Finally, since s_p and s_i satisfy the equations

$$\lim_{r \to r_w} r \frac{ds_p}{dr} = -\frac{Q_w}{2\pi T} \tag{7.315}$$

and

$$\lim_{r \to r_w} r \frac{ds_i}{dr} = 0 \tag{7.316}$$

then

$$\lim_{r \to r_w} r \frac{d(s_p + s_i)}{dr} = -\frac{Q_w}{2\pi T} \tag{7.317}$$

which shows that the sum of the drawdowns also satisfies the boundary condition given by Equation 7.313. This completes the proof that the sum of the drawdowns induced by the real and image well satisfies both the governing equation and boundary conditions of the problem.

■ Example 7.28

A well is located 100 m west of an impervious boundary that runs in the north-south direction. If the well pumps 20 L/s, the aquifer is 20 m deep, and the hydraulic conductivity is 28 m/d, estimate the drawdowns 30 m north, south, east, and west of the well. The radius of influence of the well can be taken as 600 m.

Solution

The drawdown distribution, $s(x, y)$, is the sum of the drawdowns induced by two pumping wells placed symmetrically about the impermeable boundary. Hence,

$$s(x, y) = s_p + s_i = \frac{Q_w}{2\pi T}\left[\ln\left(\frac{R}{r}\right) + \ln\left(\frac{R}{r'}\right)\right] = \frac{Q_w}{2\pi T}\ln\left(\frac{R^2}{rr'}\right)$$

where R is the radius of influence, r is the radial distance of (x, y) from the pumping well, and r' is the radial distance of (x, y) from the image well. From the given data, $Q_w = 20$ L/s $= 1{,}728$ m³/d, $K = 28$ m/d, $b = 20$ m, $T = Kb = (28)(20) = 560$ m²/d, and $R = 600$ m. Substituting the given data into the drawdown equation yields

$$s(x, y) = \frac{1728}{2\pi(560)}\ln\left(\frac{600^2}{rr'}\right) = 0.491(12.8 - \ln rr')$$

The drawdowns 30 m from the well are given in the following table:

Location	r (m)	r' (m)	s (m)
North	30	202	2.01
South	30	202	2.01
East	30	170	2.09
West	30	230	1.94

7.5.3 ■ Other Applications

Examples 7.27 and 7.28 clearly demonstrate the fundamental reasons why the method of images works, and provide sufficient guidance to apply the method of images to other cases. The linearity and homogeneity of the governing differential equation guarantee that superimposed solutions will also satisfy the governing differential equation. The selection of the location(s) of image wells is controlled by the requirement that the superimposed drawdowns must meet the boundary conditions. In the case of a constant-head boundary, an image well is placed to ensure zero drawdown at the constant-head boundary; in the case of an impermeable boundary, an image well is placed to ensure that the slope of the drawdown curve is zero at the impermeable boundary.

7.6   Saltwater Intrusion

In coastal aquifers, a transition region exists where the water in the aquifer changes from fresh water to saltwater. However, because saltwater is denser than fresh water, the saltwater tends to form a wedge beneath the fresh water as shown in Figure 7.23, for the case of an unconfined aquifer. This illustration is somewhat idealized since in reality there is not sharp interface between fresh water and saltwater, but rather a "blurred" interface resulting from diffusion and mixing caused by the relative movement of the fresh water and saltwater. This relative movement is usually associated with tides and temporal variations in aquifer stresses. The thickness of the transition zone between fresh water and saltwater can range from a few meters to over a hundred meters (Visher and Mink, 1964). The intrusion of saltwater into coastal aquifers is generally of concern because of the associated deterioration in ground-water quality. Since the recommended maximum contaminant level (MCL) for chloride in drinking water is 250 mg/L and a typical chloride level in seawater is 14,000 mg/L, then nonsaline water mixed with more than 1.8% seawater renders the mixture nonpotable. This percentage is even less if the

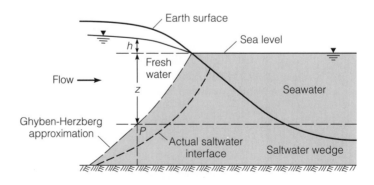

Figure 7.23 ■ Saltwater Interface in a Coastal Aquifer

fresh water contains a nonzero chloride concentration. In the United States, saltwater intrusion has resulted in the degradation of aquifers in at least 20 of the coastal states (Newport, 1977) and has been primarily caused by overpumping in sensitive portions of the aquifers. The most seriously affected states are Florida, California, Texas, New York, and Hawaii (Rail, 1989).

An approximate method for determining the location of the saltwater interface was introduced by Badon-Ghyben (1888) and Herzberg (1901) and is called the *Ghyben-Herzberg approximation.*[*] Under this approximation, the pressure distribution is assumed to be hydrostatic within any vertical section of the aquifer, which implicitly assumes that the streamlines are horizontal. Under this assumption, the hydrostatic pressure at point P in Figure 7.23 can be calculated from either the fresh-water head or the saltwater head, which means that

$$\gamma_f (h + z) = \gamma_s z \tag{7.318}$$

where γ_f is the specific weight of fresh water, γ_s is the specific weight of saltwater, h is the elevation of the water table above sea level, and z is the depth of the saltwater interface below sea level. Solving Equation 7.318 for z leads to

$$z = \frac{\gamma_f}{\gamma_s - \gamma_f} h \quad \text{or} \quad z = \frac{\rho_f}{\rho_s - \rho_f} h \tag{7.319}$$

where ρ_f is the density of fresh water and ρ_s is the density of saltwater. This equation is called the *Ghyben-Herzberg equation.* Under typical conditions, $\rho_f = 1.000$ g/cm^3 and $\rho_s = 1.025$ g/cm^3. Substituting these values into Equation 7.319 leads to

$$z \approx 40h \tag{7.320}$$

which means that the saltwater interface will typically be found at a distance below sea level equal to 40 times the elevation of the water table above sea level.

In applying the Ghyben-Herzberg approximation, it is useful to note that the assumption of horizontal flow produces acceptable results, except near the coastline where vertical flow components become significant, in which case the actual saltwater interface is expected to be found below the location predicted by the Ghyben-Herzberg equation (Bear, 1979). In the case of confined aquifers, the Ghyben-Herzberg approximation is also applicable, with the elevation of the water table replaced by the elevation of the piezometric surface. Bear and Dagan (1962) have shown that the length of saltwater intrusion into a horizontal confined aquifer of thickness b is predicted to within 5% by the Ghyben-Herzberg equation, provided that $\pi(\Delta\gamma/\gamma_f)Kb/Q > 8$, where Q is the rate of flow of fresh water per unit breadth of the aquifer, and $\Delta\gamma = \gamma_s - \gamma_f$.

Besides saltwater intrusion caused by the density difference between saltwater and fresh water, a second important mechanism for saltwater intrusion is associated with the construction of unregulated coastal drainage canals. These canals allow the inland penetration of saltwater via tidal inflow and subsequent leakage of saltwater from the canals into the aquifer. To prevent saltwater intrusion in coastal drainage canals, salinity-control gates are typically placed at the downstream end of the canal to maintain a fresh-water head (on the upstream side of the gate) over the sea elevation (on the downstream side of the gate). The fresh-water head should be sufficient to prevent saltwater intrusion in accordance with the Ghyben-Herzberg equation. During periods of high runoff and when the stages in the canals are above a prescribed level, then the canal gates are opened

---

[*]Developed independently by Badon-Ghyben (1888) and Herzberg (1901).

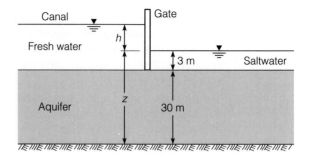

Figure 7.24 ■
Gated Canal

to permit drainage while maintaining a fresh-water head that is sufficient to prevent saltwater intrusion.

■ Example 7.29

Consider the gated canal in a coastal aquifer illustrated in Figure 7.24. If the aquifer thickness below the canal is 30 m, and at high tide the depth of seawater on the downstream side of the gate is 3 m, find the depth of fresh water on the upstream side of the gate that must be maintained to prevent saltwater intrusion.

Solution

The elevation of the fresh-water surface at the upstream side of the gate must be sufficient to maintain the saltwater interface at a depth of 33 m below sea level. According to the Ghyben-Herzberg equation (Equation 7.319), the height of the fresh-water surface above sea level, h, is given by

$$h = \frac{\rho_s - \rho_f}{\rho_f} z$$

where ρ_s and ρ_f are the densities of saltwater and fresh water, respectively, and z is the depth of the interface below sea level. Substituting $\rho_s = 1.025$ g/cm^3, $\rho_f = 1.000$ g/cm^3, and $z = 33$ m yields

$$h = \frac{1.025 - 1.000}{1.000} 33$$
$$= 0.83 \text{ m}$$

Therefore, the fresh water on the upstream side of the gate must be held at 0.83 m above the sea level on the downstream side of the gate. The depth of fresh water is 3 m + 0.83 m = 3.83 m.   ■

In addition to salinity-control gates in coastal drainage channels, other methods of controlling saltwater intrusion include modification of pumping patterns, creation of fresh-water recharge areas, and installation of extraction and injection barriers. Extraction barriers are created by maintaining a continuous pumping trough with a line of wells adjacent to the sea, and injection barriers are created by injecting high-quality fresh water into a line of recharge wells to create a high-pressure ridge. In extraction barriers, seawater flows inland toward the extraction wells and fresh water flows seaward toward the extraction wells. The pumped water is brackish and is normally discharged to the sea.

Whenever water supply wells are installed above the saltwater interface, the pumping rate from the wells must be controlled so as not to pull the saltwater up into the well. The process by which the saltwater interface rises in response to pumping is called *upconing*. This phenomenon is illustrated in Figure 7.25. Schmorak and Mercado (1969) proposed the following approximation of the rise height, z, of the saltwater interface in response to pumping

$$z = \frac{Q_w}{2\pi d K_x (\Delta\rho/\rho_f)} \tag{7.321}$$

where Q_w is the pumping rate, d is the depth of the saltwater interface below the well before pumping, K_x is the horizontal hydraulic conductivity of the aquifer, ρ_f is the density of fresh water, and $\Delta\rho$ is defined by

$$\Delta\rho = \rho_s - \rho_f \tag{7.322}$$

where ρ_s is the saltwater density. Equation 7.321 incorporates both the Dupuit and Ghyben-Herzberg approximations, and therefore care should be taken in cases where significant deviations from these approximations occur. Experiments have shown that whenever the rise height, z, exceeds a critical value, then the saltwater interface accelerates upwards toward the well. This critical rise height has been estimated to be in the range $0.3d$ to $0.5d$ (Todd, 1980). Taking the maximum allowable rise height to be $0.3d$ in Equation 7.321 corresponds to a pumping rate, Q_{max}, given by

$$Q_{max} = 0.6\pi d^2 K_x \frac{\Delta\rho}{\rho_f} \tag{7.323}$$

Therefore, as long as the pumping rate is less than or equal to Q_{max}, pumping of fresh water above a saltwater interface remains viable, although pumping rates must remain steady to avoid blurring the interface. For anisotropic aquifers in which the vertical component of the hydraulic conductivity less than the horizontal component, a max-

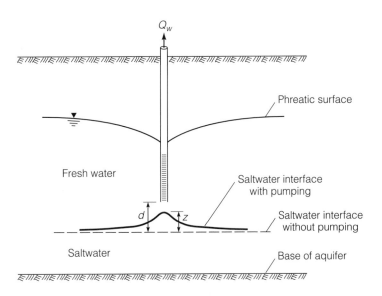

Figure 7.25 ■ Upconing Under a Partially Penetrating Well

imum well discharge larger than that given by Equation 7.323 is possible (Chandler and McWhorter, 1975).

■ **Example 7.30**

A well pumps at 5 L/s in a 30-m thick coastal aquifer that has a hydraulic conductivity of 100 m/d. How close can the saltwater wedge approach the well before the quality of the pumped water is affected?

Solution

From the given data: $Q_w = 5$ L/s $= 432$ m^3/d, $K_x = 100$ m/d, $\rho_f = 1,000$ kg/m^3, $\rho_s = 1,025$ kg/m^3, and $\Delta\rho = \rho_s - \rho_f = 1025 - 1000 = 25$ kg/m^3. Equation 7.323 gives the minimum allowable distance of the saltwater wedge from the well as

$$d = \sqrt{\frac{Q_{max}}{0.6\pi K_x \Delta\rho / \rho_f}}$$

If $Q_{max} = 432$ m^3/d, then

$$d = \sqrt{\frac{432}{0.6\pi(100)(25)/1000}} = 9.6 \text{ m}$$

Therefore, the quality of pumped water will be impacted when the saltwater interface is located 9.6 m below the pumping well.   ■

Saline ground water is a general term used to describe ground water containing more than 1,000 mg/L of total dissolved solids. There are several classification schemes for ground water based on total dissolved solids, and a widely cited classification, initially proposed by Carroll (1962), is given in Table 7.10. Intruded seawater has a total dissolved solids concentration of 35,000 mg/L and is classified as saline ground water. Other forms of saline ground water including *connate water** that was originally buried along with the aquifer material, water salinized by contact with soluble salts in the porous formation where it is situated, and water in regions with shallow water tables where evapotranspiration concentrates the salts in solution.

■ **Table 7.10**
Classification of Saline
Ground Water

Classification	Total dissolved solids (mg/L)
Fresh water	0–1,000
Brackish water	1,000–10,000
Saline water	10,000–100,000
Brine	> 100,000

Source: Carroll (1962).

---

*The word *connate* is derived from the latin word *connatus*, which means "born together."

7.7 Ground-Water Flow in the Unsaturated Zone

Porous media in which the void spaces are not completely filled with water are called *unsaturated*, and an *unsaturated zone* is generally found between the ground surface and the top of an aquifer. Understanding the movement of water in the unsaturated zone is important in describing the surface-infiltration process, the movement of water to replenish water extracted by plant transpiration, and the movement of contaminants from the ground surface into the ground water.

Within the unsaturated zone, water rises from the top of the saturated zone (water table) through the void spaces in much the same way that water rises in a capillary tube. Consider the capillary tube shown in Figure 7.26, with the *surface tension* between tube material and water given by σ. Equilibrium at the water surface in the capillary tube requires that the weight of the water column be supported by the surface tension force between the tube material and the water. Therefore

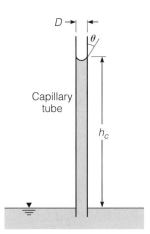

Figure 7.26 ■
Capillary Rise

$$\gamma \frac{\pi D^2}{4} h_c = \pi D \sigma \cos\theta \qquad (7.324)$$

where γ is the specific weight of water, D is the diameter of the capillary tube, h_c is the capillary rise, and θ is the *contact angle* between the water and the tube. Rearranging Equation 7.324 yields the following expression for the capillary rise, h_c,

$$\boxed{h_c = \frac{4\sigma \cos\theta}{\gamma D}} \qquad (7.325)$$

This relationship shows that the capillary rise is inversely proportional to the diameter of the tube. Since the pressure distribution within the capillary tube is hydrostatic, the pressure, p_c, at the water surface inside the capillary tube is given by

$$p_c = -\gamma h_c \qquad (7.326)$$

where the pressure is negative and therefore below atmospheric pressure.

Extending the behavior of capillary tubes to pore spaces above the saturated zone in ground water, it is easy to understand why in fine media, such as silts and clays, where the diameters of the void conduits are small, there are large capillary rises. Conversely, in coarse material such as sand and gravel, the void conduits have large diameters and the capillary rise is generally small. The capillary rise in several samples of unconsolidated materials are given in Table 7.11. Although capillary rise through porous media is very much like the rise through capillary tubes, porous media differ from capillary tubes in

■ **Table 7.11**
Capillary Rise in
Unconsolidated Materials

Material	Grain Size (mm)	Capillary rise (cm)
Fine gravel	2–5	2.5
Very coarse sand	1–2	6.5
Coarse sand	0.5–1	13.5
Medium sand	0.2–0.5	24.6
Fine sand	0.1–0.2	42.8
Silt	0.05–0.1	105.5

Source: Lohman (1972).

that the void conduits are irregular, vary in size according to the gradation of the porous matrix, and sometimes contain "dead ends." Consequently, the moisture content in the porous medium, θ_c, defined by the relation

$$\theta_c = \frac{\text{volume of water in soil sample}}{\text{sample volume}} \qquad (7.327)$$

decreases with distance above the water table. The moisture content continues to decrease with distance above the water table, until the continuity of the capillary rise is broken. When this happens, the distribution of water is no longer analogous to the rise in a capillary tube but becomes discontinuous and held by the solid matrix. The maximum moisture content in this (discontinuous) zone is called the *field capacity*, θ_f.

The distribution of moisture content above the water table is illustrated in Figure 7.27, which is commonly called the *retention curve*. In the zone between the ground surface and the level where the moisture content is controlled by capillary forces, the moisture content is determined by the ability of the soil to hold water, and the maximum (equilibrium) moisture content is equal to the field capacity. Below this zone, the soil transitions from water being held (discontinuously) by the soil to a zone where the water in the soil is continuous and rises above the water table in a manner analogous to the rise in a capillary tube. Within this region, the pressure distribution in the soil water is described by Equation 7.326, where h_c is the elevation above the water table. Water that is held discontinuously at the field capacity is called *capillary water*, while water that is continuously drawn from the water table is called *gravitational water*. The maximum moisture content in the retention curve must obviously be equal to the porosity, n.

In reality, the distribution of water within the soil column is affected by several processes. First of all, plant transpiration and surface evaporation will extract water from the unsaturated zone. If this water is extracted from the capillary water, then, because this water is discontinuous, there is no mechanism to replenish this water (other than rainfall) and the moisture content will simply decrease below the field capacity. When the moisture content is reduced to the *wilting point*, the water is held so tightly that plants cannot extract any more water, and only evaporation can further reduce the moisture content. Under these conditions, capillary forces exceed the osmotic forces and plants growing in the soil are reduced to a wilted condition. The soil can ultimately become dry as all the capillary water is removed by evaporation. At this point, a small amount of water called *hygroscopic water* is still held by the soil, and this water can be removed by oven drying the soil at a temperature of 105°C. The range of moisture contents are

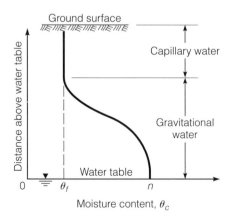

Figure 7.27 ■
Retention Curve

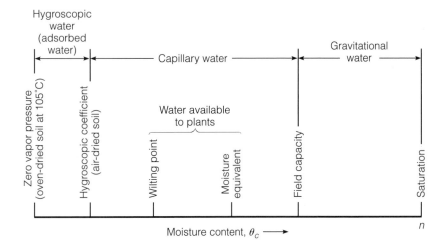

Figure 7.28 ■ Range of Moisture Contents in Porous Media
Source: Bear, Jacob. *Hydraulics of Groundwater.* The McGraw-Hill Companies, New York. Copyright © 1979. Reprinted by permission of The McGraw-Hill Companies.

illustrated in Figure 7.28. Gravitational water extracted from the unsaturated zone behaves quite differently from capillary water. Since gravitational water is continuously connected to the water table, it is replenished by capillary rise and therefore significant reductions in moisture content do not necessarily result from the extraction of gravitational water. Because of the relationship between pore geometry and capillary rise, hysteresis effects are prevalent in the distribution of gravitational water. Clearly, the moisture distribution resulting from a falling water table will be different than for a rising water table.

A useful application of the retention curve is to estimate the specific yield, S_y, of an unconfined aquifer, which is an important parameter in estimating transient effects in unconfined aquifers. Consider the case illustrated in Figure 7.29, where the water table is initially at A and falls to B. In this case, the depth of water released from storage is equal to the shaded area between the retention curves. Since the specific yield is defined as the volume of water released from storage per unit surface area per unit drop in the water table, the specific yield can be determined by dividing the depth of water released from storage by the magnitude of the drop in the water table (from A to B). A procedure for estimating the specific yield from the retention curve is illustrated in the following example.

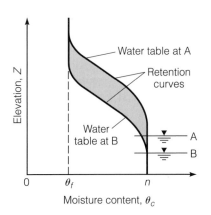

Figure 7.29 ■ Specific Yield in an Unconfined Aquifer

■ Table 7.12	Elevation above water table (m)	Moisture content
Retention Curve	0	0.20
	0.5	0.19
	1.0	0.18
	1.5	0.16
	2.0	0.13
	2.5	0.11
	3.0	0.08
	3.5	0.06
	4.0	0.05
	4.5	0.05
	5.0	0.05

■ Example 7.31

The retention curve of a soil sample from the unsaturated zone is given in Table 7.12, where the porosity of the soil is 0.2 and the field capacity of the soil is 0.05. Estimate the specific yield of the aquifer.

Solution

The initial depth of water in the unsaturated zone between the water table and 5 m above the water table, V_0, can be estimated by integrating the retention curve in Table 7.12. Therefore, using a trapezoidal rule to integrate over the 0.5-m spacing of data points yields

$$V_0 = (0.20)(0.25\ \text{m}) + (0.19 + 0.18 + 0.16 + 0.13 + 0.11 + 0.08 + 0.06$$

$$+ 0.05 + 0.05)(.5\ \text{m}) + (0.05)(0.25\ \text{m})$$

$$= 0.5563\ \text{m}$$

The initial depth of water between 1 m below the initial water table and 5 m above the initial water table, V_1, is given by

$$V_1 = 0.5563 + (0.20)(1) = 0.7563\ \text{m}$$

If the water table falls by 1 m, then the depth of water between the new water table and 6 m above the water table, V_2, is given by

$$V_2 = 0.5563 + (0.05)(1) = 0.6063\ \text{m}$$

Therefore, when the water table falls 1 m, the depth of water released from storage, $V_1 - V_2$, is

$$V_1 - V_2 = 0.7563 - 0.6063 = 0.15\ \text{m}$$

and therefore the specific yield, S_y, is given by

$$S_y = \frac{0.15}{1} = 0.15 \qquad \blacksquare$$

7.8 **Engineered Systems**

7.8.1 ■ Design of Wellfields

The objective of wellfield design is to determine the number and location of wells that are required to supply water at a specified rate. A primary constraint in wellfield design is the maximum allowable drawdown, where the wells are to be arranged in such a way that the maximum allowable drawdown is not exceeded anywhere in the aquifer. Gupta (1989) has suggested a design procedure where each well in the wellfield is initially assumed to have a drawdown (at the well) equal to one-half the allowable drawdown, with the other half of the allowable drawdown caused by interference from other wells, boundary effects, and well losses. The pumping rate, Q_w, at each well that would cause a drawdown equal to one-half the allowable drawdown is determined using the Theis equation, with a time equal to one year. The number of wells required in the wellfield is then equated to the required water-supply rate, Q, divided by the pumping rate from each well, Q_w. This ratio is rounded upward to the nearest integer. Once the number of wells and the pumping rate from each well are established, the wells are arranged in a regular pattern and spaced such that the total drawdown at each of the wells does not exceed the allowable maximum. If an analytical approach is taken, then the total-drawdown field can be derived using the principle of superposition. Selecting the relative locations of the wells will depend on the individual site characteristics, such as property boundaries and existing pipe networks, but the wells should generally be aligned parallel to and as close as possible to surface-water recharge boundaries and as far away from impermeable boundaries as possible. Typically, production wells are spaced at least 75 m apart (Walton, 1991). A wellfield design is illustrated in the following example.

■ Example 7.32

A small municipal wellfield is to be developed in an unconfined sand aquifer with a hydraulic conductivity of 50 m/d, thickness of 30 m, and specific yield of 0.2. A service demand of 7,000 m³/d is required from the wellfield, and the diameter of each well is to be 60 cm. If the drawdown in the aquifer is not to exceed 3 m when the wellfield is operational, develop a proposed layout for the wells. There are no nearby surface-water bodies.

Solution

The first step is to estimate the pumping rate from a single well that would cause a drawdown of 1.5 m, equal to one-half of the allowable drawdown of 3 m. The drawdown, s, resulting from the operation of each well can be estimated using the Theis equation, where

$$s = \frac{Q_w}{4\pi T} W(u)$$

where Q_w is the pumping rate from the well, T is the transmissivity of the aquifer, $W(u)$ is the well function, and u is defined by the relation

$$u = \frac{r^2 S_y}{4Tt}$$

where r is the distance from the well, S_y is the specific yield of the aquifer, and t is the time since the beginning of pumping. Since the drawdowns in the wellfield will vary between zero and 3 m, then the average transmissivity, T, to be used in the wellfield design is given by

$$T = KH = 50 \times \left(30 - \frac{3}{2}\right) = 1{,}425 \ \text{m}^2/\text{d}$$

For a single well with $t = 365$ days,

$$u_w = \frac{r_w^2 S_y}{4Tt} = \frac{(0.3)^2(0.2)}{(4)(1425)(365)} = 8.652 \times 10^{-9}$$

Since $u_w \leq 0.004$, then with less than 0.1% error

$$W(u_w) = -0.5772 - \ln u_w$$

and therefore

$$W(u_w) = -0.5772 - \ln(8.652 \times 10^{-9}) = 17.99$$

According to the Theis equation, the pumping rate, Q_w, required to produce a drawdown, s_w, is given by

$$Q_w = \frac{4\pi T s_w}{W(u_w)}$$

For $s_w = 1.5$ m (one-half the allowable drawdown), then

$$Q_w = \frac{4\pi(1425)(1.5)}{17.99} = 1{,}493 \ \text{m}^3/\text{d}$$

Since the service demand is 7,000 m^3/d, the number of wells required is given by

$$\text{No. of wells} = \frac{7000}{1493} = 4.69 \approx 5 \ \text{wells}$$

Consider the five-well wellfield with the arrangement shown in Figure 7.30. Each of the 5 wells are to operate at the same pumping rate, Q_w, given by

$$Q_w = \frac{7000}{5} = 1{,}400 \ \text{m}^3/\text{d}$$

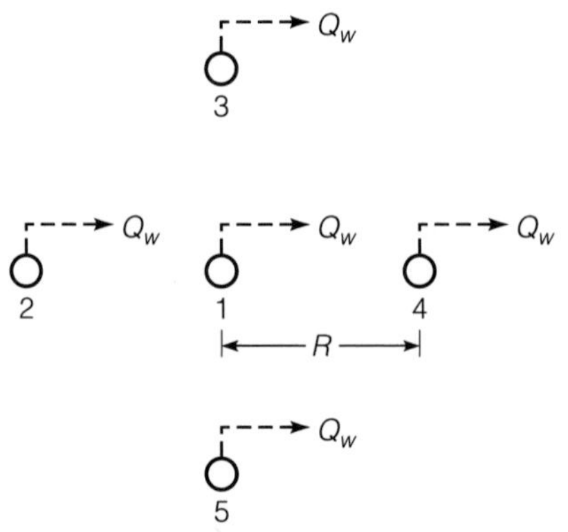

Figure 7.30 ■
Proposed Wellfield

and the arrangement consists of a central well (No. 1) surrounded by four wells (Nos. 2, 3, 4, 5) at a distance R away from the central well. Clearly, when all five wells are in operation, the drawdown will be greatest at the central well (No. 1). Denoting the drawdown at well No. 1 by s_1, then by the principle of superposition

$$s_1 = \frac{Q_w}{4\pi T} \sum_{i=1}^{5} W(u_i) \qquad (7.328)$$

where u_i are given by

$$u_1 = \frac{r_w^2 S_y}{4Tt} = \frac{(0.3)^2(0.2)}{4(1425)(365)} = 8.652 \times 10^{-9}$$

and

$$u_2 = u_3 = u_4 = u_5 = \frac{R^2 S_y}{4Tt} = \frac{R^2(0.2)}{4(1425)(365)} = 9.613 \times 10^{-8} R^2 \qquad (7.329)$$

and therefore,

$$W(u_1) = 17.99$$

$$W(u_2) = W(u_3) = W(u_4) = W(u_5)$$

Equating the drawdown at well No. 1 to the maximum allowable value of 3 m, then Equation 7.328 yields

$$3 = \frac{1400}{4\pi(1425)} [17.99 + 4W(u_2)]$$

which leads to

$$W(u_2) = 5.096$$

which, by the definition of the well function leads to

$$u_2 = 0.003437$$

Therefore, on the basis of Equation 7.329,

$$0.003437 = 9.613 \times 10^{-8} R^2$$

which leads to

$$R = 189 \text{ m}$$

Therefore, a wellfield consisting of 5 wells arranged as shown in Figure 7.30, with each well pumping 1,400 m^3/d, and spaced 189 m apart will yield the required amount of water and produce drawdowns that will not exceed the specified maximum of 3 m.   ■

7.8.2 ■ Design of Water-Supply Wells

The detailed design of a water-supply well requires specification of the production capacity (also called the *yield*), casing design, and the intake configuration of the well. A generally accepted, all-inclusive standard for designing water-supply wells is not available (USBR, 1995), however, the American Water Works Association (AWWA, 1990) and the National Water Well Association (Lehr et al., 1988) describe a number of commonly used design standards. The geology of the aquifer surrounding the well has a dominant influence on the well design, and the more certainty with which the geology surrounding the well is known the better the well design. Wells drilled in consolidated formations such as sandstone or limestone generally do not require any casing, screen, or gravel pack, while wells drilled in unconsolidated formations such as sand or gravel will generally require a casing and screen and may require a gravel pack. For major water-supply wells with yields greater than 500 L/min, it is usually advisable to first drill a pilot hole at the proposed well location to collect the relevant geologic data; if this location proves to be acceptable, the expanded pilot hole is then used as the water-supply well. If the pilot hole indicates that the proposed site is inadequate, the site can be abandoned without the major cost of drilling a producing well. For minor wells, well-drilling costs may be about the same as for a pilot hole.

Casing Design.   The well casing consists of the solid pipe above the well intake, through which ground water is pumped. The diameter of a production-well casing should be at least 5 cm larger than the nominal diameter of the pump bowls (Todd, 1980) and the flow velocity in the casing should be less than 1 m/s to prevent excessive friction losses (Lehr et al., 1988). Recommended casing diameters based on pump sizes used in the water-well industry are given in Table 7.13 (Lehr et al., 1988; Johnson Division, 1966), however, in final design the manufacturer of the selected pump system should be consulted for advice. Casing materials commonly used in practice include plastic and steel. Plastic casings, such as PVC, are used primarily in shallow, small-diameter wells, while steel casing is used in the majority of water-supply wells. Well casings should generally extend above the ground surface to prevent surface water from running down the hole and contaminating the well water.

Screen Design.   Wells driven in unconsolidated formations (such as sands and gravels) generally have screen intakes which keep the pump from being clogged with aquifer material, while wells in consolidated (rock) aquifers typically consist of the open end of the casing placed directly into the borehole. Such wells are called *open-hole wells*. In order to minimize both head losses across screens and screen clogging, entrance velocities need

■ Table 7.13 Recommended Casing Diameters Based on Pump Size	Pumping rate (L/min)	Optimum casing diameter (mm)	Minimum casing diameter (mm)
	< 400	150 mm ID*	125 mm ID
	300–660	200 mm ID	150 mm ID
	570–1,500	250 mm ID	200 mm ID
	1,300–2,500	300 mm ID	250 mm ID
	2,300–3,400	350 mm OD*	300 mm ID
	3,200–4,900	400 mm OD	350 mm OD
	4,500–6,800	500 mm OD	400 mm OD
	6,100–11,000	600 mm OD	500 mm OD

Source: Lehr et al. *Design and Construction of Water Wells*. 1988. ITP, Cincinnati.
*ID = inside diameter; OD = outside diameter.

to be controlled. The entrance velocity, v_s, can be expressed in the form (Todd, 1980)

$$v_s = \frac{Q_w}{c\pi d_s L_s P} \tag{7.330}$$

where Q_w is the pumping rate, c is the clogging coefficient (approximated as 0.5 by assuming that 50% of the open area of a screen is blocked by aquifer material), d_s is the screen diameter, L_s is the screen length, and P is the fraction of open area in the screen (which is part of the manufacturers specifications). Screen diameters should be selected based on a consideration of the desired yield from the well, and recommended minimum diameters for various well capacities are given in Table 7.14. These diameters may be increased to obtain acceptable entrance velocities, and smaller diameters are sometimes specified in the interest of economy. For screens with a given length and diameter, head losses decrease rapidly with an increase in percentage of open area up to 15%, less rapidly up to about 25%, and relatively slowly between 25% and 60%. For practical purposes, a percentage of open area of about 15% ($P = 0.15$) is acceptable and easily obtained with many commercial screens.

Walton (1962) has suggested that optimal screen entrance velocities are related to the hydraulic conductivity of the aquifer material, which is fundamentally related to the grain-size distribution of the aquifer material. The optimal screen entrance velocities for selected hydraulic conductivities are given in Table 7.15, and these limitations on entrance velocities are intended to prevent the larger grain sizes in the porous medium from being dislodged and clogging the screen. On the basis of Equation 7.330, it is clear that specification of a maximum screen entrance velocity leaves the designer the flexibility to select a variety of screen diameters and lengths. The U.S. Bureau of Reclamation (1995) recommends that the average entrance velocity generally be less than or equal to 1.8 m/min in order to limit turbulent flow and the associated head losses in the vicinity of screen intakes.

An important consideration in the selection of well screens is the size of the screen opening, commonly referred to as the *slot size*, which must be sufficiently small to prevent the screen from being clogged by dislodged portions of the aquifer matrix. The appropriate screen size can be related to the *uniformity coefficient*, U_c, of the aquifer

■ **Table 7.14** Minimum Screen Diameters	Pumping rate (L/min)	Screen diameter (mm)
	< 190	50
	190–475	100
	475–1,330	150
	1,330–3,040	200
	3,040–5,320	250
	5,320–9,500	300
	9,500–13,300	350
	13,300–19,000	400
	19,000–26,000	450
	26,600–34,200	500

Source: USBR (1995).

■ **Table 7.15** Optimal Screen Entrance Velocities	Hydraulic conductivity of aquifer (m/d)	Optimal screen entrance velocity (m/min)
	> 250	3.7
	250	3.4
	200	3.0
	160	2.7
	120	2.4
	100	2.1
	80	1.8
	60	1.5
	40	1.2
	20	0.9
	< 20	0.6

Source: Walton (1962).

matrix, defined by

$$U_c = \frac{d_{60}}{d_{10}} \tag{7.331}$$

where d_{60} and d_{10} are 60-percentile and 10-percentile particle diameters respectively of the aquifer matrix. A uniformity coefficient (U_c) of 1 indicates that the aquifer material is of uniform size, $U_c \leq 5$ indicates a *poorly graded* material, and $U_c > 5$ indicates a *well-graded* material (Kashef, 1986). Aquifers where $U_c \geq 3$ and $d_{10} > 0.25$ mm are good candidates for being *developed naturally*, meaning that initial pumping of the well will remove the fines from the surrounding aquifer material, leaving the screen surrounded by a very permeable coarse-grained annular region.

In cases where the well can be developed naturally, Todd (1980) and Lehr and colleagues (1988) suggest the criteria in Table 7.16 for selecting the screen slot size. Within the range of slot sizes indicated in Table 7.16, sizes at the low end of the range are selected whenever the well is not overlaid with a firm layer of soil, such as clay or shale, and slot sizes at the high end of the range are selected whenever firm soil layers are present (Lehr et al., 1988). By designing a screen slot size based on these criteria, the finer material in the aquifer surrounding the screen is removed during *well development*, which includes any mechanism that removes silt, fine sand, or other such material from the zone immediately surrounding a well intake. High-rate pumping and surge plungers are common methods used in well development. After well development, the screen intake is surrounded by a coarse layer with a hydraulic conductivity significantly higher than that of the undisturbed aquifer matrix. The coarse layer surrounding the screen is typi-

■ **Table 7.16** Criteria for Screen Slot Sizes in Naturally Developed Wells	Aquifer properties	Screen slot size[*]
	$U_c < 3, d_{10} > 0.25$ mm	d_{40}–d_{60}
	$3 \leq U_c \leq 5, d_{10} > 0.25$ mm	d_{40}–d_{70}
	$U_c > 5, d_{10} > 0.25$ mm	d_{50}–d_{70}

[*]Screen slot sizes are given as percentile sizes of aquifer material. For example, d_{40} is the 40-percentile size of the aquifer material.

cally about 0.5 m thick (Boonstra, 1998). If a screen is not available in the required slot size, the next smaller standard size should be selected. Slot openings in commercially available well screens typically range between 1 mm and 6 mm, and slot sizes are commonly available in multiples of 0.025 mm (0.001 in.). Hence, a No. 40 slot has a 1.0 mm (0.040 in.) slot width. In cases where the grain-size distribution varies over the depth of the aquifer, the specification of screen sections with different slot sizes is common.

Wells in confined aquifers should be screened through the entire thickness of the aquifer, while wells in unconfined aquifers should be screened in the lower one-third to one-half of the aquifer (USBR, 1995). Clearly, these specifications are not economically feasible in very deep and thick aquifers; in such aquifers, the usual practice is to penetrate a sufficient thickness of the aquifer to achieve the required discharge capacity at an acceptable pumping lift. Longer screen lengths generally result in higher pumping rates per unit drawdown.

Most screen sections are made in lengths ranging from 1.5 m to 6 m, and diameters ranging from 30 mm to 1,500 mm. Screen sections are typically joined together by welding or couplings to give almost any length of screen. The least expensive and most commonly available screens are made of low-carbon steel. Screens made of nonferrous metals and alloys, plastics, and fiberglass are used in areas of aggressive corrosion and encrustation.

Gravel-Pack Design.   In some cases, the aquifer material is so fine that selection of screen openings on the basis of the size and uniformity coefficient of the aquifer material would yield openings that are so small that the entrance velocity would be unacceptably high. Under these circumstances, *gravel packs* or (equivalently) *filter packs* are placed in the annular region between the screen and the perimeter of the borehole. In this context, the term "gravel pack" refers to any filtering media that is placed around the well screen and is not limited to a coarse gravel material as the name implies. Fine to medium sand is commonly used as gravel-pack material. According to Ahren (1957), gravel packs in unconsolidated formations are usually justified when uniformity coefficient, U_c, is less than 3 and the aquifer has a d_{10} less than 0.25 mm (0.010 in).

The thickness of the gravel pack is typically in the range of 8 to 23 cm (Todd, 1980; Ahren, 1957; Boonstra, 1998), with ideal thicknesses in the range of 10 to 15 cm (Lehr et al., 1988). It is difficult to develop a well through a gravel pack much thicker than 20 cm since the seepage velocities induced during the development procedure must be able to penetrate the gravel pack to repair the damage done by drilling, break down any residual drilling fluid on the borehole wall, and remove finer particles near the borehole (USBR, 1995). The thickness of the gravel pack should not be less than 8 cm to ensure a continuous pack will surround the entire screen (Boonstra, 1998). The specifications of the gravel pack are determined primarily by the aquifer matrix, and criteria for selecting the gravel pack and corresponding screen slot size are given in Table 7.17. Typically, a gravel pack should have a uniformity coefficient between 1 and 2.5, with a median grain size between six and nine times the median grain size of the aquifer material. The corresponding slot size of the screen opening should be between the 5- and 10-percentile grain size of the gravel pack, with the 10-percentile size usually being preferable to minimize entrance losses. The maximum grain size of a gravel pack should generally be around 10 mm, but less than 9 mm if placed through a nominal 100 mm (4 in.) tremie pipe (USBR, 1995), and the gravel pack should extend at least 3 m above the screen. Gravel packs usually consist of quartz-grained material that is sieved and washed to remove the finer material such as silt and clay. The grains of the gravel-pack material must be well rounded, since angular grains will tend to lock into adjacent grains, reducing

■ Table 7.17 Criteria for Gravel Pack Selection	Uniformity coefficient of aquifer matrix, U_c	Gravel pack criteria	Screen slot size
	< 2.5	(a) U_c between 1 and 2.5, with the 50% size not greater than six times the 50% size of the aquifer (preferable criteria)	5% to 10% passing size of the gravel pack
		(b) If (a) is not available, U_c between 2.5 and 5, with 50% size not greater than nine times the 50% size of the aquifer (alternative criteria)	
	2.5–5	(a) U_c between 1 and 2.5, with the 50% size not greater than nine times the 50% size of the formation (preferable criteria)	5% to 10% passing size of the gravel pack
		(b) If (a) is not available, U_c between 2.5 and 5, with 50% size not greater than 12 times the 50% size of the aquifer (alternative criteria)	
	> 5	Multiply the 30% passing size of the aquifer by 6 and 9 and locate the points on the grain-size distribution graph on the same horizontal line. Through these points draw two parallel lines representing materials with $U_c \leq 2.5$. Select gravel pack material that falls between the two lines.	5% to 10% passing size of the gravel pack

Source: USBR (1995).

the openings through which water and fine aquifer material can move. Gravel packs are usually considered essential in sandy aquifers.

In unconsolidated formations, whether a gravel pack is used or not, a properly developed well is surrounded by a coarse annular region that has a hydraulic conductivity much higher than the surrounding aquifer. This condition serves to increase the *effective radius* of the well, which is roughly equal to the radial extent of the coarse annular region surrounding the well screen, and drawdown calculations must be based on the effective radius.

■ Example 7.33

A water-supply well is to be installed in a 25-m thick unconfined aquifer which has a hydraulic conductivity of 30 m/d. A grain-size analysis of the aquifer indicates a uniformity coefficient of 2.7 and a 50-percentile grain size of 1 mm. If the well is to be pumped at 20 L/s, design the well screen and the gravel pack.

Solution

According to Table 7.17, since the uniformity coefficient (U_c) of the aquifer matrix is 2.7, the gravel pack should have a U_c between 1 and 2.5 and a 50% size not greater than nine times the 50-percentile size (d_{50}) of the aquifer. Hence, d_{50} for the gravel pack should be less than 9×1 mm = 9 mm. A check with a sand and gravel company will generally yield a commercial gravel-pack material that satisfies these criteria. For example, you may find "5 mm $\times$ 9" mm gravel, which contains only grain sizes between 5 mm and 9 mm. Assuming that the particle sizes are uniformly distributed between 5 mm and 9 mm, then $d_{10} \approx 5.4$ mm, $d_{60} \approx 7.4$ mm, and U_c of the gravel is approximately $7.4/5.4 = 1.4$. Hence, the 5 mm $\times$ 9 mm gravel is acceptable. Since gravel packs should be between 8 cm and 23 cm thick, a reasonable thickness would be 10 cm. This is within the range of ideal thicknesses suggested by Lehr and colleagues (1988).

According to Table 7.17, the screen slot size should be between the 5% and 10% passing size of the gravel pack. In this case, the 10% passing size is estimated to be 5.4 mm, and it is reasonable to select a slot size of 5 mm. A review of manufacturers' literature on well screens will give the commercially available slot widths and associated fractions, P, of open area. Typically, $P = 0.10$. Since the screen length, L_s, in unconfined aquifers should be between 0.3 and 0.5 times the aquifer thickness, select a screen length of 0.5 times the aquifer thickness to maximize the pumping rate per unit drawdown, in which case $L_s = 0.5 \times 25$ m = 12.5 m. For a pumping rate of 20 L/s (= 1,200 L/min), Table 7.13 indicates an optimum casing diameter of 250 mm, and Table 7.14 indicates a minimum screen diameter of 150 mm. Taking the screen and casing diameter as 250 mm, and assuming that 50% of the open area is clogged by the aquifer material, then the screen entrance velocity, v_s, is estimated by Equation 7.330 as

$$v_s = \frac{Q_w}{c\pi d_s L_s P}$$

where $Q_w = 20$ L/s = 1,728 m^3/d, $c = 0.5$, $d_s = 0.25$ m, $L_s = 12.5$ m, and $P = 0.1$, hence

$$v_s = \frac{1728}{(0.5)\pi(0.25)(12.5)(0.1)} = 3520 \text{ m/d} = 2.4 \text{ m/min}$$

According to Table 7.15, this screen entrance velocity is too high for an aquifer with a hydraulic conductivity of 30 m/d, where the desirable screen size would produce an entrance velocity on the order of 1.05 m/min. Taking $v_s = 1.05$ m/min (= 1,510 m/d), the corresponding screen diameter, d_s, is given by

$$d_s = \frac{Q_w}{c\pi v_s L_s P} = \frac{1728}{(0.5)\pi(1510)(12.5)(0.1)} = 0.583 \text{ m} \approx 600 \text{ mm}$$

In summary, a screen length of 12.5 m, diameter of 600 mm, and slot size of 5 mm is acceptable. This screen should be surrounded by a 5 mm $\times$ 9 mm gravel pack with a thickness of about 10 cm; the diameter of the well casing should be 600 mm so that it can be easily joined to the screen. ■

Sand Trap   A *sand trap* is a section of well casing installed below the screened intake section. The function of the sand trap is to store sand and silt entering the well during pumping. The length of the sand trap is usually 2 to 6 m, and the diameter is typically the same as the screen.

Well Grouting.   Well *grouting* consists of placing a cement slurry between the well casing and the borehole, and grouting is generally necessary to prevent polluted surface water or low-quality water in overlying aquifers from seeping along the outside of the casing and contaminating the well or aquifer. Grouting also serves to anchor the casing inside the borehole, protect the casing from corrosion, and prevent the caving in of land around the well. The cement slurry used in grouting is usually a mix of about 50 L of water per 100 kg of cement, with possible additives including bentonite clay, pozzolans, and perlite (Bouwer, 1978). To prevent the penetration of the cement slurry into the gravel pack, a layer of clean sand (particle size 0.3 to 0.6 mm) is placed on top of the gravel pack. This layer of sand is commonly referred to as a *sand bridge*, and cement slurries generally do not penetrate the sand layer by more than a few centimeters.

Corrosion and Encrustation.   Metal screens, casings, and pumps are subject to deterioration by corrosion, and well screens are prone to encrustation. Corrosion can be either *chemical* or *galvanic*. Chemical corrosion results in the metal going into solution, while galvanic corrosion results from the action of electrolytic cells formed between dissimilar metals. Corrosion is facilitated by certain algae, fungi, and bacteria where the byproducts of their metabolism promote chemical and galvanic corrosion. As a general rule, chemical corrosion can be anticipated under the following conditions: pH less than 7, dissolved oxygen greater than 2 mg/L, hydrogen sulfide (H_2S) greater than 1 mg/L, total dissolved solids greater than 1,000 mg/L, carbon dioxide (CO_2) greater than 50 mg/L, or chloride (Cl^-) greater than 300 mg/L. When corrosive environments are present, corrosion-resistant metals should be used for the screen and various parts of the pump. Plastic and fiberglass screens have the highest corrosion resistance, but because of their structural limitations they should be limited to use in relatively shallow, small-diameter wells of low capacity, or in unusual applications where other materials are unsuitable. Table 7.18 shows the more commonly used metallic screen materials in order of increasing cost. When all factors are considered, the most satisfactory screen materials, except under unusual conditions, are steel, stainless steel, and some of the metal alloys.

Encrustation is characterized by the accumulation of mineral deposits in and around openings in the screen, and in the voids surrounding the well. Encrustation problems are likely when the pH is greater than 7.5, carbonate hardness is greater than 300 mg/L, manganese is greater than 1 mg/L, or iron is greater than 2 mg/L (Lehr et al., 1988). The materials used in well casings and screens have little influence on the rate or character of encrustation, unless galvanic corrosion is associated with the encrustation. About 90% of encrustation problems are caused by soluble chemicals in the ground water that precipitate out of solution and are deposited in an insoluble form in the vicinity of the well screen (Lehr et al., 1988). These encrustations are mostly composed of carbonate minerals, such as $CaCO_3$, iron oxide (Fe_2O_3), and manganese oxide (MnO_2). Precipitation of these substances primarily result from large pressure drops in the immediate vicinity of well screens, causing carbon dioxide (CO_2) to come out of solution, increasing the pH, and producing favorable conditions for precipitation. The elevated oxygen levels

■ Table 7.18	Material	Corrosion resistance
Metallic Well-Screen Materials	Low carbon steel	poor
	Toncan and Armco iron	fair
	Admiralty red brass	good
	Silicon red brass	good
	304 stainless steel	very good
	Everdure bronze	very good
	Monel metal	very good
	Super nickel	very good

Source: USBR (1995).

in wells, compared to ground-water, also contribute significantly to the precipitation of iron and manganese oxides.

Pumping Rates. Several criteria for determining the optimal withdrawal rate from a well have been suggested. One of the most popular rules of thumb is that wells should be pumped at a rate that maintains a maximum drawdown of 50–60% of an unconfined aquifer's saturated thickness. This pumping rate can be estimated from the *specific capacity* of a well, which is defined as the well pumping rate per unit drawdown. The specific capacity in the case of a fully penetrating well in a homogeneous isotropic confined aquifer with negligible head losses in the developed region surrounding the intake can be derived directly from the Thiem steady-state equation (Equation 7.143) as

$$\text{Specific capacity} = \frac{Q_w}{s_w} = \frac{2\pi T}{\ln(R/r_w)} \tag{7.332}$$

or from the Theis transient equation (Equation 7.226) as

$$\text{Specific capacity} = \frac{Q_w}{s_w} = \frac{4\pi T}{W(r_w^2 S/4Tt)} \tag{7.333}$$

where Q_w is the pumping rate, s_w is the drawdown at the well, T is the transmissivity of the aquifer, R is the radius of influence (see Equation 7.144), r_w is the effective radius of the well, $W(u)$ is the well function (Equation 7.225), S is the storage coefficient, and t is the time since the beginning of pumping. Although both Equations 7.332 and 7.333 are used to estimate the specific capacity of a well, Equation 7.333 is more realistic and indicates that the specific capacity gradually decreases with time. Classifications of well productivity in terms of specific capacity are given in Table 7.19. The maximum rate at which water can be extracted from a well is defined as the *well yield*, which is equal to the specific capacity multiplied by the allowable drawdown. The well yield should not be confused with the *safe aquifer yield*, which is the maximum rate at which water can be withdrawn from an aquifer without depleting the water supply. Water managers are keenly aware that increases in ground-water withdrawals must generally be balanced by an increase in aquifer recharge, a decrease in aquifer discharge, and/or a loss of ground water storage.

■ Example 7.34

A water-supply well is to be constructed in a 25-m thick surficial aquifer in which the maximum allowable drawdown is 2 m. The well is to have an effective radius of 400 mm, and the hydraulic conductivity and specific yield of the aquifer are estimated as 20 m/d and 0.15, respectively. Estimate the specific capacity and the well yield after 3 years of service. Classify the productivity of the well.

■ Table 7.19
Classification of Well Productivity

Specific capacity, Q_w/s_w (L/min)/m	Productivity
< 0.3	negligible
0.3–3	very low
3–30	low
30–300	moderate
> 300	high

Source: Şen (1995).

Solution

From the given data: $H = 25$ m, $K = 20$ m/d, $T = KH = (20)(25) = 500$ m^2/d, $r_w = 400$ mm $= 0.4$ m, $S_y = 0.15$, and $t = 3$ years $= 3(365)$ days $= 1,095$ days. Substituting these data into Equation 7.333 (using the small-drawdown assumption for surficial aquifers) gives

$$\text{Specific capacity} = \frac{4\pi T}{W(r_w^2 S_y / 4Tt)} = \frac{4\pi(500)}{W(0.4^2 \times 0.15/4 \times 500 \times 1095)}$$

$$= 355 \text{ m}^2/\text{d} = 247 \text{ (L/min)/m}$$

Since the maximum allowable drawdown is 2 m, the well yield is given by

$$\text{well yield} = 247 \text{ (L/min)/m} \times 2 \text{ m} = 494 \text{ L/min}$$

and therefore the maximum allowable pumping rate is 494 L/min. According to Table 7.19, a well with a specific capacity of 247 (L/min)/m has a *moderate* productivity.   ■

Well Losses.   Head losses caused by turbulent flow through the coarse annular region surrounding a well, along with the head losses through the well screen, cause the water level inside the well to be less than the level in the aquifer adjacent to the well. This is illustrated in Figure 7.31. Well losses are usually added to the *formation losses* to estimate the *total drawdown*, and the water level to be expected in the well during pumping. The well loss, s_w, is typically estimated by the relation

$$\boxed{s_w = \alpha Q_w^n} \tag{7.334}$$

where α is a constant, Q_w is the pumping rate, and n is an exponent that typically varies between 1.5 and 3 (Lennox, 1966), but is commonly assumed to be equal to 2. The assumption of $n = 2$ was originally proposed by Jacob (1947) and is still widely accepted. Values of α corresponding to various states of a well were suggested by Walton (1962) and are given in Table 7.20. These data indicate that for a properly designed and functional well, α should be less than 1,800 s^2/m^5. Since *formation losses* are typically

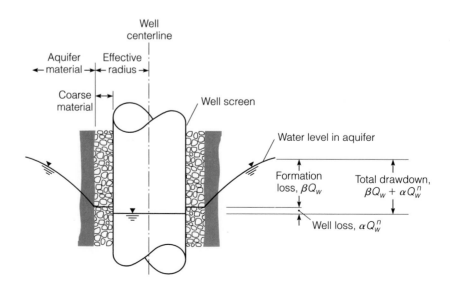

Figure 7.31 ■ Well Losses and Formation Losses

■ Table 7.20 Well Loss Coefficients	Well loss coefficient, α (s^2/m^5)	Well condition
	< 1,800	properly designed and developed
	1,800–3,600	mild deterioration or clogging
	3,600–14,400	severe deterioration or clogging
	> 14,400	difficult to restore well to original capacity

Source: Walton (1962).

proportional to the well pumping rate, Q_w, and well losses are typically proportional to Q_w^2, then the drawdown in a well, s_w, can typically be expressed in the form

$$s_w = \beta Q_w + \alpha Q_w^2 \qquad (7.335)$$

where β is the *formation loss coefficient*. Dividing both sides of Equation 7.335 by Q_w yields

$$\frac{s_w}{Q_w} = \beta + \alpha Q_w \qquad (7.336)$$

which indicates a linear relationship between s_w/Q_w and Q_w. The coefficients α and β can be determined in the field using a *step-drawdown pumping test*, where the well pumping rate is increased in a series of discrete steps, with drawdown measurements taken at a fixed time intervals (typically 1 hour) after each increase in pumping rate. The resulting data yields a relationship between s_w/Q_w and Q_w, which is fitted to the linear relationship given by Equation 7.336, and the slope and intercept of the best-fit line are taken as α and β, respectively. The *well efficiency*, e_w, is defined as the percentage of the total drawdown in the well caused by formation losses, and based on Equation 7.335 the well efficiency can be put in the form

$$e_w = \frac{\beta Q_w}{s_w} \times 100 \qquad (7.337)$$

■ Example 7.35

A step-drawdown test is conducted at a well where a pumping rate of 864 m^3/d yields a drawdown in the well of 1.54 m, and a pumping rate of 1,296 m^3/d yields a drawdown of 2.36 m. Assess the condition of the well, and estimate the well efficiency and specific capacity at a pumping rate of 2,000 m^3/d.

Solution

The given data can be put in the following tabular form:

Q (m^3/d)	s_w (m)	$\frac{s_w}{Q_w}$ (d/m^2)
864	1.54	0.00178
1,296	2.36	0.00181

According to Equation 7.336,

$$\frac{s_w}{Q_w} = \beta + \alpha Q_w$$

and combining this equation with the step-drawdown data gives

$$0.00178 = \beta + 864\alpha$$

$$0.00181 = \beta + 1296\alpha$$

Solving these equations simultaneously yields

$$\alpha = 6.9 \times 10^{-8} \text{ d}^2/\text{m}^5 = 515 \text{ s}^2/\text{m}^5 \quad \text{and} \quad \beta = 0.00172 \text{ d/m}^2$$

Comparing $\alpha = 515 \text{ s}^2/\text{m}^5$ with the guidelines in Table 7.20 indicates that the well is in good condition.

At $Q_w = 2{,}000 \text{ m}^3/\text{d}$, the expected drawdown at the well, s_w, is given by Equation 7.335 as

$$s_w = \beta Q_w + \alpha Q_w^2 = (0.00172)(2000) + (6.9 \times 10^{-8})(2000)^2 = 3.72 \text{ m}$$

The drawdown associated with formation losses is $\beta Q_w = 0.00172(2000) = 3.44 \text{ m}$, and hence the well efficiency, e_w, is given by Equation 7.337 as

$$e_w = \frac{\beta Q_w}{s_w} \times 100 = \frac{3.44}{3.72} \times 100 = 92\%$$

At $Q_w = 2{,}000 \text{ m}^3/\text{d}$, the specific capacity of the well is given by

$$\text{Specific capacity} = \frac{Q_w}{s_w} = \frac{Q_w}{\beta Q_w + \alpha Q_w^2} = \frac{1}{\beta + \alpha Q_w}$$

$$= \frac{1}{0.00172 + (6.9 \times 10^{-8})(2000)} = 538 \text{ m}^2/\text{d} = 373 \text{ (L/min)/m}$$

Comparing this result with the classification system in Table 7.19 indicates that the well is highly productive at a pumping rate of $2{,}000 \text{ m}^3/\text{d}$. ■

This analysis assumes a priori that the well losses are proportional to Q_w^2, and this assumption is justified by the fact that the well-loss coefficient used to assess the condition of the well in Table 7.20 is based on this assumption. In the more general case,

$$s_w = \beta Q_w + \alpha Q_w^n \tag{7.338}$$

which can be put in the form

$$\log\left(\frac{s_w}{Q_w} - \beta\right) = \log\alpha + (n - 1)\log Q_w \tag{7.339}$$

which shows that the relation between $\log(s_w/Q_w - \beta)$ and $\log Q_w$ is linear with a slope of $n - 1$ and an intercept of $\log\alpha$. To apply Equation 7.339 to step-drawdown data, various values of β are tried until the relation between $\log(s_w/Q_w - \beta)$ and $\log Q_w$ becomes approximately linear, then values of n and α are obtained directly from the slope and intercept of the best-fit straight line. This approach, originally proposed by Rorabaugh (1953), is sometimes referred to as the *Rorabaugh method*.

■ Example 7.36

The drawdowns, s_w (m), and corresponding pumping rates, Q_w (m^3/d), at a well during a step-drawdown test were plotted on a graph of $\log(s_w/Q_w - \beta)$ versus $\log Q_w$ for various values of β. The plotted curve is most linear when $\beta = 6.36 \times 10^{-4}$ d/m^2, and the corresponding best-fit line through the data has a slope of 1.14 and an intercept of -6.72. Determine the formation and well-loss coefficients, and estimate the efficiency and productivity of the well when the pumping rate is 900 m^3/d.

Solution

From the given data, $\beta = 6.36 \times 10^{-4}$ d/m^2; comparing the slope and intercept of the best-fit line with Equation 7.339 indicates that

$$\log \alpha = -6.72, \qquad (n-1) = 1.14$$

which gives

$$\alpha = 1.91 \times 10^{-7} \text{ d}^2/\text{m}^5, \qquad n = 2.14$$

Hence, the formation and well-loss coefficients are $\beta = 6.36 \times 10^{-4}$ d/m^2 and $\alpha = 1.91 \times 10^{-7}$ d^2/m^5 ($= 1430$ s^2/m^5) respectively, and the drawdown, s_w, is given by

$$s_w = 6.36 \times 10^{-4} Q_w + 1.91 \times 10^{-7} Q_w^{2.14}$$

The well efficiency, e_w, is equal to the percentage of total drawdown caused by formation losses, and can be expressed in the form

$$e_w = \frac{\beta Q_w}{\beta Q_w + \alpha Q_w^n} \times 100$$

When $Q_w = 900$ m^3/d,

$$e_w = \frac{(6.36 \times 10^{-4})(900)}{(6.36 \times 10^{-4})(900) + (1.91 \times 10^{-7})(900)^{2.14}} \times 100 = 59\%$$

Since the well-loss coefficient, $\alpha = 1{,}430$ s^2/m^5, indicates that the annular region surrounding the well is in fairly good condition (see Table 7.20), the low well efficiency of 59% can be attributed to formation losses being of comparable magnitude to the well losses. At $Q_w = 900$ m^3/d, the productivity of the well is measured by the specific capacity as

$$\text{Specific capacity} = \frac{Q_w}{s_w} = \frac{Q_w}{\beta Q_w + \alpha Q_w^n} = \frac{1}{\beta + \alpha Q_w^{n-1}}$$

$$= \frac{1}{6.36 \times 10^{-4} + (1.91 \times 10^{-7})(900)^{1.14}}$$

$$= 925 \text{ m}^2/\text{d} = 642 \text{ (L/min)/m}$$

Comparing this result with the classification system in Table 7.19 indicates that the well is highly productive at a pumping rate of 900 m^3/d. ■

7.8.3 ■ Wellhead Protection

Public water-supply wells must generally be protected from contaminants introduced into the aquifer from areas surrounding the well. The most effective way of protecting water-supply wells from contamination is to control land uses in the surrounding areas by zoning regulations. In the United States, the *Wellhead Protection Program* was established by the 1986 ammendments to the Safe Drinking Water Act, and all states are required to develop comprehensive wellhead protection (WHP) plans. Such plans should contain the following elements (USEPA, 1990a):

1. Specify the roles and duties of state agencies, local government entities, and public water suppliers in the development and implementation of WHP programs.
2. Delineate a *wellhead protection area* (WHPA) for each water-supply well based on reasonably available hydrogeologic information on ground-water flow, recharge and discharge, and other information the state deems necessary to adequately determine the WHPA.
3. Identify sources of contaminants within each WHPA, including all potential anthropogenic sources that may have an adverse effect on health.
4. Develop management approaches that include, as appropriate, technical assistance, financial assistance, implementation of control measures, education, training, and demonstration projects that are used to protect the water supply within the WHPA from such contaminants.
5. Develop contingency plans indicating the location and provision of alternate drinking water supplies for each public water-supply system in the event of well contamination.
6. Site new wells properly to maximize yield and minimize potential contamination.
7. Ensure public participation by establishing procedures encouraging the public to participate in developing the WHP program elements.

Much thought and effort have been put into the development of WHP plans, both in the United States and Europe, for a variety of budgets and hydrogeologic settings, and several good WHP plans have been published (e.g., ASCE, 1997; Johnson, 1997; USEPA, 1987). Wellhead protection plans developed to date indicate that there is no universal approach to wellhead protection, given the variety of political, social, economic, and technical constraints that are encountered in practice. A WHP plan should generally contain the seven key elements listed above, be tailored to local conditions, and afford a defined level of protection to the public water supply.

Delineation of Wellhead Protection Areas.   The most important and challenging component of any WHP plan, from an engineering viewpoint, is the delineation of the wellhead protection area. The Safe Drinking Water Act defines a wellhead protection area (WHPA) as "the surface and subsurface area surrounding a well or wellfield that supplies a public water system through which contaminants are likely to pass and eventually reach the water well or wellfield." A pumping well produces drawdowns in the aquifer surrounding the well, and the portion of the aquifer within which ground water moves toward the pumping well is called the *zone of contribution* (ZOC) of the well. Only pollutants introduced within the ZOC can contribute to contamination of the pumped water. Although regulators must be concerned with all pollutant sources within the ZOC, the level of protection necessarily varies with the distance from the well. In the immediate vicinity of the well, within tens of meters, protection is most stringent and all pollutant sources are typically prohibited; it is not unusual for this area to be fenced and have controlled access. This area is sometimes called the *remedial action zone*. Beyond this area, land uses that generate a limited amount of ground-water contamination, or potential for ground-water contamination, are typically permitted; however, sufficient distance to the well is provided such that the pollutant concentrations are attenuated to acceptably low levels by the time they reach the well. This area, where contaminant sources are controlled by regulatory permits, is commonly called the *attenuation zone*. Beyond the attenuation zone, but still within the ZOC, contaminant sources are unlikely to have any significant impact on the pumped water and regulatory burdens are reduced. This outer zone is commonly called the *wellfield management zone*. The primary factors to be considered in delineating WHPAs are the types and amounts of contaminants that could possibly be introduced into the ZOC and the attenuation characteristics of the contaminants in the subsurface environment.

The types of contaminants that are of most concern are organic chemicals and viruses. Organic chemicals are used as industrial solvents and pesticides, and they are the major component of gasoline, which is usually stored in underground storage tanks at gas stations. Viruses in ground water originate primarily from septic tank effluent, on-site domestic wastewater treatment systems, and broken sewer pipes. The attenuation of various contaminants in ground water occurs primarily via the processes of *dispersion*, *decay*, and *sorption*; models that describe these processes are discussed in detail in Chapter 8. Criteria commonly used for delineating WHPAs are (1) distance, (2) drawdown, (3) time of travel, (4) flow boundaries, and (5) assimilative capacity. An assessment of these criteria, and typical threshold values, are given in Table 7.21. The specific criteria selected to delineate a WHPA depends on a variety of considerations, including overall protection goals, technical considerations, and policy considerations.

The most commonly used approach is the time of travel (TOT) approach, which relates locations within the WHPA to the times of travel from those locations to the water-supply well. To define the level of protection of the water supply from a particular contaminant, the time of travel is contrasted with the time for the contaminant to undergo an acceptable amount of attenuation. The time of travel is sometimes associated with the time that is available to clean up a spill within the WHPA. In cases where decay or biotransformation is the dominant mode of attenuation, reductions in contaminant concentration are commonly described by the first-order process

$$\frac{dc}{dt} = -\lambda c \tag{7.340}$$

where c is the contaminant concentration in ground water, t is the time since release of

■ Table 7.21 WHPA Delineation Criteria	Criteria	Assessment	Typical thresholds
	Distance	Does not directly incorporate contaminant fate and transport processes. Commonly an arbitrary policy decision.	300 m–3 km
	Drawdown	Does not directly incorporate contaminant fate and transport processes. Defines areas where the influence of pumping is the same.	3 cm–30 cm
	Time of travel	Incorporates advection and decay processes, neglects dispersion. Widely used criteria.	5 y–50 y
	Flow boundaries	Highest level of protection, boundary of WHPA is taken as the ZOC boundary. Not practical in most cases.	physical and hydrologic
	Assimilative capacity	Incorporates all significant contaminant fate and transport processes. A rational approach to wellhead protection. Requires much technical expertise, rarely used.	requires drinking water standards to be met at well

Source: USEPA (1987).

the contaiminant, and λ is the decay parameter. Equation 7.340 can be solved to yield

$$\boxed{\frac{c}{c_o} = e^{-\lambda t}} \tag{7.341}$$

where c_o is the initial concentration at $t = 0$. Equation 7.341 gives the attenuation of contaminant concentration as a function of travel time. Therefore, if a pollutant source is expected to contaminate the ground water at a concentration c_o and the allowable concentration at the well is c, the minimum allowable time of travel to the well can be calculated using Equation 7.341. If there are several contaminant sources, the concentration, c, in the pumped water is equal to the sum of the concentrations contributed by each contaminant source. A useful parameter to characterize the time scale of decay is the *half-life*, T_{50}, which is the time for the concentration to decay to one-half of its original value. Equation 7.341 gives

$$T_{50} = \frac{0.693}{\lambda} \tag{7.342}$$

Time-of-travel contours surrounding single wells can be derived using the continuity relation

$$Q_w t = \pi r^2 b n \tag{7.343}$$

where Q_w is the well pumping rate, t is time, r is the distance traveled in time t, b is the aquifer thickness, and n is the porosity. Equation 7.343 can be put in the form

$$\boxed{r = \sqrt{\frac{Q_w t}{\pi b n}}} \tag{7.344}$$

which gives the radial distance from the well that has a travel time t. In the case of nonuniform aquifers and more complex multiwell scenarios, numerical ground-water models can be used to calculate travel times.

■ Example 7.37

A WHPA is to be delineated around a municipal water-supply well that pumps 0.35 m³/s from a 25-m thick aquifer with a porosity of 0.2. The contaminant of concern is viruses from residential septic tanks. Residential development is to be permitted on 2,000-m² lots, and the viral concentration under each lot resulting from septic tank effluent is expected to be 50/L. A risk analysis indicates that the maximum allowable viral concentration in the pumped water is 0.01/L, and the decay constant, λ, for viruses can be taken as 0.5 d⁻¹ (Yates, 1987). If there is to be 1 km² of residential development surrounding the water-supply well, estimate the boundary of the WHPA within which no residential development should be allowed.

Solution

The number of 2,000-m² residential lots in 1 km² is

$$\frac{1 \times 10^6}{2000} = 500 \text{ lots}$$

The allowable viral concentration at the well is 0.01/L and the maximum contribution from each lot is $0.01/500 = 2 \times 10^{-5}$/L. Since $c_o = 50$/L and $\lambda = 0.5$ d⁻¹, Equation

7.341 requires that

$$\frac{c}{c_o} = e^{-\lambda t}$$

$$\frac{2 \times 10^{-5}}{50} = e^{-0.5t}$$

which gives $t = 29.5$ days. The radial distance, r, corresponding to a travel time of 29.5 days ($= 2.55 \times 10^6$ seconds) is given by Equation 7.344 as

$$r = \sqrt{\frac{Q_w t}{\pi b n}}$$

$$= \sqrt{\frac{(0.35)(2.55 \times 10^6)}{\pi(25)(0.2)}} = 238 \text{ m}$$

Therefore, prohibiting residential development within a radius of 238 m of the well will provide a satisfactory level of protection from viral contamination. Ground-water monitoring is recommended to verify assumptions regarding viral concentrations in the ground water under septic tanks.   ■

The delineation of WHPAs based on travel time is appropriate when decay is the dominant attenuation process, contaminant dispersion is negligible, and the aquifer and pollutant source parameters are known with a reasonable degree of certainty. In cases where these conditions are not met, more comprehensive WHP models (e.g., Chin and Chittaluru, 1994) are recommended.

7.8.4 ■ Design of Aquifer Pumping Tests

Aquifer pumping tests are used to determine the hydraulic properties of aquifers. The methodology consists of matching field measurements of drawdowns caused by pumping wells with the corresponding theoretical drawdowns and determining the hydraulic properties of the aquifer that produce the best fit. The details of several analytic models for analyzing pump-test data in a variety of field scenarios have been presented in Section 7.3. Aside from analytic techniques for processing the measured data, several operational issues must be addressed to properly conduct aquifer tests. These operational procedures are described in some detail by several standards of the American Society of Testing Materials (ASTM-D5092, 1990; ASTM-D4043, 1991a; ASTM-D4050, 1991b; ASTM-D4106, 1991c), and have also been described in detail by the U.S. Environmental Protection Agency (Osbourne, 1993). Aquifer tests are usually conducted using a pumping well and at least one observation well. The key elements in designing an aquifer pumping test at any site are (1) number and location of observation wells, (2) design of observation wells, (3) approximate duration of the test, and (4) discharge rate from the pumping well. Pumping tests are usually quite expensive, and the installation of a pumping well and surrounding observation wells are only justified in cases where exploitation of the aquifer by wells at the site is being contemplated. In most cases, the pumping well is subsequently utilized as a production well.

The *lithology* of an aquifer describes the physical makeup of the aquifer, including the mineral composition, grain size, and grain packing of sediments or rocks that constitute the aquifer. Prior to conducting an aquifer test, basic data on the aquifer must be collected. These data must include, if possible, the depth, thickness, areal extent, and lithology of the aquifer, the locations of aquifer discontinuities caused by changes in lithology, the locations of surface water bodies, preliminary estimates of the transmissivity and storage coefficient of the aquifer, and preliminary estimates of leakage coefficients if semiconfining layers are present. These data facilitate the design of the discharge rate from the pumping well as well as aid in the location of observation wells. Preliminary values of the transmissivity and storage coefficient of an aquifer can be estimated by conducting slug tests (see Section 7.8.5) on wells near the site, but such tests have no more than order-of-magnitude accuracy (Osbourne, 1993). It is advisable to use existing wells to conduct aquifer tests whenever possible; however, care should be taken that existing wells are properly constructed and developed and that these wells are screened in the same aquifer zone as the one being investigated.

7.8.4.1   Design of Pumping Well

The design of the pumping well includes consideration of (1) well construction, (2) well development procedure, (3) well acccess for water level measurements, (4) a reliable power source, (5) type of pump, (6) discharge control and measurement equipment, and (7) method of waste disposal.

Well Construction.   The diameter of the pumping well must be large enough to accommodate both the test pump and space for water level measurement. Guidelines for casing diameter given in Table 7.13 are recommended. The well screen must have sufficient open area to minimize local well losses; guidelines given in Table 7.14 are recommended. If the well is located in an unconsolidated aquifer, a gravel pack should be placed in the annular region between the well screen and the perimeter of the borehole. The gravel pack should extend at least 30 cm above the top of the well screen and be designed in accordance with the guidelines given in Section 7.8.2. A seal of bentonite pellets should be placed on top of the gravel pack. A minimum of 1 m of pellets should be used, and an annulus seal of cement and/or bentonite grout should be placed on top of the bentonite pellets. The well casing should be protected at the surface with a concrete pad around the well to isolate the wellbore from surface runoff.

Well Development.   Pumping wells should be adequately developed to ensure that well losses are minimized. See Section 7.8.2 for a thorough discussion of well losses. If the well is suspected to be poorly developed or nothing is known, it is advisable to conduct a step-drawdown test to determine the magnitude of the well losses.

Water-Level Measurement Access.   It must be possible to measure the depth of water in the pumping well before, during, and after pumping. Usually, electric-sounder or pressure-transducer systems are used.

Reliable Power Source.   Power must be continuously available to the pump during the test. A power failure during the test usually requires that the test be terminated and sufficient time permitted for water levels to stabilize. This can cause expensive and time-consuming delays.

Pump Selection.   Electrically powered pumps produce the most constant discharge and are recommended for use during an aquifer test. The discharge of engine-powered (gasoline or diesel) pumps may vary greatly over a 24-h period and requires more frequent monitoring during the pump test. According to Osbourne (1993), a diesel-powered turbine pump may have more than a 10% variation in discharge as a result of daily variations in temperature.

Discharge-Control and Measurement Equipment.   Common methods of measuring well discharge include orifice plates and manometers, inline flow meters, inline calibrated pitot tubes, weirs or flumes, and (for low discharge rates) measuring the time taken to discharge a measured volume. The discharge of wells yielding less than 400 L/min can be readily measured with sufficient accuracy using a calibrated bucket or drum and a stopwatch (USBR, 1995). An important pump characteristic is that as the pump lift increases, the discharge decreases for a pump running at a constant speed. Therefore, the pump speed will usually need to be increased during the test to maintain a constant discharge.

Water Disposal.   The volume of pumped water to be produced, the storage requirements, disposal alternatives, and any treatment needs must be assessed during the planning phase. Clearly, the pumped water cannot be allowed to infiltrate back into the aquifer during the pump test. If the aquifer is unconfined and the unsaturated zone overlying the aquifer is relatively permeable, the pumped water should be transported by pipeline to a location beyond the area of influence that will develop during the pump test. If the water table is more than 30 m below the ground surface and the unsaturated zone has a low hydraulic conductivity, an open ditch may be used to transport the pumped water (USBR, 1995).

7.8.4.2   Design of Observation Wells

If existing wells are to be used as observation wells, then it should be verified that these wells are screened in the aquifer being investigated and that the screens are not clogged due to the buildup of iron compounds, carbonate compounds, sulfate compounds, or bacterial growth. The response test (Stallman, 1971; Black and Kipp, 1977) is recommended if existing wells are to be used as observation wells. In addition, the following characteristics of observation wells should be considered:

Well Diameter.   The well casing should be just large enough to allow for accurate, rapid water-level measurements. Well casings 50 mm (2 in.) in diameter are usually adequate in aquifers less than 30 m deep, but they are often difficult to develop. Well casings 100 to 150 mm (4 to 6 in.) in diameter are easier to develop and should have a better aquifer response. If a water-depth recorder is to be used, then well casings of 100 to 150 mm will usually be required. Difficulties in drilling a straight hole usually dictate that a well over 60 m deep must be at least 100 mm in diameter. Wells with diameters larger than 150 mm may cause a lag in response time due to water held in casing storage, so smaller diameters are usually preferable (Lehr et al., 1988).

Well Construction.   Observation wells should ideally have 1.5 to 6 m of perforated casing or screening near the bottom of the well. In addition, the placement of the gravel pack, bentonite pellets, cement or bentonite grout, and a concrete pad should follow the same guidelines as for the pumping well. After installation, observation wells should be developed by surging with a block, and/or submersible pump for a sufficient period (usually several hours) to meet a predetermined level of turbidity (Campbell and Lehr, 1972; Driscoll, 1986).

Distance from Pumped Well.   Single observation wells are generally located about three to five times the aquifer thickness away from the pumping well, with observation wells in unconfined aquifers located closer to the pumping well than observation wells in confined aquifers (Lehr et al., 1988). This usually works out to a distance of

10 m to 100 m (Boonstra, 1998). If the pumping well is partially penetrating, observation wells should be located at a minimum distance equal to one and a half to two times the aquifer thickness from the pumping well (USBR, 1995). At least three observation wells at different distances from the pumped well are desirable, so that results can be averaged and obviously erroneous data can be disregarded. Whenever multiple observation wells are used, they are typically placed in a straight line or along perpendicular rays originating at the pumped well. If aquifer anisotropy is expected, then observation wells should be located in a pattern based on the suspected or known anisotropic conditions at the site. If the principal directions of anisotropy are not known, then at least three wells on different rays are required. If the aquifer is vertically anisotropic, then the minimum distance, r_{min}, of an observation well from the pumping well should be taken as (USBR, 1995)

$$r_{min} = 1.5b \left(\frac{K_h}{K_v} \right)^{\frac{1}{2}} \tag{7.345}$$

where b is the saturated thickness of the aquifer, K_h is the horizontal hydraulic conductivity, and K_v is the vertical hydraulic conductivity. Observation wells should be located far enough away from geologic and hydraulic boundaries to permit recognition of drawdown trends before the boundary conditions influence the drawdown readings.

7.8.4.3   Field Procedures

Aside from the design of the pumping and observation wells, there are several operational guidelines to be considered.

Establishment of Baseline Conditions.   Prior to the initiation of a pump test, it is essential to monitor the water levels in the pumping and observation wells in addition to wells adjacent to the site. These measurements will indicate whether a measurable trend exists in the ground-water levels. In addition, the influence and scheduling of offsite pumping on the aquifer test must be assessed, and controlled if necessary. As a general rule, at least one week of observations must be available prior to the initiation of the aquifer test.

Water-Level Measurements During Test.   Immediately before pumping is to begin, static water levels in all test wells should be recorded. The recommended time intervals for recording water levels during an aquifer test are given in Table 7.22 (after ASTM Committee D-18, D 4050). After pumping is terminated, recovery measurements should be taken at the same time intervals as listed in Table 7.22, where the times are measured from the instant the pump is turned off. A check valve should be used to prevent backflow after pumping is terminated, and recovery measurements should continue until the

▪ **Table 7.22**
Recommended Time Intervals for Water-Level Measurement

Time since beginning of test	Measurement interval
0–3 min	every 30 s
3–15 min	every 1 min
15–60 min	every 5 min
60–120 min	every 10 min
120 min–10 h	every 30 min
10 h–48 h	every 4 h
48 h–shutdown	every 24 h

pre-pumping state is recovered. The drawdown, s', after pumping is terminated in homogeneous isotropic aquifers can be approximated by (temporal) superposition of the Theis equation (Equation 7.226) as

$$s' = \frac{Q_w}{4\pi T}\left[W(u) - W(u')\right] \tag{7.346}$$

where Q_w is the constant pumping rate during the pump test, T is the aquifer transmissivity, $W(u)$ is the well function, and

$$u = \frac{r^2 S}{4Tt} \quad \text{and} \quad u' = \frac{r^2 S'}{4Tt'} \tag{7.347}$$

where r is the distance of the observation well from the pumping well, S is the storage coefficient during drawdown, S' is the storage coefficient during recovery, t is the time since pumping began, and t' is the time since pumping stopped. For small values of u and u' (u, $u' < 0.004$), the Cooper-Jacob approximation to the well function (Equation 7.228) can be applied to Equation 7.346, yielding

$$s' = \frac{Q_w}{4\pi T} \ln \frac{S't}{St'} \tag{7.348}$$

or

$$\boxed{s' = \frac{Q_w}{4\pi T}\left(\ln\frac{t}{t'} + \ln\frac{S'}{S}\right)} \tag{7.349}$$

This equation is a linear relationship between s' and $\ln t/t'$. Matching Equation 7.349 to the recovery data yields a slope of $Q_w/4\pi T$ and an intercept of $Q_w/4\pi T \ln S'/S$. Knowing the pumping rate, Q_w, and the storage coefficient during the drawdown test, S, then the aquifer transmissivity, T, and recovery storage, S', can be estimated from the slope and intercept of recovery data. The estimated value of T can be used to confirm the transmissivity determined from the pump-drawdown measurements; however, no independent confirmation of the drawdown storage coefficient is obtained from this analysis. In some cases, S' is assumed to be equal to S (e.g., de Marsily, 1986), in which case the transmissivity of the aquifer is all that is extracted from the recovery data.

Discharge-Rate Measurements.   During the initial hour of the aquifer test, the discharge from the pumping well should be measured as frequently as practical, and it is important to bring the discharge rate up to the design rate as quickly as possible. Because of the variety of environmental factors that can affect the discharge rate from engine-driven pumps, the discharge should be checked four times per day, preferably early morning, mid-morning, mid-afternoon, and early evening. Osbourne (1993) has indicated that a 10% variation in discharge can result in a 100 percent variation in the estimate of aquifer transmissivity, and it is recommended that the discharge should never be allowed to vary by more than 5% during an aquifer test.

Length of Test.   The test should be of sufficient length to accurately identify the shape of the drawdown versus time curve from which the aquifer hydraulic properties are extracted. The drawdown curve should be plotted during the pump test. As a general guideline, when three or more drawdown readings taken at one-hour intervals at the most distant well fall on a straight line (on the log drawdown versus log time curve), the pump test can be terminated (Lehr et al., 1988).

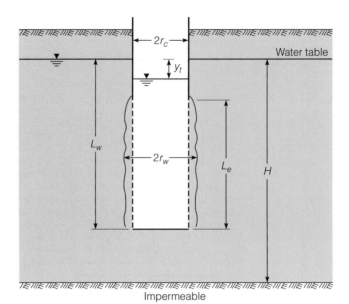

Figure 7.32 ■ Field Setup for a Slug Test

Source: Bouwer and Rice (1976).

7.8.5 ■ Slug Test

A *slug test* is commonly used to measure the hydraulic conductivity in the immediate vicinity of a well. The field setup for a slug test is illustrated in Figure 7.32, where a well with casing radius r_c extends to a depth L_w below the water table in an unconfined aquifer of thickness H. The well intake (screen) has a length L_e and effective radius r_w. The effective radius is generally greater than the actual radius of the intake and includes the high-permeability region surrounding the well caused by either well development or the placement of a gravel pack. The slug test is performed by instantaneously removing a volume V of water from the well (a "slug" of water), and relating the recovery of the water level in the well to the hydraulic conductivity of the surrounding aquifer. The instantaneous removal of water can be accomplished using a bailer (a type of bucket), or by submerging a closed cylinder in the well, letting the water reach equilibrium, and then quickly pulling the cylinder out.

A variety of analytical techniques have been proposed to relate aquifer properties to the recovery data (Cooper et al., 1967; Hvorslev, 1951; Bouwer and Rice, 1976; Nguyen and Pinder, 1984), and a comparison of these methods can be found in Herzog (1994). The most widely used slug-test procedure, applicable to fully or partially penetrating wells in unconfined aquifers, was developed by Bouwer and Rice (1976) and subsequently updated by Bouwer (1989). The basis of the Bouwer and Rice method is the Thiem equation (Equation 7.143), which can be put in the form

$$y = \frac{Q_w}{2\pi K L_e} \ln\left(\frac{R_e}{r_w}\right) \tag{7.350}$$

where y is the drawdown at the well, Q_w is the rate at which ground water is removed from the aquifer, k is the hydraulic conductivity of the aquifer, and R_e is the effective radial distance over which the head y is dissipated. Continuity requires that the flow into the well, Q_w, be equal to the rate at which the volume of water in the well increases, therefore

$$\frac{dy}{dt} = -\frac{Q_w}{\pi r_c^2} \tag{7.351}$$

Combining Equations 7.350 and 7.351 (eliminating Q_w), integrating, and solving for K yields

$$K = \frac{r_c^2 \ln(R_e/r_w)}{2L_e} \frac{1}{t} \ln \frac{y_o}{y_t} \tag{7.352}$$

where y_o is the initial drawdown in the well and y_t is the drawdown at time t. Using Equation 7.352 to estimate the hydraulic conductivity, K, requires that r_c, r_w, and L_e be known from the dimensions of the well, and t, y_o, and y_t be measured during the recovery of the water level in the well. The effective radial distance, R_e, can be estimated using the following empirical relation (Bouwer and Rice, 1976)

$$\ln \frac{R_e}{r_w} = \left\{ \frac{1.1}{\ln(L_w/r_w)} + \frac{A + B\ln[(H - L_w)/r_w]}{(L_e/r_w)} \right\}^{-1} \qquad \text{(partially penetrating well)}$$

$$\tag{7.353}$$

where A and B are dimensionless parameters that are related to L_e/r_w, as shown in Figure 7.33. Analyses by Bouwer and Rice (1976) have indicated that if $\ln[(H - L_w)/r_w] > 6$, a value of 6 should be used for this term in Equation 7.353. If $H = L_w$, the well is fully penetrating, the term $\ln[(H - L_w)/r_w]$ given in Equation 7.353 is indeterminate, and the following equation should be used to estimate R_e

$$\ln \frac{R_e}{r_w} = \left\{ \frac{1.1}{\ln(L_w/r_w)} + \frac{C}{(L_e/r_w)} \right\}^{-1} \qquad \text{(fully penetrating well)} \tag{7.354}$$

where C is a dimensionless parameter related to L_e/r_w, as shown in Figure 7.33. Values of $\ln(R_e/r_w)$ are within 10% of experimental values when $L_e/L_w > 0.4$, and within 25% of experimental values when $L_e/L_w < 0.2$ (Bouwer, 1978). It should be noted that the analysis presented here applies equally well to cases in which a "slug" of water is instantaneously added to the well at the beginning of the test (Batu, 1998). Values

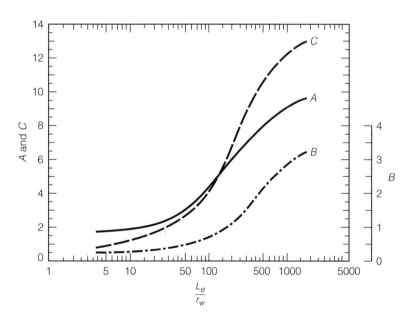

Figure 7.33 ■ Parameters in Slug Test Analysis
Source: Bouwer and Rice (1976).

of K derived from a slug test primarily measure the average horizontal component of the hydraulic conductivity over a hollow cylinder of inner radius r_w, outer radius R_e, and height slightly larger than L_e. Because of the relatively small portion of the aquifer sampled by slug tests, it is not unusual for the hydraulic conductivity values derived from slug tests at closely spaced wells to differ by several orders of magnitude (Gorelick et al., 1993). Part of the reason for this is that hydraulic conductivities derived from slug tests are highly sensitive to minor variations in well-construction details, such as the screen position and the dimensions of the gravel pack. Slug tests are particularly useful in investigating localized flow conditions, which have an important effect on contaminant transport in ground water.

Data Analysis.   Data collected during slug tests consist primarily of the drawdown, y_t, as a function of time, t, and a methodology for combining these data with Equation 7.352 to estimate the hydraulic conductivity is as follows:

1. Plot the measured data of $\ln y_t$ versus t.
2. According to Equation 7.352, the plot in Step 1 should be a straight line. If some portion of the data does not show a linear relation, then only the linear portion is used in the analysis.
3. Fit a straight line through the linear portion of the curve identified in Step 2. The equation of this line can be written as

$$\ln y_t = mt + \ln y_o \qquad (7.355)$$

Determine the slope, m, of the fitted line. Equation 7.355 also indicates that this slope is given by

$$m = -\frac{1}{t} \ln \frac{y_o}{y_t} \qquad (7.356)$$

4. Based on the known values of L_w and H, select the equation to be used for $\ln(R_e/r_w)$ from Equations 7.353 and 7.354.
5. If Equation 7.353 is to be used, determine the values of A and B from Figure 7.33. If Equation 7.354 is to be used, determine the value of C from Figure 7.33.
6. Calculate the value of $\ln(R_e/r_w)$ using the results of step 5.
7. Calculate the hydraulic conductivity, K, using Equation 7.352 with the values of m, $\ln(R_e/r_w)$, r_c, and L_e, where

$$K = -\frac{r_c^2 \ln(R_e/r_w)}{2L_e} m \qquad (7.357)$$

The application of this analysis is illustrated in the following example.

■ Example 7.38

A slug test is conducted in a monitoring well that penetrates 5 m below the water table in a 15-m thick surficial (unconfined) aquifer. The well casing has a 100-mm diameter, and the bottom 3 m of the well is screened and surrounded by a 100-mm thick gravel pack. The water level in the well is instantaneously drawn down 1,500 mm, and the observed time-drawdown relation during recovery is given in the following table:

Time (s)	0	2	4	6	8	10	12	14	16
Drawdown (mm)	1,500	645	372	195	130	67	44	23	15

Estimate the hydraulic conductivity and transmissivity of the aquifer.

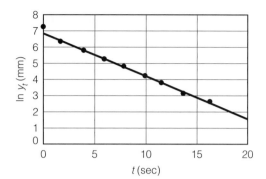

Figure 7.34 ■ Analysis of Slug Test Data

Solution

The measured data of $\ln y_t$ versus t is plotted in Figure 7.34. These data indicate that except for the first data point ($t = 0$, $y_t = 1{,}500$ mm), the relationship between $\ln y_t$ and t appears quite linear and can be fitted to the regression equation

$$\ln y_t = -0.270t + 6.976$$

which has a slope, m, of -0.270. From the given dimensions of the well and aquifer, $L_w = 5$ m, $H = 15$ m, $r_c = 50$ mm, $L_e = 3$ m, $r_w = r_c + 100$ mm $= 150$ mm, and $L_e/r_w = 3/0.15 = 20$. Equation 7.353 is appropriate for calculating $\ln(R_e/r_w)$ (since the well is partially penetrating), and the dimensionless parameters A and B are given by Figure 7.33 as

$$A = 2.1, \qquad B = 0.25$$

Substituting these data into Equation 7.353 yields

$$\ln \frac{R_e}{r_w} = \left\{ \frac{1.1}{\ln(L_w/r_w)} + \frac{A + B\ln[(H - L_w)/r_w]}{(L_e/r_w)} \right\}^{-1}$$

$$= \left\{ \frac{1.1}{\ln(5/0.15)} + \frac{2.1 + 0.25\ln[(15 - 5)/0.15]}{(3/0.15)} \right\}^{-1} = 2.12$$

and putting this result into Equation 7.357 gives

$$K = -\frac{r_c^2 \ln(R_e/r_w)}{2L_e} m$$

$$= -\frac{(0.05)^2(2.12)}{2(3)}(-0.270) = 2.39 \times 10^{-4} \text{ m/s} = 20.6 \text{ m/d}$$

This result indicates an average (horizontal) hydraulic conductivity of 20.6 m/d in the immediate vicinity of the well. The transmissivity, T, of the aquifer can be estimated by

$$T = KH = (20.6)(15) = 309 \text{ m}^2/\text{d} \qquad\blacksquare$$

Although the Bouwer and Rice (1976) slug test analysis was originally designed for unconfined aquifers, the test can also be used in confined or stratified aquifers if the top of the screen is some distance below the upper confining layer (Bedient et al., 1994).

7.8.6 ■ Design of Exfiltration Trenches

Exfiltration trenches, which are also called *french drains* and *percolation trenches*, are commonly used to discharge stormwater runoff into the subsurface and to return treated water from pump and treat systems into the ground water. A typical exfiltration trench is illustrated in Figure 7.35. The operational characteristics of exfiltration trenches are influenced by the trench hydraulic conductivity, K_t, which is defined as the flow across an aquifer interface per unit area per unit difference in piezometric head (across the surface). Trench hydraulic conductivities are commonly determined from slug tests. The functional characteristics of an exfiltration trench depend on whether the trench is discharging water at a constant rate, such as in the case of a pump and treat system, or discharging surface-water runoff from a design storm, as in the case of stormwater management applications. The design of a constant-flow exfiltration trench is described below, and the design of an exfiltration trench for stormwater management applications is described in Chapter 6, Section 6.7.2.3.

Constant-Flow Design.   Constant-flow exfiltration trenches are commonly designed assuming that the water in the trench is exfiltrated through both the bottom and the long sides of the trench. The outflow from the bottom of the trench, Q_b, is given by Darcy's law as

$$Q_b = K_t WL \tag{7.358}$$

where K_t is the trench hydraulic conductivity, W is the width of the trench, L is the length of the trench, and the hydraulic gradient is assumed equal to unity. The outflow from each side of the trench, Q_s, is given by Darcy's law as

$$Q_s = K_t A_{\text{perc}} \tag{7.359}$$

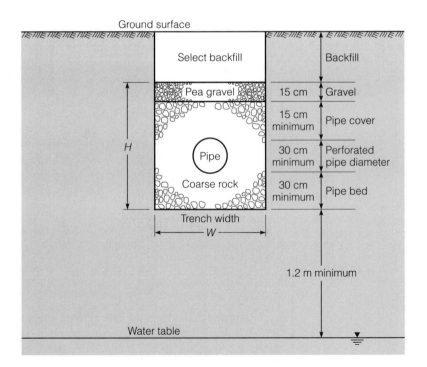

Figure 7.35 ■ Typical Exfiltration Trench

where A_{perc} is the side area through which water exfiltrates and the hydraulic gradient is taken as unity. It is common to assume that water exfiltrates through one-half of the trench height, H, in which case

$$A_{\text{perc}} = \frac{1}{2}LH \tag{7.360}$$

and Equations 7.359 and 7.360 combine to give

$$Q_s = \frac{1}{2}K_t HL \tag{7.361}$$

The total outflow rate from the exfiltration trench, Q, is equal to the sum of the flows out of the two (long) sides plus the flow out of the bottom. Therefore

$$Q = Q_b + 2Q_s$$
$$= K_t WL + 2\left(\frac{1}{2}K_t HL\right) = K_t L(W + H) \tag{7.362}$$

Solving this equation for the trench length, L, yields

$$\boxed{L = \frac{Q}{K_t(W + H)}} \tag{7.363}$$

The objective of the trench design is to select a length, L, width, W, and height, H, of the exfiltration trench, where and the trench conductivity, K_t, and the outflow rate, Q, are fixed by the site and design conditions. The maximum trench depth is limited by trench-wall stability, seasonal high ground-water level, and the depth to any impervious soil layer. Exfiltration trenches 1 m wide and 1 to 2 m deep seem to be most efficient (ASCE, 1998). Adequate ground water (quality) protection is generally obtained by providing at least 1.2 m separation between the bottom of the trench and the seasonal high water table.

After selecting the geometry of the exfiltration trench, the aggregate mix to be placed in the trench surrounding the perforated pipe must be selected such that the hydraulic conductivity of the aggregate mix significantly exceeds the hydraulic conductivity of the porous matrix surrounding the trench. The hydraulic conductivities of various aggregate mixes have been measured by Cedergren and colleagues (1972), who showed that the hydraulic conductivity of any gravel mix can be expected to exceed 4,000 m/d, which should be adequate for practically all exfiltration trench designs. Consequently, any commercially available gradation of gravel can be used as to surround the perforated pipe in the exfiltration trench. The perforated pipe used in exfiltration trenches usually consists of corrugated steel pipe with approximately 320 perforations per square meter; the diameter of the perforations are typically 0.95 mm (American Iron and Steel Institute, 1995).

After designing an exfiltration trench capable of injecting the design flow, Q, into the aquifer, the final step is to verify that the aquifer is capable of transporting the recharge water away from the trench at a sufficient rate to prevent the ground water from mounding to within a specified depth below the trench. The mounding of ground water under a rectangular recharge area of length L and width W has been investigated by Hantush (1967), who derived the following expression for the maximum aquifer thickness, h_m, under a rectangular recharge area as a function of time

$$h_m^2(t) = h_i^2 + \frac{2N}{K} v t S^* \left(\frac{W}{\sqrt{8vt}}, \frac{L}{\sqrt{8vt}} \right) \qquad (7.364)$$

where h_i is the aquifer thickness without recharge; N is the recharge rate, which can be expressed in terms of the design discharge, Q, by the relation

$$N = \frac{Q}{LW} \qquad (7.365)$$

and K (in Equation 7.364) is the hydraulic conductivity of the aquifer; v is a combination of terms given by

$$v = \frac{Kb}{S_y} \qquad (7.366)$$

where b is a mean aquifer thickness, somewhere between h_i and $h_m(t)$ and assumed to be constant; S_y is the specific yield; t is the time since the beginning of recharge; and $S^*(\alpha, \beta)$ is a function defined by

$$S^*(\alpha, \beta) = \int_0^1 \text{erf}\left(\frac{\alpha}{\sqrt{\tau}} \right) \text{erf}\left(\frac{\beta}{\sqrt{\tau}} \right) d\tau \qquad (7.367)$$

where $\text{erf}(\xi)$ is the error function. General evaluation of the function $S^*(\alpha, \beta)$ requires numerical integration of Equation 7.367; however, values of the function for a range of α and β are presented in Table 7.23. The design of an exfiltration trench is illustrated by the following example.

■ Example 7.39

Design an exfiltration trench to inject 207 m^3/d from a pump-and-treat system. Several slug tests at the site have yielded a minimum trench hydraulic conductivity of 24 m/d. The depth from the ground surface to the seasonal high-water table is 4.35 m, and the aquifer is estimated to have a hydraulic conductivity of 107 m/d, a thickness of 10.7 m, and a specific yield of 0.2. Local regulations require a factor of safety of 2 in the assumed trench hydraulic conductivity and a minimum backfill of 50 cm.

■ **Table 7.23**
S^* Function

α \ β	.00	.30	.60	.90	1.2	1.5	1.8	2.1	2.4	2.7	3.0
.00	.0000	.0000	.0000	.0000	.0000	.0000	.0000	.0000	.0000	.0000	.0000
.30	.0000	.3009	.4314	.4860	.5070	.5142	.5164	.5170	.5171	.5172	.5172
.60	.0000	.4314	.6426	.7360	.7729	.7857	.7897	.7907	.7909	.7910	.7910
.90	.0000	.4860	.7360	.8504	.8966	.9129	.9180	.9193	.9196	.9197	.9197
1.20	.0000	.5070	.7729	.8966	.9472	.9653	.9709	.9724	.9728	.9728	.9728
1.50	.0000	.5142	.7857	.9129	.9653	.9841	.9900	.9915	.9919	.9920	.9920
1.80	.0000	.5164	.7897	.9180	.9709	.9900	.9959	.9975	.9979	.9979	.9979
2.10	.0000	.5170	.7907	.9193	.9724	.9915	.9975	.9991	.9995	.9995	.9995
2.40	.0000	.5171	.7909	.9196	.9728	.9919	.9979	.9995	.9998	.9999	.9999
2.70	.0000	.5172	.7910	.9197	.9728	.9920	.9979	.9995	.9999	1.0000	1.0000
3.00	.0000	.5172	.7910	.9197	.9728	.9920	.9979	.9995	.9999	1.0000	1.0000

Solution

The length, L, of the trench is given by Equation 7.363 as

$$L = \frac{Q}{K_t(W+H)} \qquad (7.368)$$

where the dimensions H and W are illustrated in Figure 7.35. Following ASCE (1998) guidelines, specify $W = 1$ m (a typical backhoe dimension), and $H = 2$ m (a typical depth for vertical slope stability). The design injection rate, Q, is 207 m³/d, and the design trench hydraulic conductivity, K_t, is 12 m/d, equal to the minimum measured value of 24 m/d divided by the factor of safety of 2. Substituting these values for Q, K_t, W, and H into Equation 7.368 yields

$$L = \frac{Q}{K_t(W+H)} = \frac{207}{12(1+2)} = 5.75 \text{ m} \qquad (7.369)$$

A trench dimension of 6 m long by 1 m wide by 2 m deep, when filled with gravel, will be capable of transferring water into the aquifer at a rate of 207 m³/d. Since the seasonal high-water table is 4.35 m below the ground surface, there is sufficient room to install a 2-m deep trench covered with 0.5 m ($= 50$ cm) of backfill and still maintain a distance of at least 1.2 m between the bottom of the trench and the seasonal high-water table.

The next question is whether the aquifer will be able to transport the effluent away from the trench as fast as it is supplied, without causing the water table to rise to within 1.2 m of the bottom of the trench. The height of the water table above the base of the aquifer as a function of time is given by

$$h_m^2(t) = h_i^2 + \frac{2N}{K} \nu t S^* \left(\frac{W}{\sqrt{8\nu t}}, \frac{L}{\sqrt{8\nu t}} \right) \qquad (7.370)$$

where $h_i = 10.7$ m and N is given by

$$N = \frac{Q}{LW} = \frac{207}{(6)(1)} = 34.5 \text{ m/d}$$

The hydraulic conductivity of the aquifer, K, is 107 m/d, and the parameter ν in Equation 7.370 is given by

$$\nu = \frac{Kb}{S_y} = \frac{107(10.7)}{0.2} = 5{,}725 \text{ m}^2\text{/d}$$

Substituting the trench dimensions ($L = 6$ m, $W = 1$ m) and aquifer properties into Equation 7.370 yields

$$h_m^2 = 10.7^2 + \frac{2(34.5)}{107}(5725)tS^* \left(\frac{1}{\sqrt{(8)(5725)t}}, \frac{6}{\sqrt{(8)(5725)t}} \right)$$

$$= 114 + 3692tS^* \left(\frac{0.00467}{\sqrt{t}}, \frac{0.0280}{\sqrt{t}} \right) \qquad (7.371)$$

This equation relates the maximum water surface elevation below the trench to the time since the trench began operation. Values of h_m for several values of t are shown in Table 7.24, where $S^*(\alpha, \beta)$ has been estimated from the tabulated values shown in Table 7.23. The values shown in Table 7.24 indicate that the water table under the trench quickly stabilizes at an aquifer thickness of 10.8 m. This indicates that the water table below the exfiltration trench will rise (mound) by about 0.1 m ($= 10$ cm) and will

	t (days)	h_m (m)
■ **Table 7.24** h_m versus t	1	10.8
	10	10.8
	100	10.8
	1,000	10.8
	10,000	10.8

remain at an acceptable depth (> 1.2 m) below the trench. The final trench design is illustrated in Figure 7.36.   ■

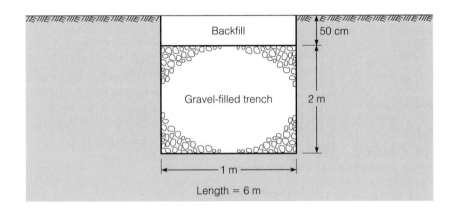

Figure 7.36 ■
Trench Design

7.9   Computer Models

Several good computer models are available for simulating ground-water flow. In their most general form, these models provide numerical solutions to the combined Darcy and continuity equations, subject to prescribed boundary conditions. Several practical applications of ground-water flow models are listed in Table 7.25. In choosing a model for a particular application, there are usually a variety of models to choose from; however, in doing work that is to be reviewed by regulatory bodies, models developed and maintained by agencies of the U.S. government have the greatest credibility and, perhaps more important, are almost universally acceptable in supporting permit applications and defending design protocols. A secondary guideline in choosing a model is that

	No.	Application
■ **Table 7.25** Practical Applications of Flow Models	1	Ground-water management and water supply
	2	Pump test evaluations
	3	Wellhead protection studies
	4	Infiltration studies
	5	Determination of velocity fields for contaminant transport models
	6	Determining barrier boundaries to stop saltwater intrusion

Source: ASCE, 1996a. *Quality of Ground Water*, p. 107. Reprinted by permission of ASCE.

the simplest model that will accomplish the design objectives should be given the most serious consideration.

By far the most widely used ground-water flow model is MODFLOW, which was developed and is maintained by the U.S. Geological Survey (McDonald and Harbaugh, 1988; Harbaugh and McDonald, 1996). MODFLOW is a three-dimensional finite-difference model that is very flexible, is applicable to numerous practical situations, has an input structure that allows for rapid and easy changes of input data, has a modular code that makes program modification simple, and is well documented. Flows can be steady or transient, aquifers confined or unconfined, and the aquifer properties can be either homogeneous or nonhomogeneous. MODFLOW has the capability of simulating a variety of aquifer stresses, including wells, areally distributed recharge, evapotranspiration, drains, and streams that penetrate the aquifer. A detailed discussion of the application of MODFLOW in ground-water modeling can be found in Anderson and Woessner (1992) as well as numerous journal publications and reports.

Other models not developed by U.S. government agencies but widely used in ground-water engineering are DYNFLOW and PLASM. DYNFLOW is a three-dimensional finite-element model capable of handling irregular boundaries and has well structured input files, and PLASM is a two-dimensional flow model that accounts for vertical leakance and has been widely taught at universities for many years. The PLASM code is simple, easy to modify and customize, and comes with a very helpful user manual.

Summary

Ground-water hydrology deals with the occurrence and movement of water below the surface of the earth. The basic equations that describe ground-water flows are the continuity and Darcy equations, where application of the Darcy equation is limited to low Reynolds number flows (Re < 10). Combining the continuity and Darcy equations yields a single governing equation in which the piezometric head is the dependent variable and space and time are the independent variables. Porous media and fluid properties are parameterized by the hydraulic conductivity tensor, K_{ij}, and the specific storage, S_s. In aquifers where vertical flows are negligible, ground-water movement can be approximated as two dimensional; the properties of the porous medium and fluid are parameterized by the transmissivity, T, and the specific yield, S_y (unconfined aquifer) or storage coefficient, S (confined aquifer). Solutions to the ground-water flow equation for a variety of initial and boundary conditions are presented. These solutions are frequently used in engineering practice, and include steady flow to a well in an unconfined aquifer with recharge, unsteady flow to a well in an unconfined aquifer, and cases of partially penetrating wells. Many of the solutions presented can be combined using the principle of superposition, where the validity of superposition is ensured by the linearity of the governing flow equation, boundary conditions, and initial conditions. Practical application of the principle of superposition is demonstrated by the method of images. In coastal areas, denser seawater intrudes into coastal aquifers (called salinity intrusion), and the application of the Ghyben-Herzberg equation to calculate the location of the saltwater interface in coastal aquifers is presented.

Engineered systems that are designed using the principles of ground-water hydrology include municipal wellfields and exfiltration trenches. In wellfield design, the objective is to design a system of wells that will produce a desired flowrate, within the constraints of available land area and allowable drawdown. A methodology for designing wellfields using fundamental solutions to the ground-water flow equation is presented as well as guidelines for designing individual water-supply wells and conducting aquifer pump tests. A methodology for obtaining localized estimates of the hydraulic conductivity using slug tests is also presented. In designing exfiltration trenches, the objective is to determine the dimensions and configuration of a trench that will exfiltrate water at a specified flowrate, within the constraint of allowable mounding of the water table. A methodology for designing exfiltration trenches and predicting mounding effects is presented. In many practical applications, a numerical ground-water flow model is required to handle complex geologies and boundary conditions. The most popular numerical model for calculating ground-water flows is MODFLOW, with other models such as DYNFLOW and PLASM also in use.

Problems

7.1. Which property is a better measure of the productivity of an aquifer: porosity or hydraulic conductivity? Explain why.

7.2. Two piezometers are installed in a confined aquifer with an estimated hydraulic conductivity of 10 m/d and a porosity of 0.2. The reference piezometer is located at point A, and the second piezometer is located at point B, 1 km from point A at an angle of 45° clockwise from true north. The water level at B is 10 cm below A. Calculate the seepage velocity along line AB. Explain why this is not the actual seepage velocity in the aquifer. A third piezometer is located in the aquifer at point C, 0.5 km from A at an angle of 140° clockwise from true north, and the water level at C is 8 cm below A. Calculate the seepage velocity and Darcy velocity (= specific discharge) in the aquifer.

7.3. If a contaminant is spilled in an aquifer, is the mean position of the contaminated cloud advected at the Darcy velocity or at the seepage velocity? Explain your answer. Why does a contaminant cloud "disperse" as it is advected through an aquifer?

7.4. A pump test in an aquifer formation indicates a hydraulic conductivity of 20 m/d when the water in the aquifer is at a temperature of 20°C. Determine the intrinsic permeability of the aquifer. If the fluid in the aquifer consists of spilled tetrachloroethylene (PERC), then determine the hydraulic conductivity of the aquifer to PERC. Under the same piezometric gradient, which fluid moves faster? (Refer to Tables B.1 and B.2 for the fluid properties.)

7.5. Hazen (1911) developed the following empirical formula for the hydraulic conductivity, K (cm/s), in terms of the effective grain size, d_{10} (cm), of a porous medium,

$$K = c d_{10}^2$$

where c is approximately equal to 100. Show that this corresponds to using $\alpha = 1.02 \times 10^{-3}$ in Equation 7.15 to estimate the intrinsic permeability. Assume that the water temperature in the porous medium is 20°C.

7.6. Show that the seepage velocity, v, induced by a pumping well as a function of the distance, r, from the well is given by

$$v = \frac{Q_w}{2\pi r H n_e} \tag{7.372}$$

where Q_w is the well pumping rate, H is the aquifer thickness, and n_e is the effective porosity.

7.7. Consider a well pumping at 0.4 m^3/s in an aquifer with an effective porosity of 0.2, a saturated thickness of 24 m, and a hydraulic conductivity of 50 m/d. If the seepage velocity as a function of distance from the well is given by Equation 7.372, then calculate the extent of the circular area surrounding the well where Darcy's law is not valid. (*Hint*: Estimate the characteristic pore diameter by first computing the intrinsic permeability and then estimating the pore diameter using Equation 7.15.)

7.8. Derive Equations 7.23 to 7.25 from Equation 7.22.

7.9. Use Equations 7.23 to 7.25 to show that the principal components of the hydraulic conductivity tensor, K_{11}' and K_{22}', can be expressed in terms of the general hydraulic conductivity tensor K_{ij} by the relations

$$K_{11}' = \frac{K_{11} + K_{22}}{2} + \left[\left(\frac{K_{11} - K_{22}}{2} \right)^2 + K_{12}^2 \right]^{\frac{1}{2}}$$

$$K_{22}' = \frac{K_{11} + K_{22}}{2} - \left[\left(\frac{K_{11} - K_{22}}{2} \right)^2 + K_{12}^2 \right]^{\frac{1}{2}}$$

7.10. The aquifer described in Problem 7.2 is found to be anisotropic, with $K_{11} = 15$ m/d, $K_{22} = 5$ m/d, $K_{12} = K_{21} = 0$ (i.e., the coordinate axes are the principal axes). If the x_1-axis is aligned with the east-west direction and the x_2-axis is aligned with the north-south direction, determine the seepage velocity and compare this velocity with the seepage velocity obtained by assuming that the aquifer is homogeneous with a hydraulic conductivity of 10 m/d.

7.11. Ground-water flow in a regional aquifer has historically been toward the east, and the hydraulic conductivity tensor is estimated to be $K_{xx} = 100$ m/d, $K_{yy} = 10$ m/d, and $K_{xy} = 0$ m/d, where the x- and y-axes are along the east/west and north/south directions, respectively. The development of a new wellfield will cause the mean head gradient to shift from the east to the southeast (a rotation of 45°). Compare the specific discharge in the direction of the head gradient for a gradient of 0.01 in the east direction with the specific discharge in the direction of the head gradient of 0.01 in the southeast direction. In either of these cases, does the ground water flow in the direction of the head gradient?

7.12. The hydraulic conductivities measured at 12 locations in a porous formation are 100, 200, 220, 250, 290, 50, 310, 400, 130, 190, 350, 500 m/d. Calculate the macroscopic hydraulic conductivites for both two- and three-dimensional flow.

7.13. If the hydraulic conductivity measurements given in Problem 7.12 were taken at 100 m intervals, then estimate the correlations between log-hydraulic conductivities at lags of 100 m and 200 m.

7.14. Write the expression for the exponential log-hydraulic conductivity correlation function along one of the Cartesian coordinate axes. (*Hint*: Take $r_1 \neq 0$, and $r_2 = r_3 = 0$.) Show that the correlation length scale in this direction is equal to the integral of the log-hydraulic conductivity correlation function in that direction.

7.15. The definition of the Laplacian operator in Cartesian coordinates is given by

$$\nabla^2 f = \frac{\partial^2 f}{\partial x^2} + \frac{\partial^2 f}{\partial y^2} + \frac{\partial^2 f}{\partial z^2}$$

If the relationship between cylindrical coordinates (r, θ, z) and Cartesian coordinates (x, y, z) is given by $x = r\cos\theta$, $y = r\sin\theta$, $z = z$, then show that the Laplacian operator in cylindrical coordinates is given by Equation 7.48. Under what circumstances would it be worthwhile to use cylindrical coordinates?

7.16. Consider the case of a three-dimensional homogeneous isotropic aquifer. Using cylindrical coordinates, give the differential equation describing the steady-state head distribution in the aquifer. Simplify this equation for a case where the head is radially symmetric.

7.17. A steady-state solution of the three-dimensional ground-water flow equation in an aquifer is found by making the simplifying assumption that the aquifer is isotropic and homogeneous, with a hydraulic conductivity of 205 m/d. The results of this simplified analysis indicate that the head, ϕ, at a location with Cartesian coordinates (100 m, 100 m, 10 m) is equal to 25 m. If the aquifer is actually anisotropic with $K_{xx} = 30$ m/d, $K_{yy} = 40$ m/d, and $K_{zz} = 13$ m/d, determine the actual location in the aquifer where the piezometric head is equal to 25 m. What is the water pressure (= pore pressure) at this location?

7.18. The hydraulic conductivity distribution in a 20-m thick stratified surficial aquifer is given by

Depth (m)	K_{xx} (m/d)	K_{yy} (m/d)
0–2	5	15
2–5	7	20
5–7	9	21
7–11	14	12
11–15	11	17
15–19	6	9
19–20	2	5

Determine the effective hydraulic conductivity when the water table is 2 m below the ground surface. Contrast this result with the effective hydraulic conductivity when the water table is 3 m below the ground surface.

7.19. Consider the case of a two-dimensional unconfined homogeneous anisotropic aquifer. Using Cartesian coordinates, give the differential equation describing the steady-state aquifer thickness. Give a real-world example of a situation in which this equation would be applicable.

7.20. Consider the case of a two-dimensional confined homogeneous isotropic aquifer. Using Cartesian coordinates, give the differential equation describing the steady-state piezometric head distribution. Give a real-world example of a situation where this equation would be applicable.

7.21. Explain why the storage coefficient is much smaller than the specific yield.

7.22. The steady-state head distribution caused by a single fully penetrating pumping well in an extensive (two-dimensional) isotropic confined aquifer can be estimated by

$$\phi(r) = \phi_o - \frac{Q_w}{2\pi T}\ln\left(\frac{R}{r}\right), \qquad r < R$$

where ϕ is the piezometric head, r is the distance from the well, ϕ_o is the head prior to pumping, Q_w is the pumping rate from the well, T is the transmissivity, and R is the radius of influence of the well. Determine the steady-state drawdown distribution for the case of an anisotropic aquifer with transmissivity components T_{xx} and T_{yy}.

7.23. Consider a two-layer stratified aquifer between two reservoirs. The water surfaces in the reservoirs are at elevations 5 m NGVD and 4 m NGVD, respectively; the ground surface between the aquifers is at elevation 10 m NGVD; the top layer of the aquifer extends from ground surface down to −10 m NGVD, and the base of the aquifer (and reservoirs) is at −20 m NGVD. The hydraulic conductivity of the top layer is 50 m/d and the hydraulic conductivity of the bottom layer is 100 m/d. If the reservoirs are 2 km apart, find the equation of the phreatic surface and the flowrate between the reservoirs. Neglect surface recharge.

7.24. Show that the flow, Q, between two reservoirs separated by a two-layer aquifer can be expressed as

$$Q = \frac{K_1}{2L}(h_L^2 - h_R^2) + (K_1 - K_2)\frac{b_2}{L}(h_R - h_L) \qquad (7.373)$$

Verify this result by using Equation 7.373 to compute the discharge in Problem 7.23.

7.25. Derive the general equation for the phreatic surface in a two-layer aquifer between two reservoirs when the recharge, $N(x)$, is not equal to zero. (*Hint*: An equation similar to Equation 7.123, but with an additional term to account for recharge.)

7.26. The equation describing the phreatic surface in a two-layer aquifer between two resevoirs has been shown to be

$$\frac{K_1}{2}h^2 + (K_2 - K_1)b_2h + \frac{Nx^2}{2} = C_1x + C_2$$

where C_1 and C_2 are constants given by

$$C_1 = \frac{K_1}{2L}(h_R^2 - h_L^2) + (K_2 - K_1)b_2\frac{(h_R - h_L)}{L} + \frac{NL}{2}$$

and

$$C_2 = \frac{K_1}{2}h_L^2 + (K_2 - K_1)b_2h_L$$

This equation describes a mounded phreatic surface, with flow to the left of the mound going toward the lefthand reservoir, and flow to the right of the mound going toward the righthand reservoir. Derive an expression for the location of the mound. It has been stated that a mound will always be located between the two reservoirs. Use your derived expression to determine whether this statement is true or false.

7.27. Derive the general equation for the phreatic surface in a three-layer aquifer between two reservoirs. Neglect surface recharge.

7.28. A well pumps at 0.4 m³/s from a confined aquifer whose thickness is 24 m. If the drawdown 50 m from the well is 1 m and the drawdown 100 m from the well is 0.5 m, then calculate the hydraulic conductivity and transmissivity of the aquifer. Do you expect the drawdowns at 50 m and 100 m from the well to approach a steady state? Explain your answer. If the radius of the pumping well is 0.5 m and the drawdown at the pumping well is measured to be 4 m, then calculate the radial distance to where the drawdown is equal to zero. Why is the steady-state drawdown equation not valid beyond this distance?

7.29. Repeat Problem 7.28 for an unconfined aquifer, assuming that the saturated thickness is 24 m prior to pumping. If actual drawdowns, rather than the modified drawdowns, are used to calculate the transmissivity of the aquifer, then what would be the percentage difference in the calculated transmissivity?

7.30. A water-supply well is located at the center of a small circular island that has a radius of 1 km. The well has a diameter of 1 m and a pumping rate of 20,000 L/min, and the (phreatic) aquifer has a hydraulic conductivity of 40 m/d, a specific yield of 0.15, and a porosity of 0.2. If the water surrounding the island is at mean sea level and the base of the aquifer is 45 m below sea level, estimate the water-table elevation at the well intake. Use the seepage velocity halfway between the well and the coastline to estimate how long a contaminant would take to travel from the perimeter of the island to the well.

7.31. A pumping well has a radius of 0.5 m and extracts water at a rate of 0.4 m³/s from a semiconfined aquifer whose thickness is 24 m. The drawdown 50 m from the well is 1 m, and the drawdown 100 m from the well is 0.5 m. Use Equation 7.184 to calculate the leakage factor, transmissivity, and hydraulic conductivity of the aquifer. Compare the transmissivity with that obtained by neglecting leakage. (See Problem 7.28.)

7.32. Repeat Problem 7.31 using the approximate relation given by Equation 7.187. Verify that it is appropriate to use Equation 7.187.

7.33. If the piezometric surface in Problem 7.31 was intially 40 m above the base of the aquifer, then plot the leakage rate into the semiconfined aquifer as a function of the radial distance from the pumping well.

7.34. The transmissivity in a confined aquifer is estimated from steady-state drawdown measurements by Equation 7.149, and in a semiconfined aquifer by Equations 7.185 and 7.186. Identify the range of leakage factors for which the leakage can be neglected in the analysis of the drawdown data. Assume that leakage can be neglected whenever the neglect of leakage results in less than 0.1% error in the calculated transmissivity.

7.35. A well of radius 0.5 m in an unconfined aquifer pumps at a rate of 0.4 m³/s. If the hydraulic conductivity of the aquifer is 30 m/d, the saturated thickness of the aquifer at the well is 24 m, and the recharge rate to the aquifer is 500 mm/y, find an expression for the distribution of aquifer thickness surrounding the well. Determine the distribution of seepage velocity surrounding the well.

7.36. Derive Equations 7.207 to 7.209 using the chain rule and Equation 7.206.

7.37. A well pumps water from a confined aquifer at a rate of 0.4 m³/s, where the radius of the well is 0.5 m and the thickness of the aquifer is 24 m. If the storage coefficient of the aquifer is 0.0012 and the hydraulic conductivity of the aquifer is estimated to be 300 m/d, then calculate the drawdown at distances of 0.5 m, 50 m, and 100 m from the well as a function of time.

7.38. It has been widely asserted that the steady-state equation for estimating the transmissivity in confined aquifers given by Equation 7.149 can be used to estimate the transmissivity in spite of the unsteadiness in the measured drawdowns.

Use the drawdown results derived in Problem 7.37 to estimate the time required for the transmissivity to be estimated within 1% accuracy by the steady-state equation.

7.39. A fully penetrating well pumps 0.2 m³/s from an unconfined aquifer where the specific yield is 0.15 and the hydraulic conductivity is 100 m/d. Prior to pumping, the aquifer thickness was uniformly equal to 28 m. Calculate the drawdowns at 50 m and 100 m from the well after 1 sec, 1 minute, 1 hour, 1 day, 1 month, and 6 months. At each time, use the steady-state (Thiem) equation to estimate the transmissivity of the aquifer based on the calculated drawdowns at 50 and 100 m. On the basis of these results, what can you say about using a steady-state equation to estimate the transmissivity from pairs of drawdowns during the transient pump test?

7.40. A confined aquifer of thickness 24 m is pumped at a rate of 0.4 m³/s, and the recorded drawdowns in a monitoring well located 50 m from the pumping well are given in Table 7.26. Estimate the hydraulic conductivity, transmissivity, and storage coefficient of the aquifer.

■ **Table 7.26** Drawdowns in Monitoring Well

Time (s)	Drawdown (cm)
1	0.00
10	9.54
100	76.39
1,000	150.03
10,000	256.85
100,000	319.72

7.41. A well of radius 0.5 m pumps water out of an unconfined aquifer at a rate of 0.4 m³/s. The aquifer thickness prior to pumping is 24 m, and the measured drawdowns 50 m from the pumping well as a function of time are given in Table 7.27. Determine the hydraulic conductivity, transmissivity, and specific yield of the aquifer.

■ **Table 7.27** Drawdowns in Monitoring Well

Time (s)	Drawdown (cm)
1	0.00
10	0.00
100	0.12
1,000	12.21
10,000	36.89
100,000	68.64

7.42. A confined aquifer of thickness 24 m is pumped at 0.27 m³/s and the recorded drawdowns in a monitoring well 50 m from the pumping well are given in Table 7.28. Use the Cooper-Jacob approximation of the well function to estimate the hydraulic conductivity, and storage coefficient of the aquifer. At what time does the Cooper-Jacob approximation become reasonable.

■ **Table 7.28** Drawdowns in Monitoring Well

Time (s)	Drawdown (cm)
1	0.00
10	6.39
100	51.18
1,000	100.52
10,000	172.09
100,000	214.21

7.43. A confined anisotropic aquifer has a hydraulic conductivity of $K_{xx} = 45$ m/d, $K_{yy} = 15$ m/d, and $K_{xy} = 0$ m/d; a thickness of 14 m; and a storage coefficient of 10^{-4}. If a well is installed in the aquifer and pumped at 1,000 L/min, estimate the drawdown at $x = 100$ m, $y = 100$ m after 1 week of pumping. How would this result change if the confining layer has a leakage factor of 2,000 m?

7.44. A fully penetrating well of radius 0.5 m pumps water at 0.4 m³/s from a semiconfined aquifer of thickness 24 m. If the hydraulic conductivity of the aquifer is 300 m/d and the storage coefficient is 0.0012, estimate the drawdown as a function of time at a distance of 50 m from the well for leakage factors of 1 m, 10 m, and 100 m. Explain your results. If the thickness of the semiconfining layer is 5 m, then what are the hydraulic conductivities of the semiconfining layers corresponding to leakage factors of 10 m and 100 m?

7.45. Consider the fully penetrating well described in Problem 7.28. What would be the drawdown in the well if only the top half of the aquifer is penetrated? Express the drawdown in the well as a function of the penetration factor. Can the drawdowns at 50 m and 100 m be reasonably estimated by assuming full penetration of the well, even if the well only penetrates the top half of the aquifer? Explain.

7.46. Consider the fully penetrating well described in Problem 7.28. Compare the additional drawdown in the well expected if the well only penetrates the top 40% of the aquifer with the additional drawdown expected if the well screen is centered in the aquifer and has a length equal to 40% of the aquifer thickness.

7.47. Three wells are located in an infinite unconfined aquifer of thickness 20 m, specific yield 0.2, and hydraulic conductivity 40 m/d. The planar coordinate locations of the wells are: point A (0 m, 0 m), point B (200 m, 200 m), and point C (200 m, −200 m). If all wells begin pumping at the same time and at a rate of 0.2 m^3/s, then calculate the drawdown as a function of time at a location (100 m, 100 m).

7.48. A well of radius 0.5 m pumps water at 0.4 m^3/s from a confined aquifer of thickness 24 m. The hydraulic conductivity of the aquifer is 300 m/d, the storage coefficient is 0.012, and the radius of influence can be taken as 1,200 m. If the planar coordinates of the well are (0 m, 0 m), and a fully penetrating river runs along the line $x = 500$ m, then calculate the steady-state drawdowns at (100 m, 0 m) and (−100 m, 0 m).

7.49. A second well located at (0 m, 200 m) is added to the wellfield described in Problem 7.48. If this well also pumps at 0.4 m^3/s and the radius of influence is 1,200 m, then calculate the drawdowns at (100 m, 0 m) and (−100 m, 0 m) and compare with those calculated in Problem 7.48. If this second well is to be placed parallel to the river and contribute no more than 1% of the total drawdown at the designated points, then determine the coordinates of the second well.

7.50. Repeat Problem 7.48, but with the fully penetrating river replaced by an impermeable barrier.

7.51. Repeat Problem 7.49, with the fully penetrating river replaced by an impermeable barrier.

7.52. A well of radius 0.5 m pumps water at 0.4 m^3/s from a confined aquifer of thickness 24 m, hydraulic conductivity 300 m/d, and storage coefficient 0.012. If a fully penetrating river is located 1 km east of the well and an impermeable boundary is located 1 km west of the well, then calculate the drawdown in the aquifer at points 200 m east of the well and 200 m west of the well. Assume that the radius of influence of the well is 1,200 m.

7.53. Show that the drawdown distribution caused by a well in an infinite strip between two fully penetrating streams is given by

$$s = \frac{Q_w}{4\pi T} \sum_{n=-\infty}^{\infty} \ln \frac{(x + x_o - 2nd)^2 + y^2}{(x - x_o - 2nd)^2 + y^2} \qquad (7.374)$$

where Q_w is the pumping rate, T is the aquifer transmissivity, x_o is the distance of the well from one stream, and d is the distance between the streams.

7.54. Show the arrangement of image and real wells that is required to calculate the drawdown induced by a well in an aquifer quadrant, where the other three quadrants contain impermeable formations.

7.55. Show that the buildup caused by injecting water at a rate Q_w into a fully penetrating well in an unconfined infinite aquifer is exactly the same as the drawdown caused by withdrawing water at a rate Q_w from the same aquifer.

7.56. The water table at a given location in a coastal aquifer is 1 m above mean sea level, and the saturated zone in the aquifer is 24 m thick. Do you expect saltwater intrusion to be a problem at this location?

7.57. A drainage canal in a coastal area terminates in a gated structure, and the thickness of the aquifer beneath the canal is 24 m. On the downstream side of the gate, the seawater fluctuates between 30 cm above and below mean sea level. If the elevation of the bottom of the canal is 3 m below sea level, determine the minimum water elevation on the upstream side of the gate to prevent saltwater intrusion.

7.58. A water-supply well is being threatened by saltwater intrusion. The well is currently pumping 0.04 m^3/s from an aquifer that has a hydraulic conductivity of 500 m/d, thickness of 50 m, porosity of 0.2, and specific yield of 0.15. If the well is screened to within 15 m of the bottom of the aquifer, what will be the thickness of the saltwater wedge under the well when the well becomes contaminated with saltwater?

7.59. Explain why a rising water table will not yield the same retention curve as a falling water table.

7.60. Consider a soil with the retention curve shown in Table 7.29. If the water table was initially 1.5 m below the ground surface and falls to 3 m below the ground surface, then determine the specific yield of the aquifer. Is the specific yield any different if the water table had fallen from 3.5 m to 5 m below the ground surface?

■ **Table 7.29**   Retention Curve

Elevation above water table (m)	Moisture content
0	0.18
0.5	0.16
1.0	0.15
1.5	0.14
2.0	0.11
2.5	0.08
3.0	0.06
3.5	0.05
4.0	0.05
4.5	0.05
5.0	0.05

7.61. A wellfield is to be developed to produce 4.44 m^3/s from an unconfined aquifer of thickness 35 m, hydraulic conductivity 500 m/d, and specific yield 0.2. If the radius of each well is to be 0.5 m, the drawdown is not to exceed 2 m, and the wells are to be arranged along a straight line, then determine the number of wells, the pumping rate from each well, and the spacing between wells.

7.62. A wellfield is to be developed such that the wells are all parallel to a stream, with each well having a diameter of 0.1 m. The aquifer is 20 m thick, with an effective porosity of 0.15 and a hydraulic conductivity of 60 m/d. If the wellfield is expected to deliver 5,000 L/min, each well is to be 100 m from the stream, and the maximum allowable drawdown is 2 m, estimate the number of wells required, the pumping rate from each well, and the spacing between the wells.

7.63. A square-shaped 50-ha parcel of land has been identified as a possible site for a wellfield to supply a population of 50,000 people. The thickness of the aquifer is 35 m, the hydraulic conductivity is 85 m/d, the porosity is 0.2, and the specific yield is 0.15. The per capita demand of the population is 580 L/person/d. Design a wellfield that can be accommodated on the parcel of land. The radius of each well can be taken as 50 cm, and the drawdown must not exceed 6 m. If the wells are screened over the bottom 20 m of the aquifer and the screen has a diameter of 50 cm with 50% open area, assess whether the screen is adequate.

7.64. A well of radius 0.5 m is to be installed in an unconfined aquifer of hydraulic conductivity 300 m/d. What is the maximum allowable screen entrance velocity, and what is the corresponding maximum pumping rate? If the uniformity coefficient of the aquifer matrix is 3.2, then write the specifications for the gravel pack and the screen slot size.

7.65. A water-supply well is to be installed in a 30-m thick sand aquifer with a uniform grain-size distribution between 0.04 mm and 2.2 mm. The hydraulic conductivity of the aquifer is estimated to be 50 m/d, the porosity is 0.15, and the specific yield is 0.2. If the pumping rate is to be 800 L/min, design the casing, screen, and gravel pack.

7.66. A 0.3-m diameter well is to be installed in a 20-m thick confined aquifer in which the allowable drawdown of the potentiometric surface at the well is 5 m. The hydraulic conductivity and storage coefficient of the aquifer are estimated as 30 m/d and 10^{-4}, respectively. Estimate the specific capacity and the well yield after 1 year, and classify the productivity of the well.

7.67. A step-drawdown test is conducted in an aquifer and yields the results shown in Table 7.30. Determine the well loss coefficient, formation loss coefficient, and the specific capacity of the well. Assess the condition and productivity of the well at a pumping rate of 5,000 m^3/s.

■ **Table 7.30** Step-Drawdown Results

Pumping rate (m^3/s)	Drawdown (cm)
0	0
0.05	20
0.10	53
0.15	86
0.20	147
0.25	210
0.30	283

7.68. The drawdowns, s_w (cm), and corresponding pumping rates, Q_w (L/min), collected at a well during a step-drawdown test were plotted on a graph of $\log(s_w/Q_w - \beta)$ versus $\log Q_w$ for various values of β. The plotted curve is most linear when $\beta = 55.2$ s/m^2, and the corresponding best-fit line through the data has a slope of 1.22. If the well efficiency is known to be 32% when the pumping rate is 620 L/min, estimate the well-loss coefficient. Assess the productivity of the well when pumping at 1,000 L/min.

7.69. Show that the half-life of a contaminant in ground water can be described by Equation 7.342.

7.70. The risk, R, of illness from viruses in pumped water is related to the viral concentration, c, by the relation

$$R = 1 - (1 + 4.76c)^{-94.9}$$

where c is in #/L. If an illness rate of 1 in 10,000 people is acceptable ($R = 10^{-4}$), determine the allowable viral concentration in the pumped water. A municipal supply well is to be installed 100 m from school that uses a septic tank for onsite wastewater disposal. The well is to have a rated capacity of 0.4 m^3/s, the aquifer is 20 m thick, and the porosity of the aquifer is 0.17. If the allowable risk of viral infection from the pumped water is 10^{-4} and the decay constant for viruses in ground water is 0.3 d^{-1}, estimate the maximum allowable virus concentration in the ground water below the school.

7.71. A 30-cm diameter water-supply well is to pump 50 L/s and be located 150 m from a fully penetrating river in a 25-m thick aquifer. The hydraulic conductivity of the aquifer is 30 m/d, the porosity is 0.2, and the radius of influence of the well is 1 km. Calculate the shortest time of travel of the ground water from the river to the well intake. If the decay constant of a contaminant is 0.01 d^{-1}, and the allowable concentration of the contaminant in the pumped water is 1 μg/L, estimate the maximum allowable concentration of the contaminant in the river by assuming the shortest time of travel. Assess the validity of this assumption.

7.72. Give an example of how you would use estimates of aquifer properties prior to conducting an aquifer pump test to aid in the design of the pump test.

7.73. Use the principle of superposition to show that Equation 7.346 can be used to describe the recovery data in a pump test.

7.74. Apply the Cooper-Jacob approximation to Equation 7.346 to show that the recovery data in a pump test can be described by Equation 7.349.

7.75. A pump test is conducted at a rate of 2,000 L/min, and pumping is terminated after 4 hours. Measurements of drawdown versus time after pumping is terminated are given in the following table:

Drawdown (m)	Time Since End of Pumping (min)
1.01	1
0.90	2
0.83	3
0.75	5
0.70	7
0.61	10
0.55	15
0.60	20
0.42	30
0.37	40
0.31	60
0.26	80
0.23	100
0.19	140
0.15	180

Time (s)	Drawdown (mm)
0	700
3	392
6	260
9	137
12	91
15	47
18	31
21	16
24	11

Estimate the hydraulic conductivity and transmissivity of the aquifer.

7.78. Repeat Problem 7.77 for the case in which the monitoring well fully penetrates the aquifer.

7.79. An exfiltration trench is to be designed to inject 500 m³/d from a pump-and-treat system. The trench hydraulic conductivity is 35 m/d, and the depth from the ground surface to the seasonal high-water table is 5.22 m. The aquifer is estimated to have a hydraulic conductivity of 70 m/d, a thickness of 15 m, and a specific yield of 0.22. Design the exfiltration trench assuming a factor of safety of 2.5 in the trench hydraulic conductivity.

7.80. Repeat Problem 7.79 assuming 1 m of backfill over the trench (so that surface vegetation can be planted).

7.81. An exfiltration trench is to be designed such that the water table remains at least 1.2 m below the trench after one year of operation. The trench hydraulic conductivity is 10 m/d, the width of the trench is to be 1 m, the depth is to be 1.5 m, and space limitations will allow a maximum trench length of 10 m. The wet-season water table is 3.1 m below ground surface, the hydraulic conductivity of the aquifer is 15 m/d, the porosity is 0.15, and the thickness of the aquifer is 7 m. What is the maximum allowable flowrate that can be exfiltrated through the trench?

If the pump-drawdown test indicates a storage coefficient of 0.0001, estimate the aquifer transmissivity and the recovery storage coefficient.

7.76. It has been stated that a 10% variation in discharge during a pump test can result in a 100% variation in the estimate of aquifer transmissivity (Osbourne, 1993). Assess the validity of this statement.

7.77. A monitoring well that penetrates 4 m below the water table in a 12-m thick unconfined aquifer is used to conduct a slug test. The well casing has a 150 mm diameter, and the bottom 2 m of the well is screened and surrounded by a 110-mm thick gravel pack. The observed time-drawdown relation during the slug test is given in the following table:

8 Hydrologic Fate and Transport Processes

8.1 Introduction

Contaminants in the water environment undergo changes in concentration resulting from physical, chemical, and/or biological processes, and a capability to understand and model these processes is at the core of water-quality management. Processes involving the transformation of substances are called *fate processes* and include both chemical and biological processes, while processes involving advection and mixing are called *transport processes*. Domestic, industrial, and agricultural activities generate a significant amount of contaminants that mix with water in the hydrologic cycle and find their way into streams, lakes, oceans, and ground waters. Water-quality management is concerned with controlling the quality of surface and ground waters, and the management objectives are typically defined by water-quality criteria and standards promulgated by regulatory agencies, based on the present and future beneficial uses of various waterbodies.

Approaches to water-quality management take a variety of forms, such as best management practices to control contaminants at their source and engineered systems to control the movement of contaminants within the hydrologic cycle. The design of engineered systems requires an understanding of contaminant fate and transport processes in the water environment, and such engineered systems include wastewater-discharge facilities, stormwater-discharge facilities, remediation systems for the cleanup of contaminated surface and ground waters, and the siting of municipal wellfields in the vicinity of existing and planned contaminant sources, commonly referred to as the delineation of wellhead protection areas. This chapter reviews the parameters commonly used to assess the quality of water in rivers, lakes, oceans, and ground waters, and presents the fundamentals of fate and transport models that are used to predict the effects of contaminant inputs on water quality.

8.2 Water Quality

The quality of various water bodies are usually assessed relative to their present and future most beneficial use. Rivers, lakes, and marine waters must be aesthetically pleasing and of acceptable quality to support healthy and diverse aquatic ecosystems as well as human recreational uses such as fishing and swimming. Furthermore, fresh-water bodies, such as rivers, lakes, and aquifers, are typically sources of drinking water, and the quality of these waters must be consistent with the level of water treatment provided prior to distribution in the public water-supply system. Sources of water pollution can be broadly

585

grouped into point sources and nonpoint sources. *Point sources* are localized discharges of contaminants, and include municipal and industrial wastewater discharges, as well as accidental and illicit spills. *Nonpoint sources* of pollution include contaminant sources that are distributed over large areas or are a composite of many point sources, including runoff from agricultural operations, the atmosphere, and urban runoff. Pollutants can be classified as either conservative or nonconservative substances. *Conservative substances* do not lose mass to chemical reactions or biochemical degradation; examples include chlorides and certain metals during times of the year when transport is in the dissolved form. *Nonconservative substances* lose mass via chemical reactions, biochemical degradation, radioactive decay, and settling of particulates out of the water column. Examples include oxidizable organic matter, nutrients, volatile chemicals, and bacteria.

Water-quality criteria can be stated either in terms of a *maximum contaminant limit* (MCL) or in terms of a concentration limit for a given frequency and duration of exposure. Water-quality criteria in terms of MCLs are usually the most stringent, while concentration limits in terms of frequency and duration tend to allow increased levels of contaminant discharges without violating ambient water-quality criteria (Chin et al., 1997). Water-quality criteria based on the frequency and duration of exposure usually state that aquatic ecosystems will not be significantly impaired if (1) the four-day (96-hour) average concentration of the pollutant does not exceed the recommended *chronic criterion* more than once every three years on average; and (2) the one-hour average concentration does not exceed the recommended *acute criterion* more than once every three years on average (USEPA, 1986b). Contaminant limits for acute exposure are generally higher than contaminant levels for chronic exposure.

8.2.1 ■ Measures of Water Quality

A wide variety of hazardous substances are found in natural and constructed water bodies. Municipal and industrial wastewaters, as well as surface-water runoff, are the most common sources of pollution. Most municipal wastewater treatment plants and stormwater outfalls discharge their effluent into rivers, lakes, or ocean waters. For river discharges, the effect of the effluent on the dissolved oxygen and nutrient levels in the river are usually of most concern, while for ocean discharges the pathogenic and heavy-metal concentrations are usually of most interest. There are a wide variety of industrial wastewaters, depending on the type of industry. Wastewaters with elevated levels of nutrients, heavy metals, heat, and toxic organic chemicals are common. Some industries provide pretreatment prior to discharging their wastewaters either directly into surface waters or into the municipal sewer system for further treatment in combination with domestic wastewater. In many countries outside the United States, industries are permitted to discharge their wastewaters without adequate pretreatment, and the resulting human and environmental impacts are usually noticeable. The impact of poor-quality surface runoff from industrial areas, construction sites, landfills, animal feedlots and agricultural areas are all of concern for surface-water bodies. Ground-water contamination from subsurface effluent originating from septic tanks, leaking underground storage tanks, and waste-injection wells is quite common and are of particular concern in aquifers used for domestic water supply.

A variety of parameters are used to assess the quality of natural water bodies, and several of these parameters are reviewed here.

Pathogenic Microorganisms.   Pathogenic microorganisms are generally found in surface-water bodies, including rivers, lakes, and oceans, and mostly result from urban stormwater runoff, domestic and municipal wastewater discharges, and combined-sewer overflows. Concern about the presence of pathogenic microorganisms in various water bodies is supported by the following facts (Tebbutt, 1998):

- Each year over five million people die from water-related diseases.
- Two million of these annual deaths are children.
- In developing countries, 80% of all illness is water related.
- At any one time, half of the population in developing countries will be suffering from one or more of the main water-related diseases.
- A quarter of the children born in developing countries will have died before the age of five, the great majority from water-related diseases.

The groups of pathogens of most concern include viruses, bacteria, and protozoans; a list of pathogens that typically find their way to natural water bodies, along with their associated diseases, is given in Table 8.1. Water uses that are impacted by pathogenic bacteria, viruses, and protozoa include bathing, fishing, and shellfish harvesting. Testing of water samples for a wide variety of pathogens is usually not practical, and the presence of nonpathogenic *fecal coliform* bacteria, which indicates the probable presence of organisms from the intestinal tract of humans and other warm-blooded animals, is used as an indicator of the presence of pathogenic microorganisms. The reliability of the test for fecal coliform (FC) bacteria as an indicator of the presence of pathogens in water depends on the persistence of the pathogens relative to fecal coliform bacteria. In surface waters, pathogenic bacteria tend to die off faster than coliforms, while viruses and protozoa tend to be more persistent. For recreational lakes and streams, FC levels of less

■ **Table 8.1** Pathogenic Microorganisms Commonly Found in Surface Waters	**Pathogen group and name**	**Associated disease**
	Virus:	
	Adenoviruses	respiratory, eye infections
	Enteroviruses	
	Polioviruses	aseptic meningitis, poliomyelitis
	Echoviruses	aseptic meningitis, diarrhea, respiratory infections
	Coxsackie viruses	aseptic meningitis, herpangina, myocarditis
	Hepatitis A virus	infectious hepatitis
	Reoviruses	not well known
	Other viruses	gastroenteritis,* diarrhea
	Bacterium:	
	Salmonella typhi	typhoid fever
	Salmonella paratyphi	paratyphoid fever
	Other salmonellae	gastroenteritis
	Shigella spp.	bacillary dysentery[†]
	Vibrio cholerae	cholera
	Other vibrios	diarrhea
	Yersinia enterocolitica	gastroenteritis
	Protozoan:	
	Entamoeba histolytica	amoebic dysentery
	Giardia lamblia	diarrhea
	Cryptosporidium spp.	diarrhea

Source: Feachem et al. *Sanitation and Disease, Health Aspects of Excreta and Wastewater Management: World Bank Studies in Water Supply and Sanitation 3.* Copyright © John Wiley & Sons, Inc. 1983. Reprinted by permission of John Wiley & Sons, Inc.
*Gastroenteritis is an inflamation of the lining membrane of the stomach and intestines.
[†]Dysentery is diarrhea with bloody stools and sometimes fever.

than 200/100 mL are usually considered acceptable. The concentration of *fecal strepto-cocci* (FS), which includes several species of streptococci that originate from humans, is also a useful indicator of the presence of pathogenic microorganisms. The ratio of FC to FS (FC/FS) has been used as a measure of whether fecal contamination is from human or animal sources. Values of FC/FS greater than 4 indicates human contamination, while FC/FS less than 4 indicates that the contamination is primarily from other warm-blooded animals. The ratio FC/FS should be used with caution, since FC and FS have different dieoff rates, and the FC/FS ratio may give misleading results far downstream of the contaminant source.

In ground waters, the fecal coliform concentration is not a reliable indicator of contamination by pathogenic microorganisms, since microorganisms larger than viruses seldom travel appreciable distances, and there is no relation between the presence of fecal coliforms and pathogenic microorganisms. The presence of viruses are of particular concern in ground waters, where they originate from septic-tank effluent and, because of their small size (typically 0.01–0.3 μm), are able to travel considerable distances in ground water. Since viruses are difficult to detect in ground water and fecal coliforms are not a reliable indicator, protection of drinking-water wells from viral contamination is usually achieved by requiring minimum setback distances for viral sources such as septic tanks and minimum disinfection requirements. In the United States, the statute that sets minimum setback distances and minimum disinfection requirements is known as the Ground Water Disinfection Rule.

Dissolved Oxygen. Dissolved oxygen (DO) is the amount of molecular oxygen dissolved in water and is a key parameter affecting the health of aquatic ecosystems, fish mortality, odors, and other aesthetic qualities of surface waters. Discharges of oxidizable organic substances into water bodies result in the consumption of oxygen and the depression of dissolved oxygen levels. Saturation levels of DO decrease with temperature, as illustrated in Table 8.2, for a standard atmospheric pressure of 101 kPa. An empirical expression for the saturation concentration of dissolved oxygen, DO_{sat}, is given by (American Public Health Association, 1992)

$$\ln DO_{sat} = -139.34411 + \frac{1.575701 \times 10^5}{T_a} - \frac{6.642308 \times 10^7}{T_a^2}$$
$$+ \frac{1.243800 \times 10^{10}}{T_a^3} - \frac{8.621949 \times 10^{11}}{T_a^4} \tag{8.1}$$

where T_a is the absolute temperature (K). The saturation concentration of oxygen in water is affected by the presence of chlorides (salt), which reduce the saturation concentration by about 0.015 mg/L per 100 mg/L chloride at low temperatures (5–10°C) and

■ **Table 8.2** Saturation Concentration of Dissolved Oxygen in Water	Temperature (°C)	Dissolved oxygen (mg/L)
	0	14.6
	5	12.8
	10	11.3
	15	10.1
	20	9.1
	25	8.2
	30	7.5
	35	6.9

by about 0.008 mg/L per 100 mg/L chloride at higher temperatures (20–30°C) (Tebbutt, 1998). Cool waters typically contain higher levels of dissolved oxygen, and consequently aquatic life in streams and lakes are usually under more oxygen stress during the warm summer months compared with the cool winter months. The minimum dissolved oxygen level needed to support a diverse aquatic ecosystem is typically on the order of 5 mg/L.

■ Example 8.1

Compare the saturation concentration of dissolved oxygen in fresh water at 20°C given by Equation 8.1 to the value given in Table 8.2. What would be the effect on the saturation concentration of dissolved oxygen if saltwater intrusion causes the chloride concentration to increase from 0 mg/L to 2,500 mg/L?

Solution

Equation 8.1 gives DO_{sat} in terms of the absolute temperature, T_a, where $T_a = 273.15 + 20 = 293.15K$. Hence, Equation 8.1 gives

$$\ln DO_{sat} = -139.34411 + \frac{1.575701 \times 10^5}{T_a} - \frac{6.642308 \times 10^7}{T_a^2}$$

$$+ \frac{1.243800 \times 10^{10}}{T_a^3} - \frac{8.621949 \times 10^{11}}{T_a^4}$$

$$= -139.34411 + \frac{1.575701 \times 10^5}{293.15} - \frac{6.642308 \times 10^7}{(293.15)^2}$$

$$+ \frac{1.243800 \times 10^{10}}{(293.15)^3} - \frac{8.621949 \times 10^{11}}{(293.15)^4}$$

$$= 2.207$$

Therefore

$$DO_{sat} = e^{2.207} = 9.1 \text{ mg/L}$$

This is the same value of DO_{sat} for fresh water given in Table 8.2.

At a temperature of 20°C, the value of DO_{sat} is reduced by 0.008 mg/L per 100 mg/L chloride; therefore, the change in the oxygen saturation concentration, ΔDO_{sat}, for a chloride concentration of 2,500 mg/L is given by

$$\Delta DO_{sat} = 25 \times 0.008 = 0.2 \text{ mg/L}$$

and the new saturation concentration at 2,500 mg/L chloride is $9.1 - 0.2 = 8.9$ mg/L.

■

Biochemical Oxygen Demand. Biochemical oxygen demand (BOD) is the amount of oxygen required to biochemically oxidize organic matter present in water. Waste discharges that contain significant amounts of biodegradable organic matter have high BOD levels and consume significant amounts of oxygen from the receiving waters, thereby reducing the level of dissolved oxygen and producing adverse impacts on aquatic ecosystems. The BOD measures the mass of oxygen consumed per liter of water, and is usually given in mg/L. A typical BOD curve is illustrated in Figure 8.1(a), where the BOD is composed of the carbonaceous BOD (CBOD) and the nitrogenous BOD (NBOD). The CBOD is exerted by heterotrophic organisms that derive their energy for

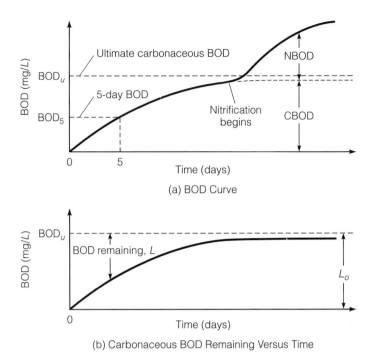

Figure 8.1 ■ Typical BOD Curve

oxidation from an organic carbon substrate, and NBOD is exerted by nitrifying bacteria that oxidize nitrogenous compounds in the wastewater. The carbonaceous demand is usually exerted first, with a lag in the growth of nitrifying bacteria. Normally, nitrogenous oxidation of raw sewage effluent is only important after 8 to 10 days of oxidation in the presence of excess oxygen; for treated effluents, however, nitrification may be important after a day or two due to the large number of nitrifying bacteria in the effluent (Tebbutt, 1998).

BOD tests are conducted using 300-mL glass bottles in which a small sample of polluted water is mixed with (clean) oxygen-saturated water containing a phosphate buffer and inorganic nutrients. The mixture is incubated at 20°C, and the dissolved oxygen in the mixture is measured as a function of time, usually for a minimum of five days. The oxygen demand of the polluted water after five days is called the *five-day BOD* and is usually written as BOD_5. The kinetics of carbonaceous BOD is illustrated in Figure 8.1(b), and can be approximated by the following first-order model

$$\frac{dL}{dt} = -k_1 L \tag{8.2}$$

where L is the carbonaceous BOD remaining at time t and k_1 is a rate constant. If L_o is the BOD remaining at time $t = 0$, equal to the *ultimate carbonaceous BOD*, then Equation 8.2 integrates to yield

$$L = L_o e^{-k_1 t} \tag{8.3}$$

Since the BOD at time t is related to L by

$$BOD = L_o - L \tag{8.4}$$

then the BOD as a function of time is given by combining Equations 8.3 and 8.4 to yield

$$BOD = L_o(1 - e^{-k_1 t}) \tag{8.5}$$

The ultimate BOD, L_o, can be expressed in terms of the five-day BOD, BOD_5, as

$$L_o = \frac{BOD_5}{1 - e^{-5k_1}} \tag{8.6}$$

where both BOD_5 and k_1 are estimated from the BOD test data. For secondary-treated municipal wastewaters, k_1 is typically in the range of 0.1 to 0.3 d^{-1} at 20°C, which gives a L_o/BOD_5 ratio of approximately 1.6. Schnoor (1996) suggests a value of 1.47 for L_o/BOD_5 in municipal wastewater. Municipal wastewater discharge permits generally allow a maximum BOD_5 of 30 mg/L, and it is recommended that communities that discharge treated domestic wastewater into lakes or pristine streams reduce their BOD_5 to below 10 mg/L to protect the natural aquatic life (Serrano, 1997).

■ Example 8.2

The results of a BOD test on secondary-treated sewage give a five-day BOD of 25 mg/L and a rate constant of 0.2 d^{-1}. Estimate the ultimate carbonaceous BOD and the time required for 90% of the carbonaceous BOD to be exerted. If the ultimate nitrogenous BOD is 20% of the ultimate carbonaceous BOD, estimate the oxygen requirement per m^3 of wastewater.

Solution

From the given data, BOD_5 = 25 mg/L and k_1 = 0.2 d^{-1}, hence the ultimate carbonaceous BOD, L_o, is given by Equation 8.6 as

$$L_o = \frac{BOD_5}{1 - e^{-5k_1}} = \frac{25}{1 - e^{-5(0.2)}} = 39.5 \text{ mg/L}$$

Letting t^* be the time for the BOD to reach 90% of its ultimate value, then Equation 8.5 gives

$$0.9(39.5) = 39.5(1 - e^{-0.2t^*})$$

which gives

$$t^* = 11.6 \text{ days}$$

Since the ultimate nitrogenous BOD is 20% of the ultimate carbonaceous BOD, the ultimate BOD is given by

$$\text{Ultimate BOD} = 1.2L_o = 1.2(39.5) = 47.4 \text{ mg/L}$$

and for 1 m^3 = 1,000 L of wastewater, the ultimate mass of oxygen consumed by biochemical reactions is

$$\text{Mass of oxygen} = 47.4(1000) = 47400 \text{ mg/m}^3 = 47.4 \text{ g/m}^3$$

If the wastewater is discharged into a surface water, this oxygen will be taken from the ambient water.   ■

Total Suspended Solids.   Total suspended solids (TSS) is the amount of suspended matter in water. TSS is measured in the laboratory by filtering a known volume of water through a 1.5 μm microfiber filter and then calculating the TSS by dividing the mass of matter retained on the filter by the volume of the water sample. TSS is normally expressed in mg/L, and suspended-solids concentrations are usually quite high in surface runoff. The sedimentation of suspended solids in receiving waters can cause a buildup of organic matter in the sediments, leading to an oxygen-demanding sludge deposit. This deposit can adversely affect fish populations by reducing their growth rate and resistance to disease, preventing the development of eggs and larvae and reducing the amount of food available on the bottom of the water body. Municipal wastewater treatment plants produce effluents with a maximum total suspended solids (TSS) of 30 mg/L, and it is recommended that communities that discharge treated domestic wastewater into lakes or pristine streams reduce their TSS to below 10 mg/L to protect the natural aquatic life (Serrano, 1997).

■ Example 8.3

An outfall discharges wastewater into a flood-control lake that is approximately 300 m long, 100 m wide, and 20 m deep. The suspended-solids concentration in the wastewater is 30 mg/L, the wastewater discharge rate is 0.05 m^3/s, and the bulk density of the settled solids is 1,600 kg/m^3. Assuming that all of the suspended solids ultimately settle out in the lake, estimate the time required for 1 cm of sediment to accumulate at the bottom of the lake.

Solution

The concentration of suspended solids in the wastewater is 30 mg/L $= 0.03$ kg/m^3, and the discharge flowrate is 0.05 m^3/s. Under steady-state condition, the rate at which suspended solids are discharged into the lake is equal to the rate of sediment accumulation at the bottom of the lake and is given by

$$\text{Sediment mass accumulation rate} = 0.03 \times 0.05 = 0.0015 \text{ kg/s}$$

Since the bulk density of the settled solids is 1,600 kg/m^3, the rate of volume accumulation is

$$\text{Sediment volume accumulation rate} = \frac{0.0015}{1600} = 9.38 \times 10^{-7} \text{ m}^3/\text{s}$$

For 1 cm ($= 0.01$ m) to cover the bottom of the 300 m $\times$ 100 m lake, the volume of sediment is $0.01 \times 300 \times 100 = 300$ m^3. Hence, the time required for 300 m^3 of sediment to accumulate is given by

$$\text{Accumulation time for 1 cm of sediment} = \frac{300}{9.38 \times 10^{-7}}$$

$$= 3.20 \times 10^8 \text{ s} = 3700 \text{ days} = 10.1 \text{ years}$$

It is interesting to note that at this rate of sediment accumulation, the lake will be completely filled in approximately 20,000 years.   ■

Nitrogen.   Nitrogen is a nutrient that stimulates the growth of algae, and the oxidation of nitrogen species can consume significant amounts of oxygen. Biological reactions generally require a minimum amount of nitrogen to proceed. There are several forms of nitrogen, including organic nitrogen (e.g., proteins, amino acids, urea), ammonia (NH$_3$), nitrite (NO$_2$), nitrate (NO$_3$), and molecular nitrogen (N$_2$). Oxidation of nitrogen compounds, termed *nitrification*, proceeds as follows

$$\text{Organic nitrogen} + O_2 \rightarrow NH_3 \text{ nitrogen} + O_2 \rightarrow NO_2 \text{ nitrogen} + O_2 \rightarrow NO_3 \text{ nitrogen}$$

Nitrate nitrogen commonly originates in runoff from agricultural areas with heavy fertilizer usage, while organic nitrogen is commonly found in municipal wastewaters.

Phosphorous.   Phosphorous is a nutrient that stimulates the growth of algae. Phosphorous commonly exists in water as phosphates and organophosphates; it often originates from household detergents and agricultural runoff. Phosphorous is usually the limiting nutrient for the growth of algae; for lakes in the northern United States to be free of algal nuisances, the generally accepted upper concentration limit is 10 μg/L of orthophosphate.

Metals.   Many trace metals, such as arsenic, chromium, and mercury, are toxic in relatively small concentrations and find their way to natural water bodies from mining and manufacturing operations, as well as by the natural processes of chemical weathering and soil leaching. At toxic levels, most metals adversely affect the internal organs of the human body. A metal of particular concern in surface waters is mercury, where the biological magnification of mercury in fresh-water food fish is a significant hazard to human health.

Synthetic Organic Chemicals.   Synthetic organic chemicals include industrial solvents, insecticides, and herbicides. Many of these chemicals are hazardous to humans in relatively small concentrations. Synthetic organic chemicals constitute many of the toxic organic substances classified as *priority pollutants*, listed by the the U.S. Environmental Protection Agency (Council on Environmental Quality, 1978).

The categories of priority pollutants are given in Table 8.3. These pollutants can enter natural waterbodies by a variety of mechanisms, such as runoff (pesticides), municipal and industrial discharges (PCBs), chlorination byproducts in municipal wastewater discharges (HAHs), and spills (PAHs). Pesticides and herbicides are frequently found in surface waters and ground waters that receive runoff and infiltration from agricultural areas, and industrial solvents are common contaminants in ground water, resulting from both accidental and illicit spills.

pH.   The pH of water is defined as the negative log of the hydrogen ion concentration (in moles/L), commonly written as

$$pH = -\log_{10}[H^+] \tag{8.7}$$

The pH of natural waters affect biological and chemical reactions, control the solubility of metal ions, and affect natural aquatic life. The desirable pH for fresh-water aquatic life is in the range 6.5–9.0, and 6.5–8.5 for marine aquatic life.

8.2.2 ■ Water-Quality Standards

The most important laws governing the quality of surface waters and ground waters in the United States are the Clean Water Act (Public Law 92-500) and the Safe Drinking Water Act (Public Law 93-523). The Clean Water Act applies to surface waters, including rivers, lakes, and oceans and focuses primarily on establishing water-quality standards and regulations to control point-source discharges into surface waters. The Safe

■ **Table 8.3**	**Category**	**Description**
Categories of Organic Priority Pollutants	Pesticides	generally chlorinated hydrocarbons
	Polychlorinated biphenyls (PCBs)	used in electrical capacitors and transformers, paints, plastics, insecticides, other industrial products
	Halogenated aliphatics (HAHs)	used in fire extinguishers, refrigerants, propellants, pesticides, solvents for oils and greases, and in dry cleaning
	Ethers	used mainly as solvents for polymer plastics
	Phthalate esters	used chiefly in production of polyvinyl chloride and thermoplastics as plasticizers
	Monocyclic aromatics (MAHs)	(excluding phenols, cresols, and pthalates) used in the manufacture of other chemicals, explosives, dyes, and pigments, and in solvents, fungicides, and herbicides
	Phenols	large-volume industrial compounds used chiefly as chemical intermediates in the production of synthetic polymers, dyestuffs, pigments, herbicides, and pesticides
	Polycyclic aromatic hydrocarbons (PAHs)	used as dyestuffs, chemical intermediates, pesticides, herbicides, motor oils, and fuels
	Nitrosamines	used in the production of organic chemicals and rubber

Source: Council on Environmental Quality (1978).

Drinking Water Act establishes water-quality standards for drinking water delivered to the public by water-supply utilities and establishes regulations to protect underground sources of drinking water. The Safe Drinking Water Act applies to public water systems that have at least 15 service connections or that serve at least 25 persons. Programs relating to the Clean Water Act and the Safe Drinking Water Act are administered by the Environmental Protection Agency Office of Ground Water and Drinking Water (http://www.epa.gov/OGWDW/).

8.2.2.1 Surface Water Standards

In accordance with the Clean Water Act, all natural water bodies in the United States are classified according to their present and future most beneficial uses. Water-quality criteria to be met by various classifications of surface waters are specified by the states, and these criteria are considered to be minimum levels that are necessary to protect the designated uses of a water body. The Environmental Protection Agency (EPA) identifies the four categories of surface waters listed in Table 8.4, and federal water-quality guidelines are specified for each category of surface water. States are generally required to classify all water bodies within their borders and promulgate water-quality criteria that are

■ **Table 8.4**	**Category**	**Description**
EPA Classifications of Surface Waters	A	water contact recreation, including swimming
	B	able to support fish and wildlife
	C	public water supply
	D	agricultural and industrial use

	Class	Description
■ **Table 8.5** Classifications of Surface Waters in Florida	I	potable water supplies
	II	shellfish propagation or harvesting
	III*	recreation, propagation, and maintenance of a healthy, well-balanced population of fish and wildlife
	IV	agricultural water supplies
	V	navigation, utility, and industrial use

*There are two sets of criteria for Class III surface waters. One set of criteria is for predominantly fresh waters (chloride < 1,500 mg/L), and the other set of criteria is for predominantly marine waters (chloride > 1500 mg/L).

at least as stringent as the federal water-quality guidelines. Several states have adopted the EPA surface-water categories and associated water-quality criteria, and some states have expanded the surface-water categories to account for local conditions. For example, Florida classifies all surface waters within the state into the five categories shown in Table 8.5, and water-quality standards have been promulgated for each classification. Surface waters are defined as either fresh waters, open waters, or coastal waters. *Fresh waters* typically include waters contained in lakes and ponds or in flowing streams above the zone in which tidal actions influence the salinity of the water and where the concentration of chloride ions is less than 1,500 mg/L. *Open waters* are typically ocean waters that are seaward of a specified depth contour (5.49 m in Florida), and *coastal waters* are waters that are between fresh waters and open waters. The quality of surface waters are controlled by state regulatory agencies that establish concentration limits on discharges into a water body and control these discharges through the issuance of permits under the National Pollutant Discharge Elimination System (NPDES). Concentration limits on discharges into a water body are called *effluent limits*. Allowable effluent limits can be either technology-based limits or based on the water-quality criteria of the receiving water body. *Technology-based limits* are determined by the performance of state-of-the-art in treatment systems. For example, secondary municipal wastewater treatment plants are capable of producing effluents with BOD and suspended solids less than 30 mg/L, and the discharge limits for municipal wastewaters are usually set at these technology limits even if the receiving water is capable of assimilating higher levels of BOD and/or suspended solids. In cases where the ambient water quality requirements are more stringent than the technology-based limits, more advanced treatment is generally required to meet the ambient water-quality criteria. Under these conditions, the allowable contaminant concentration in the discharge is *water-quality limited*.

In many cases, discharges into surface waters do not meet the water-quality criteria of the receiving water body. Whenever such discharges are deemed to be in the public interest, a *mixing zone* is delineated surrounding the discharge location to allow a sufficient amount of dilution to meet the water-quality criteria outside of the mixing zone. The state of Florida defines a mixing zone as "a volume of surface water containing the point or area of discharge and within which an opportunity for the mixture of wastes with receiving surface waters has been afforded" (Florida, 1996). Regulatory mixing zones are usually limited in relative and absolute size. For example, mixing zones in fresh-water streams in Florida are limited to 10% of the length of the stream or 800 m, whichever is less (Florida, 1995). In lakes, estuaries, bays, lagoons, bayous, sounds, and coastal waters, the mixing zones are limited to 10% of the total area of the water body or 125,600 m^2 (= area enclosed by a circle of radius 200 m), whichever is less, and in open ocean waters the area of the mixing zone must be less than or equal to 502,655 m^2 (= area enclosed by a circle of radius 400 m).

In order to establish the extent of regulatory mixing zones, engineers must be able to predict the fate and transport of contaminants discharged into receiving bodies of water and determine the regions where the contaminant concentrations can reasonably be expected to meet the ambient water-quality criteria.

8.2.2.2   Ground Water Standards

Ground waters are generally classified by the states according to their present and future most beneficial uses. The most common classifications are *potable* and *nonpotable* (Zegel, 1997); however, in states that rely heavily on ground water for their water supply, narrower classifications are usually adopted. As an example, the five classes of ground waters in Florida are shown in Table 8.6 arranged in order of degree of protection required, with Class G-I having the most stringent water-quality criteria and Class G-IV the least. Ground waters classified for potable water use are generally required to meet the *primary* and *secondary* drinking-water quality standards. Primary drinking-water standards are health-driven requirements, and secondary standards are aesthetic-driven requirements. The (federal) drinking-water standards in the United States as of January 1999, are listed in Appendix G. In cases where the quality of native ground water exceeds drinking-water standards, the quality of the (uncontaminated) native ground water serves as the water-quality standard.

The Safe Drinking Water Act has established several important programs to protect and preserve the quality of ground waters, and these programs are listed in Table 8.7.

■ Table 8.6 Classifications of Ground Waters in Florida	Class	Description
	G-I	potable water use, ground water in single source aquifers which have a total dissolved solids content of less than 3,000 mg/L
	G-II	potable water use, ground water in aquifers which has a total dissolved solids content of less than 10,000 mg/L
	G-III	nonpotable water use, ground water in unconfined aquifers which has a total dissolved solids content of 10,000 mg/L or greater
	G-IV	nonpotable water use, ground water in confined aquifers which has a total dissolved solids content of 10,000 mg/L or greater

■ Table 8.7 Programs to Protect and Preserve the Quality of Ground Waters	Program	Description
	Underground Injection Control Program	Regulates the injection of fluids into wells.
	Sole Source Aquifer Program	EPA can designate certain aquifers as the sole source or principal drinking water source for an area. No federal funds are available for projects that may contaminate these aquifers.
	Wellhead Protection Program	States are required to develop programs to protect the areas around water-supply wells from contamination. EPA provides guidance and approves the final plan.
	Ground Water Protection Program	EPA makes grants to states for developing and implementing comprehensive programs for ground water protection.

Regulations associated with ground-water protection programs generally require engineers to predict the fate and transport of contaminants released directly into the ground water or on land surfaces above the ground water. These quantitative predictions are used to assess the impact of contaminant sources on ground-water quality, and to design systems to mitigate any deleterious effects.

8.3 Fate and Transport Processes

Diffusion is the process by which a tracer spreads within a fluid. The fundamental mechanism of diffusion is the random advection of tracer molecules on scales smaller than some defined length scale. At small (microscopic) length scales, tracers diffuse primarily through Brownian motion of the tracer molecules, while at larger scales tracers are diffused by random macroscopic variations in the fluid velocity. In cases where the random macroscopic variations is velocity are caused by turbulence, the diffusion process is called *turbulent diffusion*. Where spatial variations in the macroscopic velocity are responsible for the mixing of a tracer, the process is called *dispersion*. It is common practice to use the terms diffusion and dispersion interchangeably to describe the larger-scale mixing of contaminants in natural water bodies. In open waters, spatial variations in the macroscopic velocity are usually associated with shear flow and shoreline geometry, while in ground waters macroscopic (seepage) velocity variations are caused by the complex pore geometry and spatial variations in hydraulic conductivity. Regardless of the mechanism responsible for the spatial variations in velocity, whenever these velocity variations are either truly random or spatially uncorrelated over a defined *mixing scale*, the mixing process is described by Fick's law (Fick, 1855), which can be stated in the following generalized form

$$q_i^d = -D_{ij} \frac{\partial c}{\partial x_j} \tag{8.8}$$

where q_i^d is the dispersive mass flux (M/L^2T) in the x_i direction, D_{ij} is the dispersion coefficient tensor, and c is the tracer concentration. In cases where the dispersion coefficient varies with direction, the dispersion process is called *anisotropic*; in cases where the dispersion coefficient is independent of direction, the dispersion process is called *isotropic*. Hence, for isotropic dispersion, Fick's law is given by

$$q_i^d = -D \frac{\partial c}{\partial x_i} \tag{8.9}$$

Whereas the Fickian relation given by Equation 8.9 parameterizes the mixing effect of velocity variations with correlation length scales smaller than some defined mixing scale or support scale, tracer molecules are also advected by larger-scale fluid motions. The mass flux associated with the larger-scale (advective) fluid motions is given by

$$q_i^a = V_i c \tag{8.10}$$

where q_i^a is the advective tracer mass flux (M/L^2T) in the x_i-direction, and V_i is the large-scale fluid velocity in the x_i-direction. Since tracers are transported simultaneously by both advection and dispersion, the total flux of a tracer within a fluid is the sum of the advective and diffusive fluxes, and the homogeneous isotropic dispersion is given by

$$q_i = q_i^a + q_i^d = V_i c - D \frac{\partial c}{\partial x_i} \tag{8.11}$$

where q_i is the tracer flux in the x_i-direction. Equation 8.11 can also be written in vector form as

$$\mathbf{q} = \mathbf{V}c - D\nabla c \tag{8.12}$$

where $\mathbf{q}$ is the flux vector and $\mathbf{V}$ is the larger-scale fluid velocity. Consider the finite control volume shown in Figure 8.2, where this control volume is contained within the fluid transporting the tracer. In accordance with the law of conservation of mass, the net flux (M/T) of tracer mass into the control volume is equal to the rate of change of tracer mass (M/T) within the control volume. The law of conservation of mass can be put in the form

$$\frac{\partial}{\partial t} \int_V c \, dV + \int_S \mathbf{q} \cdot \mathbf{n} \, dA = \int_V S_m \, dV \tag{8.13}$$

where V is the volume of the control volume, c is the tracer concentration, S is the surface area of the control volume, $\mathbf{q}$ is the flux vector (Equation 8.12), $\mathbf{n}$ is the unit outward normal to the control volume, and S_m is the mass flux per unit volume originating within the control volume. Equation 8.13 can be simplified using the divergence theorem, which relates a surface integral to a volume integral by the relation

$$\int_S \mathbf{q} \cdot \mathbf{n} \, dA = \int_V \nabla \cdot \mathbf{q} \, dV \tag{8.14}$$

Combining Equations 8.13 and 8.14 leads to the result

$$\frac{\partial}{\partial t} \int_V c \, dV + \int_V \nabla \cdot \mathbf{q} \, dV = \int_V S_m \, dV \tag{8.15}$$

Since the control volume is fixed in space and time, the derivative of the volume integral with respect to time is equal to the volume integral of the derivative with respect to time and Equation 8.15 can be written in the form

$$\int_V \left(\frac{\partial c}{\partial t} + \nabla \cdot \mathbf{q} - S_m \right) dV = 0 \tag{8.16}$$

This equation requires that the integral of the quantity in parentheses must be equal to zero for any arbitrary control volume, and this can only be true if the integrand itself is

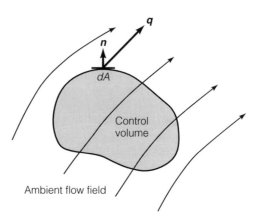

Figure 8.2 ▪ Control Volume in Fluid Transporting Tracer

equal to zero. Following this logic, Equation 8.16 requires that

$$\frac{\partial c}{\partial t} + \nabla \cdot \mathbf{q} - S_m = 0 \tag{8.17}$$

This equation can be combined with the expression for the mass flux given by Equation 8.12 and written in the expanded form

$$\frac{\partial c}{\partial t} + \nabla \cdot (\mathbf{V}c - D\nabla c) = S_m \tag{8.18}$$

which simplifies to

$$\frac{\partial c}{\partial t} + \mathbf{V} \cdot \nabla c + c(\nabla \cdot \mathbf{V}) = D\nabla^2 c + S_m \tag{8.19}$$

This equation applies to all tracers in all fluids. In the case of incompressible fluids, which is typical of the water environment, conservation of fluid mass requires that

$$\nabla \cdot \mathbf{V} = 0 \tag{8.20}$$

and combining Equations 8.19 and 8.20 yields the following diffusion equation for incompressible fluids with isotropic dispersion

$$\boxed{\frac{\partial c}{\partial t} + \mathbf{V} \cdot \nabla c = D\nabla^2 c + S_m} \tag{8.21}$$

In cases where there are no external sources or sinks of tracer mass (a conservative tracer), S_m is equal to zero and Equation 8.21 becomes

$$\boxed{\frac{\partial c}{\partial t} + \mathbf{V} \cdot \nabla c = D\nabla^2 c} \tag{8.22}$$

If the dispersion coefficient, D, is anisotropic, then the principal components of the dispersion coefficient can be written as D_i, and the diffusion equation becomes

$$\boxed{\frac{\partial c}{\partial t} + \sum_{i=1}^{3} V_i \frac{\partial c}{\partial x_i} = \sum_{i=1}^{3} D_i \frac{\partial^2 c}{\partial x_i^2} + S_m} \tag{8.23}$$

where x_i are the principal directions of the dispersion coefficient tensor. Equation 8.23 is the most commonly used relationship describing the mixing of contaminants released into natural water bodies, and it is known as either the *advection-dispersion* or the *advection-diffusion equation*, with the former term being more appropriate in most cases.

Sedimentation. The advection-dispersion equation, Equation 8.23, is appropriate for describing the fate and transport of dissolved contaminants that are advected with the same velocity as the ambient water. In the case of suspended sediments, the settling of the sediment particles is influenced by the size, shape, and density of the particles in addition to the ambient flow velocity. The process by which suspended sediments settle to the bottom of waterbodies is called *sedimentation*, and the settling velocity, v_s, of a sediment particle can be estimated by the *Stokes equation*, which is given by

Particle type	Diameter (μm)	Settling velocity (m/d)
Phytoplankton:		
Cyclotella meneghiniana	2	0.08 (0.24)*
Thalassiosira nana	4.3–5.2	0.1–0.28
Scenedesmus quadricauda	8.4	0.27 (0.89)
Asterionella formosa	25	0.2 (1.48)
Thalassiosira rotula	19–34	0.39–0.21
Coscinodiscus lineatus	50	1.9 (6.8)
Melosira agassizii	54.8	0.67 (1.87)
Rhizosolenia robusta	84	1.1 (4.7)
Particulate organic carbon	1–10	0.2
	10–64	1.5
	> 64	2.3
Clay	2–4	0.3–1
Silt	10–20	3–30

Source: Chapra, Steven. *Surface Water-Quality Monitoring.* Copyright © 1997. The McGraw-Hill Companies. Reprinted by permission of The McGraw-Hill Companies.
*Numbers in parentheses are for stationary phase of microbial growth.

$$v_s = \alpha \frac{(\rho_s/\rho_w - 1)g\phi^2}{18\nu_w} \qquad (8.24)$$

where α is a dimensionless form factor that measures the effect of particle shape ($\alpha = 1$ for spherical particles); ρ_s is the density of the sediment particle; ρ_w is the density of the ambient water; g is gravity; ϕ is the particle diameter; and ν_w is the kinematic viscosity of the ambient water. Particles in natural waters have complex shapes, and typically $\alpha < 1$. Settling velocities that are of interest in natural waters are given in Table 8.8. In cases where the ambient water moves with a horizontal velocity, V, suspended particles tend to move horizontally at the velocity, V, and vertically at the settling velocity.

Suspended solids in natural waters have two primary sources: surface runoff from drainage basins and as products of photosynthesis. The suspended-solids concentration in natural waters typically range from below 1 mg/L in clear waters to over 100 mg/L in highly turbid waters. Suspended solids derived from photosynthetic processes tend to be higher in organic matter and less dense than suspended solids derived from surface runoff. *Phytoplankton* are defined as drifting aquatic microorganisms of plant origin, and *zooplankton* are drifting aquatic microorganisms of animal origin. Caution should be used in applying the Stokes equation to calculate the settling velocity of living particles such as phytoplankton, which in their active state can become buoyant due to internal gas vacuoles.

■ Example 8.4

Analysis of water from a lake indicates a suspended-sediment concentration of 50 mg/L. The sediment particles are estimated to have an approximately spherical shape with an average diameter of 4 μm and a density of 2,650 kg/m^3. If the water temperature is 20°C, estimate the settling velocity of the suspended particles. If the suspended sediment is mostly clay, compare your estimate of the settling velocity with the data in Table 8.8. If there is 1 gram of heavy-metal ion per kilogram of sediment, determine the rate at which heavy metals are being removed from the lake by sedimentation.

Solution

From the given data, $\alpha = 1$ (spherical particles), $\rho_s = 2,650$ kg/m^3, $\rho_w = 998$ kg/m^3 at 20°C, $\phi = 4$ μm $= 4 \times 10^{-6}$ m, and $\nu_w = 1.00 \times 10^{-6}$ m^2/s. Substituting into the Stokes equation (Equation 8.24) gives

$$v_s = \alpha \frac{(\rho_s/\rho_w - 1)g\phi^2}{18\nu_w} = (1)\frac{(2650/998 - 1)(9.81)(4 \times 10^{-6})^2}{18(1.00 \times 10^{-6})}$$

$$= 1.44 \times 10^{-5} \text{ m/s} = 1.25 \text{ m/d}$$

This result is consistent with the settling velocities for clay-size particles shown in Table 8.8, which indicates that a 4 μm clay particle will have a settling velocity on the order of 1 m/d.

Since the concentration, c, of suspended sediments is 50 mg/L $= 0.05$ kg/m^3, then the rate at which sediment is accumulating on the bottom of the lake is given by

$$\text{Removal rate of sediment} = v_s c = (1.25)(0.05) = 0.0625 \text{ kg/d/m}^2$$

Since heavy metals are attached to the sediment at the rate of 1 g/kg, the removal rate of heavy metals is given by

$$\text{Removal rate of heavy metals} = (1)(0.0625) = 0.0625 \text{ g/d/m}^2 \quad ■$$

8.4　Rivers and Streams

Rivers have long been the primary source of drinking water to support human populations, and the water quality in rivers has been studied more extensively and longer than in any other bodies of water. Treated sewage effluent, industrial wastewaters, and stormwater runoff are routinely discharged into inland streams and, in most cases, the wastewater discharges do not meet the ambient water-quality standards of the stream. In such cases, *regulatory mixing zones* are usually permitted in the vicinity of the discharge location. Within these mixing zones, fate and transport processes reduce the contaminant concentrations to levels that meet the ambient water-quality criteria. The spatial extent of the mixing zone is usually restricted in size. In Florida, for example, mixing zones in streams are restricted to 800 m in length or 10% of the total length of the stream, whichever is less (Florida, 1995). Wastes can be discharged into streams either through multiport diffusers, which distribute the effluent over a finite portion of the stream width, or through single-port outfalls that discharge the effluent at a "point" in the stream.

Consider the case of a single-port discharge into a stream. The initial mixing of the pollutant is determined by the momentum and buoyancy of the discharge. As the discharge is diluted, the momentum and buoyancy of the discharge are dissipated, and further mixing of the discharge plume is dominated by ambient velocity variations in the stream. As the pollutant cloud extends over the depth and width of the stream, parts of the cloud expand into areas with significantly different longitudinal mean velocities; the pollutant cloud is "stretched" apart in the longitudinal direction in addition to diffusing in all three coordinate directions. This "stretching" of the cloud is primarily due to vertical and transverse variations in the longitudinal mean velocity. The process of "stretching" is commonly referred to as *shear dispersion* or simply *dispersion*. Contaminant discharges from point sources spread in the vertical and transverse directions by turbulent diffusion until the pollutant is well mixed across the stream cross-section, at which time almost all of the mixing will be caused by shear dispersion. Field results have shown that in most cases transverse variations in the mean velocity (across the channel) have a much greater effect on shear dispersion than vertical variations in the mean velocity.

8.4.1 ■ Initial Mixing

The turbulent velocity fluctuations in the vertical and transverse directions in rivers are on the same order of magnitude as the *shear velocity*, u_*, which is defined by

$$u_* = \sqrt{\frac{\tau_o}{\rho}} \tag{8.25}$$

where τ_o is the mean shear stress on the (wetted) perimeter of the stream and ρ is the density of the fluid. It has been shown in Chapter 4 that the boundary shear stress, τ_o, can be expressed in terms of the Darcy-Weisbach friction factor, f, and the mean (longitudinal) flow velocity, V, by

$$\tau_o = \frac{f}{8} \rho V^2 \tag{8.26}$$

Combining Equations 8.25 and 8.26 leads to the following expression for the shear velocity

$$u_* = \sqrt{\frac{f}{8}} V \tag{8.27}$$

The friction factor, f, can generally be estimated from the channel roughness, hydraulic radius, and Reynolds number using the Colebrook equation, as described in Chapter 4. Based on a theoretical analysis of turbulent mixing, Elder (1959) showed that the average value of the vertical diffusion coefficient, ε_v, in a stream can be estimated by the relation

$$\varepsilon_v = 0.067 \, du_* \tag{8.28}$$

where d is the depth of the stream. Experimental results in straight rectangular channels indicate that the transverse diffusion coefficient, ε_t, can be estimated by the relation

$$\varepsilon_t = 0.15 \, du_* \qquad \text{(straight uniform channels)} \tag{8.29}$$

where the coefficient of 0.15 can be taken to have an error bound of $\pm 50\%$ (Fischer et al., 1979). Experimental results yielded $\varepsilon_t = 0.08 du_*$ to $0.24 \, du_*$ (Lau and Krishnappan, 1977). Curves, sidewall irregularities, and variations in channel shape found in natural streams all serve to increase the transverse diffusion coefficient over that given by Equation 8.29. A more typical estimate of the transverse diffusion coefficient in natural streams is

$$\varepsilon_t = 0.6 \, du_* \qquad \text{(natural streams)} \tag{8.30}$$

where the coefficient of 0.6 can be taken to have an error bound of $\pm 50\%$.

Many mixing zone analyses assume instantaneous cross-sectional mixing and then calculate variations in cross-sectionally averaged concentrations downstream from the discharge location. Considering a stream of characteristic depth d and width w, the time scale, T_d, for mixing over the depth of the channel can be estimated by

$$T_d = \frac{d^2}{\varepsilon_v} \tag{8.31}$$

and the distance, L_d, downstream from the discharge point to where complete mixing over the depth occurs is given by

$$L_d = VT_d = \frac{Vd^2}{\varepsilon_v} \tag{8.32}$$

Similarly, the time scale, T_w, for mixing over the width, w, can be estimated by

$$T_w = \frac{w^2}{\varepsilon_t} \tag{8.33}$$

and the corresponding downstream length scale, L_w, to where the tracer is well mixed over the width can be estimated by

$$L_w = VT_w = \frac{Vw^2}{\varepsilon_t} \tag{8.34}$$

Combining Equations 8.32 and 8.34 leads to

$$\frac{L_w}{L_d} = \left(\frac{w}{d}\right)^2 \left(\frac{\varepsilon_v}{\varepsilon_t}\right) \tag{8.35}$$

According to Equations 8.28 and 8.30, it can be reasonably expected that in natural streams $\varepsilon_t \approx 10\varepsilon_v$, in which case Equation 8.35 can be approximated by

$$\boxed{\frac{L_w}{L_d} = 0.1 \left(\frac{w}{d}\right)^2} \tag{8.36}$$

Since channel widths in natural streams are usually much greater than channel depths, typically width/depth ≥ 20 (Koussis and Rodríguez-Mirasol, 1998), then for single-port discharges the downstream distance to where the tracer becomes well mixed across the stream can be expected to be at least an order of magnitude greater than the distance to where the tracer becomes well mixed over the depth. Therefore, mixing over the depth occurs well in advance of mixing over the width. The length scale to where the tracer can be expected to be well mixed over the width is given by Equation 8.34. Fischer and colleagues (1979) used field measurements to estimate the actual distance, L'_w, for a single-port discharge located on the side of a channel to mix across a stream as

$$\boxed{L'_w = 0.4 \frac{Vw^2}{\varepsilon_t}} \tag{8.37}$$

This estimate can be applied to any discharge location in the stream, where w is taken as the width over which the contaminant is to be mixed to achieve complete cross-sectional mixing. For example, if a multiport outfall of length L is placed in the center of a stream of width W, then full cross-sectional mixing occurs when the contaminant mixes over a width, $w = (W - L)/2$, and the downstream distance, L'_w, to complete cross-sectional mixing is given by

$$L'_w = 0.1 \frac{V(W - L)^2}{\varepsilon_t} \qquad (8.38)$$

Clearly, cross-sectional mixing can be accelerated by using multiport diffusers rather than single-port outlets.

■ Example 8.5

A municipality discharges wastewater from the side of a stream that is 10 m wide and 2 m deep. The average flow velocity in the stream is 1.5 m/s, and the friction factor is estimated to be 0.03 (calculated using the Colebrook equation). Estimate the time for the wastewater to become well mixed over the channel cross-section. How far downstream from the discharge location can the effluent be considered well-mixed across the stream?

Solution

From the given data, $f = 0.03$ and $V = 1.5$ m/s. Therefore, the shear velocity, u_*, is given by Equation 8.27 as

$$u_* = \sqrt{\frac{f}{8}} V = \sqrt{\frac{0.03}{8}} (1.5) = 0.092 \text{ m/s}$$

Since $d = 2$ m, the vertical and transverse diffusion coefficients are

$$\varepsilon_v = 0.067 d u_* = 0.067(2)(0.092) = 0.012 \text{ m}^2/\text{s}$$

$$\varepsilon_t = 0.6 d u_* = 0.6(2)(0.092) = 0.11 \text{ m}^2/\text{s}$$

The time scale for vertical mixing, T_d, is given by

$$T_d = \frac{d^2}{\varepsilon_v} = \frac{2^2}{0.012} = 333 \text{ s} = 5.6 \text{ min}$$

and the time scale for transverse mixing, T_w, is given by

$$T_w = \frac{w^2}{\varepsilon_t} = \frac{10^2}{0.11} = 909 \text{ s} = 15 \text{ min}$$

The discharge is well mixed over the channel cross-section when it is well mixed over both the depth and the width, which in this case occurs after about 15 minutes.

In a time interval of 15 minutes ($= 909$ s), the discharged effluent travels a distance, $V T_w$, given by

$$V T_w = (1.5)(909) = 1364 \text{ m}$$

The Fischer et al. (1979) relation given by Equation 8.37 indicates that the actual downstream distance required for complete cross-sectional mixing is $0.4 V T_w = 0.4(1364) = 546$ m.   ■

■ Example 8.6

Estimate the distance downstream to where the wastewater described in the previous example is well mixed across the stream if (a) the wastewater is discharged from the center of the stream, and (b) the wastewater is discharged through a 5-m long multiport diffuser placed in the middle of the stream.

Solution

From the previous analysis: $\varepsilon_v = 0.012$ m²/s, $\varepsilon_t = 0.11$ m²/s, and the time scale, T_s, for transverse mixing over a distance s is given by

$$T_s = \frac{s^2}{\varepsilon_t}$$

In the previous example, the wastewater was discharged from the side of the channel, so the mixing width, s, was the width of the channel, w.

(a) If the wastewater is discharged from the center of the stream, the mixing width, s, for the wastewater to become well mixed over the channel cross-section is given by

$$s = \frac{w}{2} = \frac{10}{2} = 5 \text{ m}$$

and the corresponding time scale, T_s, is given by

$$T_s = \frac{5^2}{0.11} = 227 \text{ s} = 3.8 \text{ min}$$

Since the flow velocity, V, is 1.5 m/s, the downstream distance, L, for the wastewater to become well mixed is

$$L = 0.4VT_s = 0.4(1.5)(227) = 136 \text{ m}$$

(b) If the wastewater is discharged from a 5-m long diffuser centered in the stream, the mixing width, s, for the wastewater to become well mixed over the channel cross-section is given by

$$s = \frac{w - 5}{2} = \frac{10 - 5}{2} = 2.5 \text{ m}$$

and the corresponding time scale, T_s, is given by

$$T_s = \frac{2.5^2}{0.11} = 56.8 \text{ s} = 0.95 \text{ min}$$

Since the flow velocity, V, is 1.5 m/s, the downstream distance, L, for the wastewater to become well mixed is

$$L = 0.4VT_s = 0.4(1.5)(56.8) = 34 \text{ m}$$

The results of this example illustrate that a 5-m long outfall diffuser located at the center of the stream will cause the wastewater to mix much more rapidly over the stream cross-section than a point discharge at the center of the stream.   ■

Examples 8.5 and 8.6 have illustrated why discharges from multiport outfalls generally acheive complete cross-sectional mixing more rapidly than single-port discharges. Also, at any given distance from the outfall, discharges from multiport outfalls are mixed over a larger portion of the channel width than discharges from single-port outfalls, resulting in multiport outfalls achieving larger dilutions. Consider the case where the mass flux of contaminant from an outfall (single-port or multiport) is given by $\dot{M}$, and the contaminant is mixed over a width, w, in a river, then conservation of mass requires that

$$\dot{M} = c_m V w d \tag{8.39}$$

where c_m is the mean concentration of the mixture, V is the mean velocity in the mixed portion of the river, and d is the mean depth in the mixed portion of the river. Rearranging Equation 8.39 gives the mean concentration, c_m, of the mixture by

$$\boxed{c_m = \frac{\dot{M}}{A_m V}} \tag{8.40}$$

where A_m is the area over which the contaminant is mixed ($= wd$).

■ Example 8.7

An industrial wastewater outfall discharges effluent at a rate of 3 m³/s into a river. The chromium concentration in the wastewater is 10 mg/L, the average velocity in the river is 0.5 m/s, and the average depth of the river is 3 m. Tracer tests in the river indicate that when the wastewater is discharged through a single port in the middle of the river, 100 m downstream of the outfall the plume will be mixed over a width of 4 m; if a 4-m long multiport outfall is used, the plume will be well mixed over a width of 8 m. Compare the dilution achieved by the single-port and multiport outfall at a location 100 m downstream of the discharge.

Solution

From the given data, $Q_o = 3$ m³/s, $c_o = 10$ mg/L $= 0.01$ kg/m³, and the mass flux, $\dot{M}$, of chromium released at the outfall is given by

$$\dot{M} = Q_o c_o = (3)(0.01) = 0.03 \text{ kg/s}$$

For the single-port discharge, the plume is mixed over an area, A_m, of 4 m × 3 m = 12 m². Since $V = 0.5$ m/s, the average concentration of the mixed river water 100 m downstream of the outfall is given by Equation 8.40 as

$$c_m = \frac{\dot{M}}{A_m V} = \frac{0.03}{(12)(0.5)} = 0.005 \text{ kg/m}^3 = 5 \text{ mg/L}$$

For the multiport discharge, the plume is mixed over an area, A_m, of 8 m × 3 m = 24 m², and the average concentration of the mixed river water is given by

$$c_m = \frac{\dot{M}}{A_m V} = \frac{0.03}{(24)(0.5)} = 0.0025 \text{ kg/m}^3 = 2.5 \text{ mg/L}$$

Therefore, at a location 100 m downstream of the discharge, the multiport outfall gives a dilution of $10/2.5 = 4$, compared with a dilution of $10/5 = 2$ for the single-port outfall.

■

Fate and transport models frequently assume that the discharged wasterwater is uniformly mixed over the stream cross-section in the vicinity of the discharge location.

Analyses described here demonstrate how this assertion can be assessed quantitatively. The assumption of complete cross-sectional mixing in the vicinity of a discharge location is usually justified in cases where outfall diffusers span the width of the stream and/or the wastewater flowrate is comparable to the river discharge. For single-port discharges on the sides of wide streams, where the discharge rate is small compared with the river discharge, considerable distances may be necessary for complete mixing across a river.

Consider the idealized waste discharge shown in Figure 8.3, where the river flowrate upstream of the wastewater outfall is Q_r, with a contaminant concentration, c_r, and the wastewater discharge rate is Q_w, with a contaminant concentration c_w. Assuming that the river flow and wastewater discharge are completely mixed in the *mixing zone* shown in Figure 8.3, conservation of contaminant mass requires that

$$Q_r c_r + Q_w c_w = (Q_r + Q_w)c_o \tag{8.41}$$

where c_o is the concentration of the wastewater/river water mixture downstream of the mixing zone. Rearranging Equation 8.41 gives the following expression for the concentration of the diluted wastewater after it is completely mixed with the ambient river water

$$\boxed{c_o = \frac{Q_r c_r + Q_w c_w}{Q_r + Q_w}} \tag{8.42}$$

In addition to estimating the concentrations of pollutants in mixed waters, Equation 8.42 can also be used to estimate contaminant concentrations downstream of the confluence of two or more streams, and to estimate the temperature, T_o, of mixed waters, where

$$T_o = \frac{Q_r T_r + Q_w T_w}{Q_r + Q_w} \tag{8.43}$$

and T_r and T_w are the temperatures of the river and waste discharge, respectively. The basic assumptions in applying Equation 8.43 are that the density and heat capacity of the mixed waters remain relatively constant.

In models that use the one-dimensional (along-stream) advection-dispersion equation to describe the mixing of contaminants in rivers, an initial concentration c_o is usually assumed to occur at the wastewater discharge location. It is clear from Equation 8.42 that for a given wastewater discharge, lower river flows will result in higher concentrations of the diluted wastewater. In analyzing the fate and transport of municipal and industrial discharges in rivers, the minimum seven-day average flow with a return period of 10 years is usually used as the design flow in the river (Thomann and Mueller, 1987). Such flows are typically designated using the notation aQb, where a is the number of days used in the average and b is the return period in years of the minimum a-day

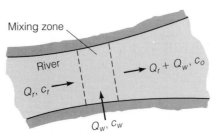

Figure 8.3 ■ Initial Mixing of Stream Discharges

average flow. Therefore, 7Q10 is the minimum seven-day average flow with a return period of 10 years.

■ **Example 8.8**

The dissolved oxygen concentration in a river upstream of a municipal wastewater out-fall is 10 mg/L. If the 7Q10 river flow upstream of the outfall is 50 m^3/s, and the out-fall discharges 2 m^3/s of wastewater with a dissolved oxygen concentration of 1 mg/L, estimate the dissolved oxygen concentration in the river after complete mixing. If the upstream river temperature is 10°C and the wastewater temperature is 20°C, estimate the temperature of the mixed river water.

Solution

From the given data, $Q_r = 50$ m^3/s, $c_r = 10$ mg/L, $Q_w = 2$ m^3/s, $c_w = 1$ mg/L, and the concentration of dissolved oxygen in the mixed river water is given by Equation 8.42 as

$$c_o = \frac{Q_r c_r + Q_w c_w}{Q_r + Q_w} = \frac{(50)(10) + (2)(1)}{50 + 2} = 9.7 \text{ mg/L}$$

Also, since $T_r = 10°C$ and $T_w = 20°C$, the temperature, T_o, of the mixture is given by Equation 8.43 as

$$T_o = \frac{Q_r T_r + Q_w T_w}{Q_r + Q_w} = \frac{(50)(10) + (2)(20)}{50 + 2} = 10.4°C$$

Therefore, in this case, the wastewater discharge has a relatively small effect on the dissolved oxygen and temperature of the river. ■

8.4.2 ■ Longitudinal Dispersion

Longitudinal mixing in streams is caused primarily by shear dispersion, which results from the "stretching effect" of both vertical and transverse variations in the longitudinal mean velocity. The *longitudinal dispersion coefficient* is used to parameterize the longitudinal mixing of a tracer that is well mixed across a stream, in which case the advection and longitudinal mixing of the tracer is described by the one-dimensional advection-dispersion equation

$$\frac{\partial c}{\partial t} + V\frac{\partial c}{\partial x} = \frac{\partial}{\partial x}\left(K_L \frac{\partial c}{\partial x}\right) + S_m \tag{8.44}$$

where c is the cross-sectionally averaged tracer concentration, V is the mean velocity in the stream, K_L is the longitudinal dispersion coefficient, x is the coordinate measured along the stream, and S_m is the net influx of tracer mass per unit volume of water per unit time. If the tracer is conservative, then $S_m = 0$. If the tracer undergoes first-order decay, then

$$S_m = -kc \tag{8.45}$$

where k is a *first-order rate constant* or *decay factor*. Combining Equations 8.44 and 8.45 gives the following equation for one-dimensional (longitudinal) advection and dispersion of a substance that undergoes first-order decay

$$\frac{\partial c}{\partial t} + V\frac{\partial c}{\partial x} = \frac{\partial}{\partial x}\left(K_L\frac{\partial c}{\partial x}\right) - kc \tag{8.46}$$

Longitudinal dispersion is caused by both vertical and transverse variations in the mean velocity. Fischer and colleagues (1979) have demonstrated that the longitudinal dispersion coefficient is proportional to the square of the distance over which the shear flow profile extends. Since natural streams typically have widths that are at least 10 times the depth, the longitudinal dispersion coefficient associated with transverse variations in the mean velocity can be expected to be on the order of 100 times larger than the longitudinal dispersion coefficient associated with vertical variations in the mean velocity. Consequently, vertical variations in the mean velocity are usually neglected in deriving expressions relating the longitudinal dispersion coefficient to the velocity distribution in natural streams. Several empirical and semiempirical formulae have been developed to estimate the longitudinal dispersion coefficient, K_L, in open-channel flow. The expressions proposed by Fischer and colleagues (1979), Liu (1977), Koussis and Rodríguez-Mirasol (1998), Iwasa and Aya (1991), and Seo and Cheong (1998) are listed in Table 8.9, where $\bar{d}$ is the mean depth of the stream, u_* is the shear velocity given by Equation 8.27, and w is the width of the stream. According to validation experiments conducted by Seo and Cheong (1998), the expressions proposed by Liu (1977), Iwasa and Aya (1991), and Seo and Cheong (1998) perform relatively well in most cases, while the expression proposed by Fischer and colleagues (1979) tends to overestimate K_L. For large rivers with channel widths larger than 200 m, both the Fischer and colleagues (1979) and the Liu (1977) expression tend to significantly overestimate K_L. Typical values of K_L are on the order of 0.05 to 0.3 m²/s for small streams (Genereux, 1991) to greater than 1,000 m²/s for large rivers (Wanner et al., 1989).

■ **Example 8.9**

Stream depths and vertically averaged velocities have been measured at 1-m intervals across a 10-m wide stream; the results are tabulated here. If the friction factor of the flow is 0.03, estimate the longitudinal dispersion coefficient using the expressions in Table 8.9.

Distance from side, y (m)	0	1	2	3	4	5	6	7	8	9	10
Depth, d (m)	0.0	0.20	0.90	1.2	2.1	3.0	2.4	1.5	0.75	0.45	0.0
Velocity, v (m/s)	0.0	0.30	0.60	0.80	1.4	2.0	1.6	1.0	0.50	0.30	0.0

■ **Table 8.9**
Estimates of Longitudinal Dispersion Coefficient in Rivers

Formula	Reference
$\frac{K_L}{\bar{d}u_*} = 0.011\left(\frac{w}{\bar{d}}\right)^2\left(\frac{V}{u_*}\right)^2$	Fischer et al. (1979)
$\frac{K_L}{\bar{d}u_*} = 0.18\left(\frac{w}{\bar{d}}\right)^2\left(\frac{V}{u_*}\right)^{0.5}$	Liu (1977)
$\frac{K_L}{\bar{d}u_*} = 0.6\left(\frac{w}{\bar{d}}\right)^2$	Koussis and Rodríguez-Mirasol (1998)
$\frac{K_L}{\bar{d}u_*} = 2.0\left(\frac{w}{\bar{d}}\right)^{1.5}$	Iwasa and Aya (1991)
$\frac{K_L}{\bar{d}u_*} = 0.64\left(\frac{w}{\bar{d}}\right)^{1.23}\left(\frac{V}{u_*}\right)^{1.25}$	Seo and Cheong (1998)

Solution

The flow area, A, can be estimated by summing the trapezoidal areas between the measurement locations which yields

$$A = (0 + 0.2 + 0.9 + 1.2 + 2.1 + 3.0 + 2.4 + 1.5 + 0.75 + 0.45 + 0)(1) = 12.5 \text{ m}^2$$

The average velocity, V, can be estimated by

$$V = \frac{1}{A} \sum_{i=1}^{10} v_i A_i$$

where v_i and A_i are the velocity and area increments measured across the channel. Therefore

$$V = \frac{1}{12.5}[(0.3)(0.2 \times 1) + (0.6)(0.9 \times 1) + (0.8)(1.2 \times 1) + (1.4)(2.1 \times 1)$$

$$+ (2.0)(3.0 \times 1) + (1.6)(2.4 \times 1) + (1.0)(1.5 \times 1) + (0.5)(0.75 \times 1)$$

$$+ (0.3)(0.45 \times 1)]$$

$$= 1.3 \text{ m/s}$$

Since $f = 0.03$ and $V = 1.3$ m/s, then the shear velocity, u_*, is given by Equation 8.27 as

$$u_* = \sqrt{\frac{f}{8}} V = \sqrt{\frac{0.03}{8}}(1.3) = 0.080 \text{ m/s}$$

and the average depth, $\bar{d}$, is given by

$$\bar{d} = \frac{A}{w} = \frac{12.5}{10} = 1.25 \text{ m}$$

The Fischer et al. (1979) expression estimates K_L by

$$\frac{K_L}{u_* \bar{d}} = 0.011 \left(\frac{w}{\bar{d}}\right)^2 \left(\frac{V}{u_*}\right)^2 = 0.011 \left(\frac{10}{1.25}\right)^2 \left(\frac{1.3}{0.08}\right)^2 = 186$$

which gives

$$K_L = 186 \bar{d} u_* = 186(1.25)(0.08) = 19 \text{ m}^2/\text{s}$$

The Liu (1977) expression estimates K_L by

$$\frac{K_L}{\bar{d} u_*} = 0.18 \left(\frac{w}{\bar{d}}\right)^2 \left(\frac{V}{u_*}\right)^{0.5} = 0.18 \left(\frac{10}{1.25}\right)^2 \left(\frac{1.3}{0.08}\right)^{0.5} = 46.4$$

which gives

$$K_L = 46.4 \bar{d} u_* = 46.4(1.25)(0.08) = 5 \text{ m}^2/\text{s}$$

The Koussis and Rodríguez-Mirasol (1998) expression estimates K_L by

$$\frac{K_L}{\bar{d} u_*} = 0.6 \left(\frac{w}{\bar{d}}\right)^2 = 0.6 \left(\frac{10}{1.25}\right)^2 = 38.4$$

which gives

$$K_L = 38.4\bar{d}u_* = 38.4(1.25)(0.08) = 4 \text{ m}^2/\text{s}$$

The Iwasa and Aya (1991) expression estimates K_L by

$$\frac{K_L}{\bar{d}u_*} = 2.0\left(\frac{w}{\bar{d}}\right)^{1.5} = 2.0\left(\frac{10}{1.25}\right)^{1.5} = 45.3$$

which gives

$$K_L = 45.3\bar{d}u_* = 45.3(1.25)(0.08) = 5 \text{ m}^2/\text{s}$$

The Seo and Cheong (1998) expression estimates K_L by

$$\frac{K_L}{\bar{d}u_*} = 0.64\left(\frac{w}{\bar{d}}\right)^{1.23}\left(\frac{V}{u_*}\right)^{1.25} = 0.64\left(\frac{10}{1.25}\right)^{1.23}\left(\frac{1.3}{0.08}\right)^{1.25} = 270$$

which gives

$$K_L = 270\bar{d}u_* = 270(1.25)(0.08) = 27 \text{ m}^2/\text{s}$$

The estimates of K_L are summarized in the following table:

Method	K_L (m^2/s)
Fischer et al. (1979)	19
Liu (1977)	5
Koussis and Rodríguez-Mirasol (1998)	4
Iwasa and Aya (1991)	5
Seo and Cheong (1998)	27

Based on these results, it should be clear that order-of-magnitude accuracy is all that can be expected in using available semiempirical and empirical equations to estimate K_L. ∎

8.4.3 ■ Spills

The governing equation for the longitudinal dispersion of contaminants that are well mixed over the cross-sections of rivers and streams and undergo first-order decay is given by Equation 8.46. The solution of Equation 8.46 for the case in which a mass, M, of contaminant is instantaneously mixed over the cross-section of the stream at time $t = 0$ is given by

$$c(x, t) = \frac{Me^{-kt}}{A\sqrt{4\pi K_L t}}\exp\left[-\frac{(x - Vt)^2}{4K_L t}\right] \tag{8.47}$$

where c is the contaminant concentration in the stream, x is the distance downstream of the spill, t is the time since the spill, A is the cross-sectional area of the stream, V is the average velocity in the stream, and k is the first-order decay factor. The concentration distribution described by Equation 8.47 has the form of a Gaussian distribution, with a variance growing with time. To see this more clearly, consider the equation of a Gaussian distribution, $f(x)$, given by

$$f(x) = \frac{M_o}{\sigma\sqrt{2\pi}} \exp\left[-\frac{1}{2}\left(\frac{x-\mu}{\sigma}\right)^2\right] \tag{8.48}$$

where μ is the mean of the distribution, σ is the standard deviation, and M_o is the total area under the curve. Note that a *normal* distribution is almost the same as a Gaussian distribution, except that the area under the curve, M_o, is equal to unity in the case of a normal distribution. Comparing the concentration distribution resulting from an instantaneous spill, Equation 8.47, to the Gaussian distribution, Equation 8.48, it is clear that the spill equation has a Gaussian distribution in the streamwise, x, direction with

$$M_o = Me^{-kt} \tag{8.49}$$

and the mean and standard deviation of the contaminant concentration distribution given by

$$\mu = Vt \tag{8.50}$$

$$\sigma = \sqrt{2K_L t} \tag{8.51}$$

Hence the concentration distribution resulting from an instantaneous spill is a symmetrical bell curve about a mean downstream location of $x = Vt$, and with a standard deviation of $\sqrt{2K_L t}$. The concentration distribution described by Equation 8.47 is illustrated in Figure 8.4.

▪ Example 8.10

Ten kilograms of a conservative contaminant ($k = 0$) are spilled in a stream that is 15 m wide, 3 m deep (on average), and has an average velocity of 35 cm/s. If the contaminant is rapidly mixed over the cross-section of the stream: (a) Derive an expression for the contaminant concentration as a function of time 500 m downstream of the spill; (b) if a peak concentration of 4 mg/L is observed 500 m downstream of the spill, estimate the longitudinal dispersion coefficient in the stream; (c) using the result in part (b), what would be the maximum contaminant concentration 1 km downstream of the spill?; and (d) if the detection limit of the contaminant is 1 μg/L, how long after release will the contaminant be detected 1 km downstream of the release point?

Solution

(a) Since the width of the stream, w, is 15 m and the average depth, $\bar{d}$, is 3 m, then the cross-sectional area, A, of the stream is given by

$$A = w\bar{d} = (15)(3) = 45 \text{ m}^2$$

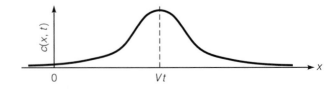

Figure 8.4 ▪ Concentration Distribution Resulting from Instantaneous Spill

The concentration distribution resulting from an instantaneous spill of a conservative contaminant ($k = 0$) is given by Equation 8.47 as

$$c(x, t) = \frac{M}{A\sqrt{4\pi K_L t}} \exp\left[-\frac{(x - Vt)^2}{4K_L t}\right]$$

In this case, $V = 35$ cm/s $= 0.35$ m/s, $M = 10$ kg, and $x = 500$ m, hence the concentration as a function of time 500 m downstream of the spill site is given by

$$c(500, t) = \frac{10}{(45)\sqrt{4\pi K_L t}} \exp\left[-\frac{(500 - 0.35t)^2}{4K_L t}\right]$$

$$= \frac{0.0627}{\sqrt{K_L t}} \exp\left[-\frac{(500 - 0.35t)^2}{4K_L t}\right] \text{ kg/m}^3$$

(b) The maximum concentration at any x occurs at a time t_o given by

$$t_o = \frac{x}{V}$$

In this case $x = 500$ m and $V = 0.35$ m/s, which gives

$$t_o = \frac{500}{0.35} = 1430 \text{ s}$$

The (maximum) concentration, c_o, at time t_o is given by

$$c_o = \frac{M}{A\sqrt{4\pi K_L t_o}} = \frac{0.0627}{\sqrt{K_L t_o}} \text{ kg/m}^3$$

When $c_o = 4$ mg/L $= 0.004$ kg/m^3 and $t_o = 1{,}430$ s, then

$$0.004 = \frac{0.0627}{\sqrt{K_L(1430)}}$$

which gives

$$K_L = 0.172 \text{ m}^2\text{/s}$$

(c) The maximum concentration 1 km downstream from the release point occurs at time t_1 where

$$t_1 = \frac{x_1}{V}$$

In this case, $x_1 = 1{,}000$ m and $V = 0.35$ m/s and therefore

$$t_1 = \frac{1000}{0.35} = 2{,}860 \text{ s}$$

The maximum concentration, c_1, occurring 1 km downstream from the release point is given by

$$c_1 = \frac{0.0627}{\sqrt{K_L t_1}} = \frac{0.0627}{\sqrt{(0.172)(2860)}} = 0.00283 \text{ kg/m}^3 = 2.83 \text{ mg/L}$$

(d) The concentration 1 km downstream of the release point as a function of time is given by

$$c(1000, t) = \frac{0.0627}{\sqrt{K_L t}} \exp\left[-\frac{(1000 - 0.35t)^2}{4K_L t}\right] \text{ kg/m}^3$$

When $c(1000, t) = 1\mu g/L = 10^{-6}\,kg/m^3$ and $K_L = 0.172\ m^2/s$,

$$10^{-6} = \frac{0.0627}{\sqrt{0.172t}} \exp\left[-\frac{(1000 - 0.35t)^2}{4(0.172)t}\right]$$

which gives

$$t = 2{,}520\ s$$

This time is relatively close to the time that the maximum concentration occurs (2,860 s, for a difference of 340 s = 5.7 min). It should be noted that there are two times when the concentration at $x = 1{,}000$ m is equa to $1\mu g/L$, once before and once after the arrival of the peak concentration.   ■

Field Measurement of K_L.   The variance of the concentration distribution resulting from an instantaneous spill is given by Equation 8.51 and can be written as

$$\sigma_x^2 = 2K_L t \tag{8.52}$$

where σ_x^2 is the variance of the concentration distribution in the streamwise, x, direction. Equation 8.52 is useful in determining K_L from field tests in which a slug of tracer is introduced into a stream, and the concentration as a function of time measured at two downstream locations, x_1 and x_2. If V is the mean flow velocity in the channel, then

$$x = Vt \tag{8.53}$$

and the variance of the temporal distribution of concentration, σ_t^2, measured at any location x can be approximately related to σ_x^2 by

$$\sigma_x^2 = V^2\sigma_t^2 \tag{8.54}$$

Combining Equations 8.52 to 8.54 gives the following expression for K_L in terms of measurable variables

$$K_L = \frac{V^3}{2}\frac{\sigma_{t2}^2 - \sigma_{t1}^2}{x_2 - x_1} \tag{8.55}$$

where σ_{t1}^2 and σ_{t2}^2 are the temporal variances of the concentration distributions measured at distances x_1 and x_2 downstream of the spill, respectively.

■ Example 8.11

A dye study is conducted to estimate the longitudinal dispersion coefficient in a river. The dye is released instantaneously from a bridge across the entire width of the river, and the dye concentrations as a function of time at locations 400 m and 700 m downstream of the bridge are tabulated here.

Time (min)	Concentration at $x = 400$ m (mg/L)	Concentration at $x = 700$ m (mg/L)
0	0	0
1	0	0
2	0	0
3	0.10	0

Time (min)	Concentration at $x = 400$ m (mg/L)	Concentration at $x = 700$ m (mg/L)
4	0.17	0
5	0.39	0
6	1.4	0
7	2.9	0
8	7.1	0
9	10.5	0.01
10	7.3	0.06
11	3.6	0.14
12	1.8	0.69
13	0.53	1.1
14	0.11	2.3
15	0.03	3.3
16	0	5.9
17	0	8.4
18	0	6.1
19	0	3.5
20	0	2.1
21	0	1.3
22	0	0.65
23	0	0.21
24	0	0.09
25	0	0.04
26	0	0
27	0	0

Estimate the longitudinal dispersion coefficient in the river and verify that mass is conserved between the upstream and downstream locations.

Solution

The longitudinal dispersion coefficient is given by Equation 8.55 as

$$K_L = \frac{V^3}{2} \frac{\sigma_{t2}^2 - \sigma_{t1}^2}{x_2 - x_1}$$

where $x_1 = 400$ m and $x_2 = 700$ m. The mean velocity, V, and the temporal variances at x_1 and x_2, σ_{t2}^2 and σ_{t1}^2, must be be estimated from the measured data. The mean velocity, V, can be estimated by

$$V = \frac{x_2 - x_1}{\bar{t}_2 - \bar{t}_1}$$

where $\bar{t}_1$ and $\bar{t}_2$ are the mean travel times to x_1 and x_2, respectively. Hence

$$\bar{t}_1 = \frac{1}{\sum_{i=1}^{27} c_{1i}} \sum_{i=1}^{27} t_i c_{1i}$$

where t_i and c_{1i} are the measurement times and concentrations at $x = 400$ m. Since

$$\sum_{i=1}^{27} c_{1i} = 0.10 + 0.17 + 0.39 + 1.4 + 2.9 + 7.1 + 10.5 + 7.3 + 3.6 + 1.8 + 0.53$$
$$+ 0.11 + 0.03$$
$$= 35.9 \text{ mg/L}$$

and

$$\sum_{i=1}^{27} t_i c_{1i} = (3)(0.10) + (4)(0.17) + (5)(0.39) + (6)(1.4) + (7)(2.9) + (8)(7.1)$$
$$+ (9)(10.5) + (10)(7.3) + (11)(3.6) + (12)(1.8) + (13)(0.53)$$
$$+ (14)(0.11) + (15)(0.03)$$
$$= 326 \text{ mg·min/L}$$

therefore

$$\bar{t}_1 = \frac{1}{35.9}(326) = 9.1 \text{ min}$$

At $x = 700$ m,

$$\bar{t}_2 = \frac{1}{\sum_{i=1}^{27} c_{2i}} \sum_{i=1}^{27} t_i c_{2i}$$

where t_i and c_{2i} are the measurement times and concentrations at $x = 700$ m. Since

$$\sum_{i=1}^{27} c_{2i} = 0.01 + 0.06 + 0.14 + 0.69 + 1.1 + 2.3 + 3.3 + 5.9 + 8.4$$
$$+ 6.1 + 3.5 + 2.1 + 1.3 + 0.65 + 0.21 + 0.09 + 0.04$$
$$= 35.9 \text{ mg/L}$$

and

$$\sum_{i=1}^{27} t_i c_{2i} = (9)(0.01) + (10)(0.06) + (11)(0.14) + (12)(0.69) + (13)(1.1)$$
$$+ (14)(2.3) + (15)(3.3) + (16)(5.9) + (17)(8.4) + (18)(6.1)$$
$$+ (19)(3.5) + (20)(2.1) + (21)(1.3) + (22)(0.65) + (23)(0.21)$$
$$+ (24)(0.09) + (25)(0.04)$$
$$= 612 \text{ mg·min/L}$$

therefore

$$\bar{t}_2 = \frac{1}{35.9}(612) = 17.0 \text{ min}$$

the mean velocity in the river is therefore given by

$$V = \frac{x_2 - x_1}{\bar{t}_2 - \bar{t}_1} = \frac{700 - 400}{17.0 - 9.1} = 38.0 \text{ m/min} = 0.63 \text{ m/s}$$

The variance of the concentration distribution at $x = 400$ m, σ_{t1}^2, is given by

$$\sigma_{t1}^2 = \frac{1}{\sum_{i=1}^{27} c_{1i}} \sum_{i=1}^{27} (t_i - \bar{t}_1)^2 c_{1i}$$

where

$$\sum_{i=1}^{27} (t_i - \bar{t}_1)^2 c_{1i} = (3 - 9.1)^2(0.1) + (4 - 9.1)^2(0.17) + (5 - 9.1)^2(0.39)$$

$$+ (6 - 9.1)^2(1.4) + (7 - 9.1)^2(2.9) + (8 - 9.1)^2(7.1)$$

$$+ (9 - 9.1)^2(10.5) + (10 - 9.1)^2(7.3) + (11 - 9.1)^2(3.6)$$

$$+ (12 - 9.1)^2(1.8) + (13 - 9.1)^2(0.53) + (14 - 9.1)^2(0.11)$$

$$+ (15 - 9.1)^2(0.03)$$

$$= 95.4 \text{ mg·min}^2/\text{L}$$

hence

$$\sigma_{t1}^2 = \frac{1}{35.9}(95.4) = 2.66 \text{ min}^2 = 9580 \text{ s}^2$$

The variance of the concentration distribution at $x = 700$ m, σ_{t2}^2, is given by

$$\sigma_{t2}^2 = \frac{1}{\sum_{i=1}^{27} c_{2i}} \sum_{i=1}^{27} (t_i - \bar{t}_2)^2 c_{2i}$$

where

$$\sum_{i=1}^{27} (t_i - \bar{t}_2)^2 c_{2i} = (9 - 17.0)^2(0.01) + (10 - 17.0)^2(0.06) + (11 - 17.0)^2(0.14)$$

$$+ (12 - 17.0)^2(0.69) + (13 - 17.0)^2(1.1) + (14 - 17.0)^2(2.3)$$

$$+ (15 - 17.0)^2(3.3) + (16 - 17.0)^2(5.9) + (17 - 17.0)^2(8.4)$$

$$+ (18 - 17.0)^2(6.1) + (19 - 17.0)^2(3.5) + (20 - 17.0)^2(2.1)$$

$$+ (21 - 17.0)^2(1.3) + (22 - 17.0)^2(0.65) + (23 - 17.0)^2(0.21)$$

$$+ (24 - 17.0)^2(0.09) + (25 - 17.0)^2(0.04)$$

$$= 175 \text{ mg·min}^2/\text{L}$$

hence

$$\sigma_{t2}^2 = \frac{1}{35.9}(175) = 4.87 \text{ min}^2 = 17500 \text{ s}^2$$

Substituting into Equation 8.55 to determine the longitudinal dispersion coefficient, K_L, gives

$$K_L = \frac{V^3}{2} \frac{\sigma_{t2}^2 - \sigma_{t1}^2}{x_2 - x_1} = \frac{(0.63)^3}{2} \frac{17500 - 9580}{700 - 400} = 3.30 \text{ m}^2/\text{s}$$

Conservation of mass requires that the areas under the concentration versus time curves at the upstream and downstream measurement locations are equal. The foregoing calculations show that the areas under the concentration versus time curves at the upstream and downstream locations are both equal to 35.9 mg·min/L; hence, mass is conserved.   ▪

8.4.4   ▪   Oxygen-Sag Model

The continuous discharge of wastewaters with high BOD into rivers and streams exerts an oxygen demand that depletes the dissolved oxygen in the ambient water and can sometimes cause severe stress on aquatic life. Aside from depleting the dissolved oxygen in streams, the oxygen demand exerted by the wastewater is partially met by oxygen transfer from the atmosphere at the surface of the stream, a process that is commonly referred to as *aeration*. The oxygen demand of wastewaters are typically measured by the BOD, and the rate of (de)oxygenation, S_1 (M/L^{3}T), is commonly described by a first-order reaction of the form

$$S_1 = -k_1 L \qquad (8.56)$$

where k_1 is a reaction-rate constant (T^{-1}), and L is the BOD remaining (M/L^3). Typical values of k_1 are shown in Table 8.10.

The rate at which oxygen is transferred from the atmosphere into the stream, defined as the *reaeration rate*, S_2 (M/L^{3}T), is commonly described by an equation of the form

$$S_2 = k_2(c_s - c) \qquad (8.57)$$

where k_2 is the *reaeration constant* (T^{-1}), c_s is the dissolved-oxygen saturation concentration (M/L^3), and c is the actual concentration of dissolved oxygen in the stream. Typical values of k_2 at 20°C are given in Table 8.11, and several empirical formulae for estimating k_2 at 20°C are given in Table 8.12, where the units of k_2 are d^{-1}, the average stream velocity, V, is in m/s, the average stream depth, $\bar{d}$, is in m, and the average channel slope, S_o, is dimensionless. In practice, the empirical formula proposed by O'Connor and Dobbins (1958) has the widest applicability and provides reasonable estimates of k_2 in most cases. Churchill and colleagues' (1962) formula applies to similar depths as the O'Connor and Dobbins model, but for faster streams. Owens and colleagues' (1964) formula is used for shallower streams; in small streams, the formula proposed by Tsivoglou and Wallace (1972) compares best with observed values (Thomann and Mueller, 1987). The formulae listed in Table 8.12 all give values of k_2 that approach zero as the depth of

---

▪ **Table 8.10**
Typical Deoxygenation
Rate Coefficients

Type of water	Ranges of k_1 at 20°C* (d^{-1})
Untreated wastewater	0.35–0.7
Treated wastewater	0.10–0.30
Polluted river	0.10–0.25

Source: Kiely (1997); Thomann and Mueller (1987).
*For temperatures other than 20°C, use (Thomann and Mueller, 1987)

$$k_{1_T} = k_{1_{20}} 1.04^{T-20}$$

where T is the temperature of the stream.

■ **Table 8.11**
Typical Reaeration
Constants

Water body	Ranges of k_2 at 20°C* (d^{-1})
Small ponds and backwaters	0.10–0.23
Sluggish streams and large lakes	0.23–0.35
Large streams of low velocity	0.35–0.46
Large streams of normal velocity	0.46–0.69
Swift streams	0.69–1.15
Rapids and waterfalls	> 1.15

Source: Adapted from *Water Quality* by George Tchobanoglous and Edward D. Schroeder. Copyright © 1985 by Addison-Wesley Publishing Company.
*For temperatures other than 20°C, use

$$k_{2_T} = k_{2_{20}} 1.024^{T-20}$$

where T is the temperature of the stream.

■ **Table 8.12**
Empirical Formulae for
Estimating Reaeration
Constant, k_2 at 20°C

Formula	Field conditions	Reference
$k_2 = 3.93 \dfrac{V^{0.5}}{\bar{d}^{1.5}}$	0.3 m < $\bar{d}$ < 9 m, 0.15 m/s < V < 0.50 m/s	O'Connor and Dobbins (1958)
$k_2 = 5.23 \dfrac{V}{\bar{d}^{1.67}}$	0.6 m < $\bar{d}$ < 3 m, 0.55 m/s < V < 1.50 m/s	Churchill et al. (1962)
$k_2 = 5.32 \dfrac{V^{0.67}}{\bar{d}^{1.85}}$	0.1 m < $\bar{d}$ < 3 m, 0.03 m/s < V < 1.50 m/s	Owens et al. (1964)
$k_2 = 3.1 \times 10^4 V S_o$	0.3 m < $\bar{d}$ < 0.9 m, 0.03 m³/s < Q < 0.3 m³/s	Tsivoglou and Wallace (1972)

the stream increases, implying that reaeration does not occur for deep bodies of water. This is certainly not the case, and the reaeration constant typically has a minimum value in the range

$$k_{2_{min}} = \frac{0.6}{\bar{d}} \quad \text{to} \quad \frac{1.0}{\bar{d}} \tag{8.58}$$

If the calculated value of k_2 falls below the range of minimum values given in Equation 8.58, then $k_2 = 0.6/\bar{d}$ should be used.

■ **Example 8.12**

A river has a width of 20 m, a mean depth of 5 m, and an estimated flowrate of 47 m³/s. Estimate the reaeration constant using the applicable equation(s) in Table 8.12. If the temperature of the river is 20°C and the dissolved oxygen concentration is 5 mg/L, estimate the reaeration rate. Determine the mass of oxygen added per day per meter along the river.

Solution

From the given data, $\bar{d} = 5$ m, $w = 20$ m, $Q = 47$ m³/s, and the average velocity, V, is given by

$$V = \frac{Q}{w\bar{d}} = \frac{47}{(20)(5)} = 0.47 \text{ m/s}$$

The O'Connor and Dobbins formula is the only applicable model and gives

$$k_2 = 3.93 \frac{V^{0.5}}{\bar{d}^{1.5}} = 3.93 \frac{(0.47)^{0.5}}{(5)^{1.5}} = 0.24 \text{ d}^{-1}$$

According to Table 8.11, this reaeration rate is typical of sluggish streams and large lakes. The calculated value of k_2 (0.24 d^{-1}) exceeds the range of minimum values of k_2 given by Equation 8.58 as 0.6/5-1.0/5 or 0.12-0.2 d^{-1}.

The reaeration rate, S_2, is given by Equation 8.57 as

$$S_2 = k_2(c_s - c)$$

where $c = 5$ mg/L, and the saturation concentration, c_s, at 20°C is given in Table 8.2 as 9.1 mg/L. Therefore, the reaeration rate, S_2, is given by

$$S_2 = (0.24)(9.1 - 5) = 0.98 \text{ mg/L·d} = 980 \text{ mg/m}^3\text{·d}$$

The volume of river water per meter is given by

$$\text{Volume of river water} = w\bar{d}(1) = (20)(5)(1) = 100 \text{ m}^3$$

Hence, the mass of oxygen added per day per meter along the river is 980(100) = 98, 000 mg/d/m = 98 g/d/m.   ∎

The total flux of oxygen into the river water, S_m (M/L^{3}T), can be estimated by adding the (de)oxygenation rate due to biodegradation, S_1, to the flux due to reaeration, S_2, to yield

$$S_m = -k_1 L + k_2(c_s - c) \tag{8.59}$$

The concentration distribution of oxygen in rivers can be described by combining the advection-dispersion equation (Equation 8.44) with the source flux given by Equation 8.59 to yield

$$\frac{\partial c}{\partial t} + V\frac{\partial c}{\partial x} = \frac{\partial}{\partial x}\left(K_L \frac{\partial c}{\partial x}\right) - k_1 L + k_2(c_s - c) \tag{8.60}$$

where the flow area, A, in the river is assumed to be constant. Assuming steady-state conditions ($\partial c/\partial t = 0$) and that the dispersive flux of oxygen is much less than oxygen fluxes due to reaeration and deoxygenation, then Equation 8.60 becomes

$$V\frac{\partial c}{\partial x} = -k_1 L + k_2(c_s - c) \tag{8.61}$$

Instead of being concerned with the oxygen concentration, c, it is convenient to deal with the *oxygen deficit*, D, defined by

$$D = c_s - c \tag{8.62}$$

Combining Equations 8.61 and 8.62 and replacing the partial derivative by the total derivative (since c is only a function of x) yields the following differential equation that describes the oxygen deficit in the river

$$\frac{dD}{dx} = \left(\frac{k_1}{V}\right)L - \left(\frac{k_2}{V}\right)D \tag{8.63}$$

The BOD remaining at any time, t, since release will follow the first-order reaction

$$\frac{dL}{dt} = -k_1 L \tag{8.64}$$

which can be solved independently of Equation 8.63 to yield

$$L = L_o \exp(-k_1 t) \tag{8.65}$$

where L_o is the BOD remaining at time $t = 0$. The time since release, t, is related to the distance traveled by

$$t = \frac{x}{V} \tag{8.66}$$

and hence the remaining BOD, L, at a distance x downstream of the wastewater discharge is derived by combining Equations 8.65 and 8.66 to give

$$L = L_o \exp\left(-k_1 \frac{x}{V}\right) \tag{8.67}$$

The differential equation describing the oxygen deficit in a river is therefore given by the combination of Equations 8.63 and 8.67, and the simultaneous solution of these equations with the boundary condition that $D = D_o$ at $x = 0$ is given by

$$D(x) = \frac{k_1 L_o}{k_2 - k_1}\left[\exp\left(-\frac{k_1 x}{V}\right) - \exp\left(-\frac{k_2 x}{V}\right)\right] + D_o \exp\left(-\frac{k_2 x}{V}\right) \tag{8.68}$$

This equation, originally derived by Streeter and Phelps (1925), is frequently referred to as the *Streeter-Phelps oxygen-sag curve*. The reason for using the term *sag curve* is apparent from a plot of the oxygen deficit, $D(x)$, as a function of distance, x, from the source as illustrated in Figure 8.5. Oxygen consumption for biodegradation begins immediately after the waste is discharged, at $x = 0$, with the oxygen deficit in the stream increasing from its initial value of D_o. Since reaeration is proportional to the oxygen deficit, the reaeration rate increases as the oxygen deficit increases, and at some point the reaeration rate becomes equal to the rate of oxygen consumption. This point is called the *critical point*, x_c, and beyond the critical point the reaeration rate exceeds the rate of oxygen consumption, resulting in a gradual decline in the oxygen deficit. The critical point, x_c, can be derived from Equation 8.68 by taking $dD/dx = 0$, which leads to

$$x_c = \frac{V}{k_2 - k_1} \ln\left[\frac{k_2}{k_1}\left(1 - \frac{D_o(k_2 - k_1)}{k_1 L_o}\right)\right] \tag{8.69}$$

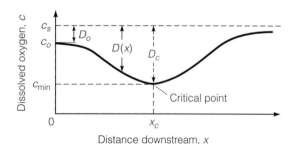

Figure 8.5 ■ Streeter-Phelps Oxygen-Sag Curve

and the corresponding critical oxygen deficit, D_c, is given by

$$\boxed{D_c = \frac{k_1}{k_2} L_o \exp\left(-\frac{k_1 x_c}{V}\right)} \tag{8.70}$$

If the value of x_c calculated using Equation 8.69 is less than or equal to zero, then the minimum oxygen deficit occurs at the waste outfall.

■ Example 8.13

A multiport outfall discharges wastewater into a slow-moving river that has a mean velocity of 3 cm/s. After initial mixing, the dissolved-oxygen concentration in the river is 9.5 mg/L and the temperature is 15°C. If the ultimate BOD of the mixed river water is 30 mg/L, the rate constant for BOD at 20°C is 0.6 d^{-1}, and the reaeration rate constant at 20°C is 0.8 d^{-1}, estimate the minimum dissolved oxygen and the critical location in the river.

Solution

At $T = 15$°C, the saturation concentration of oxygen is 10.1 mg/L (Table 8.2), and hence the initial oxygen deficit, D_o, is $10.1 - 9.5 = 0.6$ mg/L. The BOD rate constant at 15°C, $k_{1_{15}}$, is given by

$$k_{1_{15}} = k_{1_{20}} 1.04^{T-20} = (0.6)1.04^{15-20} = 0.48 \text{ d}^{-1}$$

The reaeration rate constant at 15°C, $k_{2_{15}}$, is given by

$$k_{2_{15}} = k_{2_{20}} 1.024^{T-20} = (0.8)1.024^{15-20} = 0.72 \text{ d}^{-1}$$

Since $L_o = 30$ mg/L and $V = 3$ cm/s = 2,592 m/d, Equation 8.69 gives the location, x_c, of the critical oxygen deficit as

$$\begin{aligned} x_c &= \frac{V}{k_2 - k_1} \ln\left[\frac{k_2}{k_1}\left(1 - \frac{D_o(k_2 - k_1)}{k_1 L_o}\right)\right] \\ &= \frac{2592}{0.72 - 0.48} \ln\left[\frac{0.72}{0.48}\left(1 - \frac{0.6(0.72 - 0.48)}{(0.48)(30)}\right)\right] \\ &= 4270 \text{ m} \end{aligned}$$

and Equation 8.70 gives the critical oxygen deficit, D_c, as

$$\begin{aligned} D_c &= \frac{k_1}{k_2} L_o \exp\left(-\frac{k_1 x_c}{V}\right) \\ &= \frac{0.48}{0.72}(30) \exp\left(-\frac{0.48 \times 4270}{2592}\right) = 9.0 \text{ mg/L} \end{aligned}$$

Hence the minimum dissolved oxygen level in the stream is $10.1 - 9.0 = 1.1$ mg/L. This level of dissolved oxygen will be devastating to the aquatic ecosystem. ■

Aside from neglecting longitudinal dispersion, several additional assumptions are implicit in the Streeter-Phelps oxygen-sag model, the most important of which are: (1) mixing occurs rapidly over the cross-section of the channel, and (2) biochemical oxygen demand and reaeration are the only significant oxygen sources and sinks. If mixing does not occur rapidly, as measured by the length scale for cross-sectional mixing

(Equation 8.37), significant oxygen depletion may occur prior to cross-sectional mixing, and Equation 8.42 will not give an accurate measure of the initial concentration for the one-dimensional oxygen-sag model. Under these circumstances, a more detailed two- or three-dimensional numerical model of cross-sectional mixing is necessary. Besides biochemical oxygen demand and reaeration, other sources/sinks of oxygen in rivers include photosynthesis, respiration of photosynthetic organisms, and benthic oxygen demand. Photosynthesis by phytoplankton, particularly algae, contributes oxygen to the water during daylight hours, and respiration removes oxygen at night. Since the amount of photosynthetic and respiration activity depends on the amount and intensity of sunlight, there is a diurnal and seasonal variation in dissolved oxygen amounts contributed by photosynthesis and respiration. In fact, the diurnal variation can sometimes be so extreme that the oxygen level is supersaturated during the afternoon and severely depleted just before dawn. Photosynthesis and respiration are a major source and sink of oxygen (respectively) particularly in slow-moving streams and lakes and can be expected to be significant for algal concentrations in excess of 10 g/m^3 (dry mass).

Quantification of oxygen fluxes associated with photosynthesis and respiration is difficult, and is generally dependent on such variables as temperature, nutrient concentration, sunlight, turbidity, and whether the plants are floating (phytoplankton) or on the bottom (macrophytes, periphyton). Reported photosynthetic oxygen production rates, S_p, (averaged over 24 h) range from 0.3 to 3 g/m^2·d for moderately productive surface waters up to 10 g/m^2·d for surface waters that have a significant biomass of aquatic plants (Thomann and Mueller, 1987). Reported respiration rates, S_r, have approximately the same range as photosynthetic oxygen production rates. Benthic oxygen demand primarily results from the deposition of suspended organics and native benthic organisms in the vicinity of wastewater discharges, and can be a major sink of dissolved oxygen in heavily polluted rivers and streams. Most benthic sludge undergoes anaerobic decomposition, which is a relatively slow process, however, aerobic decomposition can occur at the interface between the sludge and the flowing water. The products of anaerobic decomposition are CO_2, CH_4, and H_2S, and if gas production is especially high, floating of bottom sludge may result, leading to an aesthetic problem, as well as depletion of dissolved oxygen. Benthic oxygen demand, S_b^* (M/L^{2}T), is typically taken as a constant in most applications (Thomann, 1972), and the benthic flux of oxygen, S_b (M/L^{3}T), used in the advection-dispersion equation is derived from S_b^* using the relation

$$S_b = \frac{S_b^* A_s}{\forall} = \frac{S_b^*}{\bar{d}} \tag{8.71}$$

where A_s is the surface area of the bottom of the stream, $\forall$ is the volume of the stream section containing a benthic surface area, A_s, and $\bar{d}$ is the average depth of flow. Typical values of S_b^* at 20°C are given in Table 8.13, and calculated values of S_b at 20°C can be

■ **Table 8.13**
Typical Benthic Oxygen Demand Rates, S_b^*, at 20°C

Bottom type	Range (g/m^2·d)	Average value (g/m^2·d)
Filamentous bacteria (10 g/m^2)	—	−7
Municipal sewage sludge near outfall	−2 to −10	−4
Municipal sewage sludge downstream of outfall (aged)	−1 to −2	−1.5
Estuarine mud	−1 to −2	−1.5
Sandy bottom	−0.2 to −1.0	−0.5
Mineral soils	−0.05 to −0.1	−0.07

Source: Adapted from Thomann (1972).

converted to other temperatures using the relation

$$S_{b_T} = S_{b_{20}}(1.065)^{T-20} \tag{8.72}$$

where S_{b_T} and $S_{b_{20}}$ are the values of S_b at temperature T and 20°C, respectively. At temperatures below 10°C, S_b declines faster than indicated by Equation 8.72 and, in the range from 5°C to 0°C, S_b approaches zero (Chapra, 1997). There has been some debate about whether S_b should be taken as a constant, since it is almost certainly a function of the organic content of the sediments and the oxygen concentration in the overlying water, both of which vary with distance from the source. As more data become available, the functional approach will no doubt become the preferred formulation, however, in the meantime, a paucity of field data would support taking S_b to be a constant.

Incorporating photosynthetic, respiratory, and benthic oxygen fluxes into the oxygen-sag model yields the following modified form of Equation 8.61

$$V\frac{\partial c}{\partial x} = -k_1 L + k_2(c_s - c) + (S_p + S_r + S_b) \tag{8.73}$$

which leads to a critical point, x_c, given by

$$x_c = \frac{V}{k_2 - k_1}\ln\left[\frac{k_2}{k_1} - \frac{k_2 D_o(k_2 - k_1) + (S_p + S_r + S_b)(k_2 - k_1)}{k_1^2 L_o}\right] \tag{8.74}$$

and a corresponding critical oxygen deficit, D_c, given by

$$D_c = \frac{k_1}{k_2}L_o \exp\left(-k_1\frac{x_c}{V}\right) - \frac{S_p + S_r + S_b}{k_2} \tag{8.75}$$

In addition to the discharge of municipal wastewater at discrete locations, a spatially distributed input of BOD is typically associated with runoff from agricultural areas, stormwater discharges, and highway runoff. Therefore, even in the absence of municipal wastewater discharges, there is usually a deficit in dissolved oxygen in streams and rivers. Background deficit concentrations are typically on the order of 0.5–2 mg/L and are typically assumed as a factor of safety is assessing the assimilative capacity of streams and rivers (Schnoor, 1996).

■ Example 8.14

An outfall discharges wastewater into a slow-moving river that has a mean velocity of 3 cm/s and an average depth of 3 m. After initial mixing, the dissolved-oxygen concentration in the river is 9.5 mg/L, the saturation concentration of oxygen is 10.1 mg/L, the ultimate BOD of the mixed river water is 20 mg/L, the rate constant for BOD is 0.48 d^{-1}, and the reaeration rate constant is 0.72 d^{-1}. During the night, algal respiration exerts an oxygen demand of 2 g/m²·d, and sludge deposits downstream of the outfall exert a benthic oxygen demand of 4 g/m²·d. Estimate the minimum dissolved oxygen and the critical location in the river.

Solution

From the given data: the initial oxygen deficit, D_o, is $10.1 - 9.5 = 0.6$ mg/L, $k_1 = 0.48$ d^{-1}, $k_2 = 0.72$ d^{-1}, $S_r^* = -2$ g/m²·d, and $S_b^* = -4$ g/m²·d. Since the average depth, $\bar{d}$, of the river is 3 m, then the volumetric oxygen demand rates for respiration and benthic consumption, S_r and S_b, respectively, are

$$S_r = \frac{S_r^*}{d} = -\frac{2}{3} = -0.667 \text{ g/m}^3 \cdot \text{d} = -0.667 \text{ mg/L} \cdot \text{d}$$

and

$$S_b = \frac{S_b^*}{d} = -\frac{4}{3} = -1.33 \text{ g/m}^3 \cdot \text{d} = -1.33 \text{ mg/L} \cdot \text{d}$$

Since $L_o = 20$ mg/L and $V = 3$ cm/s $= 2592$ m/d, Equation 8.74 gives the location, x_c, of the critical oxygen deficit as

$$x_c = \frac{V}{k_2 - k_1} \ln \left[\frac{k_2}{k_1} - \frac{k_2 D_o(k_2 - k_1) + (S_p + S_r + S_b)(k_2 - k_1)}{k_1^2 L_o} \right]$$

$$= \frac{2592}{0.72 - 0.48} \ln \left[\frac{0.72}{0.48} - \frac{(0.72)(0.6)(0.72 - 0.48) + (-0.667 - 1.33)(0.72 - 0.48)}{(0.48)^2 (20)} \right]$$

$$= 4951 \text{ m}$$

and Equation 8.75 gives the critical oxygen deficit, D_c, as

$$D_c = \frac{k_1}{k_2} L_o \exp \left(-\frac{k_1 x_c}{V} \right) - \frac{S_p + S_r + S_b}{k_2}$$

$$= \frac{0.48}{0.72} (20) \exp \left(-\frac{0.48 \times 4951}{2592} \right) - \frac{-0.667 - 1.33}{0.72}$$

$$= 8.1 \text{ mg/L}$$

Hence the minimum dissolved oxygen level in the stream is $10.1 - 8.1 = 2.2$ mg/L and occurs 4,951 m downstream of the outfall location.   ■

The Streeter-Phelps model is used widely in practice, and an understanding of the assumptions incorporated in the model is essential for proper application. An important physical limitation of the Streeter-Phelps model is the neglect of longitudinal dispersion that can be important in some cases, particularly in tidal rivers and estuaries, where longitudinal dispersion causes both upstream and downstream effects on the dissolved oxygen. If longitudinal dispersion is considered, then the steady-state advection-dispersion equation for dissolved oxygen is given by the simultaneous solution of the following equations

$$K_L \frac{d^2 D}{dx^2} - V \frac{dD}{dx} - k_2 D + k_1 L = 0 \tag{8.76}$$

and

$$K_L \frac{d^2 L}{dx^2} - V \frac{dL}{dx} - k_1 L = 0 \tag{8.77}$$

where K_L is the longitudinal dispersion coefficient. Solving Equation 8.76 with the boundary conditions

$$D = 0 \text{ at } x = -\infty \qquad D = D_o \text{ at } x = 0 \qquad D = 0 \text{ at } x = +\infty \tag{8.78}$$

and solving Equation 8.77 with the boundary conditions

$$L = 0 \text{ at } x = -\infty \qquad L = L_o \text{ at } x = 0 \qquad L = 0 \text{ at } x = +\infty \qquad (8.79)$$

and combining the results gives the following expression for the oxygen deficit (Schnoor, 1996)

$$D(x) = \frac{L_o k_1 m_1}{k_2 - k_1} \left[\frac{1}{m_1} \exp\left\{ \frac{Vx}{2K_L}(1 - m_1) \right\} - \frac{1}{m_2} \exp\left\{ \frac{Vx}{2K_L}(1 - m_2) \right\} \right] \qquad (8.80)$$

where

$$m_1 = \sqrt{1 + \frac{4k_1 K_L}{V^2}} \quad \text{and} \quad m_2 = \sqrt{1 + \frac{4k_2 K_L}{V^2}} \qquad (8.81)$$

Values of m_1 and m_2 contain important dimensionless numbers that measure the relative importance of dispersion and advection, where

$$\frac{k_1 K_L}{V^2}, \frac{k_2 K_L}{V^2} > 20, \qquad \text{dispersion predominates} \qquad (8.82)$$

and

$$\frac{k_1 K_L}{V^2}, \frac{k_2 K_L}{V^2} < 0.05, \qquad \text{advection predominates} \qquad (8.83)$$

▪ Example 8.15

A tidal river has a mean velocity of 5 cm/s, a reaeration constant of 0.75 d^{-1}, and a longitudinal dispersion coefficient of 120 m^2/s. Wastewater is discharged into the river, and after initial mixing, the river has an ultimate BOD of 10 mg/L, and a BOD decay constant of 0.4 d^{-1}. If the temperature of the river is 20°C, determine the oxygen deficit 200 m downstream of the outfall. Assess whether consideration of longitudinal dispersion is important in predicting the effect of wastewater on dissolved oxygen levels in the river.

Solution

From the given data, $V = 5$ cm/s $= 4,320$ m/d, $k_1 = 0.4$ d^{-1}, $k_2 = 0.75$ d^{-1}, $K_L = 120$ m^2/s $= 1.04 \times 10^7$ m^2/d, and $L_o = 10$ mg/L. Taking longitudinal dispersion into account, the dissolved oxygen profile is given by Equation 8.80, where

$$m_1 = \sqrt{1 + \frac{4k_1 K_L}{V^2}} = \sqrt{1 + \frac{4(0.4)(1.04 \times 10^7)}{(4320)^2}} = 1.37$$

and

$$m_2 = \sqrt{1 + \frac{4k_2 K_L}{V^2}} = \sqrt{1 + \frac{4(0.75)(1.04 \times 10^7)}{(4320)^2}} = 1.63$$

Hence, Equation 8.80 gives the oxygen deficit 200 m downstream of the outfall as

$$D(200) = \frac{(10)(0.4)(1.37)}{0.75 - 0.4} \left[\frac{1}{1.37} \exp\left\{ \frac{(4320)(200)}{2(1.04 \times 10^7)} (1 - 1.37) \right\} \right.$$

$$\left. - \frac{1}{1.63} \exp\left\{ \frac{(4320)(200)}{2(1.04 \times 10^7)} (1 - 1.63) \right\} \right]$$

$$= 1.9 \text{ mg/L}$$

The dimensionless numbers given in Equations 8.82 and 8.83 are as follows

$$\frac{k_1 K_L}{V^2} = \frac{0.4(1.04 \times 10^7)}{(4320)^2} = 0.22$$

$$\frac{k_2 K_L}{V^2} = \frac{0.75(1.04 \times 10^7)}{(4320)^2} = 0.42$$

Comparing these with the limits given in Equations 8.82 and 8.83 indicates that both advection and dispersion should be considered, but neither predominates. ■

Design criteria for wastewater discharges generally require that dissolved oxygen levels in receiving waters not fall below specified water quality standards. In these cases, Equations 8.68 or 8.80 can be used to estimate the allowable BOD in the wastewater discharge.

8.5 Lakes

Lakes are large reservoirs of water in which currents are typically driven by wind rather than gravity. Natural and humanmade lakes have many uses, including recreation, municipal and industrial water supply, hydropower, and flood control. Lakes differ from rivers and streams in several important ways (James, 1993): (1) lakes rarely receive discharges of organic matter large enough to cause serious oxygen depletion; (2) lakes have significantly longer retention times than most rivers; and (3) the principal water-quality gradients are in the vertical direction rather than the longitudinal direction. Water-pollution problems in lakes can be quite persistent because of their long *hydraulic detention times*. The hydraulic detention time, t_d, is defined by the relation

$$t_d = \frac{V_L}{Q_o} \tag{8.84}$$

where V_L is the average volume of the lake, and Q_o is the average outflow rate. To give an idea of typical time scales, Baumgartner (1996) states that, on average, detention times in large lakes are on the order of years, whereas the detention times for water in rivers and streams are on the order of days. Taking the surface area, A_L, of a lake as a measure of its volume, V_L, and taking the drainage area, A_D, as a measure of the average outflow, Q_o, from the lake, the detention time, t_d (days), of a natural lake can be estimated by the relation (Bartsch and Gakstatter, 1978)

$$\log_{10} t_d = 4.077 - 1.177 \log_{10} \frac{A_D}{A_L} \qquad 1 \text{ day} < t_d < 6000 \text{ days} \tag{8.85}$$

where this equation applies only to lakes with uncontrolled discharges.

It is important to note that long detention times do not necessarily correspond to large lakes, since small lakes with small outflow rates can have comparable detention times to large lakes with large outflow rates. With low velocities and long detention times, most lake environments are favorable to sedimentation, resulting in most of the incoming sediments, and many of the organisms that grow and die in the lake, accumulating at the bottom of the lake. Over extended periods of time, these processes can permanently change the character of a lake, greatly increasing its organic content and ultimately converting it to marsh and land areas (Lamb, 1985). Lakes are either natural or humanmade; the main operational difference is that natural lakes tend to have uncontrolled outflows, while humanmade lakes have controlled outflows. Elongated and dendritic lakes are typical of artificial impoundments created by damming rivers, while natural lakes tend to be more circular. The physical parameters that most affect the water quality in lakes are detention time and depth. Detention times are typically classified as *short* if less than one year and *long* if they are more than one year; lakes are classified as *shallow* if less than 7–10 m in depth and *deep* if they are more than 10 m in depth.

■ Example 8.16

A natural lake has an estimated volume of 9×10^5 m^3, a surface area of 85,000 m^2, and an average uncontrolled outflow rate of 310 m^3/d. Estimate the hydraulic detention time in the lake and the drainage area that contributes runoff into the lake.

Solution

From the given data, $V_L = 9 \times 10^5$ m^3, $Q_o = 310$ m^3/d, and Equation 8.84 gives the hydraulic detention time, t_d, as

$$t_d = \frac{V_L}{Q_o} = \frac{9 \times 10^5}{310} = 2900 \text{ days} = 8.0 \text{ years}$$

The detention time can be empirically related to the drainage area, A_D, by Equation 8.85. Taking $A_L = 85,000$ m^2, Equation 8.85 gives

$$\log_{10} t_d = 4.077 - 1.177 \log_{10} \frac{A_D}{A_L}$$

or

$$\log_{10} 2900 = 4.077 - 1.177 \log_{10} \frac{A_D}{85000}$$

which leads to

$$A_D = 2.8 \times 10^5 \text{ m}^2$$

Hence, the estimated drainage area is about 3.3 times the size of the lake.   ■

8.5.1 ■ Near-Shore Mixing Model

Near-shore mixing models are concerned with the distribution of contaminants in the vicinity of waste discharges into large bodies of water such as lakes. Consider the wastewater discharge illustrated in Figure 8.6, where the advective currents are negligible and the steady-state advection-dispersion equation with first-order decay is given by

$$D\left(\frac{\partial^2 c}{\partial x^2} + \frac{\partial^2 c}{\partial y^2}\right) - kc = 0 \tag{8.86}$$

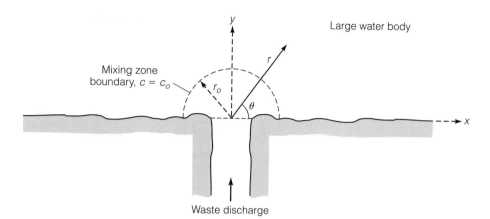

Figure 8.6 ■ Wastewater
Discharge into a Lake

where D is the dispersion coefficient, c is the contaminant concentration, and k is the first-order decay constant. Equation 8.86 can be written in polar (r, θ) coordinates as

$$\frac{\partial^2 c}{\partial r^2} + \frac{1}{r}\frac{\partial c}{\partial r} + \frac{1}{r^2}\frac{\partial^2 c}{\partial \theta^2} - \frac{k}{D}c = 0 \tag{8.87}$$

For a radially symmetric concentration distribution, $\partial c/\partial \theta$ and $\partial^2 c/\partial \theta^2$ are both equal to zero and Equation 8.87 reduces to

$$\frac{d^2 c}{dr^2} + \frac{1}{r}\frac{dc}{dr} - \frac{k}{D}c = 0 \tag{8.88}$$

which has the general solution

$$c(r) = AI_o\left(\sqrt{\frac{kr^2}{D}}\right) + BK_o\left(\sqrt{\frac{kr^2}{D}}\right) \tag{8.89}$$

where A and B are constants, and I_o and K_o are modified Bessel functions of the first and second kind, respectively (see Appendix F). Taking the boundary conditions as

$$c(r_o) = c_o \tag{8.90}$$

$$c(\infty) = 0 \tag{8.91}$$

requires that the concentration equal c_o on the boundary of a mixing zone at a distance r_o from the discharge location and that the pollutant concentration decay to zero at a large distance from the discharge location. Imposing the boundary conditions given by Equations 8.90 and 8.91 on Equation 8.89 gives the concentration distribution in the lake as (O'Connor, 1962)

$$\boxed{c = \frac{K_o\left(\sqrt{\frac{kr^2}{D}}\right)}{K_o\left(\sqrt{\frac{kr_o^2}{D}}\right)}c_o} \tag{8.92}$$

■ Example 8.17

An industrial plant discharges wastewater through a single-port outfall into the shoreline region of a lake. Field measurements indicate that the dispersion coefficient in the lake is 1 m²/s, the decay rate of the contaminant is 0.1 d⁻¹, and the effluent dilution 30 m from the outfall is 12. Estimate the distance from the outfall required to achieve a dilution of 100.

Solution

Dilution is defined as the initial concentration divided by the final concentration; hence if c_{30} is the contaminant concentration 30 m from the outfall, and c_i is the concentration at the discharge location, then

$$\frac{c_i}{c_{30}} = 12$$

and Equation 8.92 gives the concentration at a distance r from the outfall as

$$c_r = \frac{K_o\left(\sqrt{\frac{kr^2}{D}}\right)}{K_o\left(\sqrt{\frac{k30^2}{D}}\right)} c_{30} = \frac{K_o\left(\sqrt{\frac{kr^2}{D}}\right)}{K_o\left(\sqrt{\frac{k30^2}{D}}\right)}\left(\frac{c_i}{12}\right)$$

Hence the dilution, S_r, at a distance r from the outfall is given by

$$S_r = \frac{c_i}{c_r} = 12\frac{K_o\left(\sqrt{\frac{k30^2}{D}}\right)}{K_o\left(\sqrt{\frac{kr^2}{D}}\right)}$$

Since $S_r = 100$, $k = 0.1$ d⁻¹, and $D = 1$ m²/s $= 8.64 \times 10^4$ m²/d, we are looking for r such that

$$100 = 12\frac{K_o\left[\sqrt{\frac{(0.1)(30)^2}{8.64\times10^4}}\right]}{K_o\left[\sqrt{\frac{(0.1)r^2}{8.64\times10^4}}\right]}$$

or

$$\frac{K_o(0.0323)}{K_o(0.00108r)} = 8.33$$

Solving for r gives

$$r = 918 \text{ m}$$

Hence a dilution of 100 is achieved on the order of 918 m from the outfall. It is somewhat doubtful that this formulation would be applicable out to 918 m from the shoreline.   ▪

8.5.2 ▪ Eutrophication

Water bodies such as lakes and rivers can be classified in terms of their trophic state as: *oligotrophic* (poorly nourished), *mesotrophic* (moderately nourished), *eutrophic* (well nourished), and *hypereutrophic* (overnourished). In lakes, there is a natural progression from the oligotrophic state through the eutrophic state as part of the normal aging process that results from the recycling and accumulation of nutrients over a long period of time. As an example, nitrogen added to a lake is assimilated by algae; when the algae die, the bulk of the nitrogen is released and is available for assimilation by living algae. Hence the nitrogen accumulates in the lake and increases the nourishment level. This natural aging process can be accelerated by several orders of magnitude as a result of human activities. The process by which lakes become eutrophic is called *eutrophication*, and this process can have a number of deleterious effects. Negative impacts of eutrophication include: (1) the excessive growth of floating plants that decrease water clarity, clog filters at water-treatment plants, and create odors; (2) significant fluctuations in oxygen and carbon dioxide levels associated with photosynthesis and respiration, where low oxygen levels can cause the death of desirable fish species; (3) an increased sediment oxygen demand (SOD) associated with the settling of aquatic plants, resulting in low dissolved oxygen levels near the bottom of the water body; and (4) loss of diversity in aquatic ecosystems.

Aquatic plants that grow in surface waters can be broadly classified according to whether they move freely in the water (*planktonic aquatic plants*) or remain fixed in place, attached or rooted. Plants that move freely in water are called *phytoplankton* (e.g., free-floating algae), attached plants include *periphyton* (e.g., attached or benthic algae) and *macrophytes* (rooted, vascular aquatic plants). The types of fixed plant communities are varied and depend on the depth and clarity of the water. The level of eutrophication is usually measured by the *chlorophyll a* concentration (μg/L), which measures the gross level of phytoplankton.

The nutrients that are usually responsible for eutrophication are phosphorous and nitrogen. Nutrients are discharged into surface waters primarily by agricultural and urban runoff, municipal and industrial discharges, and combined-sewer overflows. Phosphorous and nitrogen usually occur in a variety of combined forms, and it is common to quantify their concentrations in terms of *total phosphorous* (TP) and *total nitrogen* (TN). The combined forms of phosphorous and nitrogen that contribute to TP and TN are illustrated in Figure 8.7. Since not all components of TP and TN are available for uptake by phytoplankton, it is important to delineate the components that are present in

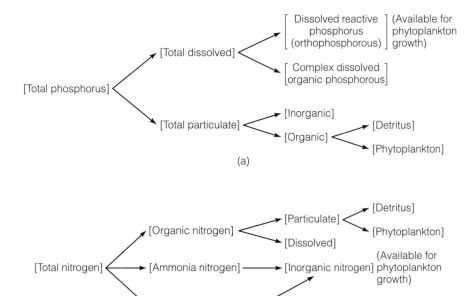

Figure 8.7 ■ Components of Total Phosphorous and Total Nitrogen

Source: Thomann, Robert and Mueller, John. *Principles of Surface Water Quality Modeling and Control.* Copyright © 1987. Scott-Foresman-Addison Wesley. Reprinted by permission.

the water. The two principal components of TP are the dissolved and particulate forms. Dissolved phosphorous includes several forms, of which dissolved reactive phosphorous (orthophosphate, PO_4^{-3}) is available for phytoplankton growth. Available phosphorous is commonly taken to be orthophosphate that passes a 0.45-μm membrane filter (Schnoor, 1996). The principal components of TN are organic, ammonia, nitrite (NO_2), and nitrate (NO_3) forms, of which ammonia, nitrite, and nitrate forms are utilized by phytoplankton for growth.

The design of systems to control eutrophication are based on identifying the limiting nutrient, and determining the allowable nutrient levels to maintain a desirable concentration of plant biomass in the water. The limiting nutrient can be identified by comparing the ratio of nitrogen to phorphorous (N/P) in the water that is available for plant growth, to the ratio of nitrogen to phosphorous required for plant growth. A consideration of the cell stoichiometry of phytoplankton shows a phosporous content of 0.5–2.0 μg/μg chlorophyll and a nitrogen content of 7–10 μg/μg chlorophyll, indicating that a N/P ratio on the order of 10 is utilized in plant growth. Hence, if the ratio of available nitrogen to available phosphorous in the water significantly exceeds 10, then phosphorous would be the limiting nutrient, whereas if the N/P ratio in the water is significantly less than 10, then nitrogen would be the limiting nutrient.

The control of eutrophication usually requires the specification of target biomass concentrations in terms of μg chlorophyll/L (μg chl/L); typical target biomass concentrations in northern temperate lakes are in the range of 5–10 μg chl/L for eutrophic lakes and 1–4 μg chl/L for oligotrophic lakes. Many lake models assume a priori that phosphorous is the limiting nutrient, in which case the biomass (chlorophyll) concentration, c_b, in a water body can be estimated based on the TP concentration. This approach implicitly assumes a stable relationship between TP and available phosphorous (orthophosphate, PO_4^{-3}) and neglects other variables that affect algal growth such as sedimentation, predation, nutrient recycling, and oxygen fluxes. Several empirical relations between c_b and TP are listed in Table 8.14, where c_b and TP are in μg/L. The variety of formulae in Table 8.14 reflect the seasonal, climatic, ecological, and hydrologic variations between lakes. Oligotrophic lakes can be defined as those with less than 10 μg TP/L; eutrophic lakes have more than 20 μg TP/L; and mesotrophic lakes have TP concentrations in the range 10–20 μg TP/L.

In municipal wastewaters with activated-sludge treatment (without phosphorous removal), nitrogen tends to be the limiting nutrient; in mixed agricultural and urban runoff, phosphorous tends to be the limiting nutrient. The limiting nutrients that are typical for several water bodies are shown in Table 8.15. According to the U.S. National Eutrophication Survey (USEPA, 1975), phosphorous is the limiting nutrient in most inland waters of the United States, where approximately 90% of the lakes studied showed phosphorous as the limiting nutrient. More detailed analyses of the impact of nutrients on lake ecosystems generally take into consideration the various ecological zones within individual lakes as listed in Table 8.16.

■ Table 8.14 Empirical Relations Between Biomass and TP Concentrations	Formula	Reference
	$\log_{10} c_b = 1.55 \log_{10} \text{TP} - 1.55 \log_{10} \left[\dfrac{6.404}{0.0204(\text{TN/TP}) + 0.334} \right]$	Smith and Shapiro (1981)
	$\log_{10} c_b = 0.807 \log_{10} \text{TP} - 0.194$	Bartsch and Gakstatter (1978)
	$\log_{10} c_b = 0.76 \log_{10} \text{TP} - 0.259$	Rast and Lee (1978)
	$\log_{10} c_b = 1.449 \log_{10} \text{TP} - 1.136$	Dillon and Rigler (1974)

Water body	Typical N/P	Limiting nutrient
Rivers and Streams		
Point source dominated:		
Without phosphorous removal	$\ll 10$	nitrogen
With phosphorous removal	$\gg 10$	phosphorous
Nonpoint-source dominated	$\gg 10$	phosphorous
Lakes		
Large, nonpoint-source dominated	$\gg 10$	phosphorous
Small, point-source dominated	$\ll 10$	nitrogen

■ **Table 8.15** Limiting Nutrients for Various Water Bodies

Source: Adapted from Thomann and Mueller (1987).

■ **Table 8.16** Ecological Zones in Lakes

Zone	Description
Littoral	shallow water, near shore
Pelagic	open water, deep
Euphotic	light penetration to 1–10% of near-surface light
Epilimnion	above the thermocline, well mixed
Hypolimnion	below the thermocline, well mixed
Sediment	diffusion of pore water, variable redox conditions

■ **Example 8.18**

Water-quality measurements in a lake indicate that available phosphorous is on the order of 60 μg/L, and available nitrogen is about 2 mg/L. Is the lake nitrogen limited or phosphorous limited? If the concentration of total phosphorous (TP) is 90 μg/L and the concentration of total nitrogen (TN) is 4 mg/L, estimate the biomass concentration and trophic state of the lake.

Solution

The ratio of available nitrogen to available phosphorous is $2 \times 10^3/60 = 33$. Since this ratio is greater than 10, the lake is phosphorous limited.

From the given data, TP = 90 μg/L and TN = 4,000 μg/L, and using the formulae in Table 8.14 gives the following results

Formula	c_b (μg/L)
Smith and Shapiro	52
Bartsch and Gakstatter	24
Rast and Lee	17
Dillon and Rigler	50

Hence the biomass concentration is estimated to be in the range 17–52 μg/L. Based on this biomass concentration (> 10 μg/L) and the concentration of total phosphorous (> 20 μg/L), the lake can be classified as eutrophic. ■

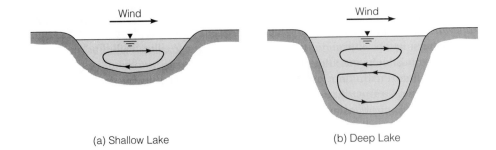

Figure 8.8 ■ Wind-
Induced Circulation in
Lakes

(a) Shallow Lake   (b) Deep Lake

8.5.3 ■ Thermal Stratification

Thermal stratification in lakes can have a pronounced effect on water quality, since temperature has a significant influence on the rates of chemical and biological reactions and strong temperature gradients can significantly limit the diffusion of dissolved oxygen from the water surface to the bottom of the lake. The fundamental processes that influence the thermal stratification of lakes are heat and momentum transfer across the lake surface and gravity forces acting on density differences within the lake.

Typical wind-induced circulation regimes in shallow and deep lakes are illustrated in Figure 8.8, where the main difference is that shallow lakes tend to have a single circulation cell, while deeper lakes tend to have more than one circulation cell. Wind-induced surface currents are called the *wind drift*, and are typically on the order of 2–3% of the wind speed. In warm summer weather, heat added at the surface of deep (> 10 m) lakes is concentrated in the top few meters, resulting in a warm, less dense layer that is well mixed, overlaying a distinctly colder and weakly mixed lower layer. The well mixed surface layer is called the *epilimnion*; the weakly mixed lower layer is called the *hypolimnion*. The warmer epilimnion is separated from the colder hypolimnion by a thin layer with a sharp temperature gradient called the *thermocline*. The turbidity of lake waters has a strong influence on the thickness of the epilimnion since surface heat attenuates rapidly in turbid waters. Waters in the epilimnion tend to be well oxygenated, while waters in the hypolimnion tend to be low in oxygen. In cool fall weather, the surface layers of deep lakes begin to become colder and more dense than the underlying water, leading to a gravity circulation, supplemented by the wind, that causes the lake waters to overturn and become better mixed. Since the maximum density of water occurs at 4°C, further cooling below 4°C under cold winter conditions causes the surface layers of the lake to become colder than 4°C, making them less dense than underlying water and the lake is again stratified. As the temperature warms during the spring, the surface waters warm to 4°C, becoming more dense than the underlying water, and causing a turnover in the lake waters. As warm summer weather returns, the lake tends to again become stratified, and the seasonal cycle is complete.

Thermal stratification is common in lakes located in temperate climates with distinct warm and cold seasons. From a water-quality viewpoint, lake turnover brings nutrients from lower layers into the surface layers, which have higher oxygen levels and more sunlight, thereby stimulating the eutrophication process. Depth and wind have significant influences on the thermal structure of lakes, with shallow lakes (< 10 m) rarely stratifying for long periods, and very deep lakes (> 30 m) generally remaining stratified on a long-term basis, either permanently in tropical climates or seasonally outside the tropics (James, 1993). The stability of a lake can be measured by a *densimetric Froude number*, Fr_D, defined by

$$\boxed{\text{Fr}_D = \frac{V}{\sqrt{\frac{\Delta\rho}{\rho_o}gd}}} \qquad (8.93)$$

where V is the average velocity in the lake, $\Delta\rho$ is the change in density over the depth, ρ_o is the average density of the water, g is the acceleration due to gravity, and d is the average depth of the lake. According to Tchobanoglous and Schroeder (1987), well-stratified lakes have $\text{Fr}_D \ll 0.1$, weakly stratified lakes have $0.1 < \text{Fr}_D < 1$, and fully mixed lakes have $\text{Fr} > 1$.

■ Example 8.19

Measurements in a 10-m deep lake show a mean velocity of 10 cm/s and a density difference between the top and bottom of the lake of 4.1 kg/m^3. If the mean density of the lake water can be taken as 998 kg/m^3, estimate the strength of the stratification.

Solution

From the given data, $d = 10$ m, $V = 10$ cm/s $= 0.1$ m/s, $\Delta\rho = 4.1$ kg/m^3, $\rho_o = 998$ kg/m^3, and hence the densimetric Froude number is given by Equation 8.93 as

$$\text{Fr}_D = \frac{V}{\sqrt{\frac{\Delta\rho}{\rho_o}gd}} = \frac{0.1}{\sqrt{\frac{4.1}{998}(9.81)(10)}} = 0.16$$

Since $0.1 < \text{Fr}_D < 1$, the lake should be classified as weakly stratified. ■

8.5.4 ■ Completely Mixed Model

The response of lakes to the input of contaminants can sometimes be estimated by assuming that the lake is well mixed. This approximation is justified when (1) wind-induced circulation is strong, and (2) the time scale of the analysis is sufficiently long (on the order of a year) that seasonal mixing processes yield a completely mixed lake. Contaminant mass fluxes into a lake can come from a variety of sources, including municipal and industrial waste discharges, inflows from polluted rivers, direct surface runoff, contaminant releases from sediments, and contaminants contained in rainfall (atmospheric sources). Denoting the rate of contaminant mass inflow to a lake by $\dot{M}$ (M/T), assuming that the lake is well mixed and assuming that the contaminant undergoes first-order decay with a decay factor, k, the law of conservation of contaminant mass requires that

$$\frac{d}{dt}(V_L c) = \dot{M} - Q_o c - kV_L c \qquad (8.94)$$

where V_L is the volume of the lake, c is the average contaminant concentration in the lake, and Q_o is the average outflow rate. Equation 8.94 states that the rate of change of contaminant mass in the lake, $d(V_L c)/dt$, is equal to the mass inflow rate, $\dot{M}$, minus the mass outflow rate, $Q_o c$, minus the rate at which mass is removed by first-order decay, $kV_L c$. Assuming that the volume of the lake, V_L, remains constant, Equation 8.94 becomes

$$V_L \frac{dc}{dt} + (Q_o + kV_L)c = \dot{M} \qquad (8.95)$$

which simplifies to the following differential equation that describes the contaminant concentration in the lake as a function of time

$$\frac{dc}{dt} + \left(\frac{Q_o}{V_L} + k\right) c = \frac{\dot{M}}{V_L} \tag{8.96}$$

Taking the mass inflow rate, $\dot{M}$, to be constant and beginning at $t = 0$, and taking the initial condition as

$$c = c_o \quad \text{at } t = 0 \tag{8.97}$$

yields the following solution to Equation 8.96 (Thomann and Mueller, 1987)

$$c(t) = \frac{\dot{M}}{Q_o + kV_L}\left\{1 - \exp\left[-\left(\frac{Q_o}{V_L} + k\right)t\right]\right\} + c_o \exp\left[-\left(\frac{Q_o}{V_L} + k\right)t\right] \tag{8.98}$$

where the first term on the right hand side of Equation 8.98 gives the buildup of concentration due to the continuous mass input, $\dot{M}$, and the second term accounts for the die-away of the initial concentration, c_o. The mass inflow rate, $\dot{M}$, and contaminant concentration, c, in Equation 8.98 as a function of time are illustrated in Figure 8.9 for cases in which the initial concentration, c_o, is less than and greater than the asymptotic concentration given by Equation 8.98. Taking $t \rightarrow \infty$ in Equation 8.98 gives the asymptotic concentration, c_∞, as

$$c_\infty = \frac{\dot{M}}{Q_o + kV_L} \tag{8.99}$$

Equation 8.98 can also be used to calculate the lake response to a mass inflow over a finite interval, Δt. This case is illustrated in Figure 8.10(a) for a mass inflow rate, $\dot{M}$, over an interval Δt. The response is described by Equation 8.98 up to $t = \Delta t$, beyond which the response is described by

$$c(t) = c_1 \exp\left[-\left(\frac{Q_o}{V_L} + k\right)(t - \Delta t)\right] \tag{8.100}$$

which derived from Equation 8.98 for a contaminant mass inflow of zero, and an initial concentration of c_1. In the case of a variable mass inflow illustrated in Figure 8.9(b), where the contaminant mass inflow rate is equal to $\dot{M}_1$ up to $t = \Delta t$, and equal to $\dot{M}_2$ thereafter, the response of the lake is described by Equation 8.98 up to $t = \Delta t$, beyond

Figure 8.9 ■ Response of a Well Mixed Lake to a Constant Contaminant Inflow

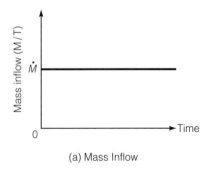

(a) Mass Inflow

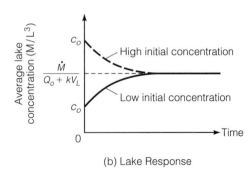

(b) Lake Response

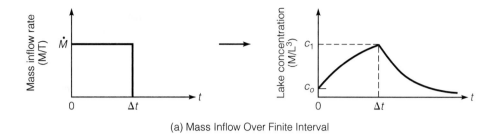

(a) Mass Inflow Over Finite Interval

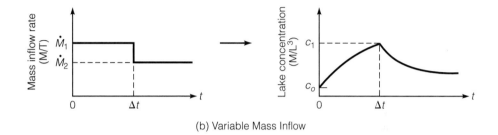

Figure 8.10 ■ Response of a Well Mixed Lake to a Variable Contaminant Inflow

(b) Variable Mass Inflow

which the concentration in the lake is described by

$$
\begin{aligned}
c(t) = {} & \frac{\dot{M}_2}{Q_o + kV_L} \left\{ 1 - \exp\left[-\left(\frac{Q_o}{V_L} + k \right)(t - \Delta t) \right] \right\} \\
& + c_1 \exp\left[-\left(\frac{Q_o}{V_L} + k \right)(t - \Delta t) \right]
\end{aligned}
\tag{8.101}
$$

which is derived from Equation 8.98 with a mass inflow rate, $\dot{M}_2$ beginning at $t = \Delta t$ with an initial concentration of c_1.

The analyses described here can also be applied to cases where the contaminants are removed by the settling of suspended solids in the lake. In this case, where the contaminants are adsorbed onto suspended solids, the removal rate due to sedimentation can be described by a settling velocity, v_s, and the conservation of mass equation can be written as

$$
\frac{d}{dt}(V_L c) = \dot{M} - Q_o c - kV_L c - v_s A_L c
\tag{8.102}
$$

where A_L is the surface area of the lake over which settling occurs. Equation 8.102 can be slightly rearranged into the form

$$
\frac{d}{dt}(V_L c) = \dot{M} - Q_o c - k' V_L c
\tag{8.103}
$$

where

$$
k' = k + \frac{v_s A_L}{V_L}
\tag{8.104}
$$

Since Equation 8.103 is identical with the conservation equation that neglects sedimentation as a removal process (Equation 8.94), with the decay coefficient, k, replaced by an effective decay coefficient, k', then all the previous results are applicable, provided that k is replaced by k'. Several variations in this model have been used in practice. In cases where there is significant vertical stratification, the lake can be considered as a well mixed epilimnion overlying a well mixed hypolimnion, with limited interaction between the two zones. Also, in large lakes ($A_L > 50–100$ km^2) it may be necessary to subdivide the lake into a number of well mixed smaller lakes in series.

■ Example 8.20

The average concentration of total phosphorous in a lake is 30 μg/L, and an attempt is to be made to reduce the phosphorous level in the lake by reducing phosphorous inflows into the lake. The target phosphorous concentration is 15 μg/L. If the discharge from the lake averages 0.09 m^3/s, the first-order decay rate for phosphorous is 0.01 d^{-1}, and the volume of the lake is 300,000 m^3, estimate the maximum allowable phosphorous inflow in kg/year. If this loading is maintained for three years but suddenly doubles in the fourth year, estimate the phosphorous concentration in the lake one month into the fourth year.

Solution

From the given data, $Q_o = 0.09$ m^3/s $= 7,776$ m^3/d, $k = 0.01$ d^{-1}, and $V_L = 300,000$ m^3. For an ultimate concentration, c_∞, of 15 μg/L ($= 15 \times 10^{-6}$ kg/m^3), Equation 8.99 gives

$$c_\infty = \frac{\dot{M}}{Q_o + kV_L}$$

which rearranges to

$$\dot{M} = [Q_o + kV_L]c_\infty = [7776 + (0.01)(300000)](15 \times 10^{-6}) = 0.16 \text{ kg/d} = 59 \text{ kg/year}$$

Maintaining a mass loading of 0.16 kg/d for 3 years ($= 1,095$ days), with $c_o = 30$ μg/L $= 30 \times 10^{-6}$ kg/m^3 yields a concentration at time t given by Equation 8.98 as

$$c(t) = \frac{\dot{M}}{Q_o + kV_L}\left\{1 - \exp\left[-\left(\frac{Q_o}{V_L} + k\right)t\right]\right\} + c_o \exp\left[-\left(\frac{Q_o}{V_L} + k\right)t\right]$$

which for $t = 1,095$ days gives

$$c(1095) = \frac{0.16}{7776 + (0.01)(300000)}\left\{1 - \exp\left[-\left(\frac{7776}{300000} + 0.01\right)(1095)\right]\right\}$$
$$+ 30 \times 10^{-6} \exp\left[-\left(\frac{7776}{300000} + 0.01\right)(1095)\right]$$
$$= 15 \text{ }\mu\text{g/L}$$

Hence after three years the phosphorous level has already decreased to the target level of 15 μg/L. In the fourth year, the mass flux doubles to $\dot{M}_2 = 2 \times 0.16 = 0.32$ kg/d, and the concentration as a function of time is given by Equation 8.101 as

$$c(t) = \frac{\dot{M}_2}{Q_o + kV_L}\left\{1 - \exp\left[-\left(\frac{Q_o}{V_L} + k\right)(t - \Delta t)\right]\right\} + c_1 \exp\left[-\left(\frac{Q_o}{V_L} + k\right)(t - \Delta t)\right]$$

where $c_1 = 15$ μg/L $= 15 \times 10^{-6}$ kg/m^3, $\Delta t = 1{,}095$ days, and after 1 month ($= 30$ days) $t = 1095 + 30 = 1{,}125$ days. The concentration in the lake is then given by

$$c(1125) = \frac{0.32}{7776 + (0.01)(300000)} \left\{ 1 - \exp\left[-\left(\frac{7776}{300000} + 0.01 \right)(1125 - 1095) \right] \right\}$$

$$+ 15 \times 10^{-6} \exp\left[-\left(\frac{7776}{300000} + 0.01 \right)(1125 - 1095) \right]$$

$$= 25 \ \mu\text{g/L}$$

Hence the lake concentration rebounds to almost the original concentration within one month. This is a reflection of the relatively short detention time in the lake. ■

8.6 Ocean Discharges

Ocean outfalls are used by many coastal communities to discharge treated sewage effluent into open-ocean waters. The wastewater discharged by an ocean outfall experiences a significant amount of mixing in the immediate vicinity of the outfall, with dilution resulting from the entrainment of ambient seawater as the buoyant fresh-water plume rises in the denser saltwater environment. As ambient seawater is entrained, the effluent plume becomes denser, rising until the plume density equals the density of the ambient seawater. If the ocean is (density) stratified, then there is the possibility that the plume will be "trapped" below the surface. On the other hand, if the ocean is unstratified, the density of the fresh-water plume can never equal the density of the ocean water (no matter how much of the ocean water is entrained) and the plume reaches the ocean surface, possibly forming a noticeable *boil*. Ambient currents advect the plume away from the outfall, and spatial variations in these currents result in further mixing of the effluent plume. The region in the immediate vicinity of the outfall, where mixing is dominated by buoyancy effects, is called the *near field*, and the region further away from the outfall, where mixing is dominated by spatial variations in ambient currents, is called the *far field*.

If the discharged effluent does not meet the ambient water-quality criteria, which is usually the case, then regulatory agencies usually allow the delineation of a *mixing zone* surrounding the outfall, within which there is sufficient dilution that the ambient water-quality criteria are met beyond the boundary of the mixing zone. There are usually statutory limits for the maximum size of a mixing zone surrounding an ocean outfall. In the state of Florida, the areas of mixing zones in open-ocean waters are required to be less than or equal to 502,655 m^2 (Florida, 1995) which is equal to the area enclosed by a circle of radius 400 m. Rapid dilution of discharged effluent within Florida mixing zones must be ensured by the use of multiport diffusers or single-port outfalls designed to achieve a 20:1 dilution of the effluent prior to reaching the surface.

The design of ocean outfalls and the analysis of plume dilution have been the subject of research for many years and are now fairly mature fields (Fischer et al., 1979; Wood et al., 1993). However, the analysis of far-field mixing is still an evolving area of research and no systematic protocol has yet emerged to analyze far-field mixing processes.

A typical ocean-outfall diffuser is illustrated in Figure 8.11. Diffusers contain multiple ports, with each port designed to discharge effluent at the same rate. The hydraulic design of diffusers can be found in Fischer and colleagues (1979). Outfall pipelines in the ocean are generally buried to the point where the water is deep enough to protect

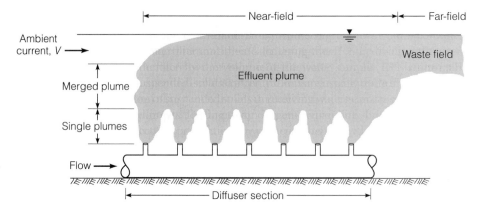

Figure 8.11 ■ Wastewater Discharge from an Ocean Outfall Diffuser

them from wave action, usually about 10 m. Beyond the buried portion, the outfall pipe rests on the bottom of the ocean, with a flanking of rock to prevent currents from undercutting it where the bottom is soft. At some outfalls, plumes originating from individual ports tend either to surface or to be trapped prior to merging, in which case dilution of the effluent is determined by the dynamics of the individual plumes discharged from each of the ports. The effects of surface waves on near-field dilution at shallow-water outfalls are usually neglected; in certain cases, however, these effects may be significant (Chin, 1987; Chin, 1988). In deep-water outfalls, plumes originating from individual ports typically merge together well in advance of reaching the surface or being trapped, and the plume dynamics are similar to a slot (or line) discharge rather than a discharge from several individual ports.

8.6.1 ■ Near-Field Mixing

The dynamics of near-field mixing or *initial dilution* depends on whether the effluent plumes originating at the diffuser ports merge prior to either being trapped or reaching the ocean surface. In cases where adjacent plumes do not merge, each plume behaves (approximately) independently and identically, and near-field mixing can be inferred from the analysis of a single plume. In cases where adjacent plumes merge well in advance of either being trapped or reaching the surface, the (merged) plume behaves as if the effluent were discharged from a long slot or line. Such plumes are called *line plumes*.

8.6.1.1   Single Plumes

Consider the case of a single effluent plume discharged through a port of diameter D at velocity u_e, where the effluent has a density ρ_e and contains a contaminant at concentration c_e. Consider further that the ambient ocean water has a density ρ_a (assuming unstratified conditions), a depth-averaged velocity u_a, and we are interested in calculating the contaminant concentration c at a distance y above the discharge port. The relationship between the contaminant concentration, c, and the parameters controlling the dilution of the effluent plume can be written in the following functional form

$$c = f_1(c_e, u_e, D, \rho_e, \rho_a, g, u_a, y) \qquad (8.105)$$

Assuming that the density differences are small compared to the absolute densities, then the kinematics of the effluent plume do not depend explicitly on the absolute densities of the effluent plume and ambient seawater, but on the difference in densities and the resultant buoyancy effect. This approximation is called the *Boussinesq approximation.* The buoyancy effect is measured by the effective gravity, g', defined by the relation

$$g' = \frac{\rho_a - \rho_e}{\rho_e} g \tag{8.106}$$

and the functional expression for the contaminant concentration in the effluent plume at a distance y above the discharge port, Equation 8.105, can be written as

$$c = f_2(c_e, u_e, D, g', u_a, y) \tag{8.107}$$

This relationship can be simplified by defining a *volume flux, Q_o; specific momentum flux, M_o;* and *specific buoyancy flux, B_o;* by the relations

$$Q_o = u_e \pi \left(\frac{D}{2}\right)^2 \tag{8.108}$$

$$M_o = Q_o u_e = u_e^2 \pi \left(\frac{D}{2}\right)^2 \tag{8.109}$$

$$B_o = Q_o g' = u_e \pi \left(\frac{D}{2}\right)^2 g' \tag{8.110}$$

The variables Q_o, M_o, and B_o involve only u_e, D, and g' and can therefore be used instead of u_e, D, and g' in Equation 8.107 to yield the following functional expression for the contaminant concentration in the effluent plume

$$c = f_3(c_e, Q_o, M_o, B_o, u_a, y) \tag{8.111}$$

Based on the Buckingham pi theorem, Equation 8.111 can be expressed as a relationship between four dimensionless groups, and the following groupings are particularly convenient

$$\frac{Q_o c_e}{u_a L_b^2 c} = f_4\left(\frac{L_M}{L_Q}, \frac{L_M}{L_b}, \frac{y}{L_b}\right) \tag{8.112}$$

where L_M, L_Q, and L_b are length scales defined by the following relations

$$L_Q = \frac{Q_o}{M_o^{1/2}} = \frac{\sqrt{\pi}}{2} D \tag{8.113}$$

$$L_M = \frac{M_o^{3/4}}{B_o^{1/2}} \tag{8.114}$$

$$L_b = \frac{B_o}{u_a^3} \tag{8.115}$$

The length scale L_Q measures the distance over which the port geometry influences the motion of the plume, the length scale L_M measures the distance to where the plume buoyancy begins to become more important than the discharge momentum in controlling the motion of the plume, and the length scale L_b measures the distance to where the ambient current begins to become more important than the plume buoyancy in controlling the motion of the plume. The length scale L_M is sometimes referred to as the *jet/plume transition length scale,* and L_b is sometimes referred to as the *plume/crossflow length scale* (Méndez-Díaz and Jirka, 1996).

Defining the plume dilution, S, by the relation

$$S = \frac{c_e}{c} \tag{8.116}$$

then the functional relationship given by Equation 8.112 can be written in the form

$$\frac{SQ_o}{u_a L_b^2} = f_4\left(\frac{L_M}{L_Q}, \frac{L_M}{L_b}, \frac{y}{L_b}\right) \tag{8.117}$$

In most sewage outfalls, the dilution of the effluent plume is dominated by the buoyancy of the discharge, with the initial momentum having a relatively minor influence. Under these circumstances, the dilution becomes insensitive to the value of L_M/L_Q, which measures the relative importance of the discharge geometry, and the functional expression for the plume dilution, Equation 8.117, can be written as

$$\frac{SQ_o}{u_a L_b^2} = f_5\left(\frac{L_M}{L_b}, \frac{y}{L_b}\right) \tag{8.118}$$

This relationship can be further reduced by considering the physical meaning of the length-scale ratio L_M/L_b. When $L_M/L_b \gg 1$, the ambient currents overwhelm the plume buoyancy before the buoyancy overwhelms the effluent momentum, and the plume motion can be expected to be dominated by ambient currents rather than buoyancy. On the other hand, when $L_M/L_b \ll 1$, the plume buoyancy becomes an important factor in plume dilution, in advance of the ambient currents dominating plume motion. According to Lee and Neville-Jones (1987), for many ocean outfalls, L_M/L_b is sufficiently small that these plumes are either buoyancy dominated over the entire depth, where $y/L_b < 5$, or influenced by both buoyancy and ambient currents, where $y/L_b > 5$. Lee and Neville-Jones (1987) termed these regimes the *buoyancy-dominated near field* (BDNF) and *buoyancy-dominated far field* (BDFF), respectively, and suggested the following formulae for horizontal plume discharges in unstratified ambient seawater

$$\text{BDNF:} \quad \frac{SQ_o}{u_a L_b^2} = 0.31\left(\frac{y}{L_b}\right)^{\frac{5}{3}} \quad \frac{y}{L_b} < 5 \tag{8.119}$$

$$\text{BDFF:} \quad \frac{SQ_o}{u_a L_b^2} = 0.32\left(\frac{y}{L_b}\right)^{2} \quad \frac{y}{L_b} \geq 5 \tag{8.120}$$

where S is the minimum dilution in the surface boil generated by the discharge and y is the depth of the discharge below the ocean surface. Equations 8.119 and 8.120 include the blocking effect of the established wastefield at the water surface and can also be used to describe the dilution of vertical discharges since, for buoyancy-dominated discharges, vertical and horizontal buoyant jets are practically the same (Huang et al., 1998). Rearranging Equations 8.119 and 8.120 reveal that the ambient current speed, u_a, is absent from the BDNF equation and the effluent buoyancy, $Q_o g'$, is absent from the BDFF equation.

Huang and colleagues (1998) have noted that there is a discontinuity in the predictions of Equations 8.119 and 8.120 at $y/L_b = 5$ and that Equation 8.119 does not accurately quantify the plume dilution in stagnant environments. Huang and colleagues (1998) suggested using the following relationship to describe the dilution in the BDNF, transition, and BDFF regimes

$$\frac{SQ_o}{u_a y^2} = 0.08 \left(\frac{y}{L_b}\right)^{-\frac{1}{3}} + \frac{0.32}{1 + 0.2\left(\frac{y}{L_b}\right)^{-0.5}} \tag{8.121}$$

In cases where the ambient current is equal to zero, the minimum dilution is more conveniently described by

$$S = 0.08 \frac{B_o^{1/3}}{Q_o} y^{5/3} \tag{8.122}$$

The application of the near-field dilution formulae is illustrated by the following example.

■ **Example 8.21**

The Central District outfall in Miami (Florida) discharges treated domestic wastewater at a depth of 28.2 m from a diffuser containing five 1.22-m diameter ports spaced 9.8 m apart. The average effluent flowrate is 5.73 m³/s, the ten percentile ambient current is 11 cm/s, and the density of the ambient seawater is 1.024 g/cm³. The density of the effluent can be assumed to be 0.998 g/cm³. Determine the length scales of the effluent plumes and calculate the minimum dilution. Neglect merging of adjacent plumes.

Solution

Calculate the basic charaterics of the effluent plume

$$\text{Effective gravity, } g' = \frac{\Delta\rho}{\rho}g = \frac{1.024 - 0.998}{0.998}(9.81) = 0.256 \text{ m/s}^2$$

$$\text{Port discharge, } Q_o = \frac{5.73}{5} = 1.15 \text{ m}^3/\text{s}$$

$$\text{Port area, } A_P = \frac{\pi}{4}(1.22)^2 = 1.169 \text{ m}^2$$

$$\text{Port velocity, } u_e = \frac{Q_o}{A_P} = \frac{1.15}{1.169} = 0.984 \text{ m/s}$$

$$\text{Momentum flux, } M_o = Q_o u_e = 1.15(0.984) = 1.13 \text{ m}^4/\text{s}^2$$

$$\text{Buoyancy flux, } B_o = Q_o g' = 1.15(0.256) = 0.294 \text{ m}^4/\text{s}^3$$

The length scales are derived from these plume characteristics as follows:

$$L_Q = \frac{Q_o}{M_o^{1/2}} = \frac{1.15}{(1.13)^{1/2}} = 1.08 \text{ m}$$

$$L_M = \frac{M_o^{3/4}}{B_o^{1/2}} = \frac{(1.13)^{3/4}}{(0.294)^{1/2}} = 2.02 \text{ m}$$

$$L_b = \frac{B_o}{u_a^3} = \frac{0.294}{(0.11)^3} = 221 \text{ m}$$

Based on these length scales, it is to be expected that port geometry will only be important within 1.08 m of the discharge port, buoyancy will be the dominant factor in plume motion after 2.02 m, and the ambient current will not dominate the plume motion before the plume surfaces ($y \ll L_b$). Using Equation 8.119 to calculate the dilution yields

$$\frac{S(1.15)}{0.11(221)^2} = 0.31\left(\frac{28.2}{221}\right)^{\frac{5}{3}} \tag{8.123}$$

which gives

$$S = 47$$

Using the equation proposed by Huang et al. (1998), Equation 8.121, yields

$$\frac{S(1.15)}{0.11(28.2)^2} = 0.08\left(\frac{28.2}{221}\right)^{-\frac{1}{3}} + \frac{0.32}{1 + 0.2\left(\frac{28.2}{221}\right)^{-0.5}}$$

which gives

$$S = 28$$

Since the Huang et al. (1998) equation is supported by a wide range of field data, accounts for transitional effects, and is asymptotically correct as the currents approach zero, then this equation is given more weight, and dilution at the Central District outfall is estimated as 28.   ■

8.6.1.2 Line Plumes

In cases where the plumes from individual ports merge together well below the rise height and the diffuser length is much greater than the water depth, the effluent plumes behave very much as if they were discharged from a slot rather than separate ports. This is a common occurrence in deep-water outfalls.

Consider the case of a buoyant jet that is discharged through a slot of width B at velocity u_e, and the effluent has a density ρ_e and contains a contaminant at concentration c_e. Consider further that the ambient ocean has a density ρ_a (assuming unstratified conditions), a depth-averaged velocity u_a, and we are interested in calculating the concentration c at a distance y above the discharge slot. The relationship between the contaminant concentration, c, and the parameters controlling the dilution of the effluent plume can be written in the following functional form

$$c = f_1'(c_e, u_e, B, \rho_e, \rho_a, g, u_a, y) \tag{8.124}$$

As in the case of a single round plume, it can be assumed that the density differences are small compared with the absolute densities (the Boussinesq assumption), and therefore the kinematics of the plume does not depend explicitly on the absolute densities of the effluent plume and ambient seawater, but on the difference in densities and the resultant buoyancy effect, which is parameterized by the effective gravity, g', defined by Equation 8.106. The functional expression for the contaminant concentration in the effluent plume at at distance y above the discharge port, Equation 8.124, can therefore be written as

$$c = f_2'(c_e, u_e, B, g', u_a, y) \tag{8.125}$$

This relationship can be simplified by defining the volume flux, q_o, specific momentum flux, m_o, and specific buoyancy flux, b_o, by the relations

$$q_o = u_e B \tag{8.126}$$

$$m_o = q_o u_e = u_e^2 B \tag{8.127}$$

$$b_o = q_o g' = u_e g' B \tag{8.128}$$

The variables q_o, m_o, and b_o involve only u_e, B, and g', and they can therefore be used instead of u_e, B, and g' in Equation 8.125 to yield the following functional expression for the contaminant concentration in the effluent plume

$$c = f_3'(c_e, q_o, m_o, b_o, u_a, y) \tag{8.129}$$

On the basis of the Buckingham pi theorem, Equation 8.129 can be expressed as a relationship between four dimensionless groups, and the following groupings are particularly convenient

$$\frac{q_o c_e}{u_a y c} = f_4'\left(\frac{l_M}{l_Q}, \frac{l_M}{l_m}, \frac{y}{l_m}\right) \tag{8.130}$$

where l_M, l_Q, and l_m are length scales defined by the following relations

$$l_Q = \frac{q_o}{b_o^{1/3}} = \left(\frac{u_e^2 B^2}{g'}\right)^{\frac{2}{3}} \tag{8.131}$$

$$l_M = \frac{m_o}{b_o^{2/3}} = \left(\frac{u_e^4 B}{g'^2}\right)^{\frac{1}{3}} \tag{8.132}$$

$$l_m = \frac{m_o}{u_a^2} = \frac{u_e^2 B}{u_a^2} \tag{8.133}$$

The length scale l_Q measures the distance over which the port geometry influences the motion of the plume, the length scale l_M measures the distance to where the plume buoyancy begins to become more important than the discharge momentum in controlling the motion of the plume, and the length scale l_m measures the distance to where the ambient current begins to become more important than the jet momentum in controlling the motion of the plume. The length scale l_Q is sometimes referred to as the *discharge/buoyancy length scale*, the length scale l_M is sometimes referred to as the *jet/plume transition length scale*, and l_m is sometimes referred to as the *jet/crossflow length scale* (Méndez-Díaz and Jirka, 1996). Defining the plume dilution, S, by the relation

$$S = \frac{c_e}{c} \tag{8.134}$$

then the functional relationship given by Equation 8.130 can be written in the form

$$\frac{S q_o}{u_a y} = f_5'\left(\frac{l_M}{l_Q}, \frac{l_M}{l_m}, \frac{y}{l_m}\right) \tag{8.135}$$

In most sewage outfalls, the dilution of the effluent plume is dominated by the buoyancy of the discharge, with the initial momentum of the discharge and port geometry having a relatively minor influence. Under these circumstances, the dilution becomes insensitive to the value of l_M/l_Q, and the functional expression for the plume dilution, Equation 8.135, can be written as

$$\frac{Sq_o}{u_a y} = f_6'\left(\frac{l_M}{l_m}, \frac{y}{l_m}\right)$$
(8.136)

This relationship can be further reduced by considering the physical meaning of the length-scale ratio l_M/l_m. Using the definitions of l_M and l_m given by Equations 8.132 and 8.133, then

$$\frac{l_M}{l_m} = \left(\frac{u_a^3}{b_o}\right)^{\frac{2}{3}}$$
(8.137)

which measures the relative importance of the ambient flow and buoyancy on the dynamics of the plume motion. An ambient/discharge Froude number, F_a, is commonly used in practice (Méndez-Díaz and Jirka, 1996; Roberts, 1977) and is defined by

$$F_a = \frac{u_a}{b_o^{1/3}}$$
(8.138)

The ratio l_M/l_m can be expressed in terms of F_a by the relation

$$\frac{l_M}{l_m} = F_a^2$$
(8.139)

and therefore the functional expression for the plume dilution becomes

$$\frac{Sq_o}{u_a y} = f_7'\left(F_a, \frac{y}{l_m}\right)$$
(8.140)

Experiments to determine the functional relationship given by Equation 8.140 were conducted by Roberts (1977) for cases where the initial momentum of the jet is negligible. Under these circumstances, the jet/crossflow length scale, l_m, drops out and Equation 8.140 can be written as

$$\frac{Sq_o}{u_a y} = f_8'(F_a)$$
(8.141)

This functional relationship, as derived from Roberts (1977), is shown in Figure 8.12, where θ indicates the direction of the current relative to the diffuser, and Roberts's Froude number, F, is related to the ambient/discharge Froude number, F_a, by the relation

$$F = F_a^3$$
(8.142)

The results shown in Figure 8.12 clearly indicate that currents perpendicular to the diffuser produce the greatest dilutions. In cases where F_a is small, the ambient flow has a minimal effect compared to the effect of buoyancy on the plume dynamics. Under these circumstances, the plume behavior is approximately the same as for a stagnant ambient, which according to Roberts (1977) is given by

$$\boxed{\frac{Sq_o}{u_a y} = 0.27 F_a^{-1}}$$
(8.143)

Experimental studies by Méndez-Díaz and Jirka (1996) indicate that ambient currents have a significant effect on the plume whenever $F_a > 0.6$ and that Roberts's results may be limited to cases where the effluent jets are located very close to the bottom of the water column.

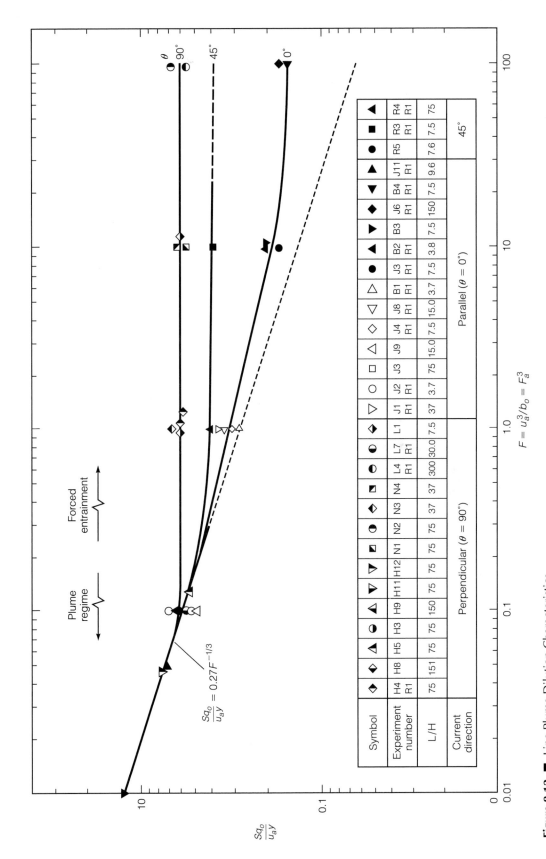

Figure 8.12 ■ Line Plume Dilution Characteristics

Source: Roberts, Philip. "Line Plume and Ocean Outfall Dispersion," *Journal of the Hydraulics Division*. vol. 15 HY4. 1977. ASCE. Reprinted by permission of ASCE.

647

■ **Example 8.22**

The Central District outfall in Miami (Florida) discharges treated domestic wastewater at a depth of 28.2 m from a diffuser containing five 1.22-m diameter ports spaced 9.8 m apart. The average effluent flowrate is 5.73 m³/s, the ten percentile ambient current is 11 cm/s, and the density of the ambient seawater is 1.024 g/cm³. The density of the effluent can be assumed to be 0.998 g/cm³. Assuming that the diffuser can be treated as a line source, calculate the expected minimum dilution for currents perpendicular and parallel to the diffuser.

Solution

For five ports spaced 9.8 m apart, the length of the diffuser, L, can be taken as

$$L = 4(9.8) = 39.2 \text{ m}$$

and the effective gravity, g', is

$$g' = \frac{\rho_a - \rho_e}{\rho_e} g = \frac{1.024 - 0.998}{0.998}(9.81) = 0.256 \text{ m/s}^2$$

The volume flux, q_o is given by

$$q_o = \frac{Q}{L} = \frac{5.73}{39.2} = 0.146 \text{ m}^2/\text{s}$$

and the buoyancy flux, b_o, is given by

$$b_o = q_o g' = (0.146)(0.256) = 0.0374 \text{ m}^3/\text{s}^3$$

From the given data, $u_a = 11$ cm/s $= 0.11$ m/s and hence the Froude number, F, in Figure 8.12 is

$$F = \frac{u_a^3}{b_o} = \frac{0.11^3}{0.0374} = 0.0356$$

Figure 8.12 indicates that for $F = 0.0356$, the minimum dilution is independent of the direction of the current, and that

$$\frac{Sq_o}{u_a y} = 0.27 F_a^{-1} = 0.27 F^{-\frac{1}{3}}$$

$$= 0.27(0.0356)^{-\frac{1}{3}} = 0.82$$

and therefore the minimum dilution, S, for currents at any angle to the diffuser is given by

$$S = 0.82 \frac{u_a y}{q_o} = 0.82 \frac{(0.11)(28.2)}{0.146} = 17.4$$

The minimum dilution in the wastefield above the diffuser is 17.4.   ■

8.6.2 ■ Far-Field Mixing

Far-field mixing is dominated by spatial and temporal variations in ocean currents, and far-field models are generally applicable when the discharge-momentum and buoyancy fluxes of the effluent plume are overwhelmed by the advection flux of the ocean currents. The dispersion coefficient in the ocean increases with the size of the contaminant plume, a fact that is observed in practically all tracer experiments in the ocean (Okubo, 1971) and is attributed to the condition that as a tracer cloud grows, the cloud experiences a wider range of velocities, which leads to increased growth rates and larger diffusion coefficients. Okubo (1971) analyzed the results of several field-scale dye experiments and derived an empirical expression for the oceanic diffusion coefficient. Within observed tracer clouds, Okubo (1971) used the area enclosed by each concentration contour to define a circle of radius r_e enclosing the same area as the irregular concentration contour, and then calculated the variance, σ_{rc}^2, of the entire tracer cloud using the relation

$$\sigma_{rc}^2 = \frac{\int_0^\infty r_e^2 c \, 2\pi r_e dr_e}{\int_0^\infty c \, 2\pi r_e dr_e} \tag{8.144}$$

where $c(r_e, t)$ is the concentration with an equivalent circular contour of radius r_e at time t. The characterization of the variance of a tracer distribution by σ_{rc}^2 can be compared with the variance characterization of bivariate Gaussian distributions, which requires specification of the variances along both the major and minor principal axes. If the tracer distribution is Gaussian and the variances along the major and minor axes are σ_x^2 and σ_y^2, respectively, then it can be shown that

$$\sigma_{rc}^2 = 2\sigma_x \sigma_y \tag{8.145}$$

Okubo (1971) plotted σ_{rc}^2 versus time for several instantaneous dye releases in the surface layers of coastal waters and showed that the following empirical relationship provided a reasonably good fit to the observed data

$$\boxed{\sigma_{rc}^2 = 0.0108 t^{2.34}} \tag{8.146}$$

where σ_{rc}^2 is measured in cm^2 and t is the time since release, measured in seconds. Okubo (1971) fitted data for times ranging from 2 hours to nearly 1 month. The variance of a tracer cloud as a function of time can be used to calculate an apparent diffusion coefficient, K_a, using the relation

$$K_a = \frac{\sigma_{rc}^2}{4t} \tag{8.147}$$

Defining the length scale, L, of the tracer cloud by

$$L = 3\sigma_{rc} \tag{8.148}$$

then Okubo (1971) used the results of field-scale dye studies to plot the apparent diffusion coefficient, K_a, versus the length scale, L, of the cloud. These results are shown in Figure 8.13. These data show a good fit to the empirical relation

$$\boxed{K_a = 0.0103 L^{1.15}} \tag{8.149}$$

where K_a is in cm^2/s and L is in cm. Equation 8.149 is widely used in practice to estimate the diffusion coefficient as a function of length scale for contaminants released into the ocean.

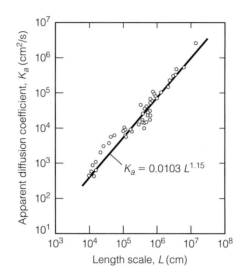

Figure 8.13 ■ Apparent Diffusion Coefficient Versus Length Scale in Coastal Waters

■ **Example 8.23**

Compare the apparent diffusion coefficient of an oil spill having a characteristic size of 50 m with an oil spill whose size is 100 m. Estimate how long it would take for a small oil spill to grow to a size of 50 m.

Solution

The apparent diffusion coefficient, K_a, is related to the length scale, L, of the oil spill by

$$K_a = 0.0103L^{1.15} \text{ cm}^2/\text{s}$$

When $L = 50$ m $= 5{,}000$ cm, then

$$K_a = 0.0103(5000)^{1.15} = 185 \text{ cm}^2/\text{s}$$

and when $L = 100$ m $= 10{,}000$ cm, then

$$K_a = 0.0103(10000)^{1.15} = 410 \text{ cm}^2/\text{s}$$

Therefore, when the oil spill doubles its size from 50 m to 100 m, the apparent diffusion coefficient more than doubles, going from 185 cm²/s to 410 cm²/s.

The variance of the oil spill as a function of time can be estimated by

$$\sigma_{rc}^2 = 0.0108t^{2.34} \text{ cm}^2$$

or

$$\sigma_{rc} = 0.104t^{1.17} \text{ cm}$$

Defining the length scale, L, by $L = 3\sigma_{rc}$, then

$$L = 0.312t^{1.17} \text{ cm}$$

When $L = 50$ m $= 5{,}000$ cm, then

$$5000 = 0.312t^{1.17} \text{ cm}$$

which leads to

$$t = 3930 \text{ s} = 65.5 \text{ min} = 1.09 \text{ h}$$

Therefore, a small oil spill will take approximately 1.09 h to grow to a size of 50 m.   ■

For multiport diffusers, far-field models typically assume that the contaminant source is a rectangular plane area perpendicular to the mean current at the trapping level of the plume, with the width of the plane area (source) equal to the length of the diffuser and the depth of the plane area equal to the plume thickness at the trapping height (e.g., Huang et al., 1996). For surfacing plumes in unstratified environments, the initial plume thickness can be taken as 30% of the total depth (Koh, 1983). Assuming that the contaminant mass flux leaving the diffuser is equal to the contaminant mass flux across the source plane at the trapping level, then

$$Qc_e = u_a Lh c_o \tag{8.150}$$

where Q is the effluent volume flux, c_e is the contaminant concentration in the effluent, u_a is the ambient velocity, L is the diffuser length, h is the height of the wastefield, and c_o is the contaminant concentration crossing the source plane. Equation 8.150 can be put in the more convenient form

$$\boxed{c_o = \frac{Qc_e}{u_a Lh}} \tag{8.151}$$

Brooks (1960) proposed a simple far-field model that assumes a constant ambient velocity, u_a, and neglects both vertical diffusion and diffusion in the flow direction. These assumptions are justified by the observations that the diffusive flux in the flow direction is usually much smaller than the advective flux, and the vertical diffusion coefficient is usually much smaller than the transverse diffusion coefficient. The governing steady-state advection-diffusion equation is therefore given by

$$\boxed{u_a \frac{\partial c}{\partial x} = \frac{\partial}{\partial y}\left(\varepsilon_y \frac{\partial c}{\partial y}\right)} \tag{8.152}$$

with boundary conditions

$$c(0, y) = \begin{cases} c_o & |y| < \frac{L}{2} \\ 0 & |y| > \frac{L}{2} \end{cases} \tag{8.153}$$

$$c(x, \pm\infty) = 0 \tag{8.154}$$

where x is in the direction of flow and y is in the transverse (horizontal) direction and ε_y is the transverse diffusion coefficient. Brooks (1960) assumed that the transverse diffusion coefficient, ε_y, increases with the size of the plume in accordance with the so-called *four-thirds law* originally proposed by Richardson (1926) and supported by the field results of Okubo (1971). According to the four-thirds law, the transverse diffusion coefficient, ε_y, is given by

$$\varepsilon_y = k\sigma_y^{4/3} \tag{8.155}$$

where σ_y is the standard deviation of the cross-sectional concentration at a distance x from the source and k is a constant. An analytic expression for the resulting concentration distribution is not available, but the maximum concentration (along the plume centerline) is given by

$$c(x, 0) = c_o \operatorname{erf} \left[\sqrt{\dfrac{3}{2\left(1 + \frac{2}{3}\frac{\beta x}{L}\right)^3 - 1}} \right] \qquad (8.156)$$

where β is defined by

$$\beta = \frac{12\varepsilon_o}{u_a L} \qquad (8.157)$$

and ε_o is the transverse diffusion coefficient at $x = 0$. The Brooks (1960) formulation given by Equation 8.156 has been widely used to predict the far-field mixing of ocean-outfall discharges, and its popularity is no doubt due to its simplicity and the fact that the maximum concentration is expressed in terms of measurable quantities.

Models of far-field mixing are still in their infancy, and it is still common practice to neglect spatial variations in the ambient velocity and parameterize the mixing process by an empirical diffusion coefficient that is unrelated to the spatial and temporal characteristics of the currents at the outfall site (e.g., the Brooks model). This pragmatic approach is usually a result of economic constraints, which typically only allow for the deployment of a single current meter in the vicinity of an ocean outfall, thereby precluding any measurement of the spatial characteristics of the velocity field. Ideally, if the ambient currents could be measured at several locations within the far field, then far-field mixing of effluent plumes could be simulated much more accurately than is presently possible using conventional methods. More comprehensive models of far-field mixing have been proposed by Chin and colleagues (1997) and Chin and Roberts (1985).

■ Example 8.24

A 39.2-m long multiport diffuser discharges treated domestic wastewater at a rate of 5.73 m³/s and at a depth of 28.2 m. If the ambient current is 11 cm/s, estimate the distance downstream of the diffuser to where the dilution is equal to 100.

Solution

Equation 8.151 estimates the initial dilution in the far-field model as

$$\frac{c_e}{c_o} = \frac{u_a L h}{Q}$$

where $u_a = 11$ cm/s $= 0.11$ m/s, $L = 39.2$ m, $h = 0.3(28.2) = 8.46$ m, $Q = 5.73$ m³/s, and hence

$$\frac{c_e}{c_o} = \frac{u_a L h}{Q} = \frac{(0.11)(39.2)(8.46)}{5.73} = 6.37$$

The initial plume dilution for the far-field model is therefore estimated as 6.37. This estimate of initial dilution is certainly not as accurate as using a near-field model to estimate the initial dilution, but this approximation is acceptable when the far-field dilution is much greater than the near-field dilution. If a near-field model is used to estimate the initial dilution, then Equation 8.151 is still valid, but the initial height of the wastefield, h, is no longer taken as 30% of the depth. Equation 8.156 gives the far-field dilution as

$$\frac{c_{max}}{c_o} = \text{erf}\left[\sqrt{\frac{3}{2\left(1 + \frac{2}{3}\frac{\beta x}{L}\right)^3 - 1}}\right]$$

and the total dilution, c_e/c_{max}, is given by

$$\frac{c_e}{c_{max}} = \frac{c_e}{c_o} \cdot \frac{c_o}{c_{max}} = 6.37\left\{\text{erf}\left[\sqrt{\frac{3}{2\left(1 + \frac{2}{3}\frac{\beta x}{L}\right)^3 - 1}}\right]\right\}^{-1} \tag{8.158}$$

where

$$\beta = \frac{12\varepsilon_o}{u_a L} \tag{8.159}$$

The diffusion coefficient at the diffuser, ε_o, can be estimated using the diffuser length, $L = 39.2$ m $= 3{,}920$ cm and the Okubo relation (Equation 8.149) as

$$\varepsilon_o = 0.0103 L^{1.15} = 0.0103(3920)^{1.15} = 140 \text{ cm}^2/\text{s} = 0.014 \text{ m}^2/\text{s}$$

The parameter, β, is therefore given by Equation 8.159 as

$$\beta = \frac{12\varepsilon_o}{u_a L} = \frac{12(0.014)}{(0.11)(39.2)} = 0.0390$$

Substituting into Equation 8.158 to find the distance, x, from the diffuser where the dilution is 100 gives

$$100 = 6.37\left\{\text{erf}\left[\sqrt{\frac{3}{2\left(1 + \frac{2}{3}\frac{0.0390x}{39.2}\right)^3 - 1}}\right]\right\}^{-1}$$

which simplifies to

$$0.0637 = \text{erf}\left[\sqrt{\frac{3}{2(1 + 0.000663x)^3 - 1}}\right]$$

Using the error function tabulated in Appendix E gives

$$x = 10200 \text{ m} = 10.2 \text{ km}$$

Hence the dilution reaches 100 at a distance of about 10.2 km downstream of the diffuser. ■

8.7 Ground Water

Ground water is a major source of drinking water in the United States, supplying approximately 40% of public water utilities and accounting for almost all of the water supply to rural households. It has been estimated that approximately 50% of the U.S. population relies on ground-water sources for drinking water (Solley et al., 1988). Many types of contaminants, from a variety of sources, have been found in ground water,

Category	Description	Examples
■ **Table 8.17** Sources of Ground-water Contamination		
I	sources designed to discharge substances	septic tanks, injection wells, land application
II	sources designed to store, treat, and/or dispose of substances; discharges through unplanned release	landfills, open dumps, underground storage tanks
III	sources designed to retain substances during transport or transmission	pipelines, materials transport and transfer
IV	sources discharging as consequence of other planned activities	pesticide/fertilizer applications, urban runoff
V	sources providing conduit or inducing discharge through altered flow patterns	production wells, construction excavation
VI	naturally occurring sources whose discharge is created and/or exacerbated by human activity	saltwater intrusion

Source: USOTA (1984).

and sources of ground-water contamination can be roughly divided into the six categories shown in Table 8.17. The most common sources of ground-water contamination encountered in practice are leaking underground storage tanks (LUSTs), landfills, surface impoundments, waste-disposal injection wells, septic tanks, hazardous chemicals in agriculture, and contamination resulting from land application of contaminated water.

Underground tanks store gasoline at service stations and are widely used by industry, agriculture, and homes to store oil, hazardous chemicals, and chemical waste products. Modern landfills are constructed with leachate prevention and treatment systems, but most older landfills are simply large holes in the ground filled with waste and covered with dirt (Bedient et al., 1994), and leaking liquids and leachate from older landfills can be a significant source of ground-water contamination. Surface impoundments are used by some municipal wastewater treatment systems for settling solids and biological oxidation, resulting in the percolation of contaminants into the ground water. Waste-disposal injection wells are used to inject contaminated water, surface runoff, and hazardous wastes deep into the ground and away from drinking water sources, but poor well design, faulty construction, inadequate understanding of the subsurface geology, and deteriorated well casings can all cause contaminants to be introduced into drinking-water sources. Septic tanks discharge pathogenic microorganisms, synthetic organic chemicals, nutrients (such as nitrogen and phosphorous), and other contaminants directly into the ground water and can cause serious problems if drinking water sources are too close to the septic tanks. The uses of pesticides and fertilizers in agricultural practice are significant sources of synthetic organic chemicals and nutrients in ground water, and the land application of waste sludges and treated wastewater (by spray irrigation) are significant sources of heavy metals, toxic chemicals, and pathogenic microorganisms. The most frequently reported contaminants in ground water are nitrates, pesticides, volatile organic compounds, petroleum products, metals, brine, and synthetic organic chemicals (USEPA, 1990b).

Contaminants in ground water undergo a variety of fate and transport processes. The fate processes that are most often considered include sorption onto the solid matrix and first-order decay, both of which affect the amount of contaminant mass in ground water.

Transport processes include advection at the mean (large-scale) seepage velocity and mixing caused by small-scale variations in the seepage velocity associated with spatial variability in the hydraulic conductivity.

8.7.1 ■ Dispersion Models

The dispersion of contaminants in ground water is adequately described by the advection-dispersion equation derived in Section 8.3 as

$$\frac{\partial c}{\partial t} + \sum_{i=1}^{3} V_i \frac{\partial c}{\partial x_i} = \sum_{i=1}^{3} D_i \frac{\partial^2 c}{\partial x_i^2} + S_m \tag{8.160}$$

where c is the contaminant concentration, equal to the mass of contaminant per unit volume of ground water; V_i are the components of the seepage velocity; D_i are the components of the dispersion coefficient, and S_m is the rate at which mass is added to the ground water per unit volume of ground water per unit time. The transport processes of advection and dispersion are parameterized by the mean seepage velocity, V_i, and the dispersion coefficient, D_i, and fate processes are parameterized by the source mass flux, S_m. For conservative substances, $S_m = 0$.

Solutions to the advection-dispersion equation are commonly referred to as *dispersion models* and can be either analytic or numerical. Numerical models provide discrete solutions to the advection-dispersion equation in space and time and are most useful in cases of complex geology and irregular boundary conditions. Analytic models provide continuous solutions to the advection-dispersion equation in space and time and are most useful in cases of simple geology and simple boundary conditions. Several useful analytic dispersion models are described in the following sections.

8.7.1.1 Instantaneous Point Source

In the case where a mass, M, of conservative contaminant is instantaneously injected over a depth, H, of a uniform aquifer with mean seepage velocity, V, the resulting concentration distribution, $c(x, y, t)$, is given by

$$\boxed{c(x, y, t) = \frac{M}{4\pi H t (D_L D_T)^{1/2}} \exp\left[-\frac{(x - Vt)^2}{4D_L t} - \frac{y^2}{4D_T t}\right]} \tag{8.161}$$

where t is the time since the injection of the contaminant, x is the coordinate measured in the direction of the seepage velocity, y is the transverse (horizontal) coordinate, the contaminant source is located at the origin of the coordinate system, and D_L and D_T are the longitudinal and transverse dispersion coefficients. Equation 8.161 is more commonly applied in cases where a contaminant is initially mixed over a depth, H, of the aquifer, not the entire depth of the aquifer, and vertical dispersion is negligible compared with longitudinal and horizontal-transverse dispersion.

■ Example 8.25

Ten kilograms of a contaminant are spilled over the top 2 m of an aquifer. If the longitudinal and horizontal-transverse dispersion coefficients are 1 m^2/d and 0.1 m^2/d, respectively; vertical mixing is negligible; and the mean seepage velocity is 0.6 m/d, estimate the maximum contaminant concentrations in the ground water 1 day, 1 week, 1 month, and 1 year after the spill. What is the contaminant concentration at the spill location after 1 week?

Solution

From the given data, $M = 10$ kg, $H = 2$ m, $D_L = 1$ m²/d, $D_T = 0.1$ m²/d, and $V = 0.6$ m/d. According to Equation 8.161, the maximum concentration, c_{max}, occurs at $x = Vt$ and $y = 0$ m, hence

$$c_{max}(t) = \frac{M}{4\pi Ht(D_L D_T)^{1/2}}$$

Substituting the given parameters gives

$$c_{max}(t) = \frac{10}{4\pi(2)t(1 \times 0.1)^{1/2}} = \frac{1.26}{t} \text{ kg/m}^3 = \frac{1260}{t} \text{ mg/L}$$

which yields the following results,

t (days)	$c_{max}(t)$ (mg/L)
1	1,260
7	180
30	42
365	3.5

The concentration at the spill location ($x = 0$ m, $y = 0$ m) as a function of time is given by Equation 8.161 as

$$c(0, 0, t) = \frac{M}{4\pi Ht(D_L D_T)^{1/2}} \exp\left[-\frac{(Vt)^2}{4D_L t}\right]$$

which at $t = 7$ days gives

$$c(0, 0, 7) = \frac{10}{4\pi(2)(7)(1 \times 0.1)^{1/2}} \exp\left[-\frac{(0.6 \times 7)^2}{4(1)(7)}\right] = 0.096^{-4} \text{ kg/m}^3 = 96 \text{ mg/L}$$

Hence, after seven days, the concentration at the site of the spill is approximately 53% of the maximum concentration of 180 mg/L. ■

8.7.1.2 Continuous Point Source

In the case where a conservative contaminant of concentration c_o is continuously injected at a rate Q (L³/T) into a uniform aquifer of depth H and mean seepage velocity V, the concentration distribution downstream of the source, $c(x, y, t)$, is given by (Fried, 1975)

$$c(x, y, t) = \frac{Qc_o}{4\pi H(D_L D_T)^{1/2}} \exp\left(\frac{Vx}{2D_L}\right)[W(0, B) - W(t, B)] \qquad (8.162)$$

where the x coordinate is in the direction of the seepage velocity; y is the transverse (horizontal) coordinate; the source is located at the origin of the coordinate system; D_L and D_T are the longitudinal and transverse dispersion coefficients, respectively, $W(t, B)$

is the well function for a leaky aquifer (Table 7.9); and B is defined by

$$B = \left[\frac{(Vx)^2}{4D_L^2} + \frac{(Vy)^2}{4D_L D_T} \right]^{\frac{1}{2}} \tag{8.163}$$

As $t \to \infty$, the concentration distribution given by Equation 8.162 approaches the steady-state solution (Bear, 1972)

$$c(x, y) = \left[\frac{Qc_0}{2\pi H (D_L D_T)^{1/2}} \right] \exp\left(\frac{Vx}{2D_L} \right) K_o \left\{ \left[\frac{V^2}{4D_L} \left(\frac{x^2}{D_L} + \frac{y^2}{D_T} \right) \right]^{\frac{1}{2}} \right\} \tag{8.164}$$

where K_o is the modified Bessel function of the second kind of order zero (described in Appendix F).

■ Example 8.26

A conservative contaminant is continuously injected through a 4-m deep perforated well into an aquifer with a mean seepage velocity of 0.8 m/d and longitudinal and transverse dispersion coefficients of 2 m²/d and 0.2 m²/d, respectively. If the injection rate of the contaminated water is 0.7 m³/d, with a contaminant concentration of 100 mg/L, estimate the steady-state contaminant concentrations at locations 1 m, 10 m, 100 m, and 1,000 m downstream of the injection well. Neglect vertical diffusion.

Solution

From the given data, $H = 4$ m, $V = 0.8$ m/d, $D_L = 2$ m²/d, $D_T = 0.2$ m²/d, $Q = 0.7$ m³/d, and $c_0 = 100$ mg/L $= 0.1$ kg/m³. The steady-state concentration is given by Equation 8.164 as

$$c(x, y) = \left[\frac{Qc_0}{2\pi H (D_L D_T)^{1/2}} \right] \exp\left[\frac{Vx}{2D_L} \right] K_o \left[\left\{ \frac{V^2}{4D_L} \left(\frac{x^2}{D_L} + \frac{y^2}{D_T} \right) \right\}^{\frac{1}{2}} \right]$$

which yields

$$c(x, 0) = \left[\frac{(0.7)(0.1)}{2\pi (4)(2 \times 0.2)^{1/2}} \right] \exp\left[\frac{(0.8)x}{2(2)} \right] K_o \left[\left\{ \frac{0.8^2}{4 \times 2} \left(\frac{x^2}{2} \right) \right\}^{\frac{1}{2}} \right]$$

$$= 0.00440 \exp(0.2x) K_o(0.2x)$$

The steady-state downstream concentrations are therefore given by

x (m)	$c(x, 0)$ (kg/m³)	$c(x, 0)$ (mg/L)
1	0.0094	9.4
10	0.0037	3.7
100	0.0012	1.2
1,000	0.00039	0.39

■

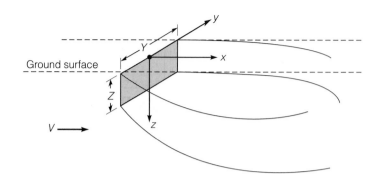

Figure 8.14 ■ Dispersion from a Continuous Finite Source

8.7.1.3   Continuous Finite Source

The case where a conservative contaminant of concentration c_o is continuously released from a finite source of dimension $Y \times Z$ is illustrated in Figure 8.14. The resulting concentration distribution, $c(x, y, z, t)$, is given by (Domenico and Robbins, 1985)

$$
\begin{aligned}
c(x, y, z, t) = \left(\frac{c_o}{8}\right) \mathrm{erfc}\left[\frac{(x - Vt)}{2(\alpha_x Vt)^{1/2}}\right] &\left\{\mathrm{erf}\left[\frac{(y + Y/2)}{2(\alpha_y x)^{1/2}}\right] - \mathrm{erf}\left[\frac{(y - Y/2)}{2(\alpha_y x)^{1/2}}\right]\right\} \\
&\left\{\mathrm{erf}\left[\frac{(z + Z)}{2(\alpha_z x)^{1/2}}\right] - \mathrm{erf}\left[\frac{(z - Z)}{2(\alpha_z x)^{1/2}}\right]\right\}
\end{aligned}
$$

$$(8.165)$$

where V is the mean seepage velocity, and α_x, α_y, and α_z are the dispersivities in the coordinate directions. The dispersivity in a porous medium is defined as the dispersion coefficient divided by the mean seepage velocity, where

$$
\alpha_x = \frac{D_x}{V}, \qquad \alpha_y = \frac{D_y}{V}, \qquad \alpha_z = \frac{D_z}{V}
\tag{8.166}
$$

and D_x, D_y, and D_z are the dispersion coefficients in the x, y, and z directions, respectively. In Equation 8.165, x is the longitudinal (flow) direction, y is the horizontal-transverse direction, and z is the vertical-transverse direction. If there is no spreading in the vertical, z, direction, then the error functions containing the z terms in Equation 8.165 are ignored, and $c_o/8$ becomes $c_o/4$ (Domenico and Schwartz, 1990). The distance, x_o, from the source to the location where the plume is well mixed over the aquifer thickness, H, can be estimated by (Domenico and Palciauskas, 1982)

$$
x_o = \frac{(H - Z)^2}{\alpha_z}
\tag{8.167}
$$

For distances less than x_o, Equation 8.165 is applicable; for distances greater than x_o, the distance x in the denominator of the error function of the z term is replaced by x_o, prohibiting further spreading for $x > x_o$. Domenico (1987) showed that for contaminants that undergo first-order decay with a decay factor λ, Equation 8.165 becomes

$$c(x, y, z, t) = \left(\frac{c_o}{8}\right) \exp\left\{\left(\frac{x}{2\alpha_x}\right)\left[1 - \left(1 + \frac{4\lambda\alpha_x}{V}\right)^{1/2}\right]\right\} \mathrm{erfc}\left[\frac{x - Vt(1 + 4\lambda\alpha_x/V)^{1/2}}{2(\alpha_x Vt)^{1/2}}\right]$$
$$\left\{\mathrm{erf}\left[\frac{(y + Y/2)}{2(\alpha_y x)^{1/2}}\right] - \mathrm{erf}\left[\frac{(y - Y/2)}{2(\alpha_y x)^{1/2}}\right]\right\}\left\{\mathrm{erf}\left[\frac{(z + Z)}{2(\alpha_z x)^{1/2}}\right] - \mathrm{erf}\left[\frac{(z - Z)}{2(\alpha_z x)^{1/2}}\right]\right\}$$

(8.168)

■ Example 8.27

A continuous contaminant source is 3 m wide by 2 m deep and contains a contaminant at a concentration of 100 mg/L. The mean seepage velocity in the aquifer is 0.4 m/d, the aquifer is 7 m deep, and the longitudinal, horizontal-transverse, and vertical-transverse dispersivities are 3 m, 0.3 m, and 0.03 m, respectively. Assuming that the contaminant is conservative, determine the downstream location at which the contaminant plume will be fully mixed over the depth of the aquifer. Estimate the contaminant concentrations at the water table at locations 10 m, 100 m, and 1,000 m downstream of the source after 10 years. If the contaminant undergoes biodegradation with a decay rate of 0.01 d^{-1}, estimate the effect on the concentrations downstream of the source.

Solution

From the given data, $Y = 3$ m, $Z = 2$ m, $c_o = 100$ mg/L $= 0.1$ kg/m^3, $V = 0.4$ m/d, $H = 7$ m, $\alpha_x = 3$ m, $\alpha_y = 0.3$ m, and $\alpha_z = 0.03$ m. The contaminant plume becomes well mixed at a distance x_o downstream, where x_o is given by Equation 8.167 as

$$x_o = \frac{(H - Z)^2}{\alpha_z} = \frac{(7 - 2)^2}{0.03} = 833 \text{ m}$$

The concentration along the line $y = 0$ m, $z = 0$ m is given by Equation 8.165 as

$$c(x, 0, 0, t) = \left(\frac{c_o}{8}\right) \mathrm{erfc}\left[\frac{(x - Vt)}{2(\alpha_x Vt)^{1/2}}\right]\left\{\mathrm{erf}\left[\frac{(Y/2)}{2(\alpha_y x)^{1/2}}\right] - \mathrm{erf}\left[\frac{(-Y/2)}{2(\alpha_y x)^{1/2}}\right]\right\}$$
$$\left\{\mathrm{erf}\left[\frac{(Z)}{2(\alpha_z x)^{1/2}}\right] - \mathrm{erf}\left[\frac{(-Z)}{2(\alpha_z x)^{1/2}}\right]\right\}$$

and therefore at $t = 10$ years $= 3{,}650$ days,

$$c(x, 0, 0, 3650) = \left(\frac{0.1}{8}\right) \mathrm{erfc}\left[\frac{(x - 0.4 \times 3650)}{2(3 \times 0.4 \times 3650)^{1/2}}\right]\left\{\mathrm{erf}\left[\frac{(3/2)}{2(0.3x)^{1/2}}\right]\right.$$
$$\left. - \mathrm{erf}\left[\frac{(-3/2)}{2(0.3x)^{1/2}}\right]\right\}\left\{\mathrm{erf}\left[\frac{(2)}{2(0.03x)^{1/2}}\right] - \mathrm{erf}\left[\frac{(-2)}{2(0.03x)^{1/2}}\right]\right\}$$

which simplifies to*

$$c(x, 0, 0, 3650) = 0.05\, \mathrm{erfc}\left[\frac{x - 1460}{132}\right] \mathrm{erf}\left(\frac{1.37}{\sqrt{x}}\right) \mathrm{erf}\left(\frac{5.77}{\sqrt{x}}\right)$$

---

*This simplification uses the identity

$$\mathrm{erf}(-x) = -\mathrm{erf}(x)$$

Since the contaminant becomes well mixed at $x_o = 833$ m, this formulation can only be used for calculating the concentrations at $x \leq 833$ m. At $x = 10$ m and $x = 100$ m, the equation yields

x (m)	$c(x, 0, 0, 3650)$ (kg/m^3)	$c(x, 0, 0, 3650)$ (mg/L)
10	0.046	46
100	0.0090	9.0

At $x = 1{,}000$ m, the plume is well mixed over the vertical, and the contaminant concentration is calculated by replacing x by x_o ($= 833$ m) in the denominator of the error function in the z term to yield

$$c(x, 0, 0, 3650) = 0.05\,\text{erfc}\left[\frac{x - 1460}{132}\right]\text{erf}\left(\frac{1.37}{\sqrt{x}}\right)\text{erf}\left(\frac{5.77}{\sqrt{833}}\right)$$

$$= 0.0111\,\text{erfc}\left[\frac{x - 1460}{132}\right]\text{erf}\left(\frac{1.37}{\sqrt{x}}\right)$$

which gives $c(1000, 0, 0, 3650) = 0.0011$ kg/m^3 $= 1.1$ mg/L.

If the contaminant undergoes first-order decay with $\lambda = 0.01$ d^{-1}, the concentration profile along the line $y = 0$ m, $z = 0$ m, is given by Equation 8.168 as

$$c(x, 0, 0, 3650) = \left(\frac{0.1}{8}\right)\exp\left\{\left(\frac{x}{2(3)}\right)\left[1 - \left(1 + \frac{4(0.01)(3)}{0.4}\right)^{1/2}\right]\right\}$$

$$\text{erfc}\left[\frac{x - 0.4(3650)(1 + 4 \times 0.01 \times 3/0.4)^{1/2}}{2(3 \times 0.4 \times 3650)^{1/2}}\right]$$

$$\left\{\text{erf}\left[\frac{(3/2)}{2(0.3x)^{1/2}}\right] - \text{erf}\left[\frac{(-3/2)}{2(0.3x)^{1/2}}\right]\right\}\left\{\text{erf}\left[\frac{(2)}{2(0.03x)^{1/2}}\right]\right.$$

$$\left.- \text{erf}\left[\frac{(-2)}{2(0.03x)^{1/2}}\right]\right\}$$

which simplifies to

$$c(x, 0, 0, 3650) = 0.05\exp(-0.0234x)\text{erfc}\left[\frac{x - 1665}{132}\right]\text{erf}\left(\frac{1.37}{\sqrt{x}}\right)\text{erf}\left(\frac{5.77}{\sqrt{x}}\right)$$

and at $x = 10$ m and $x = 100$ m yields the following results

x (m)	$c(x, 0, 0, 3650)$ (kg/m^3)	$c(x, 0, 0, 3650)$ (mg/L)
10	0.036	36
100	0.00087	0.87

Replacing x by x_o ($= 833$ m) in the z term yields $c(1000, 0, 0, 3650) = 7.48 \times 10^{-14}$ kg/m^3 ≈ 0 mg/L.

The results of this example show that biodegradation will have a significant effect on the contaminant concentrations downstream of the source. Beyond $x = 100$ m, the biodegraded contaminant concentrations are negligible.   ■

8.7.2 ■ Transport Processes

Dispersion of contaminants in ground water is caused by spatial variations in hydraulic conductivity and, to a much smaller extent, pore-scale mixing and molecular diffusion. Pore-scale mixing results from the differential movement of ground water through pores of various sizes and shapes, a process called *mechanical dispersion*; the combination of mechanical dispersion and molecular diffusion is called *hydrodynamic dispersion*. Dispersion caused by large-scale variations in hydraulic conductivity is called *macrodispersion*. Consider a porous medium in which several samples of characteristic size L are tested for their hydraulic conductivity, K. The hydraulic conductivity (K) is then a random space function (RSF) with support scale L. Assuming that K is log normally distributed, it is convenient to work with the variable Y defined as

$$Y = \ln K \tag{8.169}$$

where Y is a normally distributed random space function, characterized by a mean, $\langle Y \rangle$; variance, σ_Y^2; and correlation length scales, λ_i, in the x_i coordinate directions. The geometric mean hydraulic conductivity, K_G, is related to $\langle Y \rangle$ by

$$K_G = e^{\langle Y \rangle} \tag{8.170}$$

Freeze (1975) analyzed data from a variety of geologic cores, and the statistics of the measured hydraulic conductivities are tabulated in Table 8.18. These data indicate relatively high values of σ_Y, which reflect a significant degree of variability about the mean hydraulic conductivity. The variance of the hydraulic conductivity is inversely proportional to the magnitude of the support scale, with larger support scales resulting in smaller variances in the hydraulic conductivity. Consequently, whenever values of σ_Y are cited, it is sound practice also to state the corresponding support scale. The support scale of the data shown in Table 8.18 is on the order of 10 cm. The spatial covariance of Y must also be associated with a stated support scale, since both σ_Y and the correlation length scale, λ_i, depend on the support scale. Larger support scales generally yield larger correlation length scales. Porous media in which the correlation length scales of the hydraulic conductivity in the principal directions differ from each other are called *anisotropic* media, and porous media where the correlation length scales of the hydraulic conductivity in the principal directions are all equal are called *isotropic* media.

■ **Table 8.18**
Hydraulic Conductivity Statistics

Formation	$\langle Y \rangle = \langle \ln K \rangle$ (K in m/d)	K_G (m/d)	σ_Y
Sandstone	−2.0	0.13	0.92
Sandstone	−0.98	0.38	0.46
Sand and gravel	—	—	1.01
Sand and gravel	—	—	1.24
Sand and gravel	—	—	1.66
Silty clay	−0.15	0.86	2.14
Loamy sand	0.59	1.81	1.98

Source: Freeze (1975).

Detailed discussions of dispersion in both isotropic and anisotropic media can be found in Dagan (1989), Chin and Wang (1992), Gelhar (1993), and Chin (1997). The mean seepage velocity, V_i, in isotropic porous media is given the the Darcy equation

$$V_i = -\frac{K_{\text{eff}}}{n_e} J_i \qquad (8.171)$$

where K_{eff} is the effective hydraulic conductivity, n_e is the effective porosity, and J_i is the slope of the piezometric surface in the i direction. The effective hydraulic conductivity in isotropic media can be expressed in terms of statistics of the hydraulic conductivity field by the relations (Dagan, 1989)

$$\begin{aligned}
\text{1-D flow} &: \quad K_{\text{eff}} = K_G\left(1 - \frac{\sigma_Y^2}{2}\right) \\
\text{2-D flow} &: \quad K_{\text{eff}} = K_G \\
\text{3-D flow} &: \quad K_{\text{eff}} = K_G\left(1 + \frac{\sigma_Y^2}{6}\right)
\end{aligned} \qquad (8.172)$$

The dispersion coefficient in porous media can be stated generally as a tensor quantity, D_{ij}, which is typically expressed in terms of the magnitude of the mean seepage velocity, V, by the relation (Bear, 1979)

$$D_{ij} = \alpha_{ij} V \qquad (8.173)$$

where α_{ij} is called the *dispersivity* of the porous medium. In general porous media, α_{ij} is a symmetric tensor with six independent components, and can be written in the form

$$\alpha_{ij} = \begin{bmatrix} \alpha_{11} & \alpha_{12} & \alpha_{13} \\ \alpha_{21} & \alpha_{22} & \alpha_{23} \\ \alpha_{31} & \alpha_{32} & \alpha_{33} \end{bmatrix} \qquad (8.174)$$

where $\alpha_{ij} = \alpha_{ji}$. In cases where the flow direction coincides with one of the principal directions of the hydraulic conductivity, then the off-diagonal terms in the dispersivity tensor are equal to zero, and α_{ij} can be written in the form

$$\alpha_{ij} = \begin{bmatrix} \alpha_{11} & 0 & 0 \\ 0 & \alpha_{22} & 0 \\ 0 & 0 & \alpha_{33} \end{bmatrix} \qquad (8.175)$$

where α_{11} is generally taken as the dispersivity in the flow direction, and α_{22} and α_{33} are the dispersivities in the horizontal and vertical transverse principal directions of the hydraulic conductivity. The component of the dispersivity in the direction of flow is called the *longitudinal dispersivity*, and the other components of the dispersivity are called the *transverse dispersivities*.

The dispersivites used to describe the transport of contaminants in porous media cannot be taken as constant unless the contaminant cloud has traversed several correlation length scales of the hydraulic conductivity, or the contaminant cloud is sufficiently large to encompass several correlation length scales. If either of these conditions is violated, then the dispersivity increases as the contaminant cloud moves through the porous medium, includes an expanding range of hydraulic conductivity variations and ultimately approaches a constant value called the *asymptotic macrodispersivity* or simply the *macrodispersivity*. In isotropic media, the correlation length scale, λ, of the hydraulic conductivity is the same in all directions, and the components of the macrodispersivity can be estimated using the approximate relations (Dagan, 1989; Chin and Wang, 1992)

$$\alpha_{11} = \sigma_Y^2 \lambda, \quad \alpha_{22} = \alpha_{33} = 0 \tag{8.176}$$

The components of the macrodispersion tensor estimated by Equation 8.176 can be taken to be approximately valid up to $\sigma_Y = 1.5$. In cases where the porous medium is stratified, isotropic in the horizontal plane, and anisotropic in the vertical plane, then the correlation length scale of the hydraulic conductivity in the horizontal plane can be denoted by λ_h, and the correlation length scale in the vertical direction denoted by λ_v. The *anisotropy ratio*, e, is then defined by

$$e = \frac{\lambda_v}{\lambda_h} \tag{8.177}$$

and is typically on the order of 0.1 in most stratified media. Gelhar and Axness (1983) have derived approximate relations to estimate the components of the macrodispersivity in the case that the flow is in the plane of isotropy. In this case, the longitudinal and transverse components of the macrodispersivity tensor can be estimated by

$$\alpha_{11} = \sigma_Y^2 \lambda_h, \quad \alpha_{22} = \alpha_{33} = 0 \tag{8.178}$$

The relationships given in Equation 8.178 are approximately valid for $\sigma_Y < 1$, but the exact range of validity has not yet been established (Chin, 1997). Typical values of σ_Y, λ_h, and λ_v in several formations are listed in Table 8.19. It is important to note that even though the hydraulic-conductivity statistics given in Table 8.19 depend on the support scale of the samples used to derive the statistics, the macrodispersivities calculated using these statistics are (theoretically) independent of the support scale of the samples. In estimating the (total) dispersivity in porous media, the macrodispersivities calculated using Equations 8.176, and 8.178 are additive to the dispersivities associated with hydrodynamic dispersion, which result from pore-scale mixing and molecular diffusion.

■ Example 8.28

Several hydraulic conductivity measurements in an isotropic aquifer indicate that the spatial covariance, C_Y, of the log-hydraulic conductivity can be approximated by the equation

■ **Table 8.19**
Variances and Correlation Length Scales of Hydraulic Conductivity

Formation	σ_Y	λ_h (m)	λ_v (m)	Reference
Sandstone	1.5–2.2	—	0.3–1.0	Bakr (1976)
Sandstone	0.4	8	3	Goggin et al. (1988)
Sand	0.9	> 3	0.1	Byers and Stephens (1983)
Sand	0.6	3	0.12	Sudicky (1986)
Sand	0.5	5	0.26	Hess (1989)
Sand	0.4	8	0.34	Woodbury and Sudicky (1991)
Sand	0.4	4	0.2	Robin et al. (1991)
Sand	0.2	5	0.21	Woodbury and Sudicky (1991)
Sand and gravel	5.0	12	1.5	Boggs et al. (1990)
Sand and gravel	2.1	13	1.5	Rehfeldt et al. (1989)
Sand and gravel	1.9	20	0.5	Hufschmied (1986)
Sand and gravel	0.8	5	0.4	Smith (1978); Smith (1981)

$$C_Y = \sigma_Y^2 \exp\left[-\frac{r_1^2}{\lambda^2} - \frac{r_2^2}{\lambda^2} - \frac{r_3^2}{\lambda^2} \right]$$

where $\sigma_Y = 0.5$, $\lambda = 5$ m, the spatial lags r_1 and r_2 are measured in the horizontal plane, and r_3 is measured in the vertical plane. The mean hydraulic gradient is 0.001, the effective porosity is 0.2, and the mean log-hydraulic conductivity is 2.5 (where the hydraulic conductivity is in m/d). Estimate the effective hydraulic conductivity and the macrodispersion coefficient in the aquifer.

Solution

From the given data, the hydraulic conductivity field is described statistically by $\langle Y \rangle = 2.5$, $\sigma_Y = 0.5$, and $\lambda = 5$ m. The geometric mean hydraulic conductivity, K_G, is given by Equation 8.170 as

$$K_G = e^{\langle Y \rangle} = e^{2.5} = 12 \text{ m/d}$$

and the effective hydraulic conductivity, for three-dimensional flow, is given by Equation 8.172 as

$$K_{\text{eff}} = K_G \left(1 + \frac{\sigma_Y^2}{6} \right) = (12) \left(1 + \frac{0.5^2}{6} \right) = 12.5 \text{ m/d}$$

The mean seepage velocity, V, in the aquifer is given by Equation 8.171, as

$$V = -\frac{K_{\text{eff}}}{n_e} J$$

where $J = -0.001$ and $n_e = 0.2$, hence

$$V = -\frac{12.5}{0.2}(-0.001) = 0.063 \text{ m/d}$$

Since $\sigma_Y = 0.5$ and $\lambda = 5$ m, the longitudinal macrodispersivity, α_{11}, can be estimated by Equation 8.176 as

$$\alpha_{11} = \sigma_Y^2 \lambda = (0.5)^2(5) = 1.25 \text{ m}$$

and, according to Equation 8.176, the theoretical transverse macrodispersivities are both zero. The longitudinal dispersion coefficient, D_{11}, is given by

$$D_{11} = \alpha_{11} V = (1.25)(0.063) = 0.079 \text{ m}^2/\text{d}$$ ■

The relative importance of advective transport to dispersive transport can be measured by the *Peclet number*, Pe, defined as

$$\text{Pe} = \frac{VL}{D_L} \tag{8.179}$$

where V is the mean seepage velocity, L is the characteristic length scale, and D_L is the characteristic longitudinal dispersion coefficient. In municipal wellfields, values of Pe within several meters of the well tend to be high, indicating that contaminant transport is advection-dominated and dispersion effects can be neglected. A Peclet number, Pe_m, can be defined in terms of the molecular diffusion coefficient where

$$\text{Pe}_m = \frac{Vd}{D_m} \tag{8.180}$$

where d is the characteristic pore size, and D_m is the molecular diffusion coefficient. Previous investigations have shown that the pore-scale longitudinal dispersion coefficient is much greater than the molecular diffusion coefficient when $\text{Pe}_m > 10$, and the transverse dispersion coefficient is much greater than the molecular diffusion coefficient when $\text{Pe}_m > 100$ (Perkins and Johnson, 1963).

■ Example 8.29

The mean seepage velocity in an aquifer is 1 m/d, the mean pore size is 1 mm, and the molecular diffusion coefficient of a certain toxic contaminant in water is 10^{-9} m²/s. Determine whether molecular diffusion should be considered in a contaminant transport model.

Solution

From the given data, $V = 1$ m/d, $d = 1$ mm $= 0.001$ m, and $D_m = 10^{-9}$ m²/s $= 8.64 \times 10^{-5}$ m²/d. The Peclet number, Pe_m, is given by

$$\text{Pe}_m = \frac{Vd}{D_m} = \frac{(1)(0.001)}{8.64 \times 10^{-5}} = 12$$

Since $\text{Pe}_m > 10$, molecular diffusion has a negligible contribution to longitudinal dispersion, but since $\text{Pe}_m < 100$ molecular diffusion will contribute significantly to transverse dispersion. ■

In most practical cases longitudinal dispersion is dominated by macrodispersion, vertical dispersion is dominated by local hydrodynamic dispersion and the horizontal-transverse dispersion is significantly influenced by temporal variations in the seepage velocity (Rehfeldt and Gelhar, 1992). Field studies indicate that horizontal-transverse dispersivities can be related to longitudinal dispersivities using a ratio of longitudinal to horizontal-transverse dispersivity in the range of 6 to 20 (Anderson, 1979; Klotz et al., 1980). Common practice is to estimate the longitudinal dispersivity using a theoretical or empirical relation such as Equation 8.176 and estimate the horizontal-transverse dispersivity as one-tenth of the longitudinal dispersivity. The horizontal-transverse dispersivity is usually much larger than the vertical-transverse dispersivity, which tends to be very small, and primarily associated with pore-scale mixing and molecular diffusion.

Longitudinal dispersivities derived from 55 field experiments around the world have been collated by Gelhar and colleagues (1992) and are shown in Figure 8.15. Based on these results, it is clear that the longitudinal dispersivity coefficient increases with the distance traveled by the contaminant cloud, indicating that field formations are seldom homogeneous, and that the variability in hydraulic conductivity increases with scale. Of all the experiments reviewed by Gelhar and colleagues (1992), only five studies were considered to provide reliable estimates of the dispersivity, a further 18 values were considered of intermediate reliability, and the most reliable dispersivity estimates were at the lower end of the length scale. In the absence of field measurements of the hydraulic conductivity, from which the spatial statistics are parameterized by $\langle Y \rangle$, σ_Y, and λ, Figure 8.15 provides a useful basis for estimating the dispersivity in porous formations. Analyses by Neuman (1990) indicate that the longitudinal macrodispersivity, α_{11}, can

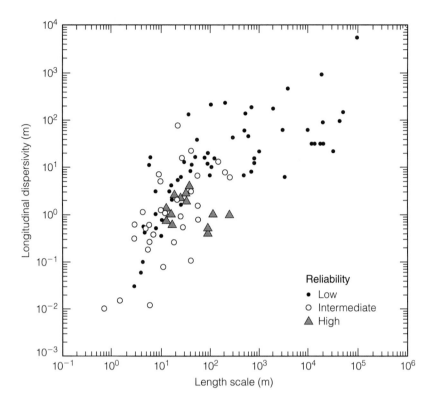

Figure 8.15 ■ Longitudinal Dispersivity Versus Length Scale in Ground Water

be related to the travel distance, L, by the relation

$$\alpha_{11} = 0.0175 L^{1.46}, \qquad 100 \text{ m} < L < 3500 \text{ m} \tag{8.181}$$

where α_{11} and L are both measured in meters. For $L < 100$ m, a better match with field data was obtained using

$$\alpha_{11} = 0.0169 L^{1.53}, \qquad L < 100 \text{ m} \tag{8.182}$$

Recent analyses by Al-Suwaiyan (1998) demonstrate that the observed macrodispersivities reported by Gelhar and colleagues (1992) are scattered about the mean (approximated by Equations 8.181 and 8.182), with the upper limit of the scatter at about 5 times the mean, and the lower limit at about 1/5 of the mean. These uncertainty limits should be accounted for whenever Equations 8.181 and 8.182 are used in contaminant-transport predictions. For very small scales, on the order of the pore size, dispersion is caused primarily by pore-scale mechanical dispersion and molecular diffusion, where the longitudinal and transverse dispersivities can be estimated by the relations

$$\begin{aligned} \alpha_L &= \alpha_L^* + \frac{D_m}{\tau V} \\ \alpha_T &= \alpha_T^* + \frac{D_m}{\tau V} \end{aligned} \tag{8.183}$$

where α_L^* and α_T^* are pore-scale longitudinal and transverse dispersivities respectively, D_m is the molecular diffusion coefficient in water, τ is the tortuosity (which accounts for the effect of the solid matrix on diffusion), and V is the mean seepage velocity. The molecular diffusion coefficient divided by the tortuosity represents the effective molecular diffusion coefficient in porous media and is sometimes called the *bulk diffusion coefficient*. Values of α_L^* are typically on the order of the pore size of the porous medium, α_T^* is typically on the order of $0.1\alpha_L^*$ to $0.01\alpha_L^*$ (Delleur, 1998), τ is typically in the range of 2 to 100 (lower values are associated with coarse material such as sands; higher values are associated with finer material such as clays), and typical values of the molecular diffusion coefficient are in the range of 1×10^{-9} to 2×10^{-9} m^2/s.

For travel distances longer than 3,500 m, the longitudinal dispersivity tends to asymptote to an upper limit that is consistent with a finite variability in the hydraulic conductivity. In cases where the dispersivity increases with travel distance, the Fickian assumption of a constant dispersion coefficient is not supported and the dispersion is termed *non-Fickian*. However, a Fickian approximation to the mixing process is obtained by adjusting the dispersion coefficient with length scale, and the advection-dispersion equation can be used to approximate the dispersion process. Typical values of the longitutinal dispersivity for various ranges of length scales are shown in Table 8.20.

■ Example 8.30

A contaminant plume in an aquifer is approximately 50 m long, 10 m wide, and 3 m deep. The characteristic pore size in the aquifer is 3 mm, the molecular diffusion coefficient is 2×10^{-9} m^2/s, the tortuosity is 1.5, and the mean seepage velocity is 0.5 m/d. Estimate the components of the dispersion coefficient.

Solution

From the given data, $L_x = 50$ m, $L_y = 10$ m, $L_z = 3$ m, $D_m = 2 \times 10^{-9}$ m^2/s $= 1.73 \times 10^{-4}$ m^2/d, $\tau = 1.5$, and $V = 0.5$ m/d. The length scale, L, of the contaminant plume can be approximated by the relation

$$L = \sqrt{L_x L_y} = \sqrt{(50)(10)} = 22 \text{ m}$$

and since $L < 100$ m, the longitudinal macrodispersivity can be estimated by Equation 8.182 as

$$\alpha_{11} = 0.0169L^{1.53} = 0.0169(22)^{1.53} = 1.9 \text{ m}$$

The horizontal-transverse macrodispersivity, α_{22}, can be estimated as $0.1\alpha_{11}$, which gives

$$\alpha_{22} = 0.1\alpha_{11} = 0.1(1.9) = 0.19 \text{ m}$$

■ Table 8.20
Typical Longitudinal
Dispersivities for Various
Length Scales

Scale (m)	Longitudinal dispersivity (m)	
	Average	Range
< 1	0.001–0.01	0.0001–0.01
1–10	0.1–1.0	0.001–1.0
10–100	25	1–100

Source: Schnoor, Jerald L. *Environmental Modeling.* Copyright © 1996. John Wiley & Sons, Inc. New York. Reprinted by permission of John Wiley & Sons, Inc.

The vertical-transverse macrodispersivity can be taken as zero, hence

$$\alpha_{33} = 0 \text{ m}$$

The local longitudinal dispersivity, α_L, is given by Equation 8.183, where α_L^* is on the order of the pore size (0.003 m), and hence

$$\alpha_L = \alpha_L^* + \frac{D_m}{\tau V} = 0.003 + \frac{1.73 \times 10^{-4}}{(1.5)(0.5)} = 0.0032 \text{ m}$$

and taking α_T^* as $0.1\alpha_L^* = 0.1(0.003) = 0.0003$ m, then Equation 8.183 gives

$$\alpha_T = \alpha_T^* + \frac{D_m}{\tau V} = 0.0003 + \frac{1.73 \times 10^{-4}}{(1.5)(0.5)} = 0.00053 \text{ m}$$

The principal components of the dispersion coefficient are then given by

$$D_{11} = (\alpha_{11} + \alpha_L)V = (1.9 + 0.0032)(0.5) = 0.95 \text{ m}^2/\text{d}$$

$$D_{22} = (\alpha_{22} + \alpha_T)V = (0.19 + 0.00053)(0.5) = 0.095 \text{ m}^2/\text{d}$$

$$D_{33} = (\alpha_{33} + \alpha_T)V = (0 + 0.00053)(0.5) = 0.00027 \text{ m}^2/\text{d}$$

It must be emphasized that these estimates of the dispersion coefficient are order-of-magnitude estimates only, and it would be entirely appropriate to take $D_{11} = 1$ m^2/d, $D_{22} = 0.1$ m^2/d, and $D_{33} = 0.0003$ m^2/d.   ■

8.7.3 ■ Fate Processes

Fate processes include all mechanisms that remove tracer mass from the water environment. These processes include chemical reactions, decay, and sorption. *Sorption* processes include *adsorption*, *chemisorption*, and *absorption*, where adsorption is the process by which a solute attaches itself to a solid surface, chemisorption occurs when the solute is incorporated onto a solid surface by ion exchange, and absorption occurs when the solute diffuses into the aquifer particles and is sorbed onto the interior surfaces of the particles. Fate processes in the environment are complex and difficult to study at the field scale, and usually they are studied under idealized laboratory conditions, with the results fitted to idealized models. In engineering applications, the most commonly considered fate processes are sorption and decay.

8.7.3.1   Sorption

Models that describe the partitioning of dissolved mass onto solid surfaces are called *sorption isotherms*, since they describe sorption at a constant temperature. The most widely used isotherm is the *Freundlich isotherm* given by

$$F = K_F c_{\text{aq}}^{1/m} \tag{8.184}$$

where F is the mass of tracer sorbed per unit mass of solid phase, c_{aq} is the concentration of the tracer dissolved in the water (aqueous concentration), and K_F and m are constants. In most cases, the constant m is approximately equal to unity, in which case Equation 8.184 is linear and is written in the form

$$F = K_d c_{\text{aq}} \tag{8.185}$$

where K_d is called the *distribution coefficient* (L^3/M), and is defined as the ratio of the sorbed tracer mass per unit mass of solid matrix to the aqueous concentration.

There are two important limitations in applying linear isotherms such as Equation 8.185 (Fetter, 1992): (1) Such isotherms do not limit the amount of solute that can be sorbed onto the solid matrix (there must be an upper limit); and (2) the actual isotherm is usually *piecewise linear*, and therefore extrapolation beyond the range of experimental conditions used to estimate K_d is not recommended. Values of K_d in Equation 8.185 range from near zero to 10^3 mL/g or greater. In the case of organic compounds, the mass of the organic compound sorbed per unit mass of solid matrix has been observed to depend primarily on the amount of organic carbon in the solid matrix (Karickhoff et al., 1979), and it is more appropriate to deal with the sorption coefficient K_{oc}, which is defined as the ratio of sorbed mass of organic compound per unit mass of organic carbon to the aqueous concentration. Therefore, the distribution coefficient, K_d, is related to the organic carbon sorption coefficient, K_{oc}, by

$$K_d = f_{oc} K_{oc} \tag{8.186}$$

where f_{oc} is the fraction of organic carbon in the porous medium [M/M]. The formulation described by Equation 8.186 is appropiate whenever the organic fraction exceeds 1%, and values of K_{oc} for several organic compounds typically found in contaminated groundwater are given in Appendix B.2. When the organic fraction is less than 1%, it is not automatic that the soil or aquifer organic carbon will be the primary surface onto which the organic compounds will partition (Fetter, 1993). The octanol-water partition coefficient, K_{ow}, is a widely available and easily measured parameter that gives the distribution of a chemical between *n*-octanol and water in contact with each other. This coefficient is defined by

$$K_{ow} = \frac{c_o}{c_w} \tag{8.187}$$

where c_o is the concentration in the octanol, and c_w is the concentration in the water.

There have been many experimental studies of the relationship between K_{oc} and K_{ow}. Several proposed empirical relationships are shown in Table 8.21, where K_{oc} is in cm^3/g and K_{ow} is dimensionless. Clearly there is no universal relation for deriving K_{oc} from the easily measured K_{ow}, although Fetter (1993) has shown that most estimates of K_{oc}

■ **Table 8.21** Empirical Relationships Between K_{oc} and K_{ow}	**Equation**	**Chemicals**	**Reference**
	$\log K_{oc} = 1.00 \log K_{ow} - 0.21$	10 polyaromatic hydrocarbons	Karickhoff et al. (1979)
	$K_{oc} = 0.63 K_{ow}$	miscellaneous organics	Karickhoff et al. (1979)
	$\log K_{oc} = 0.544 \log K_{ow} + 1.377$	45 organics, mostly pesticides	Kenaga and Goring (1980)
	$\log K_{oc} = 1.029 \log K_{ow} - 0.18$	13 pesticides	Rao and Davidson (1980)
	$\log K_{oc} = 0.94 \log K_{ow} + 0.22$	s-trizines and dinitroanalines	Rao and Davidson (1980)
	$\log K_{oc} = 0.989 \log K_{ow} - 0.346$	5 polyaromatic hydrocarbons	Karickhoff (1981)
	$\log K_{oc} = 0.937 \log K_{ow} - 0.006$	aromatics, polyaromatics, triazines	Lyman (1982)
	$\ln K_{oc} = \ln K_{ow} - 0.7301$	DDT, tetrachlorobiphenyl, lindane, 2,4-D and dichloropropane	McCall et al. (1983)
	$\log K_{oc} = 0.72 \log K_{ow} + 0.49$	methylated and chlorinated benzenes	Schwarzenbach and Westall (1981)
	$\log K_{oc} = 1.00 \log K_{ow} - 0.317$	22 polynuclear aromatics	Hassett et al. (1980)

■ Table 8.22
Values of K_{ow} for Selected
Organic Compounds

Compound	log K_{ow}	Reference
Acetone	−0.24	Schwarzenbach et al. (1993)
Atrazine	2.56	Schwarzenbach et al. (1993)
Benzene	2.01–2.13*	Hansch and Leo (1979), MacKay (1991)
Carbon tetrachloride	2.64–2.72	Schnoor (1996), Hansch and Leo (1979)
Chlorobenzene	2.49–2.84*	Hansch and Leo (1979), MacKay (1991)
Chloroform	1.95–1.97*	Hansch and Leo (1979)
DDT	4.98–6.91	Hansch and Leo (1979), Schnoor (1996)
Dieldrin	5.48	Schwarzenbach et al. (1993)
Lindane	3.78	Schwarzenbach et al. (1993)
Malathion	2.89	Schwarzenbach et al. (1993)
Naphthalene	3.29–3.35*	MacKay (1991), Schnoor (1996)
n-Octane	5.18	Schwarzenbach et al. (1993)
Parathion	3.81	Schwarzenbach et al. (1993)
Phenol	1.46*–1.49	Hansch and Leo (1979), MacKay (1991)
Polychlorinated biphenyls	4.09–8.23	Schwarzenbach et al. (1993)
2,3,7,8-Tetrachlorodibenzo-p-dioxin	6.64	Schwarzenbach et al. (1993)
Toluene	2.69*	MacKay (1991)
1,1,1-Trichloroethane	2.47*–2.51	MacKay (1991), Schnoor (1996)
Trichloroethylene (TCE)	2.29*	MacKay (1991)
p-Xylene	3.12–3.18	Schwarzenbach et al. (1993)

*@ 25°C.

derived using any equation in Table 8.21 are likely to fall within one standard deviation of the geometric mean of K_{oc} estimated from the combined predictions of all the equations listed in Table 8.21. Values of K_{ow} for selected organic compounds are listed in Table 8.22, and it is best to estimate K_{oc} from these values using an empirical relation derived for similar chemicals. The amount of sorbed mass per unit volume of the porous medium, c_s, is related to the mass of tracer sorbed per unit mass of solid phase, F, by the expression

$$c_s = \rho_b F \tag{8.188}$$

where n is the porosity of the medium, and ρ_s is the bulk density of the solid (= oven-dried mass of a soil sample divided by the sample volume), which is typically in the range of 1,600–2,100 kg/m^3. Combining Equations 8.185 and 8.188 leads to the following linear relationship between the concentration of the sorbed mass, c_s, and the concentration of the dissolved mass, c_{aq},

$$\boxed{c_s = \beta c_{aq}} \tag{8.189}$$

where β is a dimensionless constant given by

$$\boxed{\beta = \rho_b K_d} \tag{8.190}$$

The fate term, S_m, in the dispersion equation, Equation 8.160, is equal to the rate at which tracer mass is added to the water in a unit volume of water. In the case of sorption, the rate at which tracer mass is added to the aqueous phase is equal to the rate at which tracer mass is lost from the solid phase. Therefore, the rate at which tracer mass is added to the water per unit volume of water, S_m, is given by

$$S_m = -\frac{1}{n}\frac{\partial c_s}{\partial t} = -\frac{\beta}{n}\frac{\partial c_{aq}}{\partial t} \tag{8.191}$$

In applications to flow in porous media, the contaminant concentration, c, used in the advection-dispersion equation is equal to the mass of contaminant per unit volume of porous medium, and c is related to the aqueous concentration, c_{aq}, by

$$c_{aq} = c \tag{8.192}$$

Combining Equations 8.191 and 8.192, the tracer mass flux, S_m, given by Equation 8.191 can be written as

$$S_m = -\frac{\beta}{n}\frac{\partial c}{\partial t} \tag{8.193}$$

Substituting this sorption model into the advection-dispersion equation, Equation 8.160, yields

$$\left(1+\frac{\beta}{n}\right)\frac{\partial c}{\partial t} + \sum_{i=1}^{3} V_i \frac{\partial c}{\partial x_i} = \sum_{i=1}^{3} D_i \frac{\partial^2 c}{\partial x_i^2} \tag{8.194}$$

The term $(1+\beta/n)$ is commonly referred to as the *retardation factor*, R_d, where

$$\boxed{R_d = 1 + \frac{\beta}{n}} \tag{8.195}$$

Dividing both sides of Equation 8.194 by R_d yields the following form of the dispersion equation

$$\boxed{\frac{\partial c}{\partial t} + \sum_{i=1}^{3}\left(\frac{V_i}{R_d}\right)\frac{\partial c}{\partial x_i} = \sum_{i=1}^{3}\left(\frac{D_i}{R_d}\right)\frac{\partial^2 c}{\partial x_i^2}} \tag{8.196}$$

Comparing Equation 8.196, which accounts for the sorbing of contaminants onto porous media, to the advection-dispersion equation for conservative contaminants, Equation 8.160, it is clear that both equations have the same form, with sorption being accounted for by reducing the mean seepage velocity and dispersion coefficients by a factor $1/R_d$. In other words, the fate and transport of a sorbing tracer can be modeled by neglecting sorption but reducing the mean velocity and dispersion coefficients by a factor $1/R_d$.

■ Example 8.31

One kilogram of a contaminant is spilled over a 1-m depth of ground water and spreads laterally as the ground water moves with an average velocity of 0.1 m/d. The longitudinal and transverse dispersion coefficients are 0.03 m²/d and 0.003 m²/d, respectively; the porosity is 0.2; the density of the aquifer matrix is 2.65 g/cm³; log K_{oc} is 1.72 (K_{oc} in cm³/g); and the organic fraction in the aquifer is 5%. Calculate the concentration at the spill location after 1 hour, 1 day, and 1 week. Compare these values with the concentration obtained by neglecting sorption.

Solution

The distribution coefficient, K_d, is given by

$$K_d = f_{oc} K_{oc}$$

where $f_{oc} = 0.05$ and $K_{oc} = 10^{1.72} = 52.5$ cm^3/g. Therefore

$$K_d = (0.05)(52.5) = 2.63 \text{ cm}^3/\text{g}$$

The dimensionless constant β is given by

$$\beta = \rho_b K_d = (1 - n)\rho_s K_d$$

where $n = 0.2$, $\rho_s = 2.65$ g/cm^3, and therefore

$$\beta = (1 - 0.2)(2.65)(2.63) = 5.58$$

The retardation factor, R_d, is then given by

$$R_d = 1 + \frac{\beta}{n} = 1 + \frac{5.58}{0.2} = 29$$

For an instantaneous release, the resulting concentration distribution is given by

$$c(x, y, t) = \frac{M}{4\pi H t \sqrt{D_L D_T}} \exp\left[-\frac{(x - Vt)^2}{4D_L t} - \frac{y^2}{4D_T t}\right]$$

where $M = 1$ kg, $H = 1$ m, $D_L = 0.03/R_d$ m^2/d, $D_T = 0.003/R_d$ m^2/d, $V = 0.1/R_d$ m/d, $x = 0$ m, and $y = 0$ m. Substituting these values into the above expression for the concentration distribution yields

$$c(0, 0, t) = \frac{(1)R_d}{4\pi (1)t\sqrt{(0.03)(0.003)}} \exp\left[-\frac{(-0.1/R_d t)^2}{4(0.03/R_d)t}\right]$$

$$= \frac{8.39 R_d}{t} \exp\left[-\frac{0.083t}{R_d}\right] \text{ kg/m}^3$$

In the absence of sorption, $R_d = 1$, for a sorbing contaminant $R_d = 29$, and the concentrations at $t = 1$ h, 1 day, and 1 week are given in the following table

Time	Without sorption ($R_d = 1$) (kg/m^3)	With sorption ($R_d = 29$) (kg/m^3)
1 hour	199	5,792
1 day	7.7	243
1 week	0.67	34

Sorption results in higher contaminant concentrations in the ground water near the spill. This is a result of the requirement that higher water concentrations are necessary to maintain an equilibrium with the sorbed mass. Unrealistically high concentrations calculated at early times are a result of the model assumption that the spill occurs over an infinitesimally small volume. To be realistic, the calculated concentrations should be less than the solubility of the contaminant.   ■

8.7.3.2 First-Order Decay

Many chemical compounds in the environment ultimately decompose into other compounds, usually through chemical reactions such as hydrolysis or biodegradation. The most frequently used model of decomposition is the following first-order decay model

$$S_m = -\lambda c \qquad (8.197)$$

where S_m is the rate at which tracer mass is added to ground water per unit volume of ground water, c is the concentration of tracer in the ground water, and λ is the first-order decay coefficient. Substituting this decay model into the dispersion equation, Equation 8.160, leads to

$$\frac{\partial c}{\partial t} + \sum_{i=1}^{3} V_i \frac{\partial c}{\partial x_i} = \sum_{i=1}^{3} D_i \frac{\partial^2 c}{\partial x_i^2} - \lambda c \qquad (8.198)$$

which can be modified by changing variables from c to c_*, where

$$c = c_* e^{-\lambda t} \qquad (8.199)$$

Substituting Equation 8.199 into Equation 8.198 and dividing both sides by $e^{-\lambda t}$ yields

$$\frac{\partial c_*}{\partial t} + \sum_{i=1}^{3} V_i \frac{\partial c_*}{\partial x_i} = \sum_{i=1}^{3} D_i \frac{\partial^2 c_*}{\partial x_i^2} \qquad (8.200)$$

which is exactly the same as the advection-dispersion equation for a conservative tracer. The practical implication of this result is that the fate and transport of a tracer undergoing first-order decay is the same as if the tracer is initially assumed to be conservative, and the resulting concentration distribution reduced by a factor $e^{-\lambda t}$, where t is the time since release of the tracer mass.

■ Example 8.32

Ten kilograms of a contaminant is spilled into the ground water and is well mixed over a 1-m depth. The mean seepage velocity in the aquifer is 0.5 m/d, the longitudinal dispersion coefficient is 1 m²/d, the horizontal-transverse dispersion coefficient is 0.1 m²/d, vertical mixing is negligible, and the first-order decay constant of the contaminant is 0.01 d⁻¹. Determine the maximum concentration in the ground water after 1, 10, 100, and 1,000 days. Compare these concentrations to the maximum concentration without decay.

Solution

From the given data, $M = 10$ kg, $H = 1$ m, $V = 0.5$ m/d, $D_L = 1$ m²/d, $D_T = 0.1$ m²/d, and $\lambda = 0.01$ d⁻¹. Neglecting decay, the contaminant concentration downstream of the spill is given by

$$c^*(x, y, t) = \frac{M}{4\pi H t \sqrt{D_L D_T}} \exp\left[-\frac{(x - Vt)^2}{4D_L t} - \frac{y^2}{4D_T t} \right]$$

and the maximum concentration, at $x = Vt$, is given by

$$c_{\max}^*(t) = \frac{M}{4\pi H t \sqrt{D_L D_T}}$$

which yields

$$c_{\max}^*(t) = \frac{10}{4\pi(1)t\sqrt{(1)(0.1)}} = \frac{2.52}{t} \text{ kg/m}^3 = \frac{2520}{t} \text{ mg/L}$$

Accounting for first-order decay, the maximum concentration, $c_{\max}(t)$, is given by

$$c_{\max}(t) = c_{\max}^*(t)e^{-\lambda t} = \frac{2520}{t}e^{-0.01t}$$

Hence the maximum concentrations at 1, 10, 100, and 1,000 days are given by

t (d)	$c_{\max}^*(t)$ (mg/L)	$c_{\max}(t)$ (mg/L)
1	2,520	2,490
10	252	228
100	25.2	9.3
1000	2.52	0.0001

Therefore, as time increases, the decay effect becomes more pronounced.   ■

8.7.3.3 Combined Processes

In some cases, both sorption and first-order decay processes occur simultateously. Assuming that sorption is described by the linear sorption isotherm, Equation 8.189, and that the sorbed mass decays as a first-order process, described by

$$\frac{\partial c_s}{\partial t} = -\lambda c_s \tag{8.201}$$

where c_s is the sorbed mass per unit volume of the porous medium, the mass flux per unit volume of ground water into the aqueous phase due to desorption, S_m^1, is given by

$$S_m^1 = -\frac{\beta}{n}\frac{\partial c_{\text{aq}}}{\partial t} - \frac{\lambda c_s}{n} = -\frac{\beta}{n}\frac{\partial c_{\text{aq}}}{\partial t} - \frac{\lambda \beta c_{\text{aq}}}{n} \tag{8.202}$$

where c_{aq} is the aqueous concentration of the contaminant. In applications to flow in porous media, the contaminant concentration, c, used in the advection-dispersion equation is equal to the aqueous concentration, c_{aq}, hence

$$c_{\text{aq}} = c \tag{8.203}$$

Combining Equations 8.202 and 8.203 gives the tracer mass flux, S_m^1, as

$$S_m^1 = -\frac{\beta}{n}\frac{\partial c}{\partial t} - \lambda\frac{\beta}{n}c \tag{8.204}$$

In addition to the mass flux into the aqueous phase due to desorption, there is the additional mass, S_m^2, being removed from the ground water due to decay of the dissolved contaminant, where

$$S_m^2 = -\lambda c \qquad (8.205)$$

The total rate at which mass is added to the ground water, S_m, is equal to the mass flux into the ground water due to desorption plus the mass flux due to first-order decay of the dissolved contaminant, therefore

$$
\begin{aligned}
S_m &= S_m^1 + S_m^2 \\
&= -\frac{\beta}{n}\frac{\partial c}{\partial t} - \lambda\frac{\beta}{n}c - \lambda c \\
&= -\frac{\beta}{n}\frac{\partial c}{\partial t} - \left(1 + \frac{\beta}{n}\right)\lambda c \qquad (8.206)
\end{aligned}
$$

Substituting this fate model into the dispersion equation, Equation 8.160, and simplifying yields

$$\frac{\partial c}{\partial t} + \sum_{i=1}^{3}\left(\frac{V_i}{R_d}\right)\frac{\partial c}{\partial x_i} = \sum_{i=1}^{3}\left(\frac{D_i}{R_d}\right)\frac{\partial^2 c}{\partial x_i^2} - \lambda c \qquad (8.207)$$

where R_d is the retardation factor defined by Equation 8.195. Equation 8.207 indicates that the fate and transport of a tracer that is undergoing both sorption and first-order decay is the same as if sorption is neglected, but the mean fluid velocity and dispersion coefficients are reduced by a factor $1/R_d$. Equation 8.207 can be further simplified by changing variables from c to c_* where

$$c = c_* e^{-\lambda t} \qquad (8.208)$$

Substituting Equation 8.208 into Equation 8.207 and simplifying yields

$$\boxed{\frac{\partial c_*}{\partial t} + \sum_{i=1}^{3}\left(\frac{V_i}{R_d}\right)\frac{\partial c_*}{\partial x_i} = \sum_{i=1}^{3}\left(\frac{D_i}{R_d}\right)\frac{\partial^2 c_*}{\partial x_i^2}} \qquad (8.209)$$

This is the dispersion equation for a conservative contaminant and demonstrates that the fate and transport of an sorbing tracer undergoing first-order decay can be modeled by: (1) reducing the fluid velocity and dispersion coefficients by the factor $1/R_d$; (2) neglecting both sorption and decay; and (3) reducing the resulting concentration distribution by the factor $e^{-\lambda t}$, where t is the time since the release of the tracer mass.

■ Example 8.33

Three kilograms of a contaminant are spilled over a 1-m depth of ground water that is moving with an average seepage velocity of 0.1 m/d. Vertical diffusion is negligible, and the longitudinal and horizontal-transverse dispersion coefficients are 0.05 m^2/d and 0.005 m^2/d, respectively. If the retardation factor is equal to 20 and the first-order decay factor is 2 d^{-1}, calculate the concentration at the spill location after 1 day and 1 week.

Solution

The concentration distribution (accountng for sorption but prior to correction for decay) is given by

$$c_*(x, y, t) = \frac{M}{4\pi H t\sqrt{D_L D_T}}\exp\left[-\frac{(x - Vt)^2}{4D_L t} - \frac{y^2}{4D_T t}\right]$$

where $M = 3$ kg, $H = 1$ m, $D_L = 0.05/R_d = 0.05/20 = 0.0025$ m²/d, $D_T = 0.005/R_d = 0.005/20 = 0.00025$ m²/d, $x = 0$ m, $y = 0$ m, and $V = 0.1/R_d = 0.1/20 = 0.005$ m/d. Substituting these values into the previous equation yields

$$c_*(0, 0, t) = \frac{3}{4\pi(1)t\sqrt{(0.0025)(0.00025)}} \exp\left[-\frac{(0.005t)^2}{4(0.0025)t}\right] \text{kg/m}^3$$

$$= \frac{302}{t} \exp[-0.0025t] \text{ kg/m}^3$$

Correcting for decay requires multiplying by $e^{-\lambda t}$, where $\lambda = 2$ d⁻¹, and therefore the actual concentration as a function of time is given by

$$c(0, 0, t) = c_*(0, 0, t)e^{-2t}$$

$$= \frac{302}{t} \exp[-0.0025t - 2t] \text{ kg/m}^3$$

$$= \frac{302}{t} \exp[-2.0025t] \text{ kg/m}^3$$

Therefore at $t = 1$ day, $c(0, 0, t) = 41$ kg/m³ $= 41,000$ mg/L, and at $t = 7$ days $c(0, 0, t) = 3.5 \times 10^{-5}$ kg/m³ $= 0.035$ mg/L.   ■

8.7.4 ■ Nonaqueous-Phase Liquids

Many organic compounds are only slightly soluble in water and exist in both the dissolved and insoluble (pure) phase in ground water. Pure liquids that are not dissolved are called *nonaqueous-phase liquids* (NAPLs). NAPLs have been identified at four out of five hazardous waste sites in the United States (Plumb and Pitchford, 1985), and are typically composed of either a single chemical or a mixture of several chemicals. NAPLs are further classified as light NAPLs (LNAPLs) that are less dense than water and tend to float on the water table, and dense NAPLs (DNAPLs) that are denser than water and tend to sink to the bottom of the aquifer. Spills of petroleum products such as gasoline, kerosene, and diesel fuels are common sources of LNAPLs in ground water, while DNAPLs are primarily chlorinated solvents such as trichloroethylene (TCE) and perchloroethylene (PCE) originating from a variety of industrial activities, such as degreasing and metal stripping. The densities and solubilities of several NAPLs are given in Table 8.23.

Movement of ground water past NAPLs trapped in the porous medium results in the dissolution of soluble compounds and an associated downstream plume. In some cases, the dissolved concentrations are sufficient to significantly affect the density of the water, inducing a vertical ground-water velocity, v_z, given by (Frind, 1982)

$$\boxed{v_z = -\frac{K_z}{n_e}\left(\frac{\rho}{\rho_o} - 1\right)} \tag{8.210}$$

where K_z is the hydraulic conductivity in the vertical direction, n_e is the effective porosity, ρ is the density of the dissolved mixture, and ρ_o is the density of the native ground water. The relative magnitude of v_z to the horizontal seepage velocity will give an

Liquid	Density @ 15°C (kg/m³)	Solubility @ 10°C (mg/L)
LNAPLs		
Medium distillates (fuel oil)	820–860	3–8
Petroleum distillates (jet fuel)	770–830	10–150
Gasoline	720–780	150–300
Crude oil	800–880	3–25
DNAPLs		
Trichloroethlene (TCE)	1,460	1,070
Tetrachloroethylene (PCE)	1,620	160
1,1,1-Trichloroethane (TCA)	1,320	1,700
Dichloromethane (CH_2Cl_2)	1,330	13,200
Chloroform ($CHCl_3$)	1,490	8,200
Carbon tetrachloride (CCl_4)	1,590	785
Creosote	1,110	20

■ **Table 8.23**
Densities and Solubilities of NAPLs

Source: Schnoor, Jerald L. *Environmental Modeling*. Copyright © 1996. John Wiley & Sons, Inc. New York. Reprinted by permission of John Wiley & Sons, Inc.

indication of the extent to which the contaminant plume affects the ground-water flow. Since LNAPLs do not penetrate very deeply into the water table and are relatively biodegradable under natural conditions, they are generally thought to be a more manageable environmental problem than DNAPLs, which tend to be trapped deep in the aquifer (Bedient et al., 1994). Other factors that make DNAPL contamination harder to remediate are the following: (1) chlorinated solvents do not biodegrade very rapidly and persist for long periods of time in ground water (in fact, products of microbial degradation of halogenated solvents are sometimes more toxic than the parent compounds [Parsons et al., 1984]); and (2) chlorinated solvents have physical properties, such as small viscosities, that allow movement through very small fractures and downward penetration to great distances. The pattern of DNAPL penetration in aquifers is commonly referred to as *viscous fingering*. DNAPL pools on impermeable boundaries are often difficult to locate and remediate using existing technologies (Blatchley and Thompson, 1998). A detailed account of the fate and transport of DNAPLs in ground water can be found in Pankow and Cherry (1996).

The movement of NAPLs in ground water is governed primarily by gravity, buoyancy, and capillary forces. At low concentrations, NAPLs tend to become discontinuous and immobilized by capillary forces, and they end up trapped in the pores of aquifers. Under these conditions, the concentration of the NAPL is termed the *residual saturation*, which is defined as the fraction of total pore volume occupied by residual NAPL under ambient ground-water flow conditions. In the unsaturated zone, residual saturation values are typically in the range of 5% to 20%, while in the saturated zone this range is typically on the order of 15% to 50% (Mercer and Cohen, 1990; Schwille, 1988). Even at residual saturation levels, NAPLs are capable of contaminating large volumes of water, and cannot be easily removed except by dissolution in flowing ground water.

■ **Example 8.34**

A cubic meter of aquifer has a porosity of 0.3 and contains TCE at a residual saturation of 20%. If the density of TCE is 1,470 kg/m³, the solubility of TCE in water is 1,100 mg/L, and the mean seepage velocity of the ground water is 0.02 m/d, estimate the time it would take for the TCE to be removed by dissolution.

Solution

From the given data, $n = 0.3$, and the residual saturation, S_r, is 0.20, hence the residual volume of TCE in 1 m^3 of aquifer is given by

$$\text{Volume of TCE} = (0.20)(0.3)(1) = 0.06 \text{ m}^3$$

Since the density of TCE is 1,470 kg/m^3, then 0.06 m^3 corresponds to $(0.06)(1470) = 88.2$ kg of TCE. With a solubility of 1100 mg/L $= 1.1$ kg/m^3, the volume of water required to dissolve the 88.2 kg of TCE is given by

$$\text{Dissolution water required} = \frac{88.2}{1.1} = 80.2 \text{ m}^3$$

Since the seepage velocity of the ground water is 0.02 m/d, assuming that the contaminated volume is 1 m $\times$ 1 m $\times$ 1 m block of aquifer, the time required for 80.2 m^3 of water to flow through the 1 m^3 of contaminated aquifer is given by

$$\text{Time} = \frac{80.2}{0.02n(1 \times 1)} = \frac{80.2}{0.02(0.3)(1)} = 13,367 \text{ days} = 36.6 \text{ years}$$

Hence the residual NAPL will generate a contaminant plume at saturation level (1,100 mg/L) for 36.6 years! This result should be considered as somewhat approximate, since dissolution rates are highly dependent on the range and size distribution of NAPL blobs (Schnoor, 1996). ▪

8.8 Computer Models

Hundreds of computer models are available for simulating the water quality in rivers, lakes, oceans, and ground water. These models typically provide numerical solutions to the advection-dispersion equation, or some other form of the law of conservation of mass, at discrete locations and times for multiple interacting constituents, complex boundary conditions, spatially and temporally distributed contaminant sources and sinks, multiple fate processes, and variable flow and dispersion conditions. In engineering practice, the use of computer models to apply the fundamental principles covered in this chapter is sometimes essential. In choosing a model for a particular application, there is usually a variety of models to choose from. However, in doing work that is to be reviewed by regulatory agencies, or where professional liability is a concern, models developed and maintained by the U.S. government have the greatest credibility and, perhaps more important, are almost universally acceptable in developing permit applications and defending design protocols on liability issues. Several of the more widely used models that have been developed and endorsed by U.S. government agencies are described briefly here.

▪ Rivers

▪ **QUAL-II.** QUAL-II (USEPA, 1987) is presently the most widely used computer model for simulating streamwater quality. This model is capable of simulating several interacting water-quality constituents in connected (dendritic) streams that are well mixed across their cross-section. Water-quality constituents that can be simulated include dissolved oxygen, temperature, BOD, available nitrogen, available phosphorous, and algae as chlorophyll a. The latest version of the program, known as the *enhanced QUAL-II* *model* or QUAL-2E, is currently maintained by the EPA Center for Water Quality Modeling in Athens, Georgia.

■ **WASP 4.** WASP 4, an acronym for Water Quality Analysis Simulation Program, (USEPA, 1988) simulates contaminant transport in streams in one, two, or three dimensions. The contaminants that can be simulated include BOD, dissolved oxygen, and nutrients. This model is similar to QUAL-2E.

■ **HEC-6.** HEC-6 (USACE, 1991) is a one-dimensional transient numerical model developed by the U.S. Army Corps of Engineers that simulates sediment transport in river and reservoir systems with tributary inflows. This model can perform continuous simulation of aggradation and degradation in streams, and deposition in reservoirs.

■ Lakes and Reservoirs

■ **CE-QUAL-R1.** CE-QUAL-R1 (USAWES, 1986) is one-dimensional (in vertical) horizontally averaged continuous simulation model. This model conceptualizes the lake as a vertical sequence of horizontal layers, where contaminants are uniformly distributed in each layer. The model simulates inflows, outflows, vertical diffusion, and interactions of a number of water-quality constituents.

■ **CE-QUAL-W2.** CE-QUAL-W2 (USAWES, 1986) is a two-dimensional, laterally averaged, continuous-simulation, finite-difference model for reservoir water-quality simulation. This model accounts for vertical and longitudinal diffusion as well as interaction among various water-quality constituents. CE-QUAL-W2 is more complex and sophisticated than CE-QUAL-R1.

■ Oceans

■ **PLUMES.** PLUMES (Baumgartner et al., 1994) is a computer model used for simulating the dilution of fresh-water discharges into ocean waters. The PLUMES model contains four well-established submodels for near-field and far-field dilution. The near-field models are applicable to both single-port and multiport discharges, and the far-field models are implementations of the Brooks far-field dispersion equation (Equation 8.156). The PLUMES model is useful in cases of complex stratification, variable currents, and transient outfall discharges. The PLUMES model is currently maintained by the EPA Pacific Ecosystems Branch in Newport, Oregon.

■ Ground Water

■ **MOC.** MOC (Method of Characteristics) is the most popular model used for contaminant transport in ground water (ASCE, 1996c). This model was originally developed by the U.S. Geological Survey (Konikow and Bredehoeft, 1978) to simulate contaminant fate and transport in two-dimensional single-layer heterogeneous aquifers for both steady and unsteady flow. The model has evolved through several updates and improvements (Goode and Konikow, 1989; Konikow et al., 1994). The processes simulated by the model include advection, hydrodynamic dispersion, sorption, and a variety of chemical reactions (Konikow and Reilly, 1998). MOC has many limitations that may preclude its use in certain applications, such as in the case of complex three-dimensional geologies. A three-dimensional version of the model (MOC3D) has been developed to use MODFLOW (see Section 7.9) to simulate the ground-water flow (Konikow et al.,

1996). A detailed review of the capabilities and usage of several widely used ground-water fate and transport models can be found in ASCE (1996c).

Only a few of the more widely used computer models have been cited here. Certainly there are many other good models that are capable of performing the same tasks.

Summary

Rivers, lakes, oceans, and ground waters are depositories of much of the contaminated water that circulates through the hydrologic cycle, and the quality of these waters has a significant effect on the health and well-being of human populations. In the United States, water-quality criteria are promulgated by states for all natural water bodies, and these criteria are generally based on the most beneficial uses for the present and the future. To preserve or improve water quality, discharge permits are generally required for domestic and industrial wastewater discharges as well as municipal stormwater discharges. The allowable quality of discharges is usually based on available technology for water treatment, with more advanced treatment required for discharges into pristine waters. Measures of water quality generally include pathogenic microorganisms, dissolved oxygen, biochemical oxygen demand, total suspended solids, nitrogen, phosphorous, heavy metals, synthetic organic chemicals, and pH.

The fate and transport of contaminants discharged into natural water bodies are mostly modeled using the advection-dispersion equation, which is an analytic statement of the law of conservation of mass. Contaminant advection by large-scale velocities is parameterized by a velocity field, mixing by smaller-scale velocities is parameterized by a dispersion coefficient, and fate processes are parameterized by the rate at which contaminant mass is added to the water. Analyses of contaminant transport in rivers and streams mostly assume that the contaminant is well mixed across the cross-section, and only one-dimensional mixing along the river is usually modeled. Models have been presented to determine the along-stream distance required for complete cross-sectional mixing, the initial concentration after complete cross-sectional mixing, and the longitudinal dispersion coefficient in terms of channel geometry and flow condition. Using these results, analytic solutions to the advection-dispersion equation are presented for instantaneous spills, and continuous discharges of high-BOD wastewaters into rivers.

Lakes differ from rivers in that they typically have much lower velocities, longer detention times, and the water-quality gradients of concern are usually in the vertical direction. A major cause for concern is that lakes tend to accumulate nutrients and ultimately become eutrophic, in which case the respiration and photosynthesis of phytoplankton can cause significant and deleterious effects on the dissolved oxygen in the lake. The design of systems to control eutrophication involve the identification of the limiting nutrient, usually phosphorous or nitrogen, and control of nutrient fluxes into the lake. The effect of contaminant inputs on the water quality in lakes can sometimes be adequately analyzed using a completely mixed model, which is an appropriate formulation when the wind-induced circulation is strong and the time scale of the analysis is sufficiently long that seasonal mixing processes yield a completely mixed lake.

Coastal communities commonly discharge treated domestic wastewater into the ocean using submerged outfalls. Discharged effluent generally undergoes two distinct phases of mixing, termed near-field mixing and far-field mixing. Near-field mixing occurs in the immediate vicinity of the outfall and is influenced primarily by the buoyancy of the fresh-water discharge; far-field mixing occurs when mixing is dominated by the ambient ocean currents. Near-field mixing models are fairly mature, and formulations are presented that yield fairly accurate predictions of initial dilution. Far-field models are still in the development stage, however, an approximate far-field model is presented.

Ground water is a major source of drinking water in the United States, supplying approximately 50% of the population. Contamination of ground water occurs from a variety of sources, including leaking underground storage tanks, septic systems, agricultural applications of hazardous chemicals, and accidental spills. Dispersion models that simulate the fate and transport of spills, continuous point sources, and continuous finite sources are useful in preliminary analyses of exposures of human populations to contaminated ground water. Aside from the application of fate and transport models in ground water, the determination of model parameters is usually quite challenging. The dispersion coefficient of large-scale plumes in ground water is generally determined by the statistics of the spatial variations in hydraulic conductivity, which are difficult to measure in the field, however, the dispersion coefficient can also be related empircally to the length scale of the plume. Fate processes commonly accounted for in ground-water dispersion models include sorption and decay. Sorption is usually parameterized by the distribution coefficient, decay by a first-order decay factor, and these processes can be modeled by appropriate modification of conservative-contaminant models. Nonaqueous phase liquids are of special concern in ground waters, since they tend to have low solubilities, high residual concentrations, and can persist in the pores of aquifers for many years, providing a continuous source of contamination.

In conclusion, it must be emphasized that much of this chapter is concerned with the fundamental processes affecting water quality in natural water bodies. Practical application of these fundamentals to complex real-world cases frequently require computer models to account for complex geometries and spatial and temporal variabilities. A few of the more common models that are used in water-quality modeling have been briefly described.

Problems

8.1. Saltwater intrusion into a river has increased the average chloride concentration to 3,000 mg/L. If the summer water temperature is around $25°C$ and the winter temperature is around $15°C$, compare the saturated dissolved oxygen level in the summer with the level in the winter.

8.2. At what temperature does the saturation concentration of dissolved oxygen in water fall below the minimum desirable level of 5 mg/L?

8.3. A BOD test on an industrial wastewater indicates that the five-day BOD is 49 mg/L and that the ultimate carbonaceous BOD is 75 mg/L. Estimate the decay factor.

8.4. Analyses of an industrial wastewater indicate that nitrogenous BOD (NBOD) begins after 10 days of incubation and can be described by the same exponential function as the carbonaceous BOD (CBOD). If the rate constant is 0.1 d^{-1} and the five-day BOD is 20 mg/L, estimate the total BOD after 20 days and the ultimate BOD.

8.5. The suspended solids in a 200 m × 200 m lake is measured to be 45 mg/L, and the average settling velocity is estimated as 0.1 m/d. Estimate the rate at which sediment mass is accumulating on the bottom of the lake. If the suspended-solids concentration remains fairly steady and the water leaving the lake does not have a significant suspended sediment content, at what rate is sediment mass entering the lake?

8.6. The phytoplankton *Coscinodiscus lineatus* has a typical diameter of 50 μm and an estimated density of 1,600 kg/m^3. Assuming that the phytoplankton is approximately spherical and the water temperature is $20°C$, estimate the settling velocity using the Stokes equation. Compare your result with the settling velocity given in Table 8.8, and provide possible reasons for any discrepancy.

8.7. A stormwater outfall discharges runoff into a pristine river (with negligible dissolved solids) such that the suspended-solids concentration of the combined water just downstream of the outfall is 100 mg/L. The settling velocity of the sediment is estimated to be 2 m/d, the flow velocity in the river is 0.4 m/s, and the river is 10 m wide and 2 m deep. How far downstream from the outfall will it be before the suspended sediment all settles out? Estimate the rate at which sediment is accumulating downstream of the outfall.

8.8. Consider the (common) case in which an outfall discharges treated domestic wastewater at a rate of 80 L/s with a suspended-solids concentration of 30 mg/L into a river. The suspended solids are composed of predominantly silt particles, and the river has a trapezoidal shape with a flow depth of 4 m, a bottom width of 6 m, and side slopes of 2:1 (H:V). The mean velocity in the river is 3 cm/s. Estimate the distance from the outfall within which most of the suspended particles are deposited on the bottom. Estimate the rate at which sediment is accumulating on the bottom within 500 m of the outfall.

8.9. A natural river has a top width of 18 m, a flow area of 75 m^2, a wetted perimeter of 25 m, and the roughness elements on the wetted perimeter have a characteristic height of 7 mm. If the flowrate in the river is 100 m^3/s, and the temperature of the water is 20°C, estimate the friction factor and the turbulent diffusion coefficients in the vertical and transverse directions. (*Hint:* Use the hydraulic depth [= flow area/top width] as the characteristic depth of the channel.)

8.10. If the river described in Problem 8.8 has a characteristic roughness of 8 mm, estimate the vertical and transverse mixing coefficients. If the outfall discharges the effluent across the entire bottom width of 6 m, estimate the distance downstream to where the effluent is completely mixed across the channel cross-section.

8.11. A single-port outfall is located on the side of a stream that is 15 m wide and 3 m deep, and the flow velocity in the stream is 2 m/s. If the friction factor is estimated to be 0.035 (calculated using the Colebrook equation), how far downstream from the discharge location can the effluent be considered well mixed across the stream? How is this mixing distance affected if the single-port outfall is replaced by a 5-m long multiport outfall located in the center of the stream?

8.12. A regulatory mixing zone in a river extends 200 m downstream of a 5-m long industrial multiport outfall. The river has an average depth of 3 m, an average width of 30 m, and an average velocity of 0.8 m/s; the outfall discharges 10 m^3/s of wastewater containing 5 mg/L of a toxic contaminant. It is estimated that the plume width on the downstream boundary of the mixing zone is 15 m. Estimate the plume dilution on the (downstream) boundary of the mixing zone.

8.13. A relatively clear river containing 5 mg/L of suspended solids has a temperature of 15°C and intersects a turbid river with a suspended-solids concentration of 35 mg/L and a temperature of 20°C. If the discharge in the clear river is 100 m^3/s and the discharge in the turbid river is 20 m^3/s, estimate the suspended-solids concentration and temperature downstream of the confluence of the two rivers.

8.14. Stream depths and vertically averaged velocities at 1-m intervals across a 10-m wide stream are given in the following table. If the friction factor is 0.04, estimate the longitudinal dispersion coefficient across the channel using the formulae in Table 8.9.

Distance from side, y (m)	Depth, d (m)	Velocity, v (m/s)
0	0.0	0.0
1	0.30	0.45
2	1.30	0.90
3	1.8	1.20
4	3.1	2.1
5	4.5	3.0
6	3.6	2.4
7	2.2	1.5
8	1.20	0.75
9	0.70	0.45
10	0.0	0.0

8.15. Fifteen kilograms of a contaminant are spilled into a stream 4 m wide and 2 m deep. If the average velocity in the stream is 0.8 m/s, the longitudinal dispersion coefficient is 0.2 m^2/s, and the first-order decay constant of the contaminant is 0.05 h^{-1}, determine the maximum concentration in the stream at a drinking-water intake 1 km downstream from the spill. How would this concentration be affected if the decay constant is actually one-half of the estimated value?

8.16. The plume resulting from a spill of a conservative contaminant passes an observation point 1 km downstream of the spill location, where the concentration as a function of time is measured. If the river has a mean velocity of 25 cm/s, and the concentration distribution is observed to be Gaussian with a maximum concentration of 5 mg/L, estimate the maximum concentration 1.5 km downstream of the spill location.

8.17. Dye is released instantaneously from a bridge across a river, and the dye concentrations as a function of time are measured at locations 500 m and 1,000 m downstream of the release location. The measured concentrations at the downstream locations are:

Time (min)	Concentration at $x = 500$ m (mg/L)	Concentration at $x = 1,000$ m (mg/L)
0	0	0
1	0	0
2	0	0
3	0	0
4	0	0
5	0	0
6	0.03	0
7	0.05	0
8	0.12	0
9	0.45	0
10	0.92	0
11	2.3	0
12	3.3	0
13	2.3	0.02

Time (min)	Concentration at $x = 500$ m (mg/L)	Concentration at $x = 1,000$ m (mg/L)
14	1.1	0.04
15	0.57	0.22
16	0.17	0.35
17	0.04	0.73
18	0.01	1.1
19	0	1.9
20	0	2.7
21	0	1.9
22	0	1.1
23	0	0.67
24	0	0.41
25	0	0.21
26	0	0.07
27	0	0.03
28	0	0.01
29	0	0
30	0	0

Estimate the longitudinal dispersion coefficient in the river.

8.18. If the river described in Problem 8.17 is 20 m wide and 5 m deep, estimate the mass of dye that was used in the tracer study. Verify that the assumption of a conservative dye is reasonable.

8.19. Estimate the reaeration rate in a stream that is 10 m wide, 2 m deep (on average), and has a flowrate of 2 m^3/s. The stream temperature is 20°C and the dissolved oxygen concentration is 7 mg/L. How would the reaeration rate be affected if the temperature in the river dropped to 15°C?

8.20. A small stream has a mean depth of 0.3 m, a mean velocity of 0.5 m/s, a discharge rate of 0.3 m^3/s, and a slope of 0.1%. Compare the values of the reaeration rate constants estimated using the formulae given in Table 8.12. Comment on your results.

8.21. After initial mixing of a wastewater discharge in a 5-m deep river, the dissolved oxygen concentration in the river is 7 mg/L and the temperature is 22°C. The average flow velocity in the river is 6 cm/s. If the ultimate BOD of the mixed river water is 15 mg/L, the rate constant for BOD at 20°C is 0.5 d^{-1}, and the reaeration rate constant at 20°C is 0.7 d^{-1}, estimate the minimum dissolved oxygen concentration in the river. How far downstream of the outfall location will the minimum dissolved oxygen concentration occur?

8.22. After inital mixing of the wastewater discharged in Problem 8.8, the five-day BOD of the river water is 15 mg/L and the disolved oxygen concentration is 2 mg/L. If the temperature of the river water is 25°C, estimate the dissolved oxygen concentration 500 m downstream of the discharge. Does the minimum dissolved oxygen concentration occur within 500 m of the outfall?

8.23. Measurements in a river indicate that the BOD and reaeration rate constants are 0.3 d^{-1} and 0.5 d^{-1}, and the ultimate BOD of the river water after mixing with a wastewater discharge is 20 mg/L. If the average velocity in the river is 5 cm/s, and the saturation concentration of dissolved oxygen is 12.8 mg/L, determine the initial dissolved oxygen concentration of the mixed river water at which the minimum dissolved oxygen concentration will occur at the outfall.

8.24. Repeat Problem 8.21 accounting for a respiration oxygen demand of 3 g/m^2·d and a benthic oxygen demand of 5 g/m^2·d. Determine the initial dissolved oxygen deficit that will cause the critical oxygen level to occur at the outfall.

8.25. A wastewater is discharged into a large tidal river that has a mean velocity of 5 cm/s, a reaeration rate constant of 0.6 d^{-1}, and a longitudinal dispersion coefficient of 30 m^2/s. After initial mixing of the wastewater, the river has an ultimate BOD of 15 mg/L, and a BOD decay constant of 0.3 d^{-1}. Account for longitudinal dispersion in determining the oxygen deficit 1 km downstream of the outfall. Assess whether it is important to consider longitudinal dispersion in this case.

8.26. A rectangular flood-control lake is 100 m × 70 m and has an average depth of 5 m. If the average inflow and outflow rate is 0.05 m^3/s, estimate the detention time in the lake. Compare this detention time with a larger lake that is 200 m × 140 m and 10 m deep, with an average inflow/outflow rate of 0.1 m^3/s.

8.27. A large natural lake has a surface area of 2.5×10^6 m^2 and the catchment for the lake has an area of 2.0×10^7 m^2. Estimate the hydraulic detention time of the lake.

8.28. A near-shore discharge of a contaminant from a single-port outfall into a lake results in a contaminant concentration of 10 mg/L at a distance of 10 m from the discharge location. The lake currents are negligible, the dispersion coefficient in the lake is 5 m^2/s, and the decay rate of the contaminant is 0.01 d^{-1}. Estimate the contaminant concentration 100 m from the discharge location.

8.29. Measurements in the vicinity of a near-shore outfall in a lake indicate that the contaminant concentration 20 m from the outfall is approximately twice the concentration at a distance 40 m from the outfall. Observations also indicate that the currents in the lake are negligible, and the decay constant of the contaminant is 0.05 d^{-1}. Estimate the dispersion coefficient.

8.30. The orthophosphate concentration in a lake is measured as 30 μg/L and the available nitrogen is measured at 0.2 mg/L. Determine the limiting nutrient for algal growth. If the biomass concentration in the lake is too high, suggest a method to limit the growth of biomass.

8.31. The total-phosphorous concentration is a lake is measured as 15 μg/L and the concentration of total nitrogen is estimated as 0.17 mg/L. Estimate the biomass concentration and trophic state of the lake.

8.32. The density difference between the top and bottom of a 7-m deep lake is measured as 3.5 kg/m^3, and the mean density of the lake water is 998 kg/m^3. If the velocity in the lake is typically on the order of 2% of the wind speed, estimate the maximum wind speed for the lake to remain strongly stratified.

8.33. The currents in a 7-m deep lake are on the order of 5 cm/s. If the mean density of the water is 998 kg/m^3, estimate the density difference between the top and the bottom of the lake for the lake to be strongly stratified. What temperature difference could be responsible for such a density variation?

8.34. You have been appointed to be the project engineer to direct the clean up of a polluted lake in an urban development. Congratulations! The lake is approximately circular with a radius of 100 m and an average depth of 5 m. The target biomass concentration in the lake is 5 μg chl/L, and the current biomass concentration is estimated to be 15 μg chl/L. If the average inflow and outflow from the lake is 5 L/s, estimate the allowable concentration of total phosphorous in the lake inflow such that the target biomass concentration is reached in six months. The decay rate for phosphorus can be taken as 0.01 d^{-1}.

8.35. The average concentration of total phosphorous (TP) in a flood-control lake is 25 μg/L, and daily flows into and out of the lake are typically on the order of 0.13 m^3/s. If the decay rate of TP is 0.2 d^{-1} and the volume of the lake is 2.8 × 10^5 m^3, estimate the mass loading that must be maintained to ultimately bring the TP concentration down to 15 μg/L. If this mass loading is maintained for one week, estimate the TP concentration at the end of each day.

8.36. If the mass loading of TP in Problem 8.35 suddenly drops to zero at the end of the first week, estimate the daily TP concentration during the second week.

8.37. If the mass loading of TP in Problem 8.35 doubles at the end of the first week, estimate the daily TP concentration during the second week.

8.38. The surface area of the lake described in Problem 8.35 is 28,000 m^2. If the effective settling velocity of TP is 0.1 m/d, repeat Problem 8.35 to assess the effect of sedimentation on the TP concentrations.

8.39. A single-port outfall discharges treated municipal wastewater into an unstratified stagnant ocean at a depth of 15 m. The diameter of the outfall port is 0.7 m, the discharge velocity is 3 m/s, the density of the wastewater is 998 kg/m^3, and the density of the ambient seawater is 1,024 kg/m^3. Estimate the plume dilution.

8.40. Repeat Problem 8.39 for the case in which the ambient current is 15 cm/s. Determine the approximate current speed at which the plume changes from the BDNF to BDFF regime.

8.41. An ocean outfall discharges at a depth of 31 m from a diffuser containing seven 1-m diameter ports spaced 10 m apart. The average effluent flow rate is 7.5 m^3/s, the ambient current is 8 cm/s, the density of the ambient seawater is

1.024 g/cm^3, and the density of the effluent is 0.998 g/cm^3. Determine the length scales of the effluent plumes, and calculate the minimum dilution. Neglect merging of adjacent plumes.

8.42. Consider an outfall that is required to discharge treated domestic wastewater at 3 m^3/s in 20 m of water, where the ten percentile current is 5 cm/s. What would the wastewater dilution be if the wastewater is simply discharged out of the end of the 900-mm diameter outfall pipe? What would the wastewater dilution be if the wastewater were discharged through closely spaced ports along a 12-m (long) diffuser? Assume typical values for the wastewater and seawater densities.

8.43. Repeat Problem 8.41 treating the diffuser as a line source. Calculate the dilution for currents perpendicular and parallel to the diffuser. Determine the effluent discharge rate at which the plume dilution becomes independent of the current direction.

8.44. Use Equation 8.143 to derive an expression for the dilution of a slot plume in a stagnant environment. Express the dilution in terms of y, g', L, and Q.

8.45. A 15-kg slug of Rhodamine WT dye is released instantaneously at one point into the ocean, and the concentration distribution of the dye is measured every 3 h for the 12 h duration of daylight when the dye can be seen. The horizontal variance of the dye cloud as a function of time is given in the following table. Estimate the apparent diffusion coefficient as a function of time.

Time, t (h)	$\sigma_{x'}^2$ (cm^2)	$\sigma_{y'}^2$ (cm^2)
0	1.1 × 10^4	1.2 × 10^4
3	3.3 × 10^7	3.0 × 10^7
6	1.5 × 10^8	1.6 × 10^8
9	4.1 × 10^8	3.9 × 10^8
12	7.9 × 10^8	7.7 × 10^8

Compare your results to the Okubo relation (Equation 8.149).

8.46. Compare the apparent diffusion coefficient of a contaminant cloud with a characteristic size of 100 m with the apparent diffusion coefficient of a contaminant cloud whose size is 200 m. Estimate how long it would take for a small contaminant spill to grow to a size of 100 m.

8.47. A 50-m long multiport diffuser discharges effluent at a rate of 6.5 m^3/s and at a depth of 32 m. If the ambient current is 20 cm/s, estimate the distance downstream of the diffuser to where the dilution is equal to 150. Assume that the initial wastefield thickness is 30% of the depth.

8.48. If the near-field mixing at the outfall described in Problem 8.47 is analyzed more closely using the Roberts line-plume model, compare the assumed wastefield thickness (30%) with that derived from using the near-field dilution model. Assume that the current is perpendicular to the outfall, the density of the discharge is 988 kg/m^3, the density of the seawater is 1,025 kg/m^3, and the average dilution in the plume is $\sqrt{2}$ times the minimum dilution.

8.49. Five kilograms of a conservative contaminant is spilled into the ground water and is well mixed over the top one meter. The longitudinal and (horizontal) transverse dispersion coefficients are 0.5 m^2/d and 0.05 m^2/d, respectively; vertical mixing is negligible; and the mean seepage velocity is 0.3 m/d. Determine the concentrations at the spill location for the first seven days after the spill. How are your calculated concentrations affected by the assumed depth of the spill?

8.50. Determine the maximum contaminant concentrations in Problem 8.49 for the first seven days after the spill.

8.51. A contaminant is continuously injected over a depth of 3 m into an aquifer with a mean seepage velocity of 0.45 m/d, and longitudinal and transverse dispersion coefficients of 1 m^2/d and 0.1 m^2/d, respectively. The injection rate is 0.4 m^3/d with a concentration of 130 mg/L. Estimate the steady-state contaminant concentration 30 m downstream of the injection location.

8.52. For the case described in Problem 8.51, determine the distance from the injection location to the point where the contaminant concentration is 1% on the injection concentration.

8.53. A contaminant source is 5 m wide by 2 m deep and continuously releases a conservative contaminant at a concentration of 70 mg/L. The mean seepage velocity in the aquifer is 0.1 m/d, the aquifer is 10 m deep, and the longitudinal, horizontal transverse, and vertical dispersivities are 1 m, 0.1 m, and 0.01 m, respectively. Determine the downstream location at which the plume will be fully mixed over the depth of the aquifer. Estimate the contaminant concentrations at the water table 200 m downstream of the source after five years of operation.

8.54. Repeat Problem 8.53 for the case in which the contaminant undergoes biodegradation with a decay rate of 0.01 d^{-1}. Assess whether biodegradation has a significant effect on the downstream concentration.

8.55. Hydraulic conductivity measurements (in m/d) on 50-cm core samples taken from an isotropic aquifer indicate a variable log-hydraulic conductivity that can be described by a mean of 3.2, variance of 1.6, and a correlation length scale of 1.3 m. Estimate the effective hydraulic conductivity and macrodispersivity of the aquifer. If the mean hydraulic gradient in the aquifer is 0.005 and the effective porosity is 0.15, estimate the components of the macrodispersion coefficient.

8.56. How would your dispersivity estimates change in Problem 8.55 if the aquifer were anisotropic with a horizontal correlation length scale of 1.3 m and an anisotropy ratio of 0.1?

8.57. The mean seepage velocity in an aquifer is 1 m/d, the mean pore size is 3 mm, and the molecular diffusion coefficient of a toxic contaminant in the groundwater is 2×10^{-9} m^2/s. Should molecular diffusion be considered in a contaminant transport model?

8.58. Seepage velocities surrounding a municipal wellfield are on the order of 5 m/d, the wellfield is approximately circular with a radius of 70 m, and the longitudinal dispersion coefficient is on the order of 50 m^2/d. Estimate whether contaminant transport from the boundary of the wellfield is advection or dispersion dominated.

8.59. A contaminant spill in an aquifer has resulted in a pollutant cloud that is 11 m long, 5 m wide, and 2 m deep. The pore sizes in the aquifer are on the order of 2 mm, the molecular diffusion coefficient is 10^{-9} m^2/s, the tortuosity is 1.3, and the mean seepage velocity is 0.1 m/d. Estimate the components of the dispersion coefficient that should be used in modeling plume transport.

8.60. The seepage velocity, v, surrounding a well is described by the relation

$$v = \frac{Q}{2\pi r b n}$$

where Q is the pumping rate at the well, r is the radial distance from the well, b is the aquifer thickness, and n is the effective porosity of the aquifer. At a particular well, the pumping rate is 20,000 L/min, the aquifer thickness is 15 m, and the effective porosity is 0.15. Estimate the extent of the zone surrounding the well where advection transport dominates macrodispersion. (*Hint:* Use the Peclet number vr/D_L as a basis for your analysis.)

8.61. Determine whether the dispersivities given by Equation 8.182 are consistent with the values given in Table 8.20.

8.62. The concentration of trichloroethylene (TCE) in a ground water is measured as 100 mg/L, the fraction of organic carbon in the solid matrix is estimated as 1.5%, and the bulk density of the aquifer is approximately 1,800 kg/m^3. Estimate the mass of TCE sorbed per unit volume of the aquifer.

8.63. Three kilograms of tetrachloroethylene is spilled over a 1.2-m depth of ground water and spreads laterally as the ground water moves with an average velocity of 0.2 m/d. The longitudinal and transverse dispersion coefficients are 0.05 m^2/d and 0.005 m^2/d, respectively; the porosity is 0.15; the density of the soil matrix is 2.65 g/cm^3; $\log K_{oc}$ is 2.42 (K_{oc} in g/cm^3); and the organic fraction in the soil is 8%. Calculate the concentration at the spill location after 1 hour, 1 day, and 1 week. Compare these values with the concentration obtained by neglecting sorption.

8.64. Use all the empirical relationships in Table 8.21 to estimate the organic carbon sorption coefficient of TCE. Assume that $\log K_{ow}$ of TCE is 2.29 (as shown in Table 8.22). Verify the claim that the actual value of $\log K_{oc}$, given in Appendix B, is within one standard deviation of the of the mean of the predictions given by the empirical equations listed in Table 8.21.

8.65. A buried drum containing 10 kg of a contaminant suddenly ruptures and spills all of its contents into the ground water over a 1-m depth. If the mean seepage velocity in the aquifer is 0.5 m/d, the longitudinal dispersion coefficient is 1 m^2/d, the horizontal-transverse dispersion coefficient is 0.1 m^2/d, vertical mixing is negligible, and the first-order decay constant of the contaminant is 0.02 d^{-1}. Determine the maximum concentration in the ground water after 100 days. Compare this concentration to the maximum concentration without decay.

8.66. If the decay factor in Problem 8.65 could be increased by adding nutrients to the ground water, determine the required decay rate for the calculated maximum concentration to be reduced by 90%.

8.67. Five kilograms of TCE are spilled over a 0.8-m depth of ground water that moves with an average seepage velocity of 0.2 m/d. Vertical dispersion is negligible, and the longitudinal and transverse dispersion coefficients are 0.1 m^2/d and 0.01 m^2/d, respectively. If the retardation factor is equal to 15 and the first-order decay factor is 1 d^{-1}, calculate the concentration at the spill location after 1 day and 1 week.

8.68. A 2 m × 2 m × 3 m (deep) portion of an aquifer contains chlorobenzene at a residual saturation of 15%. If the porosity of the contaminated portion of the aquifer is 0.17, the density of chlorobenzene is 1,110 kg/m^3, the solubility of chlorobenzene in water is 500 mg/L, and the mean seepage velocity of the ambient ground water is 0.05 m/d, estimate the time it would take for the chlorobenzene to be removed by dissolution.

Units and Conversion Factors

A.1 Units

The Système International d'Unités (International System of Units, or SI system) was adopted by the 11th General Conference on Weights and Measures (CGPM) in 1960 and is now used almost worldwide. In the SI system, all quantities are expressed in terms of seven base (fundamental) units. These base units and their standard abbreviations are as follows (Dean, 1985; Szirtes, 1977):

meter (m): distance light travels in a vacuum during 1/299 792 458 of a second.*

kilogram (kg): mass of a cylinder of platinum-iridium alloy kept at the International Bureau of Weights and Measures in Sèvres, France.

second (s): duration of 9,192,631,770 cycles of the radiation corresponding to the transition between two hyper fine levels of the ground state of cesium-133 atom.

ampere (A): magnitude of the current that, when flowing through each of two long parallel wires separated by 1 meter in free space, results in a force between the two wires of 2×10^{-7} newtons for each meter of length.

kelvin (K): defined in the thermodynamic scale by assigning 273.16 K to the triple point of water (freezing point, 273.16 K = 0°C).

candela (cd): luminous intensity in a given direction, of a source that emits monochromatic radiation of frequency 5.4×10^{14} hertz and that has a radiant intensity in that direction of $\frac{1}{683}$ watt per steradian.

mole (mol): amount of substance that contains as many specified entities (molecules, atoms, ions, electrons, photons, etc.) as there are atoms of carbon-12 in exactly 0.012 kg of that nuclide.

In addition to the seven base units of the SI system, there are two supplementary units: the radian and the steradian. The *radian* (rad) is defined as the angle at the center of a circle subtended by an arc equal in length to the radius, and the *steradian* (sr) is defined as the solid angle with its vertex at the center of a sphere that is subtended by an area of the spherical surface equal to the radius squared. SI units should not be confused with the now obsolete *metric units* which were developed in Napoleonic France approximately 200 years ago. The primary difference between the metric and SI units is that the former uses centimeters and grams to measure length and mass, while these quantities are measured in meters and kilograms in SI units. The United States is gradually moving

---

*Whereas *meter* is the accepted U.S. spelling, the rest of the world uses the British spelling *metre*.

toward the SI standard, but there is still widespread use of the English system of units, also referred to as "U.S. Customary" or "British Gravitational" units.

The SI system also includes several derived units that are given special names. These derived units are listed in Table A.1 (Wandmacher and Johnson, 1995).

■ **Table A.1**
SI Derived Units

Quantity	Unit name	Symbol	In terms of base units
Frequency	hertz	Hz	s^{-1}
Force	newton	N	$kg \cdot m/s^2$
Pressure, stress	pascal	Pa	N/m^2
Energy, work, heat	joule	J	$N \cdot m$
Power, radiant flux	watt	W	J/s
Quantity of electricity, electric charge	coulomb	C	$A \cdot s$
Electric potential, potential difference, electromotive force	volt	V	W/A
Capacitance	farad	F	C/V
Electric resistance	ohm	Ω	V/A
Conductance	siemens*	S	A/V
Magnetic flux	weber	Wb	$V \cdot s$
Magnetic flux density	telsa	T	Wb/m^2
Inductance	henry	H	Wb/A
Luminous flux	lumen	lm	$cd \cdot sr$
Illuminance	lux	lx	lm/m^2
Activity of a radionuclide	becquerel	Bq	s^{-1}
Adsorbed dose	gray	Gy	J/kg

*Previously called the mho.

A.2 Conversion Factors

■ **Table A.2**
Multiplicative Factors for Unit Conversion

Quantity	Convert from	Convert to	Multiply by
Area	ac	ha	0.404687
Energy	Btu	J	1054.350 264
	cal	J	4.184*
Energy/area	ly[†]	kJ/m^2	41.84*
Flowrate	cfs	m^3/s	0.02831685
	gpm[†]	L/s	0.06309
	mgd[†]	m^3/s	0.04381
Force	lbf	N	4.448 221 615 260 5*
Length	ft	m	0.3048*
	in.	m	0.0254*
	mi (U.S. statute)	km	1.609 344*
	mi (U.S. nautical)	km	1.852 000*
	yd	m	0.9144*

Quantity	Convert from	Convert to	Multiply by
Mass	g	kg	0.001*
	lbm	kg	0.453 592 37*
Permeability	darcy	m^2	0.987×10^{-12}
Power	hp	W	745.699 87
Pressure	atm	kPa	101.325*
	bar	kPa	100.000*
	psi	kPa	6.894757
Speed	knot	m/s	0.514 444 444
Viscosity (dynamic)	cp	Pa·s	0.001*
Viscosity (kinematic)	cs	m^2/s	10^{-6}*
Volume	gal	L	3.785 411 784*

*Exact conversion
[†] ly ≡ langley; gpm ≡ gallons per minute; mgd ≡ million gallons per day.

Fluid Properties

B.1   Water*

Temperature (°C)	Density (kg/m³)	Dynamic viscosity (mPa·s)	Heat of vaporization (MJ/kg)	Saturation vapor pressure (kPa)	Surface tension (mN/m)	Bulk modulus (10^6 kPa)
0	999.8	1.781	2.499	0.611	75.6	2.02
5	1000.0	1.518	2.487	0.872	74.9	2.06
10	999.7	1.307	2.476	1.227	74.2	2.10
15	999.1	1.139	2.464	1.704	73.5	2.14
20	998.2	1.002	2.452	2.337	72.8	2.18
25	997.0	0.890	2.440	3.167	72.0	2.22
30	995.7	0.798	2.428	4.243	71.2	2.25
40	992.2	0.653	2.405	7.378	69.6	2.28
50	988.0	0.547	2.381	12.340	67.9	2.29
60	983.2	0.466	2.356	19.926	66.2	2.28
70	977.8	0.404	2.332	31.169	64.4	2.25
80	971.8	0.354	2.307	47.367	62.6	2.20
90	965.3	0.315	2.282	70.113	60.8	2.14
100	958.4	0.282	2.256	101.325	58.9	2.07

*Density, viscosity, surface tension, and bulk modulus are from Franzini and Finnemore (1997); heat of vaporization and saturation vapor pressure are from Viessman and Lewis (1996).

B.2   Organic Compounds Found in Contaminated Water

Contaminant	Density @ 20°C (kg/m³)	Dynamic viscosity @ 20°C (mN·s/m²)	Solubility (mg/L)	Sorption Coefficient, log K_{oc} (log cm³/g)	Source
Acetone					‡
Benzene	880*		1780*	1.92[†]	‡
Bis(2-ethylhexyl)phthalate (DEHP)					‡
Chlorobenzene	1110*		472*–500[†]	2.63[†]	‡
Chloroethane			5,740[†]	1.57[†]	‡
Trichloromethane (Chloroform)	1480*		8,200[§]	1.92[†]	‡
1,1-Dichloroethane	1180*		400[†]	1.80[†]	‡
1,2-Dichloroethane	1240*		8,000[†]–8,426[¶]	1.57[†]	‡
Ethlybenzene	870*		152*	3.13[†]	‡
Gasoline[‖]	680	0.31			**
Methyl tert-butyl ether (MTBE)					‡
Dichloromethane (Methylene chloride)	1330*		13,000*		‡
Naphthalene	1030*		31[§]–33*	3.08[†]	‡
Phenol			82,000[§]–93,000[†]	1.27[†]	‡
Tetrachloroethene[††]	1623.0	1.932	200[†]	2.57[†]	‡‡
Methylbenzene (Toluene)	870*		515*–535[†]	2.48[†]	‡
1,1,1-Trichloroethane	1340*		4,400[†]	2.30[†]	‡
Trichloroethene (TCE)	1460*		1,000*–1,100[§]	2.15[†]	‡
Chloroethene (Vinyl chloride)	910*		90*	0.93[†]	‡
1,2 Dimethylbenzine (o-Xylene)	880*		175*		‡
1,4 Dimethylbenzine (p-Xylene)	860		185[§]		

*Hemond and Fechner (1993). Solubility is at 20°C.

[†] Schnoor (1996). Solubility at 20°C.

[‡] Bazzazieh (1996).

[§] Mackay (1991). Solubility is at 25°C.

[¶] Schwarzenbach et al. (1993). Solubility is at 25°C.

[‖] Typical properties at 15.6°C. Properties of petroleum products vary.

** Munson et al. (1990).

[††] Also known as: perchloroethylene, PCE, PERC, and 1,1,2,2-tetrachloroethene.

[‡‡] Dean (1985).

C Geometric Properties of Plane Surfaces

Shape	Sketch	Area, A	Location of centroid	I_c or I [a]*
Rectangle		bh	$y_c = \dfrac{h}{2}$	$I_c = \dfrac{bh^3}{12}$
Triangle		$\dfrac{bh}{2}$	$y_c = \dfrac{h}{3}$	$I_c = \dfrac{bh^3}{36}$
Circle		$\dfrac{\pi D^2}{4}$	$y_c = \dfrac{D}{3}$	$I_c = \dfrac{\pi D^4}{64}$
Semicircle		$\dfrac{\pi D^2}{8}$	$y_c = \dfrac{4r}{3\pi}$	$I = \dfrac{\pi D^4}{128}$
Ellipse		$\dfrac{\pi bh}{4}$	$y_c = \dfrac{h}{2}$	$I_c = \dfrac{\pi bh^3}{64}$
Semiellipse		$\dfrac{\pi bh}{4}$	$y_c = \dfrac{4h}{3\pi}$	$I_c = \dfrac{\pi bh^3}{16}$
Parabola		$\dfrac{2bh}{3}$	$x_c = \dfrac{3b}{8}$ $y_c = \dfrac{3b}{5}$	$I = \dfrac{2bh^3}{7}$

*Parallel axis theorem: $I = I_c + A y_c^2$.

Figure C.1 ■ Geometric Properties of Various Shapes

D Statistical Tables

Areas Under Standard Normal Curve

z	0.00	0.01	0.02	0.03	0.04	0.05	0.06	0.07	0.08	0.09
−4.0	.0000	.0000	.0000	.0000	.0000	.0000	.0000	.0000	.0000	.0000
−3.9	.0000	.0000	.0000	.0000	.0000	.0000	.0000	.0000	.0000	.0000
−3.8	.0001	.0001	.0001	.0001	.0001	.0001	.0001	.0001	.0001	.0001
−3.7	.0001	.0001	.0001	.0001	.0001	.0001	.0001	.0001	.0001	.0001
−3.6	.0002	.0002	.0001	.0001	.0001	.0001	.0001	.0001	.0001	.0001
−3.5	.0002	.0002	.0002	.0002	.0002	.0002	.0002	.0002	.0002	.0002
−3.4	.0003	.0003	.0003	.0003	.0003	.0003	.0003	.0003	.0003	.0002
−3.3	.0005	.0005	.0005	.0004	.0004	.0004	.0004	.0004	.0004	.0003
−3.2	.0007	.0007	.0006	.0006	.0006	.0006	.0006	.0005	.0005	.0005
−3.1	.0010	.0009	.0009	.0009	.0008	.0008	.0008	.0008	.0007	.0007
−3.0	.0013	.0013	.0013	.0012	.0012	.0011	.0011	.0011	.0010	.0010
−2.9	.0019	.0018	.0018	.0017	.0016	.0016	.0015	.0015	.0014	.0014
−2.8	.0026	.0025	.0024	.0023	.0023	.0022	.0021	.0021	.0020	.0019
−2.7	.0035	.0034	.0033	.0032	.0031	.0030	.0029	.0028	.0027	.0026
−2.6	.0047	.0045	.0044	.0043	.0041	.0040	.0039	.0038	.0037	.0036
−2.5	.0062	.0060	.0059	.0057	.0055	.0054	.0052	.0051	.0049	.0048
−2.4	.0082	.0080	.0078	.0075	.0073	.0071	.0069	.0068	.0066	.0064
−2.3	.0107	.0104	.0102	.0099	.0096	.0094	.0091	.0089	.0087	.0084
−2.2	.0139	.0136	.0132	.0129	.0125	.0122	.0119	.0116	.0113	.0110
−2.1	.0179	.0174	.0170	.0166	.0162	.0158	.0154	.0150	.0146	.0143
−2.0	.0228	.0222	.0217	.0212	.0207	.0202	.0197	.0192	.0188	.0183
−1.9	.0287	.0281	.0274	.0268	.0262	.0256	.0250	.0244	.0239	.0233
−1.8	.0359	.0351	.0344	.0336	.0329	.0322	.0314	.0307	.0301	.0294
−1.7	.0446	.0436	.0427	.0418	.0409	.0401	.0392	.0384	.0375	.0367
−1.6	.0548	.0537	.0526	.0516	.0505	.0495	.0485	.0475	.0465	.0455
−1.5	.0668	.0655	.0643	.0630	.0618	.0606	.0594	.0582	.0571	.0559
−1.4	.0808	.0793	.0778	.0764	.0749	.0735	.0721	.0708	.0694	.0681
−1.3	.0968	.0951	.0934	.0918	.0901	.0885	.0869	.0853	.0838	.0823
−1.2	.1151	.1131	.1112	.1093	.1075	.1056	.1038	.1020	.1003	.0985
−1.1	.1357	.1335	.1314	.1292	.1271	.1251	.1230	.1210	.1190	.1170
−1.0	.1587	.1562	.1539	.1515	.1492	.1469	.1446	.1423	.1401	.1379
−.9	.1841	.1814	.1788	.1762	.1736	.1711	.1685	.1660	.1635	.1611
−.8	.2119	.2090	.2061	.2033	.2005	.1977	.1949	.1922	.1894	.1867
−.7	.2420	.2389	.2358	.2327	.2297	.2266	.2236	.2206	.2177	.2148

z	0.00	0.01	0.02	0.03	0.04	0.05	0.06	0.07	0.08	0.09
−.6	.2743	.2709	.2676	.2643	.2611	.2578	.2546	.2514	.2483	.2451
−.5	.3085	.3050	.3015	.2981	.2946	.2912	.2877	.2843	.2810	.2776
−.4	.3446	.3409	.3372	.3336	.3300	.3264	.3228	.3192	.3156	.3121
−.3	.3821	.3783	.3745	.3707	.3669	.3632	.3594	.3557	.3520	.3483
−.2	.4207	.4168	.4129	.4090	.4052	.4013	.3974	.3936	.3897	.3859
−.1	.4602	.4562	.4522	.4483	.4443	.4404	.4364	.4325	.4286	.4247
.0	.5000	.5040	.5080	.5120	.5160	.5199	.5239	.5279	.5319	.5359
.1	.5398	.5438	.5478	.5517	.5557	.5596	.5636	.5675	.5714	.5753
.2	.5793	.5832	.5871	.5910	.5948	.5987	.6026	.6064	.6103	.6141
.3	.6179	.6217	.6255	.6293	.6331	.6368	.6406	.6443	.6480	.6517
.4	.6554	.6591	.6628	.6664	.6700	.6736	.6772	.6808	.6844	.6879
.5	.6915	.6950	.6985	.7019	.7054	.7088	.7123	.7157	.7190	.7224
.6	.7257	.7291	.7324	.7357	.7389	.7422	.7454	.7486	.7517	.7549
.7	.7580	.7611	.7642	.7673	.7704	.7734	.7764	.7794	.7823	.7852
.8	.7881	.7910	.7939	.7967	.7995	.8023	.8051	.8078	.8106	.8133
.9	.8159	.8186	.8212	.8238	.8264	.8289	.8315	.8340	.8365	.8389
1.0	.8413	.8438	.8461	.8485	.8508	.8531	.8554	.8577	.8599	.8621
1.1	.8643	.8665	.8686	.8708	.8729	.8749	.8770	.8790	.8810	.8830
1.2	.8849	.8869	.8888	.8907	.8925	.8944	.8962	.8980	.8997	.9015
1.3	.9032	.9049	.9066	.9082	.9099	.9115	.9131	.9147	.9162	.9177
1.4	.9192	.9207	.9222	.9236	.9251	.9265	.9279	.9292	.9306	.9319
1.5	.9332	.9345	.9357	.9370	.9382	.9394	.9406	.9418	.9429	.9441
1.6	.9452	.9463	.9474	.9484	.9495	.9505	.9515	.9525	.9535	.9545
1.7	.9554	.9564	.9573	.9582	.9591	.9599	.9608	.9616	.9625	.9633
1.8	.9641	.9649	.9656	.9664	.9671	.9678	.9686	.9693	.9699	.9706
1.9	.9713	.9719	.9726	.9732	.9738	.9744	.9750	.9756	.9761	.9767
2.0	.9772	.9778	.9783	.9788	.9793	.9798	.9803	.9808	.9812	.9817
2.1	.9821	.9826	.9830	.9834	.9838	.9842	.9846	.9850	.9854	.9857
2.2	.9861	.9864	.9868	.9871	.9875	.9878	.9881	.9884	.9887	.9890
2.3	.9893	.9896	.9898	.9901	.9904	.9906	.9909	.9911	.9913	.9916
2.4	.9918	.9920	.9922	.9925	.9927	.9929	.9931	.9932	.9934	.9936
2.5	.9938	.9940	.9941	.9943	.9945	.9946	.9948	.9949	.9951	.9952
2.6	.9953	.9955	.9956	.9957	.9959	.9960	.9961	.9962	.9963	.9964
2.7	.9965	.9966	.9967	.9968	.9969	.9970	.9971	.9972	.9973	.9974
2.8	.9974	.9975	.9976	.9977	.9977	.9978	.9979	.9979	.9980	.9981
2.9	.9981	.9982	.9982	.9983	.9984	.9984	.9985	.9985	.9986	.9986
3.0	.9987	.9987	.9987	.9988	.9988	.9989	.9989	.9989	.9990	.9990
3.1	.9990	.9991	.9991	.9991	.9992	.9992	.9992	.9992	.9993	.9993
3.2	.9993	.9993	.9994	.9994	.9994	.9994	.9994	.9995	.9995	.9995
3.3	.9995	.9995	.9995	.9996	.9996	.9996	.9996	.9996	.9996	.9997
3.4	.9997	.9997	.9997	.9997	.9997	.9997	.9997	.9997	.9997	.9998
3.5	.9998	.9998	.9998	.9998	.9998	.9998	.9998	.9998	.9998	.9998
3.6	.9998	.9998	.9999	.9999	.9999	.9999	.9999	.9999	.9999	.9999
3.7	.9999	.9999	.9999	.9999	.9999	.9999	.9999	.9999	.9999	.9999
3.8	.9999	.9999	.9999	.9999	.9999	.9999	.9999	.9999	.9999	.9999
3.9	1.0000	1.0000	1.0000	1.0000	1.0000	1.0000	1.0000	1.0000	1.0000	1.0000
4.0	1.0000	1.0000	1.0000	1.0000	1.0000	1.0000	1.0000	1.0000	1.0000	1.0000

D.2 Critical Values of the Chi-Square Distribution

v	\multicolumn{8}{c}{α}
	0.995	0.990	0.975	0.950	0.050	0.025	0.010	0.005
1	.000	.000	.001	.004	3.841	5.024	6.635	7.880
2	.010	.020	.051	.103	5.991	7.378	9.210	10.597
3	.072	.115	.216	.352	7.815	9.348	11.345	12.838
4	.207	.297	.484	.711	9.488	11.143	13.277	14.861
5	.412	.554	.831	1.145	11.071	12.833	15.086	16.750
6	.676	.872	1.237	1.635	12.592	14.449	16.812	18.548
7	.989	1.239	1.690	2.167	14.067	16.013	18.476	20.279
8	1.344	1.646	2.180	2.733	15.507	17.535	20.090	21.956
9	1.735	2.088	2.700	3.325	16.919	19.023	21.666	23.590
10	2.156	2.558	3.247	3.940	18.307	20.483	23.210	25.189
11	2.603	3.053	3.816	4.575	19.675	21.920	24.725	26.757
12	3.074	3.571	4.404	5.226	21.026	23.337	26.217	28.300
13	3.565	4.107	5.009	5.892	22.362	24.736	27.688	29.819
14	4.075	4.660	5.629	6.571	23.685	26.119	29.141	31.319
15	4.601	5.229	6.262	7.261	24.996	27.488	30.578	32.801
16	5.142	5.812	6.908	7.962	26.296	28.845	32.000	34.267
17	5.697	6.408	7.564	8.672	27.587	30.191	33.409	35.718
18	6.265	7.015	8.231	9.390	28.869	31.526	34.805	37.156
19	6.844	7.633	8.907	10.117	30.144	32.852	36.191	38.582
20	7.434	8.260	9.591	10.851	31.410	34.170	37.566	39.997
21	8.034	8.897	10.283	11.591	32.671	35.479	38.932	41.401
22	8.643	9.542	10.982	12.338	33.924	36.781	40.289	42.796
23	9.260	10.196	11.689	13.091	35.172	38.076	41.638	44.181
24	9.886	10.856	12.401	13.848	36.415	39.364	42.980	45.559
25	10.520	11.524	13.120	14.611	37.652	40.646	44.314	46.928
26	11.160	12.198	13.844	15.379	38.885	41.923	45.642	48.290
27	11.808	12.879	14.573	16.151	40.113	43.195	46.963	49.645
28	12.461	13.565	15.308	16.928	41.337	44.461	48.278	50.993
29	13.121	14.256	16.047	17.708	42.557	45.722	49.588	52.336
30	13.787	14.953	16.791	18.493	43.773	46.979	50.892	53.672
40	20.707	22.164	24.433	26.509	55.758	59.342	63.691	66.766
50	27.991	29.707	32.357	34.764	67.505	71.420	76.154	79.490
60	35.534	37.485	40.482	43.188	79.082	83.298	88.379	91.952
70	43.275	45.442	48.758	51.739	90.531	95.023	100.425	104.215
80	51.172	53.540	57.153	60.391	101.879	106.629	112.329	116.321
90	59.196	61.754	65.647	69.126	113.145	118.136	124.116	128.299
100	67.328	70.065	74.222	77.929	124.342	129.561	135.807	140.170

D.3 Critical Values for the Kolmogorov-Smirnov Test Statistic

Sample size (n)	Significance level				
	0.20	0.15	0.10	0.05	0.01
1	0.900	0.925	0.950	0.975	0.995
2	0.684	0.726	0.776	0.842	0.929
3	0.585	0.597	0.642	0.708	0.829
4	0.494	0.525	0.564	0.624	0.734
5	0.446	0.474	0.510	0.563	0.669
6	0.410	0.436	0.470	0.521	0.618
7	0.381	0.405	0.438	0.486	0.577
8	0.358	0.381	0.411	0.457	0.543
9	0.339	0.360	0.388	0.432	0.514
10	0.322	0.342	0.368	0.409	0.486
11	0.307	0.326	0.352	0.391	0.468
12	0.295	0.313	0.338	0.375	0.450
13	0.284	0.302	0.325	0.361	0.433
14	0.274	0.292	0.314	0.349	0.418
15	0.266	0.283	0.304	0.338	0.404
16	0.258	0.274	0.295	0.328	0.391
17	0.250	0.266	0.286	0.318	0.380
18	0.244	0.259	0.278	0.309	0.370
19	0.237	0.252	0.272	0.301	0.361
20	0.231	0.246	0.264	0.294	0.352
25	0.210	0.220	0.240	0.264	0.320
30	0.190	0.200	0.220	0.242	0.290
35	0.180	0.190	0.210	0.230	0.270
40				0.210	0.250
50				0.190	0.230
60				0.170	0.210
70				0.160	0.190
80				0.150	0.180
90				0.140	
100				0.140	
Asymptotic formula:	$\dfrac{1.07}{\sqrt{n}}$	$\dfrac{1.14}{\sqrt{n}}$	$\dfrac{1.22}{\sqrt{n}}$	$\dfrac{1.36}{\sqrt{n}}$	$\dfrac{1.63}{\sqrt{n}}$

Special Functions

E.1 Error Function

z	erf(z)
.0	.00000
.1	.11246
.2	.22270
.3	.32863
.4	.42839
.5	.52050
.6	.60386
.7	.67780
.8	.74210
.9	.79691
1.0	.84270
1.1	.88021
1.2	.91031
1.3	.93401
1.4	.95229
1.5	.96611
1.6	.97635
1.7	.98379
1.8	.98909
1.9	.99279
2.0	.99532
2.1	.99702
2.2	.99814
2.3	.99886
2.4	.99931
2.5	.99959
2.6	.99976
2.7	.99987
2.8	.99992
2.9	.99996
3.0	.99998
∞	1.00000

E.2   Gamma Function

z	$\Gamma(z)$	z	$\Gamma(z)$	z	$\Gamma(z)$
1.00	1.00000	1.34	0.89222	1.68	0.90500
1.01	0.99433	1.35	0.89115	1.69	0.90678
1.02	0.98884	1.36	0.89018	1.70	0.90864
1.03	0.98355	1.37	0.88931	1.71	0.91057
1.04	0.97844	1.38	0.88854	1.72	0.91258
1.05	0.97350	1.39	0.88785	1.73	0.91467
1.06	0.96874	1.40	0.88726	1.74	0.91683
1.07	0.96415	1.41	0.88676	1.75	0.91906
1.08	0.95973	1.42	0.88636	1.76	0.92137
1.09	0.95546	1.43	0.88604	1.77	0.92376
1.10	0.95135	1.44	0.88581	1.78	0.92623
1.11	0.94740	1.45	0.88566	1.79	0.92877
1.12	0.94359	1.46	0.88560	1.80	0.93138
1.13	0.93993	1.47	0.88563	1.81	0.93408
1.14	0.93642	1.48	0.88575	1.82	0.93685
1.15	0.93304	1.49	0.88595	1.83	0.93969
1.16	0.92980	1.50	0.88623	1.84	0.94261
1.17	0.92670	1.51	0.88659	1.85	0.94561
1.18	0.92373	1.52	0.88704	1.86	0.94869
1.19	0.92089	1.53	0.88757	1.87	0.95184
1.20	0.91817	1.54	0.88818	1.88	0.95507
1.21	0.91558	1.55	0.88887	1.89	0.95838
1.22	0.91311	1.56	0.88964	1.90	0.96177
1.23	0.91075	1.57	0.89049	1.91	0.96523
1.24	0.90852	1.58	0.89142	1.92	0.96877
1.25	0.90640	1.59	0.89243	1.93	0.97240
1.26	0.90440	1.60	0.89352	1.94	0.97610
1.27	0.90250	1.61	0.89468	1.95	0.97988
1.28	0.90072	1.62	0.89592	1.96	0.98374
1.29	0.89904	1.63	0.89724	1.97	0.98768
1.30	0.89747	1.64	0.89864	1.98	0.99171
1.31	0.89600	1.65	0.90012	1.99	0.99581
1.32	0.89464	1.66	0.90167	2.00	1.00000
1.33	0.89338	1.67	0.90330		

F

Bessel Functions

Definition

A second-order linear homogeneous differential equation of the form

$$x^2 \frac{d^2 y}{dx^2} + x \frac{dy}{dx} + (x^2 - n^2)y = 0, \qquad n \geq 0 \tag{F.1}$$

is called *Bessel's equation*. The general solutions of Bessel's equation are

$$y = AJ_n(x) + BJ_{-n}(x), \qquad n \neq 1, 2, \ldots \tag{F.2}$$

$$y = AJ_n(x) + BY_n(x), \qquad \text{all } n \tag{F.3}$$

where $J_n(x)$ is called the *Bessel function of the first kind of order n*, and $Y_n(x)$ is called the *Bessel function of the second kind of order n*.

If Bessel's equation (Equation F.1) is slightly modified and written in the form

$$x^2 \frac{d^2 y}{dx^2} + x \frac{dy}{dx} - (x^2 + n^2)y = 0, \qquad n \geq 0 \tag{F.4}$$

then this equation is called the *modified Bessel's equation*. The general solutions of the modified Bessel's equation are

$$y = AI_n(x) + BI_{-n}(x), \qquad n \neq 1, 2, \ldots \tag{F.5}$$

$$y = AI_n(x) + BK_n(x), \qquad \text{all } n \tag{F.6}$$

where $I_n(x)$ is called the *modified Bessel function of the first kind of order n*, and $K_n(x)$ is called the *modified Bessel function of the second kind of order n*.

F.2 Evaluation of Bessel Functions

The solution of Bessel's equations can be found in most calculus texts, such as Hildebrand (1976). The Bessel functions cannot generally be expressed in closed form and are usually presented as an infinite series.

F.2.1 ■ Bessel Function of the First Kind of Order n

This function is given by

$$J_n(x) = \frac{x^n}{2^n \Gamma(n+1)} \left\{ 1 - \frac{x^2}{2(2n+2)} + \frac{x^4}{2 \cdot 4(2n+2)(2n+4)} - \cdots \right\} \tag{F.7}$$

$$= \sum_{k=0}^{\infty} \frac{(-1)^k (x/2)^{n+2k}}{k! \Gamma(n+k+1)} \tag{F.8}$$

The Bessel function $J_{-n}(x)$ can be derived from Equation F.8 by simply replacing n by $-n$ in the formula. A convenient relationship to note is

$$J_{-n}(x) = (-1)^n J_n(x), \qquad n = 0, 1, 2, \ldots \tag{F.9}$$

F.2.2 ■ Bessel Function of the Second Kind of Order n

This function is given by

$$Y_n(x) = \begin{cases} \frac{J_n(x)\cos n\pi - J_{-n}(x)}{\sin n\pi}, & n \neq 0, 1, 2, \ldots \\ \lim_{p \to n} \frac{J_p(x)\cos p\pi - J_{-p}(x)}{\sin p\pi}, & n = 0, 1, 2, \ldots \end{cases} \tag{F.10}$$

F.2.3 ■ Modified Bessel Function of the First Kind of Order n

This function is given by

$$I_n(x) = \frac{x^n}{2^n \Gamma(n+1)} \left\{ 1 + \frac{x^2}{2(2n+2)} + \frac{x^4}{2 \cdot 4(2n+2)(2n+4)} + \cdots \right\} \tag{F.11}$$

$$= \sum_{k=0}^{\infty} \frac{(x/2)^{n+2k}}{k! \Gamma(n+k+1)} \tag{F.12}$$

The Bessel function $I_{-n}(x)$ can be derived from Equation F.12 by simply replacing n by $-n$ in the formula. A convenient relationship to note is

$$I_{-n}(x) = I_n(x), \qquad n = 0, 1, 2, \ldots \tag{F.13}$$

F.2.4 ■ Modified Bessel Function of the Second Kind of Order n

This function is given by

$$K_n(x) = \begin{cases} \frac{\pi}{2\sin n\pi}[I_{-n}(x) - I_n(x)], & n \neq 0, 1, 2, \ldots \\ \lim_{p \to n} \frac{\pi}{2\sin p\pi}[I_{-p}(x) - I_p(x)], & n = 0, 1, 2, \ldots \end{cases} \tag{F.14}$$

F.3   Tabulated Bessel Functions

F.3.1   ■ $I_0(x)$, $K_0(x)$, $I_1(x)$, and $K_1(x)$

x	$I_0(x)$	$K_0(x)$	$I_1(x)$	$K_1(x)$
.001	1.0000	7.0237	.0005	999.9962
.002	1.0000	6.3305	.0010	499.9932
.003	1.0000	5.9251	.0015	333.3237
.004	1.0000	5.6374	.0020	249.9877
.005	1.0000	5.4143	.0025	199.9852
.006	1.0000	5.2320	.0030	166.6495
.007	1.0000	5.0779	.0035	142.8376
.008	1.0000	4.9443	.0040	124.9782
.009	1.0000	4.8266	.0045	111.0871
.010	1.0000	4.7212	.0050	99.9739
.020	1.0001	4.0285	.0100	49.9547
.030	1.0002	3.6235	.0150	33.2715
.040	1.0004	3.3365	.0200	24.9233
.050	1.0006	3.1142	.0250	19.9097
.060	1.0009	2.9329	.0300	16.5637
.070	1.0012	2.7798	.0350	14.1710
.080	1.0016	2.6475	.0400	12.3742
.090	1.0020	2.5310	.0450	10.9749
.100	1.0025	2.4271	.0501	9.8538
.110	1.0030	2.3333	.0551	8.9353
.120	1.0036	2.2479	.0601	8.1688
.130	1.0042	2.1695	.0651	7.5192
.140	1.0049	2.0972	.0702	6.9615
.150	1.0056	2.0300	.0752	6.4775
.160	1.0064	1.9674	.0803	6.0533
.170	1.0072	1.9088	.0853	5.6784
.180	1.0081	1.8537	.0904	5.3447
.190	1.0090	1.8018	.0954	5.0456
.200	1.0100	1.7527	.1005	4.7760
.210	1.0111	1.7062	.1056	4.5317
.220	1.0121	1.6620	.1107	4.3092
.230	1.0133	1.6199	.1158	4.1058
.240	1.0145	1.5798	.1209	3.9191
.250	1.0157	1.5415	.1260	3.7470
.260	1.0170	1.5048	.1311	3.5880
.270	1.0183	1.4697	.1362	3.4405
.280	1.0197	1.4360	.1414	3.3033
.290	1.0211	1.4036	.1465	3.1755
.300	1.0226	1.3725	.1517	3.0560
.310	1.0242	1.3425	.1569	2.9441
.320	1.0258	1.3136	.1621	2.8390
.330	1.0274	1.2857	.1673	2.7402
.340	1.0291	1.2587	.1725	2.6470
.350	1.0309	1.2327	.1777	2.5591
.360	1.0327	1.2075	.1829	2.4760
.370	1.0345	1.1832	.1882	2.3973
.380	1.0364	1.1596	.1935	2.3227

x	$I_0(x)$	$K_0(x)$	$I_1(x)$	$K_1(x)$
.390	1.0384	1.1367	.1987	2.2518
.400	1.0404	1.1145	.2040	2.1844
.410	1.0425	1.0930	.2093	2.1202
.420	1.0446	1.0721	.2147	2.0590
.430	1.0468	1.0518	.2200	2.0006
.440	1.0490	1.0321	.2254	1.9449
.450	1.0513	1.0129	.2307	1.8915
.460	1.0536	.9943	.2361	1.8405
.470	1.0560	.9761	.2415	1.7916
.480	1.0584	.9584	.2470	1.7447
.490	1.0609	.9412	.2524	1.6997
.500	1.0635	.9244	.2579	1.6564
.510	1.0661	.9081	.2634	1.6149
.520	1.0688	.8921	.2689	1.5749
.530	1.0715	.8766	.2744	1.5364
.540	1.0742	.8614	.2800	1.4994
.550	1.0771	.8466	.2855	1.4637
.560	1.0800	.8321	.2911	1.4292
.570	1.0829	.8180	.2967	1.3960
.580	1.0859	.8042	.3024	1.3638
.590	1.0889	.7907	.3080	1.3328
.600	1.0920	.7775	.3137	1.3028
.610	1.0952	.7646	.3194	1.2738
.620	1.0984	.7520	.3251	1.2458
.630	1.1017	.7397	.3309	1.2186
.640	1.1051	.7277	.3367	1.1923
.650	1.1084	.7159	.3425	1.1668
.660	1.1119	.7043	.3483	1.1420
.670	1.1154	.6930	.3542	1.1181
.680	1.1190	.6820	.3600	1.0948
.690	1.1226	.6711	.3659	1.0722
.700	1.1263	.6605	.3719	1.0503
.710	1.1301	.6501	.3778	1.0290
.720	1.1339	.6399	.3838	1.0083
.730	1.1377	.6300	.3899	.9882
.740	1.1417	.6202	.3959	.9686
.750	1.1456	.6106	.4020	.9496
.760	1.1497	.6012	.4081	.9311
.770	1.1538	.5920	.4142	.9130
.780	1.1580	.5829	.4204	.8955
.790	1.1622	.5740	.4266	.8784
.800	1.1665	.5653	.4329	.8618
.810	1.1709	.5568	.4391	.8456
.820	1.1753	.5484	.4454	.8298
.830	1.1798	.5402	.4518	.8144
.840	1.1843	.5321	.4581	.7993
.850	1.1889	.5242	.4646	.7847
.860	1.1936	.5165	.4710	.7704
.870	1.1984	.5088	.4775	.7564
.880	1.2032	.5013	.4840	.7428
.890	1.2080	.4940	.4905	.7295
.900	1.2130	.4867	.4971	.7165

x	$I_0(x)$	$K_0(x)$	$I_1(x)$	$K_1(x)$
.910	1.2180	.4796	.5038	.7039
.920	1.2231	.4727	.5104	.6915
.930	1.2282	.4658	.5171	.6794
.940	1.2334	.4591	.5239	.6675
.950	1.2387	.4524	.5306	.6560
.960	1.2440	.4459	.5375	.6447
.970	1.2494	.4396	.5443	.6336
.980	1.2549	.4333	.5512	.6228
.990	1.2604	.4271	.5582	.6122
1.000	1.2661	.4210	.5652	.6019
1.100	1.3262	.3656	.6375	.5098
1.200	1.3937	.3185	.7147	.4346
1.300	1.4693	.2782	.7973	.3725
1.400	1.5534	.2437	.8861	.3208
1.500	1.6467	.2138	.9817	.2774
1.600	1.7500	.1880	1.0848	.2406
1.700	1.8640	.1655	1.1963	.2094
1.800	1.9896	.1459	1.3172	.1826
1.900	2.1277	.1288	1.4482	.1597
2.000	2.2796	.1139	1.5906	.1399
2.100	2.4463	.1008	1.7455	.1227
2.200	2.6291	.0893	1.9141	.1079
2.300	2.8296	.0791	2.0978	.0950
2.400	3.0493	.0702	2.2981	.0837
2.500	3.2898	.0623	2.5167	.0739
2.600	3.5533	.0554	2.7554	.0653
2.700	3.8417	.0493	3.0161	.0577
2.800	4.1573	.0438	3.3011	.0511
2.900	4.5027	.0390	3.6126	.0453
3.000	4.8808	.0347	3.9534	.0402
3.100	5.2945	.0310	4.3262	.0356
3.200	5.7472	.0276	4.7343	.0316
3.300	6.2426	.0246	5.1810	.0281
3.400	6.7848	.0220	5.6701	.0250
3.500	7.3782	.0196	6.2058	.0222
3.600	8.0277	.0175	6.7927	.0198
3.700	8.7386	.0156	7.4357	.0176
3.800	9.5169	.0140	8.1404	.0157
3.900	10.3690	.0125	8.9128	.0140
4.000	11.3019	.0112	9.7595	.0125
5.000	27.2399	.0037	24.3356	.0040
6.000	67.2344	.0012	61.3419	.0013
7.000	168.5939	.0004	156.0391	.0005
8.000	427.5641	.0001	399.8731	.0002
9.000	1093.5880	.0001	1030.9150	.0001
10.000	2815.7170	.0000	2670.9880	.0000

G   Drinking-Water Standards

G.1   Primary Drinking-Water Standards

The following primary drinking-water standards are current as of January 1999:

■ Table G.1
National Primary
Drinking-Water Standards

Contaminants	MCLG* (mg/L)	MCL† (mg/L)
Inorganic Chemicals:		
Antimony	0.006	0.006
Arsenic	—	0.05
Asbestos	7 MFL‡	7 MFL
Barium	2	2
Beryllium	0.004	0.004
Cadmium	0.005	0.005
Chromium (total)	0.1	0.1
Copper	1.3	TT£
Cyanide (as free cyanide)	0.2	0.2
Fluoride	4.0	4.0
Lead	0	TT£
Mercury (inorganic)	0.002	0.002
Nitrate (as N)	10	10
Nitrite (as N)	1	1
Selenium	0.05	0.05
Thallium	0.0005	0.002
Organic Chemicals:		
Acrylamide	0	TT£
Alachlor	0	0.002
Atrazine	0.003	0.003
Benzene	0	0.005
Benzo(a)pyrene	0	0.0002
Carbofuran	0.04	0.04
Carbon tetrachloride	0	0.005
Chlordane	0	0.002
Chlorobenzene	0.1	0.1
2,4-D	0.07	0.07
Dalapon	0.2	0.2
1,2-Dibromo-3-chloropropane (DBCP)	0	0.0002

▪ Table G.1
(*Continued*)

Contaminants	MCLG* (mg/L)	MCL† (mg/L)
o-Dichlorobenzene	0.6	0.6
p-Dichlorobenzene	0.075	0.075
1,2-Dichloroethane	0	0.005
1,1-Dichloroethylene	0.007	0.007
cis-1,2-Dichloroethylene	0.07	0.07
trans-1,2 Dichloroethylene	0.1	0.1
Dichloromethane	0	0.005
1,2-Dichloropropane	0	0.005
Di(2-ethylhexyl)adipate	0.4	0.4
Di(2-ethylhexyl)phthalate	0	0.006
Dinoseb	0.007	0.007
Dioxin (2,3,7,8-TCDD)	0	0.00000003
Diquat	0.02	0.02
Endothall	0.1	0.1
Endrin	0.002	0.002
Epichlorohydrin	0	TT£
Ethylbenzene	0.7	0.7
Ethlyene dibromide	0	0.00005
Glyphosate	0.7	0.7
Heptachlor	0	0.0004
Heptachlor epoxide	0	0.0002
Hexachlorobenzene	0	0.001
Hexachlorocyclopentadiene	0.05	0.05
Lindane	0.0002	0.0002
Methoxychlor	0.04	0.04
Oxamyl (Vydate)	0.2	0.2
Polychlorinated biphenlys (PCBs)	0	0.0005
Pentachlorophenol	0	0.001
Picloram	0.5	0.5
Simazine	0.004	0.004
Styrene	0.1	0.1
Tetrachloroethylene	0	0.005
Toluene	1	1
Total trihalomethanes (TTHMs)	—	0.10
Toxaphene	0	0.003
2,4,5-TP Silvex	0.05	0.05
1,2,4-Trichlorobenzene	0.07	0.07
1,1,1-Trichloroethane	0.20	0.2
1,1,2-Trichloroethane	0.003	0.005
Trichloroethylene	0	0.005
Vinyl chloride	0	0.002
Xylenes (total)	10	10
Radionuclides:		
Beta particles and photon emitters	—	4 mrem/y
Gross alpha particle activity	—	15 pCi/L
Radium 226 and 228	—	5 pCi/L

■ **Table G.1**
(*Continued*)

Contaminants	MCLG* (mg/L)	MCL† (mg/L)
Microorganisms:		
Giardia lamblia	0	TT$
Heterotrophic plate count	N/A	TT$
Legionella	0	TT$
Total coliforms	0	< 5/100 mL
Turbidity	N/A	TT$
Viruses	0	TT$

*MCLG = maximum contaminant level goal.

†MCL = maximum contaminant level.

‡MFL = million fibers per liter longer than 10 μm.

$TT = treatment technique requirement. Action level for copper is 1.3 mg/L; lead is 0.015 mg/L; acrylamide is 0.05% dosed at 1 mg/L; epichlorohydrin is 0.01% dosed at 20 mg/L; *Giardia lamblia* is 99.9% killed/inactivated; viruses is 99.99% killed/inactivated; *Legionella* is *Giardia* and virus inactivation; turbidity is 1 NTU; and heterotrophic plate count is 500 bacterial colonies per milliliter.

G.2 Secondary Drinking-Water Standards

■ **Table G.2**
Secondary Drinking-Water
Standards

Contaminants	Secondary standard
Aluminum	0.05 to 0.2 mg/L
Chloride	250 mg/L
Color	15 (color units)
Copper	1.0 mg/L
Corrosivity	noncorrosive
Fluoride	2.0 mg/L
Foaming agents	0.5 mg/L
Iron	0.3 mg/L
Manganese	0.05 mg/L
Odor	3 threshold odor number
pH	6.5–8.5
Silver	0.10 mg/L
Sulfate	250 mg/L
Total dissolved solids	500 mg/L
Zinc	5 mg/L

Bibliography

M. Abramowitz and I. A. Stegun. *Handbook of Mathematical Functions*. Dover, New York, 1965.

P. Ackers, W. R. White, J. A. Perkins, and A. J. M. Harrison. *Weirs and Flumes for Flow Measurement*. Wiley, Chichester, 1978.

N. Ahmed and D. K. Sunada. Nonlinear flow in porous media. *Journal of the Hydraulics Division, ASCE*, 95(HY6) (1969): 1847–57.

T. P. Ahrens. Well design criteria. *Water Well Journal*, (September–November 1957).

A. J. Aisenbrey, Jr., R. B. Hayes, H. J. Warren, D. L. Winsett, and R. B. Young. *Design of Small Canal Structures*. U. S. Department of the Interior, Bureau of Reclamation, Denver, CO, 1974.

A. O. Akan. *Urban Stormwater Hydrology*. Technomic Publishing Co., Lancaster, PA, 1993.

M. S. Al-Suwaiyan. Nondeterministic evaluation of field-scale dispersivity relation. *Journal of Hydrologic Engineering*, 3, no. 3 (1998): 215–17.

R. Aldridge and R. J. Jackson. Interception of rainfall by hard beech. *New Zealand Journal of Science*, 16(1973): 185–98.

R. G. Allen. A Penman for all seasons. *Journal of Irrigation and Drainage Engineering*, 112, no. 4 (1986): 348–68.

R. G. Allen, M. E. Jensen, J. L. Wright, and R. D. Burman. Operational estimates of reference evapotranspiration. *Journal of Agronomy*, 81 (1989): 650–62.

R. G. Allen and W. O. Pruitt. Rational use of the FAO Blaney-Criddle formula. *Journal of Irrigation and Drainage Engineering*, 112(IR2)(1986): 139–55.

M. Amein. Stream flow routing on computer by characteristics. *Water Resources Research*, 2, no. 1(1966): 123–30.

M. Amein and C. S. Fang. Stream flow routing (with applications to North Carolina rivers). Report 17, Water Resources Research Institute of the University of North Carolina, Raleigh, NC, 1969.

American Concrete Pipe Association. *Concrete Pipe Handbook*. ACPA, Vienna, VA, 1981.

American Concrete Pipe Association. *Concrete Pipe Design Manual*. ACPA, Vienna, VA, 1985.

American Iron and Steel Institute. *Modern Sewer Design*, 3rd ed. AISI, Washington, DC, 1995.

American Public Health Association. *Standard Methods for the Examination of Water and Wastewater*. APHA, Washington, DC, 18th ed., 1992.

American Society of Civil Engineers. Report of ASCE Task Force on Friction Factors in Open Channels. *Proceedings of the American Society of Civil Engineers*, 89(HY2) (March 1963): 97.

American Society of Civil Engineers. *Gravity Sanitary Sewer Design and Construction*. ASCE Manuals and Reports on Engineering Practice, no. 60. ASCE, New York, 1982.

American Society of Civil Engineers. *Evapotranspiration and Irrigation Water Requirements*. Manual of Practice No. 70. ASCE, New York, 1990.

American Society of Civil Engineers. *Design and Construction of Urban Stormwater Management Systems*. ASCE, New York, 1992.

American Society of Civil Engineers. *Hydrology Handbook*, 2nd ed. ASCE Manual of Practice No. 28. ASCE, New York, 1996a.

American Society of Civil Engineers. *Flood-Runoff Analysis.* Technical Engineering and Design Guides as Adapted from the U. S. Army Corps of Engineers, No. 19. ASCE, New York, 1996b.

American Society of Civil Engineers. *Quality of Ground Water, Guidelines for Selection and Application of Frequently Used Models.* Manual of Practice No. 85. ASCE, New York, 1996,

American Society of Civil Engineers. *Urban Runoff Quality Management.* Manual of Practice No. 23. ASCE, New York, 1998.

American Society of Heating, Refrigerating and Air Conditioning Engineers. *ASHRAE Handbook, 1981. Fundamentals.* ASHRAE, New York, 1981.

American Water Works Association. *1984 Water Utility Operating Data.* AWWA, Denver, CO, 1986.

American Water Works Association. *Distribution System Requirements for Fire Protection.* Manual of Water Supply Practices M31. AWWA, Denver, CO, 1992.

W. F. Ames. *Numerical Methods for Partial Differential Equations,* 2nd ed. Academic Press, New York, 1977.

M. P. Anderson. Using models to simulate the movement of contaminants through groundwater flow systems. *Critical Reviews in Environmental Controls,* 9, no. 2 (1979): 97–156.

M. P. Anderson and W. W. Woessner. *Applied Groundwater Modeling, Simulation of Flow and Transport.* Academic Press, New York, 1992.

American Society of Testing Materials, Committee D-18 on Soil and Rock. *Standard Practice for Design and Installation of Ground Water Monitoring Wells in Aquifers.* Technical Report D5092-1990. ASTM, 1990.

American Society of Testing Materials, Committee D-18 on Soil and Rock. *Standard Guide for Selection of Aquifer Test Method in Determining of Hydraulic Properties by Well Techniques.* Technical Report D4043-1991. ASTM, 1991a.

American Society of Testing Materials, Committee D-18 on Soil and Rock. *Standard Test Method (Field Procedure) for Withdrawal and Injection Well Tests for Determining Hydraulic Properties of Aquifer Systems.* Technical Report D4050-1991. ASTM, 1991b.

American Society of Testing Materials, Committee D-18 on Soil and Rock. *Standard Test Method (Analytical Procedure) for Determining Transmissivity and Storage Coefficient of Nonleaky Confined Aquifers by Overdamped Well Response to Instantaneous Change in Head (Slug Test).* Technical Report D4106-1991. ASTM, 1991c.

G. Aussenac and C. Boulangeat. Interception des precipitation et evapotranspiration reelle dans des peuplements de feuillu (*fagus silvatica* L.) et de resineux [*pseudotsuga menziesii* (Mirb.) Franco]. *Annales des Sciences Forestieres,* 37, no. 2 (1980): 91–107.

W. Badon-Ghyben. *Nota in Verband met de Voorgenomen Putboring Nabij Amsterdam* [Notes on the probable results of well drilling near Amsterdam]. Technical Report 1888/9, Tijdschr. Kon. Inst. Ing., The Hague, 1888.

A. A. Bakr. Stochastic analysis of the effects of spatial variations of hydraulic conductivity on groundwater flow. Ph.D thesis. New Mexico Institute of Mining and Technology, Socorro, NM, 1976.

A. F. Bartsch and J. H. Gakstatter. Management decisions for lake systems on a survey of trophic status, limiting nutrients, and nutrient loadings. In *American-Soviet Symposium on Use of Mathematical Models to Optimize Water Quality Management,* 1975. Technical Report EPA-600/9-78-024, U.S. Environmental Protection Agency, Office of Research and Development, Environmental Research Laboratory, Gulf Breeze, FL, 1978.

V. Batu. *Aquifer Hydraulics.* Wiley, New York, 1998.

D. J. Baumgartner. Surface water pollution. In I. L. Pepper, C. P. Gerba, and M. L. Brusseau, eds., *Pollution Science,* pp. 189–209. Academic Press, New York, 1996.

D. J. Baumgartner, W. E. Frick, and P. J. W. Roberts. *Dilution Models for Effluent Discharges,* 3rd ed. Technical Report EPA/600/R-94/086, U.S. Environmental Protection Agency, ERL-N, Newport, OR, 1994.

N. Bazzazieh. Groundwater treatment technology. *Water Environment and Technology,* 8, no. 12, (December 1996): 61–65.

J. Bear. *Dynamics of Fluids in Porous Media*. Dover, New York, 1972.

J. Bear. *Hydraulics of Groundwater*. McGraw-Hill, New York, 1979.

J. Bear and G. Dagan. The transition zone between fresh and salt waters in a coastal aquifer. Progress Report 1: The steady interface between two immiscible fluids in a two-dimensional field of flow. Technical Report. Hydraulic Lab, Technion, Haifa, Israel, 1962.

J. S. Beard. Rainfall interception by grass. *South African Forestry Journal*, 42 (1962): 12–25.

P. B. Bedient and W. C. Huber. *Hydrology and Floodplain Analysis*. Prentice-Hall, Englewood Cliffs, NJ, 1992.

P. B. Bedient, H. S. Rifai, and C. J. Newell. *Ground Water Contamination*. Prentice-Hall, Englewood Cliffs, NJ, 1994.

J. R. Benjamin and C. Allin Cornell. *Probability, Statistics, and Decision for Civil Engineers*. McGraw-Hill, New York, 1970.

J. H. Black and K. L. Kipp. Observation well response time and its effect upon aquifer test results. *Journal of Hydrology*, 34 (1977): 297–306.

P. E. Black. *Watershed Hydrology*, 2nd ed. Ann Arbor Press, Chelsea, MI, 1996.

H. Blasius. Das Ähnlichkeitsgesetz bei Reibungsvorgängen in Flüssigkeiten. *Forsch. Gebiete Ingenieurw.*, 131 (1913).

E. R. Blatchley III and J. E. Thompson. Groundwater contaminants. In J. W. Delleur, ed., *The Handbook of Groundwater Engineering*, pp. 13.1–13.30. CRC Press, Boca Raton, FL, 1998.

R. D. Blevins. *Applied Fluid Dynamics Handbook*. Van Nostrand Reinhold, New York, 1984.

F. E. Blow. Quantity and hydrologic characteristics of litter upon upland oak forests in eastern Tennessee. *Journal of Forestry*, 53: 190–195, 1955.

J. M. Boggs, S. C. Young, D. J. Benton, and Y. C. Chung. Hydrogeologic characterization of the MADE site. Interim Report EPRI EN-6915. Electric Power Research Institute, Palo Alto, CA, 1990. Project 2485-5.

J. Boonstra. Well hydraulics and aquifer tests. In J. W. Delleur, ed., *The Handbook of Groundwater Engineering*, pp. 8.1–8.34. CRC Press, Boca Raton, FL, 1998.

L. E. Borgman. Risk criteria. *Journal of Waterways and Harbors Division, ASCE*, 89(WW3) (August 1963): 1–35.

J. F. Bosen. A formula for approximation of the saturation vapor pressure over water. *Monthly Weather Review*, 88, no. 8 (1960): 275.

Boulder County, CO. *Drainage Criteria Manual*. Boulder County, CO, 1984.

H. Bouwer. *Groundwater Hydrology*. McGraw-Hill, New York, 1978.

H. Bouwer. Elements of soil science and groundwater hydrology. In G. Bitton and C. P. Gerba, eds., *Groundwater Pollution Microbiology*, pp. 9–38. Krieger, Malabar, FL, 1984.

H. Bouwer. The Bouwer and Rice slug test: An update. *Ground Water*, 27 (1989): 304–309.

H. Bouwer and R. C. Rice. A slug test for determining hydraulic conductivity of unconfined aquifers with completely or partially penetrating wells. *Water Resources Research*, 12(1976): 423–428.

E. F. Brater, H. W. King, J. E. Lindell, and C. Y. Wei. *Handbook of Hydraulics*, 2nd ed. McGraw-Hill, New York, 1996.

K. N. Brooks, P. F. Folliott, H. M. Gregersen, and J. L. Thames. *Hydrology and the Management of Watersheds*. Iowa State University Press, Ames, IA, 1991.

N. H. Brooks. Diffusion of sewage effluent in an ocean current. In E. A. Pearson, ed., *Proceedings, First International Conference on Waste Disposal in the Marine Environment*, pp. 246–67. Pergamon Press, New York, 1960.

D. S. Brookshire and D. Whittington. Water resources issues in developing countries. *Water Resources Research*, 29 no. 7 (July 1993): 1883–88.

D. Brunt. Notes on radiation in the atmosphere. *Quarterly Journal of Royal Meteorological Society*, 58 (1932): 389–418.

E. Buckingham. Model experiments and the forms of empirical equations. *Trans. ASME*, 37 (1915): 263.

N. M. Burns and F. Rosa. In situ measurement of settling velocity of organic carbon particles and 10 species of phytoplankton. *Limnol. Oceanogr.*, 25 (1980): 855–64.

F. L. Burton. Wastewater-Collection Systems. In L. W. Mays, ed., *Water Resources Handbook*, pp. 19.1–19.53. McGraw-Hill, New York, 1996.

J. A. Businger. Some remarks on Penman's equations for the evapotranspiration. *Netherlands J. Agric. Sci.*, 4(1956): 77.

E. Byers and D. B. Stephens. Statistical and stochastic analyses of hydraulic conductivity and particle size in a fluvial sand. *Soil Science Society of America Journal*, 47 (1983): 1072–81.

I. R. Calder. Hydrologic effects of land-use change. In D. R. Maidment, ed., *Handbook of Hydrology*, pp. 13.1–13.50. McGraw-Hill, New York, 1993.

T. R. Camp. Design of sewers to facilitate flow. *Sewer Works Journal*, 18 (1946): 3.

M. D. Campbell and J. H. Lehr. *Water Well Technology*. McGraw-Hill, New York, 1972.

P. C. Carman. Fluid flow through a granular bed. *Transaction of the Institutions of Chemical Engineers*, 15(1937): 150–56.

P. C. Carman. *Flow of Gasses Through Porous Media*. Butterworths, London, 1956.

D. Carroll. Rainwater as a chemical agent and geologic processes- A review. Water-Supply Paper 1535-G. U.S. Geological Survey, 1962.

H. R. Cedergren, K. H. O'Brien, and J. A. Arman. Guidelines for the design of subsurface drainage systems for highway structural sections. Technical Report FHWA-RD-72-30. Federal Highway Administration, June 1972.

A. Chadwick and J. Morfett. *Hydraulics in Civil and Environmental Engineering*. E. and F. N. Spon, London, 1993.

R. A. Chandler and D. B. McWhorter. Upconing of the salt-water–fresh-water interface beneath a pumping well. *Ground Water*, 13(1975): 354–59.

S. C. Chapra. *Surface Water-Quality Modeling*. McGraw-Hill, New York, 1997.

M. H. Chaudhry. *Open-Channel Flow*. Prentice-Hall, Englewood Cliffs, NJ, 1993.

C. Chen. Rainfall intensity-duration-frequency formulas. *Journal of Hydraulic Engineering*, 109, no. 12 (1983): 1603–21.

C.-N. Chen and T. S. W. Wong. Critical rainfall duration for maximum discharge from overland plane. *Journal of Hydraulic Engineering*, 119, no. 9 (September 1993): 1040–45.

D. A. Chin. Influence of surface waves on outfall dilution. *Journal of Hydraulic Engineering*, 113, no. 8 (1987): 1005–17.

D. A. Chin. Model of buoyant jet-surface wave interaction. *Journal of Waterway, Port, Coastal and Ocean Engineering*, 114, no. 3 (1988): 331–45.

D. A. Chin. Leakage of clogged channels that partially penetrate surficial aquifers. *Journal of Hydraulic Engineering*, 117, no. 4 (April 1991): 467–88.

D. A. Chin. *A Risk Management Strategy for Wellhead Protection*. Technical Report CEN93-3, University of Miami, Department of Civil Engineering, Coral Gables, FL, 1993a.

D. A. Chin. *Analysis and Prediction of South-Florida Rainfall*. Technical Report CEN-93-2, University of Miami, Coral Gables, FL, 1993b.

D. A. Chin. An assessment of first-order stochastic dispersion theories in porous media. *Journal of Hydrology*, 199 (1997): 53–73.

D. A. Chin and P. V. K. Chittaluru. Risk management in wellhead protection. *Journal of Water Resources Planning and Management*, 120, no. 3 (1994): 294–315.

D. A. Chin, L. Ding, and H. Huang. Ocean-outfall mixing zone delineation using Doppler radar. *Journal of Environmental Engineering*, 123, no. 12 (December 1997): 1217–26.

D. A. Chin and P. J. W. Roberts. Time series modeling of coastal currents. *Journal of Waterway, Port, Coastal and Ocean Engineering*, 111, no. 6 (November 1985): 954–72.

D. A. Chin and T. Wang. An investigation of the validity of first-order stochastic dispersion theories in isotropic porous media. *Water Resources Research*, 28 no. 6 (June 1992): 1531–42.

V. T. Chow. *Frequency Analysis of Hydrologic Data with Special Application to Rainfall Intensities*. Bulletin 414. University of Illinois Engineering Experiment Station, 1953.

V. T. Chow. The log-probability law and its emerging applications. *Proceedings of the American Society of Civil Engineers*, 80 (1954): 536-1–536-25.

V. T. Chow. *Open-Channel Hydraulics*. McGraw-Hill, New York, 1959.

V. T. Chow, D. R. Maidment, and L. W. Mays. *Applied Hydrology*. McGraw-Hill, New York, 1988.

J. E. Christiansen. Pan evaporation and evapotranspiration from climatic data. *Journal of Irrigation and Drainage Engineering*, 94 (1968): 243–65.

J. E. Christiansen and G. H. Hargreaves. Irrigation requirements from evaporation. *Trans. Int. Comm. on Irrig. and Drain.*, 3(1969): 23.569–23.596.

M. A. Churchill, H. L. Elmore, and R. A. Buckingham. Prediction of stream reaeration rates. *Journal of the Sanitary Engineering Division*, SA4 (1962): 1.

R. M. Clark. Water supply. In R. A. Corbitt, ed., *Standard Handbook of Environmental Engineering*, pp. 5.1–5.225. McGraw-Hill, New York, 1990.

C. F. Colebrook. Turbulent flow in pipes with particular reference to the transition between smooth and rough pipe laws. *J. Inst. Civ. Eng. Lond.*, 11(1939).

H. H. Cooper, J. D. Bredehoeft, and I. S. Papadopulos. Response of a finite-diameter well to an instantaneous change of water. *Water Resources Research*, 3, no. 1 (1967): 263–69.

H. H. Cooper and C. E. Jacob. A generalized graphical method for evaluating formation constants and summarizing well-field history. *Transactions of the American Geophysical Union*, 27 (1946): 526–34.

R. A. Corbitt. Wastewater Disposal. In R. A. Corbitt, ed., *Standard Handbook of Environmental Engineering*, pp. 6.1–6.274. McGraw-Hill, New York, 1990.

Council on Environmental Quality. *Environmental Quality, The Ninth Annual Report of the Council on Environmental Quality*. U.S. Government Printing Office, Washington, DC, 1978.

J. J. Coyle. *Grassed Waterways and Outlets*. Engineering Field Manual. U.S. Soil Conservation Service, April 1975.

H. Cross. *Analysis of Flow in Networks of Conduits and Conductors*. Bulletin 286. University of Illinois Engineering Experiment Station, 1936.

J. A. Cunge. On the subject of flood propagation method. *Journal of Hydraulic Research, IAHR*, 7, no. 2 (1967): 205–30.

C. Cunnane. Methods and merits of regional flood frequency analysis. *Journal of Hydrology*, 100(1988): 269–90.

G. Dagan. *Flow and Transport in Porous Formations*. Springer-Verlag, New York, 1989.

M. Dagg. Evaporation pans in East Africa. *Proceedings Fourth Specialist Meeting on Applied Meteorology in East Africa*, 1968.

J. W. Daily and D. R. F. Harleman. *Fluid Dynamics*. Addison-Wesley, Reading, MA, 1966.

H. Darcy. *Les Fontaines Publiques de la Ville de Dijon*. Victor Dalmont, Paris, 1856.

G. de Marsily. *Quantitative Hydrogeology, Groundwater Hydrology for Engineers*. Academic Press, San Diego, CA, 1986.

John A. Dean, ed. *Lange's Handbook of Chemistry*, 13th ed. McGraw-Hill, New York, 1985.

T. N. Debo and A. J. Reese. *Municipal Storm Water Management*. Lewis Publishers, Boca Raton, FL, 1995.

J. W. Delleur. Elementary groundwater flow and transport processes. In J. W. Delleur, ed., *The Handbook of Groundwater Engineering*, pp. 2.1–2.40. CRC Press, Boca Raton, FL, 1998.

P. J. Dillon and F. H. Rigler. The phosphorous-chlorophyll relationship in lakes. *Limnol. Oceanogr.*, 19, no. 4 (1974): 767–73.

South Florida Water Management District. *Management and Storage of Surface Waters, Permit Information Manual*, Vol. 4. West Palm Beach, FL, May 1994.

P. A. Domenico. An analytical model for multidimensional transport of a decaying contaminant species. *Journal of Hydrology*, 91 (1987): 49–58.

P. A. Domenico and V. V. Palciaukas. Alternative boundaries in solid waste management. *Ground Water*, 20 (1982): 303–11.

P. A. Domenico and G. A. Robbins. A new method of contaminant plume analysis. *Ground Water*, 23, no. 4 (1985): 476–85.

P. A. Domenico and F. W. Schwartz. *Physical and Chemical Hydrogeology*. Wiley, New York, 1990.

J. Doorenbos and W. O. Pruitt. *Guidelines for the Prediction of Crop Water Requirements*. Irrigation and Drainage Paper 24. UN Food and Agriculture Organization, Rome, 1975.

J. Doorenbos and W. O. Pruitt. *Guidelines for the Prediction of Crop Water Requirements*, 2nd ed. Irrigation and Drainage Paper 24. UN Food and Agriculture Organization, Rome, 1977.

F. G. Driscoll. *Ground Water and Wells*, 2nd ed. Johnson Division, St. Paul, MN, 1986.

N. E. Driver and G. D. Tasker. *Techniques for Estimation of Storm-Runoff Loads, Volumes, and Selected Constituent Concentrations in Urban Watersheds in the United States*. Open File Report 88-191. U.S. Geological Survey, Denver, CO, 1988.

N. E. Driver and G. D. Tasker. *Techniques for Estimation of Storm-Runoff Loads, Volumes, and Selected Constituent Concentrations in Urban Watersheds in the United States*. Water-Supply Paper 2363. U.S. Geological Survey, Washington, DC, 1990.

J. A. Duffie and W. A. Beckman. *Solar Engineering of Thermal Processes*. Wiley, New York, 1980.

J. Dupuit. *Études Théoriques et Pratiques sur le Mouvement des Eaux dans les Canaux Decouverts et á Travers les Terrains Perméables*, 2nd ed. Dunod, Paris, 1863.

H. B. Dwight. *Tables of Integrals and Other Mathematical Data*, 4th ed. Macmillan, New York, 1961.

B. Dziegielewski, E. M. Opitz, and D. Maidment. Water Demand Analysis. In L. W. Mays, ed., *Water Resources Handbook*, pp. 23.1–23.62. McGraw-Hill, New York, 1996.

S. M. Easa. Geometric Design. In W. F. Chen, ed., *The Civil Engineering Handbook*, pp. 2287–19. CRC Press, Boca Raton, FL, 1995.

G. Echávez. Increase in losses coefficient with age for small diameter pipes. *Journal of Hydraulic Engineering*, 123, no. 2(February 1997): 157–59.

H. A. Einstein. Der Hydraulische oder Profil-radius. *Schweizerische Bauzeitung*, 103, no. 8 (February 1934): 89–91.

H. A. Einstein and R. B. Banks. Fluid resistance of composite roughness. *Transactions of the American Geophysical Union*, 31, no. 4 (1951): 603–10.

N. El-Jabi and S. Sarraf. Effect of maximum rainfall position on rainfall-runoff relationship. *Journal of Hydraulic Engineering*, 117, no. 5 (May 1991): 681–85.

J. W. Elder. The dispersion of marked fluid in turbulent shear flow. *Journal of Fluid Mechanics*, 5(1959): 544–60.

W. J. Elliot. Precipitation. In A. D. Ward and W. J. Elliot, eds., *Environmental Hydrology*, pp. 19–49. Lewis Publishers, Boca Raton, FL, 1995.

W. H. Espey and D. G. Altman. Nomographs for ten-minute unit hydrographs for small urban watersheds. Addendum 3 of *Urban Runoff Control Planning*, EPA-600/9-78-035. USEPA, Washington, DC, 1978.

G. M. Fair and L. P. Hatch. Fundamental factors governing the streamline flow of water through sand. *Journal of the American Water Works Association*, 25 (1933): 1551–65.

J. A. Fay. *Introduction to Fluid Mechanics*. MIT Press, Cambridge, MA, 1994.

R. G. Feachem, D. J. Bradley, H. Garelick, and D. D. Mara. *Sanitation and Disease, Health Aspects of Excreta and Wastewater Management: World Bank Studies in Water Supply and Sanitation 3*. Wiley, New York, 1983.

M. C. Feller. Water balances in *eucalyptus regnans, e. obliqua*, and *pinus radiata* forests in Victoria. *Australian Forestry*, 44, no. 3 (1981): 153–61.

C. W. Fetter. *Contaminant Hydrogeology*. Macmillan, New York, 1992.

C. W. Fetter. *Applied Hydrogeology*, 3rd ed. Macmillan, New York, 1993.

A. Fick. On liquid diffusion. *Phils. Mag.*, 4, no. 10 (1855): 30–39.

H. B. Fischer, E. J. List, R. C. Y. Koh, J. Imberger, and N. H. Brooks. *Mixing in Inland and Coastal Waters*. Academic Press, New York, 1979.

R. A. Fisher and L. H. C. Tippett. Limiting forms of the frequency distribution of the largest or smallest member of a sample. *Proc. Cambridge Phil. Soc.*, 24, part II (1928): 180–91.

Florida Department of Environmental Protection. *Permits*. Florida Administrative Code Chapter 62-4. Tallahassee, FL, 1995.

Florida Department of Environmental Protection. *Surface Water Quality Standards*. Florida Administrative Code Chapter 62-302. Tallahassee, FL, 1996.

P. Forchheimer. *Hydraulik*. Teubner Verlagsgesellschaft, Stuttgart, Germany, 1930.

S. Fortier and F. C. Scobey. Permissible canal velocities. *Transactions of the American Society of Civil Engineers*, 89 (1926): 940–84.

H. A. Foster. Theoretical frequency curves and their application to engineering problems. *Transactions of the American Society of Civil Engineers*, 87 (1924): 142–73.

R. W. Fox and A. T. McDonald. *Introduction to Fluid Mechanics*, 4th ed. Wiley, New York, 1992.

J. B. Franzini and E. J. Finnemore. *Fluid Mechanics with Engineering Applications*, 9th ed. McGraw-Hill, New York, 1997.

R. H. Frederick, V. A. Myers, and E. P. Auciello. Five- to 60-minute precipitation frequency for the eastern and central United States. NOAA Technical Memorandum NWS HYDRO-35, National Oceanic and Atmospheric Administration, National Weather Service, Silver Spring, MD, June 1977.

R. A. Freeze. A stochastic-conceptual analysis of one-dimensional groundwater flow in non-uniform homogeneous media. *Water Resources Research*, 11(1975): 725–41.

R. A. Freeze and J. A. Cherry. *Groundwater*. Prentice-Hall, Englewood Cliffs, NJ, 1979.

R. H. French. *Open-Channel Hydraulics*. McGraw-Hill, New York, 1985.

M. Frére and G. F. Popov. Agrometeorological crop monitoring and forecasting. FAO Plant Production and Protection Paper 17, pp. 38–43. FAO, Rome, 1979.

J. J. Fried. *Groundwater Pollution*. Elsevier Scientific Publishing Co., Amsterdam, 1975.

E. O. Frind. Simulation of long-term transient density-dependent transport in groundwater. *Advances in Water Resources*, 5(1982): 73–88.

E. Ganguillet and W. R. Kutter. Versuch zur Aufstellung einer neuen allegemeinen Formel für die gleichförmige Bewegung des Wassers in Canälen und Flüssen [An investigation to establish a new general formula for uniform flow of water in canals and rivers]. *Zeitschrift des Oesterreichischen Ingenieur- und Architekten Vereines*, 21, no. 1, 2–3 (1869): 6–25,46–59. [Published in Bern, Switzerland, 1877; trans. Rudolph Hering and John C. Trautwine, Jr., as *A general Formula for the Uniform Flow of Water in Rivers and Other Channels*, Wiley, New York, 1st ed., 1888; 2d ed., 1891 and 1901].

L. W. Gelhar. *Stochastic Subsurface Hydrology*. Prentice-Hall, Englewood Cliffs, NJ, 1993.

L. W. Gelhar and C. L. Axness. Three-dimensional stochastic analysis of dispersion in aquifers. *Water Resources Research*, 19, no. 1 (January 1983): 161–80.

L. W. Gelhar, C. Welty, and K. R. Rehfeldt. A critical review of data on field-scale dispersion in aquifers. *Water Resources Research*, 28, no. 7 (1992): 1955–74.

D. P. Genereux. Field studies of streamflow generation using natural and injected tracers on Bicford and Walker branch watersheds. Ph.D thesis. MIT, Boston, MA, 1991.

P. M. Gerhart, R. J. Gross, and J. I. Hochstein. *Fundamentals of Fluid Mechanics*, 2d ed. Addison-Wesley, Reading, MA, 1992.

P. H. Gleick. An introduction to global fresh water issues. In P. H. Gleick, ed., *Water in Crisis*, pp. 3–12. Oxford University Press, New York, 1993.

GLUMB. *Recommended Standards for Water Works, 1987*. Great Lakes Upper Mississippi River Board of State Public Health and Environmental Managers, Health Research, 1987.

D. J. Goggin, M. A. Chandler, G. Kacurek, and L. W. Lake. Patterns of permeability in Eolian deposits: Page sandstone (Jurassic), Northeastern Arizona. *SPE Formation Evaluation* (June 1988): 297–306.

D. J. Goode and L. F. Konikow. *Modification of a Method-of-Characteristics Solute-Transport Model to Incorporate Decay and Equilibrium Controlled Sorption or Ion Exchange.* Water-Resources Investigation 89-4030. U.S. Geological Survey, 1989.

S. M. Gorelick, R. A. Freeze, D. Donohue, and J. F. Keely. *Groundwater Contamination.* Lewis, Boca Raton, FL, 1993.

R. A. Granger. *Fluid Mechanics.* Holt, Rinehart and Winston, New York, 1985.

D. M. Gray, ed. *Handbook on the Principles of Hydrology.* National Research Council, Water Information Center, Port Washington, Canada, 1973.

W. H. Green and G. Ampt. Studies of soil physics, Part I: The flow of air and water through soils. *J. Agric. Sci.*, 4, no. 1 (1911): 1–24.

J. E. Gribbin. *Hydraulics and Hydrology for Stormwater Management.* Delmar, Albany, NY, 1997.

I. I. Gringorten. A plotting rule for extreme probability paper. *Journal of Geophysical Research*, 68, no. 3 (1963): 813–14.

T. L. Grizzard, C. W. Randall, B. L. Weand, and K. L. Ellis. Effectiveness of extended detention ponds. In *Urban Runoff Quality.* ASCE, NY, 1986.

E. J. Gumbel. *Statistical Theory of Extreme Values and Some Practical Applications.* Applied Mathematics Series 33, U.S. National Bureau of Standards, Washington, DC, 1954.

E. J. Gumbel. *Statistics of Extremes.* Columbia University Press, New York, 1958.

R. S. Gupta. *Hydrology and Hydraulic Systems.* Waveland Press, Prospect Heights, IL, 1989.

C. T. Haan. *Statistical Methods in Hydrology.* The Iowa State University Press, Ames, IA, 1977.

C. T. Haan, B. J. Barfield, and J. C. Hayes. *Design Hydrology and Sedimentology for Small Catchments.* Academic Press, New York, 1994.

Haestad Methods. *1997 Practical Guide to Hydraulics and Hydrology.* Haestad Press, Waterbury, CT, 1997a.

Haestad Methods. *Computer Applications in Hydraulic Engineering.* Haestad Press, Waterbury, CT, 1997b.

W. H. Hager. *Energy Dissipators and Hydraulic Jump.* Kluwer Academic Publishers, Dordrecht, Netherlands, 1991.

C. Hansch and A. Leo. *Substitute Constants for Correlation Analysis in Chemistry and Biology.* Wiley, New York, 1979.

E. M. Hansen, L. C. Schreiner, and J. F. Miller. *Application of Probable Maximum Precipitation Estimates—United States East of the 105th meridian.* NOAA Hydrometeorological Report 52. National Weather Service, Washington, DC, August 1982.

M. S. Hantush. Analysis of data from pumping tests in leaky aquifers. *Transactions of the American Geophysical Union*, 37, no. 6 (1956): 702–14.

M. S. Hantush. Growth and decay of groundwater-mounds in response to uniform percolation. *Water Resources Research*, 3, no. 1 (1967): 227–34.

M. S. Hantush and C. E. Jacob. Non-steady radial flow in an infinite leaky aquifer. *Transactions of the American Geophysical Union*, 36, no. 1 (1955): 95–100.

A. W. Harbaugh and M. G. McDonald. *User's Documentation for MODFLOW-96, An Update to the U.S. Geological Survey Modular Finite-Difference Ground-Water Flow Model.* Open File Report 96-485. U.S. Geological Survey, 1996.

G. L. Hargreaves, G. H. Hargreaves, and J. P. Riley. Agric. benefits for Senegal River Basin. *Journal of Irrigation and Drainage Engineering*, 111, no. 2 (1995): 113–24.

G. L. Hargreaves and Z. A. Samani. Reference-crop evapotranspiration from temperature. *Applied Engrg. in Agric.*, 1, no. 2 (1985): 96–99.

D. R. F. Harleman, P. F. Melhorn, and R. R. Rumer. Dispersion-permeability correlation in porous media. *Journal of the Hydraulics Division*, 89(1963): 67–85.

B. W. Harrington. Design and construction of infiltration trenches. In L. A. Roesner, B. Urbonas, and M. B. Sonnen, eds., *Design of Urban Runoff Quality Controls*, pp. 291–306. ASCE, New York, 1989.

L. P. Harrison. Fundamental concepts and definitions relating to humidity. In A. Wexler, ed., *Humidity and Moisture*. Reinhold Publishing Co., New York, 1963.

J. P. Hartigan. Basis for design of wet detention basin BMP's. In L. A. Roesner, B. Urbonas, and M. B. Sonnen, eds., *Design of Urban Runoff Quality Controls*, pp. 122–44. ASCE, New York, 1989.

J. J. Hassett, J. C. Means, W. L. Banwart, and S. G. Wood. Sorption properties of sediments and energy-related pollutants. Technical Report EPA-600/3-80-041. U.S. Environmental Protection Agency, 1980.

A. Hazen. Discussion of "Dams on Sand Foundations, by A. C. Koenig". *Transactions of the American Society of Civil Engineers*, 73(1911): 199.

J. F. Heany, W. C. Huber, H. Sheikh, M. A. Medina, J. R. Doyle, W. A. Peltz, and J. E. Darling. *Nationwide Evaluation of Combined Sewer Overflows and Urban Stormwater Discharges*, Vol. 2, *Cost Assessment and Impacts*. Report EPA-600/2-77-064. USEPA, Washington, DC, 1977.

J. D. Helvey. *Rainfall Interception by Hardwood Forest Litter in the Southern Appalachians*. Research Paper SE-8, U.S. Department of Agriculture, Forest Service, Southeastern Forest Experiment Station, Asheville, NC, 1964.

J. D. Helvey. Interception by eastern white pine. *Water Resources Research*, 3, no. 3 (1967): 723–29.

H. F. Hemond and E. J. Fechner. *Chemical Fate and Transport in the Environment*. Academic Press, New York, 1993.

F. M. Henderson. *Open Channel Flow*. Macmillan Publishing Co., New York, 1966.

D. M. Hershfield. *Rainfall Frequency Atlas of the United States for Durations from 30 Minutes to 24 Hours and Return Periods from 1 to 100 Years*. Technical Paper 40. U.S. Department of Commerce, Weather Bureau, Washington, DC, May 1961.

A. Herzberg. Die Wasserversorgung einiger Nordseebaden [The water supply on parts of the North Sea coast in Germany]. *Z. Gasbeleucht. Wasserversorg.*, 44(1901): 815–19, 824–44.

B. L. Herzog. Slug tests for determining hydraulic conductivity of natural geologic deposits. In D. E. Daniel and S. J. Trautwein, eds., *Hydraulic Conductivity and Waste Contaminant Transport in Soil*, pp. 95–110. ASTM, Philadelphia, PA, 1994.

K. M. Hess. Use of a borehole flowmeter to determine spatial homogeneity of hydraulic conductivity and macrodispersion in a sand and gravel aquifer, Cape Cod, Massachusetts. In F. J. Moltz, J. G. Melville, and O. Guven, eds., *Proceedings of the Conference on New Field Techniques for Quantifying the Physical and Chemical Properties of Heterogeneous Aquifers*, pp. 497–508. National Water Well Association, Dublin, Ohio, 1989.

W. I. Hicks. A method of computing urban runoff. *Transactions of the American Society of Civil Engineers*, 109(1944): 1217.

F. Hildebrand. *Advanced Calculus for Applications*, 2d ed. Prentice-Hall, Englewood Cliffs, NJ, 1976.

A. T. Hjelmfelt, Jr. Investigation of curve number procedure. *Journal of Hydraulic Engineering*, 117, no. 6 (1991): 725–37.

R. A. Horton. Separate roughness coefficients for channel bottom and sides. *Engineering News Record*, 111, no. 22 (November 1933): 652–53.

R. E. Horton. Rainfall interception. *Monthly Weather Review*, 47(1919): 603–23.

R. E. Horton. Analysis of runoff plot experiments with varying infiltration capacity. *Trans. American Geophysical Union*, 20(1939): 693–711.

R. E. Horton. An approach toward a physical interpretation of infiltration capacity. *Soil Science Society of America Proceedings*, 5(1940): 399–417.

H. Huang, R. E. Fergen, J. R. Proni, and J. J. Tsai. Probabilistic analysis of ocean outfall mixing zones. *Journal of Environmental Engineering*, 122, no. 5 (1996): 359–67.

H. Huang, R. E. Fergen, J. R. Proni, and J. J. Tsai. Initial dilution equations for buoyancy-dominated jets in current. *Journal of Hydraulic Engineering*, 124, no. 1 (January 1998): 105–108.

W. C. Huber and R. E. Dickinson. *Storm Water Management Model, Version 4: User's Manual, with Addendums*. Technical Report EPA-600/3-88/001a, USEPA, Athens, GA, 1988.

P. Hufschmied. Estimation of three-dimensional, statistically anisotropic hydraulic conductivity field by means of single well pumping tests combined with flow meter measurements. *Hydrogeologie*, 2(1986): 163–174.

M. J. Hvorslev. Time lag and soil permeability in ground-water observation. Bulletin 36, Waterways Experiment Station. U.S. Army Corps of Engineers, Vicksburg, MI, 1951.

Insurance Services Office. *Fire Suppression Rating Schedule*. ISO, New York, 1980.

Y. Iwasa and S. Aya. Predicting longitudinal dispersion coefficient in open-channel flows. In *Proceedings of the International Symposium on Environmental Hydraulics*, pp. 505–10, Hong Kong, 1991.

C. F. Izzard. The surface-profile of overland flow. *Transactions of the American Geophysical Union*, 25(1944): 959–69.

I. J. Jackson. Problems of throughfall and interception assessment under tropical forest. *Journal of Hydrology*, 12(1971): 234–54.

C. E. Jacob. Drawdown test to determine effective radius of artesian well. Paper 2321. *Transactions of the American Society of Civil Engineers*, 112(1947): 1047–64.

A. K. Jain. Accurate explicit equation for friction factor. *Journal of the Hydraulics Division*, 102(HY5)(May 1976): 674–77.

A. K. Jain, D. M. Mohan, and P. Khanna. Modified Hazen-Williams formula. *Journal of the Environmental Engineering Division*, 104(EE1)(February 1978): 137–46.

A. James. Modelling water quality in lakes and reservoirs. In A. James, ed., *An Introduction to Water Quality Modelling*, 2d ed., pp. 233–60. Wiley, New York, 1993.

W. S. Janna. *Introduction to Fluid Mechanics*, 3d ed. PWS-KENT Publishing Co., Boston, MA, 1993.

S. W. Jens. Design of urban highway drainage—The state of the art. Report FHWA-TS-79-225. U.S. Department of Transportation, Federal Highway Administration, Washington, DC, 1979.

M. E. Jensen. Empirical methods of estimating or predicting evapotranspiration using radiation. In *Proceedings of Conference on Evapotranspiration*, pp. 57–61, 64. ASAE, Chicago, IL, December 1966.

M. E. Jensen and H. R. Haise. Estimating evapotranspiration from solar radiation. *Journal of Irrigation and Drainage Engineering*, 89(1963): 15–41.

M. E. Jensen, J. L. Wright, and B. J. Pratt. Estimating soil moisture depletion from climate, crop, and soil data. *Transactions of the American Society of Agricultural Engineers*, 14(1971): 954–59.

J. L. Johnson. *Local Wellhead Protection in Florida*. Technical Report. City of Tallahassee, Water Quality Division, Aquifer Protection Section, Tallahassee, FL, 1997.

P. A. Johnson. Uncertainty in hydraulic parameters. *Journal of Hydraulic Engineering*, 22, no. 2 (February 1996): 112–14.

Johnson Division, Universal Oil Products Company. *Ground Water and Wells*, St. Paul MN; Universal Oil Products Company, 1966.

A. G. Journel. *Fundamentals of Geostatistics in Five Lessons*. Vol. 8, *Short Course in Geology*. American Geophysical Union, Washington, DC, 1989.

S. W. Karickhoff. Estimation of sorption of hydrophobic semi-empirical pollutants on natural sediments and soils. *Chemosphere*, 10, no. 8 (1981): 833–46.

S. W. Karickhoff, D. S. Brown, and T. A. Scott. Sorption of hydrophobic pollutants on natural sediments. *Water Resources*, 13(1979): 241–48.

A. I. Kashef. *Groundwater Engineering*. McGraw-Hill, New York, 1986.

E. E. Kenaga and C. A. I. Goring. Relationship between water solubility, soil sorption, octanol-water partitioning, and bioconcentration of chemicals in biota. In *Third Aquatic Toxicology Symposium, Proceedings of the American Society of Testing and Materials*, no. 707, pp. 78–115, 1980.

W. S. Kerby. Time of concentration for overland flow. *Civil Engineering*, 29, no. 3 (1959): 174.

D. F. Kibler. Desk-top runoff methods. In D. F. Kibler, ed., *Urban Stormwater Hydrology*, pp. 87–135. American Geophysical Union, Water Resources Monograph Series, Washington, DC, 1982.

G. Kiely. *Environmental Engineering*. McGraw-Hill, New York, 1997.

P. Z. Kirpich. Time of concentration of small agricultural watersheds. *Civil Engineering*, 10, no. 6 (1940): 362.

G. W. Kite. *Frequency and Risk Analysis in Hydrology*. Water Resources Publications, Fort Collins, CO, 1977.

D. Klotz, K. P. Seiler, H. Moser, and F. Neumaier. Dispersivity and velocity relationship from laboratory and field relationships. *Journal of Hydrology*, 45, no. 3 (1980): 169–84.

R. C. Y. Koh. Wastewater field thickness and initial dilution. *Journal of Hydraulic Engineering*, 109 (September 1983): 1232–40.

L. F. Konikow and J. D. Bredehoeft. *Computer Model of Two-Dimensional Solute Transport and Dispersion in Ground Water*. Techniques of Water-Resources Investigations Book 7, Chapter C2. U.S. Geological Survey, 1978.

L. F. Konikow, D. J. Goode, and G. Z. Hornberger. *A Three-Dimensional Method-of-Characteristics Solute-Transport Model (MOC3D)*. Water-Resources Investigation 96-4267. U.S. Geological Survey, 1996.

L. F. Konikow, G. E. Granato, and G. Z. Hornberger. *User's Guide to Revised Method-of-Characteristics Solute-Transport Model (MOC—Version 3.1)*. Water-Resources Investigation 94-4115. U.S. Geological Survey, 1994.

L. F. Konikow and T. E. Reilly. Groundwater modeling. In J. W. Delleur, ed., *The Handbook of Groundwater Engineering*, pp. 20.1–20.40. CRC Press, Boca Raton, FL, 1998.

A. Koussis and J. Rodríguez-Mirasol. Hydraulic estimation of dispersion coefficient for streams. *Journal of Hydraulic Engineering*, 124, no. 3 (March 1998): 317–20.

J. Kozeny. Über Kapillare Leitung des Wassers im Boden. *Sitzungsber. Akad. Wiss.*, 36(1927): 271–306.

M. Krishnamurthy and B. A. Christensen. Equavalent roughness for shallow channels. *Journal of the Hydraulics Division*, 98, no. 12 (1972): 2257–63.

W. C. Krumbein and G. D. Monk. Permeability as a function of the size parameters of unconsolidated sand. *Am. Inst. Min. & Met. Eng. Tech. Pub.*, 150, no. 11 (1942): 1492.

W. C. Krumbein and G. D. Monk. Permeability as a function of the size parameters of unconsolidated sand. *Am. Inst. Min. & Met. Eng. Tech. Pub.*, 151(1943): 153–63.

E. Kuichling. The relation between the rainfall and the discharge of sewers in populous districts. *ASCE Transactions*, 20(1889): 1–56.

J. C. Lamb. *Water Quality and Its Control*. Wiley, New York, 1985.

E. W. Lane. Design of stable channels. *Transactions of the American Society of Civil Engineers*, 120(1955): 1234–79.

Y. L. Lau and B. G. Krishnappan. Transverse dispersion in rectangular channels. *Journal of the Hydraulics Division*, 103(HY10)(1977):1173–89.

G. Leclerc and J. C. Schaake. *Derivation of Hydrologic Frequency Curves*. Report 142. R. M. Parsons Laboratory of Hydrodynamics and Water Resources, MIT, Cambridge, MA, 1972.

J. H. W. Lee and P. Neville-Jones. Initial dilution of horizontal jet in crossflow. *Journal of Hydraulic Engineering*, 113, no. 5 (May 1987): 615–29.

J. Lehr, S. Hurlburt, B. Gallagher, and J. Voytek. *Design and Construction of Water Wells, A Guide for Engineers*. Van Nostrand Reinhold, New York, 1988.

D. H. Lennox. Analysis of step-drawdown test. *Journal of the Hydraulics Division*, 92(HY6) (1966): 25–48.

L. Leyton, E. R. C. Reynolds, and F. B. Thompson. Rainfall interception in forest and moorland. In W. E. Sopper and H. W. Lull, eds., *International Symposium on Forest Hydrology*, pp. 163–78. Pergamon Press, Oxford, 1967.

J. A. Liggett. *Fluid Mechanics*. McGraw-Hill, New York, 1994.

R. K. Linsley, J. B. Franzini, D. L. Freyberg, and G. Tchobanoglous. *Water-Resources Engineering*, 4th ed. McGraw-Hill, New York, 1992.

R. K. Linsley, M. A. Kohler, and J. L. H. Paulhus. *Hydrology for Engineers*, 3d ed. McGraw-Hill, New York, 1982.

C. P. Liou. Limitations and proper use of the Hazen-Williams equation. *Journal of Hydraulic Engineering*, 124, no. 9 (September 1998): 951–54.

H. Liu. Prediction dispersion coefficient of stream. *Journal of the Environmental Engineering Division*, 103, no. 1 (1977): 59–69.

D. E. Lloyd-Davies. The elimination of storm water from sewerage systems. *Proc. Inst. Civ. Eng.*, 164(1906): 41–67.

G.V. Loganathan, D. F. Kibler, and T. J. Grizzard. Urban Stormwater Management. In L. W. Mays, ed., *Water Resources Handbook*, pp. 26.1–26.35. McGraw-Hill, New York, 1996.

S. W. Lohman. *Ground-Water Hydraulics*. Professional Paper 708, U.S. Geological Survey, 1972.

G. K. Lotter. Considerations of hydraulic design of channels with different roughness of walls. *Trans. All Union Scientific Research, Institute of Hydraulic Engineering*, 9(1933): 238–41.

W. J. Lyman. Adsorption coefficient for soils and sediment. In W. J. Lyman et al., ed., *Handbook of Chemical Property Estimation Methods*, pp. 4.1–4.33. McGraw-Hill, New York, 1982.

MacKay, D. *Multimedia Environmental Models: The Fugacity Approach*. Lewis Publishers, Chelsea, MI, 1991.

R. Manning. Flow of water in open channels and pipes. *Trans. Inst. Civil Engrs. (Ireland)*, 20, 1890.

L. W. Mays. Water resources: An introduction. In L. W. Mays, ed., *Water Resources Handbook*, pp. 1.3–1.35. McGraw-Hill, New York, 1996.

P. J. McCall, R. L. Swann, and D. A. Laskowski. Partition models for equilibrium distribution of chemicals in environmental compartments. In R. L. Swann and A. Eschenroder, eds., *Fate of Chemicals in the Environment*, pp. 105–23. American Chemical Society, 1983.

G. T. McCarthy. The unit-hydrograph and flood routing. Unpublished paper presented at U.S. Army Corps of Engineers North Atlantic Division Conference, 1938.

R. H. McCuen. *Hydrologic Analysis and Design*. Prentice-Hall, Englewood Cliffs, NJ, 1989.

R. H. McCuen, S. L. Wong, and W. J. Rawls. Estimating urban time of concentration. *Journal of Hydraulic Engineering*, 110, no. 7 (1984): 887–904.

M. G. McDonald and A. W. Harbaugh. A modular three-dimensional finite-difference groundwater flow model. Techniques of Water Resources Investigations of the United States Geological Survey, Book 6, Chapter A1. U.S. Geological Survey, Reston, VA, 1988.

B. M. McEnroe. Characteristics of intense storms in Kansas. In *World Water Issues in Evolution*, pp. 933–40. ASCE, New York, 1986.

T. J. McGhee. *Water Supply and Sewerage*, 6th ed. McGraw-Hill, New York, 1991.

R. G. Mein and C. L. Larson. Modeling infiltration during a steady rain. *Water Resources Research*, 9, no. 2 (1973): 384–94.

M. M. Méndez-Díaz and G. H. Jirka. Buoyant plumes from multiport diffuser discharge in deep coflowing water. *Journal of Hydraulic Engineering*, 122, no. 8 (August 1996): 428–35.

J. W. Mercer and R. M. Cohen. A review of immiscible fluids in the subsurface: Properties, models, characterization, and remediation. *Journal of Contaminant Hydrology*, 6(1990): 107–63.

R. A. Meriam. A note on the interception loss equation. *Journal of Geophysical Research*, 65(1960): 3850–51.

D. W. Miller, J. J. Geraghty, and R. S. Collins. *Water Atlas of the United States*. Water Information Center, Port Washington, NY, 1962.

J. F. Miller, R. H. Frederick, and R. J. Tracey. *Precipitation Frequency Atlas of the Western United States*. NOAA Atlas 2. U.S. National Weather Service, Silver Springs, MD, 1973.

J. L. Monteith. Evaporation and the environment. *Symp. Soc. Expl. Biol.*, 19(1965): 205–34.

J. L. Monteith. Evaporation and surface temperature. *Quarterly Journal of Royal Meteorological Society*, 107(1981): 1–27.

S. Montes. *Hydraulics of Open Channel Flow*. ASCE Press, Reston, VA, 1998.

L. F. Moody. Friction factors for pipe flow. *Trans. ASME*, 66, no. 8 (1944).

L. F. Moody. Some pipe characteristics of engineering interest. *Houille Blanche*, May-June 1950.

H. J. Morel-Seytoux and J. P. Verdin. *Extension of Soil Conservation Service Rainfall Runoff Methodology for Ungaged Watersheds*. Report FHWA/RD-81/060. U.S. Federal Highway Administration, Washington, DC, 1981.

D. A. Morris and A. I. Johnson. *Summary of Hydrologic and Physical Properties of Rock and Soil Materials, as Analysed by the Hydrologic Laboratory of the U.S. Geologic Survey, 1948–1960*. Water-Supply Paper 1839-D. U.S. Geological Survey, 1967.

A. K. Motayed and M. Krishnamurthy. Composite roughness of natural channels. *Journal of the Hydraulics Division*, 106, no. 6 (1980): 1111–16.

R. L. Mott. *Applied Fluid Mechanics*. Merrill, New York, 1994.

L. Muhlhofer. Rauhigkeitsuntersuchungen in einem Stollen mit betonierter Sohle und unverkleideten Wanden. *Wasserkraft und Wasserwirtschaft*, 28, no. 8 (1933): 85–88.

T. J. Mulvaney. On the use of self-registering rain and flood gauges. *Trans. Inst. Civ. Eng. Ireland*, 4, no. 2 (1850): 1–8.

B. R. Munson, D. F. Young, and T. H. Okiishi. *Fundamentals of Fluid Mechanics*, 2d ed. Wiley, New York, 1994.

W. R. C. Myers. Influence of geometry on discharge capacity of open channels. *Journal of Hydraulic Engineering*, 117, no. 5 (May 1991): 676–80.

National Research Council. *Opportunities in the Hydrologic Sciences*. National Academy Press, Washington, DC, 1991.

L. C. Neale and R. E. Price. Flow characteristics of PVC sewer pipe. *Journal of the Sanitary Engineering Division*, 90(SA3)(1964): 109.

S. P. Neuman. Theory of flow in unconfined aquifers considering delayed response of the water table. *Water Resources Research*, 8, no. 4 (August 1972): 1031–45.

S. P. Neuman. Universal scaling of hydraulic conductivities and dispersivities in geologic media. *Water Resources Research*, 26, no. 8 (1990): 1749–58.

S. P. Neuman and P. A. Witherspoon. Theory of flow in aquicludes adjacent to slightly leaky aquifers. *Water Resources Research*, 4, no. 1 (1968): 103–12.

S. P. Neuman and P. A. Witherspoon. Theory of flow in a confined two aquifers system. *Water Resources Research*, 5, no. 2 (1969a): 803–16.

S. P. Neuman and P. A. Witherspoon. Applicability of current theories of flow in leaky aquifers. *Water Resources Research*, 5, no. 4 (1969b): 817–29.

S. P. Neuman and P. A. Witherspoon. Field determination of the hydraulic properties of leaky multiple aquifer systems. *Water Resources Research*, 8, no. 5 (1972): 1284–98.

B. D. Newport. Salt water intrusion in the United States. Technical Report 600/8-77-011, USEPA, Washington, DC, 1977.

V. Nguyen and G. F. Pinder. Direct calculation of aquifer parameters in slug test analysis. In J. Rosenshein and G. D. Bennett, eds., *Groundwater Hydraulics*, pp. 222–239. Water Resources Monograph 9. American Geophysical Union, 1984.

National Academy of Sciences. *Safety of Existing Dams: Evaluation and Improvement*. National Academy Press, Washington, DC, 1983.

J. Nikuradse. Gesetzmässigkeiten der turbulenten Strömung in glatten Röhren. *VDI-Firschungush*, (1932): 356.

J. Nikuradse. Strömungsgesetze in rauhen Röhren. *VDI-Forschungsh*, (1933): 361.

V. Novotny, K. R. Imhoff, M. Olthof, and P. A. Krenkel. *Karl Imhoff's Handbook of Urban Drainage and Wastewater Disposal*. Wiley, New York, 1989.

D. J. O'Connor. The bacterial distribution in a lake in the vicinity of a sewage discharge. In *Proceedings of the 2nd Purdue Industrial Waste Conference*, West Lafayette, IN, 1962.

D. J. O'Connor and W. E. Dobbins. Mechanism of reaeration in natural streams. *Transactions of the American Society of Civil Engineers*, 123(1958): 655.

A. Okubo. Ocean diffusion diagrams. *Deep-Sea Research*, 18(1971): 789–802.

I. C. Olmsted. *Stomatal Resistance and Water Stress in Melaleuca*. Final Report. Contract with USDA, Forest Service, August 1978.

R. M. Olson and S. J. Wright. *Essentials of Engineering Fluid Mechanics*, 5th ed. Harper & Row, New York, 1990.

P. S. Osbourne. *Suggested Operating Procedures for Aquifer Pumping Tests*. EPA Ground Water Issue EPA/540/S-93/503, USEPA, February 1993.

D. E. Overton and M. E. Meadows. *Storm Water Modeling*. Academic Press, New York, 1976.

M. Owens, R. Edwards, and J. Gibbs. Some reaeration studies in streams. *International Journal of Air and Water Pollution*, 8(1964): 469–86.

J. F. Pankow and J. A. Cherry. *Dense Chlorinated Solvents and other DNAPLs in Groundwater*. Waterloo Press, Portland, OR, 1996.

I. S. Papadopulos and H. H. Cooper, Jr. Drawdown in a well of large diameter. *Water Resources Research*, 3, no. 1 (1967): 241–44.

F. Parsons, P. R. Wood, and J. DeMarco. Transformation of tetrachloroethylene and trichloroethylene in microcosms and groundwater. *Journal of the American Water Works Association*, 26, no. 2 (1984): 56f.

R. Pecher. The runoff coefficient and its dependence on rain duration. *Berichte aus dem Institut fur Wasserwirtschaft und Gesundheitsingenieurwesen*, no. 2, TU, Munich, 1969.

H. L. Penman. Evaporation: An introductory survey. *Netherlands J. Agric. Sci.*, 1(1956): 9–29, 87–97, 151–53.

H. L. Penman. *Vegetation and Hydrology*. Technical Communication 53, Commonwealth Bureau of Soils, Harpenden, England, 1963.

T. K. Perkins and O. C. Johnson. A review of diffusion and dispersion in porous media. *Society of Petroleum Engineers Journal*, 3(1963): 70–84.

J. R. Philip. An infiltration equation with physical significance. *Soil Science*, 77(1954): 153–57.

J. R. Philip. Variable-head ponded infiltration under constant or variable rainfall. *Water Resources Research*, 29, no. 7 (July 1993): 2155–65.

N. N. Pillai. Effect of shape on uniform flow through smooth rectangular open channels. *Journal of Hydraulic Engineering*, 123, no. 7 (July 1997): 656–58.

R. H. Plumb, Jr. and A. M. Pitchford. Volatile organic scans: Implication for ground-water monitoring. In *Proceedings Petroleum Hydrocarbons and Organic Chemicals in Ground Water*, pp. 207–22. National Water Well Association, Houston, TX, November 13–15, 1985.

R. D. Pomeroy and J. D. Parkhurst. The forecasting of sulfide buildup rates in sewers. *Progress in Water Technology*, 9, 1977.

V. M. Ponce. *Engineering Hydrology, Principles and Practices*. Prentice-Hall, Englewood Cliffs, NJ, 1989.

V. M. Ponce and R. H. Hawkins. Runoff curve number: Has it reached maturity? *Journal of Hydrologic Engineering*, 1, no. 1 (January 1996): 11–19.

V. M. Ponce, A. K. Lohani, and P. T. Huston. Surface albedo and water resources: Hydroclimatological impact of human activities. *Journal of Hydrologic Engineering*, 2, no. 4 (October 1997): 197–203.

M. C. Potter and D. C. Wiggert. *Mechanics of Fluids*. Prentice-Hall, Englewood Cliffs, NJ, 1991.

C. H. B. Priestly and R. J. Taylor. On the assessment of surface heat flux and evaporation using large-scale parameters. *Mon. Weath. Rev.*, 100(1972): 81–92.

L. G. Puls. *Construction of Flood Flow Routing Curves, 1928*. U.S. 70th Congress, first session, House Document 185, pp. 46–52, and U.S. 71st Congress, second session, House Document 328, pp. 190–191.

V. L. Quisenberry and R. E. Phillips. Percolation of surface-applied water in the field. *Soil Sci. Soc. Am. J.*, 40(1976): 484–89.

M. Radojkovic and C. Maksimovic. On standardization of computational models for overland flow. In B.C. Yen, ed., *Proceedings of the Fourth International Conference on Urban Storm Drainage*, pp. 100–105. International Association for Hydraulic Research, Lausanne, Switzerland, 1987.

C. D. Rail. *Groundwater Contamination*. Technomic Publishing Co., Lancaster, PA, 1989.

P. S. C. Rao and J. M. Davidson. Estimation of pesticide retention and transformation parameters required in nonpoint source pollution models. In M. R. Overcash and J. M. Davidson, eds., *Environmental Impact of Nonpoint Source Pollution*, pp. 23–67. Ann Arbor Science Publishers, Ann Arbor, MI, 1980.

W. Rast and G. F. Lee. *Summary Analysis of the North American (U. S. Portion) OECD Eutrophication Project: Nutrient Loading-Lake Response Relationships and Trophic State Indices*. Technical Report EPA-600/3-78-008. USEPA, Corvallis, OR, 1978.

A. J. Raudkivi and R. A. Callander. *Analysis of Groundwater Flow*. Wiley, New York, 1976.

W. Rawls, P. Yates, and L. Asmussen. *Calibration of Selected Infiltration Equations for the Georgia Coastal Plain*. Technical Report ARS-S-113. U.S. Department of Agriculture, Agricultural Research Service, Washington, DC, 1976.

W. J. Rawls and D. L. Brakensiek. Estimating soil water retention from soil porosities. *Journal of Irrigation and Drainage Engineering*, 108, no. 2 (1982): 166–71.

W. J. Rawls, D. L. Brakensiek, and N. Miller. Green-Ampt infiltration parameters from soils data. *Journal of Hydraulic Engineering*, 109, no. 1 (1983): 1316–20.

K. R. Rehfeldt and L. W. Gelhar. Stochastic analysis of dispersion in unsteady flow in heterogeneous aquifers. *Water Resources Research*, 28, no. 8 (1992): 2085–99.

K. R. Rehfeldt, L. W. Gelhar, J. B. Southard, and A. M. Dasinger. *Estimates of Macrodispersivity Based on Analyses of Hydraulic Conductivity Variability at the MADE Site*. Technical Report EPRI EN-6405. Electric Power Research Institute, Palo Alto, CA, 1989. Project 2485-5.

L. F. Richardson. Atmospheric diffusion shown on a distance-neighbour graph. *Proceedings of the Royal Society*, A110(1926): 709.

J. A. Roberson, J. J. Cassidy, and M. H. Chaudhry. *Hydraulic Engineering*, 2d ed. Wiley, New York, 1998.

J. A. Roberson and C. T. Crowe. *Engineering Fluid Mechanics*, 6th ed. Wiley, New York, 1993.

P. J. W. Roberts. Line plume and ocean outfall dispersion. *Journal of the Hydraulics Division*, 105(HY4)(April 1977): 313–31.

M. J. L. Robin, E. A. Sudicky, R. W. Gillham, and R. G. Kachanaski. Spatial variability of strontium distribution coefficients and their correlation with hydraulic conductivity in the Canadian forces base Borden aquifer. *Water Resources Research*, 27, no. 10 (1991): 2619–32.

M. I. Rorabaugh. Graphical and theoretical analysis of step-drawdown test of artesian well. *Proceedings of the American Society of Civil Enginers*, 79, no. 362 (1953).

R. L. Rossmiller. The rational formula revisited. In *Proceedings of the International Symposium on Urban Storm Runoff*, University of Kentucky, Lexington, KY, 1980.

J. Rothacher. Net precipitation under a Douglas-fir forest. *Forest Science*, 9, no. 4 (1963): 423–29.

H. Rouse. *Elementary Fluid Mechanics*. Wiley, New York, 1946.

L. K. Rowe. Rainfall interception by a beech-podocarp-hardwood forest near Reefton, North Westland, New Zealand. *Journal of Hydrology (N.Z.)*, 18, no. 2 (1979): 63–72.

Barre de Saint-Venant. Theory of unsteady water flow, with application to river floods and to propagation of tides in river channels. *French Academy of Science*, 73 (1871): 148–54, 237–40, 1871.

G. D. Salvucci and D. Entekhabi. Explicit expressions for Green-Ampt (delta function diffusivity) infiltration rate and cumulative storage. *Water Resources Research*, 30, no. 9 (September 1994): 2661–63.

C. N. Sawyer, P. L. McCarty, and G. F. Parkin. *Chemistry for Environmental Engineering*, 4th ed. McGraw-Hill, New York, 1994.

S. Schmorak and A. Mercado. Upconing of fresh water-sea water interface below pumping wells, field study. *Water Resources Research*, 5(1969): 1290–1311.

J. L. Schnoor. *Environmental Modeling: Fate and Transport of Pollutants in Water, Air, and Soil.* Wiley, New York, 1996.

R. P. Schwarzenbach, P. M. Gschwend, and D. M. Imboden. *Environmental Organic Chemistry.* Wiley, New York, 1993.

R. P. Schwarzenbach and J. Westall. Transport of non-polar organic compounds from surface water to groundwater: Laboratory sorption studies. *Environmental Science and Technology*, 15, no. 11 (1981): 1360–67.

F. Schwille. *Dense Chlorinated Solvents in Porous and Fractured Media: Model Experiments (English Translation).* Lewis Publishers, Boca Raton, FL, 1988.

C. K. Sehgal. Design guidelines for spillway gates. *Journal of Hydraulic Engineering*, 122, no. 3 (March 1996): 155–65.

Z. Şen. *Applied Hydrogeology for Scientists and Engineers.* Lewis Publishers, Boca Raton, FL, 1995.

I. W. Seo and T.S. Cheong. Predicting longitudinal dispersion coefficient in natural streams. *Journal of Hydraulic Engineering*, 124, no. 1 (January 1998): 25–32.

S. E. Serrano. *Hydrology for Engineers, Geologists, and Environmental Professionals.* HydroScience, Lexington, KY, 1997.

I. H. Shames. *Mechanics of Fluids*, 3d ed. McGraw-Hill, New York, 1992.

R. M. Shane and W. R. Lynn. Mathematical model for flood risk evaluation. *Journal of the Hydraulics Division, ASCE*, 90(HY6)(November 1964): 1–20.

L. K. Sherman. Stream-flow from rainfall by the unit graph method. *Engineering News Record*, 108(1932): 501–505.

W. J. Shuttleworth. Evaporation. In D. R. Maidment, ed., *Handbook of Hydrology*, pp. 4.1–4.53. McGraw-Hill, New York, 1993.

V. Sichard. Das Fassungsvermögen von Bohrbrunnen und Eine Bedeutung für die Grundwasser-ersenkung inbesondere für grossere Absentiefen. Ph.D thesis. Tech. Hochschule, Berlin, 1927.

J. R. Simanton and H. B. Osborn. Reciprocal-distance estimate of point rainfall. *Journal of the Hydraulic Engineering Division*, 106(HY7), July 1980.

V. P. Singh. *Hydrologic Systems, Rainfall-Runoff Modeling, vol. I.* Prentice-Hall, Englewood Cliffs, NJ, 1989.

V. P. Singh. *Elementary Hydrology.* Prentice-Hall, Englewood Cliffs, NJ, 1992.

L. Smith. A stochastic analysis of steady state groundwater flow in a bounded domain. Ph.D thesis. University of British Columbia, Vancouver, 1978.

L. Smith. Spatial variability of flow parameters in a stratified sand. *Mathematical Geology*, 13, no. 1 (1981): 1–21.

M. K. Smith. Throughfall, stemflow, and interception in pine and eucalypt forests. *Australian Forestry*, 36(1974): 190–97.

R. E. Smith. Discussion of "Runoff curve number: Has it reached maturity?" by V. M. Ponce and R. H. Hawkins. *Journal of Hydrologic Engineering*, 2, no. 3 (1997): 145–47.

V. H. Smith and J. Shapiro. *A Retrospective Look at the Effects of Phosphorous Removal in Lakes, in Restoration of Lakes and Inland Waters.* Technical Report EPA-440/5-81-010. USEPA, Office of Water Regulations and Standards, Washington, DC, 1981.

Soil Conservation Service. *Computer Program for Project Formulation Hydrology [draft].* Technical Release 20. SCS, U.S. Department of Agriculture, Washington, DC, 1983.

Soil Conservation Service. *Urban Hydrology for Small Watersheds.* Technical Release 55. SCS, U.S. Department of Agriculture, Washington, DC, 1986.

Soil Conservation Service. *SCS National Engineering Handbook, Section 4: Hydrology.* U.S. Department of Agriculture, Washington, DC, 1993.

W. B. Solley, C. F. Merk, and R. R. Pierce. *Estimated Use of Water in the United States in 1985.* Circular 1004. U.S. Geological Survey, Reston, VA, 1988.

South Carolina Land Resources Conservation Commission. *South Carolina Stormwater Management and Sediment Reduction Regulations*. Technical Report. Division of Engineering, Columbia, SC, 1992.

M. S. Sperry and J. J. Pierce. A model for estimating the hydraulic conductivity of granular material based on grain shape, grain size, and porosity. *Ground Water*, 33, no. 6 (November–December 1995): 892–98.

P. Stahre and B. Urbonas. Swedish approach to infiltration and percolation design. In L. A. Roesner, B. Urbonas, and M. B. Sonnen, eds., *Design of Urban Runoff Quality Controls*, pp. 307–23. ASCE, New York, 1989.

R. W. Stallman. Aquifer-test design, observation and data analysis. *Techniques of Water Resources Investigations*, Book 3, Chapter B1. U.S. Geological Survey, 1971.

G. Stanhill. The control of field irrigation practice from measurements of evaporation. *Israel J. Agric. Res.*, 12(1962): 51–62.

T. E. Stanton and J. R. Pannell. Similarity of motion in relation to surface friction of fluids. *Philosophical Transactions, Royal Society of London*, 214A(1914): 199–224.

R. L. Street, G. Z. Watters, and J. K. Vennard. *Elementary Fluid Mechanics*, 7th ed. Wiley, New York, 1996.

H. W. Streeter and E. B. Phelps. *A Study of the Pollution and Natural Purification of the Ohio River*, iii. *Factors Concerned in the Phenomena of Oxidation and Reaeration*. Bulletin 146. U.S. Public Health Service, 1925.

V. L. Streeter and E. B. Wylie. *Fluid Mechanics*, 8th ed. McGraw-Hill, New York, 1985.

V. L. Streeter, E. B. Wylie, and K. W. Bedford. *Fluid Mechanics*, 9th ed. McGraw-Hill, New York, 1998.

J. M. Stubchaer. *The Santa Barbara Urban Hydrograph Method*. Proceedings, National Symposium of Hydrology and Sediment Control. University of Kentucky, Lexington, KY, 1980.

E. A. Sudicky. A natural gradient experiment on solute transport in a sand aquifer: Spatial variability of hydraulic conductivity and its role in the dispersion process. *Water Resources Research*, 22, no. 13 (1986): 2069–82.

P. K. Swamee and A. K. Jain. Explicit equations for pipe-flow problems. *ASCE, Journal of the Hydraulics Division*, 102(HY5)(May 1976): 657–64.

W. T. Swank, N. B. Goebel, and J. D. Helvey. Interception loss in loblolly pine stands in the piedmont of South Carolina. *Journal of Soil and Water Conservation*, 27, no. 4 (1972): 160–64.

R. M. Sykes. Water and wastewater planning. In W. F. Chen, ed., *The Civil Engineering Handbook*, pp. 169–223. CRC Press, Boca Raton, FL, 1995.

T. Szirtes. *Applied Dimensional Analysis and Modeling*. McGraw-Hill, New York, 1977.

D. W. Taylor. *Fundamentals of Soil Mechanics*. Wiley, New York, 1948.

G. Tchobanoglous and E.D. Schroeder. *Water Quality*. Addison-Wesley, Reading, MA, 1987.

T. H. Y. Tebbutt. *Principles of Water Quality Control*, 5th ed. Butterworth-Heinemann, Oxford, England, 1998.

C. V. Theis. The relation between lowering of the piezometric surface and the rate and duration of discharge of a well using ground water storage. In *Transactions of the American Geophysical Union, 16th Annual Meeting*, Part 2, pp. 519–24, 1935.

H. J. Thiébaux. *Statistical Data Analysis for Ocean and Atmospheric Sciences*. Academic Press, San Diego, CA, 1994.

G. Thiem. *Hydrologische*. Gebhardt, Leipzig, 1906.

A. H. Thiessen. Precipitation for large areas. *Monthly Weather Review*, 39(July 1911): 1082–84.

A. L. Tholin and C. J. Keifer. The hydrology of urban runoff. *Transactions of the American Society of Civil Engineers*, 125(1960): 1308.

A. S. Thom and H. R. Oliver. On Penman's equation for estimating regional evaporation. *Quarterly Journal of Royal Meteorological Society*, 193(1977): 345–57.

R. V. Thomann. *Systems Analysis and Water Quality Management*. Technical Report. Environmental Research Applications, New York, 1972.

R. V. Thomann and J. A. Mueller. *Principles of Surface Water Quality Modeling and Control*. Harper & Row, New York, 1987.

G. W. Thomas and R. E. Phillips. Consequences of water movement in macropores. *J. Environ. Qual.*, 8, no. 2 (1979): 149–52.

C. W. Thornthwaite. An approach toward a rational classification of climate. *Geograph. Rev*, 38(1948): 55.

C. W. Thornthwaite and J. R. Mather. The water balance. *Climatology*, 8(1). Lab. of Climat., Centeron, NJ, 1955.

D. K. Todd. *Groundwater Hydrology*, 2d ed. Wiley, New York, 1980.

E. C. Tsivoglou and S. R. Wallace. *Characterization of Stream Reaeration Capacity*. Technical Report EPA-R3-72-012. USEPA, 1972.

L. Turc. Evaluation des besoins en eau d'irrigation, evapotranspiration potentielle, formule climatique simplifice et mise a jour. [Estimation of irrigation water requirements, potential evapotranspiration: A simple climatic formula evolved up to date]. *Ann. Agron.*, 12(1961): 13–49.

Urban Drainage and Flood Control District. *Urban Storm Drainage Criteria Manual*. Denver Regional Council of Governments, Denver, CO, 1984.

B. Urbonas and P. Stahre. *Stormwater: Best Management Practices and Detention for Water Quality, Drainage, and CSO Management*. Prentice-Hall, Englewood Cliffs, NJ, 1993.

U.S. Army Corps of Engineers. *Scour and deposition in rivers and reservoirs, HEC-6 User's Manual*. Technical Report. The Hydrologic Engineering Center, Davis, CA, 1991.

U.S. Army Corps of Engineers. *Hydraulic Design of Flood Control Channels*. ASCE, New York, 1995.

U.S. Army Waterways Experiment Station. *A Numerical One-Dimensional Model of Reservoir Water Quality*. CE-QUAL-R1. Technical Report. USAWES, Vicksburg, MI, 1986.

U.S. Army Waterways Experiment Station. *A Numerical Two-Dimensional Laterally-Averaged Model of Hydrodynamics and Water Quality*. CE-QUAL-W2. Technical Report. USAWES, Vicksburg, MI, 1986.

U.S. Bureau of Reclamation. *Design of Small Canal Structures*. U.S. Department of Interior, Denver, CO, 1978.

U.S. Bureau of Reclamation. *Ground Water Manual*. U.S. Department of the Interior, 1977.

U.S. Bureau of Reclamation. *Ground Water Manual*, 2nd ed. U.S. Department of the Interior, 1995.

U.S. Department of Agriculture. *Irrigation Water Requirements*. Technical Release 21 (rev). Soil Conservation Service, 1970.

U.S. Environmental Protection Agency. *A Compendium of Lake and Reservoir Data Collected by the National Eutrophication Survey in the Northeast and North-Central United States*. Working Paper 474. USEPA, Corvalis, OR, 1975.

U.S. Environmental Protection Agency. *Results of the Nationwide Urban Runoff Program, Final Report*. Technical Report Accession No. PB84-185552. NTIS, Washington, DC, December 1983.

U.S. Environmental Protection Agency. *Methodology for Analysis of Detention Basins for Control of Urban Runoff Quality*. Technical Report EPA 440/5-87-001. USEPA, Office of Water Regulations and Standards, Washington, DC, 1986a.

U.S. Environmental Protection Agency. *Quality Criteria for Water*. Technical Report EPA-440/5-86-001. Office of Water Regulations and Standards, Washington, DC, 1986b.

U.S. Environmental Protection Agency. *Guidelines for Delineation of Wellhead Protection Areas*. Technical Report EPA 440/6-87-010. Office of Ground-Water Protection, Washington, DC, June 1987.

U.S. Environmental Protection Agency. *WASP4, A Hydrodynamic and Water Quality Model, Model Theory, User's Manual, and Programmer's Guide*. Technical Report EPA/600/3-86-034. Athens, GA: USEPA, 1988.

U.S. Environmental Protection Agency. *Guide to Ground-Water Supply Contingency Planning for Local and State Governments*. Technical Assistance Document EPA 440/6-90-003. Office of Water, Washington, DC, 1990a.

U.S. Environmental Protection Agency. *National Water Quality Inventory: 1988 Report to Congress*. Technical Report EPA-440-4-90-003. USEPA, 1990b.

U.S. Environmental Protection Agency. *Office of Water. National Primary Drinking Water Standards*. Technical Report EPA 810-F-94-001A. February 1994.

U.S. Federal Highway Administration. *Hydraulic Design of Highway Culverts*. Hydraulic Design Series No. 5 FHWA-IP-85-15. Washington, DC: FHWA, 1984a.

U.S. Federal Highway Administration. HEC–12. U.S. Department of Transportation, McLean, VA: FHWA, 1984b.

U.S. Federal Highway Administration. *Hydrology* HEC-19. Technical Report FHWA-IP-95. Washington, DC: FHWA, 1995.

U.S. Interagency Advisory Committee on Water Data. *Guidelines for Determining Flood Flow Frequency*. Bulletin 17B of the Hydrology Committee, U.S. Geological Survey, Reston, VA, 1982.

U.S. Office of Technology Assessment. *Protecting the Nation's Groundwater from Contamination*. Technical Report. USOTA, Washington, DC, 1984.

U.S. Weather Bureau. *Two-to-Ten-Day Precipitation for a Return Period of 2 to 100 Years in the Contiguous United States*. Technical Paper 49. U.S. Department of Commerce, Washington, DC, 1964.

USSR National Committee for the International Hydrological Decade. World water balance and water resources of the earth. In *Studies and Reports in Hydrology*, vol. 25. UNESCO, Paris, 1978.

C. H. M. van Bavel. Potential evaporation: The combination concept and its experimental verification. *Water Resources Research*, 2, no. 3 (1966): 455–67.

J. A. Van Mullem. Runoff and peak discharges using Green-Ampt infiltration model. *Journal of Hydraulic Engineering*, 117, no. 3 (March 1991): 354–70.

V. A. Vanoni. Velocity distribution in open channels. *Civil Engineering*, 11(1941): 356–57.

J. P. Velon and T. J. Johnson. Water Distribution and Treatment. In V.J. Zipparro and H. Hasen, eds., *Davis's Handbook of Applied Hydraulics*, 4th ed., pp. 27.1–27.50. McGraw-Hill, New York, 1993.

P. Venkataraman and P. R. M. Rao. Darcian, transitional, and turbulent flow through porous media. *Journal of Hydraulic Engineering*, 124, no. 8 (August 1998): 840–46.

W. Viessman and M. J. Hammer. *Water Supply and Pollution Control*, 6th ed. Addison-Wesley, Reading, MA, 1998.

W. Viessman, J. Knapp, and G. L. Lewis. *Introduction to Hydrology*, 2d ed. Harper & Row, New York, 1977.

W. Viessman and G. L. Lewis. *Introducton to Hydrology*, 4th ed. HarperCollins, New York, 1996.

W. Viessman and C. Welty. *Water Management Technology and Institutions*. Harper & Row, New York, 1985.

J. R. Villemonte. Submerged weir discharge studies. *Engineering News Record*, December 25, 1947.

F. N. Visher and J. F. Mink. *Ground-Water Resources in Southern Oahu, Hawaii*. Water Supply Paper 1778, U.S. Geological Survey, 1964.

G. K. Voigt. Distribution of rainfall under forest stands. *Forest Science*, 6(1960): 2–10.

T. M. Walski. Water Distribution. In L. W. Mays, ed., *Water Resources Handbook*, pp. 18.1–18.45. McGraw-Hill, New York, 1996.

W. C. Walton. *Selected Analytical Methods for Well and Aquifer Evaluation*. Bulletin 49. Illinois State Water Survey, Urbana, IL, 1962.

W. C. Walton. *Groundwater Resources Evaluation*. McGraw-Hill, New York, 1970.

W. C. Walton. *Principles of Groundwater Engineering*. CRC Press, Boca Raton, FL, 1991.

C. Wandmacher and A. I. Johnson. *Metric Units in Engineering, Going SI*. ASCE Press, New York, 1995.

M. Wanielista, R. Kersten, and R. Eaglin. *Hydrology: Water Quantity and Quality Control*, 2d ed. Wiley, New York, 1997.

M. P. Wanielista and Y. A. Yousef. *Stormwater Management*. Wiley, New York, 1993.

O. Wanner, T. Egli, T. Fleischmann, K. Lanz, P. Reichert, and R. P. Schwarzenbach. Behavior of the insecticides Disulfoton and Thiometon in the Rhine river: A chemodynamic study. *Environmental Science and Technology*, 23, no. 10 (1989): 1232–42.

A. D. Ward. Surface runoff and subsurface drainage. In A. D. Ward and W. J. Elliot, eds., *Environmental Hydrology*, pp. 133–75. Lewis Publishers, Boca Raton, FL, 1995.

A. D. Ward and J. Dorsey. Infiltration and soil water procsses. In A. D. Ward and W. J. Elliot, eds., *Environmental Hydrology*, pp. 51–90. Lewis Publishers, Boca Raton, FL, 1995.

D. L. Warner and J. H. Lehr. *Subsurface Wastewater Injection*. Premier Press, Berkeley, CA, 1981.

Water Pollution Control Federation. *Gravity Sanitary Sewer Design and Construction, WPCF Manual of Practice No.FD-5*. WPCF, Washington, DC, 1982.

T. C. Wei and J. L. McGuinness. *Reciprocal Distance Squared Method, a Computer Technique for Estimating Area Precipitation*. Technical Report ARS-NC-8. U.S. Agricultural Research Service, North Central Region, Coshocton, OH, 1973.

W. Weibull. A statistical theory of the strength of materials. *Ing. Vetenskapsakad. Handl. [Stockh.]*, 151(1939): 15.

H. G. Wenzel. The effect of raindrop impact and surface roughness on sheet flow. WRC Research Report 34. Water Resources Centre, University of Illinois, Urbana, IL, 1970.

H. G. Wenzel, Jr. Rainfall for urban design. In D. F. Kibler, ed., *Urban Stormwater Hydrology*, pp. 35–67. American Geophysical Union, Water Resources Monograph Series, Washington, DC, 1982.

R. G. Wetzel. *Limnology*. Saunders, Philadelphia, PA, 1975.

W. Whipple and C. W. Randall. *Detention and Flow Retardation Devices*. Prentice-Hall, Englewood Cliffs, NJ, 1983. Stormwater Management in Urbanizing Areas.

F. M. White. *Fluid Mechanics*, 3d ed. McGraw-Hill, New York, 1994.

E. R. Wilcox. *A Comparative Test of the Flow of Water in 8-inch Concrete and Vitrified Clay Sewer Pipe*. Experiment Station Series Bulletin 27. University of Washington, 1924.

Willeke. Discussion of "Runoff Curve Number: Has it Reached Maturity?" by V. M. Ponce and R. H. Hawkins. *Journal of Hydrologic Engineering*, 2, no. 3 (1997): 147.

G. S. Williams and A. H. Hazen. *Hydraulic Tables*, 3d ed. Wiley, New York, 1933.

J. Williamson. The laws of motion in rough pipes. *La Houille Blanche*, 6, no. 5 (September–October 1951): 738.

M. D. Winsberg. *Florida Weather*. University of Central Florida Press, Orlando, FL, 1990.

T. S. W. Wong and C.-N. Chen. Time of concentrtion formula for sheet flow of varying flow regime. *Journal of Hydrologic Engineering*, 2, no. 3 (July 1997): 136–39.

E. F. Wood. *Global Scale Hydrology: Advances in Land Surface Modeling, U.S. National Report to International Union of Geodesy and Geophysics 1987–1990*. Technical Report. American Geophysical Union, Washington, DC, 1991.

I. R. Wood, R. G. Bell, and D. L. Wilkinson. *Ocean Disposal of Wastewater*. World Scientific, Singapore, 1993.

S. L. Woodall. *Rainfall Interception Losses from Melaleuca Forest in Florida*. Research Note SE-323, U. S. Department of Agriculture, Southeastern Forest Experiment Station, Forest Resources Laboratory, Lehigh Acres, FL, February 1984.

A. D. Woodbury and E. A. Sudicky. The Geostatistical Characteristics of the Borden Aquifer. *Water Resources Research*, 27, no. 4 (1991): 533–46.

D. A. Woolhiser, R. E. Smith, and J.-V. Giraldez. Effects of spatial variability of saturated hydraulic conductivity on Hortonian overland flow. *Water Resources Research*, 32, no. 3 (March 1996): 671–78.

World Meteorological Organization. *Guide to Hydrological Practices*, Vol. 2, *Analysis, Forecasting and Other Applications*, 4th ed. Technical Report. WMO, Geneva, Switzerland, 1983.

J. L. Wright. New evapotranspiration crop coefficients. *Journal of Irrigation and Drainage Engineering*, 108(IR2)(1982): 57–74.

J. L. Wright and M. E. Jensen. Peak water requirements of crops in southern Idaho. *Journal of Irrigation and Drainage Engineering*, 96(IR1)(1972): 193–201.

S. Wu and N. Rajaratnam. Transition from hydraulic jump to open channel flow. *Journal of Hydraulic Engineering*, 122, no. 9 (September 1996): 526–28.

D. L. Yarnell and S. M. Woodward. *The Flow of Water in Drain Tile*. Bulletin 854. U.S. Department of Agriculture, Washington, DC, 1924.

M. V. Yates. *Septic Tank Siting to Minimize the Contamination of Ground Water by Microorganisms*. Technical Report EPA 440/6-87-007. USEPA, Office of Ground-Water Protection, Washington, DC, 1987.

B. C. Yen and V. T. Chow. Design hyetographs for small drainage structures. *Journal of Hydraulic Engineering*, 106(HY6), 1980.

V. Yevjevich. *Probability and Statistics in Hydrology*. Water Resources Publications, Littleton, CO, 1972.

D. F. Young, B. R. Munson, and T. H. Okiishi. *A Brief Introduction to Fluid Mechanics*. Wiley, New York, 1997.

Y. A. Yousef and M. P. Wanielista. Efficiency optimization of wet-detention ponds for urban stormwater management. State of Florida Department of Environmental Regulation, Tallahassee, FL, 1989.

W. C. Zegel. Standards. In D. H. F. Liu and B. G. Lipták, eds., *Environmental Engineers' Handbook*, pp. 185–226. Lewis Publishers, Boca Raton, FL, 1997.

V. J. Zipparro and H. Hasen. *Davis Handbook of Applied Hydraulics*, 4th ed. McGraw-Hill, New York, 1993.

Index